# Detection, Estimation, and Modulation Theory

# Detection, Estimation, and Modulation Theory

## Radar-Sonar Processing and Gaussian Signals in Noise

HARRY L. VAN TREES
George Mason University

A Wiley-Interscience Publication
JOHN WILEY & SONS, INC.
New York • Chichester • Weinheim • Brisbane • Singapore • Toronto

This text is printed on acid-free paper. ♾



Published simultaneously in Canada.



For ordering and customer service, call 1-800-CALL-WILEY.

***Library of Congress Cataloging in Publication Data is available.***

ISBN 0-471-10793-X

10 9 8 7 6

*To Diane*

and Stephen, Mark, Kathleen, Patricia,
Eileen, Harry, and Julia

and the next generation—
Brittany, Erin, Thomas, Elizabeth, Emily,
Dillon, Bryan, Julia, Robert, Margaret,
Peter, Emma, Sarah, Harry, Rebecca,
and Molly

# *Preface for Paperback Edition*

In 1968, Part I of *Detection, Estimation, and Modulation Theory* [VT68] was published. It turned out to be a reasonably successful book that has been widely used by several generations of engineers. There were thirty printings, but the last printing was in 1996. Volumes II and III ([VT71a], [VT71b]) were published in 1971 and focused on specific application areas such as analog modulation, Gaussian signals and noise, and the radar–sonar problem. Volume II had a short life span due to the shift from analog modulation to digital modulation. Volume III is still widely used as a reference and as a supplementary text. In a moment of youthful optimism, I indicated in the the Preface to Volume III and in Chapter III-14 that a short monograph on optimum array processing would be published in 1971. The bibliography lists it as a reference, *Optimum Array Processing,* Wiley, 1971, which has been subsequently cited by several authors. After a 30-year delay, *Optimum Array Processing,* Part IV of *Detection, Estimation, and Modulation Theory* will be published this year.

A few comments on my career may help explain the long delay. In 1972, MIT loaned me to the Defense Communication Agency in Washington, D.C. where I spent three years as the Chief Scientist and the Associate Director of Technology. At the end of the tour, I decided, for personal reasons, to stay in the Washington, D.C. area. I spent three years as an Assistant Vice-President at COMSAT where my group did the advanced planning for the INTELSAT satellites. In 1978, I became the Chief Scientist of the United States Air Force. In 1979, Dr. Gerald Dinneen, the former Director of Lincoln Laboratories, was serving as Assistant Secretary of Defense for C3I. He asked me to become his Principal Deputy and I spent two years in that position. In 1981, I joined M/A-COM Linkabit. Linkabit is the company that Irwin Jacobs and Andrew Viterbi had started in 1969 and sold to M/A-COM in 1979. I started an Eastern operation which grew to about 200 people in three years. After Irwin and Andy left M/A-COM and started Qualcomm, I was responsible for the government operations in San Diego as well as Washington, D.C. In 1988, M/A-COM sold the division. At that point I decided to return to the academic world.

I joined George Mason University in September of 1988. One of my priorities was to finish the book on optimum array processing. However, I found that I needed to build up a research center in order to attract young research-oriented faculty and doctoral students. The process took about six years. The Center for Excellence in Command, Control, Communications, and Intelligence has been very successful and has generated over $300 million in research funding during its existence. During this growth period, I spent some time on array processing but a concentrated effort was not possible. In 1995, I started a serious effort to write the Array Processing book.

Throughout the *Optimum Array Processing* text there are references to Parts I and III of *Detection, Estimation, and Modulation Theory*. The referenced material is available in several other books, but I am most familiar with my own work. Wiley agreed to publish Part I and III in paperback so the material will be readily available. In addition to providing background for Part IV, Part I is still useful as a text for a graduate course in Detection and Estimation Theory. Part III is suitable for a second level graduate course dealing with more specialized topics.

In the 30-year period, there has been a dramatic change in the signal processing area. Advances in computational capability have allowed the implementation of complex algorithms that were only of theoretical interest in the past. In many applications, algorithms can be implemented that reach the theoretical bounds.

The advances in computational capability have also changed how the material is taught. In Parts I and III, there is an emphasis on compact analytical solutions to problems. In Part IV, there is a much greater emphasis on efficient iterative solutions and simulations. All of the material in parts I and III is still relevant. The books use continuous time processes but the transition to discrete time processes is straightforward. Integrals that were difficult to do analytically can be done easily in Matlab®. The various detection and estimation algorithms can be simulated and their performance compared to the theoretical bounds. We still use most of the problems in the text but supplement them with problems that require Matlab® solutions.

We hope that a new generation of students and readers find these reprinted editions to be useful.

HARRY L. VAN TREES

*Fairfax, Virginia*
*June 2001*

# Preface

In this book I continue the study of detection, estimation, and modulation theory begun in Part I [1]. I assume that the reader is familiar with the background of the overall project that was discussed in the preface of Part I. In the preface to Part II [2] I outlined the revised organization of the material. As I pointed out there, Part III can be read directly after Part I. Thus, some persons will be reading this volume without having seen Part II. Many of the comments in the preface to Part II are also appropriate here, so I shall repeat the pertinent ones.

At the time Part I was published, in January 1968, I had completed the "final" draft for Part II. During the spring term of 1968, I used this draft as a text for an advanced graduate course at M.I.T. and in the summer of 1968, I started to revise the manuscript to incorporate student comments and include some new research results. In September 1968, I became involved in a television project in the Center for Advanced Engineering Study at M.I.T. During this project, I made fifty hours of videotaped lectures on applied probability and random processes for distribution to industry and universities as part of a self-study package. The net result of this involvement was that the revision of the manuscript was not resumed until April 1969. In the intervening period, my students and I had obtained more research results that I felt should be included. As I began the final revision, two observations were apparent. The first observation was that the manuscript has become so large that it was economically impractical to publish it as a single volume. The second observation was that since I was treating four major topics in detail, it was unlikely that many readers would actually use all of the book. Because several of the topics can be studied independently, with only Part I as background, I decided to divide the material into three sections: Part II, Part III, and a short monograph on *Optimum Array Processing* [3]. This division involved some further editing, but I felt it was warranted in view of increased flexibility it gives both readers and instructors.

In Part II, I treated nonlinear modulation theory. In this part, I treat the random signal problem and radar/sonar. Finally, in the monograph, I discuss optimum array processing. The interdependence of the various parts is shown graphically in the following table. It can be seen that Part II is completely separate from Part III and *Optimum Array Processing*. The first half of *Optimum Array Processing* can be studied directly after Part I, but the second half requires some background from Part III. Although the division of the material has several advantages, it has one major disadvantage. One of my primary objectives is to present a unified treatment that enables the reader to solve problems from widely diverse physical situations. Unless the reader sees the widespread applicability of the basic ideas he may fail to appreciate their importance. Thus, I strongly encourage all serious students to read at least the more basic results in all three parts.

| | Prerequisites |
|---|---|
| Part II | Chaps. I-5, I-6 |
| Part III | |
| Chaps. III-1 to III-5 | Chaps. I-4, I-6 |
| Chaps. III-6 to III-7 | Chaps. I-4 |
| Chaps. III-8-end | Chaps. I-4, I-6, III-1 to III-7 |
| Array Processing | |
| Chaps. IV-1, IV-2 | Chaps. I-4 |
| Chaps. IV-3-end | Chaps. III-1 to III-5, AP-1 to AP-2 |

The character of this book is appreciably different that that of Part I. It can perhaps be best described as a mixture of a research monograph and a graduate level text. It has the characteristics of a research monograph in that it studies particular questions in detail and develops a number of new research results in the course of this study. In many cases it explores topics which are still subjects of active research and is forced to leave some questions unanswered. It has the characteristics of a graduate level text in that it presents the material in an orderly fashion and develops almost all of the necessary results internally.

The book should appeal to three classes of readers. The first class consists of graduate students. The random signal problem, discussed in Chapters 2 to 7, is a logical extension of our earlier work with deterministic signals and completes the hierarchy of problems we set out to solve. The last half of the book studies the radar/sonar problem and some facets of the digital communication problem in detail. It is a thorough study of how one applies statistical theory to an important problem area. I feel that it provides a useful educational experience, even for students who have no ultimate interest in radar, sonar, or communications, because it demonstrates system design techniques which will be useful in other fields.

The second class consists of researchers in this field. Within the areas studied, the results are close to the current research frontiers. In many places, specific research problems are suggested that are suitable for thesis or industrial research.

The third class consists of practicing engineers. In the course of the development, a number of problems of system design and analysis are carried out. The techniques used and results obtained are directly applicable to many current problems. The material is in a form that is suitable for presentation in a short course or industrial course for practicing engineers. I have used preliminary versions in such courses for several years.

The problems deserve some mention. As in Part I, there are a large number of problems because I feel that problem solving is an essential part of the learning process. The problems cover a wide range of difficulty and are designed to both augment and extend the discussion in the text. Some of the problems require outside reading, or require the use of engineering judgement to make approximations or ask for discussion of some issues. These problems are sometimes frustrating to the student but I feel that they serve a useful purpose. In a few of the problems I had to use numerical calculations to get the answer. I strongly urge instructors to work a particular problem before assigning it. Solutions to the problems will be available in the near future.

As in Part I, I have tried to make the notation mnemonic. All of the notation is summarized in the glossary at the end of the book. I have tried to make my list of references as complete as possible and acknowledge any ideas due to other people.

Several people have contributed to the development of this book. Professors Arthur Baggeroer, Estil Hoversten, and Donald Snyder of the M.I.T. faculty, and Lewis Collins of Lincoln Laboratory, carefully read and criticized the entire book. Their suggestions were invaluable. R. R. Kurth read several chapters and offered useful suggestions. A number of graduate students offered comments which improved the text. My secretary, Miss Camille Tortorici, typed the entire manuscript several times.

My research at M.I.T. was partly supported by the Joint Services and by the National Aeronautics and Space Administration under the auspices of the Research Laboratory of Electronics. I did the final editing while on Sabbatical Leave at Trinity College, Dublin. Professor Brendan Scaife of the Engineering School provided me office facilities during this period, and M.I.T. provided financial assistance. I am thankful for all of the above support.

Harry L. Van Trees

Dublin, Ireland,

## REFERENCES

[1] Harry L. Van Trees, *Detection, Estimation, and Modulation Theory, Pt. I*, Wiley, New York, 1968.

[2] Harry L. Van Trees, *Detection, Estimation, and Modulation Theory, Pt. II*, Wiley, New York, 1971.

[3] Harry L. Van Trees, *Optimum Array Processing*, Wiley, New York, 1971.

# *Contents*

# 1

# *Introduction*

This book is the third in a set of four volumes. The purpose of these four volumes is to present a unified approach to the solution of detection, estimation, and modulation theory problems. In this volume we study two major problem areas. The first area is the detection of random signals in noise and the estimation of random process parameters. The second area is signal processing in radar and sonar systems. As we pointed out in the Preface, Part III does not use the material in Part II and can be read directly after Part I.

In this chapter we discuss three topics briefly. In Section 1.1, we review Parts I and II so that we can see where the material in Part III fits into the over-all development. In Section 1.2, we introduce the first problem area and outline the organization of Chapters 2 through 7. In Section 1.3, we introduce the radar–sonar problem and outline the organization of Chapters 8 through 14.

## 1.1 REVIEW OF PARTS I AND II

In the introduction to Part I [1], we outlined a hierarchy of problems in the areas of detection, estimation, and modulation theory and discussed a number of physical situations in which these problems are encountered.

We began our technical discussion in Part I with a detailed study of classical detection and estimation theory. In the classical problem the observation space is finite-dimensional, whereas in most problems of interest to us the observation is a waveform and must be represented in an infinite-dimensional space. All of the basic ideas of detection and parameter estimation were developed in the classical context.

In Chapter I-3, we discussed the representation of waveforms in terms of series expansions. This representation enabled us to bridge the gap between the classical problem and the waveform problem in a straightforward manner. With these two chapters as background, we began our study of the hierarchy of problems that we had outlined in Chapter I-1.

In the first part of Chapter I-4, we studied the detection of known signals in Gaussian noise. A typical problem was the binary detection problem in which the received waveforms on the two hypotheses were

$$r(t) = s_1(t) + n(t), \qquad T_i \le t \le T_f : H_1, \tag{1}$$

$$r(t) = s_0(t) + n(t), \qquad T_i \le t \le T_f : H_0, \tag{2}$$

where $s_1(t)$ and $s_0(t)$ were known functions. The noise $n(t)$ was a sample function of a Gaussian random process.

We then studied the parameter-estimation problem. Here, the received waveform was

$$r(t) = s(t, \mathbf{A}) + n(t), \qquad T_i \le t \le T_f. \tag{3}$$

The signal $s(t, \mathbf{A})$ was a known function of $t$ and $\mathbf{A}$. The parameter $\mathbf{A}$ was a vector, either random or nonrandom, that we wanted to estimate.

We referred to all of these problems as known signal-in-noise problems, and they were in the first level in the hierarchy of problems that we outlined in Chapter I-1. The common characteristic of first-level problems is the presence of a *deterministic signal* at the receiver. In the binary detection problem, the receiver decides which of the two deterministic waveforms is present in the received waveform. In the estimation problem, the receiver estimates the value of a parameter contained in the signal. In all cases it is the additive noise that limits the performance of the receiver.

We then generalized the model by allowing the signal component to depend on a finite set of unknown parameters (either random or nonrandom). In this case, the received waveforms in the binary detection problem were

$$\begin{aligned} r(t) &= s_1(t, \boldsymbol{\theta}) + n(t), \qquad T_i \le t \le T_f : H_1, \\ r(t) &= s_0(t, \boldsymbol{\theta}) + n(t), \qquad T_i \le t \le T_f : H_0. \end{aligned} \tag{4}$$

In the estimation problem the received waveform was

$$r(t) = s(t, \mathbf{A}, \boldsymbol{\theta}) + n(t), \qquad T_i \le t \le T_f. \tag{5}$$

The vector $\boldsymbol{\theta}$ denoted a set of unknown and unwanted parameters whose presence introduced a new uncertainty into the problem. These problems were in the second level of the hierarchy. The additional degree of freedom in the second-level model allowed us to study several important physical channels such as the random-phase channel, the Rayleigh channel, and the Rician channel.

In Chapter I-5, we began our discussion of modulation theory and continuous waveform estimation. After formulating a model for the problem, we derived a set of integral equations that specify the optimum demodulator.

In Chapter I-6, we studied the linear estimation problem in detail. Our analysis led to an integral equation,

$$K_{dr}(t, u) = \int_{T_i}^{T_f} h_o(t, \tau) K_r(\tau, u) \, d\tau, \qquad T_i < t, u < T_f, \tag{6}$$

that specified the optimum receiver. We first studied the case in which the observation interval was infinite and the processes were stationary. Here, the spectrum-factorization techniques of Wiener enabled us to solve the problem completely. For finite observation intervals and nonstationary processes, the state-variable formulation of Kalman and Bucy led to a complete solution. We shall find that the integral equation (6) arises frequently in our development in this book. Thus, many of the results in Chapter I-6 will play an important role in our current discussion.

In Part II, we studied nonlinear modulation theory [2]. Because the subject matter in Part II is essentially disjoint from that in Part III, we shall not review the contents in detail. The material in Chapters I-4 through Part II is a detailed study of the first and second levels of our hierarchy of detection, estimation, and modulation theory problems.

There are a large number of physical situations in which the models in the first and second level do not adequately describe the problem. In the next section we discuss several of these physical situations and indicate a more appropriate model.

## 1.2 RANDOM SIGNALS IN NOISE

We begin our discussion by considering several physical situations in which our previous models are not adequate. Consider the problem of detecting the presence of a submarine using a *passive* sonar system. The engines, propellers, and other elements in the submarine generate acoustic signals that travel through the ocean to the hydrophones in the detection system. This signal can best be characterized as a sample function from a random process. In addition, a hydrophone generates self-noise and picks up sea noise. Thus a suitable model for the detection problem might be

$$r(t) = s(t) + n(t), \qquad T_i \le t \le T_f : H_1, \tag{7}$$

$$r(t) = n(t), \qquad T_i \le t \le T_f : H_0. \tag{8}$$

Now $s(t)$ is a sample function from a random process. The new feature in this problem is that the mapping from the hypothesis (or source output) to the signal $s(t)$ is no longer deterministic. The detection problem is to decide whether $r(t)$ is a sample function from a signal plus noise process or from the noise process alone.

A second area in which we decide which of two processes is present is the digital communications area. A large number of digital systems operate over channels in which randomness is inherent in the transmission characteristics. For example, tropospheric scatter links, orbiting dipole links, chaff systems, atmospheric channels for optical systems, and underwater acoustic channels all exhibit random behavior. We discuss channel models in detail in Chapters 9–13. We shall find that a typical method of communicating digital data over channels of this type is to transmit one of two signals that are separated in frequency. (We denote these two frequencies as $\omega_1$ and $\omega_0$). The resulting received signal is

$$\begin{aligned} r(t) &= s_1(t) + n(t), \qquad T_i \le t \le T_f : H_1, \\ r(t) &= s_0(t) + n(t), \qquad T_i \le t \le T_f : H_0. \end{aligned} \tag{9}$$

Now $s_1(t)$ is a sample function from a random process whose spectrum is centered at $\omega_1$, and $s_0(t)$ is a sample function from a random process whose spectrum is centered at $\omega_0$. We want to build a receiver that will decide between $H_1$ and $H_0$.

Problems in which we want to estimate the parameters of random processes are plentiful. Usually when we model a physical phenomenon using a stationary random process we assume that the power spectrum is known. In practice, we frequently have a sample function available and must determine the spectrum by observing it. One procedure is to parameterize the spectrum and estimate the parameters. For example, we assume

$$S(\omega, A) = \frac{A_1}{\omega^2 + A_2{}^2}, \qquad -\infty < \omega < \infty, \tag{10}$$

and try to estimate $A_1$ and $A_2$ by observing a sample function of $s(t)$ corrupted by measurement noise. A second procedure is to consider a small frequency interval and try to estimate the average height of spectrum over that interval.

A second example of estimation of process parameters arises in such diverse areas as radio astronomy, spectroscopy, and passive sonar. The source generates a narrow-band random process whose center frequency identifies the source. Here we want to estimate the center frequency of the spectrum.

A closely related problem arises in the radio astronomy area. Various sources in our galaxy generate a narrow-band process that would be centered at some known frequency if the source were not moving. By estimating the center frequency of the received process, the velocity of the source can be determined. The received waveform may be written as

$$r(t) = s(t, v) + n(t), \qquad T_i \leq t \leq T_f, \tag{11}$$

where $s(t, v)$ is a sample function of a random process whose statistical properties depend on the velocity $v$.

These examples of detection and estimation theory problems correspond to the third level in the hierarchy that we outlined in Chapter I-1. They have the common characteristic that the information of interest is imbedded in a random process. Any detection or estimation procedure must be based on how the statistics of $r(t)$ vary as a function of the hypothesis or the parameter value.

In Chapter 2, we formulate a quantitative model of the simple binary detection problem in which the received waveform consists of a white Gaussian noise process on one hypothesis and the sum of a Gaussian signal process and the white Gaussian noise process on the other hypothesis. In Chapter 3, we study the general problem in which the received signal is a sample function from one of two Gaussian random processes. In both sections we derive optimum receiver structures and investigate the resulting performance.

In Chapter 4, we study four special categories of detection problems for which complete solutions can be obtained. In Chapter 5, we consider the $M$-ary problem, the performance of suboptimum receivers for the binary problem, and summarize our detection theory results.

In Chapters 6 and 7, we treat the parameter estimation problem. In Chapter 6, we develop the model for the single-parameter estimation problem, derive the optimum estimator, and discuss performance analysis techniques. In Chapter 7, we study four categories of estimation problems in which reasonably complete solutions can be obtained. We also extend our results to include multiple-parameter estimation and summarize our estimation theory discussion.

The first half of the book is long, and several of the discussions include a fair amount of detail. This detailed discussion is necessary in order to develop an ability actually to solve practical problems. Strictly speaking, there are no new concepts. We are simply applying decision theory and estimation theory to a more general class of problems. It turns out that the transition from the concept to actual receiver design requires a significant amount of effort.

The development in Chapters 2 through 7 completes our study of the hierarchy of problems that were outlined in Chapter I-1. The remainder of the book applies these ideas to signal processing in radar and sonar systems.

## 1.3 SIGNAL PROCESSING IN RADAR–SONAR SYSTEMS

In a conventional active radar system we transmit a pulsed sinusoid. If a target is present, the signal is reflected. The received waveform consists of the reflected signal plus interfering noises. In the simplest case, the only source of interference is an additive Gaussian receiver noise. In the more general case, there is interference due to external noise sources or reflections from other targets. In the detection problem, the receiver processes the signal to decide whether or not a target is present at a particular location. In the parameter estimation problem, the receiver processes the signal to measure some characteristics of the target such as range, velocity, or acceleration. We are interested in the signal-processing aspects of this problem.

There are a number of issues that arise in the signal-processing problem.

1. We must describe the reflective characteristics of the target. In other words, if the transmitted signal is $s_t(t)$, what is the reflected signal?
2. We must describe the effect of the transmission channels on the signals.
3. We must characterize the interference. In addition to the receiver noise, there may be other targets, external noise generators, or clutter.
4. After we develop a quantitative model for the environment, we must design an optimum (or suboptimum) receiver and evaluate its performance.

In the second half of the book we study these issues. In Chapter 8, we discuss the radar–sonar problem qualitatively. In Chapter 9, we discuss the problem of detecting a slowly fluctuating point target at a particular range and velocity. First we assume that the only interference is additive white Gaussian noise, and we develop the optimum receiver and evaluate its performance. We then consider nonwhite Gaussian noise and find the optimum receiver and its performance. We use complex state-variable theory to obtain complete solutions for the nonwhite noise case.

In Chapter 10, we consider the problem of estimating the parameters of a slowly fluctuating point target. Initially, we consider the problem of estimating the range and velocity of a single target when the interference is additive white Gaussian noise. Starting with the likelihood function, we develop the structure of the optimum receiver. We then investigate the performance of the receiver and see how the signal characteristics affect the estimation accuracy. Finally, we consider the problem of detecting a target in the presence of other interfering targets.

The work in Chapters 9 and 10 deals with the simplest type of target and models the received signal as a known signal with unknown random parameters. The background for this problem was developed in Section I-4.4, and Chapters 9 and 10 can be read directly after Chapter I-4.

In Chapter 11, we consider a point target that fluctuates during the time during which the transmitted pulse is being reflected. Now we must model the received signal as a sample function of a random process.

In Chapter 12, we consider a slowly fluctuating target that is distributed in range. Once again we model the received signal as a sample function of a random process. In both cases, the necessary background for solving the problem has been developed in Chapters III-2 through III-4.

In Chapter 13, we consider fluctuating, distributed targets. This model is useful in the study of clutter in radar systems and reverberation in sonar systems. It is also appropriate in radar astronomy and scatter communications problems. As in Chapters 11 and 12, the received signal is modeled as a sample function of a random process. In all three of these chapters we are able to find the optimum receivers and analyze their performance.

Throughout our discussion we emphasize the similarity between the radar problem and the digital communications problem. Imbedded in various chapters are detailed discussions of digital communication over fluctuating channels. Thus, the material will be of interest to communications engineers as well as radar/sonar signal processors.

Finally, in Chapter 14, we summarize the major results of the radar–sonar discussion and outline the contents of the subsequent book on *Array Processing* [3]. In addition to the body of the text, there is an Appendix on the complex representation of signals, systems, and processes.

## REFERENCES

[1] H. L. Van Trees, *Detection, Estimation, and Modulation Theory, Part I*, Wiley, New York, 1968.

[2] H. L. Van Trees, *Detection, Estimation, and Modulation Theory, Part II*, Wiley New York, 1971.

[3] H. L. Van Trees, *Array Processing*, Wiley, New York (to be published).

# 2

# *Detection of Gaussian Signals in White Gaussian Noise*

In this chapter we consider the problem of detecting a sample function from a Gaussian random process in the presence of additive white Gaussian noise. This problem is a special case of the general Gaussian problem described in Chapter 1. It is characterized by the property that on both hypotheses, the received waveform contains an additive noise component $w(t)$, which is a sample function from a zero-mean white Gaussian process with spectral height $N_0/2$. When $H_1$ is true, the received waveform also contains a signal $s(t)$, which is a sample function from a Gaussian random process whose mean and covariance function are known. Thus,

$$r(t) = s(t) + w(t), \qquad T_i \leq t \leq T_f : H_1 \tag{1}$$

and

$$r(t) = w(t), \qquad T_i \leq t \leq T_f : H_0. \tag{2}$$

The signal process has a mean value function $m(t)$,

$$E[s(t)] = m(t), \qquad T_i \leq t \leq T_f, \tag{3}$$

and a covariance function $K_s(t, u)$,

$$E[s(t) - m(t))(s(u) - m(u))] \triangleq K_s(t, u), \qquad T_i \leq t, u \leq T_f. \tag{4}$$

Both $m(t)$ and $K_s(t, u)$ are known. We assume that the signal process has a finite mean-square value and is statistically independent of the additive noise. Thus, the covariance function of $r(t)$ on $H_1$ is

$$E[(r(t) - m(t))(r(u) - m(u)) \mid H_1] \triangleq K_1(t, u) = K_s(t, u) + \frac{N_0}{2}\,\delta(t - u), \qquad T_i \leq t, u \leq T_f. \tag{5}$$

We refer to $r(t)$ as a *conditionally Gaussian* random process. The term "conditionally Gaussian" is used because $r(t)$, given $H_1$ is true, and $r(t)$, given $H_0$ is true, are the two Gaussian processes in the model.

We observe that the mean value function can be viewed as a deterministic component in the input. When we want to emphasize this we write

$$\begin{aligned} r(t) &= m(t) + [s(t) - m(t)] + w(t) \\ &= m(t) + s_R(t) + w(t), \qquad T_i \le t \le T_f : H_1. \end{aligned} \tag{6}$$

(The subscript $R$ denotes the random component of the signal process.) Now the waveform on $H_1$ consists of a known signal corrupted by two independent zero-mean Gaussian processes. If $K_s(t, u)$ is identically zero, the problem degenerates into the known signal in white noise problem of Chapter I-4. As we proceed, we shall find that all of the results in Chapter I-4 except for the random phase case in Section I-4.4.1 can be viewed as special cases of various problems in Chapters 2 and 3.

In Section 2.1, we derive the optimum receiver and discuss various procedures for implementing it. In Section 2.2, we analyze the performance of the optimum receiver. Finally, in Section 2.3, we summarize our results.

Most of the original work on the detection of Gaussian signals is due to Price [1]–[4] and Middleton [17]–[20]. Other references are cited at various points in the Chapter.

## 2.1 OPTIMUM RECEIVERS

Our approach to designing the optimum receiver is analogous to the approach in the deterministic signal case (see pages I-250–I-253). The essential steps are the following:

1. We expand $r(t)$ in a series, using the eigenfunctions of the signal process as coordinate functions. The noise term $w(t)$ is white, and so the coefficients of the expansion will be conditionally uncorrelated on both hypotheses. Because the input $r(t)$ is Gaussian on both hypotheses, the coefficients are conditionally statistically independent.
2. We truncate the expansion at the $K$th term and denote the first $K$ coefficients by the vector $\mathbf{r}$. The waveform corresponding to the sum of the first $K$ terms in the series is $r_K(t)$.
3. We then construct the likelihood ratio,

$$\Lambda(r_K(t)) = \Lambda(\mathbf{R}) = \frac{p_{\mathbf{r}|H_1}(\mathbf{R} \mid H_1)}{p_{\mathbf{r}|H_0}(\mathbf{R} \mid H_0)}, \tag{7}$$

and manipulate it into a form so that we can let $K \to \infty$.

4. We denote the limit of $\Lambda(r_K(t))$ as $\Lambda(r(t))$. The test consists of comparing the likelihood ratio with a threshold $\eta$,

$$\Lambda[r(t)] \underset{H_0}{\overset{H_1}{\gtrless}} \eta. \tag{8}$$

As before, the threshold $\eta$ is determined by the costs and a-priori probabilities in a Bayes test and the desired $P_F$ in a Neyman-Pearson test.

We now carry out these steps in detail and then investigate the properties of the resulting tests.

The orthonormal functions for the series expansion are the eigenfunctions of the integral equation†

$$\lambda_i^s \phi_i(t) = \int_{T_i}^{T_f} K_s(t, u)\phi_i(u)\, du, \qquad T_i \le t \le T_f. \tag{9}$$

We shall assume that the orthonormal functions form a complete set. This will occur naturally if $K_s(t, u)$ is positive-definite. If $K_s(t, u)$ is only non-negative-definite, we augment the set to make it complete.

The coefficients in the series expansion are

$$r_i \triangleq \int_{T_i}^{T_f} r(t)\phi_i(t)\, dt. \tag{10}$$

The $K$-term approximation is

$$r_K(t) = \sum_{i=1}^{K} r_i \phi_i(t), \qquad T_i \le t \le T_f \tag{11}$$

and

$$r(t) = \underset{K\to\infty}{\text{l.i.m.}}\; r_K(t), \qquad T_i \le t \le T_f. \tag{12}$$

The statistical properties of the coefficients on the two hypotheses follow easily.

$$E[r_i \mid H_0] = E\left[\int_{T_i}^{T_f} w(t)\phi_i(t)\, dt\right] = 0. \tag{13}$$

$$E[r_i r_j \mid H_0] = \frac{N_0}{2}\delta_{ij}. \tag{14}$$

$$\begin{aligned} E[r_i \mid H_1] &= E\left[\int_{T_i}^{T_f} s(t)\phi_i(t)\, dt + \int_{T_i}^{T_f} w(t)\phi_i(t)\, dt\right] \\ &= \int_{T_i}^{T_f} m(t)\phi_i(t)\, dt \triangleq m_i. \end{aligned} \tag{15}$$

† Series expansions were developed in detail in Chapter I-3.

Notice that (15) implies that the $m_i$ are the coefficients of an orthogonal expansion of the mean-value function; that is,

$$m(t) = \sum_{i=1}^{\infty} m_i \phi_i(t), \qquad T_i \le t \le T_f. \tag{16}$$

The covariance between coefficients is

$$E[(r_i - m_i)(r_j - m_j) \mid H_1] = \left(\lambda_i^s + \frac{N_0}{2}\right) \delta_{ij}, \tag{17}$$

where $\lambda_i^s$ is the $i$th eigenvalue of (9). The superscript $s$ emphasizes that it is an eigenvalue of the signal process, $s(t)$.

Under both hypotheses, the coefficients $r_i$ are statistically independent Gaussian random variables. The probability density of $\mathbf{r}$ is just the product of the densities of the coefficients. Thus,

$$\Lambda(\mathbf{R}) \triangleq \frac{p_{\mathbf{r}|H_1}(\mathbf{R} \mid H_1)}{p_{\mathbf{r}|H_0}(\mathbf{R} \mid H_0)}$$
$$= \frac{\left(\prod_{i=1}^{K} \frac{1}{[2\pi(N_0/2 + \lambda_i^s)]^{1/2}}\right) \exp\left(-\frac{1}{2} \sum_{i=1}^{K} \frac{(R_i - m_i)^2}{\lambda_i^s + (N_0/2)}\right)}{\left(\prod_{i=1}^{K} \frac{1}{[2\pi(N_0/2)]^{1/2}}\right) \exp\left(-\frac{1}{2} \sum_{i=1}^{K} \frac{R_i^2}{N_0/2}\right)}. \tag{18}$$

Multiplying out each term in the exponent, canceling common factors, taking the logarithm, and rearranging the results, we have

$$\ln \Lambda(\mathbf{R}) = \frac{1}{N_0} \sum_{i=1}^{K} \left(\frac{\lambda_i^s}{\lambda_i^s + N_0/2}\right) R_i^2 + \sum_{i=1}^{K} \left(\frac{1}{\lambda_i^s + N_0/2}\right) m_i R_i$$
$$- \frac{1}{2} \sum_{i=1}^{K} \left(\frac{1}{\lambda_i^s + N_0/2}\right) m_i^2 - \frac{1}{2} \sum_{i=1}^{K} \ln\left(1 + \frac{2\lambda_i^s}{N_0}\right). \tag{19}$$

The final step is to obtain closed form expressions for the various terms when $K \to \infty$. To do this, we need the inverse kernel that was first introduced in Chapter I-4 [see (I-4.152)]. The covariance function of the *entire* input $r(t)$ on $H_1$ is $K_1(t, u)$. The corresponding inverse kernel is defined by the relation

$$\int_{T_i}^{T_f} K_1(t, u) Q_1(u, z)\, du = \delta(t - z), \qquad T_i < t, z < T_f. \tag{20}$$

In terms of eigenfunctions and eigenvalues,

$$Q_1(t, u) = \sum_{i=1}^{\infty} \frac{1}{\lambda_i^s + N_0/2} \phi_i(t) \phi_i(u), \qquad T_i < t, z < T_f. \tag{21}$$

We also saw in Chapter I-4 (I-4.162) that we could write $Q_1(t, u)$ as a sum of an impulse component and a well-behaved function,

$$Q_1(t, u) = \frac{2}{N_0}(\delta(t-u) - h_1(t, u)), \qquad T_i < t, u < T_f, \tag{22}$$

where the function $h_1(t, u)$ satisfies the integral equation

$$\boxed{\frac{N_0}{2} h_1(t, u) + \int_{T_i}^{T_f} h_1(t, z) K_s(z, u)\, dz = K_s(t, u), \qquad T_i \le t, u \le T_f.} \tag{23}$$

The endpoint values of $h_1(t, u)$ are defined as a limit of the open-interval values because we assume that $h_1(t, u)$ is continuous. (Recall the discussion on page I-296.) We also recall that we could write the solution to (23) in terms of eigenfunctions and eigenvalues.

$$h_1(t, u) = \sum_{i=1}^{\infty} \frac{\lambda_i^s}{\lambda_i^s + N_0/2} \phi_i(t)\phi_i(u), \qquad T_i \le t, u \le T_f. \tag{24}$$

We now rewrite the first three terms in (19) by using (10) and (15) to obtain

$$\begin{aligned}
\ln \Lambda(r_K(t)) = {} & \frac{1}{N_0} \iint\limits_{T_i}^{T_f} r(t) \left[ \sum_{i=1}^{K} \left( \frac{\lambda_i^s}{\lambda_i^s + N_0/2} \right) \phi_i(t)\phi_i(u) \right] r(u)\, dt\, du \\
& + \iint\limits_{T_i}^{T_f} m(t) \left[ \sum_{i=1}^{K} \left( \frac{1}{\lambda_i^s + N_0/2} \right) \phi_i(t)\phi_i(u) \right] r(u)\, dt\, du \\
& - \tfrac{1}{2} \iint\limits_{T_i}^{T_f} m(t) \left[ \sum_{i=1}^{K} \left( \frac{1}{\lambda_i^s + N_0/2} \right) \phi_i(t)\phi_i(u) \right] m(u)\, dt\, du \\
& - \frac{1}{2} \sum_{i=1}^{K} \ln \left( 1 + \frac{2\lambda_i^s}{N_0} \right)
\end{aligned} \tag{25}$$

We now let $K \to \infty$ in (25) and use (21) and (24) to evaluate the first three terms in (25). The result is

$$\begin{aligned}
\ln \Lambda(r(t)) = {} & \frac{1}{N_0} \iint\limits_{T_i}^{T_f} r(t) h_1(t, u) r(u)\, dt\, du + \iint\limits_{T_i}^{T_f} m(t) Q_1(t, u) r(u)\, dt\, du \\
& - \tfrac{1}{2} \iint\limits_{T_i}^{T_f} m(t) Q_1(t, u) m(u)\, dt\, du - \frac{1}{2} \sum_{i=1}^{\infty} \ln \left( 1 + \frac{2\lambda_i^s}{N_0} \right).
\end{aligned} \tag{26}$$

We can further simplify the second and third terms on the right side of (26) by recalling the definition of $g(u)$ in (I-4.168),

$$g_1(u) \triangleq \int_{T_i}^{T_f} m(t)Q_1(t, u)\, dt, \qquad T_i < u < T_f. \tag{27}$$

Notice that $m(t)$ plays the role of the known signal [which was denoted by $s(t)$ in Chapter I-4]. We also observe that the third and fourth term are *not* functions of $r(t)$ and may be absorbed in the threshold. Thus, the likelihood ratio test (LRT) is,

$$\boxed{\frac{1}{N_0} \iint_{T_i}^{T_f} r(t)h_1(t, u)r(u)\, dt\, du + \int_{T_i}^{T_f} g_1(u)r(u)\, du \underset{H_0}{\overset{H_1}{\gtrless}} \gamma_*,} \tag{28}$$

where

$$\boxed{\gamma_* \triangleq \ln \eta + \tfrac{1}{2} \int_{T_i}^{T_f} g_1(u)m(u)\, du + \tfrac{1}{2} \sum_{i=1}^{\infty} \ln \left(1 + \frac{2\lambda_i^s}{N_0}\right).} \tag{29}$$

If we are using a Bayes test, we must evaluate the infinite sum on the right side in order to set the threshold. On page 22 we develop a convenient closed-form expression for this sum. For the Neyman-Pearson test we adjust $\gamma_*$ directly to obtain the desired $P_F$ so that the exact value of the sum is not needed as long as we know the sum converges. The convergence follows easily.

$$\sum_{i=1}^{\infty} \ln \left(1 + \frac{2\lambda_i^s}{N_0}\right) \le \sum_{i=1}^{\infty} \frac{2\lambda_i^s}{N_0} = \frac{2}{N_0} \int_{T_i}^{T_f} K_s(t, t)\, dt. \tag{30}$$

The integral is just the expected value of the energy in the process, which was assumed to be finite.

The first term on the left side of (28) is a quadratic operation on $r(t)$ and arises because the signal is random. If $K_s(t, u)$ is zero (i.e., the signal is deterministic), this term disappears. We denote the first term by $l_R$. (The subscript $R$ denotes random.) The second term on the left side is a linear operation on $r(t)$ and arises because of the mean value $m(t)$. Whenever the signal is a zero-mean process, this term disappears. We denote the second term by $l_D$. (The subscript $D$ denotes deterministic.) It is also convenient to denote the last two terms on the right side of (29) as

$(-l_B^{[2]})$ and $(-l_B^{[1]})$. Thus, we have the definitions

$$l_R \triangleq \frac{1}{N_0} \iint_{T_i}^{T_f} r(t) h_1(t, u) r(u)\, dt\, du, \tag{31}$$

$$l_D \triangleq \int_{T_i}^{T_f} g_1(u) r(u)\, du, \tag{32}$$

$$l_B^{[1]} \triangleq -\tfrac{1}{2} \sum_{i=1}^{\infty} \ln\left(1 + \frac{2\lambda_i^s}{N_0}\right), \tag{33}$$

$$l_B^{[2]} \triangleq -\tfrac{1}{2} \int_{T_i}^{T_f} g_1(u) m(u)\, du. \tag{34}$$

In this notation, the LRT is

$$l_R + l_D \underset{H_0}{\overset{H_1}{\gtrless}} \ln \eta - l_B^{[1]} - l_B^{[2]} = \gamma - l_B^{[1]} - l_B^{[2]} \triangleq \gamma_*. \tag{35}$$

The second term on the left side of (35) is generated physically by either a cross-correlation or a matched filter operation, as shown in Fig. 2.1. The impulse response of the matched filter in Fig. 2.1*b* is

$$h(\tau) = \begin{cases} g_1(T_f - \tau), & 0 \le \tau \le T_f - T_i, \\ 0, & \text{elsewhere.} \end{cases} \tag{36}$$

We previously encountered these operations in the colored noise detection problem discussed in Section I-4.3. Thus, the only new component in the optimum receiver is a device to generate $l_R$. In the next several paragraphs we develop a number of methods of generating $l_R$.

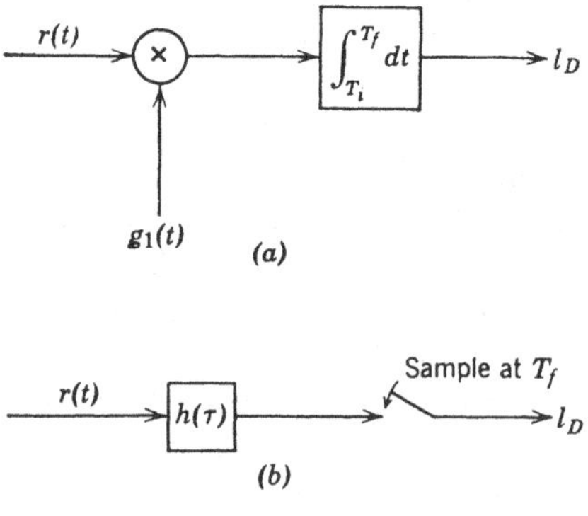


**Fig. 2.1 Generation of $l_D$.**

### 2.1.1 Canonical Realization No. 1: Estimator-Correlator

We want to generate $l_R$, where

$$l_R = \frac{1}{N_0} \iint_{T_i}^{T_f} r(t)\, h_1(t, u)\, r(u)\, dt\, du, \tag{37}$$

and $h_1(t, u)$ satisfies (23). An obvious realization is shown in Fig. 2.2*a*. Notice that $h_1(t, u)$ is an unrealizable filter. Therefore, in order actually to build it, we would have to allow a delay in the filter in the system in Fig. 2.2*a*. This is done by defining a new filter whose output is a delayed version of the output of $h_1(t, u)$,

$$h_{1d}(t, u) = \begin{cases} h_1(t - T, u), & T_i + T \le t \le T_f + T,\ T_i \le u \le T_f, \\ 0, & \text{elsewhere}, \end{cases} \tag{38}$$

where

$$T \triangleq T_f - T_i \tag{39}$$

is the length of the observation interval. Adding a corresponding delay in the upper path and the integrator gives the system in Fig. 2.2*b*.

This realization has an interesting interpretation. We first assume that $m(t)$ is zero and then recall that we have previously encountered (23) in the

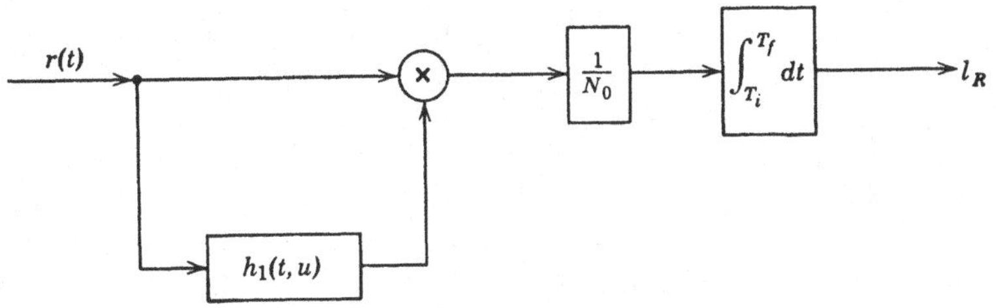


(*a*) Unrealizable filter

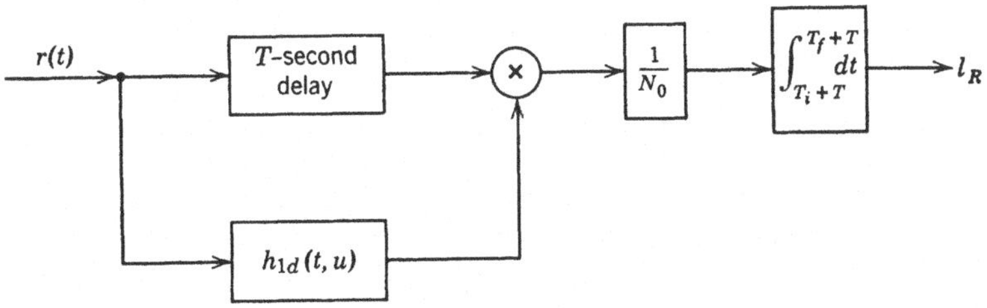


(*b*) Realization with delay

**Fig. 2.2 Generation of $l_R$.**

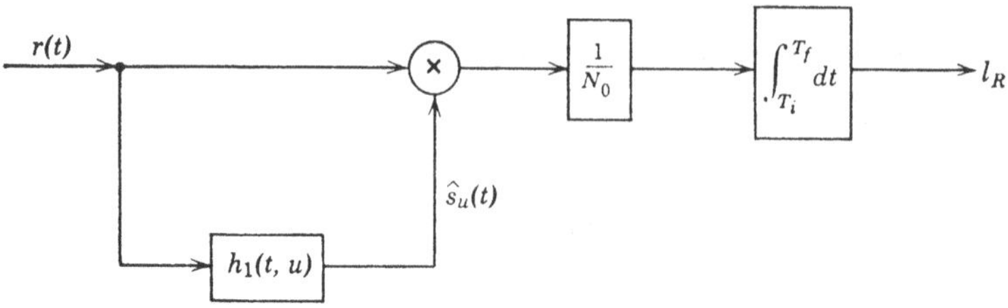


**Fig. 2.3 Estimator-correlator (zero-mean case).**

linear filter context. Specifically, *if* we had available a waveform

$$r(t) = s(t) + w(t), \qquad T_i \leq t \leq T_f \tag{40}$$

and wanted to estimate $s(t)$ using a minimum mean-square error (MMSE) or maximum a-posteriori probability (MAP) criterion, then, from (I-6.16), we know that the resulting estimate $\hat{s}_u(t)$ would be obtained by passing $r(t)$ through $h_1(t, u)$.

$$\hat{s}_u(t) = \int_{T_i}^{T_f} h_1(t, u) r(u)\, du, \qquad T_i \leq t \leq T_f, \tag{41}$$

where $h_1(t, u)$ satisfies (23) and the subscript $u$ emphasizes that the estimate is unrealizable. Looking at Fig. 2.3, we see that the receiver is correlating $r(t)$ with the MMSE estimate of $s(t)$. For this reason, the realization in Fig. 2.3 is frequently referred to as an estimator-correlator receiver. This is an intuitively pleasing interpretation. (This result is due to Price [1]–[4].)

Notice that the interpretation of the left side of (41) as the MMSE estimate is only valid when $r(t)$ is zero-mean. However, the output of the receiver in Fig. 2.3 is $l_R$ for either the zero-mean or the non-zero-mean case. We also obtain an estimator-correlator interpretation in the non-zero-mean case by a straightforward modification of the above discussion (see Problem 2.1.1).

Up to this point all of the filters except the one in Fig. 2.2*b* are unrealizable and are obtained by solving (23). The next configuration eliminates the unrealizability problem.

### 2.1.2 Canonical Realization No. 2: Filter-Correlator Receiver

The realization follows directly from (37). We see that because of the symmetry of the kernel $h_1(t, u)$, (37) can be rewritten as

$$l_R = \frac{2}{N_0} \int_{T_i}^{T_f} r(t) \left[ \int_{T_i}^{t} h_1(t, u) r(u)\, du \right] dt. \tag{42}$$

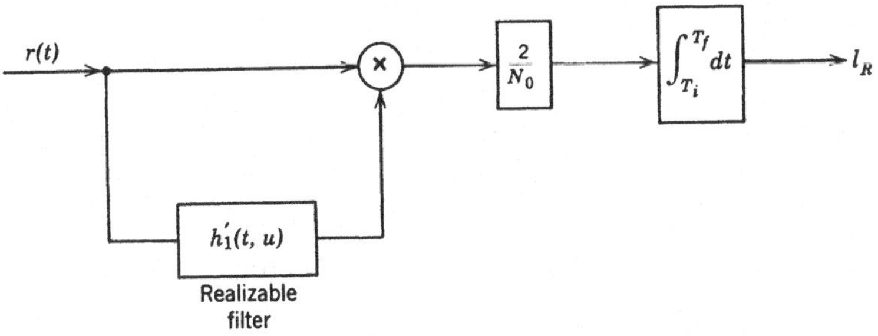


**Fig. 2.4 Filter-correlator receiver.**

In this form, the inner integral represents a *realizable* operation. Thus, we can build the receiver using a realizable filter,

$$h_1'(t, u) = \begin{cases} h_1(t, u), & t \geq u, \\ 0, & t < u, \end{cases} \tag{43}$$

This realization is shown in Fig. 2.4. Observe that the output of the realizable filter $h_1'(t, u)$ is *not* the realizable MMSE estimate of $s(t)$. The impulse response of the optimum realizable linear filter for estimating $s(t)$ is $h_{or}(t, u)$ and its satisfies the equation

$$\frac{N_0}{2} h_{or}(t, u) + \int_{T_i}^{t} h_{or}(t, z) K_s(z, u)\, dz = K_s(t, u), \qquad T_i \leq u \leq t, \tag{44}$$

which is not the same filter specified by (23) plus (43). (This canonical realization is also due to Price [1].) The receiver in Fig. 2.4 is referred to as a filter-correlator receiver. We have included it for completeness. It is used infrequently in practice and we shall not use it in any subsequent discussions.

### 2.1.3 Canonical Realization No. 3: Filter-Squarer-Integrator (FSI) Receiver

A third canonical form can be derived by factoring $h_1(t, u)$. We define $h_f(z, t)$ by the relation

$$h_1(t, u) = \int_{T_i}^{T_f} h_f(z, t) h_f(z, u)\, dz, \qquad T_i \leq t, u \leq T_f. \tag{45}$$

If we do not require that $h_f(z, t)$ be realizable, we can find an infinite number of solutions to (45). From (24), we recall that

$$h_1(t, u) = \sum_{i=1}^{\infty} h_i \phi_i(t) \phi_i(u), \qquad T_i \leq t, u \leq T_f, \tag{46}$$

where

$$h_i = \frac{\lambda_i^s}{\lambda_i^s + N_0/2}. \tag{47}$$

We see that

$$h_{fu}(z, t) = \sum_{i=1}^{\infty} \pm\sqrt{h_i}\, \phi_i(z)\phi_i(t), \qquad T_i \le z,\ t \le T_f \tag{48}$$

is a solution to (45) for any assignment of plus and minus signs in the series.

Using (45) in (37), $l_R$ becomes

$$l_R = \frac{1}{N_0}\int_{T_i}^{T_f} dz\left[\int_{T_i}^{T_f} h_{fu}(z, t)r(t)\, dt\right]^2. \tag{49}$$

This can be realized by a cascade of an unrealizable filter, a square-law device, and an integrator as shown in Fig. 2.5.

Alternatively, we can require that $h_1(t, u)$ be factored using realizable filters. In other words, we must find a solution $h_{fr}(z, t)$ to (45) that is zero for $t > z$. Then,

$$l_R = \frac{1}{N_0}\int_{T_i}^{T_f} dz\left[\int_{T_i}^{z} h_{fr}(z, t)r(t)\, dt\right]^2, \tag{50}$$

and the resulting receiver is shown in Fig. 2.6. If the time interval is finite, a realizable solution to (45) is difficult to find for arbitrary signal processes. Later we shall encounter several special situations that lead to simple solutions.

The integral equation (45) is a functional relationship somewhat analogous to the square-root relation. Thus, we refer to $h_f(z, t)$ as the *functional square root* of $h_1(t, u)$. We shall only define functional square roots for symmetric two-variable functions that can be expanded as in (46) with non-negative coefficients. We frequently use the notation

$$h_1(t, u) = \int_{T_i}^{T_f} h_1^{[1/2]}(z, t)h_1^{[1/2]}(z, u)\, dz. \tag{51}$$

Any solution to (51) is called a functional square root. Notice that the solutions are not necessarily symmetric.

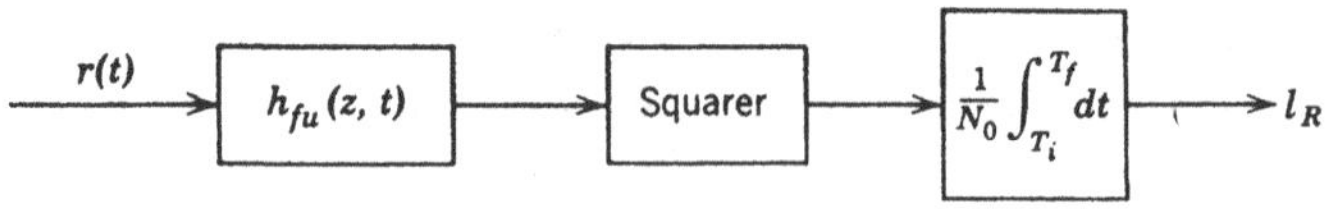


**Fig. 2.5 Filter-squarer receiver (unrealizable).**

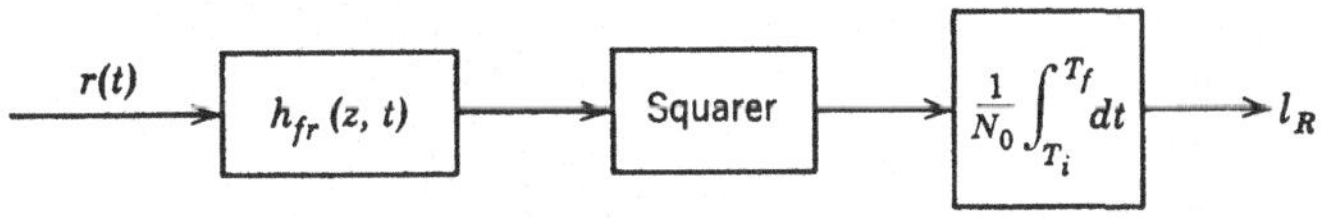


**Fig. 2.6 Filter-squarer receiver (realizable).**

The difficulty with all of the configurations that we have derived up to this point is that to actually implement them we must solve (23). From our experience in Chapter I-4 we know that we can do this for certain classes of kernels and certain conditions on $T_i$ and $T_f$. We explore problems of this type in Chapter 4. On the other hand, in Section I-6.3 we saw that whenever the processes could be generated by exciting a linear finite-dimensional dynamic system with white noise, we had an effective procedure for solving (44). Fortunately, many of the processes (both nonstationary and stationary) that we encounter in practice have a finite-dimensional state representation.

In order to exploit the effective computation procedures that we have developed, we now modify our results to obtain an expression for $l_R$ in which the optimum realizable linear filter specified by (44) is the only filter that we must find.

### 2.1.4 Canonical Realization No. 4: Optimum Realizable Filter Receiver

The basic concept involved in this realization is that of generating the likelihood ratio in real time as the output of a nonlinear dynamic system.† The derivation is of interest because the basic technique is applicable to many problems. For notational simplicity, we let $T_i = 0$ and $T_f = T$ in this section. Initially we shall assume that $m(t) = 0$ and consider only $l_R$.

Clearly, $l_R$ is a function of the length of the observation interval $T$. To emphasize this, we can write

$$l_R(T \mid r(u), 0 \le u \le T) \triangleq l_R(T). \tag{52}$$

More generally, we could define a likelihood function for any value of time $t$.

$$l_R(t \mid r(u), 0 \le u \le t) \triangleq l_R(t), \tag{53}$$

where $l_R(0) = 0$. We can write $l_R(T)$ as

$$l_R(T) = \int_0^T \frac{dl_R(t)}{dt}\, dt = \int_0^T \dot{l}_R(t)\, dt. \tag{54}$$

† The original derivation of (66) was done by Schweppe [5]. The technique is a modification of the linear filter derivation in [6].

Now we want to find an easy method for generating $\dot{l}_R(t)$. Replacing $T$ by $t$ in (31), we have

$$l_R(t) = \frac{1}{N_0}\int_0^t d\tau\, r(\tau)\int_0^t du\, h_1(\tau, u:t)r(u), \tag{55}$$

where $h_1(\tau, u:t)$ satisfies the integral equation

$$\frac{N_0}{2}h_1(\tau, u:t) + \int_0^t h_1(\tau, z:t)K_s(z, u)\, dz = K_s(\tau, u), \qquad 0 \le \tau, u \le t. \tag{56}$$

[Observe that the solution to (56) depends on $t$. We emphasize this with the notation $h_1(\cdot, \cdot : t)$.] Differentiating (55), we obtain

$$\dot{l}_R(t) = \frac{1}{N_0}\Bigg[r(t)\int_0^t du\, h_1(t, u:t)r(u) + \int_0^t d\tau\, r(\tau)\left(h_1(\tau, t:t)r(t) + \int_0^t \frac{\partial h_1(\tau, u:t)}{\partial t}\, r(u)\, du\right)\Bigg]. \tag{57}$$

We see that the first two terms in (57) depend on $h_1(t, u:t)$. For this case, (56) reduces to

$$\frac{N_0}{2}h_1(t, u:t) + \int_0^t h_1(t, z:t)K_s(z, u)\, dz = K_s(t, u), \qquad 0 \le u \le t. \tag{58}†$$

We know from our previous work in Chapter I-6 that

$$\hat{s}_r(t) = \int_0^t h_1(t, u:t)r(u)\, du \tag{59}$$

or

$$\hat{s}_r(t) = \int_0^t h_1(u, t:t)r(u)\, du. \tag{60}$$

[The subscript $r$ means that the operation in (59) can be implemented with a realizable filter.] The result in (60) follows from the symmetry of the solution to (56). Using (59) and (60) in (57) gives

$$\dot{l}_R(t) = \frac{1}{N_0}\left[2r(t)\hat{s}_r(t) + \int_0^t d\tau \int_0^t du\, r(\tau)\frac{\partial h_1(\tau, u:t)}{\partial t}\, r(u)\right]. \tag{61}$$

In Problem I-4.3.3, we proved that

$$\frac{\partial h_1(\tau, u:t)}{\partial t} = -h_1(\tau, t:t)h_1(t, u:t), \qquad 0 \le \tau, u \le t. \tag{62}$$

Because the result is the key step, we include the proof (from [7]).

† Notice that $h_1(t, u:t) = h_{or}(t,u)$ [compare (44) and (58)].

*Proof of (62).* Differentiating (56) gives

$$\frac{N_0}{2}\frac{\partial h_1(\tau, u:t)}{\partial t} + \int_0^t \frac{\partial h_1(\tau, z:t)}{\partial t} K_s(z, u)\, dz + h_1(\tau, t:t)K_s(t,u) = 0, \qquad 0 \le \tau, u \le t. \tag{63}$$

Now replace $K_s(t, u)$ with the left side of (58) and rearrange terms. This gives

$$-\frac{N_0}{2}\left\{\frac{\partial h_1(\tau, u:t)}{\partial t} + h_1(\tau, t:t)h_1(t, u:t)\right\} = \int_0^t \left\{\frac{\partial h_1(\tau, z:t)}{\partial t} + h_1(\tau, t:t)h_1(t, z:t)\right\} \times K_s(z, u)\, dz, \qquad 0 \le \tau, u \le t. \tag{64}$$

We see that the terms in braces play the role of an eigenfunction with an eigenvalue of $(-N_0/2)$. However, $K_s(z, u)$ is non-negative definite, and so it cannot have a negative eigenvalue. Thus, the term in braces must be identically zero in order for (64) to hold. This is the desired result.

Substituting (62) into (61) and using (58), we obtain the desired result,

$$l_R(t) = \frac{1}{N_0}\,[2r(t)\hat{s}_r(t) - \hat{s}_r^{\,2}(t)]. \tag{65}$$

Then

$$\boxed{l_R = l_R(T) = \frac{1}{N_0}\int_0^T [2r(t)\hat{s}_r(t) - \hat{s}_r^{\,2}(t)]\, dt.} \qquad (66)\dagger$$

Before looking at the optimum receiver configuration and some examples, it is appropriate to digress briefly and demonstrate an algorithm for computing the infinite sum $\sum_{i=1}^{\infty} \ln(1 + 2\lambda_i^s/N_0)$ that is needed to evaluate the bias in the Bayes test. We do this now because the derivation is analogous to the one we just completed. Two notational comments are necessary:

1. The eigenvalues in the sum depend on the length of the interval. We emphasize this with the notation $\lambda_i^s(T)$.

2. The eigenfunctions also depend on the length of the interval, and so we use the notation $\phi_i(t:T)$.

This notation was used previously in Chapter I-3 (page I-204).

† A result equivalent to that in (66) was derived independently by Stratonovich and Sosulin [21]–[24]. The integral in (66) is a stochastic integral, and some care must be used when one is dealing with arbitrary (not necessarily Gaussian) random processes. For Gaussian processes it can be interpreted as a Stratonovich integral and used rigorously [25]. For arbitrary processes an Itô integral formulation is preferable [26]–[28]. Interested readers should consult these references or [29]–[30]. For our purposes, it is adequate to treat (66) as an ordinary integral and manipulate it using the normal rules of calculus

We write

$$\sum_{i=1}^{\infty} \ln\left(1 + \frac{2}{N_0}\lambda_i^s(T)\right) = \int_0^T dt\left[\frac{d}{dt}\sum_{i=1}^{\infty} \ln\left(1 + \frac{2}{N_0}\lambda_i^s(t)\right)\right]. \tag{67}$$

Performing the indicated differentiation, we have

$$\frac{d}{dt}\sum_{i=1}^{\infty} \ln\left(1 + \frac{2}{N_0}\lambda_i^s(t)\right) = \frac{2}{N_0}\sum_{i=1}^{\infty}\frac{[d\lambda_i^s(t)]/dt}{1 + (2/N_0)\lambda_i^s(t)}. \tag{68}$$

In Chapter I-3 (page I-3.163), we proved that

$$\frac{d\lambda_i^s(t)}{dt} = \lambda_i^s(t)\phi_i^2(t:t), \tag{69}$$

and we showed that (I-3.154),

$$h_1(t, t:t) = \sum_{i=1}^{\infty}\frac{\lambda_i^s(t)}{\lambda_i^s(t) + N_0/2}\phi_i^2(t:t), \tag{70}$$

where $h_1(t, t:t)$ is the optimum MMSE realizable linear filter specified by (58). From (I-3.155), (44), and (58), the minimum mean-square realizable estimation error $\xi_{Ps}(t)$ is

$$\xi_{Ps}(t) = \frac{N_0}{2}h_1(t, t:t) \triangleq \frac{N_0}{2}h_{or}(t, t). \tag{71}$$

Thus

$$\sum_{i=1}^{\infty} \ln\left(1 + \frac{2\lambda_i^s(T)}{N_0}\right) = \int_0^T h_{or}(t, t)\, dt = \frac{2}{N_0}\int_0^T \xi_{Ps}(t)\, dt. \tag{72}$$

From (33),

$$\boxed{l_B^{[1]} = -\frac{1}{2}\sum_{i=1}^{\infty} \ln\left(1 + \frac{2\lambda_i^s}{N_0}\right) = -\frac{1}{N_0}\int_0^T \xi_{Ps}(t)\, dt.} \tag{73}$$

We see that whenever we use Canonical Realization No. 4, we obtain the first bias term needed for the Bayes test as a by-product. The second bias term [see (34)] is due to the mean, and its computation will be discussed shortly. A block diagram of Realization No. 4 for generating $l_R$ and $l_B^{[1]}$ is shown in Fig. 2.7.

Before leaving our discussion of the bias term, some additional comments are in order. The infinite sum of the left side of (72) will appear in several different contexts, so that an efficient procedure for evaluating it is important. It can also be written as the logarithm of the Fredholm

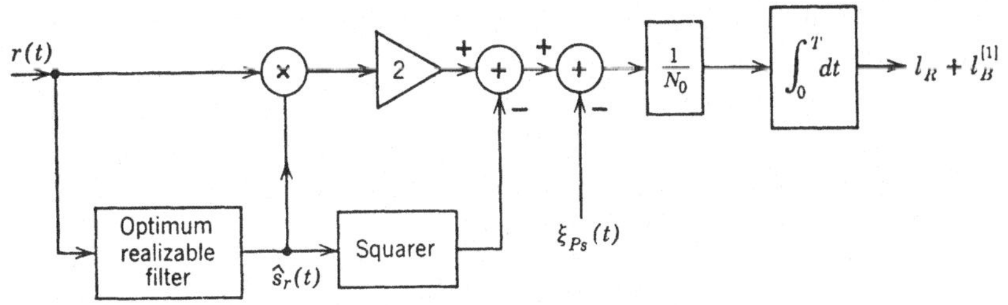


**Fig. 2.7 Optimum realizable filter realization (Canonical Realization No. 4).**

determinant [8],

$$\sum_{i=1}^{\infty} \ln\left(1 + \frac{2\lambda_i^s}{N_0}\right) = \left[\ln\left\{\prod_{i=1}^{\infty}(1 + z\lambda_i^s)\right\}\right]_{z=2/N_0} \triangleq \ln D_{\mathscr{F}}\left(\frac{2}{N_0}\right). \tag{74}$$

Now, unless we can find $D_{\mathscr{F}}(2/N_0)$ effectively, we have not made any progress. One procedure is to evaluate $\xi_{P_s}(t)$ and use the integral expression on the right side of (73). A second procedure for evaluating $D_{\mathscr{F}}(\cdot)$ is a by-product of our solution procedure for Fredholm equations for certain signal processes (see the Appendix in Part II). A third procedure is to use the relation

$$\sum_{i=1}^{\infty} \ln\left(1 + \frac{2\lambda_i^s}{N_0}\right) = \int_0^{2/N_0} dz \int_{T_i}^{T_f} h_1(t, t \mid z)\, dt, \tag{75}$$

where $h_1(t, t \mid z)$ is the solution to (23) when $N_0/2$ equals $z$. Notice that this is the optimum unrealizable filter. This result is derived in Problem 2.1.2. The choice of which procedure to use depends on the specific problem.

Up to this point in our discussion we have not made any detailed assumptions about the signal process. We now look at Realization No. 4 for signal processes that can be generated by exciting a finite-dimensional linear system with white noise. We refer to the corresponding receiver as Realization No. 4S ("S" denotes "state").

### 2.1.5 Canonical Realization No. 4S: State-variable Realization

The class of signal processes of interest was described in detail in Section I-6.3 (see pages I-516–I-538). The process is described by a state equation,

$$\dot{\mathbf{x}}(t) = \mathbf{F}(t)\mathbf{x}(t) + \mathbf{G}(t)\mathbf{u}(t), \tag{76}$$

where $\mathbf{F}(t)$ and $\mathbf{G}(t)$ are possibly time-varying matrices, and by an observation equation,

$$s(t) = \mathbf{C}(t)\mathbf{x}(t), \tag{77}$$

where $\mathbf{C}(t)$ is the modulation matrix. The input, $\mathbf{u}(t)$, is a sample function from a zero-mean vector white noise process,

$$E[\mathbf{u}(t)\mathbf{u}^T(\tau)] = \mathbf{Q}\,\delta(t-\tau), \tag{78}$$

and the initial conditions are

$$E[\mathbf{x}(0)] = \mathbf{0}, \tag{79}$$

$$E[\mathbf{x}(0)\mathbf{x}^T(0)] \triangleq \mathbf{P}_0. \tag{80}$$

From Section I-6.3.2 we know that the MMSE realizable estimate of $s(t)$ is given by the equations

$$\hat{s}_r(t) = \mathbf{C}(t)\hat{\mathbf{x}}(t), \tag{81}$$

$$\dot{\hat{\mathbf{x}}}(t) = \mathbf{F}(t)\hat{\mathbf{x}}(t) + \boldsymbol{\xi}_P(t)\mathbf{C}^T(t)\frac{2}{N_0}[r(t) - \mathbf{C}(t)\hat{\mathbf{x}}(t)]. \tag{82}$$

The matrix $\boldsymbol{\xi}_P(t)$ is the error covariance matrix of $\mathbf{x}(t) - \hat{\mathbf{x}}(t)$.

$$\boldsymbol{\xi}_P(t) \triangleq E[(\mathbf{x}(t) - \hat{\mathbf{x}}(t))(\mathbf{x}^T(t) - \hat{\mathbf{x}}^T(t))]. \tag{83}$$

It satisfies the nonlinear matrix differential equations,

$$\dot{\boldsymbol{\xi}}_P(t) = \mathbf{F}(t)\boldsymbol{\xi}_P(t) + \boldsymbol{\xi}_P(t)\mathbf{F}^T(t) - \boldsymbol{\xi}_P(t)\mathbf{C}^T(t)\frac{2}{N_0}\mathbf{C}(t)\boldsymbol{\xi}_P(t) + \mathbf{G}(t)\mathbf{Q}\mathbf{G}^T(t). \tag{84}$$

The mean-square error in estimating $s(t)$ is

$$\xi_{Ps}(t) = \mathbf{C}(t)\boldsymbol{\xi}_P(t)\mathbf{C}^T(t). \tag{85}$$

Notice that $\boldsymbol{\xi}_P(t)$ is the error covariance matrix for the state vector and $\xi_{Ps}(t)$ is the scalar mean-square error in estimating $s(t)$. Both (84) and (85) can be computed either before $r(t)$ is received *or* simultaneously with the computation of $\hat{\mathbf{x}}(t)$.

The system needed to generate $l_R$ and $l_B^{[1]}$ follows easily and is shown in Fig. 2.8. The state equation describing $l_R$ is obtained from (65),

$$\dot{l}_R(t) = \frac{1}{N_0}[2r(t)\hat{s}_r(t) - \hat{s}_r^2(t)], \tag{86}$$

where $\hat{s}_r(t)$ is defined by (81)–(84) and

$$l_R \triangleq l_R(T). \tag{87}$$

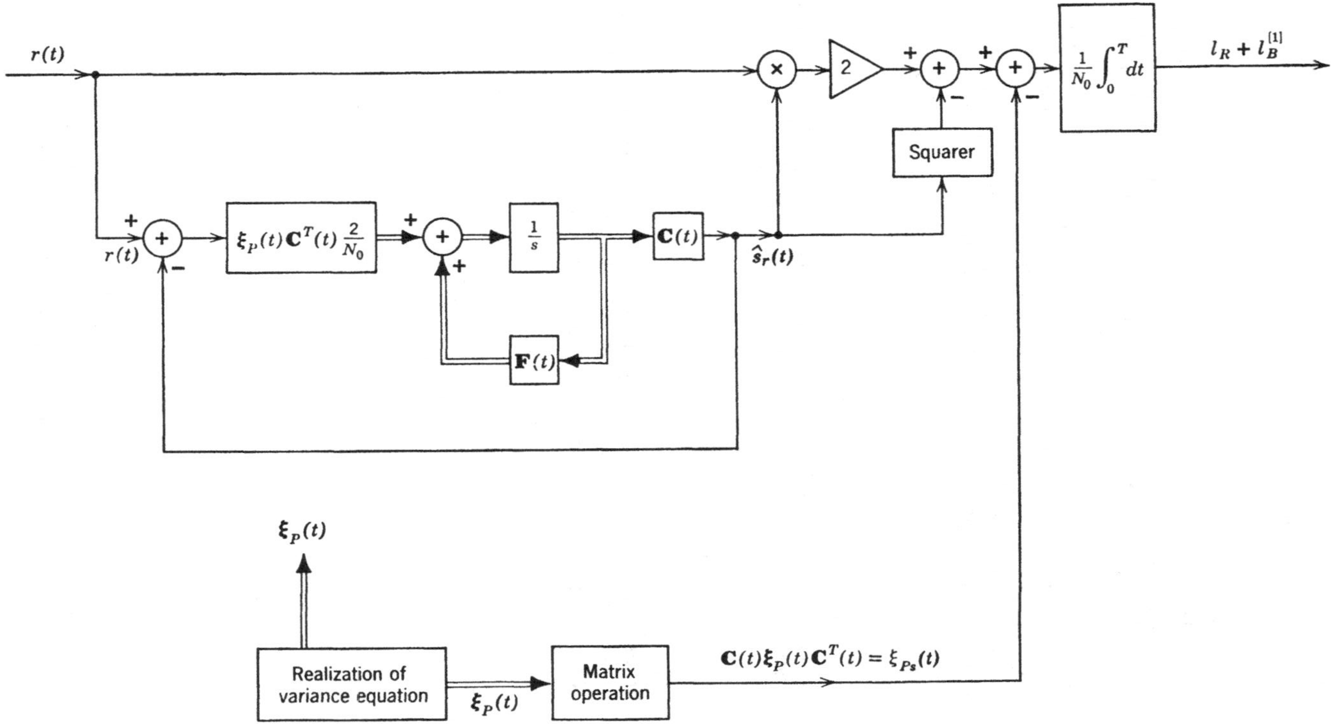


**Fig. 2.8 State-variable receiver (Canonical Realization No. 4S).**

The important feature of this realization is that there are *no* integral equations to solve. The likelihood ratio is generated as the output of a dynamic system. We now consider a simple example to illustrate the application of these ideas.

**Example.** In Fig. 2.9 we show a hypothetical communication system that illustrates many of the important features encountered in actual systems operating over fading channels. In Chapter 10, we shall develop models for fading channels and find that the models are generalizations of the system in this example. When $H_1$ is true, we transmit a *deterministic* signal $f(t)$. When $H_0$ is true, we transmit nothing. The channel affects the received signal in two ways. The transmitted signal is *multiplied* by a sample function of a Gaussian random process $b(t)$. In many cases, this channel process will be stationary over the time intervals of interest. The output of the multiplicative part of the channel is corrupted by additive white Gaussian noise $w(t)$, which is statistically independent of $b(t)$. Thus the received waveforms on the two hypotheses are

$$\begin{aligned} r(t) &= f(t)b(t) + w(t), \qquad 0 \le t \le T: H_1, \\ r(t) &= w(t), \qquad 0 \le t \le T: H_0. \end{aligned} \tag{88}$$

We assume that the channel process has a state representation

$$\dot{\mathbf{x}}(t) = \mathbf{F}(t)\mathbf{x}(t) + \mathbf{G}(t)\mathbf{u}(t), \tag{89}$$

where $\mathbf{u}(t)$ satisfies (78) and

$$b(t) = \mathbf{C}_b(t)\mathbf{x}(t). \tag{90}$$

The signal process on $H_1$ is $s(t)$, where

$$s(t) \triangleq f(t)b(t). \tag{91}$$

Notice that, unless $f(t)$ is constant over the interval $[0, T]$, the process, $s(t)$, will be nonstationary even though $b(t)$ is stationary. Clearly, $s(t)$ has the same state equation as $b(t)$, (89). Combining (90) and (91) gives the observation equation,

$$s(t) = f(t)\mathbf{C}_b(t)\mathbf{x}(t) \triangleq \mathbf{C}(t)\mathbf{x}(t). \tag{92}$$

We see that the transmitted signal $f(t)$ appears only in the modulation matrix, $\mathbf{C}(t)$.

It is instructive to draw the receiver for the simple case in which $b(t)$ has a one-dimensional state equation with constant coefficients. We let

$$\mathbf{F}(t) = -k_b, \tag{93}$$

$$\mathbf{G}(t) = 1, \tag{94}$$

$$Q = 2k_b\sigma_b^2, \tag{95}$$

$$\mathbf{C}_b(t) = 1, \tag{96}$$

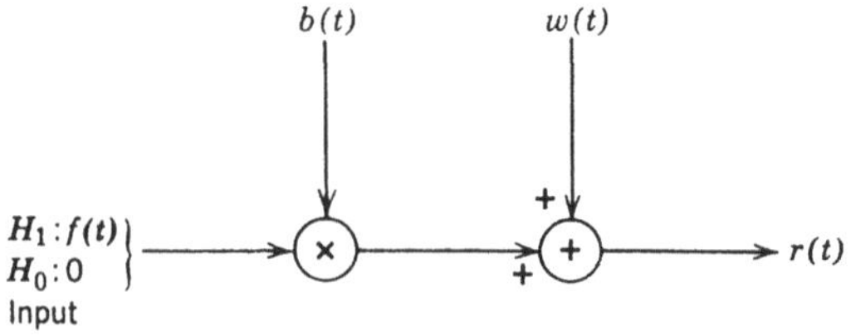


**Fig. 2.9 A simple multiplicative channel.**

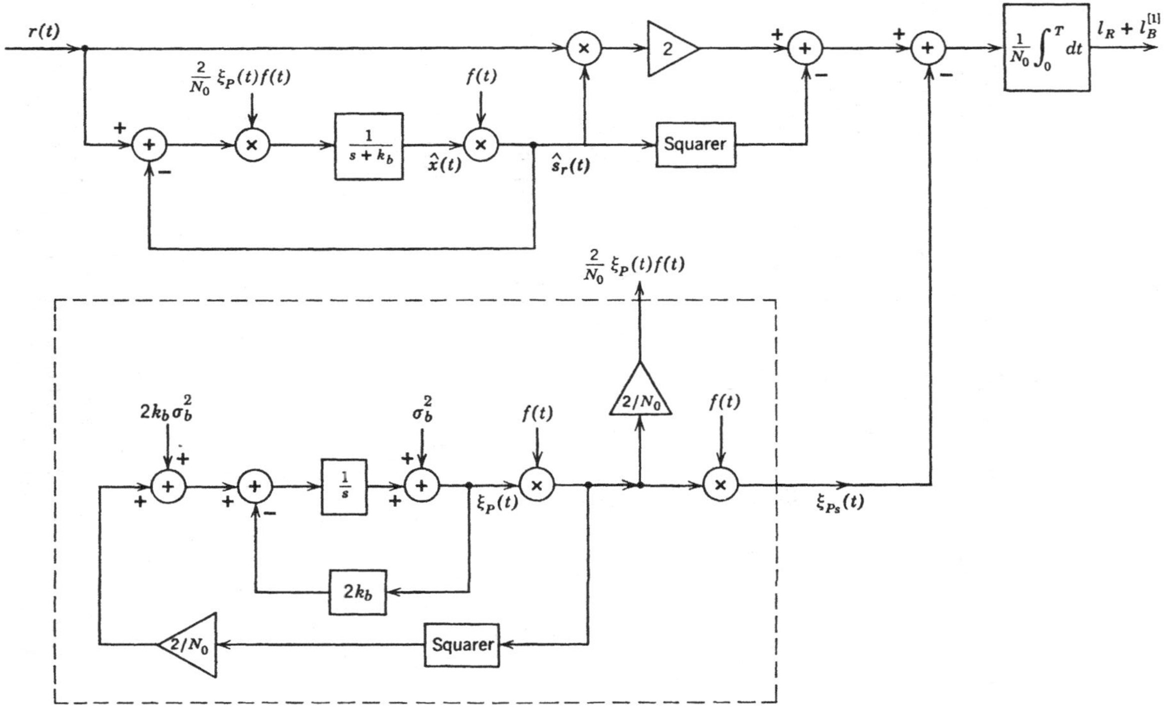


Realization of variance equation (either in real time or precomputed)

**Fig. 2.10 Feedback realization for generating $l_R + l_B^{[1]}$.**

and

$$\xi_P(0) = \sigma_b^2. \tag{97}$$

Then (82) and (84) reduce to

$$\dot{\hat{x}}(t) = -k_b\hat{x}(t) + \frac{2}{N_0}\,\xi_P(t)f(t)[r(t) - f(t)\hat{x}(t)], \tag{98}$$

$$\dot{\xi}_P(t) = -2k_b\xi_P(t) - \frac{2}{N_0}f^2(t)\xi_P^2(t) + 2k_b\sigma_b^2, \tag{99}$$

and

$$\xi_{Ps}(t) = f^2(t)\xi_P(t). \tag{100}$$

The resulting receiver structure is shown in Fig. 2.10.

We shall encounter other examples of Canonical Realization No. 4S as we proceed. Before leaving this realization, it is worthwhile commenting on the generation of $l_D$, the component in the likelihood ratio that arises because of the mean value in the signal process. If the process has a finite state representation, it is usually easier to generate $l_D$ using the optimum realizable filter. The derivation is identical with that in (54)–(66). From (22) and (26)–(28) we have

$$l_D(T) = \frac{2}{N_0}\int_0^T \left[m(\tau) - \int_0^T h_1(\tau, u:T)m(u)\,du\right] r(\tau)\,d\tau. \tag{101}$$

As before,

$$l_D(T) = \int_0^T \frac{dl_D(t)}{dt}\,dt, \tag{102}$$

and

$$\frac{dl_D(t)}{dt} = \frac{2}{N_0}\left[r(t)\left(m(t) - \int_0^t h_1(t, u:t)m(u)\,du\right) - m(t)\int_0^t h_1(\tau, t:t)r(\tau)\,d\tau \right.$$
$$\left. + \iint_0^t h_1(\tau, t:t)h_1(t, u:t)m(u)r(\tau)\,d\tau\,du\right]. \tag{103}$$

The resulting block diagram is shown in Fig. 2.11. The output of the bottom path is just a deterministic function, which we denote by $K(t)$,

$$K(t) \triangleq m(t) - \int_0^t h_1(t, u:t)m(u)\,du, \qquad 0 \le t \le T. \tag{104}$$

Because $K(t)$ does not depend on $r(t)$, we can generate it before any data are received. This suggests the two equivalent realizations in Fig. 2.12.

Notice that (101) (and therefore Figs. 2.11 and 2.12) does not require that the processes be state-representable. If the processes have a finite state, the optimum realizable linear filter can be derived easily using state-variable techniques. Using the state representation in (76)–(80) gives

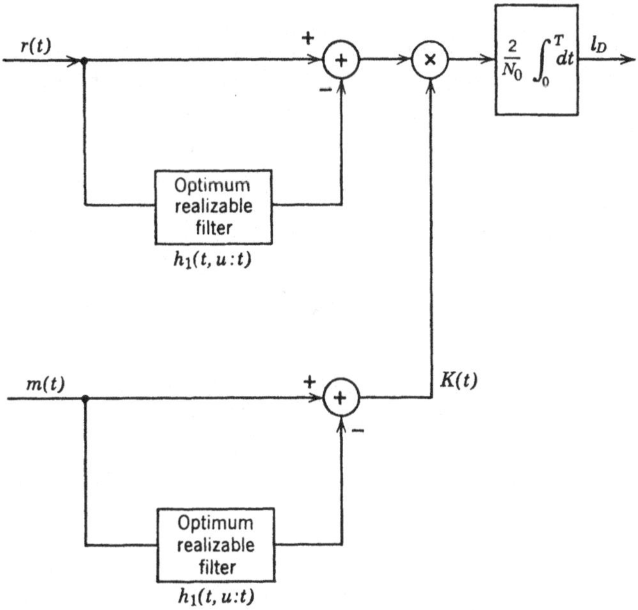


Fig. 2.11 Generation of $l_D$ using optimum realizable filters.

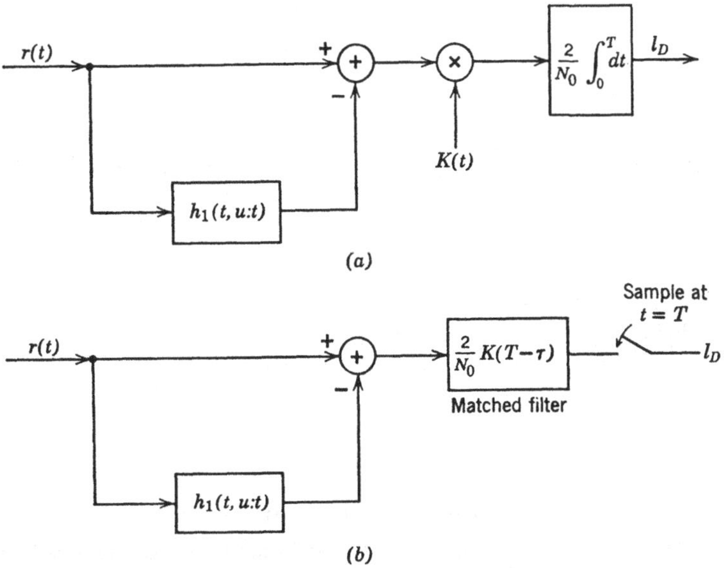


Fig. 2.12 Generation of $l_D$.

the realization in Fig. 2.13*a*.† Notice that the state vector in Fig. 2.13*a* is *not* $\hat{\mathbf{x}}(t)$, because $r(t)$ has a nonzero mean. We denote it by $\check{\mathbf{x}}(t)$.

The block diagram in Fig. 2.13*a* can be simplified as shown in Fig. 2.13b. We can also write $l_D(t)$ in a canonical state-variable form:

$$\begin{bmatrix} \dot{l}_D(t) \\ \dot{\check{\mathbf{x}}}(t) \end{bmatrix} = \left[\begin{array}{c|c} 0 & -\dfrac{2}{N_0} K(t)\mathbf{C}(t) \\ \hline \mathbf{0} & \mathbf{F}(t) - \dfrac{2}{N_0} \boldsymbol{\xi}_P(t)\mathbf{C}^T(t)\mathbf{C}(t) \end{array}\right] \begin{bmatrix} l_D(t) \\ \check{\mathbf{x}}(t) \end{bmatrix} + \left[\begin{array}{c} \dfrac{2}{N_0} K(t) \\ \hline \dfrac{2}{N_0} \boldsymbol{\xi}_P(t)\mathbf{C}^T(t) \end{array}\right] r(t), \tag{105}$$

where $K(t)$ is defined in Fig. 2.11 and $\boldsymbol{\xi}_P(t)$ satisfies (84).

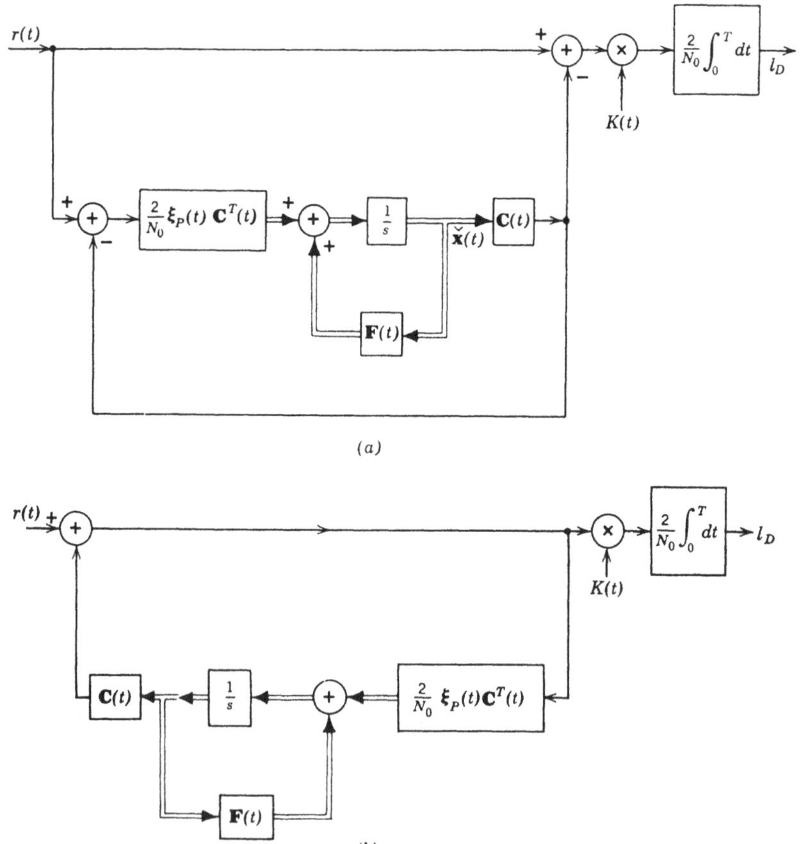


(b)

**Fig. 2.13 State-variable realizations to generate $l_D$.**

† As we would expect, the system in Fig. 3.12 is identical with that obtained using a whitening approach (e.g., Collins [9] or Problem 2.1.3).

Looking at (32) and (34), we see that their structure is identical. Thus, we can generate $l_B^{[2]}$ by driving the dynamic system in (105) with $(-m(t)/2)$ instead of $r(t)$.

It is important to emphasize that the presence of $m(t)$ does not affect the generation of $l_R(t)$ in (86). The only difference is that $\hat{s}_r(t)$ and $\hat{\mathbf{x}}(t)$ are no longer MMSE estimates, and so we denote them by $\check{s}_r(t)$ and $\check{\mathbf{x}}(t)$, respectively. The complete set of equations for the non-zero-mean case may be summarized as follows:

$$\dot{l}_R(t) = \frac{1}{N_0}[-\check{s}_r^{\,2}(t) + 2r(t)\check{s}_r(t)], \tag{106}$$

$$\dot{l}_D(t) = \left(-\frac{2}{N_0}K(t)\mathbf{C}(t)\right)l_D(t) + \frac{2}{N_0}K(t)r(t), \tag{107}$$

$$\check{s}_r(t) = \mathbf{C}(t)\check{\mathbf{x}}(t), \tag{108}$$

$$\dot{\check{\mathbf{x}}}(t) = \mathbf{F}(t)\check{\mathbf{x}}(t) + \frac{2}{N_0}\boldsymbol{\xi}_P(t)\mathbf{C}^T(t)[r(t) - \mathbf{C}(t)\check{\mathbf{x}}(t)], \tag{109}$$

with initial conditions

$$\check{\mathbf{x}}(0) = \mathbf{0} \tag{110}$$

and

$$l_D(0) = 0. \tag{111}$$

The matrix $\boldsymbol{\xi}_P(t)$ is specified by (84). The biases are described in (73) and a modified version of (105).

This completes our discussion of state-variable realizations of the optimum receiver for the Gaussian signal problem. We have emphasized structures based on realizable estimators. An alternative approach based on unrealizable estimator structures can also be developed (see Problem I-6.6.4 and Problem 2.1.4). Before discussing the performance of the optimum receiver, we briefly summarize our results concerning receiver structures.

### 2.1.6 Summary: Receiver Structures

In this section we derived the likelihood ratio test for the simple binary detection problem in which the received waveforms on the two hypotheses were

$$\begin{aligned} r(t) &= w(t), & T_i \le t \le T_f: H_0,\\ r(t) &= s(t) + w(t), & T_i \le t \le T_f: H_1. \end{aligned} \tag{112}$$

The result was the test

$$l_R + l_D \underset{H_0}{\overset{H_1}{\gtrless}} \ln \eta - l_B^{[1]} - l_B^{[2]}, \tag{113}$$

where the various terms were defined by (31)–(34).

We then looked at four receivers that could be used to implement the likelihood ratio test. The first three configurations were based on the optimum unrealizable filter and required the solution of a Fredholm integral equation (23). In Chapter 4 we shall consider problems where this equation can be easily solved. The fourth configuration was based on an optimum *realizable* filter. For this realization we had to solve (44). For a large class of processes, specifically those with a finite state representation, we have already developed an efficient technique for solving this problem (the Kalman-Bucy technique). It is important to re-emphasize that all of the receivers implement the likelihood ratio test and therefore must have identical error probabilities. By having alternative configurations available, we may choose the one easiest to implement.† In the next section we investigate the performance of the likelihood ratio test.

## 2.2 PERFORMANCE

In this section we analyze the performance of the optimum receivers that we developed in Section 2.1. All of these receivers perform the test indicated in (35) as

$$l = l_R + l_D + l_B \underset{H_0}{\overset{H_1}{\gtrless}} \ln \eta = \gamma, \tag{114}$$

where

$$l_R = \frac{1}{N_0} \iint_{T_i}^{T_f} r(t) h_1(t, u) r(u)\, dt\, du, \tag{115}$$

$$l_D = \int_{T_i}^{T_f} g_1(u) r(u)\, du, \tag{116}$$

and

$$l_B \triangleq l_B^{[1]} + l_B^{[2]}. \tag{117}$$

† The reader may view the availability of alternative configurations as a mixed blessing, because it requires some mental bookkeeping to maintain the divisions between realizable and unrealizable filters, the zero-mean and non-zero-mean cases, and similar separations. The problems at the end of Chapter 4 will help in remembering the various divisions.

From (33) and (34), we recall that

$$l_B^{[1]} \triangleq -\frac{1}{2}\sum_{i=1}^{\infty} \ln\left(1 + \frac{2\lambda_i^s}{N_0}\right), \tag{118}$$

$$l_B^{[2]} \triangleq -\frac{1}{2}\int_{T_i}^{T_f} g_1(u)m(u)\,du. \tag{119}$$

To compute $P_D$ and $P_F$, we must find the probability that $l$ will exceed $\gamma$ on $H_1$ and $H_0$, respectively. These probabilities are

$$P_D = \int_{\gamma}^{\infty} p_{l|H_1}(L \mid H_1)\,dL \tag{120}$$

and

$$P_F = \int_{\gamma}^{\infty} p_{l|H_0}(L \mid H_0)\,dL. \tag{121}$$

The $l_D$ component is a linear function of a Gaussian process, so that it is a Gaussian random variable whose mean and variance can be computed easily. However, $l_R$ is obtained by a nonlinear operation on $r(t)$, and so its probability density is difficult to obtain. To illustrate the difficulty, we look at the first term in (25). Because this term corresponds to $l_R$ *before* we let $K \to \infty$, we denote it by $l_R{}^K$,

$$l_R{}^K \triangleq \frac{1}{N_0}\sum_{i=1}^{K} \frac{\lambda_i^s}{\lambda_i^s + N_0/2} R_i^2. \tag{122}$$

We see that $l_R{}^K$ is a weighted sum of *squared* Gaussian random variables. The expression in (122) is familiar from our work on the general Gaussian problem, Section I-2.6. In fact, if the $R_i$ were zero-mean, (122) would be identical with (I-2.420). At that point we observed that if the $\lambda_i^s$ were all equal, $l_R{}^K$ had a chi-square density with $K$ degrees of freedom (e.g., I-2.406). On the other hand, for unequal $\lambda_i^s$, we could write an expression for the probability density but it was intractable for large $K$. Because of the independence of the $R_i$, the characteristic function and moment-generating function of $l_R{}^K$ followed easily (e.g., Problem I-4.4.2). Given the characteristic function, we could, in principle at least, find the probability density by computing the Fourier transform numerically. In practice, we are usually interested in small error probabilities, and so we must know the tails of $p_{l|H_i}(L \mid H_i)$ accurately. This requirement causes the amount of computation required for accurate numerical inversion to be prohibitive. This motivated our discussion of performance bounds and approximations in Section I-2.7. In this section we carry out an analogous discussion for the case in which $K \to \infty$.

We recall† that the function $\mu_K(s)$ played the central role in our discussion. From (I-2.444),

$$\mu_K(s) \triangleq \ln\,[\phi_{l(\mathbf{R})|H_0}(s)], \tag{123}$$

[The subscript $K$ is added to emphasize that we are dealing with $K$-term approximation to $r(t)$.] Where $l(\mathbf{R})$ is the logarithm of the likelihood ratio

$$l(\mathbf{R}) = \ln \Lambda(\mathbf{R}) = \ln\left(\frac{p_{\mathbf{r}|H_1}(\mathbf{R}\mid H_1)}{p_{\mathbf{r}|H_0}(\mathbf{R}\mid H_0)}\right), \tag{124}$$

and $\phi_{l(\mathbf{R})|H_0}(s)$ is its moment-generating function,

$$\phi_{l(\mathbf{R})|H_0}(s) = E[e^{sl(\mathbf{R})}\mid H_0], \tag{125}$$

for real $s$. Using the definition of $l(\mathbf{R})$ in (124),

$$\mu_K(s) = \ln \int_{-\infty}^{\infty}\cdots\int_{-\infty}^{\infty}[p_{\mathbf{r}|H_1}(\mathbf{R}\mid H_1)]^s[p_{\mathbf{r}|H_0}(\mathbf{R}\mid H_0)]^{1-s}\,d\mathbf{R}. \tag{126}$$

We then developed upper bounds on $P_F$ and $P_M$.‡

$$\begin{aligned} P_F &\leq \exp\,[\mu_K(s) - s\dot{\mu}_K(s)], \\ P_M &\leq \exp\,[\mu_K(s) + (1-s)\dot{\mu}_K(s)], \end{aligned} \qquad 0 \leq s \leq 1, \tag{127}$$

where $\dot{\mu}_K(s) = \gamma_K$, the threshold in the LRT. By varying the parameter $s$, we could study threshold settings anywhere between $E[l \mid H_1]$ and $E[l \mid H_0]$. The definition of $l(\mathbf{R})$ in (124) guaranteed that $\mu_K(s)$ existed for $0 \leq s \leq 1$.

We now define a function $\mu(s)$,

$$\mu(s) \triangleq \lim_{K\to\infty} \mu_K(s). \tag{128}$$

If we can demonstrate that the limit exists, our bounds in (127) will still be valid. However, in order to be useful, the expression for $\mu(s)$ must be in a form that is practical to evaluate. Thus, our first goal in this section is to find a convenient closed-form expression for $\mu(s)$.

The second useful set of results in Section I-2.7 was the approximate error expressions in (I-2.480) and (I-2.483),

$$P_F \simeq \frac{1}{\sqrt{2\pi s^2\ddot{\mu}(s)}}\,e^{\mu(s)-s\dot{\mu}(s)}, \qquad s \geq 0, \tag{129}$$

and

$$P_M \simeq \frac{1}{\sqrt{2\pi(1-s)^2\ddot{\mu}(s)}}\,e^{\mu(s)+(1-s)\dot{\mu}(s)}, \qquad s \leq 1. \tag{130}$$

† Our discussion assumes a thorough understanding of Section I-2.7, so that a review of that section may be appropriate at this point.

‡ Pr ($\varepsilon$) bounds were also developed. Because they are more appropriate to the general binary problem in the next section, we shall review them then.

As we pointed out on page I-124, the exponents in these expressions were identical with the Chernoff bounds in (127), but the multiplicative factor was significant in many applications of interest to us. In order to derive (129) and (130), we used a central limit theorem argument. For the problems considered in Section I-2.7 (e.g., Examples 2, 3, and 3A on pages I-127–I-132), it was easy to verify that the central limit theorem is applicable. However, for the case of interest in most of this chapter, the sum defining $l_R$ in (122) violates a necessary condition for the validity of the central limit theorem. Thus, we must use a new approach in order to find an approximate error expression. This is the second goal of this section.

In addition to these two topics, we develop an alternative expression for computing $\mu(s)$ and analyze a typical example in detail. Thus, there are four subsections:

2.2.1. Closed-form expressions for $\mu(s)$.
2.2.2. Approximate error expressions.
2.2.3. An alternative expression for $\mu(s)$.
2.2.4. Performance for a typical example.

### 2.2.1 Closed-form Expression for $\mu(s)$

We first evaluate $\mu_K(s)$ for finite $K$. Substituting (18) into (126) gives

$$\mu_K(s) = \ln \int_{-\infty}^{\infty} \cdots \int_{-\infty}^{\infty} \left\{ \left[ \prod_{i=1}^{K} \frac{1}{\sqrt{2\pi(N_0/2 + \lambda_i^s)}} \right] \exp\left( -\frac{1}{2} \sum_{i=1}^{K} \frac{(R_i - m_i)^2}{(\lambda_i^s + N_0/2)} \right) \right\}^s$$
$$\times \left\{ \left[ \prod_{i=1}^{K} \frac{1}{\sqrt{2\pi(N_0/2)}} \right] \exp\left( -\frac{1}{2} \sum_{i=1}^{K} \frac{R_i^2}{N_0/2} \right) \right\}^{1-s} dR_1 \cdots dR_K. \quad (131)$$

Performing the integration, we have

$$\mu_K(s) = \tfrac{1}{2} \sum_{i=1}^{K} \left[ (1 - s) \ln \left( 1 + \frac{2\lambda_i^s}{N_0} \right) - \ln \left( 1 + \frac{2(1 - s)\lambda_i^s}{N_0} \right) \right]$$
$$- \frac{s}{2} \sum_{i=1}^{K} \left( \frac{m_i^2}{N_0/2(1 - s) + \lambda_i^s} \right), \qquad 0 \le s \le 1. \quad (132)$$

From our discussion on page 13, we know the first sum on the right side of (132) is well behaved as $K \to \infty$. The convergence of the second sum follows easily.

$$\sum_{i=1}^{K} \frac{m_i^2}{(N_0/2(1 - s) + \lambda_i^s)} \le \sum_{i=1}^{K} \frac{m_i^2}{N_0/2(1 - s)} \le \frac{2(1 - s)}{N_0} \int_{T_i}^{T_f} m^2(t)\, dt. \quad (133)$$

We now take the limit of (132) as $K \to \infty$. The first sum is due to the randomness in $s(t)$, and so we denote it by $\mu_R(s)$. The second sum is due to the deterministic component in $r(t)$, and so we denote it by $\mu_D(s)$.

$$\mu_R(s) \triangleq \frac{1}{2} \sum_{i=1}^{\infty} \left[ (1-s) \ln \left(1 + \frac{2\lambda_i^s}{N_0}\right) - \ln \left(1 + \frac{2(1-s)\lambda_i^s}{N_0}\right) \right]. \tag{134}$$

$$\mu_D(s) \triangleq -\frac{s}{2} \sum_{i=1}^{\infty} \frac{m_i^2}{N_0/2(1-s) + \lambda_i^s}. \tag{135}$$

We now find a closed-form expression for the sums in (134) and (135). First we consider $\mu_R(s)$. Both of the sums in (134) are related to realizable linear filtering errors. To illustrate this, we consider the linear filtering problem in which

$$r(u) = s(u) + w(u), \qquad T_i \le u \le t, \tag{136}$$

where $s(u)$ is a zero-mean message process with covariance function $K_s(t, u)$ and the white noise has spectral height $N_0/2$. Using our results in Chapter I-6, we can find the linear filter whose output is the MMSE point estimate of $s(\cdot)$ and evaluate the resulting mean-square error. We denote this error as $\xi_P(t \mid s(\cdot), N_0/2)$. (The reason for the seemingly awkward notation will be apparent in a moment.) Using (72), we can write the mean-square error in terms of a sum of eigenvalues.

$$\sum_{i=1}^{\infty} \ln \left(1 + \frac{2\lambda_i^s}{N_0}\right) = \frac{2}{N_0} \int_{T_i}^{T_f} \xi_P\left(t \mid s(\cdot), \frac{N_0}{2}\right) dt. \tag{137}$$

Comparing (134) and (137) leads to the desired result.

$$\mu_R(s) = \frac{1-s}{N_0} \int_{T_i}^{T_f} dt \left[ \xi_P\left(t \mid s(\cdot), \frac{N_0}{2}\right) - \xi_P\left(t \mid s(\cdot), \frac{N_0}{2(1-s)}\right) \right]. \tag{138}$$

Thus, to find $\mu_R(s)$, we must find the mean-square error for two realizable linear filtering problems. In the first, the signal is $s(\cdot)$ and the noise is white with spectral height $N_0/2$. In the second, the signal is $s(\cdot)$ and the noise is white with spectral height $N_0/2(1 - s)$. An alternative expression for $\mu_R(s)$ also follows easily.

$$\mu_R(s) = \frac{1}{N_0} \int_{T_i}^{T_f} \left[ (1-s)\xi_P\left(t \mid s(\cdot), \frac{N_0}{2}\right) - \xi_P\left(t \mid \sqrt{1-s}\, s(\cdot), \frac{N_0}{2}\right) \right] dt. \tag{139}$$

Here the noise level is the same in both calculations, but the amplitude of the signal process is changed. These equations are the first key results in our performance analysis. Whenever we have a signal process such that we can calculate the realizable mean-square filtering error for the problem of estimating $s(t)$ in the presence of white noise, then we can find $\mu_R(s)$.

The next step is to find a convenient expression for $\mu_D(s)$. To evaluate the sum in (135), we recall the problem of detecting a *known* signal in colored noise, which we discussed in detail in Section I-4.3. The received waveforms on the two hypotheses are

$$\begin{aligned} r(t) &= m(t) + n_c(t) + w(t), \qquad T_i \le t \le T_f : H_1, \\ r(t) &= n_c(t) + w(t), \qquad T_i \le t \le T_f : H_0. \end{aligned} \tag{140}$$

By choosing the covariance function of $n_c(t)$ and $w(t)$ appropriately, we can obtain the desired interpretation. Specifically, we let

$$E[n_c(t)n_c(u)] = K_s(t, u), \qquad T_i \le t, u \le T_f \tag{141}$$

and

$$E[w(t)w(u)] = \frac{N_0}{2(1-s)}\,\delta(t-u), \qquad T_i \le t,\ u \le T_f. \tag{142}$$

Then, from Chapter I-4 (page I-296), we know that the optimum receiver correlates $r(t)$ with a function $g(t \mid N_0/2(1-s))$, which satisfies the equation†

$$m(t) = \int_0^T \left[K_s(t, u) + \frac{N_0}{2(1-s)}\,\delta(t-u)\right] g\left(u \,\middle|\, \frac{N_0}{2(1-s)}\right) du. \tag{143}$$

We also recall that we can write $g(t \mid \cdot)$ explicitly in terms of the eigenfunctions and eigenvalues of $K_s(t, u)$. Writing

$$g\left(u \,\middle|\, \frac{N_0}{2(1-s)}\right) = \sum_{i=1}^{\infty} g_i \phi_i(u), \qquad T_i \le u \le T_f, \tag{144}$$

substituting into (143), and solving for the $g_i$ gives

$$g_i = \frac{m_i}{\lambda_i^s + N_0/2(1-s)}, \tag{145}$$

where

$$m_i \triangleq \int_{T_i}^{T_f} m(t)\phi_i(t)\,dt. \tag{146}$$

† This notation is used to emphasize that $g(t \mid \cdot)$ depends on both $N_0$ and $s$, in addition to $K_s(t, u)$.

Substituting (145) and (146) into (135) and using Parseval's theorem, we have

$$\mu_D(s) = -\frac{s}{2}\int_{T_i}^{T_f} m(t)g\left(t \,\Big|\, \frac{N_0}{2(1-s)}\right) dt. \tag{147}$$

We observe that the integral in (147) is just $d^2$ for the known signal in colored noise problem described in (140) [see (I-4.198)]. We shall encounter several equivalent expressions for $\mu_D(s)$ later.

We denote the limit of the right side of (132) as $K \to \infty$ as $\mu(s)$. Thus,

$$\mu(s) = \mu_R(s) + \mu_D(s). \tag{148}$$

Using (138) and (147) in (148) gives a closed-form expression for $\mu(s)$. This enables us to evaluate the Chernoff bounds in (127) when $K \to \infty$. In the next section we develop approximate error expressions similar to those in (129) and (130).

### 2.2.2 Approximate Error Expressions

In order to derive an approximate error expression, we return to our derivation in Section I-2.7 (page I-123). After tilting the density and standardizing the tilted variable, we have the expression for $P_F$ given in (I-2.477). The result is

$$P_F = e^{\mu(s) - s\dot{\mu}(s)} \int_0^\infty e^{-s\sqrt{\ddot{\mu}(s)}\,Y} p_y(Y)\, dY, \tag{149}$$

where $Y$ is a zero-mean, unit-variance, random variable and we assume that $\dot{\mu}(s)$ equals $\gamma$. Recall that

$$y = \frac{x_s - \dot{\mu}(s)}{\sqrt{\ddot{\mu}(s)}}, \tag{150}$$

where

$$p_{x_s}(X) = e^{sX - \mu(s)} p_{l|H_0}(X \mid H_0), \tag{151}$$

and $l$ is the log likelihood ratio which can be written as

$$l = l_R + l_D + l_B^{[1]} + l_B^{[2]}. \tag{152}$$

[Notice that the threshold is $\gamma$ as defined in (35).] The quantity $l$ is also the limit of the sum in (19) as $K \to \infty$. If the weighted variables in the first sum in (19) were identically distributed, then, as $K \to \infty$, $p_y(Y)$ would approach a Gaussian density. An example of a case of this type was given in Example 2 on page I-127. In that problem,

$$\lambda_i^s = \sigma_s^2, \qquad i = 1, 2, \ldots, N, \tag{153}$$

so that the weighting in the first term of (19) was uniform and the variables were identically distributed. In the model of this chapter, we assume that

$s(t)$ has finite average power [see the sentence below (4)]. Thus $\sum_{i=1}^{\infty} \lambda_{i_s}$ is finite. Whenever the sum of the variances of the component random variables is finite, the central limit theorem cannot hold (see [10]). This means that we must use some other argument to get an approximate expression for $P_F$ and $P_M$.

A logical approach is to expand $p_y(Y)$ in an Edgeworth series. The first term in the expansion is a Gaussian density. The remaining terms take into account the non-Gaussian nature of the density. On the next few pages we carry out the details of the analysis. The major results are approximations to $P_F$ and $P_M$,

$$P_F \simeq \frac{1}{\sqrt{2\pi s^2 \ddot{\mu}(s)}} e^{\mu(s) - s\dot{\mu}(s)}, \qquad 0 \leq s \leq 1 \tag{154}$$

and

$$P_M \simeq \frac{1}{\sqrt{2\pi(1-s)^2 \ddot{\mu}(s)}} e^{\mu(s) + (1-s)\dot{\mu}(s)}, \qquad 0 \leq s \leq 1. \tag{155}$$

We see that (154) and (155) are identical with (129) and (130). Thus, our derivation leads us to the same result as before. The important difference is that we get to (154) and (155) without using the central limit theorem.

***Derivation of Error Approximations*†** The first term in the Edgeworth series is the Gaussian density,

$$\phi(Y) \triangleq \frac{1}{\sqrt{2\pi}} e^{-Y^2/2}. \tag{156}$$

The construction of the remaining terms in the series and the ordering of terms are discussed in detail on pages 221–231 of Cramèr [12]). The basic functions are

$$\phi^{(k)}(Y) \triangleq \frac{d^k}{dY^k}\left[\frac{1}{\sqrt{2\pi}} e^{-Y^2/2}\right]. \tag{157}$$

We write

$$\begin{aligned}
p_y(Y) = \phi(Y) &- \left[\frac{\gamma_3}{6}\phi^{(3)}(Y)\right] \\
&+ \left[\frac{\gamma_4}{4!}\phi^{(4)}(Y) + \frac{10\gamma_3^{\,2}}{6!}\phi^{(6)}(Y)\right] \\
&- \left[\frac{\gamma_5}{5!}\phi^{(5)}(Y) + \frac{35\gamma_3\gamma_4}{7!}\phi^{(7)}(Y) + \frac{280\gamma_3^{\,3}}{9!}\phi^{(9)}(Y)\right] \\
&+ \left[\frac{\gamma_6}{720}\phi^{(6)}(Y) + \left(\frac{\gamma_4^{\,2}}{1152} + \frac{\gamma_3\gamma_5}{720}\right)\phi^{(8)}(Y)\right. \\
&\left. + \frac{\gamma_3^{\,2}\gamma_4}{1728}\phi^{(10)}(Y) + \frac{\gamma_3^{\,4}}{31104}\phi^{(12)}(Y)\right] + \cdots,
\end{aligned} \tag{158}$$

† This derivation was done originally in [11].

where

$$\gamma_n \triangleq \frac{d^n/ds^n[\mu(s)]}{[\ddot{\mu}(s)]^{n/2}}, \qquad n = 3, 4, \ldots. \tag{159}$$

We see that all of the coefficients can be expressed in terms of $\mu(s)$ and its derivatives. We now substitute (158) into the integral in (149). The result is a sum of integrals of the form

$$I_k(\alpha) = \int_0^\infty \phi^{(k)}(Y)e^{-\alpha Y}\,dY, \tag{160}$$

where

$$\alpha \triangleq s(\ddot{\mu}(s))^{1/2}. \tag{161}$$

Repeated integration by parts gives an expression for $I_k(\alpha)$ in terms of erfc*($\alpha$). The integrals are

$$I_0(\alpha) = \text{erfc}_*\,(\alpha)\,e^{\alpha^2/2} \tag{162}$$

and

$$I_k(\alpha) = \alpha I_{k-1}(\alpha) - \phi^{(k-1)}(0), \qquad k \geq 1. \tag{163}$$

If we use just the first term in the series,

$$\boxed{P_F \simeq P_F^{[1]} \triangleq \exp\left(\mu(s) - s\dot{\mu}(s) + \frac{s^2\ddot{\mu}(s)}{2}\right)\text{erfc}_*\left(s\sqrt{\ddot{\mu}(s)}\right)} \tag{164}$$

For large $s(\ddot{\mu}(s)^{1/2}(\geq 2)$, we may use the approximation to erfc$_*$ $(X)$ given in Fig. 2.10 of Part I.

$$\text{erfc}_*\,(X) \simeq \frac{1}{\sqrt{2\pi}\,X}\,e^{-X^2/2}, \qquad X \geq 2. \tag{165}$$

Then (164) reduces to

$$\boxed{P_F \simeq P_{F_*}^{[1]} \triangleq \frac{1}{\sqrt{2\pi s^2\ddot{\mu}(s)}}\,e^{\mu(s)-s\dot{\mu}(s)}.} \tag{166}$$

This, of course, is the same answer we obtained when the central limit theorem was valid. The second term in the approximation is obtained by using $I_3(\alpha)$ from (163).

$$P_F^{[2]} = -\frac{\gamma_3}{6}\,e^{\mu(s)-s\dot{\mu}(s)}\left[(s\sqrt{\ddot{\mu}(s)})^3 I_0(s\sqrt{\ddot{\mu}(s)}) + \frac{1}{\sqrt{2\pi}}\,(1 - s^2\ddot{\mu}(s))\right]. \tag{167}$$

Now,

$$I_0(s\sqrt{\ddot{\mu}(s)}) = \text{erfc}_*(s\sqrt{\ddot{\mu}(s)})\exp\left(\frac{s^2\ddot{\mu}(s)}{2}\right). \tag{168}$$

In Problem I-2.2.15 on page I-137, we showed that

$$\frac{1}{\sqrt{2\pi}\,X}\left(1 - \frac{1}{X^2}\right)e^{-X^2/2} < \text{erfc}_*\,(X) < \frac{1}{\sqrt{2\pi}\,X}\left(1 - \frac{1}{X^2} + \frac{3}{X^4}\right)e^{-X^2/2}. \tag{169}$$

We can now place an upper bound on the magnitude of $P_F^{[2]}$.

$$|P_F^{[2]}| \leq \left|\frac{\gamma_3}{6} e^{\mu(s)-s\dot{\mu}(s)}\left[(s\sqrt{\ddot{\mu}(s)})^3 \frac{1}{\sqrt{2\pi}\, s\sqrt{\ddot{\mu}(s)}}\left(1 - \frac{1}{s^2\ddot{\mu}(s)} + \frac{3}{s^4(\ddot{\mu}(s))^2}\right) + \frac{1}{\sqrt{2\pi}}(1 - s^2\ddot{\mu}(s)\right]\right|$$

$$= \left|\frac{\gamma_3}{2} \frac{1}{\sqrt{2\pi}\, s^2\ddot{\mu}(s)} e^{\mu(s)-s\dot{\mu}(s)}\right|$$

$$= \left|\frac{\gamma_3}{2} \frac{1}{s\sqrt{\ddot{\mu}(s)}}\right| P_{F*}^{[1]}. \tag{170}$$

Using (159),

$$|P_F^{[2]}| \leq \left|\frac{\mu^{(3)}(s)}{2s[\ddot{\mu}(s)]^2}\right| P_{F*}^{[1]}. \tag{171}$$

Thus, for any particular $\mu(s)$, we can calculate a bound on the size of the second term in relation to a bound on the first term. By using more terms in the series in (169), we can obtain bounds on the other terms in (158). Notice that this is not a bound on the percentage error in $P_F$; it is just a bound on the magnitude of the successive terms. In most of our calculations we shall use just the first-order term $P_F^{[1]}$. We calculated $P_F^{[2]}$ for a number of examples, and it was usually small compared to $P_F^{[1]}$. The bound on $P_F^{[2]}$ is computed for several typical systems in the problems.

To derive an approximate expression for $P_M$, we go through a similar argument. The starting point is (172), which is obtained from (I-2.465) by a change of variables.

$$P_M = e^{\mu(s)+(1-s)\dot{\mu}(s)} \int_{-\infty}^{0} e^{(1-s)\sqrt{\ddot{\mu}(s)}\,Y} p_y(Y)\, dY. \tag{172}$$

The first-term approximation is

$$P_M \simeq P_M^{[1]} = \left[\exp\left[\mu(s) + (1-s)\dot{\mu}(s) + \frac{(1-s)^2}{2}\ddot{\mu}(s)\right]\right] \text{erfc}_*\,[(1-s)\sqrt{\ddot{\mu}(s)}], \quad 0 \leq s \leq 1. \tag{173}$$

Using the approximation in (165) gives

$$P_M \simeq P_{M*}^{[1]} \triangleq \frac{1}{\sqrt{2\pi(1-s)^2\ddot{\mu}(s)}} e^{\mu(s)+(1-s)\dot{\mu}(s)}, \qquad 0 \leq s \leq 1. \tag{174}$$

The higher-order terms are derived exactly as in the $P_F$ case.

The results in (164), (166), (173), and (174), coupled with the closed-form expression for $\mu(s)$ in (138) and (147), give us the ability to calculate the approximate performance of the optimum test in an efficient manner. A disadvantage of our approach is that for the general case we cannot bound the error in our approximation. Later, we shall obtain bounds for some special cases and shall see that our first-order approximation is accurate in those cases.

We now return to the problem of calculating $\mu(s)$ and develop an alternative procedure.

### 2.2.3 An Alternative Expression for $\mu_R(s)$†

The expressions in (138) and (139) depend on the realizable mean-square estimation error. If we are going to build the optimum receiver using a state-variable realization, we will have $\xi_P(t \mid s(\cdot), N_0/2)$ available. On the other hand, there are many cases in which we want to compute the performance for a number of systems in order to select one to build. In this case we want an expression for $\mu_R(s)$ that requires the *least* amount of computation. Specifically, we would like to find an expression for $\mu(s)$ that does not require the computation of $\xi_P(t \mid s(\cdot), N_0/2)$ at *each* point in $[T_i, T_f]$. Whenever the random process has a finite-dimensional state representation, we can find a much simpler expression for $\mu(s)$. The new expression is based on an alternative computation of the integral‡

$$\frac{2}{N_0}\int_{T_i}^{T_f} \xi_P\left(t \,\middle|\, s(\cdot), \frac{N_0}{2}\right) dt. \tag{175}$$

**Derivation.** We use the state model in (76)–(80),

$$\dot{\mathbf{x}}(t) = \mathbf{F}(t)\mathbf{x}(t) + \mathbf{G}(t)u(t), \tag{176}$$

$$s(t) = \mathbf{C}(t)\mathbf{x}(t), \tag{177}$$

and the initial conditions

$$E[\mathbf{x}(T_i)] = \mathbf{0}, \tag{178}$$

$$E[\mathbf{x}(T_i)\mathbf{x}^T(T_i)] = \boldsymbol{\xi}_P(T_i) \triangleq \mathbf{P}_0. \tag{179}$$

Recall that the error covariance matrix is

$$\boldsymbol{\xi}_P(t) = E[(\mathbf{x}(t) - \hat{\mathbf{x}}(t))(\mathbf{x}^T(t) - \hat{\mathbf{x}}^T(t))]. \tag{180}$$

Using (177),

$$\xi_P\left(t \,\middle|\, s(\cdot), \frac{N_0}{2}\right) = \mathbf{C}(t)\boldsymbol{\xi}_P(t)\mathbf{C}^T(t). \tag{181}$$

We first recall several results from Chapter I-6 and introduce some simplifying notation. From Property 16 on page I-545, we know that the variance equation (84) can be related to two simultaneous linear equations (I-6.335 or I-6.336),

$$\frac{d}{dt}\begin{bmatrix} \mathbf{v}_1(t) \\ \mathbf{v}_2(t) \end{bmatrix} = \left[\begin{array}{c:c} \mathbf{F}(t) & \mathbf{G}(t)\mathbf{Q}\mathbf{G}^T(t) \\ \hdashline \mathbf{C}^T(t)\dfrac{2}{N_0}\mathbf{C}(t) & -\mathbf{F}^T(t) \end{array}\right]\begin{bmatrix} \mathbf{v}_1(t) \\ \mathbf{v}_2(t) \end{bmatrix}. \tag{182}$$

† This section may be omitted on the first reading.
‡ This derivation is due to Collins [13].

The transition matrix of (182), $\mathbf{T}(t, T_i)$, satisfies the differential equation

$$\frac{d}{dt}[\mathbf{T}(t, T_i)] = \left[\begin{array}{c|c} \mathbf{F}(t) & \mathbf{G}(t)\mathbf{Q}\mathbf{G}^T(t) \\ \hline \mathbf{C}^T(t)\frac{2}{N_0}\mathbf{C}(t) & -\mathbf{F}^T(t) \end{array}\right] \mathbf{T}(t, T_1), \tag{183}$$

with initial conditions $\mathbf{T}(T_i, T_i) = \mathbf{I}$. In addition, from (I-6.338), the error covariance matrix is given by

$$\boldsymbol{\xi}_P(t) = [\mathbf{T}_{11}(t, T_i)\boldsymbol{\xi}_P(T_i) + \mathbf{T}_{12}(t, T_i)][\mathbf{T}_{21}(t, T_i)\boldsymbol{\xi}_P(T_i) + \mathbf{T}_{22}(t, T_i)]^{-1}. \tag{184}$$

The inverse of the second matrix always exists because it is the transition matrix of a linear dynamical system. For simplicity, we define two new matrices,

$$\begin{aligned} \boldsymbol{\Gamma}_1(t) &= \mathbf{T}_{11}(t, T_i)\boldsymbol{\xi}_P(T_i) + \mathbf{T}_{12}(t, T_i), \\ \boldsymbol{\Gamma}_2(t) &= \mathbf{T}_{21}(t, T_i)\boldsymbol{\xi}_P(T_i) + \mathbf{T}_{22}(t, T_i). \end{aligned} \tag{185}$$

Thus,

$$\boldsymbol{\xi}_P(t) = \boldsymbol{\Gamma}_1(t)\boldsymbol{\Gamma}_2^{-1}(t), \tag{186}$$

and $\boldsymbol{\Gamma}_1(t)$ and $\boldsymbol{\Gamma}_2(t)$ satisfy

$$\frac{d}{dt}\begin{bmatrix} \boldsymbol{\Gamma}_1(t) \\ \boldsymbol{\Gamma}_2(t) \end{bmatrix} = \left[\begin{array}{c|c} \mathbf{F}(t) & \mathbf{G}(t)\mathbf{Q}\mathbf{G}^T(t) \\ \hline \mathbf{C}^T(t)\frac{2}{N_0}\mathbf{C}(t) & -\mathbf{F}^T(t) \end{array}\right] \begin{bmatrix} \boldsymbol{\Gamma}_1(t) \\ \boldsymbol{\Gamma}_2(t) \end{bmatrix}, \tag{187}$$

with initial conditions

$$\boldsymbol{\Gamma}_1(T_i) = \xi_P(T_i) \triangleq \mathbf{P}_0 \tag{188}$$

and

$$\boldsymbol{\Gamma}_2(T_i) = \mathbf{I}. \tag{189}$$

We now proceed with the derivation. Multiplying both sides of (181) by $2/N_0$ and integrating gives

$$\begin{aligned} \frac{2}{N_0}\int_{T_i}^{T_f} \xi_P\left(t \,\middle|\, s(\cdot), \frac{N_0}{2}\right) dt &= \frac{2}{N_0}\int_{T_i}^{T_f} \mathbf{C}(t)\boldsymbol{\xi}_P(t)\mathbf{C}^T(t)\, dt \\ &= \frac{2}{N_0}\int_{T_i}^{T_f} \mathbf{C}(t)[\boldsymbol{\Gamma}_1(t)\boldsymbol{\Gamma}_2^{-1}(t)]\mathbf{C}^T(t)\, dt. \end{aligned} \tag{190}$$

Now recall that

$$\mathbf{x}^T\mathbf{B}\mathbf{x} = \operatorname{Tr}[\mathbf{x}\mathbf{x}^T\mathbf{B}] \tag{191}$$

for any vector $\mathbf{x}$. Thus,

$$\frac{2}{N_0}\int_{T_i}^{T_f} \xi_P\left(t \,\middle|\, s(\cdot), \frac{N_0}{2}\right) dt = \int_{T_i}^{T_f} \operatorname{Tr}\left[\left(\mathbf{C}^T(t)\frac{2}{N_0}\mathbf{C}(t)\boldsymbol{\Gamma}_1(t)\boldsymbol{\Gamma}_2^{-1}(t)\right)\right] dt \tag{192}$$

Using (187) to eliminate $\mathbf{\Gamma}_1(t)$, we have

$$\frac{2}{N_0}\int_{T_i}^{T_f}\xi_P\left(t \mid s(\cdot), \frac{N_0}{2}\right)dt = \int_{T_i}^{T_f}\mathrm{Tr}\left[\left(\frac{d\mathbf{\Gamma}_2(t)}{dt} + \mathbf{F}^T(t)\mathbf{\Gamma}_2(t)\right)\mathbf{\Gamma}_2^{-1}(t)\right]dt$$
$$= \int_{T_i}^{T_f}\mathrm{Tr}\left[\frac{d\mathbf{\Gamma}_2(t)}{dt}\mathbf{\Gamma}_2^{-1}(t)\right]dt + \int_{T_i}^{T_f}\mathrm{Tr}\,[\mathbf{F}^T(t)]\,dt$$
$$= \int_{T_i}^{T_f}\mathrm{Tr}\,[\mathbf{\Gamma}_2^{-1}(t)\,d\mathbf{\Gamma}_2(t)] + \int_{T_i}^{T_f}\mathrm{Tr}\,[\mathbf{F}(t)]\,dt. \tag{193}$$

From (9.31) of [14],

$$\int_{T_i}^{T_f}\mathrm{Tr}\,[\mathbf{\Gamma}_2^{-1}(t)\,d\mathbf{\Gamma}_2(t)] = \int_{T_i}^{T_f} d[\ln\det\mathbf{\Gamma}_2(t)]$$
$$= \ln\det\mathbf{\Gamma}_2(T_f) - \ln\det\mathbf{\Gamma}_2(T_i)$$
$$= \ln\det\mathbf{\Gamma}_2(T_f). \tag{194}$$

Thus,

$$\boxed{\frac{2}{N_0}\int_{T_i}^{T_f}\xi_P\left(t \mid s(\cdot), \frac{N_0}{2}\right)dt = \ln\det\mathbf{\Gamma}_2(T_f) + \int_{T_i}^{T_f}\mathrm{Tr}\,[\mathbf{F}(t)]\,dt,} \tag{195}$$

which is the desired result.†

We see that we have to compute $\mathbf{\Gamma}_2(T_f)$ at only one point rather than over an entire interval. This is particularly important when an analytic expression for $\mathbf{\Gamma}_2(T_f)$ is available. If we have to find $\mathbf{\Gamma}_2(T_f)$ by numerically integrating (187), there is no significant saving in computation.

The expression in (195) is the desired result. In the next section we consider a simple example to illustrate the application of the result we have derived.

### 2.2.4 Performance for a Typical System

In this section we analyze the performance of the system described in the example of Section 2.1.5. It provides an immediate application of the performance results we have just developed. In Chapter 4, we shall consider the performance for a variety of problems.

**Example.** We consider the system described in the example on page 26. We assume that the channel process $b(t)$ is a stationary zero-mean Gaussian process with a spectrum

$$S_b(\omega) = \frac{2k\sigma_b^2}{\omega^2 + k^2}. \tag{196}$$

† This result was first obtained by Baggeroer as a by-product of his integral equation work [15]. See Siegert [16] for a related result.

We assume that the transmitted signal is a rectangular pulse,

$$f(t) = \begin{cases} \sqrt{\dfrac{E_t}{T}}, & 0 \le t \le T, \\ 0, & \text{elsewhere} \end{cases} \tag{197}$$

As we pointed out in our earlier discussion, this channel model has many of the characteristics of models of actual channels that we shall study in detail in Chapter 10. The optimum receiver is shown in Fig. 2.10. To illustrate the techniques involved, we calculate $\mu(s)$ using both (138) and (195). [Notice that $\mu_D(s)$ is zero.] To use (138), we need the realizable mean-square filtering error. The result for this particular spectrum was derived in Example 1 on pages I-546–I-548. From (I-6.353),

$$\xi_P\left(t \,\middle|\, s(\cdot), \frac{N_0}{2}\right) = \frac{2\bar{E}_r}{T} \frac{1}{(1+\alpha)} \left\{\frac{1 - [(1-\alpha)/(1+\alpha)]e^{-2k\alpha t}}{1 - [(1-\alpha)^2/(1+\alpha)^2]e^{-2k\alpha t}}\right\}, \qquad 0 \le t \le T, \tag{198}$$

where

$$\bar{E}_r \triangleq \sigma_b^2 E_t \tag{199}$$

is the average received energy and

$$\alpha \triangleq \sqrt{1 + \frac{4\bar{E}_r}{kTN_0}}. \tag{200}$$

Integrating, we obtain

$$\int_{T_i}^{T_f} \xi_P\left(t \,\middle|\, s(\cdot), \frac{N_0}{2}\right) dt = \frac{N_0}{2}\left\{\ln\left[\frac{(1+\alpha)^2 e^{2k\alpha T} - (1-\alpha)^2}{4\alpha}\right] - (\alpha + 1)kT\right\}. \tag{201}$$

We now derive (201) using the expression in (195). The necessary quantities are

$$\begin{aligned} \mathbf{F}(t) &= -k, \\ \mathbf{G}(t)\mathbf{Q}\mathbf{G}^T(t) &= 2k\sigma_b^2, \\ \mathbf{C}(t) &= 1, \\ \mathbf{P}_0 &= \sigma_b^2. \end{aligned} \tag{202}$$

The transition matrix is given in (I-6.351) as

$$\mathbf{T}(T + T_i, T_i) = \left[\begin{array}{c:c} \cosh(\gamma T) - \dfrac{k}{\gamma}\sinh(\gamma T) & \dfrac{2k\sigma_b^2}{\gamma}\sinh(\gamma T) \\ \hdashline \dfrac{2}{N_0\gamma}\sinh(\gamma T) & \cosh(\gamma T) + \dfrac{k}{\gamma}\sinh(\gamma T) \end{array}\right], \tag{203}$$

where

$$\gamma = k\sqrt{1 + \frac{4\sigma_b^2 E_t}{kN_0}} = k\alpha \tag{204}$$

From the definition in (185),

$$\Gamma_2(T_f) = \frac{2\sigma_b^2}{N_0\gamma} \sinh(\gamma T) + \cosh(\gamma T) + \frac{k}{\gamma} \sinh(\gamma T)$$
$$= \cosh(\gamma T) + \frac{k}{\gamma}\left(1 + \frac{2\sigma_b^2}{kN_0}\right) \sinh(\gamma T) \tag{205}$$
$$= e^{-k\alpha T}\left[\frac{1 - [(\alpha+1)^2/(\alpha-1)^2]\, e^{2k\alpha T}}{1 - (\alpha+1)^2/(\alpha-1)^2}\right].$$

Using (202) and (205) in (195), we have

$$\frac{2}{N_0}\int_{T_i}^{T_f} \xi_P\left(t \mid s(\cdot), \frac{N_0}{2}\, dt\right) = \ln\left[\frac{(\alpha-1)^2 - (\alpha+1)^2 e^{2k\alpha T}}{(\alpha-1)^2 - (\alpha+1)^2}\right] - k(\alpha+1)T, \tag{206}$$

which is identical with (201). To get the second term in (138), we define

$$\alpha_s \triangleq \sqrt{1 + \frac{4\bar{E}_r(1-s)}{kTN_0}} \tag{207}$$

and replace $\alpha$ by $\alpha_s$ in (201). Then

$$\mu(s) = \frac{1-s}{2}\left\{\ln\left[\frac{[(1+\alpha)^2 e^{2kT\alpha} - (1-\alpha)^2]\alpha_s}{[(1+\alpha_s)^2 e^{2kT\alpha_s} - (1-\alpha_s)^2]\alpha}\right] - \frac{4\bar{E}_r}{N_0}\left[\frac{1}{\alpha-1} - \frac{1}{\alpha_s-1}\right]\right\}. \tag{208}$$

We see that $\mu(s)$ (and therefore the error expression) is a function of two quantities. $\bar{E}_r/N_0$, the average energy divided by the noise spectral height and the $kT$ product, The 3-db-bandwidth of the spectrum is $k$ radians per second, so that $kT$ is a time-bandwidth product.

To use the approximate error expressions in (154) and (155), we find $\dot{\mu}(s)$ and $\ddot{\mu}(s)$ from (208). The simplest way to display the results is to fix $P_F$ and plot $P_M$ versus $kT$ for various values of $2\bar{E}_r/N_0$. We shall not carry out this calculation at this point. In Example 1 of Chapter 4, we study this problem again from a different viewpoint. At that time we plot a detailed set of performance curves (see Figs. 4.7–4.9 and Problem 4.1.21).

This example illustrates the application of our results to a typical problem of interest. Other interesting cases are developed in the problems. We now summarize the results of the Chapter.

## 2.3 SUMMARY: SIMPLE BINARY DETECTION

In Sections 2.1 and 2.2 we considered in detail the problem of detecting a sample function of a Gaussian random process in the presence of additive white Gaussian noise. In Section 2.1 we derived the likelihood ratio test and discussed various receiver configurations that could be used to implement the test. The test is

$$l_R + l_D + l_B^{[1]} + l_B^{[2]} \underset{H_0}{\overset{H_1}{\gtrless}} \ln \eta, \tag{209}$$

where

$$l_R = \frac{1}{N_0} \iint\limits_{T_i}^{T_f} r(t) h_1(t, u) r(u)\, dt\, du, \tag{210}$$

$$l_D = \int_{T_i}^{T_f} g_1(u) r(u)\, du, \tag{211}$$

$$l_B^{[1]} = -\frac{1}{N_0} \int_{T_i}^{T_f} \xi_{P_s}(t)\, dt, \tag{212}$$

$$l_B^{[2]} = -\tfrac{1}{2} \int_{T_i}^{T_f} g_1(u) m(u)\, du. \tag{213}$$

The operation needed to generate $l_R$ was a quadratic operation. The receiver structures illustrated different schemes for computing $l_R$. The three receivers of most importance in practice are the following:

1. The estimator-correlator receiver (Canonical Realization No. 1).
2. The filter-squarer receiver (Canonical Realization No. 3).
3. The optimum realizable filter receiver (Canonical Realizations Nos. 4 and 4S).

The most practical realization will depend on the particular problem of interest.

In Section 2.2 we considered the performance of the optimum receiver. In general, it was not possible to find the probability density of $l_R$ on the two hypotheses. By extending the techniques of Chapter I-2, we were able to find good approximations to the error probabilities. The key function in this analysis was $\mu(s)$.

$$\mu(s) = \mu_R(s) + \mu_D(s), \tag{214}$$

where

$$\mu_R(s) = \frac{1-s}{N_0} \int_{T_i}^{T_f} dt \left[ \xi_P\left(t \,\middle|\, s(\cdot), \frac{N_0}{2}\right) - \xi_P\left(t \,\middle|\, s(\cdot), \frac{N_0}{2(1-s)}\right)\right] \tag{215}$$

and

$$\mu_D(s) = -\frac{s}{2} \int_{T_i}^{T_f} m(t) g\left(t \,\middle|\, \frac{N_0}{2(1-s)}\right) dt. \tag{216}$$

The performance was related to $\mu(s)$ through the Chernoff bounds,

$$P_F \le e^{\mu(s) - s\dot{\mu}(s)},$$
$$P_M \le e^{\mu(s) + (1-s)\dot{\mu}(s)}, \qquad 0 \le s \le 1, \tag{217}$$

where

$$\dot{\mu}(s) = \gamma = \ln \eta. \tag{218}$$

An approximation to the performance was obtained by an Edgeworth series expansion,

$$P_F \simeq \frac{1}{\sqrt{2\pi s^2 \ddot{\mu}(s)}} e^{\mu(s) - s\dot{\mu}(s)}, \tag{219}$$

$$P_M \simeq \frac{1}{\sqrt{2\pi(1-s)^2 \ddot{\mu}(s)}} e^{\mu(s) + (1-s)\dot{\mu}(s)}, \qquad 0 \le s \le 1. \tag{220}$$

By varying $s$, we could obtain a segment of an approximate receiver operating characteristic.

We see that both the receiver structure and performance are closely related to the optimum linear filtering results of Chapter I-6. This close connection is important because it means that all of our detailed studies of optimum linear filters are useful for the Gaussian detection problem.

At this point, we have developed a set of important results but have not yet applied them to specific physical problems. We continue this development in Chapter 4, where we consider three important classes of physical problems and obtain specific results for a number of interesting examples. Many readers will find it helpful to study Section 4.1.1 before reading Chapter 3 in detail.

## 2.4 PROBLEMS

### P.2.1 Optimum Receivers

**Problem 2.1.1.** Consider the model described by (1)–(6). Assume that $m(t)$ is not zero. Derive an estimator-correlator receiver analogous to that in Fig. 2.3 for this case.

**Problem 2.1.2** Consider the function $h_1(t, t \mid z)$, which is specified by the equation

$$zh_1(t, u \mid z) + \int_{T_i}^{T_f} h_1(t, y \mid z) K_s(y, u)\, dy = K_s(t, u), \qquad T_i \le t, u \le T_f.$$

Verify that (75) is true. [*Hint:* Recall (I-3.154).]

**Problem 2.1.3.**

1. Consider the waveform

$$r(\tau) = n_c(\tau) + w(\tau), \qquad T_i \le \tau \le t,$$

where $n_c(\tau)$ can be generated as the output of a dynamic system,

$$\dot{\mathbf{x}}(t) = \mathbf{F}(t)\mathbf{x}(t) + \mathbf{G}(t)u(t),$$

$$n_c(t) = \mathbf{C}(t)\mathbf{x}(t),$$

driven by a statistically independent white noise $u(t)$. Denote the MMSE realizable estimate of $n_c(\tau)$ as $\hat{n}_c(\tau)$. Prove that the process

$$r_*(t) \triangleq r(t) - \hat{n}_c(t) = r(t) - \mathbf{C}(t)\hat{\mathbf{x}}(t)$$

is white.

2. Use the result of part 1 to derive the receiver in Fig. 2.11 by inspection.

**Problem 2.1.4.** Read Problem I-6.6.4 and the Appendix to Part II (sect. A.4–A.6). With this background derive a procedure for generating $l_R$ using unrealizable filters expressed in terms of vector-differential equations. For simplicity, assume zero means.

**Problem 2.1.5.** The received waveforms on the two hypotheses are

$$r(t) = s(t) + w(t), \qquad 0 \le t \le T: H_1,$$

$$r(t) = w(t), \qquad 0 \le t \le T: H_0.$$

The process $w(t)$ is a sample function of a white Gaussian random process with spectral height $N_0/2$. The process $s(t)$ is a Wiener process that is statistically independent of $w(t)$.

$$s(0) = 0,$$

$$E[s^2(t)] = \sigma^2 t.$$

1. Find the likelihood ratio test.
2. Draw a realization of the optimum receiver. Specify all components completely.

**Problem 2.1.6.** The received waveforms on the two hypotheses are

$$r(t) = s(t) + w(t), \qquad 0 \le t \le T: H_1,$$

$$r(t) = w(t), \qquad 0 \le t \le T: H_0.$$

The process $w(t)$ is a sample function of a white Gaussian random process with spectral height $N_0/2$. The signal $s(t)$ is a sample function of a Gaussian random process and can be written as

$$s(t) = at, \qquad 0 \le t,$$

where $a$ is a zero-mean Gaussian random variable with variance $\sigma_a^2$. Find the optimum receiver. Specify all components completely.

**Problem 2.1.7.** Repeat Problem 2.1.6 for the case in which

$$s(t) = at + b, \qquad 0 \le t,$$

where $a$ and $b$ are statistically independent, zero-mean Gaussian random variables with variances $\sigma_a^2$ and $\sigma_b^2$, respectively.

**Problem 2.1.8.**

1. Repeat Problem 2.1.7 for the case in which $a$ and $b$ are statistically independent Gaussian random variables with means $m_a$ and $m_b$ and variances $\sigma_a^2$ and $\sigma_b^2$, respectively.
2. Consider four special cases of part 1:

(i) $m_a = 0$,

(ii) $m_b = 0$,

(iii) $\sigma_a^2 = 0$,

(iv) $\sigma_b^2 = 0$.

Verify that the receiver for each of these special cases reduces to the correct structure.

**Problem 2.1.9.** Consider the model in Problem 2.1.6. Assume that $s(t)$ is a piecewise constant waveform,

$$s(t) = \begin{cases} b_1, & 0 < t \le T_0, \\ b_2, & T_0 < t \le 2T_0, \\ b_3, & 2T_0 < t \le 3T_0, \\ \vdots & \\ b_n, & (n-1)T_0 < t \le nT_0, \end{cases}$$

The $b_i$ are statistically independent, zero-mean Gaussian random variables with variances equal to $\sigma_b^2$. Find the optimum receiver.

**Problem 2.1.10.** Consider the model in Problem 2.1.6. Assume

$$s(t) = \sum_{i=1}^{K} a_i t^i, \qquad 0 \le t,$$

where the $a_i$ are statistically independent random variables with variances $\sigma_i^2$. Find the optimum receiver.

**Problem 2.1.11.** Re-examine Problems 2.1.6 through 2.1.10. If you implemented the optimum receiver using Canonical Realization No. 4S, go back and find an easier procedure.

**Problem 2.1.12** Consider the model in Problem 2.1.5. Assume that $s(t)$ is a segment of a stationary zero-mean Gaussian process with an $n$th-order Butterworth spectrum

$$S_s(\omega: n) = \frac{2nP}{k} \frac{\sin(\pi/2n)}{(\omega/k)^{2n} + 1}, \qquad n = 1, 2, \ldots.$$

1. Review the state representation for these processes in Example 2 on page I-548 Make certain that you understand the choice of initial conditions.

2. Draw a block diagram of the optimum receiver.

**Problem 2.1.13.** From (31), we have

$$l_R = \frac{1}{N_0} \iint_{T_i}^{T_f} r(t) h(t, u) r(u)\, dt\, du.$$

One possible factoring of $h(t, u)$ was given in (45). An alternative factoring is

$$h(t, u) = \int_{\Omega_T} g_1(z, t) g_1(z, u)\, dz, \qquad T_i \le t, u \le T_f \tag{P.1}$$

where

$$[T_i, T_f] \subset \Omega_T.$$

1. Explain the physical significance of this operation. Remember that our model assumes that $r(t)$ is only observed over the interval $[T_i, T_f]$.

2. Give an example in which the factoring indicated in (P.1) is easier than that in the text.

**Problem 2.1.14.** Consider the expression for $l_R$ in (31). We want to decompose $h_1(t, u)$ in terms of two new functions, $k_1(T_f, z)$ and $k_2(z, t)$, that satisfy the equation

$$h_1(t, u) = \int_{T_i}^{T_f} k_1(T_f, z)k_2(z, t)k_2(z, u)\, dz, \qquad T_i \le t, u \le T_f.$$

1. Draw a block diagram of the optimum receiver in terms of these new functions.
2. Give an example in which this realization would be easier to find than Canonical Realization No. 3.
3. Discuss the decomposition

$$h_1(t, u) = \int_{\Omega_T} k_1(T_f, z)k_2(z, t)k_2(z, u)\, dz, \qquad T_i \le t, u \le T_f, \ [T_i, T_f] \subset \Omega_T.$$

**Problem 2.1.15.** From (86) and (87),

$$l_R = \frac{1}{N_0}\int_0^T [2r(t)\hat{s}(t) - \hat{s}^2(t)]\, dt. \tag{P.1}$$

Consider the case in which

$$\dot{\hat{s}}(t) + a_0\hat{s}(t) = b_0 r(t), \qquad 0 \le t$$

and

$$\hat{s}(0) = 0. \tag{P.2}$$

1. Implement the optimum receiver in the form shown in Fig. P.2.1. Specify the time-invariant filter completely.
2. Discuss the case in which

$$\ddot{\hat{s}}(t) + a_1\dot{\hat{s}}(t) + a_0\hat{s}(t) = b_0 r(t).$$

Suggest some possible modifications to the structure in Fig. P.2.1.

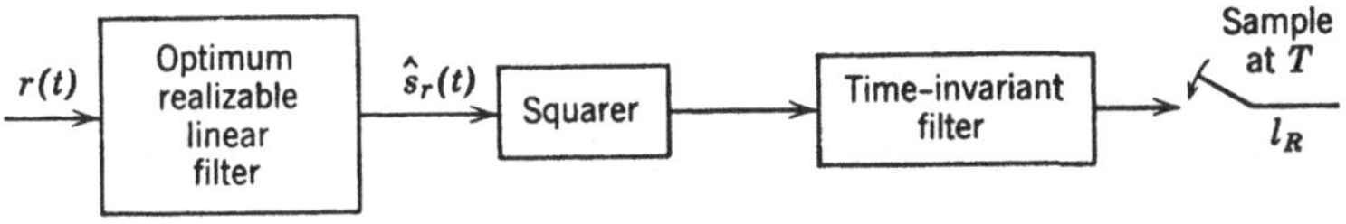


**Fig. P.2.1**

3. Extend your discussion to the general case in which the estimate $\hat{s}(t)$ is described by an $n$th-order differential equation with constant coefficients.

**Problem 2.1.16.** On both hypotheses there is a sample function of a zero-mean Gaussian white noise process with spectral height $N_0/2$. On $H_1$, the signal is equally likely to be a sample function from any one of $M$ zero-mean Gaussian processes. We denote the covariance function of the $i$th process as $K_{s_i}(t, u)$, $i = 1, \ldots, M$. Thus,

$$r(t) = s_i(t) + w(t), \quad T_i \le t \le T_f, \quad \text{with probability } \frac{1}{M} : H_1, \qquad i = 1, \ldots, M.$$

$$r(t) = w(t), \qquad T_i \le t \le T_f : H_0.$$

Find the optimum Bayes receiver to decide which hypothesis is true.

**Problem 2.1.17.** Consider the vector version of the simple binary detection problem. The received waveforms on the two hypotheses are

$$\begin{aligned} \mathbf{r}(t) &= \mathbf{s}(t) + \mathbf{w}(t), \qquad T_i \le t \le T_f : H_1, \\ &= \mathbf{w}(t), \qquad T_i \le t \le T_f : H_0, \end{aligned} \tag{P.1}$$

where $\mathbf{s}(t)$ and $\mathbf{w}(t)$ are sample functions of zero-mean, statistically independent, $N$-dimensional, vector Gaussian processes with covariance matrices

$$\mathbf{K}_s(t, u) \triangleq E[\mathbf{s}(t)\mathbf{s}^T(u)] \tag{P.2}$$

and

$$\mathbf{K}_w(t, u) \triangleq E[\mathbf{w}(t)\mathbf{w}^T(u)] = \frac{N_0}{2}\,\delta(t - u)\mathbf{I}. \tag{P.3}$$

1. Derive the optimum receiver for this problem. (*Hint:* Review Sections I-3.7 and I-4.5.)
2. Derive the equations specifying the four canonical realizations. Draw a block diagram of the four realizations.
3. Consider the special case in which

$$\mathbf{K}_s(t, u) = K_s(t, u)\mathbf{I}. \tag{P.4}$$

Explain what the condition in (P.4) means. Give a physical situation that would lead to this condition. Simplify the optimum receiver in part 1.

4. Consider the special case in which

$$\mathbf{K}_s(t, u) = K_s(t, u)\begin{bmatrix} 1 & 1 & \cdots & 1 \\ 1 & 1 & & 1 \\ \cdot & & \cdot & \cdot \\ \cdot & & \cdot & \cdot \\ \cdot & & \cdot & \cdot \\ 1 & & & 1 \end{bmatrix}. \tag{P.5}$$

Repeat part 3.

**Problem 2.1.18.** Consider the model in Problem 2.1.17. The covariance of $\mathbf{w}(t)$ is

$$\mathbf{K}_w(t, u) = \mathbf{N}\,\delta(t - u)\mathbf{I},$$

where $\mathbf{N}$ is a nonsingular matrix.

1. Repeat parts 1 and 2 of Problem 2.1.17. (*Hint:* Review Problem I-4.5.2 on page I-408.)
2. Why do we assume that $\mathbf{N}$ is nonsingular?
3. Consider the special case in which

$$\mathbf{K}_s(t, u) = K_s(t, u)\mathbf{I}$$

and $\mathbf{N}$ is diagonal. Simplify the results in part 1.

**Problem 2.1.19.** Consider the model in Problem 2.1.17. Assume

$$E[\mathbf{s}(t)] = \mathbf{m}(t).$$

All of the other assumptions in Problem 2.1.17 are still valid. Repeat Problem 2.1.17

**Problem 2.1.20.** In Section 2.1.5 we considered a simple multiplicative channel. A more realistic channel model is the Rayleigh channel model that we encountered previously in Section I-4.4.2 and Chapter II-8. We shall study it in detail in Chapter 10.

On $H_1$ we transmit a bandpass signal,

$$s_t(t) \triangleq \sqrt{2P}\, f(t) \cos \omega_c t,$$

where $f(t)$ is a slowly varying function (the envelope of the signal). The received signal is

$$r(t) = \sqrt{2P}\, b_1(t) f(t) \cos \omega_c t + \sqrt{2P}\, b_2(t) f(t) \sin \omega_c t + w(t), \qquad T_i \le t \le T_f : H_1.$$

The channel processes $b_1(t)$ and $b_2(t)$ are statistically independent, zero-mean Gaussian processes whose covariance functions are $K_b(t, u)$. The additive noise $w(t)$ is a sample function of a statistically independent, zero-mean Gaussian process with spectral height $N_0/2$. The channel processes vary slowly compared to $\omega_c$. On $H_0$, only white noise is present.

1. Derive the optimum receiver for this model of the Rayleigh channel.
2. Draw a filter-squarer realization for the optimum receiver.
3. Draw a state-variable realization of the optimum receiver. Assume that

$$S_b(\omega) = \frac{2k\sigma_b^2}{\omega^2 + k^2}.$$

**Problem 2.1.21.** The model for a Rician channel is the same as that in Problem 2.1.19, except that

$$E[b_1(t)] = m$$

instead of zero. Repeat Problem 2.1.19 for this case.

## P.2.2. Performance

**Problem 2.2.1.** Consider the problem of evaluating $\mu_D(s)$, which is given by (135) or (147). Assume that $s(t)$ has a finite-dimensional state representation. Define

$$\mu_D(s, T) = -\frac{s}{2} \int_0^T m(x) g\left(x \,\middle|\, \frac{N_0}{2(1-s)}\right) dx.$$

Find a finite-dimensional dynamic system whose output is $\mu_D(s, T)$.

**Problem 2.2.2.** Consider the model in the example in Section 2.2.4. Assume that

$$E[b(t)] = m$$

instead of zero. Evaluate $\mu_D(s)$ for this problem. [*Hint:* If you use (147), review pages I-320 and I-390.]

**Problem 2.2.3.**

1. Consider the model in Problem 2.1.5. Evaluate $\mu(s)$ for this system.
2. Define

$$\gamma \triangleq \sqrt{\frac{2\sigma^2}{N_0}}.$$

Simplify the expression in part 1 for the case in which $\gamma T \gg 1$.

**Problem 2.2.4 (continuation).** Use the expression for $\mu(s)$ in part 2 of Problem 2.2.3. Evaluate $P_F^{[2]}$ and $P_M^{[2]}$ [see (167)]. Compare their magnitude with that of $P_F^{[1]}$ and $P_M^{[1]}$.

**Problem 2.2.5.** Consider the model in Problem 2.1.5. Assume that

$$r(t) = s(t) + m(t) + w(t), \qquad 0 \le t \le T,$$

$$r(t) = w(t), \qquad 0 \le t \le T,$$

where $m(t)$ is a deterministic function. The processes $s(t)$ and $w(t)$ are as described in Problem 2.1.5. Evaluate $\mu_D(s)$ for this model.

**Problem 2.2.6.**

1. Evaluate $\mu(s)$ for the system in Problem 2.1.6.
2. Plot the result as a function of $s$.
3. Find $P_F$ and $P_D$.

**Problem 2.2.7.** Evaluate $\mu(s)$ for the system in Problem 2.1.7.

**Problem 2.2.8.** Evaluate $\mu(s)$ for the system in Problem 2.1.8.

**Problem 2.2.9.**

1. Evaluate $\mu(s)$ for the system in Problem 2.1.9.
2. Evaluate $P_F$ and $P_D$.

**Problem 2.2.10.** Consider the system in Problem 2.1.17.

1. Assume that (P.4) in part 3 is valid. Find $\mu(s)$ for this special case.
2. Assume that (P.5) in part 4 is valid. Find $\mu(s)$ for this special case.
3. Derive an expression for $\mu(s)$ for the general case.

**Problem 2.2.11.** Consider the system in Problem 2.1.19. Find an expression for $\mu_D(s)$ for this system.

**Problem 2.2.12.** Find $\mu(s)$ for the Rayleigh channel model in Problem 2.1.20.

**Problem 2.2.13.** Find $\mu(s)$ for the Rician channel model in Problem 2.1.21.

## REFERENCES


[1] R. Price, "Statistical Theory Applied to Communication through Multipath Disturbances," Massachusetts Institute of Technology Research Laboratory of Electronics, Tech. Rept. 266, September 3, 1953.

[2] R. Price, "The Detection of Signals Perturbed by Scatter and Noise," IRE Trans. **PGIT-4,** 163–170 (Sept. 1954).

[3] R. Price, "Notes on Ideal Receivers for Scatter Multipath," Group Rept. 34–39, Lincoln Laboratory, Massachusetts Institute of Technology, May 12, 1955.

[4] R. Price, "Optimum Detection of Random Signals in Noise, with Application to Scatter-Multipath Communication. I," IRE Trans. **PGIT-6,** 125–135 (Dec. 1956).

[5] F. Schweppe, "Evaluation of Likelihood Functions for Gaussian Signals," IEEE Trans. **IT-11,** No. 1, 61–70 (Jan. 1965).

[6] R. E. Kalman and R. S. Bucy, "New Results in Linear Filtering and Prediction Theory," ASME J. Basic Eng., **83,** 95–108 (March 1961).

[7] L. D. Collins, "An Expression for $\partial h_0(s, \tau:t)/\partial t$," Detection and Estimation Theory Group Internal Memorandum IM-LDC-6, Massachusetts Institute of Technology, April 1966.

[8] W. Lovitt, *Linear Integral Equations*, Dover Publications, New York, 1924.
[9] L. D. Collins, "Realizable Whitening Filters and State-Variable Realizations," IEEE Proc. **56**, No. 1, 100–101 (Jan. 1968).
[10] W. Feller, *An Introduction to Probability Theory and Its Applications*, Vol. II, John Wiley, New York, 1966.
[11] L. D. Collins, "Asymptotic Approximations to the Error Probability for Detecting Gaussian Signals," Massachusetts Institute of Technology, Department of Electrical Engineering, Sc.D. Thesis Proposal, January 1968.
[12] H. Cramèr, *Mathematical Methods in Statistics*, Princeton University Press, Princeton, N.J., 1946.
[13] L. D. Collins, "Closed-Form Expressions for the Fredholm Determinant for State-Variable Covariance Functions," IEEE Proc. **56**, No. 4 (April 1968).
[14] M. Athans and F. C. Schweppe, "Gradient Matrices and Matrix Calculations," Technical Note 1965–53, Lincoln Laboratory, Massachusetts Institute of Technology, 1965.
[15] A. B. Baggeroer, "A State-Variable Approach to the Solution of Fredholm Integral Equations," November 15, 1967.
[16] A. J. F. Siegert, "A Systematic Approach to a Class of Problems in the Theory of Noise and Other Random Phenomena. II. Examples," IRE Trans. **IT-3**, No. 1, 38–43 (March 1957).
[17] D. Middleton, "On the Detection of Stochastic Signals in Additive Normal Noise. I," IRE Trans. Information Theory **IT-3**, 86–121 (June 1957).
[18] D. Middleton, "On the Detection of Stochastic Signals in Additive Normal Noise. II," IRE Trans. Information Theory **IT-6**, 349–360 (June 1960).
[19] D. Middleton, "On Singular and Nonsingular Optimum (Bayes) Tests for the Detection of Normal Stochastic Signals in Normal Noise," IRE Trans. Information Theory **IT-7**, 105–113 (April 1961).
[20] D. Middleton, *Introduction to Statistical Communication Theory*, McGraw-Hill, New York, 1960.
[21] R. L. Stratonovich and Y. G. Sosulin, "Optimal Detection of a Markov Process in Noise," Eng. Cybernet. **6**, 7–19 (Oct. 1964).
[22] R. L. Stratonovich and Y. G. Sosulin, "Optimal Detection of a Diffusion Process in White Noise," Radio Eng. Electron. Phys. **10**, 704–713 (May 1965).
[23] R. L. Stratonovich and Y. G. Sosulin, "Optimum Reception of Signals in Non-Gaussian Noise," Radio Eng. Electron. Phys. **11**, 497–507 (April 1966).
[24] Y. G. Sosulin, "Optimum Extraction of Non-Gaussian Signals in Noise," Radio Eng. Electron. Phys. **12**, 89–97 (Jan. 1967).
[25] R. L. Stratonovich, "A New Representation for Stochastic Integrals," J. SIAM Control **4**, 362–371 (1966).
[26] J. L. Doob, *Stochastic Processes*, Wiley, New York, 1953.
[27] K. Itô, *Lectures on Stochastic Processes*, Tata Institute for Fundamental Research, Bombay, 1961.
[28] T. Duncan, "Probability Densities for Diffusion Processes with Applications to Nonlinear Filtering Theory and Detection Theory," Information Control **13**, 62–74 (July 1968).
[29] T. Kailath and P. A. Frost, "Mathematical Modeling of Stochastic Processes," JACC Control Symposium (1969).
[30] T. Kailath, "A General Likelihood-Ratio Formula for Random Signals in Noise," IEEE Trans. Information Theory **IT-5**, No. 3, 350–361 (May 1969).

# 3

# General Binary Detection: Gaussian Processes

In this Chapter we generalize the model of Chapter 2 to include other Gaussian problems that we encounter frequently in practice. After developing the generalized model in Section 3.1, we study the optimum receiver and its performance for the remainder of the chapter.

## 3.1 MODEL AND PROBLEM CLASSIFICATION

An obvious generalization is suggested by the digital communication system on page 26. In this case we transmit a different signal on each hypothesis. Typically we transmit

$$\sqrt{2P}\sin(\omega_1 t), \qquad T_i \le t \le T_f : H_1 \tag{1}$$

and

$$\sqrt{2P}\sin(\omega_0 t), \qquad T_i \le t \le T_f : H_0. \tag{2}$$

If the channel is the simple multiplicative channel shown in Fig. 2.9, the received waveforms on the two hypotheses are

$$r(t) = \sqrt{2P}\, b(t)\sin(\omega_1 t) + w(t), \qquad T_i \le t \le T_f : H_1, \tag{3}$$

$$r(t) = \sqrt{2P}\, b(t)\sin(\omega_0 t) + w(t), \qquad T_i \le t \le T_f : H_0, \tag{4}$$

where $b(t)$ is a sample function of Gaussian random process. This is just a special case of the general problem in which the received waveforms on the two hypotheses are

$$\begin{aligned} r(t) &= s_1(t) + w(t), \qquad T_i \le t \le T_f : H_1, \\ r(t) &= s_0(t) + w(t), \qquad T_i \le t \le T_f : H_0, \end{aligned} \tag{5}$$

where $s_1(t)$ and $s_0(t)$ are Gaussian processes with mean-value functions $m_1(t)$ and $m_0(t)$ and covariance functions $K_1(t, u)$ and $K_0(t, u)$, respectively. In many cases, we also have a colored noise term, $n_c(t)$, present on both hypotheses. Then

$$\begin{aligned} r(t) &= s_1(t) + n_c(t) + w(t), \qquad T_i \le t \le T_f : H_1, \\ r(t) &= s_0(t) + n_c(t) + w(t), \qquad T_i \le t \le T_f : H_0. \end{aligned} \tag{6}$$

We can include both these problems and many others in the general formulation,

$$\begin{aligned} r(t) &= r_1(t), \qquad T_f \le t \le T_i : H_1, \\ r(t) &= r_0(t), \qquad T_i \le t \le T_i : H_0. \end{aligned} \tag{7}$$

On $H_1$, $r(t)$ is a sample function from a Gaussian random process with mean-value function $m_1(t)$ and covariance function $K_{H_1}(t, u)$. On $H_0$, $r(t)$ is a sample function from a Gaussian random process with mean-value function $m_0(t)$ and covariance function $K_{H_0}(t, u)$. For algebraic simplicity, we assume that $r(t)$ is zero-mean on both hypotheses in our initial discussion. The results regarding mean-value functions in Chapter 2 generalize in an obvious manner and are developed in Section 3.4.

Some of our discussion will be for the general problem in (7). On the other hand, many results are true only for subclasses of this problem. For bookkeeping purposes we define these classes by the table in Fig. 3.1. In all cases, the various processes are statistically independent. The subscript $w$ implies that the same white noise component is present on both hypotheses. There may also be other processes present on both $H_1$ and $H_0$. The absence of the subscript means that a white noise component is not necessarily present. The class inclusions are indicated by solid lines. Thus,

$$B_w \subset A_w \subset A \subset GB, \tag{8}$$

$$B_w \subset B. \tag{9}$$

Two additional subscripts may be applied to any of the above classes. The additional subscript $s$ means that *all* of the processes involved have a finite-dimensional state representation. The additional subscript $m$ means that *some* of the processes involved have a nonzero mean. The absence of the subscript $m$ implies that all processes are zero-mean. We see that the simple binary problem in Chapter 2 is the special case of class $B_w$, in which $n_c(t)$ is not present. This class structure may seem cumbersome, but it enables us to organize our results in a clear manner.

As in the simple binary problem, we want to find the optimum receiver and evaluate its performance. The reason the calculation of the likelihood ratio was easy in the simple binary case was that only white noise was

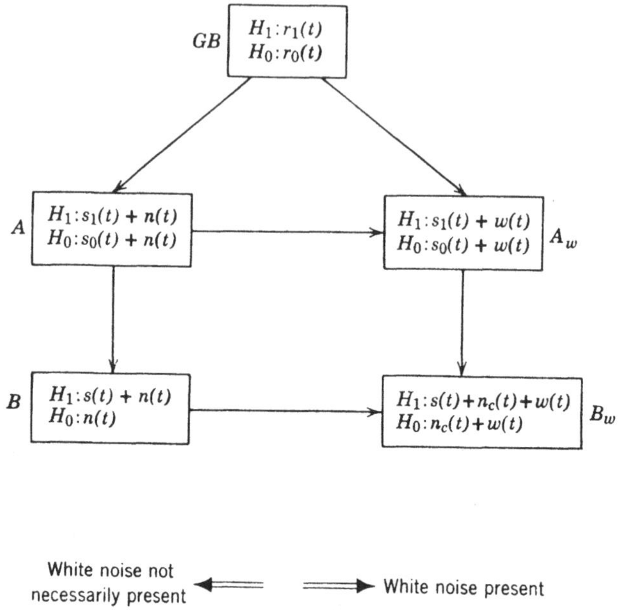


**Fig. 3.1 Classification of Gaussian detection problems.**

present on $H_0$. Thus, we could choose our coordinate system based on the covariance function of the signal process on $H_1$. As a result of this choice, we had statistically independent coefficients on both hypotheses. Now the received waveform may have a nonwhite component on both hypotheses. Therefore, except for the trivial case in which the nonwhite components have the same eigenfunctions on both hypotheses, the technique in Section 2.1 will give correlated coefficients. There are several ways around this difficulty. An intuitively appealing method is the whitening approach, which we encountered originally in Chapter I-4 (page I-290). We shall use this approach in the text.

In Section 3.2 we derive the likelihood ratio test and develop various receiver structures for the class $A_w$ problem. In Section 3.3 we study the performance for the class $A_w$ problem. In Section 3.4 we discuss four important special situations: the binary symmetric problem, the non-zero-mean problem, the bandpass problem, and the binary symmetric bandpass problem. In Section 3.5 we look at class $GB$ problems and discuss the singularity problem briefly. We have deliberately postponed our discussion of the general case because almost all physical situations can be modeled by a class $A_w$ system. Finally, in Section 3.6, we summarize our results and discuss some related issues.

## 3.2 RECEIVER STRUCTURES

In this section, we derive the likelihood ratio test for problems in class $A_w$ and develop several receiver configurations. Looking at Fig. 3.1, we see that class $A_w$ implies that the same white noise process is present on both hypotheses. Thus,

$$
\begin{aligned}
r(t) &= s_1(t) + w(t), \qquad T_i \le t \le T_f : H_1, \\
r(t) &= s_0(t) + w(t), \qquad T_i \le t \le T_f : H_0.
\end{aligned} \tag{10}
$$

In addition, we assume that both $s_1(t)$ and $s_0(t)$ are zero-mean Gaussian processes with finite mean-square values. They are statistically independent of $w(t)$ and have continuous covariance functions $K_1(t, u)$ and $K_0(t, u)$, respectively. The spectral height of the Gaussian white noise is $N_0/2$. Therefore, the covariance functions of $r(t)$ on the two hypotheses are

$$E[r(t)r(u) \mid H_1] \triangleq K_{H_1}(t, u) = K_1(t, u) + \frac{N_0}{2}\,\delta(t - u), \tag{11}$$

$$E[r(t)r(u) \mid H_0] \triangleq K_{H_0}(t, u) = K_0(t, u) + \frac{N_0}{2}\,\delta(t - u). \tag{12}$$

We now derive the likelihood ratio test by a whitening approach.

### 3.2.1 Whitening Approach

The basic idea of the derivation is straightforward. We whiten $r(t)$ on one hypothesis and then operate on the whitened waveform using the techniques of Section 2.1. As long as the whitening filter is reversible, we know that the over-all system is optimum (see page I-289). (Notice that realizability is *not* an issue.)

The whitening filter is shown in Fig. 3.2. We choose $h_{w_0}(t, u)$ so that $r_*(t)$ is white on $H_0$ and has a unity spectral height. Thus,

$$E[r_*(t)r_*(u) \mid H_0] = \delta(t - u), \qquad T_i \le t, u \le T_f. \tag{13}$$

On pages I-290–I-297 we discussed construction of the whitening filter.

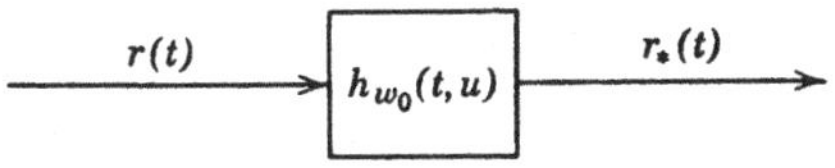


**Fig. 3.2 Whitening filter.**

From that discussion we know that we can always find a filter such that (13) is satisfied. Because

$$r_*(t) = \int_{T_i}^{T_f} h_{w_0}(t, u) r(u)\, du, \tag{14}$$

(13) implies that

$$\iint_{T_i}^{T_f} h_{w_0}(t, \alpha) h_{w_0}(u, \beta) K_{H_0}(\alpha, \beta)\, d\alpha\, d\beta = \delta(t - u). \tag{15}$$

The covariance function of $r_*(t)$ on $H_1$ is

$$E[r_*(t) r_*(u) \mid H_1] = \iint_{T_i}^{T_f} h_{w_0}(t, \alpha) h_{w_0}(u, \beta) K_{H_1}(\alpha, \beta)\, d\alpha\, d\beta \triangleq K_1^*(t, u). \tag{16}$$

We now expand $r_*(t)$ using the eigenfunctions of $K_1^*(t, u)$, which are specified by the equation

$$\lambda_i^* \varphi_i(t) = \int_{T_i}^{T_f} K_1^*(t, u) \varphi_i(u)\, du, \qquad T_i \le t \le T_f. \tag{17}$$

Proceeding as in Section 2.1, we find that

$$l_R = -\frac{1}{2} \sum_{i=1}^{\infty} \left(\frac{1}{\lambda_i^*} - 1\right) r_i^2. \tag{18}$$

(Remember that the whitened noise on $H_0$ has unity spectral height.) As before we define an inverse kernel, $Q_1^*(t, u)$,

$$\int_{T_i}^{T_f} Q_1^*(t, u) K_1^*(u, z)\, du = \delta(t - z), \qquad T_i < t, z < T_f. \tag{19}$$

Then we can write

$$l_R = -\tfrac{1}{2} \iint_{T_i}^{T_f} dt\, du\, r_*(t) [Q_1^*(t, u) - \delta(t - u)] r_*(u). \tag{20a}$$

It is straightforward to verify that the kernel in (20*a*) is always square-integrable (see Problem 3.2.11). Using (14), we can write this in terms of $r(t)$.

$$l_R = -\tfrac{1}{2} \iint_{T_i}^{T_f} r(\alpha) \left\{ \iint_{T_i}^{T_f} h_{w_0}(t, \alpha) [Q_1^*(t, u) - \delta(t - u)] h_{w_0}(u, \beta)\, dt\, du \right\} r(\beta)\, d\alpha\, d\beta. \tag{20b}$$

We want to examine the term in the braces. The term contributed by the impulse is just $Q_{H_0}(\alpha, \beta)$, the inverse kernel of $K_{H_0}(\alpha, \beta)$ [see (I-4.152)]. We now show that the remaining term is $Q_{H_1}(\alpha, \beta)$. We must show that

$$Q_{H_1}(\alpha, \beta) = \iint_{T_i}^{T_f} h_{w_0}(t, \alpha) Q_1^*(t, u) h_{w_0}(u, \beta)\, dt\, du. \tag{21}$$

This result is intuitively obvious from the relationship between $K_{H_1}(\alpha, \beta)$ and $K_1^*(t, u)$ expressed in (16). It can be verified by a few straightforward manipulations. [Multiply both sides of (16) by

$$Q_1^*(z_1, t) h_{w_0}^{-1}(z_2, u) Q_{H_1}(z_2, z_3) h_{w_0}(z_1, z_4).$$

Integrate the left side with respect to $u$, $\beta$, $z_2$, and $\alpha$, in that order. Integrate the right side with respect to $t$, $z_1$, $u$, and $z_2$, in that order. At each step simplify by using known relations.] The likelihood function in (19) can now be written as

$$l_R = -\frac{1}{2} \iint_{T_i}^{T_f} d\alpha\, d\beta\, r(\alpha) r(\beta) [Q_{H_1}(\alpha, \beta) - Q_{H_0}(\alpha, \beta)]. \tag{22}$$

In a moment we shall see that the impulses in the inverse kernels cancel, so that kernel is a square-integrable function. This can also be written formally as a difference of two quadratic forms,

$$\boxed{l_R = \frac{1}{2} \iint_{T_i}^{T_f} d\alpha\, d\beta\, r(\alpha) Q_{H_0}(\alpha, \beta) r(\beta) - \frac{1}{2} \iint_{T_i}^{T_f} d\alpha\, d\beta\, r(\alpha) Q_{H_1}(\alpha, \beta) r(\beta).} \tag{23}$$

The reader should note the similarity between (23) and the LRT for the finite-dimensional general Gaussian problem in (I-2.327). This similarity enables one to guess both the form of the test for nonzero means and the form of the bias terms. Several equivalent forms of (22) are also useful.

### 3.2.2 Various Implementations of the Likelihood Ratio Test

To obtain the first equivalent form, we write $Q_{H_1}(\alpha, \beta)$ and $Q_{H_0}(\alpha, \beta)$ in terms of an impulse and a well-behaved function,

$$Q_{H_i}(\alpha, \beta) = \frac{2}{N_0} [\delta(\alpha - \beta) - h_i(\alpha, \beta)], \qquad i = 0, 1, \tag{24}$$

where $h_i(\alpha, \beta)$ satisfies the equation

$$\frac{N_0}{2} h_i(\alpha, \beta) + \int_{T_i}^{T_f} h_i(\alpha, x) K_i(x, \beta)\, dx = K_i(\alpha, \beta),$$

$$T_i \leq \alpha, \beta \leq T_f, \quad i = 0, 1. \quad (25)$$

Using (24) in (22) gives

$$\boxed{l_R = l_{R_1} - l_{R_0},} \quad (26)$$

where

$$\boxed{l_{R_i} = \frac{1}{N_0} \iint_{T_i}^{T_f} r(\alpha) r(\beta) h_i(\alpha, \beta)\, d\alpha\, d\beta, \qquad i = 0, 1.} \quad (27)$$

It is easy to verify (see Problem 3.2.1) that the bias term can be written as

$$l_B = l_{B_1} - l_{B_0}, \quad (28)$$

where [by analogy with (2.73)]

$$l_{B_i} = -\frac{1}{N_0} \int_{T_i}^{T_f} \xi_{P_{s_i}}(t)\, dt, \qquad i = 0, 1. \quad (29)$$

The complete LRT is

$$l_{R_1} + l_{B_1} - l_{R_0} - l_{B_0} \underset{H_0}{\overset{H_1}{\gtrless}} \ln \eta. \quad (30)$$

We see that the receiver can be realized as two *simple* binary receivers in parallel, with their outputs subtracted. Thus, any of the four canonic realizations developed in Section 2.1 (Figs. 2.2–2.7) can be used in each path. A typical structure using Realization No. 1 is shown in Fig. 3.3. This parallel processing structure is frequently used in practice.

A second equivalent form of (22) is also useful. We define a function

$$h_\Delta(t, u) \triangleq Q_{H_0}(t, u) - Q_{H_1}(t, u). \quad (31)$$

Then,

$$\boxed{l_R = \tfrac{1}{2} \iint_{T_i}^{T_f} r(t) h_\Delta(t, u) r(u)\, dt\, du.} \quad (32)$$

To eliminate the inverse kernels in (31), we multiply by the two covariance

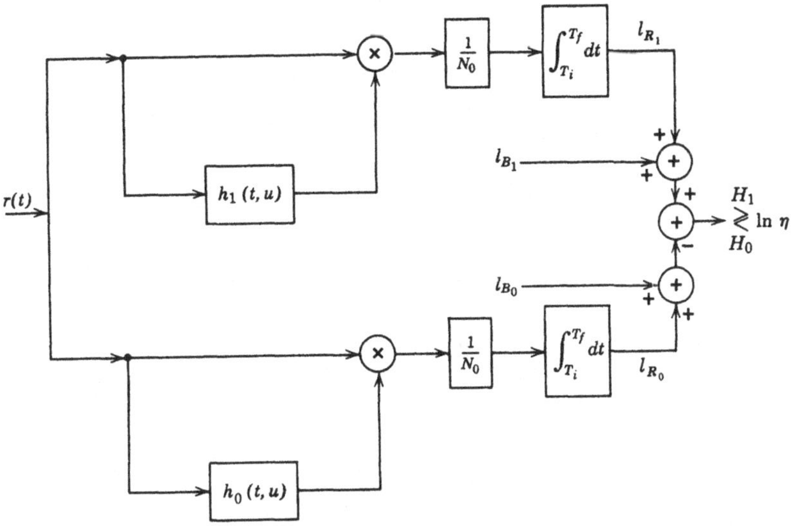


**Fig. 3.3 Parallel processing realization of general binary receiver (class $A_w$).**

functions and integrate. The result is an integral equation specifying $h_\Delta(t, u)$.

$$\iint_{T_i}^{T_f} dt\, du K_{H_0}(x, t) h_\Delta(t, u) K_{H_1}(u, z) = K_{H_1}(x, z) - K_{H_0}(x, z), \quad T_i \leq x, z \leq T_f. \tag{33}$$

This form of the receiver is of interest because the white noise level does not appear *explicitly*. Later we shall see that (32) and (33) specify the receiver for class *GB* problems. The receiver is shown in Fig. 3.4.

Two other forms of the receiver are useful for class $B_w$ problems. In this case, the received waveform contains the same noise process on both hypotheses and an additional signal process on $H_1$. Thus,

$$\begin{aligned} r(t) &= s(t) + n_c(t) + w(t), \qquad T_i \leq t \leq T_f : H_1, \\ r(t) &= n_c(t) + w(t), \qquad T_i \leq t \leq T_f : H_0, \end{aligned} \tag{34}$$

where $s(t)$, $n_c(t)$, and $w(t)$ are zero-mean, statistically independent Gaussian random processes with covariance functions $K_s(t, u)$, $K_c(t, u)$, and $(N_0/2)\delta(t - u)$, respectively. On $H_0$,

$$K_{H_0}(t, u) = K_c(t, u) + \frac{N_0}{2}\delta(t - u) \triangleq K_n(t, u), \qquad T_i \leq t, u \leq T_f. \tag{35}$$

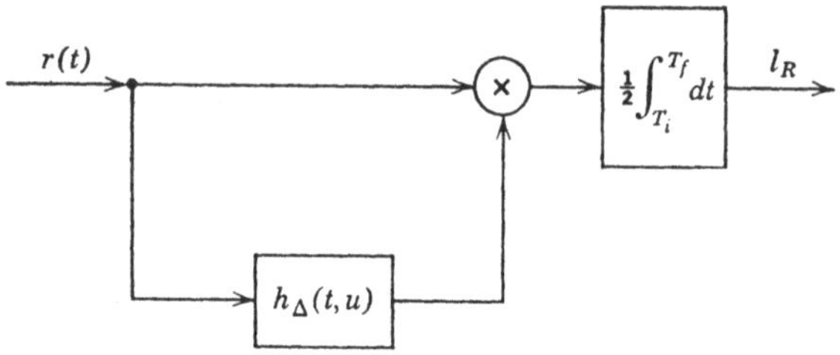


**Fig. 3.4 Receiver for class $A_w$ problem.**

For this particular case the first form is an alternative realization that corresponds to the estimator-correlator of Fig. 2.3. We define a new function $h_o(t, z)$ by the relation

$$h_o(x, u) = \int_{T_i}^{T_f} h_\Delta(t, u) K_{H_0}(x, t)\, dt. \tag{36}$$

Using (36) and the definition of $h_\Delta(t, u)$ in (33), we have

$$\int_{T_i}^{T_f} h_o(t, x)[K_s(x, u) + K_n(x, u)]\, dx = K_s(t, u), \qquad T_i \le x, z \le T_f. \tag{37}$$

This equation is familiar from Chapter I-6 as the equation specifying the optimum linear filter for estimating $s(t)$ from an observation $r(t)$ assuming that $H_1$ is true. Thus,

$$\hat{s}(t) = \int_{T_i}^{T_f} h_o(t, u) r(u)\, du. \tag{38}$$

We now implicitly define a function $r_g(t)$,

$$r(t) = \int_{T_i}^{T_f} K_{H_0}(t, x) r_g(x)\, dx, \qquad T_i \le t \le T_f. \tag{39}$$

Equivalently,

$$r_g(t) = \int_{T_i}^{T_f} Q_{H_0}(t, x) r(x)\, dx, \qquad T_i \le t \le T_f. \tag{40}$$

This type of function is familiar from Chapter I-5 (I-5.32). Then, from (36) and (40), we have

$$l_R = \tfrac{1}{2}\int_{T_i}^{T_f} \hat{s}(t) r_g(t)\, dt. \tag{41}$$

The resulting receiver structure is shown in Fig. 3.5. We see that this has the same structure as the optimum receiver for known signals in colored noise (Fig. I-4.38c) except that a MMSE estimate $\hat{s}(t)$ has replaced the

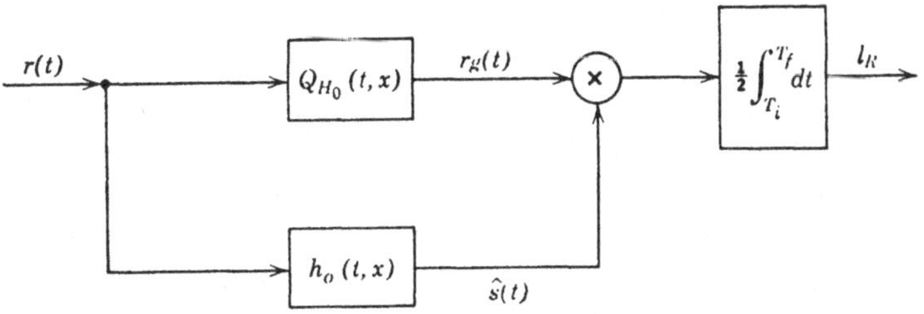


**Fig. 3.5 Estimator-correlator realization for class $B_w$ problems.**

known signal in the correlation operation. This configuration is analogous to the estimator-correlator in Fig. 2.3.

The second receiver form of interest for class $B_w$ is the filter-squarer realization. For this class a functional square root exists,

$$h_\Delta(t, u) = \int_{T_i}^{T_f} h_\Delta^{[1/2]}(z, t) h_\Delta^{[1/2]}(z, u)\, dz, \qquad T_i \leq t, u \leq T_f. \tag{42}$$

The existence can be shown by verifying that one solution to (42) is

$$h_\Delta^{[1/2]}(t, u) = \int_{T_i}^{T_f} h_1^{*[1/2]}(t, z) h_{w_0}(z, u)\, dz, \qquad T_i \leq t, u \leq T_f, \tag{43}$$

since both functions in the integrand exist (see Problem 3.2.10). This filter-squarer realization is shown in Fig. 3.6. For class $A_w$ problems a functional square root of $h_\Delta(t, u)$ may not exist, and so a filter-squarer realization is not always possible (see Problem 3.2.10).

### 3.2.3 Summary: Receiver Structures

In this section we have derived the likelihood ratio test for the class $A_w$ problem. The LRT was given in (23). We then looked at various receiver configurations. The parallel processing configuration is the one most commonly used. All of the canonical receiver configuration developed for the simple binary problem can be used in each path. For class $B_w$ problems, the filter-squarer realization shown in Fig. 3.6 is frequently used.

The next problem of interest is the performance of the optimum receiver

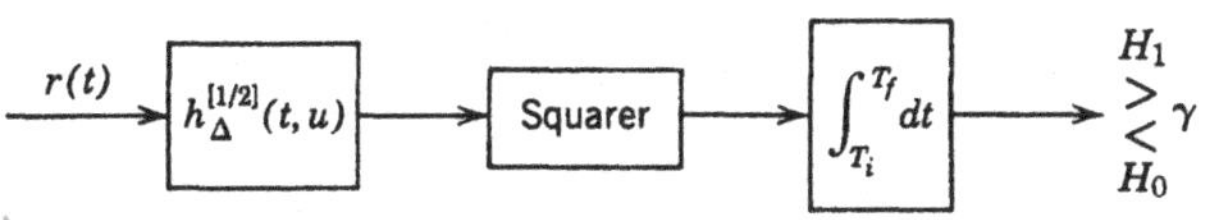


**Fig. 3.6 Filter-squarer realization for class $B_w$ problems.**

## 3.3 PERFORMANCE

All of our performance discussion in Section 2.2 is valid for class $A_w$ problems with the exception of the closed-form expressions for $\mu(s)$. In this section we derive an expression for $\mu(s)$. Just as in the derivation of the optimum receiver, there is a problem due to the nonwhite process that is present on both hypotheses. As before, one way to avoid this is to prewhiten the received signal on $H_0$. It is possible to carry out this derivation, but it is too tedious to have much pedagogical appeal. Of the various alternatives available at this point the sampling approach seems to be the simplest. In Section 3.5, we study the performance question again. The derivation of $\mu(s)$ at that point is much neater.

In the problems on pages I-231–233 of Chapter I-3, we discussed how many of the continuous waveform results could be derived easily using a sampling approach. The received waveforms on the two hypotheses are given by (5). We sample $r(t)$ every $T/N$ seconds. This gives us an $N$-dimensional vector $\mathbf{r}$ whose mean and covariance matrix are sampled versions of the mean-value function and covariance function of the process. We can then use the $\mu(s)$ expression derived in Section I-2.7. Finally, we let $N \to \infty$ to get the desired result. For algebraic simplicity, we go through the details for the zero-mean case.

Denote the sample at $t_i$ as $r_i$. The covariances between the samples are

$$E[r_i r_j \mid H_\alpha] = K_{H_\alpha}(t_i, t_j) = K_{\alpha,ij}, \qquad i, j = 1, \dots, N, \alpha = 0, 1. \tag{44}$$

The set of samples is denoted by the vector $\mathbf{r}$. The covariance matrix of $\mathbf{r}$ is

$$E[\mathbf{r}\mathbf{r}^T \mid H_\alpha] = \mathbf{K}_\alpha, \qquad \alpha = 0, 1. \tag{45}$$

The matrices in (45) are $N \times N$ covariance matrices. The elements of the matrices on the two hypotheses are

$$K_{1,ij} = K_{s_1,ij} + \frac{N_0}{2}\,\delta_{ij}, \tag{46}$$

$$K_{0,ij} = K_{s_0,ij} + \frac{N_0}{2}\,\delta_{ij}. \tag{47}$$

Notice that

$$K_{s_1,ij} = K_{s_1}(t_i, t_j) \tag{48}$$

and

$$K_{s_0,ij} = K_{s_0}(t_i, t_j). \tag{49}$$

We can write (46) and (47) in matrix notation as

$$\mathbf{K}_1 \triangleq \mathbf{K}_{s_1} + \frac{N_0}{2}\mathbf{I}, \tag{50}$$

$$\mathbf{K}_0 \triangleq \mathbf{K}_{s_0} + \frac{N_0}{2}\mathbf{I}. \tag{51}$$

We can now use the $\mu(s)$ expression derived in Chapter I-2. From the solution to Problem I-2.7.3,

$$\mu_N(s) = -\tfrac{1}{2}\ln\left(|\mathbf{K}_1|^{s-1}\,|\mathbf{K}_0|^{-s}\,|\mathbf{K}_0 s + \mathbf{K}_1(1-s)|\right), \qquad 0 \le s \le 1. \tag{52}$$

Notice that $|\cdot|$ denotes the determinant of a matrix. Substituting (50) and (51) into (52), we have

$$\mu_N(s) = -\tfrac{1}{2}\ln\left\{\left|\frac{N_0}{2}\mathbf{I} + \mathbf{K}_{s_1}\right|^{s-1}\left|\frac{N_0}{2}\mathbf{I} + \mathbf{K}_{s_0}\right|^{-s} \times \left|\left(\frac{N_0}{2}\mathbf{I} + \mathbf{K}_{s_0}\right)s + \left(\frac{N_0}{2}\mathbf{I} + \mathbf{K}_{s_1}\right)(1-s)\right|\right\}. \tag{53}$$

The matrices in (53) cannot be singular, and so all of the indicated operations are valid. Collecting $N_0/2$ from the various terms and rewriting (53) as a sum of logarithms, we have

$$\mu_N(s) = \frac{1}{2}\left\{(1-s)\ln\left|\mathbf{I} + \frac{2}{N_0}\mathbf{K}_{s_1}\right| + s\ln\left|\mathbf{I} + \frac{2}{N_0}\mathbf{K}_{s_0}\right| - \ln\left|\mathbf{I} + \frac{2}{N_0}(s\mathbf{K}_{s_0} + (1-s)\mathbf{K}_{s_1})\right|\right\}. \tag{54}$$

Now each term is the logarithm of the determinant of a matrix and can be rewritten as the sum of the logarithms of the eigenvalues of the matrix by using the Cayley-Hamilton theorem. For example,

$$\ln\left|\mathbf{I} + \frac{2}{N_0}\mathbf{K}_{s_1}\right| = \sum_{i=1}^{N}\ln\left(1 + \frac{2}{N_0}\lambda_{s_1,i}\right), \tag{55}$$

where $\lambda_{s_1,i}$ is the $i$th eigenvalue of $\mathbf{K}_{s_1}$. As $N \to \infty$, this function of the eigenvalues of the matrix, $\mathbf{K}_{s_1}$, will approach the same function of the eigenvalues of the kernel, $K_{s_1}(t, u)$.† We denote the eigenvalues of $K_{s_1}(t, u)$ by $\lambda_i^{s_1}$. Thus,

$$\lim_{N\to\infty}\sum_{i=1}^{N}\ln\left(1 + \frac{2}{N_0}\lambda_{s_1,i}\right) = \sum_{i=1}^{\infty}\ln\left(1 + \frac{2}{N_0}\lambda_i^{s_1}\right). \tag{56}$$

† We have not proved that this statement is true. It is shown in various integral equation texts (e.g., Lovitt [1, Chapter III]).

The sum on the right side is familiar from (2.73) as

$$\sum_{i=1}^{\infty} \ln\left(1 + \frac{2\lambda_i^{s}}{N_0}\right) = \frac{2}{N_0} \int_{T_i}^{T_f} \xi_P\left(t \mid s_1(\cdot), \frac{N_0}{2}\right) dt. \tag{57}$$

Thus, the first term in $\mu(s)$ can be expressed in terms of the realizable mean-square error for the problem of filtering $s_1(t)$ in the presence of additive white noise. A similar interpretation follows for the second term. To interpret the third term we define a new composite signal process,

$$s_{\text{com}}(t, s) \triangleq \sqrt{s}\, s_0(t) + \sqrt{(1 - s)}\, s_1(t). \tag{58}$$

This is a fictitious process constructed by generating two sample functions $s_0(t)$ and $s_1(t)$ from statistically independent random processes with covariances $K_0(t, u)$ and $K_1(t, u)$ and then forming a weighted sum. The resulting composite process has a covariance function

$$K_{\text{com}}(t, u{:}s) = sK_0(t, u) + (1 - s)K_1(t, u), \qquad T_i \le t, u \le T_f. \tag{59}$$

We denote the realizable mean-square filtering error in the presence of white noise as $\xi_P(t \mid s_{\text{com}}(\cdot), N_0/2)$. The resulting expression for $\mu(s)$ is

$$\boxed{\begin{aligned}\mu(s) = \frac{1}{N_0} \int_{T_i}^{T_f} dt \Bigg[ (1 - s)\xi_P\left(t \mid s_1(\cdot), \frac{N_0}{2}\right) \\ + s\xi_P\left(t \mid s_0(\cdot), \frac{N_0}{2}\right) - \xi_P\left(t \mid s_{\text{com}}(\cdot), \frac{N_0}{2}\right)\Bigg].\end{aligned}} \tag{60}$$

We see that for the general binary problem, we can express $\mu(s)$ in terms of three different realizable filtering errors.

To evaluate the performance, we use the expression for $\mu(s)$ in (60) in the Chernoff bounds in (2.127), or the approximate error expressions in (2.164), (2.166), (2.173), and (2.174). We shall look at some specific examples in Chapter 4. We now look at four special situations.

## 3.4 FOUR SPECIAL SITUATIONS

In this section, we discuss four special situations that arise in practice:

1. The binary symmetric problem.
2. The non-zero-mean problem.
3. The stationary independent bandpass problem.
4. The binary symmetric bandpass problem.

We define each of these problems in detail in the appropriate subsection.

### 3.4.1 Binary Symmetric Case

In this case the received waveforms on the two hypotheses are

$$r(t) = s_1(t) + w(t), \qquad T_i \le t \le T_f : H_1,$$
$$r(t) = s_0(t) + w(t), \qquad T_i \le t \le T_f : H_0. \tag{61}$$

We assume that the signal processes $s_1(t)$ and $s_0(t)$ have identical eigenvalues and that their eigenfunctions are essentially disjoint. For stationary processes, this has the simple interpretation illustrated by the spectra in Fig. 3.7. The two processes have spectra that are essentially disjoint in frequency and are identical except for a frequency shift. The additive noise $w(t)$ is white with spectral height $N_0/2$. This class of problems is encountered frequently in binary communications over a fading channel and is just the waveform version of Case 2 on page I-114. We shall discuss the physical channel in more detail in Chapter 11 and see how this mathematical model arises. The receiver structure is just a special case of Fig. 3.3. We can obtain $\mu_{BS}(s)$ from (60) by the following observations (the subscript denotes binary symmetric):

1. The minimum mean-square filtering error only depends on the *eigenvalues* of the process. Therefore,

$$\xi_P\left(t \mid s_1(\cdot), \frac{N_0}{2}\right) = \xi_P\left(t \mid s_0(\cdot), \frac{N_0}{2}\right). \tag{62}$$

2. If two processes have *no* eigenfunctions in common, then the minimum mean-square error in filtering their sum is the sum of the minimum mean-square errors for filtering the processes individually. Therefore,

$$\xi_P\left(t \mid s_{\text{com}}(\cdot), \frac{N_0}{2}\right) = \xi_P\left(t \mid \sqrt{s}\, s_0(\cdot), \frac{N_0}{2}\right) + \xi_P\left(t \mid \sqrt{1-s}\, s_0(\cdot), \frac{N_0}{2}\right). \tag{63}$$

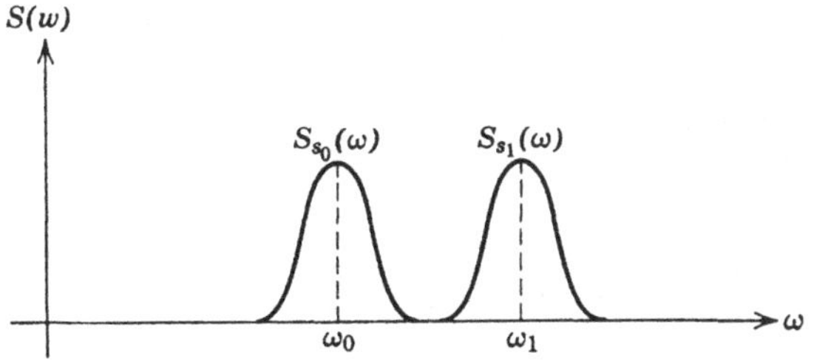


**Fig. 3.7 Disjoint processes (spectrum is symmetric around ω = 0; only positive frequencies are shown).**

Using (62) and (63) in (64), we have

$$\mu_{BS}(s) = \frac{1}{N_0}\int_{T_i}^{T_f} dt\left[(1-s)\xi_P\left(t \mid s_0(\cdot), \frac{N_0}{2}\right) - \xi_P\left(t \mid \sqrt{1-s}\, s_0(\cdot), \frac{N_0}{2}\right)\right.$$
$$\left. + s\xi_P\left(t \mid s_0(\cdot), \frac{N_0}{2}\right) - \xi_P\left(t \mid \sqrt{s}\, s_0(\cdot), \frac{N_0}{2}\right)\right]. \quad (64)$$

This can be rewritten in two different forms. Looking at the expression for $\mu(s)$ in (2.139) for the simple binary problem, we see that (64) can be written as

$$\boxed{\mu_{BS}(s) = \mu_{SIB}(s) + \mu_{SIB}(1-s),} \quad (65)\dagger$$

where the subscript $SIB$ denotes simple binary, and, from (2.139),

$$\mu_{SIB}(s) = \frac{1}{N_0}\int_{T_i}^{T_f} dt\left[(1-s)\xi_P\left(t \mid s_0(\cdot), \frac{N_0}{2}\right) - \xi_P\left(t \mid \sqrt{1-s}\, s_0(\cdot), \frac{N_0}{2}\right)\right]. \quad (66)$$

From (65), it is clear that $\mu_{BS}(s)$ is symmetric about $s = 1/2$. A second form of $\mu_{BS}(s)$ that is frequently convenient is

$$\mu_{BS}(s) = \frac{1}{N_0}\int_{T_i}^{T_f} dt\left[\xi_P\left(t \mid s_0(\cdot), \frac{N_0}{2}\right)\right.$$
$$\left. - (1-s)\xi_P\left(t \mid s_0(\cdot), \frac{N_0}{2(1-s)}\right) - s\xi_P\left(t \mid s_0(\cdot), \frac{N_0}{2s}\right)\right]. \quad (67)$$

The binary symmetric model is frequently encountered in communication systems. In most cases the a-priori probabilities of the two hypotheses are equal,

$$\Pr[H_0] = \Pr[H_1] = \tfrac{1}{2}, \quad (68)$$

and the criterion is minimum $\Pr(\epsilon)$,

$$\Pr(\epsilon) = \tfrac{1}{2}P_F + \tfrac{1}{2}P_M. \quad (69)$$

Under these conditions the threshold, $\ln\eta$, equals zero. All of our bounds and performance expressions require finding the value of $s$ where

$$\dot{\mu}(s) = \ln\eta. \quad (70)$$

In this case, we want the value of $s$ where

$$\dot{\mu}_{BS}(s) = 0. \quad (71)$$

† This particular form was derived in [2].

From the symmetry it is clear that

$$\dot{\mu}_{BS}(s)\Big|_{s=1/2} = 0. \tag{72}$$

Thus, the important quantity is $\mu_{BS}(1/2)$. From (65),

$$\boxed{\mu_{BS}(\tfrac{1}{2}) = 2\mu_{SIB}(\tfrac{1}{2}).} \tag{73}$$

From (66),

$$\boxed{\mu_{BS}(\tfrac{1}{2}) = \frac{1}{N_0}\int_{T_i}^{T_f} dt\left[\xi_P\left(t \mid s_0(\cdot), \frac{N_0}{2}\right) - \xi_P(t \mid s_0(\cdot), N_0)\right].} \tag{74}$$

Using (I-2.473), we have a bound on $\Pr(\epsilon)$,

$$\boxed{\Pr(\epsilon) \le \tfrac{1}{2}\exp\left[\mu_{BS}(\tfrac{1}{2})\right].} \tag{75}$$

In order to get an approximate error expression, we proceed in exactly the same manner as in (2.164) and (2.173). The one-term approximation is

$$\boxed{\Pr(\epsilon) \simeq \left[\operatorname{erfc}_*\left(\frac{\sqrt{\ddot{\mu}_{BS}(\frac{1}{2})}}{2}\right)\right]\exp\left(\mu_{BS}(\tfrac{1}{2}) + \frac{\ddot{\mu}_{BS}(\frac{1}{2})}{8}\right).} \tag{76}$$

When the argument of $\operatorname{erfc}_*(\cdot)$ is greater than two, this can be approximated as

$$\Pr(\epsilon) \simeq \left[\frac{2}{\pi\ddot{\mu}_{BS}(\frac{1}{2})}\right]^{1/2}\exp\left(\mu_{BS}(\tfrac{1}{2})\right). \tag{77}$$

As before, the coefficient is frequently needed in order to get a good estimate of the $\Pr(\epsilon)$. On page 79 we shall revisit this problem and investigate the accuracy of (77) in more detail.

Two other observations are appropriate:

1. From our results in (2.72) and (2.74), we know that $\mu_{BS}(s)$ can be written in terms of Fredholm determinants. Using these equations, we have

$$\mu_{BS}(s) = \tfrac{1}{2}\ln\left[\frac{D_{\mathscr{F}}(2/N_0)}{D_{\mathscr{F}}([2(1-s)]/N_0)D_{\mathscr{F}}(2s/N_0)}\right] \tag{78}$$

and

$$\mu_{BS}(\tfrac{1}{2}) = \tfrac{1}{2}\ln\left[\frac{D_{\mathscr{F}}([2(1-s)]/N_0)}{D_{\mathscr{F}}^{2}(1/N_0)}\right]. \tag{79}$$

2. The negative of $\mu(\frac{1}{2})$ has been used as criterion for judging the quality of a test by a number of people. It was apparently first introduced by Hellinger [3] in 1909. It is frequently referred to as the Bhattacharyya distance [4]. (Another name used less frequently is the Kakutani distance [5].) It is essential to observe that the importance of $\mu(\frac{1}{2})$ arises from both the symmetry of the problem *and* the choice of the threshold. If either of these elements is changed, $\mu(s)$ for some $s \neq \frac{1}{2}$ will provide a better measure of performance. It is easy to demonstrate cases in which ordering tests by their $\mu(\frac{1}{2})$ value or designing signals to minimize $\mu(\frac{1}{2})$ gives incorrect results because the model is asymmetric.

The formulas derived in this section are essential in the analysis of binary symmetric communication systems. In Chapter 5 we shall derive corresponding results for $M$-ary systems. The next topic of interest is the effect of nonzero means.

### 3.4.2 Non-zero Means

All of our discussion of the general binary problem up to this point has assumed that the processes were zero-mean on both hypotheses. In this section we consider a class $A_{wm}$ problem and show how nonzero means affect the optimum receiver structure and the system performance. The received waveforms on the two hypotheses are

$$
\begin{aligned}
r(t) &= s_1(t) + w(t), \qquad T_i \le t \le T_f : H_1, \\
r(t) &= s_0(t) + w(t), \qquad T_i \le t \le T_f : H_0,
\end{aligned}
\tag{80}
$$

where

$$E[s_1(t)] = m_1(t) \tag{81}$$

and

$$E[s_0(t)] = m_0(t). \tag{82}$$

The covariance functions of $s_1(t)$ and $s_0(t)$ are $K_{s_1}(t, u)$ and $K_{s_0}(t, u)$, respectively. The additive zero-mean white Gaussian noise is independent of the signal processes and has spectral height $N_0/2$. As in the simple binary problem, we want to obtain an expression for $l_D$ and $\mu_D(s)$. [Recall the definition of these quantities in (2.32) and (2.147).] Because of the similarity of both the derivation and the results to the simple binary case, we simply state the answers and leave the derivations as an exercise (see Problem 3.4.1).

Modifying (23), we obtain

$$l_D = \int_{T_i}^{T_f} r(u)\, du \left[ \int_{T_i}^{T_f} [m\ (t) Q_{H_1}(t, u) - m_0(t) Q_{H_0}(t, u)]\, dt \right]. \tag{83}$$

This can be written as

$$\boxed{l_D = \int_{T_i}^{T_f} r(u)[g_1(u) - g_0(u)]\,du,} \tag{84}$$

where

$$g_1(u) \triangleq \int_{T_i}^{T_f} m_1(t)Q_{H_1}(t, u)\,dt, \qquad T_i < u < T_f \tag{85}$$

and

$$g_0(u) \triangleq \int_{T_i}^{T_f} m_0(t)Q_{H_0}(t, u)\,dt, \qquad T_i < u < T_f. \tag{86}$$

The functions $g_1(u)$ and $g_0(u)$ can also be defined implicitly by the relations

$$m_1(t) = \int_{T_i}^{T_f} K_{H_1}(t, u)g_1(u)\,du, \qquad T_i \le t \le T_f \tag{87}$$

and

$$m_0(t) = \int_{T_i}^{T_f} K_{H_0}(t, u)g_0(u)\,du, \qquad T_i \le t \le T_f. \tag{88}$$

The resulting test is

$$l_R + l_D \underset{H_0}{\overset{H_1}{\gtrless}} \gamma', \tag{89}$$

where $l_R$ is given by (23) or (32) and $\gamma'$ is the threshold that includes the bias terms. An alternative expression for the test derived in Problem 3.4.1 is

$$l \triangleq \int_{T_i}^{T_f} r(t)g(t)\,dt + \frac{1}{2}\iint_{T_i}^{T_f} [r(t) - m_1(t)]h_\Delta(t, u)[r(u) - m_1(u)]\,dt\,du \underset{H_0}{\overset{H_1}{\gtrless}} \gamma'', \tag{90}$$

where $g(t)$ satisfies the equation

$$\int_{T_i}^{T_f} K_{H_0}(t, u)g(u)\,du = m_1(t) - m_0(t), \qquad T_i \le t \le T_f \tag{91}$$

and $h_\Delta(t, u)$ satisfies (33). The advantage of the form in (90) is that it requires solving two integral equations rather than three.

The derivation of $\mu_D(s)$ is a little more involved (see Problem 3.4.2). We define a function

$$m_\Delta(t) = m_0(t) - m_1(t) \tag{92}$$

and a composite signal process

$$s_{\text{com}}(t, s) = \sqrt{s}\,s_0(t) + \sqrt{1 - s}\,s_1(t), \tag{93}$$

whose covariance is denoted by $K_{com}(t, u)$. In (93), $s_0(t)$ and $s_1(t)$ are assumed to be statistically independent processes. Thus,

$$K_{com}(t, u) = sK_0(t, u) + (1 - s)K_1(t, u). \tag{94}$$

This process was encountered previously in (58). Finally we define $g_{\Delta com}(t \mid N_0/2)$ implicitly by the integral equation

$$m_\Delta(t) = \int_{T_i}^{T_f} \left[K_{com}(t, u) + \frac{N_0}{2}\delta(t - u)\right] g_{\Delta com}\left(u \,\middle|\, \frac{N_0}{2}\right) du. \tag{95}$$

Then we can show that

$$\boxed{\mu_D(s) = -\frac{s(1 - s)}{2}\int_{T_i}^{T_f} m_\Delta(t) g_{\Delta com}\left(t \,\middle|\, \frac{N_0}{2}\right) dt.} \tag{96}$$

To get the $\mu(s)$ for the entire problem, we add the $\mu(s)$ from $l_R$ [denoted now by $\mu_R(s)$ and defined in (60)] and $\mu_D(s)$.

$$\mu(s) = \mu_R(s) + \mu_D(s). \tag{97}$$

The results in (84), (90), and (96) specify the non-zero-mean problem. Some typical examples are developed in the problems.

### 3.4.3 Stationary "Carrier-symmetric" Bandpass Problems

Many of the processes that we encounter in practice are bandpass processes centered around a carrier frequency. In Chapter 11, we shall explore this class of problem in detail. By introducing suitable notation we shall be able to study the general bandpass process efficiently. In this section we consider a special class of bandpass problems that can be related easily to the corresponding low-pass problem. We introduce this special class at this point because it occurs frequently in practice. Thus, it is a good vehicle for discussing some of the solution techniques in Chapter 4.

The received waveforms on the two hypotheses are

$$\begin{aligned} r(t) &= s_1(t) + w(t), \qquad T_i \le t \le T_f : H_1, \\ r(t) &= s_0(t) + w(t), \qquad T_i \le t \le T_f : H_0. \end{aligned} \tag{98}$$

The signal $s_1(t)$ is a segment of a sample function of a zero-mean stationary Gaussian process whose spectrum is narrow-band and symmetric about a carrier $\omega_1$. The signal $s_0(t)$ is a segment of a sample function of a zero-mean stationary Gaussian process whose spectrum is narrow-band and symmetric about a carrier $\omega_0$. The two spectra are essentially disjoint, as illustrated

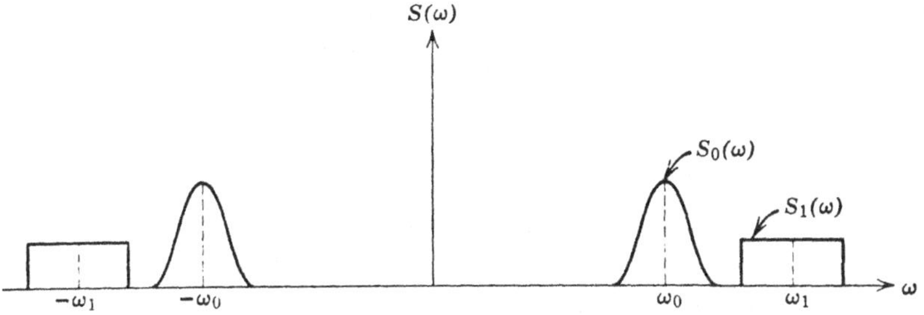


**Fig. 3.8 Disjoint bandpass spectra.**

in Fig. 3.8. This problem differs from that in Section 3.4.1 in that we do *not* require that the two processes have the same eigenvalues in the present problem.

To develop the receiver structure, we multiply $r(t)$ by the four carriers shown in Fig. 3.9 and pass the resulting outputs through ideal low-pass filters. These low-pass filters pass the frequency-shifted versions of $s_1(t)$ and $s_0(t)$ without distortion. We now have four waveforms, $r_{c_1}(t)$, $r_{s_1}(t)$, $r_{c_0}(t)$, and $r_{s_0}(t)$, to use as inputs for our likelihood ratio test. The four waveforms on the two hypotheses are

$$\left.\begin{aligned} r_{c_1}(t) &= s_{c_1}(t) + w_{c_1}(t), & T_i \le t \le T_f \\ r_{s_1}(t) &= s_{s_1}(t) + w_{s_1}(t), & T_i \le t \le T_f \\ r_{c_0}(t) &= w_{c_0}(t), & T_i \le t \le T_f \\ r_{s_0}(t) &= w_{s_0}(t), & T_i \le t \le T_f \end{aligned}\right\} H_1,$$

$$\left.\begin{aligned} r_{c_1}(t) &= w_{c_1}(t), & T_i \le t \le T_f \\ r_{s_1}(t) &= w_{s_1}(t), & T_i \le t \le T_f \\ r_{c_0}(t) &= s_{c_0}(t) + w_{c_0}(t), & T_i \le t \le T_f \\ r_{s_0}(t) &= s_{s_0}(t) + w_{s_0}(t), & T_i \le t \le T_f \end{aligned}\right\} H_0. \tag{99}$$

Because of the assumed symmetry of the spectra, *all* of the processes are statistically independent (e.g., Appendix A.3.1). The processes $s_{c_1}(t)$ and $s_{s_1}(t)$ have identical spectra, which we denote by $S_{L_1}(\omega)$. It is just the low-pass component of the bandpass spectrum after it has been shifted to the origin. Similarly, $s_{c_0}(t)$ and $s_{s_0}(t)$ have identical spectra, which we denote by $S_{L_0}(\omega)$. In view of the statistical independence, we can write the LRT by inspection. By analogy with (30), the LRT is

$$l_{R_{c_1}} + l_{R_{s_1}} + l_{B_{c_1}} + l_{B_{s_1}} - l_{R_{c_0}} - l_{R_{s_0}} - l_{B_{c_0}} - l_{B_{s_0}} \underset{H_0}{\overset{H_1}{\gtrless}} \ln \eta, \tag{100}$$

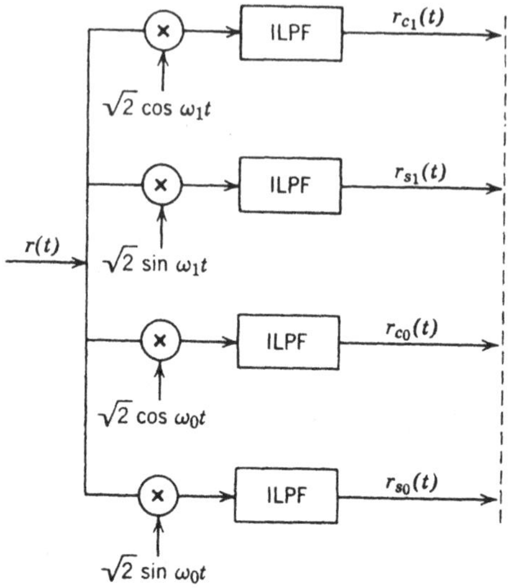


**Fig. 3.9 Generation of low-pass waveforms.**

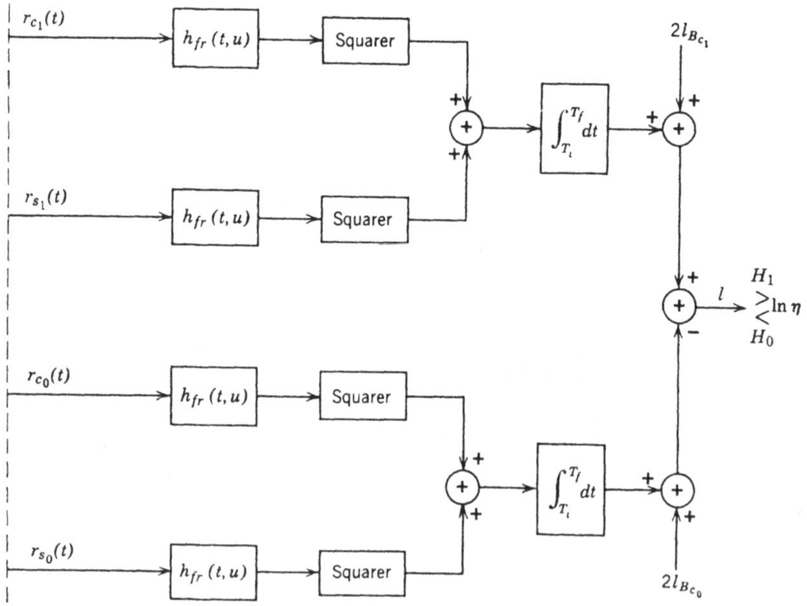


**Fig. 3.10 Optimum processing of low-pass waveforms.**

where the definitions of the various terms are parallel to (27) and (29). A filter-squarer version of the optimum receiver is shown in Fig. 3.10. (Notice that $l_{B_{c1}} = l_{B_{s1}}$.) In most cases, the filters before the squarer are low-pass, so that the ideal low-pass filters in Fig. 3.9 can be omitted. In Chapter 11, we develop a more efficient realization using bandpass filters and square-law envelope detectors.

To evaluate the performance, we observe that the sine components provide exactly the same amount of information as the cosine components. Thus, we would expect that

$$\boxed{\mu_{BP}(s) = 2\mu_{LP}(s),} \tag{101}$$

where the subscript *BP* denotes the actual bandpass problem and the subscript *LP* denotes a low-pass problem with inputs $r_{c_1}(t)$ and $r_{c_0}(t)$. Notice that the power (or energy) in the low-pass problem is one-half the power (or energy) in the bandpass problem.

$$P_{LP} = \frac{P_{BP}}{2}, \tag{102}$$

$$E_{r_{LP}} = \frac{E_{r_{BP}}}{2}. \tag{103}$$

It is straightforward to verify that (101)–(103) are correct (see Problem 3.4.8). Notice that since the bandpass process generates two statistically independent low-pass processes, we can show that the eigenvalues of the bandpass process occur in pairs.

The important conclusion is that, for this *special class* of bandpass problems, there is an equivalent low-pass problem that can be obtained by a simple scale change. Notice that three assumptions were made:

1. The signal processes are stationary.
2. The signal spectra on the two hypotheses are essentially disjoint.
3. The signal spectra are symmetric about their respective carriers.

Later we shall consider asymmetric spectra and nonstationary spectra. In those cases the transition will be more involved and it will be necessary to develop a more efficient notation.

### 3.4.4 Error Probability for the Binary Symmetric Bandpass Problem

In this section we consider the binary symmetric bandpass problem. The model for this problem satisfies the assumptions of both Sections

3.4.1 and 3.4.3. We shall derive tight upper and lower bounds for the Pr ($\epsilon$).†

Because we have assumed equally likely hypotheses and a minimum total probability of error criterion, we have

$$\begin{aligned}\Pr[\epsilon] &= \Pr[\epsilon \mid H_0]\\ &= \int_0^{\infty} p_{l|H_0}(L)\, dL\\ &= \int_0^{\infty} \frac{1}{2\pi j}\int_{\sigma-j\infty}^{\sigma+j\infty} M_{l|H_0}(w)e^{-wL}\, dw\, dL \quad \text{for} \quad 0 \le \sigma \le 1. \end{aligned}\tag{104}$$

Notice that $w$ is a complex variable,‡

$$w = \sigma + jv. \tag{105}$$

Interchanging the order of integration and evaluating the results in the integral, we obtain

$$\begin{aligned}\Pr[\epsilon] &= \frac{1}{2\pi j}\int_{\sigma-j\infty}^{\sigma+j\infty} M_{l|H_0}(w)\int_0^{\infty} e^{-wL}\, dL\, dw\\ &= \frac{1}{2\pi j}\int_{\sigma-j\infty}^{\sigma+j\infty} \frac{1}{w} M_{l|H_0}(w)\, dw\\ &= \frac{1}{2\pi j}\int_{\sigma-j\infty}^{\sigma+j\infty} \frac{e^{\mu(w)}}{w}\, dw, \qquad 0 < \sigma \le 1. \end{aligned}\tag{106}$$

For our specific problem, $\mu(w)$ follows immediately from (57) and (67),

$$\mu(w) = \sum_{i=1}^{\infty}\left[\ln\left(1 + \frac{2\lambda_i}{N_0}\right) - \ln\left(1 + \frac{2w\lambda_i}{N_0}\right) - \ln\left(1 + \frac{2(1-w)\lambda_i}{N_0}\right)\right]. \tag{107}$$

Notice that we have used (101) to eliminate the one-half factor in $\mu(s)$. As pointed out earlier, this is because the eigenvalues appear in pairs in the bandpass problem. From (107),

$$e^{\mu(w)} = \prod_{i=1}^{\infty} \frac{(1 + (2\lambda_i/N_0))}{(1 + (2w\lambda_i/N_0))(1 + (2(1-w)\lambda_i/N_0))}. \tag{108}$$

Thus,

$$\Pr[\epsilon] = \frac{1}{2\pi j}\int_{\sigma-j\infty}^{\sigma+j\infty} \frac{1}{w}\prod_{i=1}^{\infty} \frac{(1 + (2\lambda_i/N_0))\, dw}{(1 + (2w\lambda_i/N_0))(1 + (2(1-w)\lambda_i/N_0))} \quad \text{for} \quad 0 < \sigma \le 1. \tag{109}$$

† Our discussion in this section follows [2]. The original results are due to Pierce [6].
‡ All of our previous discussions assumed that the argument of $M_{l|H_0}(\cdot)$ was real. The necessary properties are also valid for complex arguments with the restriction $0 \le \text{Re}[w] \le 1$.

The result in (109) is due to Turin ([7], Eq. 27). Pierce [6] started with (109) and derived tight upper and lower bounds on Pr ($\epsilon$). Since his derivation is readily available, we omit it and merely state the result. (A simple derivation using our notation is given in [2].) We can show that

$$\operatorname{erfc}_*\left(-\frac{\sqrt{\ddot{\mu}(\frac{1}{2})}}{2}\right)\exp\left[\mu(\tfrac{1}{2})+\frac{\ddot{\mu}(\frac{1}{2})}{8}\right] \le \Pr[\epsilon] \le \frac{\exp[\mu(\frac{1}{2})]}{2(1+\sqrt{\ddot{\mu}(\frac{1}{2})/8})}. \tag{110}$$

The lower bound can be further relaxed to yield

$$\boxed{\frac{\exp[\mu(\frac{1}{2})]}{2(1+\sqrt{(\frac{\pi}{8})\ddot{\mu}(\frac{1}{2})})} \le \Pr[\epsilon] \le \frac{\exp[\mu(\frac{1}{2})]}{2(1+\sqrt{(\frac{1}{8})\ddot{\mu}(\frac{1}{2})})},} \tag{111}$$

which is Pierce's result. Notice that the upper and lower bounds differ at most by $\sqrt{\pi}$. From (76), we see that our first-order approximation to Pr ($\epsilon$) is identical with the lower bound in (110). Thus, for the *binary-symmetric* bandpass case, our approximate error expression is always within a factor of $\sqrt{\pi}$ of the exact Pr ($\epsilon$). Notice that our result assumes that the spectra are symmetric about their carriers. The results in (110) and (111) are also valid for asymmetric spectra. We shall prove this in Chapter 11.

We have not been able to extent Pierce's derivation to the asymmetric case in order to obtain tight bounds on $P_D$ and $P_F$. However several specific examples indicate that our approximate error expressions give accurate results.

In this section we have examined four special models of the Gaussian problem. In the next section we return to the general problem and look at the effect of removing the white noise assumption.

## 3.5 GENERAL BINARY CASE: WHITE NOISE NOT NECESSARILY PRESENT: SINGULAR TESTS

In this section, we discuss the general binary problem briefly. The received waveforms on the two hypotheses are

$$\begin{aligned} r(t) &= r_1(t), \qquad T_i \le t \le T_f : H_1,\\ r(t) &= r_0(t), \qquad T_i \le t \le T_f : H_0. \end{aligned} \tag{112}$$

The processes are zero-mean Gaussian processes with covariance functions $K_{H_1}(t, u)$ and $K_{H_0}(t, u)$, respectively. We assume that both processes are (strictly) positive-definite.

### 3.5.1 Receiver Derivation

To solve the problem we first pass $r(t)$ through a whitening filter to generate an output $r_*(t)$, which is white on $H_0$. Previously we have denoted the whitening filter by $h_w(t, z)$. In our present discussion we shall denote it by $K_{H_0}^{[-1/2]}(t, z)$. The reason for this notation will become obvious shortly.

$$r_*(t) = \int_{T_i}^{T_f} K_{H_0}^{[-1/2]}(t, z) r(z)\, dz. \tag{113}$$

The whitening requirement implies

$$\begin{aligned} E[r_*(t)r_*(u) \mid H_0] &= \delta(t - u) \\ &= \iint_{T_i}^{T_f} K_{H_0}^{[-1/2]}(t, z) K_{H_0}(z, y) K_{H_0}^{[-1/2]}(u, y)\, dz\, dy. \end{aligned} \tag{114}$$

On $H_1$, the covariance of $r_*(t)$ is

$$E[r_*(t)r_*(u) \mid H_1] = \iint_{T_i}^{T_f} K_{H_0}^{[-1/2]}(t, z) K_{H_1}(z, y) K_{H_0}^{[-1/2]}(u, y)\, dz\, dy \triangleq K_1^*(t, u). \tag{115}$$

We can now expand $r_*(t)$ using the eigenfunctions of $K_1{}^*(t, u)$,

$$\lambda_i^* \phi_i(t) = \int_{T_i}^{T_f} K_1^*(t, u) \phi_i(u)\, du, \qquad T_i \le t \le T_f. \tag{116}$$

The coefficients are

$$r_i \triangleq \int_{T_i}^{T_f} r_*(t) \phi_i(t)\, dt, \tag{117}$$

and the waveform is

$$r_*(t) = \underset{K \to \infty}{\text{l.i.m.}} \sum_{i=1}^{K} r_i \phi_i(t), \qquad T_i \le t \le T_f. \tag{118}$$

The coefficients are zero-mean. Their covariances are

$$E[r_i r_j \mid H_1] = \lambda_i{}^* \delta_{ij} \tag{119}$$

and

$$E[r_i r_j \mid H_0] = \delta_{ij}. \tag{120}$$

Notice that we could also write (116) as

$$\lambda_i^* \phi_i(t) = \int_{T_i}^{T_f} \left[ \iint_{T_i}^{T_f} K_{H_0}^{[-1/2]}(t, z) K_{H_1}(z, y) K_{H_0}^{[-1/2]}(u, y)\, dz\, dy \right] \phi_i(u)\, du, \qquad T_i \le t \le T_f. \tag{121}$$

Now we define the function $K_{H_0}^{[1/2]}(z, t)$ implicitly by the relation

$$K_{H_0}(t, u) = \int_{T_i}^{T_f} K_{H_0}^{[1/2]}(z, t) K_{H_0}^{[1/2]}(z, u)\, dz. \tag{122}$$

We see that $K_{H_0}^{[1/2]}(z,t)$ is just the functional square root of the covariance function $K_{H_0}(t, u)$. Observe that

$$\delta(t - u) = \int_{T_i}^{T_f} K_{H_0}^{[1/2]}(t, z) K_{H_0}^{[-1/2]}(z, u)\, dz. \tag{123}$$

The result in (123) can be verified by writing each term in an orthogonal series expansion. Multiplying both sides of (121) by $K_{H_0}^{[-1/2]}(\cdot, \cdot)$, integrating, and using (123), we obtain

$$\lambda_i^* \int_{T_i}^{T_f} K_{H_0}(t, u)\, du \left[ \int_{T_i}^{T_f} K_{H_0}^{[-1/2]}(u, z) \phi_i(z)\, dz \right] = \int_{T_i}^{T_f} K_{H_1}(t, u)\, du \left[ \int_{T_i}^{T_f} K_{H_0}^{[-1/2]}(u, z) \phi_i(z)\, dz \right]. \tag{124}$$

If we define

$$\psi_i(u) \triangleq \int_{T_i}^{T_f} K_{H_0}^{[-1/2]}(u, z) \phi_i(z)\, dz, \tag{125}$$

(124) becomes

$$\boxed{\lambda_i^* \int_{T_f}^{T_f} K_{H_0}(t, u) \psi_i(u)\, du = \int_{T_i}^{T_f} K_{H_1}(t, u) \psi_i(u)\, du.} \tag{126}$$

Notice that we could also write the original waveform $r(t)$ as

$$r(t) = \underset{K \to \infty}{\text{l.i.m.}} \sum_{i=1}^{K} r_i \left\{ \int_{T_i}^{T_f} K_{H_0}(t, u) \psi_i(u)\, du \right\}. \tag{127}$$

Thus we have available a decomposition that gives statistically independent coefficients on both hypotheses. The likelihood ratio test is

$$\Lambda(\mathbf{R}) = \frac{\displaystyle\prod_{i=1}^{K} \frac{1}{(2\pi)^{K/2} (\lambda_i^*)^{1/2}} \exp\left(-\frac{1}{2}\frac{R_i^2}{\lambda_i^*}\right)}{\displaystyle\prod_{i=1}^{K} \frac{1}{(2\pi)^{K/2}} \exp\left(-\tfrac{1}{2} R_i^2\right)} \underset{H_0}{\overset{H_1}{\gtrless}} \eta. \tag{128}$$

If we let $K \to \infty$, this reduces to

$$l_R \triangleq \frac{1}{2} \sum_{i=1}^{\infty} R_i^2 \left( \frac{\lambda_i^* - 1}{\lambda_i^*} \right) \underset{H_0}{\overset{H_1}{\gtrless}} \ln \eta + \frac{1}{2} \sum_{i=1}^{\infty} \ln \lambda_i^* \triangleq \gamma. \tag{129}$$

We now define a kernel,

$$h_*(t, u) \triangleq \sum_{i=1}^{\infty} \left( \frac{\lambda_i^* - 1}{\lambda_i^*} \right) \phi_i(t) \phi_i(u), \qquad T_i \le t, u \le T_f, \tag{130}$$

that satisfies the integral equation

$$\int_{T_i}^{T_f} h_*(t, z)K_1^*(z, u)\, dz = K_1^*(t, u) - \delta(t - u), \qquad T_i \le t, u \le T_f. \tag{131}$$

Then

$$l_R = \tfrac{1}{2} \iint_{T_i}^{T_f} r_*(t)h_*(t, u)r_*(u)\, dt\, du. \tag{132}$$

Using (113), we have

$$l_R = \tfrac{1}{2} \iint_{T_i}^{T_f} dx\, dy\, r(x)\left\{\iint_{T_i}^{T_f} K_{H_0}^{[-\frac{1}{2}]}(t, x)h_*(t, u)K_{H_0}^{[-\frac{1}{2}]}(u, y)\, dt\, du\right\}r(y). \tag{133}$$

Defining

$$h_\Delta(x, y) \triangleq \iint_{T_i}^{T_f} K_{H_0}^{[-\frac{1}{2}]}(t, x)h_*(t, u)K_{H_0}^{[-\frac{1}{2}]}(u, y)\, dt\, du, \tag{134}$$

we have

$$\boxed{l_R = \tfrac{1}{2} \iint_{T_i}^{T_f} dx\, dy\, r(x)h_\Delta(x, y)r(y).} \tag{135}$$

Starting with (134), it is straightforward to show that $h_\Delta(x, y)$ satisfies the equation

$$\boxed{\iint_{T_i}^{T_f} K_{H_0}(t, x)h_\Delta(x, y)K_{H_1}(y, u)\, dx\, dy = K_{H_1}(t, u) - K_{H_0}(t, u), \qquad T_i \le t, u \le T_f.} \tag{136}$$

As we would expect, the result in (136) is identical with that in (33). The next step is to evaluate the performance of the optimum receiver.

### 3.5.2 Performance: General Binary Case

To evaluate the performance in the general binary case, we use (2.126) to evaluate $\mu(s)$.

$$\mu(s) = \lim_{K\to\infty} \sum_{i=1}^{K} \ln \left[\int_{-\infty}^{\infty} \frac{1}{\sqrt{2\pi}\,(\lambda_i^*)^{s/2}} \exp\left(-\frac{sR_i^2}{2\lambda_i^*}\right) \exp\left(-\frac{(1-s)R_i^2}{2}\right) dR_i\right]. \tag{137}$$

Evaluating the integral, we have

$$\boxed{\mu(s) = \sum_{i=1}^{\infty} \ln \left\{ \frac{(\lambda_i^*)^{(1-s)/2}}{(s + (1-s)\lambda_i^*)^{1/2}} \right\},} \tag{138}$$

where the $\lambda_i^*$ are the eigenvalues of the kernel,

$$\boxed{K_1^*(t, u) = \int_{T_i}^{T_f} K_{H_0}^{[-1/2]}(t, z) K_{H_1}(z, x) K_{H_0}^{[-1/2]}(u, x)\, dz\, dx.} \tag{139}$$

In our discussion of performance for the case of a known signal in Gaussian noise, we saw that when there was no white noise present it was possible to make perfect decisions under some circumstances (see pages I-303–I-306). We now consider the analogous issue for the general Gaussian problem.

### 3.5.3 Singularity

The purpose of our singularity discussion is to obtain a necessary and sufficient condition for the test to be nonsingular. The derivation is a sequence of lemmas. As before, we say that a test is singular if $\Pr(\epsilon) = 0$. Notice that we do not assume equally likely hypotheses, and

$$\Pr(\epsilon) = P_1 \Pr(\epsilon \mid H_1) + P_0 \Pr(\epsilon \mid H_0). \tag{140}$$

The steps in the development are the following:

1. We show that the $\Pr(\epsilon)$ is greater than zero iff $\mu(\frac{1}{2})$ is finite.
2. We then derive a necessary and sufficient condition for $\mu(\frac{1}{2})$ to be finite.

Finally we consider two simple examples of singular tests.

**Lemma 1.** The $\Pr(\epsilon)$ can be bounded by

$$\tfrac{1}{2}\{\min[P_{H_1}, P_{H_0}]\} e^{2\mu(1/2)} \le \Pr(\epsilon) \le \tfrac{1}{2} e^{\mu(1/2)} \tag{141}$$

Therefore the $\Pr(\epsilon)$ will equal zero if $P_0$ or $P_1$ or $e^{\mu(1/2)}$ equals zero. If we assume that $P_1$ and $P_0$ are positive then $\Pr(\epsilon)$ will be greater than zero *iff* $\mu(\frac{1}{2})$ is finite. In other words, a singular test will occur *iff* $\mu(\frac{1}{2})$ diverges.

The upper bound is familiar. The proof of the lower bound is straightforward.

*Proof.*† Let

$$\alpha \triangleq e^{\mu(1/2)} = \int_{-\infty}^{\infty} [p_{\mathbf{r}|H_1}(\mathbf{R} \mid H_1) p_{\mathbf{r}|H_0}(\mathbf{R} \mid H_0)]^{1/2}\, d\mathbf{R}. \tag{142a}$$

† This result is similar to that in [8].

Now observe, from the Schwarz inequality, that for any set $S$,

$$\int_S [p_{\mathbf{r}|H_1}(\mathbf{R} \mid H_1) p_{\mathbf{r}|H_0}(\mathbf{R} \mid H_0)]^{1/2}\, d\mathbf{R} \le \left[\int_S p_{\mathbf{r}|H_1}(\mathbf{R} \mid H_1)\, d\mathbf{R} \int_S p_{\mathbf{r}|H_0}(\mathbf{R} \mid H_0)\, d\mathbf{R}\right]^{1/2}$$

$$\le \left[\int_S p_{\mathbf{r}|H_m}(\mathbf{R} \mid H_m)\, d\mathbf{R}\right]^{1/2}, \qquad m = 0, 1. \tag{142b}$$

We recall from page I-30 that the probability of error using the optimum test is

$$\Pr(\epsilon) = P_1 \int_{Z_0} p_{\mathbf{r}|H_1}(\mathbf{R} \mid H_1)\, d\mathbf{R} + P_0 \int_{Z_1} p_{\mathbf{r}|H_0}(\mathbf{R} \mid H_0)\, d\mathbf{R}$$

$$\ge \min\,[P_1, P_0]\left\{\int_{Z_0} p_{\mathbf{r}|H_1}(\mathbf{R} \mid H_1)\, d\mathbf{R} + \int_{Z_1} p_{\mathbf{r}|H_0}(\mathbf{R} \mid H_0)\, d\mathbf{R}\right\} \tag{143a}$$

Using the result in (142b) on each integral in (143a) gives

$$\Pr(\epsilon) \ge \min\,[P_1, P_0]\left\{\left(\int_{Z_0} [p_{\mathbf{r}|H_1}(\mathbf{R} \mid H_1) p_{\mathbf{r}|H_0}(\mathbf{R} \mid H_0)]^{1/2}\, d\mathbf{R}\right)^2 + \left(\int_{Z_1} [p_{\mathbf{r}|H_1}(\mathbf{R} \mid H_1) p_{\mathbf{r}|H_0}(\mathbf{R} \mid H_0)]^{1/2}\, d\mathbf{R}\right)^2\right\}$$

$$= \min\,[P_1, P_0]\left\{\left(\int_{Z_0} [p_{\mathbf{r}|H_1}(\mathbf{R} \mid H_1) p_{\mathbf{r}|H_0}(\mathbf{R} \mid H_0)]^{1/2}\, d\mathbf{R}\right)^2 + \left(\alpha - \int_{Z_0} [p_{\mathbf{r}|H_1}(\mathbf{R} \mid H_1) p_{\mathbf{r}|H_0}(\mathbf{R} \mid H_0)]^{1/2}\, d\mathbf{R}\right)^2\right\}$$

$$= \min\,[P_1, P_0]\{x^2 + (\alpha - x)^2\}, \tag{143b}$$

where

$$x \triangleq \int_{Z_0} [p_{\mathbf{r}|H_1}(\mathbf{R} \mid H_1) p_{\mathbf{r}|H_0}(\mathbf{R} \mid H_0)]^{1/2}\, d\mathbf{R}, \tag{143c}$$

and $x$ will lie somewhere in the range $[0, \alpha]$.

The term in the brackets in (143b) could be minimized by setting

$$x = \frac{\alpha}{2} \tag{143d}$$

and the minimum value is

$$\frac{\alpha^2}{2} = \frac{1}{2} e^{2\mu(1/2)} \tag{143e}$$

Thus,

$$\Pr(\epsilon) \ge \min\,[P_1, P_0] \cdot \frac{1}{2} e^{2\mu(1/2)} \tag{144}$$

which is the desired result. We should emphasize that the lower bound in (141) is used for the purpose of our singularity discussion and so it does not need to be a tight bound.

**Lemma 2.** From (138),

$$\mu(\tfrac{1}{2}) = \frac{1}{4}\sum_{i=1}^{\infty} \ln\left(\frac{4\lambda_i^*}{(1+\lambda_i^*)^2}\right). \tag{145}$$

In order for $\mu(\frac{1}{2})$ to be finite, all $\lambda_i^*$ must be greater than zero. If this is true, then, in order for $\mu(\frac{1}{2})$ to be finite, it is necessary and sufficient that

$$\sum_{i=1}^{\infty} (1-\lambda_i^*)^2 < \infty. \tag{146}$$

*Proof* (from [9]). The convergence properties of the following sums can be demonstrated.

$$\sum_{i=1}^{\infty} \ln\left(\frac{4\lambda_i^*}{(1+\lambda_i^*)^2}\right) < \infty \tag{147}$$

iff

$$\sum_{i=1}^{\infty}\left(1-\frac{4\lambda_i^*}{(1+\lambda_i^*)^2}\right) = \sum_{i=1}^{\infty}\frac{(1-\lambda_i^*)^2}{(1+\lambda_i^*)^2} < \infty \tag{148}$$

iff

$$\sum_{i=1}^{\infty} (1-\lambda_i^*)^2 < \infty. \tag{149}$$

These equivalences can be verified easily.

**Lemma 3.** Define

$$\lambda_i^{**} \triangleq \lambda_i^* - 1 \tag{150}$$

and a kernel,

$$Y(t,u) \triangleq \iint_{T_i}^{T_f} K_{H_0}^{[-1/2]}(t,x)K_{H_1}(x,z)K_{H_0}^{[-1/2]}(u,z)\,dx\,dz - \delta(t-u), \qquad T_i \le t, u \le T_f. \tag{151}$$

The $\lambda_i^{**}$ are the eigenvalues of $Y(t, u)$. Notice that $Y(t, u)$ is not necesarily positive-definite (i.e., some of the $\lambda_i^{**}$ may be negative).

**Lemma 4.** The value of $\mu(\frac{1}{2})$ will be finite iff:

(i) All $\lambda_i^{**} > -1$,

(ii) The sum $\sum_{i=1}^{\infty} (\lambda_i^{**})^2$ is finite.

Assuming the first condition is satisfied, then, in order for

$$\sum_{i=1}^{\infty} (\lambda_i^{**})^2 < \infty \tag{152}$$

it is necessary and sufficient that

$$\iint_{T_i}^{T_f} Y^2(t, u)\, dt\, du < \infty. \tag{153}$$

The equation (151) can also be written as

$$K_\Delta(t, u) \triangleq K_{H_1}(t, u) - K_{H_0}(t, u) = \int_{T_i}^{T_f} K_{H_0}(t, x) Y(x, u)\, dx, \qquad T_i \leq t, u \leq T_f. \tag{154}$$

This equation must have a square-integrable solution.

Summarizing, a necessary and sufficient condition for a nonsingular test is that the function $Y(t, u)$ defined by (151) or (154) be square-integrable and not have $-1$ as an eigenvalue.

Several observations are useful.

1. The result in (150)–(154) has a simple physical interpretation. The covariance function of the whitened waveform $r_*(t)$ on $H_1$ must consist of an impulse with unit area and a positive-definite square-integrable component.

2. The problem is symmetric, so that the entire discussion is valid with the subscripts 0 and 1 interchanged. Thus we can check the conditions given in (151) and (153) for whichever case is the simplest. Notice that it is not necessary to check both.

3. The function $\mu(s)$ can be written in terms of the eigenvalues of $Y(t, u)$. Using (138) and (150),

$$\mu(s) = \sum_{i=1}^{\infty} \ln \left[ \frac{(1 + \lambda_i^{**})^{(1-s)/2}}{(1 + (1 - s)\lambda_i^{**})^{1/2}} \right], \tag{155}$$

where the $\lambda_i^{**}$ are the eigenvalues of $Y(t, u)$, which may be either positive or negative. Notice that in order for $\mu(s)$ to be finite, it is sufficient, but not necessary, for the logarithm of the numerator and denominator of (155) to converge individually (see Problem 3.5.11).

We now consider two simple examples of singular tests.

**Example 1.** Let

$$K_{H_0}(t, u) = K(t, u) \tag{156}$$

and

$$K_{H_1}(t, u) = \alpha K(t, u). \tag{157}$$

Then

$$\int\!\!\int_{T_i}^{T_f} K_{H_0}^{[-1/2]}(t, x) K_{H_1}(x, y) K_{H_0}^{[-1/2]}(u, y)\, dx\, dy = \alpha\delta(t-u) \tag{158}$$

and

$$Y(t, u) = (\alpha - 1)\delta(t-u), \tag{159}$$

which is not square-integrable unless $\alpha = 1$.

Thus, when the covariance functions on the two hypotheses are identical except for an amplitude factor, the test is singular.

**Example 2.** Let

$$K_{H_0}(t, u) = P_0 \exp(-\alpha|t-u|) \tag{160}$$

and

$$K_{H_1}(t, u) = P_1 \exp(-\beta|t-u|). \tag{161}$$

For this particular example, the simplest procedure is to construct the whitening filter From page I-312,

$$r_*(t) = \frac{1}{\sqrt{2\alpha P_0}}[\dot{r}(t) + \alpha r(t)], \tag{162}$$

or

$$K_{H_0}^{[-1/2]}(t, u) = \frac{1}{\sqrt{2\alpha P_0}}[\delta^{[1]}(t-u) + \alpha\delta(t-u).] \tag{163†}$$

The covariance function of $r_*(t)$ on $H_1$ is

$$K_1^*(t, u) = \frac{1}{2\alpha P_0}\left\{\frac{\partial^2 K_{H_1}(t, u)}{\partial t\, \partial u} + \alpha\frac{\partial K_{H_1}(t, u)}{\partial t} + \alpha\frac{\partial K_{H_1}(t, u)}{\partial u} + \alpha^2 K_{H_1}(t, u)\right\}. \tag{164}$$

Only the first term contains an impulse,

$$\frac{\partial^2 K_{H_1}(t, u)}{\partial t\, \partial u} = 2\beta P_1\, \delta(t-u) - \beta^2 P_1 \exp(-\beta|t-u|). \tag{165}$$

In order for the test to be nonsingular, we require

$$\frac{\beta P_1}{\alpha P_0} = 1. \tag{166}$$

Otherwise (153) cannot be satisfied.

Example 2 suggests a simple test for singularity that can be used when the random processes on the two hypotheses are stationary with rational spectra. In this case, a necessary and sufficient condition for a nonsingular test is

$$\boxed{\lim_{\omega \to \infty}\left(\frac{S_{H_1}(\omega)}{S_{H_0}(\omega)}\right) = 1} \tag{167}$$

(see Problem 3.5.12).

† The symbol $\delta^{[1]}(\tau)$ denotes a doublet at $\tau = 0$.

Several other examples of singular tests are discussed in the problems. For a rigorous and more detailed discussion, the interested reader should consult Root [9], Feldman [10]–[11], Hájek [12] or Shepp [15]. As we commented in our discussion of the class $A_w$ problem, we can guarantee that the test is nonsingular by including the same white noise component on both hypotheses. Since the inclusion of the white noise component usually can be justified on physical grounds, we can avoid the singularity problem in this manner. We now summarize our results for the general binary detection problem.

## 3.6 SUMMARY: GENERAL BINARY PROBLEM

In this chapter we have discussed the general binary problem. In our initial discussion we considered class $A_w$ problems. In this class, the same white noise process was present on both hypotheses. The likelihood ratio test can be implemented as a parallel processing receiver that computes $l_{R_i}$ and $l_{B_i}$,

$$l_{R_i} = \frac{1}{N_0} \iint_{T_i}^{T_f} r(\alpha) h_i(\alpha, \beta) r(\beta)\, d\alpha\, d\beta, \qquad i = 0, 1, \tag{168}$$

where $h_i(\alpha, \beta)$ is defined by (25), and

$$l_{B_i} = -\frac{1}{N_0} \int_{T_i}^{T_f} \xi_{P_i}\left(t \,\middle|\, \frac{N_0}{2}\right) dt, \qquad i = 0, 1, \tag{169}$$

where $\xi_{P_i}(t \mid N_0/2)$ is a MMSE defined on page 22. The receiver then performs the test

$$l_{R_1} + l_{B_1} - l_{R_0} - l_{B_0} \underset{H_0}{\overset{H_1}{\gtrless}} \ln \eta. \tag{170}$$

The processing indicated in (168) can be implemented using the one of the canonical realizations developed in Section 2.1. The performance was calculated by computing $\mu(s)$ defined in (60).

$$\mu(s) = \frac{1}{N_0} \int_{T_i}^{T_f} dt \left[ (1 - s) \xi_P\left(t \,\middle|\, s_1(\cdot), \frac{N_0}{2}\right) + s \xi_P\left(t \,\middle|\, s_0(\cdot), \frac{N_0}{2}\right) - \xi_P\left(t \,\middle|\, s_{\text{com}}(\cdot), \frac{N_0}{2}\right) \right], \tag{171}$$

where the individual terms are defined on page 68.

For the general binary case, the test is

$$l_R + l_B \underset{H_0}{\overset{H_1}{\gtrless}} \ln \eta, \tag{172}$$

where

$$l_R = \frac{1}{2}\iint_{T_i}^{T_f} dx\, dy\, r(x)h_\Delta(x, y)r(y) \tag{173}$$

and

$$l_B = -\frac{1}{2}\sum_{i=1}^{\infty} \ln(\lambda_i^*). \tag{174}$$

The kernel $h_\Delta(x, y)$ satisfies the equation

$$\iint_{T_i}^{T_f} K_{H_0}(t, x)h_\Delta(x, y)K_{H_1}(y, u)\, dx\, dy = K_{H_1}(t, u) - K_{H_0}(t, u),$$

$$T_i \le t, u \le T_f. \tag{175}$$

The $\lambda_i^*$ are the eigenvalues of the kernel

$$K_1^*(t, u) = \iint_{T_i}^{T_f} K_{H_0}^{[-1/2]}(t, z)K_{H_1}(z, x)K_{H_0}^{[-1/2]}(u, x)\, dz\, dx. \tag{176}$$

When we remove the white noise assumption, we have to be careful that our model does not lead to a singular test. We demonstrated that a necessary and sufficient condition for a nonsingular test is that

$$Y(t, u) \triangleq K_1^*(t, u) - \delta(t - u), \qquad T_i \le t, u \le T_f \tag{177}$$

be a square-integrable function which does not have $-1$ as an eigenvalue. The performance could be evaluated by computing $\mu(s)$:

$$\mu(s) = \sum_{i=1}^{\infty} \ln\left\{\frac{(1 + \lambda_i^{**})^{(1-s)/2}}{(1 + (1 - s)\lambda_i^{**})^{1/2}}\right\}, \tag{178}$$

where the $\lambda_i^{**}$ are the eigenvalues of $Y(t, u)$.

In addition to our general discussion, we considered several special situations. The first was the binary symmetric problem. The most important result was the relationship of $\mu_{BS}(s)$ to $\mu_{SIB}(s)$ for the simple binary problem of Section 3.4,

$$\mu_{BS}(s) = \mu_{SIB}(s) + \mu_{SIB}(1 - s). \tag{179}$$

We also observed that when $\ln \eta = 0$, $\mu_{BS}(\frac{1}{2})$ was the appropriate quantity for the Pr ($\epsilon$) bounds.

The second situation was the non-zero-mean case. This resulted in two new terms,

$$l_D = \int_{T_i}^{T_f} r(u)[g_1(u) - g_0(u)]\, du, \tag{180}$$

in the LRT. The functions $g_i(u)$ were specified by

$$m_i(t) = \int_{T_i}^{T_f} K_{H_i}(t, u) g_i(u)\, du, \qquad T_i \le t \le T_f, \quad i = 0, 1. \tag{181}$$

In the performance calculation we added a term $\mu_D(s)$, which was specified by (96).

We then observed that for bandpass processes whose spectra are symmetric about the carrier there is a simple relationship between the actual bandpass problem and an equivalent low-pass problem. Finally, for the binary symmetric bandpass problem, a tight bound on the Pr $(\epsilon)$ was derived.

$$\frac{\exp[\mu(\frac{1}{2})]}{2(1 + \sqrt{(\pi/8)\ddot{\mu}(\frac{1}{2})})} \le \Pr(\epsilon) \le \frac{\exp[\mu(\frac{1}{2})]}{2(1 + \sqrt{(1/8)\ddot{\mu}(\frac{1}{2})})}. \tag{182}$$

This bound was useful for this particular problem. In addition, it provided a good estimate of the accuracy of our approximate expression. There are large numbers of problems in which we can evaluate the approximate expression but have not been able to find tight bounds.

Throughout our discussion in Chapters 2 and 3, we have encountered linear filters, estimates of random processes, and mean-square error expressions that we had to find in order to specify the optimum receiver and its performance completely. In many cases we used processes with finite state representations as examples, because the procedure for finding the necessary quantities was easy to demonstrate. In the next chapter we consider three other categories of problems for which we can obtain a complete solution.

## 3.7 PROBLEMS

### P.3.2 Receiver Structures

**Problem 3.2.1.**

1. Verify the result in (21) by following the suggested approach.
2. Verify that the bias term can be written as in (28).

*Comment*: In many of the problems we refer to a particular class of problems. These classes were defined in Fig. 3.1.

**Problem 3.2.2.** Consider the class $A_w$ problem in which both $s_0(t)$ and $s_1(t)$ are Wiener processes,

$$E[s_0^2(t)] = \sigma_0^2 t, \qquad 0 \le t$$

and

$$E[s_1^2(t)] = \sigma_1^2 t, \qquad 0 \le t.$$

Find the optimum receiver. Use a parallel processing configuration initially and then simplify the result. Describe the filters using state equations.

**Problem 3.2.3.** Consider the class $A_w$ problem in which

$$K_{s_1}(t, u) = \alpha K_{s_0}(t, u). \tag{P.1}$$

1. Use the condition in (P.1) to simplify the optimum receiver.
2. Derive the optimum receiver directly for the case in (P.1). (Do not use the results of Chapter 3. You may use the results of Chapter 2.)

**Problem 3.2.4.** Consider the class $A_w$ problem in which $s_0(t)$ is a Wiener process and $s_1(t)$ is a sample function from a stationary Gaussian random process whose spectrum is

$$S_{s_1}(\omega) = \frac{2kP}{\omega^2 + k^2}.$$

Find the optimum receiver.

**Problem 3.2.5.** Consider the class $B_w$ problem in which both $s(t)$ and $n_c(t)$ have finite-dimensional state representations. Derive a state-variable realization for the optimum receiver. The receiver should contain $\hat{s}_r(t)$, the MMSE realizable estimate, as one of the internal waveforms. (Notice that the parallel processing receiver in Fig. 3.3 will satisfy this requirement if we use Canonical Realization No. 4S in each path. The desired receiver is analogous to that in Fig. 3.5.)

**Problem 3.2.6 (continuation).** Consider the special case of Problem 3.2.5 in which $n_c(t)$ is a Wiener process and $s(t)$ is a stationary process whose spectrum is

$$S_s(\omega) = \frac{2kP}{\omega^2 + k^2}.$$

Specify the optimum receiver in Problem 3.2.5 completely.

**Problem 3.2.7.** In the vector version of the class $A_w$ problem, the received waveforms are

$$\mathbf{r}(t) = \mathbf{s}_1(t) + \mathbf{w}(t), \qquad T_i \le t \le T_f : H_1,$$
$$\mathbf{r}(t) = \mathbf{s}_0(t) + \mathbf{w}(t), \qquad T_i \le t \le T_f : H_0.$$

The signal processes are sample functions from $N$-dimensional, zero-mean, vector Gaussian random processes with covariance function matrices $\mathbf{K}_{\mathbf{s}_1}(t, u)$ and $\mathbf{K}_{\mathbf{s}_0}(t, u)$. The additive noise process $\mathbf{w}(t)$ is a sample function from a statistically independent, zero-mean, vector Gaussian random process whose covariance function matrix is $(N_0/2)\delta(t - u)\mathbf{I}$.

1. Find the optimum receiver.
2. Derive the vector versions of (32) and (33).
3. Consider the special case in which

$$\mathbf{K}_{\mathbf{s}_1}(t, u) = K_{s_1}(t, u)\mathbf{I}$$

and

$$\mathbf{K}_{\mathbf{s}_0}(t, u) = K_{s_0}(t, u)\mathbf{I}.$$

Simplify the optimum receiver.

**Problem 3.2.8.** Consider the model in Problem 3.2.7. Assume

$$E[\mathbf{w}(t)\mathbf{w}^T(u)] = \mathbf{N}\delta(t-u),$$

where $\mathbf{N}$ is a nonsingular matrix. Repeat Problem 3.2.7.

**Problem 3.2.9 (continuation).** Consider the vector version of the class $B_w$ problem. Derive the vector analog to (41).

**Problem 3.2.10.**

1. Prove the result in (43).
2. Consider the functional square root defined in (42). Give an example of a class $A_w$ problem in which $h_\Delta^{[1/2]}(t, u)$ does not exist.

**Problem 3.2.11.** Consider the development in (16)–(23). The output of the whitening filter is a waveform $r_*(t)$, whose covariance function on $H_1$ is $K_1^*(t, u)$. Suppose that we write

$$K_1^*(t, u) = \delta(t-u) + Y(t, u).$$

1. Show that $Y(t, u)$ is not necessarily non-negative-definite.
2. Prove that $Y(t, u)$ is a square-integrable function. [*Hint*: Write $\iint_{T_i}^{T_f}(K_1^*(t, u)) \, dt\, du$ as a 6-fold integral using (16). Simplify the result by using the fact that the same white noise is present on both hypotheses.]

## P.3.3 Performance

**Problem 3.3.1.** Derive the result in (60) by using a whitening approach.

**Problem 3.3.2.** Consider the composite process defined in (58). Assume that both $s_1(t)$ and $s_0(t)$ have finite-dimensional state representations. Write the state equations for $s_{com}(t)$. What is the dimension of the resulting system?

**Problem 3.3.3 (continuation).** Specialize your results in Problem 3.3.2 to the case in which

$$K_1(t, u) = \alpha K_0(t, u).$$

**Problem 3.3.4.** Consider the class $A_w$ problem in which both $s_0(t)$ and $s_1(t)$ are Wiener processes, where

$$E[s_0^2(t)] = \sigma_0^2 t, \qquad t \geq 0$$

and

$$E[s_1^2(t)] = \sigma_1^2 t, \qquad t \geq 0.$$

Evaluate $\mu(s)$.

**Problem 3.3.5.** Define

$$\mu(s, t) = \frac{1}{N_0}\int_{T_i}^{t} du\left[(1-s)\xi_P\left(u \mid s_1(\cdot), \frac{N_0}{2}\right) + s\xi_P\left(u \mid s_0(\cdot), \frac{N_0}{2}\right) - \xi_P\left(u \mid s_{com}(\cdot), \frac{N_0}{2}\right)\right].$$

Assume that $s_0(t)$ and $s_1(t)$ have finite-dimensional state representations.

1. Write a differential equation for $\mu(s, t)$.
2. Define the Bhattacharyya distance as

$$B(T_f) = -\mu(\tfrac{1}{2}, T_f).$$

Write a differential equation for $B(t)$.

## P.3.4 Special Situations

NON-ZERO MEANS

**Problem 3.4.1.** In the class $A_{wm}$ problem, the received waveforms on the two hypotheses are

$$r(t) = s_1(t) + w(t), \qquad T_i \le t \le T_f{:}H_1$$

and

$$r(t) = s_0(t) + w(t), \qquad T_i \le t \le T_f{:}H_0,$$

where

$$E[s_1(t)] = m_1(t)$$

and

$$E[s_0(t)] = m_0(t).$$

1. Derive (83)–(86).
2. Assume that a Bayes test with threshold $\eta$ is desired. Evaluate the threshold $\gamma'$ in (89).
3. Derive (90).
4. Find the threshold $\gamma''$ in (90) in terms of $\eta$.
5. Check your results in parts 1–4 for the case in which

$$K_{H_1}(t, u) = K_{H_0}(t, u).$$

**Problem 3.4.2.** Consider the model in Problem 3.4.1. Derive the expression for $\mu_D(s)$ in (96).

**Problem 3.4.3.** Consider the class $A_{wm}$ problem in which $s_1(t)$ and $s_0(t)$ have finite dimensional state representations.

1. Derive a state-variable realization for $l_D$.
2. Derive a state equation for $\mu_D(s)$.

**Problem 3.4.4.** Consider the class $A_{wm}$ problem in which

$$m_1(t) = +m, \qquad T_i \le t \le T_f,$$

$$m_0(t) = -m, \qquad T_i \le t \le T_f,$$

$$\mathscr{F}[K_{s_1}(\tau)] = \frac{2\beta P_1}{\omega^2 + \beta^2},$$

$$\mathscr{F}[K_{s_0}(\tau)] = \frac{2\beta P_0}{\omega^2 + \beta^2},$$

and

$$E[w(t)w(u)] = \frac{N_0}{2}\,\delta(t - u).$$

1. Find the optimum receiver for this problem.
2. Find $\mu_D(s)$ and $\mu_R(s)$.

**Problem 3.4.5.** Consider the modification of Problem 3.4.4 in which

$$\mathscr{F}[K_{s_0}(\tau)] = \frac{2\alpha P_0}{\omega^2 + \alpha^2}$$

where

$$\frac{\beta P_1}{\alpha P_0} = 1.$$

1. Evaluate $\mu_D(s)$ and $\mu_R(s)$ for the case in which $N_0 = 0$.
2. Find the optimum receiver for this case.

**Problem 3.4.6.** Consider the class $A$ problem in which $s_0(t)$ and $s_1(t)$ are sample functions from stationary random processes whose spectra are

$$S_1(\omega) = S_x(\omega) + \frac{\beta^2}{2} u_0(\omega - \omega_1) + \frac{\beta^2}{2} u_0(\omega + \omega_1) \tag{P.1}$$

and

$$S_0(\omega) = S_y(\omega), \tag{P.2}$$

where $S_x(\omega)$ and $S_y(\omega)$ are rational functions of $\omega$.

1. Find the optimum receiver.
2. Find $\mu(s)$.
3. How does the model in (P.1) differ from the case in which

$$E[s_1(t)] = m_1(t) = \sqrt{2}\,\beta \cos(\omega_1 t)?$$

BANDPASS PROBLEMS

**Problem 3.4.7.** Consider the model described in (98).

1. Verify that a necessary and sufficient condition for $r_{c_1}(t)$ and $r_{s_1}(t)$ to be statistically independent is that $S_1(\omega)$ be symmetric around the carrier.
2. Verify the result in (102) and (103).

**Problem 3.4.8.** Consider the model in (99). This is a four-dimensional vector problem that is a special case of the model in Problem 3.2.8.

1. Use the results of Problem 3.2.8 to verify that (100) is correct. Write out the terms on the left side of (100).
2. Verify that (101) is correct.

**Problem 3.4.9.** Whenever the spectra are not symmetric around the carrier, the low-pass processes are not statistically independent. The most efficient way to study this problem is to introduce a complex signal. We use this technique extensively, starting in Chapter 9.

In this problem we carry out the analysis using vector techniques. Perhaps the prime benefit of doing the problem will be an appreciation for the value of the complex representation when we reach Chapter 9.

1. Consider the model in (98). Initially, we assume

$$s_0(t) = 0,$$

so that we have the simple binary problem. Evaluate the cross-correlation function between $s_{c_1}(t)$ and $s_{s_1}(t)$. Evaluate the corresponding cross-spectrum. Notice that we do not assume that $S_{s_1}(\omega)$ is symmetric about $\omega_1$. Check your answer with (A.67) and (A.70).

2. Use the results of Problem 3.2.8 to find the optimum receiver.

3. Derive an expression for $\mu(s)$.

4. Generalize your results to include the original model in (98). Allow $s_0(t)$ to have an asymmetric spectrum about $\omega_0$.

**Problem 3.4.10.**

1. Read [6] and verify that (110) is correct.

2. Discuss the difficulties that arise when the criterion is not minimum Pr $(\epsilon)$ (i.e., the threshold changes).

## P.3.5 Singularity

**Problem 3.5.1.** Draw a block diagram of the receiver operations needed to generate the $r_i$ in (117).

**Problem 3.5.2.** Consider the integral equation in (126). Assume

$$K_{H_0}(t, u) = \sigma^2 \min [t, u]$$

and

$$K_{H_1}(t, u) = Pe^{-\alpha|t-u|}.$$

Find the eigenfunctions and eigenvalues of (126).

**Problem 3.5.3.** Consider the integral equation in (126). Assume

$$K_{H_0}(t, u) = e^{-\beta|t-u|}$$

and

$$K_{H_1}(t, u) = e^{-\alpha|t-u|}.$$

Find the eigenfunctions and eigenvalues of (126).

**Problem 3.5.4.** Assume that

$$K_{H_0}(t, u) = K_0(t, u) + \frac{N_0}{2}\delta(t-u) \tag{P.1}$$

and

$$K_{H_1}(t, u) = K_1(t, u) + \frac{N_0}{2}\delta(t-u). \tag{P.2}$$

How does this assumption affect the eigenvalues and eigenfunctions of (126)?

**Problem 3.5.5.** Assume that $r_0(t)$ and $r_1(t)$ in (112) have finite-dimensional state representations. Extend the technique in the Appendix to Part II to find the solution to (126).

**Problem 3.5.6.** Assume that $K_{H_0}(t, u)$ and $K_{H_1}(t, u)$ are both separable:

$$K_{H_1}(t, u) = \sum_{i=1}^{N_1} \sigma_{1i}^2 f_i(t) f_i(u)$$

and

$$K_{H_0}(t, u) = \sum_{j=1}^{N_2} \sigma_{0j}^2 g_j(t) g_j(u),$$

where

$$\int_{T_i}^{T_f} f_i(t) f_j(t)\, dt = \int_{T_i}^{T_f} g_i(t) g_j(t)\, dt = \delta_{ij}.$$

1. Assume

$$\int_{T_i}^{T_f} f_i(t) g_j(t)\, dt = 0, \qquad \text{all } i, j. \tag{P.1}$$

Solve (126).

2. Assume that $f_i(t)$ and $g_j(t)$ do not necessarily satisfy (P.1) for all $i$ and $j$. Solve (126). How many eigenvalues does (126) have?

**Problem 3.5.7.** Assume

$$K_{H_0}(t, u) = \begin{cases} 1 - |t - u|, & |t - u| < 1, \\ 0, & \text{elsewhere}, \end{cases}$$

and

$$K_{H_1}(t, u) = \sigma^2 \min [t, u].$$

Solve (126).

**Problem 3.5.8.** Consider the definition of $h_\Delta(x, y)$ in (134). Verify that (136) is valid.

**Problem 3.5.9.** Verify the equivalences in (147)–(149).

**Problem 3.5.10.**

1. Can a class $A$ problem be singular? Prove your answer.
2. Can a class $B$ problem be singular? Prove your answer.

**Problem 3.5.11.** Give an example of a case in which the logarithm of neither the numerator nor the denominator of (155) converges but the sum in (155) does.

**Problem 3.5.12.** Verify the result in (167). Is the result also true for nonrational spectra?

**Problem 3.5.13.** Assume that

$$S_{H_0}(\omega) = \frac{1}{c_n \omega^{2n} + c_{n-1}\omega^{2(n-1)} + \cdots + c_0}.$$

Assume that $r_1(t)$ has a finite-dimensional state representation *and* that the detection problem is nonsingular.

1. Find a state-variable realization of the optimum receiver.
2. Find a differential equation specifying $\mu(s)$.

**Problem 3.5.14 (continuation).** Assume that

$$S_{H_0}(\omega) = \frac{1}{c_1 \omega^2 + c_0}$$

and that $r_1(t)$ is a segment of a stationary process with a finite-dimensional state representation. Assume that the detection problem is nonsingular.

1. Draw a block diagram of the optimum receiver. Specify all components completely.
2. Evaluate $\mu(s)$.

**Problem 3.5.15.**

1. Generalize the result in Problem 3.5.14 to the case in which

$$S_{H_0}(\omega) = \frac{2nP}{k} \frac{\sin(\pi/2n),}{1 + (\omega/k)^{2n}} \qquad n = 1, 2, \ldots.$$

2. How must $S_{H_1}(\omega)$ behave as $\omega$ approaches infinity in order for the test to be nonsingular?

**Problem 3.5.16.** Assume that both $r_1(t)$ and $r_0(t)$ are sample functions from stationary processes with flat bandlimited spectra. Under what conditions will the test be nonsingular?

**Problem 3.5.17.** In Section I-4.3.7 we discussed the sensitivity problem for the known signal case. Read [13, page 420] and discuss the sensitivity problem for the general binary case.

**Problem 3.5.18.** Extend the discussion in Section 3.5 to the general vector case. Specifically, find the vector versions of (126), (135), (136), (138), (139), (151), (154), and (167).

**Problem 3.5.19** [14]. Consider the integral equation in (126). Assume

$$K_{H_0}(t, u) = 1 - \frac{|t - u|}{2T}, \quad -T \le t,u \le T$$

and

$$K_{H_1}(t, u) = e^{-|t-u|/T}.$$

Let $T_i = -T$ and $T_f = +T$. Find the eigenvalues and eigenfunctions of (126).

## REFERENCES


[1] W. V. Lovitt, *Linear Integral Equations*, McGraw-Hill, New York, 1924.

[2] L. D. Collins, "Asymptotic Approximation to the Error Probability for Detecting Gaussian Signals," Sc.D Thesis, Massachusetts Institute of Technology, June 1968.

[3] E. Hellinger, "Neue Begrundung der Theorie quadràtischer Formen von unendlichvielen Verànderlichen," J. Reine Angew. Math. **49**, 214–224 (1909).

[4] A. Bhattacharyya, "On a Measure of Divergence between Two Statistical Populations Defined by Their Probability Distributions," Bull. Calcutta Math. Soc. **35**, 99–109 (1943).

[5] S. Kakutani, "On Equivalence of Infinite Product Measures," Ann. of Math. **49**, 214–224 (1948).

[6] J. N. Pierce, "Approximate Error Probabilities for Optimal Diversity Combining, "IEEE Trans. Commun. Syst. **CS-11**, No. 3, 352–354 (Sept. 1963).

[7] G. L. Turin, "Error Probabilities for Binary Symmetric Ideal Reception through Nonselective Slow Fading and Noise," Proc. IRE **46**, 1603–1619 (Sept. 1958).

[8] C. H. Kraft, "Some Conditions for Consistency and Uniform Consistency of Statistical Procedures," University of California Publications in Statistics, 1955.

[9] W. L. Root, "Singular Gaussian Measures in Detection Theory," *Time Series Analysis*, M. Rosenblatt, Ed., Wiley, New York, 1963, pp. 292–315.

[10] J. Feldman, "Equivalence and Perpendicularity of Gaussian Processes," Pacific J. Math. **8,** No. 4, 699–708 (1958).

[11] J. Feldman, "Some Classes of Equivalent Gaussian Processes on an Interval," Pacific J. Math. **10,** No. 4, 1211–1220 (1960).

[12] J. Hájek, "On a Property of Normal Distribution of Any Stochastic Process," Čy. Math. J. **8,** 610–617 (1958).

[13] W. L. Root, "Stability in Signal Detection Problems," Stochastic Processes in Mathematical Physics and Engineering, Proc. Symp. Appl. Math. **16,** (1964).

[14] T. T. Kadota, "Simultaneously Orthogonal Expansion of Two Stationary Gaussian Processes—Examples," Bell Syst. Tech. J. **45,** No. 7, 1071–1096 (Sept. 1966).

[15] L. A. Shepp, "Radon-Nikodym derivative of Gaussian Measures," Annals of Math. Stat. **37,** 321–354 (1966).

# 4

# *Special Categories of Detection Problems*

In Chapters 2 and 3, we studied the simple binary detection problem and the general binary detection problem. Most of our examples dealt with state-representable processes, because we could obtain a complete solution for this class of problem. In this chapter, we discuss three categories of problems for which we can also obtain a complete solution. The three categories are the following:

1. The stationary-processes, long-observation-time (SPLOT) problem.
2. The separable-kernel (SK) problem.
3. The low-energy-coherence (LEC) problem.

We shall explain the categories in detail in the appropriate sections. The discussion is important for two reasons. First, almost all physical situations fall into one of these four categories (the above three categories plus finite-state processes). Second, we can obtain a complete solution for problems in these categories.

## 4.1 STATIONARY PROCESSES: LONG OBSERVATION TIME

In many physical situations of interest, the received waveforms under both hypotheses are segments of stationary processes. Thus, we can characterize the processes by their power density spectra. If the spectra are rational, they will have a finite-dimensional state representation and we can solve the problem using state-variable techniques. In our previous work with state variables we saw that when the input was a stationary process the gains in the optimum system approached constant values and the system approached a time-invariant system. In this section, we consider cases in which the observation time is *long* compared with the time necessary for the system transients to decay. By ignoring the transient, we can obtain much simpler solutions. If desired, we can always check the validity of the approximation by solving the problem with state-variable techniques. We refer to the results obtained by ignoring the transients as asymptotic results and add a subscript $\infty$ to the various expressions. As in the general case, we are interested in optimum receiver structures and their performance. We begin our discussion with the simple binary problem.

### 4.1.1 Simple Binary Problem

The model for the simple binary problem was given in Section 2.1. For algebraic simplicity we discuss only the zero-mean case in the text. The received waveforms are

$$\begin{aligned} r(t) &= s(t) + w(t), \qquad && T_i \le t \le T_f : H_1, \\ r(t) &= w(t), && T_i \le t \le T_f : H_0, \end{aligned} \tag{1}$$

We assume that $s(t)$ is a zero-mean Gaussian process with spectrum $S_s(\omega)$. The noise $w(t)$ is a white, zero-mean Gaussian process that is statistically independent of $s(t)$ and has a spectral height $N_0/2$. The LRT is

$$l_R + l_B \underset{H_0}{\overset{H_1}{\gtrless}} \ln \eta. \tag{2}$$

We first examine various receiver realizations for computing $l_R$. Next we derive a formula for $l_B$. Finally, we compute the performance.

If we use Canonical Realization No. 1 (pages 15–16),

$$l_R = \frac{1}{N_0} \iint\limits_{T_i}^{T_f} r(t) h_1(t, u) r(u)\, dt\, du, \tag{3}$$

where $h_1(t, u)$ is a solution to (4),

$$\frac{N_0}{2} h_1(t, u) + \int_{T_i}^{T_f} h_1(t, z) K_s(z - u)\, dz = K_s(t - u), \qquad T_i \le t, u \le T_f. \tag{4}$$

From our work in Chapter I-4 (page I-321), we know that the total solution is made up of a particular solution that does not depend on the limits and a weighted sum of bounded homogeneous solutions that give the correct endpoint conditions. These homogeneous solutions decay as we move into the interior of the interval. If the time interval is large, the particular solution will exert the most influence on $l_R$, so that we can obtain a good approximation to the solution by neglecting the homogeneous solutions. To accomplish this, we let $T_i = -\infty$ and $T_f = \infty$ in (4). With the infinite limits, we would assume that we could find a solution to (4) that corresponded to a time-invariant filter. To verify this, we let

$$h_1(t, u) = h_{1\infty}(t - u) \tag{5}$$

in (4) and try to find a solution. Rewriting (4), we have

$$\frac{N_0}{2} h_{1\infty}(t - u) + \int_{-\infty}^{\infty} h_{1\infty}(t - z) K_s(z - u)\, dz = K_s(t - u), \qquad -\infty < t, u < \infty, \tag{6}$$

which can be solved by using Fourier transforms. Transforming, we have

$$\boxed{H_{1\infty}(j\omega) = \frac{S_s(\omega)}{S_s(\omega) + (N_0/2)},} \tag{7}$$

which is the desired result. This filter is familiar from our work with unrealizable MMSE estimators in Section I-6.2.3. The resulting receiver is shown in Fig. 4.1. Notice that we have used only the infinite limits to solve the integral equation. The receiver still operates on $r(t)$ over $[T_i, T_f]$.

To implement Canonical Realization No. 3, we must solve (2.45).

$$h_1(t, u) = \int_{T_i}^{T_f} h_f(z, t) h_f(z, u)\, dz, \qquad T_i \le t,\ u \le T_f. \tag{8}$$

To find the asymptotic solution, we let $T_i = -\infty$ and $T_f = \infty$, use (5), and assume that a time-invariant solution exists. The resulting equation is

$$h_{1\infty}(t - u) = \int_{-\infty}^{\infty} h_{f\infty}(z - t) h_{f\infty}(z - u)\, dz, \qquad -\infty < t, u < \infty. \tag{9}$$

Transforming, we have

$$H_{1\infty}(j\omega) = |H_{f\infty}(j\omega)|^2. \tag{10}$$

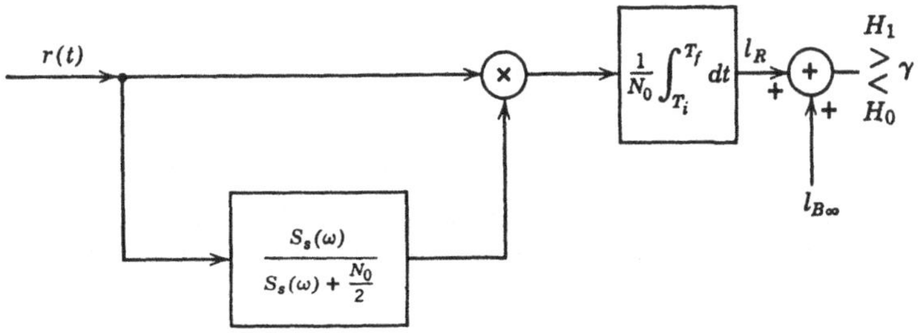


**Fig. 4.1 Canonical Receiver No. 1: stationary process, long observation time.**

This type of equation is familiar from our spectrum factorization work in Section I-6.2. Because $H_{1\infty}(j\omega)$ has all the properties of a power density spectrum, we can obtain a realizable solution easily.

$$\boxed{H_{fr\infty}(j\omega) = [H_{1\infty}(j\omega)]^{+}.} \tag{11}$$

We recall that the superscript + means that we assign all the poles and zeros of $H_{1\infty}(s)$ that lie in the left half of the complex $s$-plane to $H_{fr\infty}(s)$. Notice that this assignment of zeros is somewhat arbitrary. (Recall the discussion on page I-311.) Thus the solution to (10) that we have indicated in (11) is not unique. The resulting receiver is shown in Fig. 4.2. Notice that we can also choose an unrealizable solution to (10). An example is

$$H_{fu\infty}(j\omega) = |H_{1\infty}(j\omega)|^{1/2}. \tag{12}$$

To implement Canonical Realization No. 4, we must solve the realizable filtering problem. By letting $T_i = -\infty$ and assuming stationarity, we obtain the Wiener filtering problem. The solution is given by (I-6.78),

$$H_{or\infty}(j\omega) = \frac{1}{[S_s(\omega) + (N_0/2)]^{+}}\left[\frac{S_s(\omega)}{[S_s(\omega) + (N_0/2)]^{-}}\right]_{+}. \tag{13}$$

The receiver is shown in Fig. 4.3. Comparing Figs. 4.1, 4.2, and 4.3, we see that Canonical Realization No. 3 in Fig. 4.2 is the simplest to implement.

To evaluate the bias $l_B$, we begin with (2.73).

$$l_B = -\frac{1}{N_0}\int_{T_i}^{T_f} \xi_{Ps}(t)\, dt, \tag{14}$$

where $\xi_{Ps}(t)$ is the realizable mean-square error in estimating $s(t)$, assuming that $H_1$ is true. In our work in Section I-6.3 (particularly Examples 1 and 2 on pages I-546–I-555), we saw that $\xi_{Ps}(t)$ approached the steady-state, mean-square error, $\xi_{P\infty}$, reasonably quickly. Thus, if $T_f - T_i \triangleq T$ is long compared to the length of this initial transient, we can obtain a good approximation to $l_B$ by replacing $\xi_{Ps}(t)$ with $\xi_{P\infty}$.

$$l_{B\infty} \triangleq -\frac{1}{N_0}\int_{T_i}^{T_f} \xi_{P\infty}\, dt = -\frac{T}{N_0}\,\xi_{P\infty}. \tag{15}$$

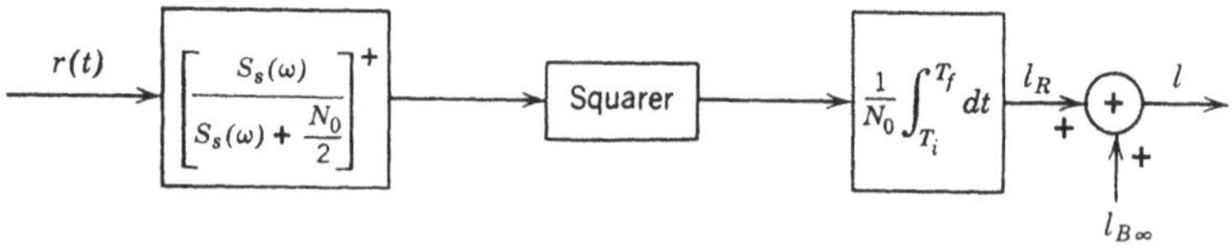


**Fig. 4.2 Canonical Receiver No. 3: stationary process, long observation time.**

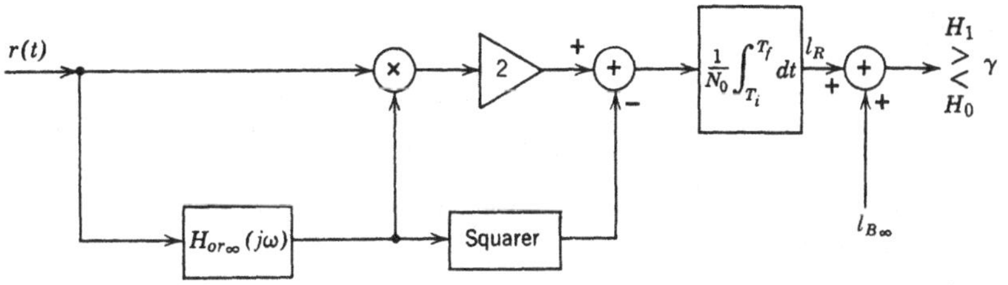


**Fig. 4.3 Canonical Realization No. 4: stationary process, long observation time.**

In Section I-6.2.4 we derived a closed-form expression for $\xi_{P\infty}$. From (I-6.152),

$$\xi_{P\infty} = \frac{N_0}{2}\int_{-\infty}^{\infty} \ln\left[1 + \frac{S_s(\omega)}{N_0/2}\right]\frac{d\omega}{2\pi}. \tag{16}$$

Using (16) in (15) gives

$$l_{B\infty} = -\frac{T}{2}\int_{-\infty}^{\infty} \ln\left[1 + \frac{S_s(\omega)}{N_0/2}\right]\frac{d\omega}{2\pi}, \tag{17}$$

where

$$T \triangleq T_f - T_i. \tag{18}$$

The result in (17) can also be obtained directly from the asymptotic value of the logarithm of the Fredholm determinant in (2.74) [e.g., page I-207 (I-3.182)].

An identical argument gives the asymptotic form of $\mu(s)$, which we denote by $\mu_\infty(s)$. From (2.138),

$$\mu(s) = \frac{1-s}{N_0}\int_{T_i}^{T_f} dt\left\{\xi_P\left(t \mid s(\cdot), \frac{N_0}{2}\right) - \xi_P\left(t \mid s(\cdot), \frac{N_0}{2(1-s)}\right)\right\}. \tag{19}$$

(Notice that $\mu(s) = \mu_R(s)$ because of the zero mean assumption.) Replacing $\xi_P(t \mid s(\cdot), \cdot)$ by $\xi_{P\infty}(s(\cdot), \cdot)$, we have

$$\mu_\infty(s) = \frac{(1-s)T}{N_0}\left\{\xi_{P\infty}\left(s(\cdot), \frac{N_0}{2}\right) - \xi_{P\infty}\left(s(\cdot), \frac{N_0}{2(1-s)}\right)\right\}. \tag{20}$$

Using (16), we obtain

$$\mu_\infty(s) = \frac{T}{2}\left\{(1-s)\int_{-\infty}^{\infty} \ln\left[1 + \frac{2S_s(\omega)}{N_0}\right]\frac{d\omega}{2\pi} - \int_{-\infty}^{\infty} \ln\left[1 + \frac{2(1-s)S_s(\omega)}{N_0}\right]\frac{d\omega}{2\pi}\right\}. \tag{21}$$

An equivalent form is

$$\mu_\infty(s) = \frac{T}{2}\int_{-\infty}^{\infty} \ln\left[\frac{[1 + (2S_s(\omega)/N_0)]^{1-s}}{[1 + (2(1-s)S_s(\omega)/N_0)]}\right]\frac{d\omega}{2\pi} \tag{22}$$

To illustrate the application of these asymptotic results, we consider two simple examples.

**Example 1. First-Order Butterworth Spectrum.** The received waveforms on the two hypotheses are

$$\begin{aligned} r(t) &= s(t) + w(t), & T_i \le t \le T_f : H_1,\\ r(t) &= w(t), & T_i \le t \le T_f : H_0. \end{aligned} \tag{23}$$

The signal process, $s(t)$, is a sample function from a stationary, zero-mean, Gaussian random process with spectrum $S_s(\omega)$,

$$S_s(\omega) = \frac{2kP}{\omega^2 + k^2}, \quad -\infty < \omega < \infty. \tag{24}$$

The noise process is a statistically independent, zero-mean white Gaussian random process with spectral height $N_0/2$.

We shall use Canonical Realization No. 3 (Fig. 4.2) for the receiver. Using (24) in (7), we obtain

$$H_{1\infty}(j\omega) = \frac{2kP/(\omega^2 + k^2)}{(2kP/(\omega^2 + k^2)) + N_0/2} = \frac{k^2\Lambda_1}{[\omega^2 + k^2(1 + \Lambda_1)]}, \tag{25}$$

where

$$\Lambda_1 = \frac{4P}{kN_0} \tag{26}$$

is the signal-to-noise ratio in the message bandwidth. From (11),

$$H_{f\infty}(j\omega) = [H_{1\infty}(j\omega)]^+ = \frac{k\Lambda_1^{1/2}}{j\omega + k\sqrt{1 + \Lambda_1}}. \tag{27}$$

We obtain the bias term from (15). The mean-square error $\xi_{P\infty}$ was evaluated for the first-order Butterworth spectrum in Example 3 on page I-495. From (I-6.112),

$$\xi_{P\infty} = \frac{2P}{1 + \sqrt{1 + \Lambda_1}}. \tag{28}$$

Using (28) in (15), we have

$$l_{B\infty} = -\frac{2PT}{N_0[1 + \sqrt{1 + \Lambda_1}]}. \tag{29}$$

The resulting receiver is shown in Fig. 4.4. By incorporating part of the filter gain in the integrator, we can implement the filter as a simple resistor-capacitor circuit. Notice that the location of the pole of the filter depends on $\Lambda_1$. As $\Lambda_1$ decreases, the filter pole approaches the pole of the message spectrum. As $\Lambda_1$ increases, the bandwidth of the filter increases.

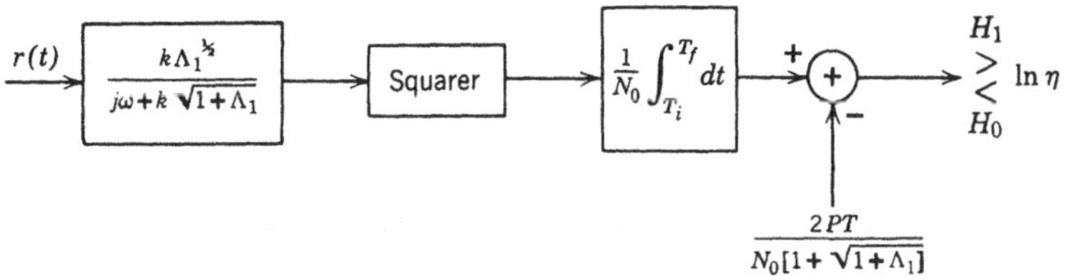


**Fig. 4.4 Filter-squarer receiver: first-order Butterworth spectrum, long observation time.**

To evaluate the performance, we find $\mu_\infty(s)$ by using (28) in (20).

$$\mu_\infty(s) = \frac{(1-s)T}{N_0}\left\{\frac{2P}{1+\sqrt{1+\Lambda_1}} - \frac{2P}{1+\sqrt{1+(1-s)\Lambda_1}}\right\}. \tag{30}$$

At this point it is useful to introduce an efficient notation to emphasize the important parameters in the performance expression.

We introduce several quantities,

$$\bar{E}_r \triangleq PT, \tag{31}$$

which is the average energy in the signal process, and

$$D_1 \triangleq \frac{kT}{2}, \tag{32}$$

which is a measure of the time-bandwidth product of the signal process. Notice that

$$\Lambda_1 = \frac{2\bar{E}_r/N_0}{D_1}. \tag{33}$$

Using (31) in (30), we obtain

$$\mu_\infty(s) = -\left(\frac{2\bar{E}_r}{N_0}\right) g_1(s, \Lambda_1), \tag{34}$$

where

$$g_1(s, \Lambda_1) \triangleq -(1-s)\{(1+\sqrt{1+\Lambda_1})^{-1} - (1+\sqrt{1+(1-s)\Lambda_1})^{-1}\}. \tag{35}$$

The first factor in (34) is the average signal energy-to-noise ratio and appears in all detection problems. The second term includes the effect of the spectral shape, the signal-to-noise ratio in the message bandwidth, and the threshold. It is this term that will vary in different examples. To evaluate the approximate expressions for $P_F$ and $P_D$, we need $\dot{\mu}_\infty(s)$ and $\ddot{\mu}_\infty(s)$. Then, from (2.166) and (2.174),

$$P_F \simeq \frac{1}{\sqrt{2\pi s^2 \ddot{\mu}_\infty(s)}} \exp\left[\mu_\infty(s) - s\dot{\mu}_\infty(s)\right] \tag{36}$$

and

$$P_M \simeq \frac{1}{\sqrt{2\pi(1-s)^2 \ddot{\mu}_\infty(s)}} \exp\left[\mu_\infty(s) + (1-s)\dot{\mu}_\infty(s)\right] \tag{37}$$

From (34) and (35) we can obtain the necessary quantities to substitute into (36) and (37). In Figs. 4.5–4.7 we have plotted the approximate performance characteristics

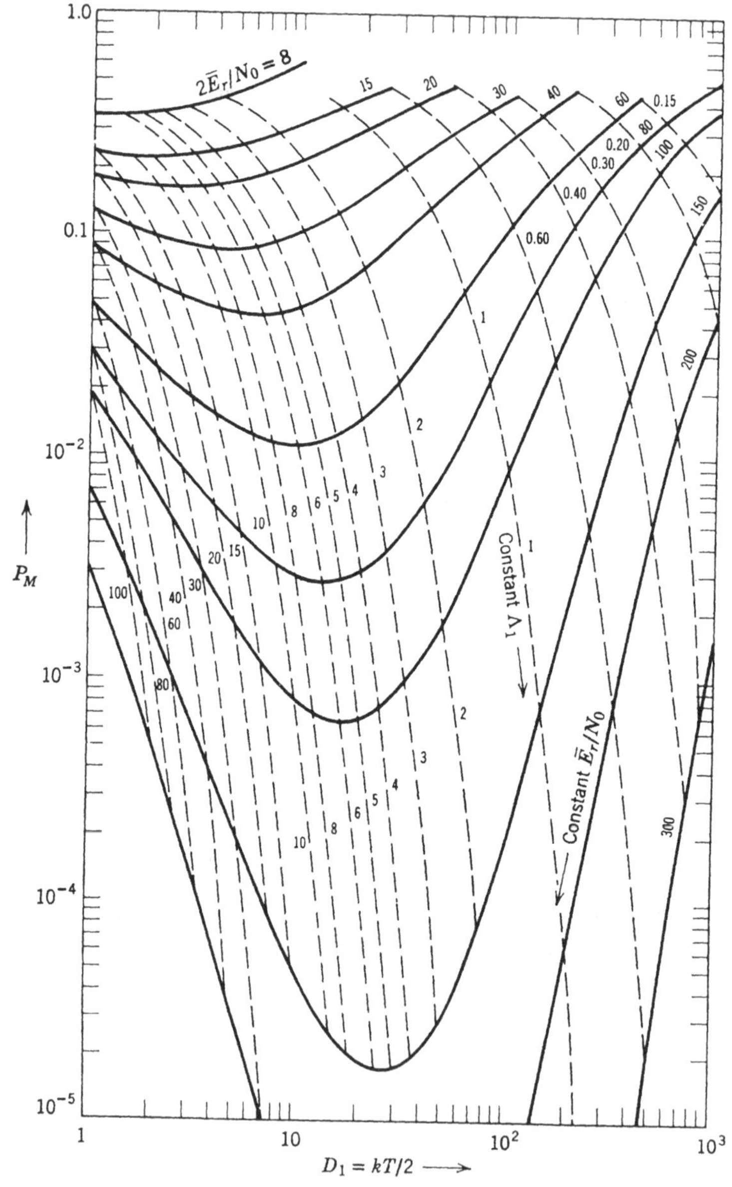


**Fig. 4.5 Probability of miss versus time-bandwidth product: first-order Butterworth spectrum, $P_F = 10^{-1}$.**

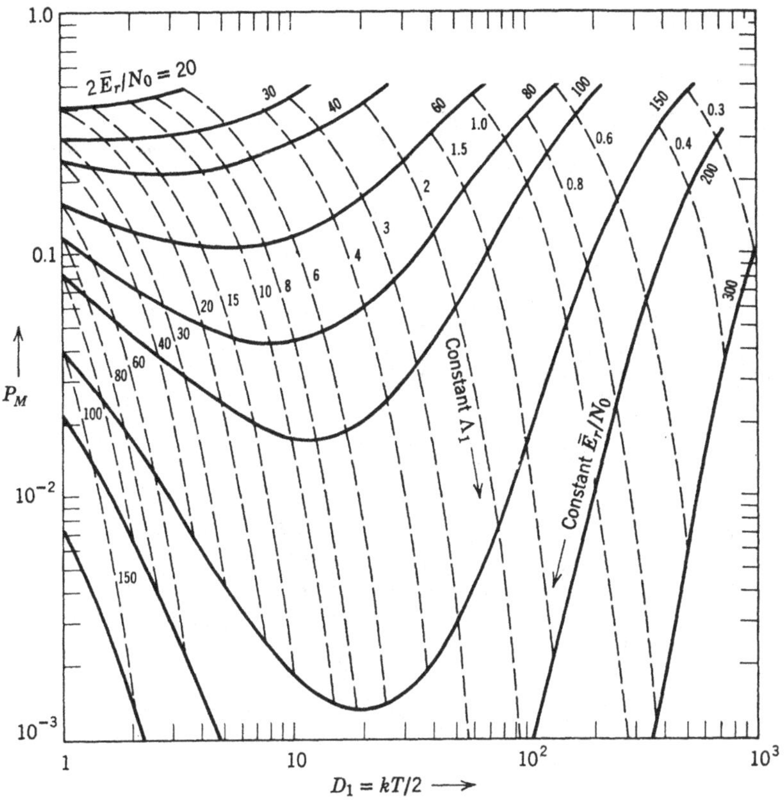


**Fig. 4.6 Probability of miss versus time-bandwidth product for first-order Butterworth spectrum, $P_F = 10^{-3}$.**

indicated by (36) and (37). In Fig. 4.5 we have constrained $P_F$ to equal $10^{-1}$. The horizontal axis is $D_1$ ($= kT/2$). The vertical axis is $P_M$. The solid curves correspond to constant values of $2\bar{E}_r/N_0$. We see that the performance is strongly dependent on the time-bandwidth product of the signal process. Notice that there is an optimum value of $\Lambda_1$ for each value of $2\bar{E}_r/N_0$. This optimum value is in the vicinity of $\Lambda_1 = 6$. (We shall find the exact minimum in a later example.) The dashed curves correspond to constant values of $\Lambda_1$. Moving to the right on a constant $\Lambda_1$ curve corresponds physically to increasing the observation time. Similar results are shown for $P_F = 10^{-3}$ and $P_F = 10^{-5}$ in Figs. 4.6 and 4.7, respectively.

For small values of $D_1$ (say, $D_1 < 2$), the curves should be checked using state-variable techniques, because the SPLOT approximation may not be valid.

For larger time-bandwidth products our performance calculations give good results, for two reasons:

1. The error resulting from the large time-interval approximation decreases rapidly as $kT$ increases. We shall make some quantitative statements about the error on page 142.

2. The error resulting from truncating the Edgeworth series at the first term decreases as $kT$ increases, because there are more significant eigenvalues. As the number of

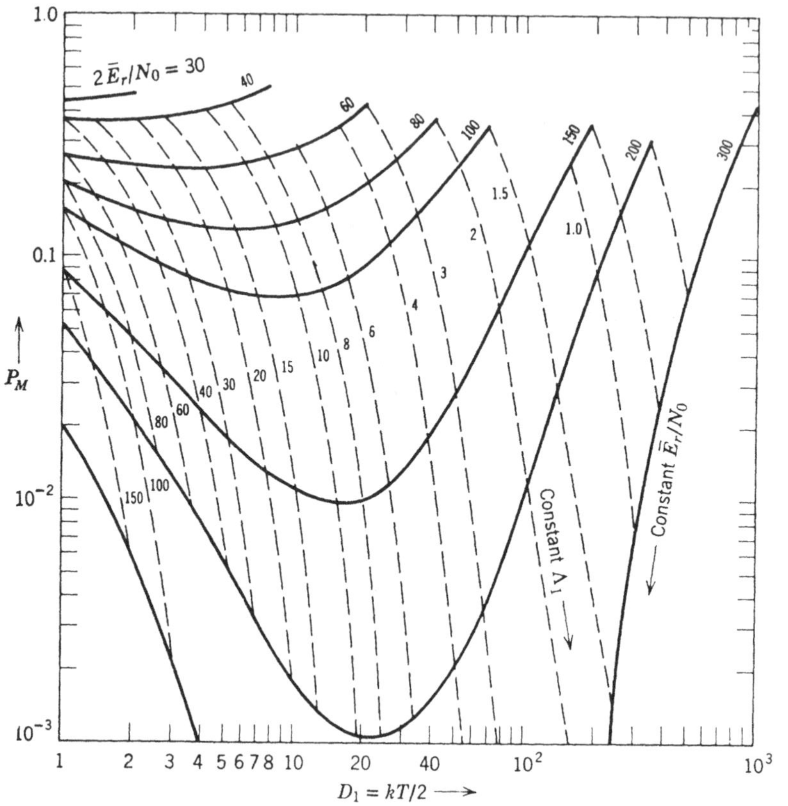


**Fig. 4.7 Probability of miss versus time-bandwidth product for first-order Butterworth spectrum, $P_F = 10^{-5}$.**

significant eigenvalues increases, the tilted density becomes closer to a Gaussian density.

Notice that if the system is operating close to the optimum value of $\Lambda_1$, $D_1$ will be large enough to make the SPLOT approximation valid.

Similar results for higher-order Butterworth spectra can be obtained easily (see Problem 4.1.3). In the next example we consider the case in which the signal has an ideal bandlimited message spectrum. This is a problem that is difficult to treat using state-variable techniques but is straightforward when the SPLOT condition is valid.

**Example 2.** In this example, we assume that $S_s(\omega)$ has a bandlimited spectrum

$$S_s(\omega) = \begin{cases} \dfrac{P}{2W}, & -2\pi W \le \omega \le 2\pi W, \\ 0, & \text{elsewhere}. \end{cases} \tag{38}$$

The most practical receiver configuration is No. 3. From (38) and (7),

$$H_{1\infty}(j\omega) = \begin{cases} \dfrac{P}{P + N_0 W}, & -2\pi W \le \omega \le 2\pi W, \\ 0, & \text{elsewhere}. \end{cases} \tag{39}$$

Thus,

$$H_{fu\infty}(j\omega) = \begin{cases} \dfrac{1}{(1 + N_0 W/P)^{1/2}}, & -2\pi W \le \omega \le 2\pi W, \\ 0, & \text{elsewhere}. \end{cases} \tag{40}$$

The bias term is obtained by using (38) in (17).

$$l_{B\infty} = -WT \ln\left(1 + \frac{P}{N_0 W}\right). \tag{41}$$

The resulting receiver is shown in Fig. 4.8. Notice that we cannot realize the filter in (40) exactly. We can approximate it arbitrarily closely by using an $n$th-order Butterworth filter, where $n$ is chosen large enough to obtain the desired approximation accuracy.

To calculate the performance, we find $\mu_\infty(s)$ from (21). The result is

$$\mu_\infty(s) = \frac{T(1-s)}{N_0}\left\{N_0 W \ln\left(1 + \frac{P}{N_0 W}\right) - \frac{N_0 W}{(1-s)} \ln\left(1 + \frac{P(1-s)}{N_0 W}\right)\right\}. \tag{42}$$

This can be written as

$$\mu_\infty(s) = -\frac{2\bar{E}_r}{N_0} g_\infty(s, \Lambda_\infty), \tag{43}$$

where

$$\begin{aligned} g_\infty(s, \Lambda_\infty) &= -\frac{1}{2\Lambda_\infty}\left[(1-s)\ln(1+\Lambda_\infty) - \ln(1 + (1-s)\Lambda_\infty)\right] \\ &= \frac{-1}{2\Lambda_\infty}\left[\ln\left[\frac{(1+\Lambda_\infty)^{1-s}}{(1 + (1-s)\Lambda_\infty)}\right]\right] \end{aligned} \tag{44}$$

and

$$\Lambda_\infty = \frac{P}{N_0 W}. \tag{45}$$

Notice that the $\infty$ subscript of $\Lambda_\infty$ and $g_\infty(\cdot, \cdot)$ denotes an infinite-order Butterworth spectrum. In Figs. 4.9 and 4.10, we plot the same results as in Example 1.

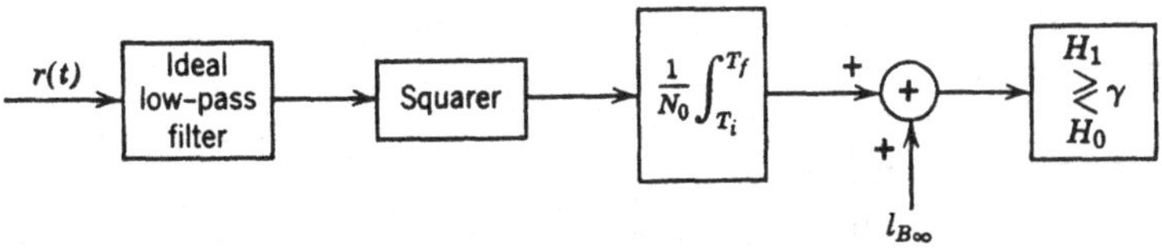


**Fig. 4.8 Optimum receiver: ideal low-pass spectrum, long observation time.**

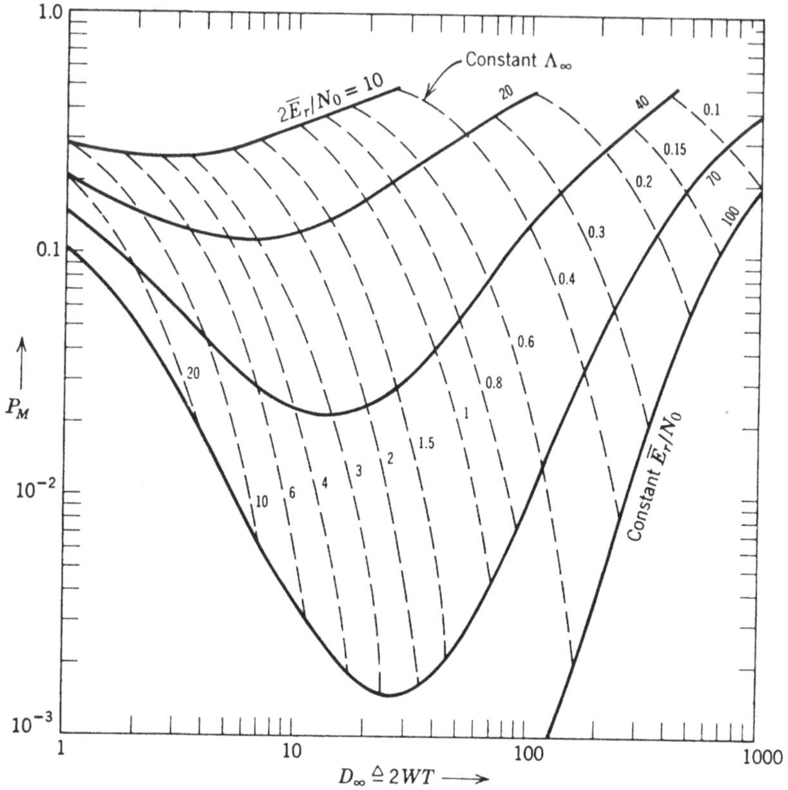


**Fig. 4.9 Probability of miss versus time-bandwidth product, ideal bandlimited spectrum, $P_F = 10^{-1}$.**

In this section we have considered the simple binary problem, developed the appropriate asymptotic formulas, and analyzed two typical examples. The next problem of interest is the general binary problem.

### 4.1.2 General Binary Problem

In Chapter 3 we extended the results from the simple binary case to the general binary case. Because of the strong similarities, we can simply summarize some of the appropriate asymptotic formulas for the general case. In Table 4.1, we have listed the transfer functions of the filters in the optimum receiver. In Table 4.2, we have listed the asymptotic formulas for $\mu_\infty(s)$. In Table 4.3, we have listed various relationships that are useful in general binary problems.

To illustrate the application of some of these results, we consider two examples.

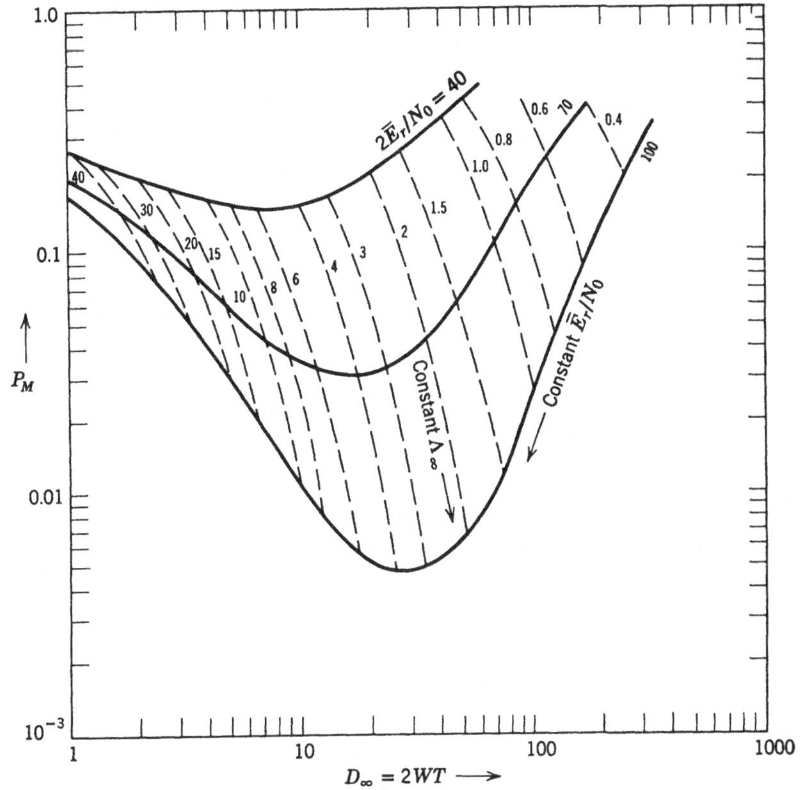


**Fig. 4.10 Probability of miss versus time-bandwidth product, ideal bandlimited spectrum, $P_F = 10^{-3}$.**

**Example 3.** In this example we consider a binary symmetric problem. The transmitted signals on the two hypotheses are

$$s_t(t) = \begin{cases} \sqrt{\dfrac{2E_t}{T}} \sin \omega_1 t, & T_i \le t \le T_f : H_1, \\ \sqrt{\dfrac{2E_t}{T}} \sin \omega_0 t, & T_i \le t \le T_f : H_0, \end{cases} \tag{46}$$

where

$$T \triangleq T_f - T_i. \tag{47}$$

The signal passes over a fluctuating Rayleigh channel.† The received waveforms are

$$\begin{aligned} r(t) &= s_1(t) + w(t), \qquad T_i \le t \le T_f : H_1, \\ r(t) &= s_0(t) + w(t), \qquad T_i \le t \le T_f : H_0. \end{aligned} \tag{48}$$

† We previously encountered fluctuating Rayleigh channels in Chapter II-8 and in the problem section of Chapter II-2. We discuss the model in more detail in Chapter 9.

**Table 4.1 Asymptotic formulas for filters in optimum receivers**

| No. | Problem | Reference | SPLOT formula |
|---|---|---|---|
| 1 | General binary | (3.33) | $H_\Delta(j\omega) = \dfrac{S_{H_1}(\omega) - S_{H_0}(\omega)}{S_{H_1}(\omega)S_{H_0}(\omega)}$ |
| 2 | Class $A_w$ | (3.33) | $H_\Delta(j\omega) = \dfrac{S_{s_1}(\omega) - S_{s_0}(\omega)}{\left(S_{s_1}(\omega) + \dfrac{N_0}{2}\right)\left(S_{s_0}(\omega) + \dfrac{N_0}{2}\right)}$ |
| 3 | Class $B_w$ | (3.42) | $[H_\Delta(j\omega)]^+ = \left[\dfrac{S_s(\omega)}{(S_s(\omega) + S_n(\omega))S_n(\omega)}\right]^+$ |

**Table 4.2 Asymptotic formulas for $\mu_\infty(s)$**

| No. | Problem | Reference | $\mu_\infty(s)$ |
|---|---|---|---|
| 1 | General binary: nonsingular | (3.178) | $\mu_\infty(s) = T\displaystyle\int_{-\infty}^{\infty} \ln\left\{\frac{(S_{H_1}(\omega)/S_{H_0}(\omega))^{(s-1)/2}}{s + (1-s)[S_{H_1}(\omega)/S_{H_0}(\omega)]}\right\}\frac{d\omega}{2\pi}$ |
| 2 | Class $A_w$ | (3.60) | $\mu_\infty(s) = \dfrac{T}{2}\displaystyle\int_{-\infty}^{\infty}\left[(1-s)\ln\left(1 + \frac{2S_{s_1}(\omega)}{N_0}\right) + s\ln\left(1 + \frac{2S_{s_0}(\omega)}{N_0}\right) - \ln\left(1 + \frac{2(sS_{s_0}(\omega) + (1-s)S_{s_1}(\omega))}{N_0}\right)\right]\frac{d\omega}{2\pi}$ |
| 3 | Class $B$ | (3.60) | $\mu_\infty(s) = \dfrac{T}{2}\displaystyle\int_{-\infty}^{\infty} \ln\left\{\frac{[S_s(\omega) + S_n(\omega)]^{1-s}[S_n(\omega)]^s}{[(1-s)S_s(\omega) + S_n(\omega)]}\right\}\frac{d\omega}{2\pi}$ |
| 4 | Class $B_w$ | (3.60) | $\mu_\infty(s) = \dfrac{T}{2}\left\{(1-s)\displaystyle\int_{-\infty}^{\infty}\ln\left(1 + \frac{S_s(\omega)}{S_n(\omega) + (N_0/2)}\right)\frac{d\omega}{2\pi} - \int_{-\infty}^{\infty}\ln\left(1 + \frac{(1-s)S_s(\omega)}{S_n(\omega) + (N_0/2)}\right)\frac{d\omega}{2\pi}\right.$ |

**Table 4.3 Relationship among various $\mu_\infty(s)$ results**

| No. | Reference | Relation |
|---|---|---|
| 1 | (3.101) | $\mu_{BP,\infty}(s) = 2\mu_{LP,\infty}(s)$ |
| 2 | (3.65) | $\mu_{BS,\infty}(s) = \mu_{SIB,\infty}(s) + \mu_{SIB,\infty}(1-s)$ |
| 3 | (3.73) | $\mu_{BS,BP,\infty}(\frac{1}{2}) = 4\mu_{SIB,LP,\infty}(\frac{1}{2})$ |

We assume that $s_0(t)$ and $s_1(t)$ are bandpass processes centered at $\omega_0$ and $\omega_1$, respectively, which are essentially disjoint and are symmetric about their respective carriers (see Figs. 3.7 and 3.8). The low-pass spectrum of the signal processes is $S_{s,LP}(\omega)$, where

$$S_{s,LP}(\omega) = \frac{2kP_{LP}}{\omega^2 + k^2}. \tag{49}$$

The power in the received process depends on the transmitted power and the channel attenuation,

$$P_{LP} \triangleq \frac{E_t\sigma_b^2}{T}, \tag{50}$$

where $\sigma_b^2$ is a measure of the mean-square channel strength. Notice that the total average received power is

$$P_r = 2P_{LP}, \tag{51}$$

and that the total average received signal energy is

$$\bar{E}_r = 2\bar{E}_{r,LP} = 2E_t\sigma_b^2 = 2P_L\ T\sigma_b^2. \tag{52}$$

We assume that the hypotheses are equally likely and that the criterion is minimum Pr ($\epsilon$).

The receiver structure follows easily by combining the results from the bandpass discussion (pages 74–77) with the results in Example 1 (page 104). The receiver is shown in Fig. 4.11. The four low-pass filters are identical:

$$H_{fr} = \frac{k\sqrt{1+\Lambda_1}}{j\omega + k\sqrt{1+\Lambda_1}}. \tag{53}$$

We have eliminated the ideal low-pass filters included in Fig. 3.9 because $H_{fr}(j\omega)$ is low-pass. We have also eliminated the gain in the integrator because it is the same in each path and the threshold is zero.

We can evaluate the performance by suitably modifying the results of Example 1. The first step to go from the low-pass asymmetric problem to the bandpass asymmetric problem. Recall from Table 4.3 that

$$\mu_{BP,\infty}(s) = 2\mu_{LP,\infty}(s). \tag{54}$$

Using (30) in (54) gives

$$\mu_{BP,\infty}(s) = \frac{2\bar{E}_r(1-s)}{N_0}\left\{\frac{1}{1+\sqrt{1+\Lambda_1}} - \frac{1}{1+\sqrt{1+(1-s)\Lambda_1}}\right\}, \tag{55}$$

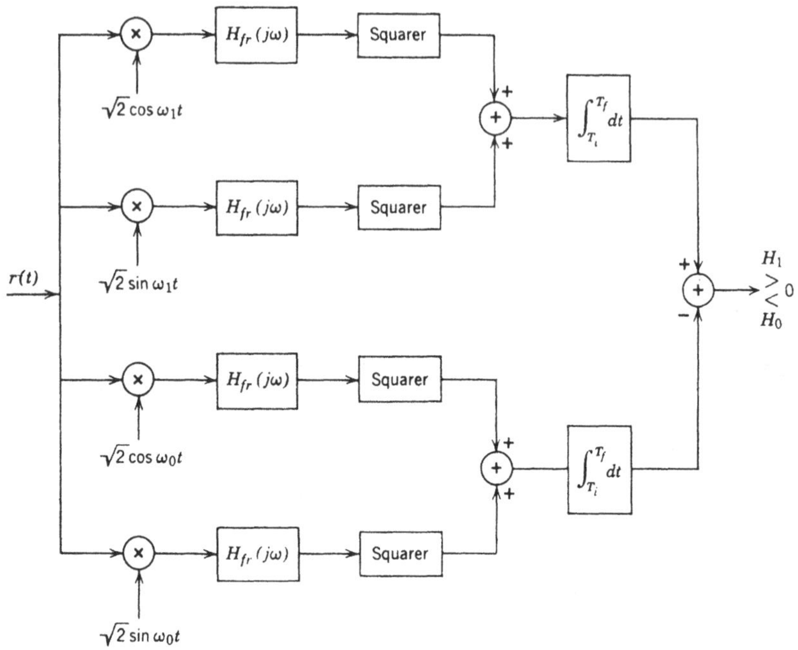


**Fig. 4.11 Optimum receiver, binary symmetric bandpass problem, long observation time.**

where

$$\Lambda_1 \triangleq \frac{4P_{LP}}{kN_0}. \tag{56}$$

The next step is go from the asymmetric (or simple binary) problem to the binary symmetric problem. We recall that

$$\mu_{BS}(s) = \mu_{SIB}(s) + \mu_{SIB}(1 - s). \tag{57}$$

Using (55) in (57) gives

$$\mu_{BS,BP,\infty}(s) = \frac{2\bar{E}_r}{N_0}\left\{\frac{1}{1 + \sqrt{1 + \Lambda_1}} - \frac{(1 - s)}{1 + \sqrt{1 + (1 - s)\Lambda_1}} - \frac{s}{1 + \sqrt{1 + s\Lambda_1}}\right\}. \tag{58}$$

This reduces to

$$\mu_{BS,BP,\infty}(s) = \frac{2\bar{E}_r}{N_0}\left\{\frac{1}{\Lambda_1}[(1 + \Lambda_1)^{1/2} - (1 + \Lambda_1(1 - s))^{1/2} - (1 + \Lambda_1 s)^{1/2} + 1]\right\}. \tag{59}$$

The important quantity in the probability of error expressions is $\mu_{BS,BP,\infty}(\frac{1}{2})$ Letting $s = \frac{1}{2}$ in (59) gives

$$\mu_{BS,BP,\infty}(\tfrac{1}{2}) = \frac{2\bar{E}_r}{N_0}\left\{\frac{1}{\Lambda_1}\left[(1 + \Lambda_1)^{1/2} - 2\left(1 + \frac{\Lambda_1}{2}\right)^{1/2} + 1\right]\right\}. \tag{60}$$

If we define

$$g_{BP,1}(\Lambda_1) \triangleq \frac{-1}{2\Lambda_1}\left\{(1+\Lambda_1)^{1/2} - 2\left(1+\frac{\Lambda_1}{2}\right)^{1/2} + 1\right\}, \tag{61}$$

then we can write

$$\mu_{BS,BP,\infty}(\tfrac{1}{2}) = -\frac{\bar{E}_r}{N_0}\cdot 4g_{BP,1}(\Lambda_1). \tag{62}$$

We refer to $4g_{BP,1}(\Lambda_1)$ as the efficiency function of the binary communication system.

To find the approximate Pr ($\epsilon$) we need $\ddot{\mu}_{BS,BP,\infty}(\frac{1}{2})$. Differentiating (59) twice with respect to $s$ and evaluating the result at $s = \frac{1}{2}$, we have

$$\ddot{\mu}_{BS,BP,\infty}(\tfrac{1}{2}) = \frac{2\bar{E}_r}{N_0}\left\{\frac{\Lambda_1}{2}\left(1+\frac{\Lambda_1}{2}\right)^{-3/2}\right\}. \tag{63}$$

The approximate Pr ($\epsilon$) follows from (3.77) as

$$\Pr(\epsilon) \simeq 2\left(\frac{N_0}{2\bar{E}_r}\right)^{1/2}\frac{[1+(\Lambda_1/2)]^{3/4}}{\sqrt{\pi\Lambda_1}}\exp\left[-4g_{BP,1}(\Lambda_1)\cdot\frac{\bar{E}_r}{N_0}\right]. \tag{64}$$

We see that the Pr ($\epsilon$) depends on $\Lambda_1$, the signal-to-noise ratio in the signal process bandwidth, and $2\bar{E}_r/N_0$, the ratio of the average received signal energy to the noise spectral height.

The next step in the analysis depends on the transmitter constraints. If it is completely specified, we simply evaluate Pr ($\epsilon$). If the signals are specified to be a segment of sine waves, as in (46), and the transmitter is peak-power-limited (i.e., $E_t/T$ is limited), the performance is monotonic with $T$. On the other hand, if the transmitter is energy-limited, we may be able to optimize the performance by choosing $T$ appropriately. This is an elementary version of the signal design problem. Later we shall look at the effect of different signal shapes.

We assume that $\sigma_b^2$, $N_0$, and $k$ are fixed. Then, if we fix $E_t$, this fixes $\bar{E}_r$. The only remaining parameter is $T$ (or, equivalently, $\Lambda_1$). We could choose $\Lambda_1$ to minimize the Pr ($\epsilon$). A slightly easier procedure is to choose it to minimize $\mu_{BS,BP,\infty}(\frac{1}{2})$. From (60), we see that this is equivalent to maximizing the efficiency factor. In Fig. 4.12 we have plotted $g_{BP,1}(\Lambda_1)$ as a function of $\Lambda_1$. We see that the maximum occurs in the vicinity of $\Lambda_1 = 7$. We refer to this point as $\Lambda_{1,OPT}$,

$$\Lambda_{1,OPT} = 6.88, \tag{65}$$

and

$$g_{BP,1}(\Lambda_{1,OPT}) = 0.0592. \tag{66}$$

We observe that the curve is very flat near the maximum, so that a precise adjustment of $\Lambda_1$ is not necessary. Using (66) in (64) gives an expression for the probability of error when the optimum value of $\Lambda_1$ is used.

$$\Pr_o(\epsilon) \simeq 1.32\left(\frac{N_0}{2\bar{E}_r}\right)\exp\left(-0.118\,\frac{\bar{E}_r}{N_0}\right). \tag{67}$$

We see that when the system uses the optimum value of $\Lambda_1$, the probability of error decreases exponentially with increasing $\bar{E}_r/N_0$.

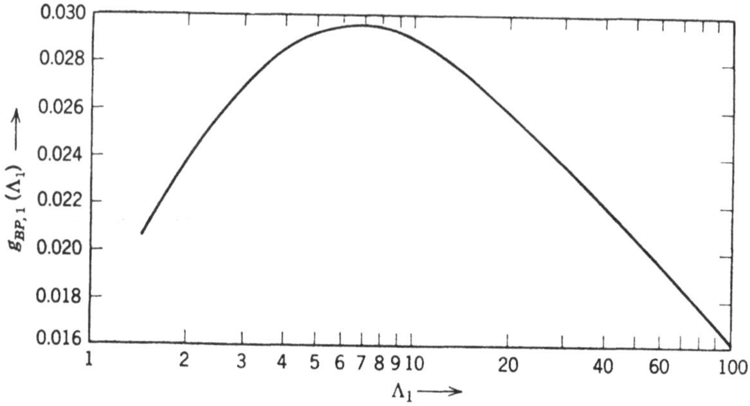


**Fig. 4.12** $g_{\text{BP},1}(\Lambda_1)$ **versus** $\Lambda_1$.

We now consider another example. All of the techniques are identical with Example 3. The reason for including the example is to derive some specific numerical results that we shall use later.

**Example 4. Symmetric Hypotheses, Bandlimited Spectrum.** The basic model is the same as in the preceding example [see (46) and (48)]. Now we assume that the low-pass signal process has a bandlimited spectrum

$$S_s(\omega) = \begin{cases} \dfrac{P_{\text{LP}}}{2W}, & -2\pi W \leq \omega \leq 2\pi W, \\ 0, & \text{elsewhere}. \end{cases} \tag{68}$$

The receiver structure is an obvious combination of the structures in Figs. 4.8 and 4.11. As pointed out on page 109, we cannot realize the filters exactly but can approximate them arbitrarily closely. For the present we are concerned with the system performance. Using (42)–(45), Tables 4.2 and 4.3, and (68), we obtain

$$\mu_{\text{BS,BP},\infty}(s) = -\frac{2\bar{E}_r}{N_0}\left\{\frac{1}{2\Lambda_\infty}\ln\left[\frac{[1+(1-s)\Lambda_\infty][1+s\Lambda_\infty]}{[1+\Lambda_\infty]}\right]\right\}, \tag{69}$$

where

$$\Lambda_\infty \triangleq \frac{P_{\text{LP}}}{N_0 W}. \tag{70}$$

Letting $s = \frac{1}{2}$ in (69) gives

$$\mu_{\text{BS,BP},\infty}(\tfrac{1}{2}) = -\frac{\bar{E}_r}{N_0}\cdot g_{\text{BP},\infty}(\Lambda_\infty), \tag{71}$$

where we have defined

$$g_{\text{BP},\infty}(\Lambda_\infty) \triangleq \frac{1}{\Lambda_\infty}\ln\left[\frac{[1+(\Lambda_\infty/2)]^2}{(1+\Lambda_\infty)}\right]. \tag{72}$$

Thus,

$$\exp\left[\mu_{BS,BP,\infty}(\tfrac{1}{2})\right] = \left\{\frac{[1 + (\Lambda_\infty/2)]^2}{(1 + \Lambda_\infty)}\right\}^{-\bar{E}_r/N_0\Lambda_\infty} \tag{73}$$

To get the coefficient for the approximate Pr $(\epsilon)$ expression, we differentiate (71) twice and evaluate the result at $s = \frac{1}{2}$. The result is

$$\ddot{\mu}_{BS,BP,\infty}(\tfrac{1}{2}) = \frac{2\bar{E}_r}{N_0}\frac{\Lambda_\infty}{[1 + (\Lambda_\infty/2)]^2}. \tag{74}$$

Then, using (3.76),

$$\Pr(\epsilon) \simeq \frac{(1 + \Lambda_\infty)^{WT}}{\sqrt{\pi WT}\,\Lambda_\infty[1 + (\Lambda_\infty/2)]^{2WT-1}} \tag{75}$$

As before, we can find an optimum value of $\Lambda_\infty$ by maximizing $g_{BP,\infty}(\Lambda_\infty)$. The result is

$$\Lambda_{\infty,OPT} = \frac{\bar{E}_r/N_0}{2WT} = 3.07. \tag{76}$$

Substituting (76) into (75), we obtain

$$\Pr_o(\epsilon) \simeq \sqrt{\frac{0.815}{\bar{E}_r/N_0}}\exp\left(-0.1488\left[\frac{\bar{E}_r}{N_0}\right]\right). \tag{77}$$

We see that the magnitude of the coefficient in the exponent of (77) is slightly larger than in the one-pole case [recall (67)].

The communication systems in Examples 3 and 4 have illustrated the application of long-time approximations to particular problems. In addition, they have given us some interesting results for binary FSK communication over fluctuating symmetric Rayleigh channels. It is interesting to compare these results with those we obtained in Chapter I-4 for binary PSK and FSK systems operating over an additive noise channel. From (I-4.40) and (I-4.36) we have

$$\Pr_{FSK}(\epsilon) = \operatorname{erfc}_*\left(\sqrt{\frac{E_r}{N_0}}\right), \tag{78}$$

or

$$\Pr_{FSK}(\epsilon) \simeq \left(\frac{N_0}{2\pi E_r}\right)^{1/2}\exp\left(-\frac{E_r}{2N_0}\right), \qquad \sqrt{\frac{E_r}{N_0}} \geq 2. \tag{79}$$

Similarly,

$$\Pr_{PSK}(\epsilon) \simeq \left(\frac{N_0}{\pi E_r}\right)^{1/2}\exp\left(-\frac{E_r}{N_0}\right), \qquad \sqrt{\frac{E_r}{N_0}} \geq 2. \tag{80}$$

Recall that the received signal energy is fixed in an additive noise channel. The results from (67), (77), (79), and (80) are summarized in Table 4.4.

**Table 4.4 Efficiency Factors for Various Binary Communication Systems (Large $\bar{E}_r/N_0$)**

| System | Signals | Channel | Efficiency Factor | Loss in db (relative to system 1) |
|---|---|---|---|---|
| 1 | PSK | Additive white Gaussian noise | 1.0 | 0 |
| 2 | FSK | Additive white Gaussian noise | 0.5 | 3 |
| 3 | FSK | Rayleigh channel: ideal bandlimited spectrum | 0.149 | 8.28 |
| 4 | FSK | Rayleigh channel: one-pole spectrum | 0.118 | 9.30 |

We denote the coefficient of $\bar{E}_r/N_0$ as the *efficiency factor* of a particular communication scheme. Comparing the exponents, we see that a band-limited Rayleigh channel requires about 5.28 db more average energy than the binary FSK system to obtain the same error exponent. A Rayleigh channel with a first-order Butterworth spectrum requires about 6.30 db more average energy to obtain the same error exponent. We have assumed that $\bar{E}_r/N_0$ is large.

There are several restrictions to our analysis that should be emphasized:

1. We assumed that a rectangular pulse was transmitted. In Chapter 11, we shall prove that the efficiency factor for *any* Rayleigh channel and any signal shape is bounded by 0.1488. We shall see that for certain channels the system in Example 4 corresponds to the optimum binary orthogonal signaling scheme.

2. We used long-time-interval approximations. If $\bar{E}_r/N_0$ is large and we use the optimum time-bandwidth product, the approximation will always be valid.

3. We detected each signal individually and did not try to exploit the continuity of the channel from baud to baud by performing a continuous measurement. In Section 5.1.3, we shall discuss this type of system briefly.

4. We considered only Rayleigh channels whose fading spectra were symmetric about the carrier. In Chapter 11, we shall analyze more general channels.

We now summarize briefly the results for the long time interval–stationary process case.

### 4.1.3 Summary: SPLOT Problem

In this section we studied the case in which the received waveform is a sample function of a stationary random process *and* the observation interval is long. By neglecting the transient effects at the ends of the observation interval, we were able to implement the receiver using time-invariant filters. The resulting receiver is suboptimum but approaches the optimum receiver rapidly as the time-bandwidth product of the signal process increases.

We have not discussed how long the observation interval must be in order for the SPLOT approximation to be valid. Whenever the processes have rational spectra, we can compute the performance of both the optimum receiver and the SPLOT receiver using state-variable techniques. Thus, in any particular situation we can check the validity of the approximation quantitatively. A conservative requirement for using the approximation is to check the time-bandwidth product at the input to the squarer in Canonical Realization No. 3. If the product is greater than 5, the approximation is almost always valid. In many cases, the SPLOT receiver is essentially optimum for products as low as 2.

The performance expressions for the SPLOT case were simplified because we could use the asymptotic expressions for the Fredholm determinant. Thus, the calculation of $\mu_\infty(s)$ always reduced to finding the mean-square filtering error in some realizable Wiener filtering problem. This reduction meant that many of the detailed results in Section I-6.2 were directly applicable to the Gaussian detection problem. In many situations we can exploit this similarity to obtain answers efficiently.

In addition to considering the general SPLOT problem, we considered the problem of binary communication over a Rayleigh channel. We found that if we were allowed to control the time-bandwidth product of the receiver signal process, we could achieve a Pr $(\epsilon)$ that decreased exponentially with $\bar{E}_r/N_0$. This behavior was in contrast to the nonfluctuating Rayleigh channel discussed in Section I-4.4.2, in which the Pr $(\epsilon)$ decreased linearly with $\bar{E}_r/N_0$.

This completes our discussion of the SPLOT problem. There are a number of problems in Section 4.5 that illustrate the application of the results to specific situations.

## 4.2 SEPARABLE KERNELS

In this section we consider a class of signal covariance functions that lead to a straightforward solution for the optimum receiver and its performance. In Section 4.2.1, we consider the separable kernel model and derive the necessary equations that specify the optimum receiver and its performance. In Sections 4.2.2 and 4.2.3, we consider physical situations in which the separable kernel model is valid. Finally, in Section 4.2.4, we summarize our results.

### 4.2.1 Separable Kernel Model

Our initial discussion is in the context of the simple binary problem with zero-mean processes. The received waveforms on the two hypotheses are

$$\begin{aligned} r(t) &= s(t) + w(t), \qquad && T_i \le t \le T_f : H_1, \\ r(t) &= w(t), && T_i \le t \le T_f : H_0. \end{aligned} \tag{81}$$

The noise $w(t)$ is a sample function from a zero-mean white Gaussian random process with spectral height $N_0/2$. The signal $s(t)$ is a sample function from a zero-mean Gaussian random process with covariance function $K_s(t, u)$.

From (2.28) the LRT is

$$l_R = \frac{1}{N_0} \iint_{T_i}^{T_f} r(t) h_1(t, u) r(u)\, dt\, du \underset{H_0}{\overset{H_1}{\gtrless}} \gamma, \tag{82a}$$

where $h_1(t, u)$ is specified by the integral equation

$$\frac{N_0}{2} h_1(t, u) + \int_{T_i}^{T_f} h_1(t, z) K_s(z, u)\, dz = K_s(t, u), \qquad T_i \le t, u \le T_f. \tag{82b}$$

In Section I-4.3.6 we studied solution techniques for this integral equation. On page I-322, we observed that whenever the kernel of the integral equation [i.e., the signal covariance function $K_s(t, u)$] was separable, the solution to (82*b*) followed by inspection. A separable kernel corresponds to a signal process with a finite number of eigenvalues. Thus, we can write

$$\boxed{K_s(t, u) = \sum_{i=1}^{K} \lambda_i^s \phi_i(t) \phi_i(u), \qquad T_i \le t, u \le T_f,} \tag{83}$$

where $\phi_i(t)$ and $\lambda_i^s$ are the eigenfunctions and eigenvalues, respectively, of the signal process. In this case the solution to (82*b*) is

$$h_1(t, u) = \sum_{i=1}^{K} h_i \phi_i(t) \phi_i(u) = \sum_{i=1}^{K} \frac{\lambda_i^s}{(N_0/2) + \lambda_i^s} \phi_i(t) \phi_i(u), \qquad T_i \le t, u \le T_f. \tag{84}$$

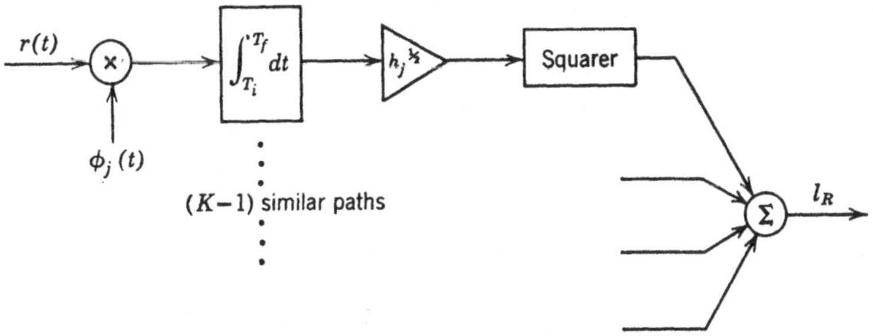


**Fig. 4.13 Correlator realization of separable kernel receiver.**

This can be verified by substituting (84) into (82*b*). For separable kernels, the simplest realization is Canonical Realization No. 3 (the filter-squarer receiver). From (2.45),

$$h_1(t, u) = \int_{T_i}^{T_f} h_f(z, t) h_f(z, u)\, dz, \tag{85}$$

whose solution is

$$h_{fu}(z, t) = \sum_{i=1}^{K} h_i^{1/2} \phi_i(z) \phi_i(t), \qquad T_i \le t, z \le T_f. \tag{86}$$

Using (85) and (86) in (82*a*) and performing the integration on $z$ we obtain

$$\boxed{l_R = \frac{1}{N_0} \sum_{i=1}^{K} h_i \left[ \int_{T_i}^{T_f} r(t) \phi_i(t)\, dt \right]^2.} \tag{87}$$

The operations on $r(t)$ can be realized using either correlators or matched filters. These realizations are shown in Figs. 4.13 and 4.14. These receiver structures are familiar from Fig. I-4.66 on page I-353. Looking at

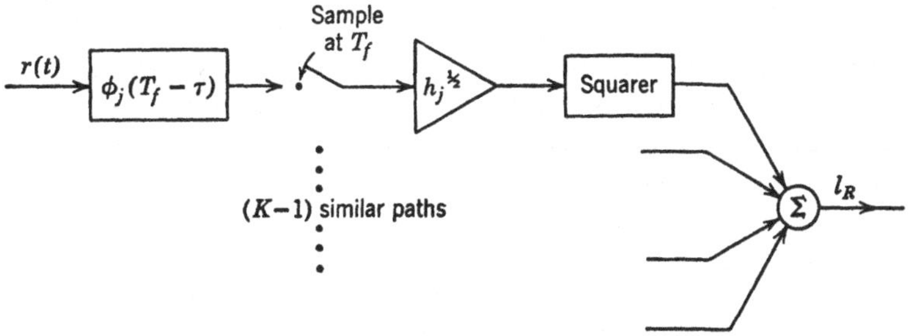


**Fig. 4.14 Matched filter realization of separable kernel receiver.**

(I-4.399), we see that the received signal on $H_1$ was

$$r(t) = \sum_{i=1}^{K} a_i s_i(t) + w(t), \qquad 0 \le t \le T, \tag{88}$$

where the $a_i$ are $N(0, \sigma_{a_i})$ and the $s_i(t)$ are orthonormal. The total signal is

$$s(t) = \sum_{i=1}^{K} a_i s_i(t), \tag{89}$$

which is a zero-mean Gaussian process with covariance function

$$K_s(t, u) = \sum_{i=1}^{K} \sigma_{a_i}^2 s_i(t) s_i(u), \qquad 0 < t, u \le T. \tag{90}$$

Comparing (90) and (83), we see that the separable kernel problem is identical with the problem that we solved in Section I-4.4.2, in the context of an unwanted parameter problem. As we observed on page I-353, the problem is also identical with the general Gaussian problem that we solved in Section I-2.6. The reason for this simplification is that the signal has only a finite number of eigenvalues. Thus, we can immediately map $r(t)$ into a $K$-dimensional vector $\mathbf{r}$ that is a sufficient statistic. Therefore, all of the examples in Sections I-2.6 and I-4.4 correspond to separable kernel Gaussian process problems, and we have a collection of results that are useful here.

The approximate performance of the optimum receiver is obtained by calculating $\mu(s)$ and using the approximate expressions in (2.166) and (2.174). From the first term in (2.132), we have

$$\boxed{\mu_R(s) = \frac{1}{2} \sum_{i=1}^{K} \ln \left\{ \frac{(1 + 2\lambda_i^s/N_0)^{1-s}}{1 + [2(1-s)/N_0]\lambda_i^s} \right\}.} \tag{91}$$

Using (91) in (2.166) and (2.174) gives an approximate expression for $P_F$ and $P_M$. We recall that when the $K$ eigenvalues were equal we could obtain an exact expression. Even in this case the approximate expressions are easier to use and give accurate answers for moderate $K$ (see Fig. I-2.42).

At this point we have established that the separable kernel problem is identical with problems that we have already solved. The next step is to discuss several important physical situations in which the signal processes have separable kernels.

### 4.2.2 Time Diversity

Historically, the first place that this type of problem arose was in pulsed radar systems. The transmitted signal is a sequence of pulsed sinusoids at a carrier frequency $\omega_c = 2n\pi/T$, where $n$ is a large integer. The sequence is shown in Fig. 4.15. The $i$th signal is

$$s_i(t) \triangleq \begin{cases} \sqrt{\dfrac{2}{T}} \sin \omega_c t, & (i-1)T_p \le t \le (i-1)T_p + T, \\ 0, & \text{elsewhere.} \end{cases} \tag{92}$$

If a target is present, the pulses are reflected. We shall discuss target reflection models in detail in Chapter 9. There we shall see that for many targets the reflection from the $i$th pulse can be modeled as

$$s_{ri}(t) = \begin{cases} v_i \sqrt{\dfrac{2}{T}} \sin (\omega_c t + \phi_i), & (i-1)T_p \le t \le (i-1)T_p + T, \\ 0, & \text{elsewhere,} \end{cases} \tag{93}$$

where the $v_i$ are Rayleigh random variables and the $\phi_i$ are uniform random variables. (Notice that we have put the target at zero range for simplicity.) As in Section I-4.4.2, we write $s_{ri}(t)$ in terms of two quadrature components,

$$s_{ri}(t) \triangleq \begin{cases} b_{si} \sqrt{\dfrac{2}{T}} \sin \omega_c t + b_{ci} \sqrt{\dfrac{2}{T}} \cos \omega_c t, \\ \qquad\qquad (i-1)T_p \le t \le (i-1)T_p + T, \\ 0, \quad \text{elsewhere.} \end{cases} \tag{94a}$$

Equivalently, we can write

$$s_{ri}(t) \triangleq b_{si}\phi_{si}(t) + b_{ci}\phi_{ci}(t), \qquad -\infty < t < \infty, \tag{94b}$$

where $\phi_{si}(t)$ and $\phi_{ci}(t)$ include the time interval in their definition. The $b_{si}$ and $b_{ci}$ are statistically independent Gaussian random variables with variances $\sigma_b^2$. The average received energy per pulse is

$$\bar{E}_{r1} \triangleq 2\sigma_b^2. \tag{95}$$

The received waveform consists of the signal reflected from the sequence of pulses plus a white noise component,

$$r(t) = \sum_{i=1}^{K} [b_{si}\phi_{si}(t) + b_{ci}\phi_{ci}(t)] + w(t), \qquad T_i \le t \le T_f. \tag{96}$$

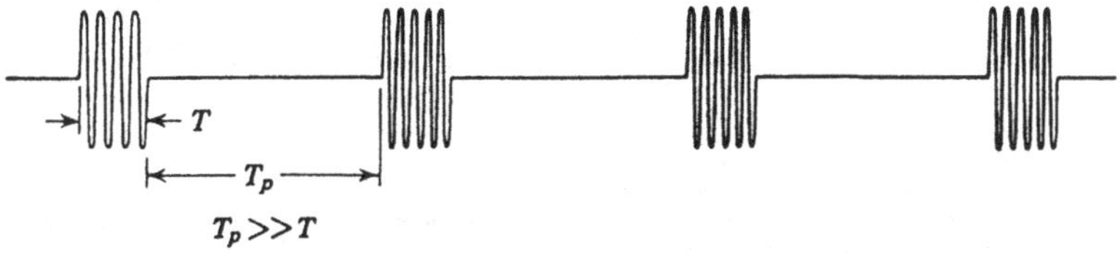


**Fig. 4.15 Transmitted pulse sequence.**

The observation interval includes all of the reflected pulses completely. Now, when we let $v_i$ and $\phi_i$ be random variables, we are assuming that the target reflection is essentially constant over the pulse duration. In general, $T_p$ is much larger than the pulse duration. Thus, if the target is fluctuating, it is plausible to assume that the $v_i$ and $\phi_i$ are independent random variables for different $i$. This means that the $b_{ci}$ and $b_{si}$ are independent for different $i$. The covariance function of the signal process is

$$K_s(t, u) = \sum_{i=1}^{K} \{\sigma_b^2 \phi_{ci}(t)\phi_{ci}(u) + \sigma_b^2 \phi_{si}(t)\phi_{si}(u)\}, \qquad T_i \le t, u \le T_f. \tag{97}$$

Thus we have a separable kernel with $2K$ equal eigenvalues. Using Figs. 4.13 and I-4.68, we obtain the receiver structure shown in Fig. 4.16. Here the orthogonality arises because the signals are nonoverlapping in time. We refer to this as the time-diversity case.

We have already computed the performance for this problem (Case 1A on page I-108). By letting

$$N = 2K, \tag{98}$$

$$\sigma_s^2 = \sigma_b^2 = \frac{\bar{E}_{r1}}{2}, \tag{99}$$

$$\sigma_n^2 = \frac{N_0}{2}, \tag{100}$$

the results in Fig. I-2.35 apply directly. Notice that

$$\mu_{\text{BP,SK}}(s) = K \ln \left[\frac{(1 + \bar{E}_{r1}/N_0)^{1-s}}{1 + (1-s)(\bar{E}_{r1}/N_0)}\right] \tag{101}$$

[use either (I-2.501) or (91)].

The ROC is shown in Fig. 4.17. The average received signal energy per pulse is $\bar{E}_{r1}$, and the total average received energy is $\bar{E}_r$, where

$$\bar{E}_r = K\bar{E}_{r1}. \tag{102}$$

In Fig. 4.18, we fix $\bar{E}_r$ and $P_F$ and plot $P_M$ as a function of $K$. (These are Figs. I-2.35$b$ and $c$ relabeled.) This shows us how to optimize the number

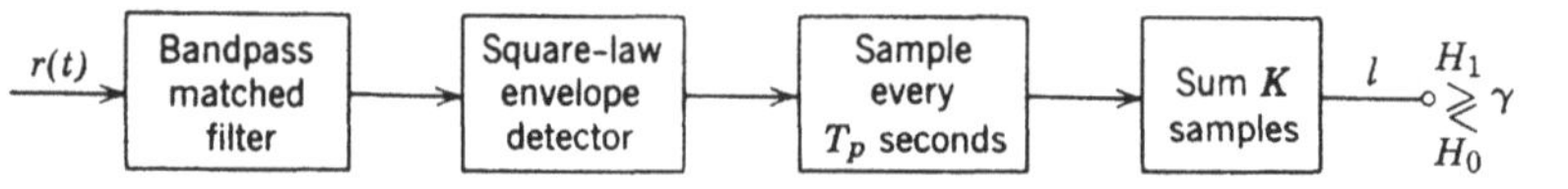


**Fig. 4.16 Optimum receiver for pulsed radar problem.**

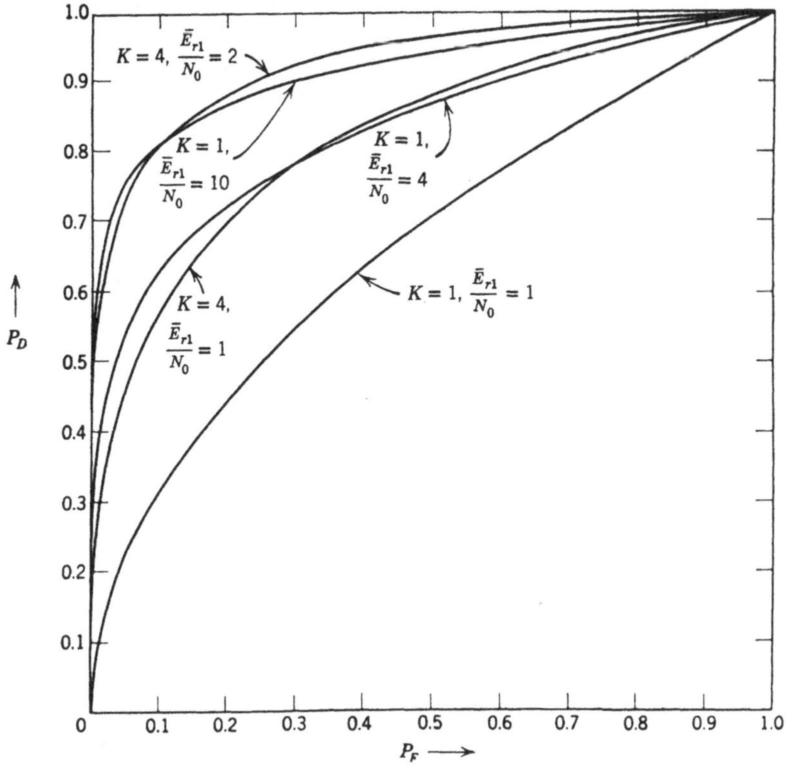


**Fig. 4.17 Receiver operating characteristic: pulsed radar, Rayleigh target.**

of transmitted pulses in various situations. Notice that Fig. I-2.35 was based on an exact calculation. As shown in Fig. I-2.42, an approximate calculation gives a similar results.

A second place that time-diversity occurs is in ionospheric communication. In the HF frequency range, long-range communication schemes frequently rely on waves reflected from the ionosphere. As a result of multiple paths, a single transmitted pulse may cause a sequence of pulses to appear at the receiver. Having traveled by separate paths, the amplitudes and phases of the different pulses are usually not related. A typical situation is shown in Fig. 4.19. If the output pulses do not overlap, this is commonly referred to as a resolvable multipath problem. If the path lengths are known (we discuss the problem of unknown path lengths in a later section), this is identical with the time diversity problem above.

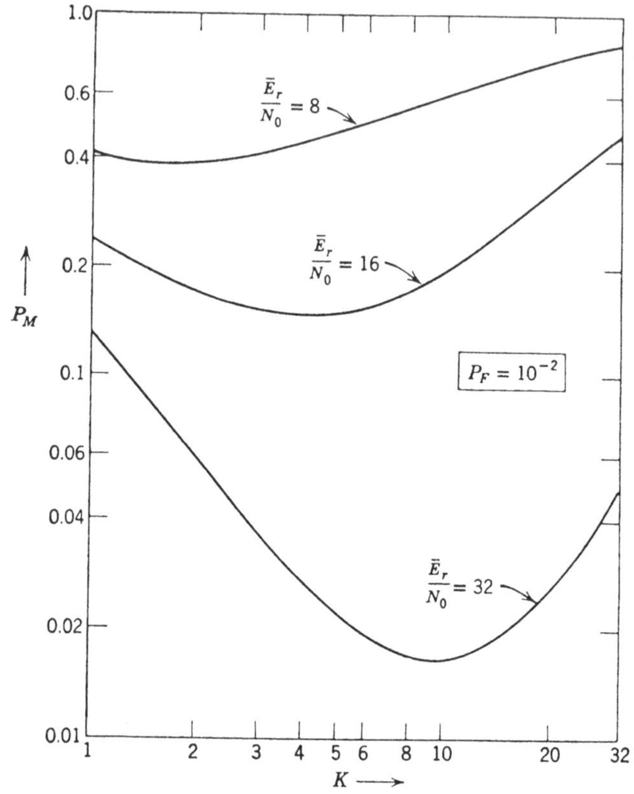


**Fig. 4.18** $P_M$ **as a function of the number of transmitted pulses (total energy fixed).**

### 4.2.3 Frequency Diversity

The obvious dual to the time problem occurs when we transmit $K$ pulses at different frequencies but at the same time. A typical application would be a frequency diversity communication system operating over $K$ nonfluctuating Rayleigh channels. On $H_1$, we transmit $K$ signals in disjoint frequency bands,

$$s_{t1}(t) = \begin{cases} \sum_{i=1}^{K} \sqrt{\frac{2}{T}} \sin \omega_{1_i} t, & 0 \le t \le T, \\ 0, & \text{elsewhere}. \end{cases} \tag{103}$$

On $H_0$, we transmit $K$ signals in a different set of disjoint frequency bands,

$$s_{t0}(t) = \begin{cases} \sum_{j=1}^{K} \sqrt{\frac{2}{T}} \sin \omega_{0_j} t, & 0 \le t \le T, \\ 0, & \text{elsewhere}. \end{cases} \tag{104}$$

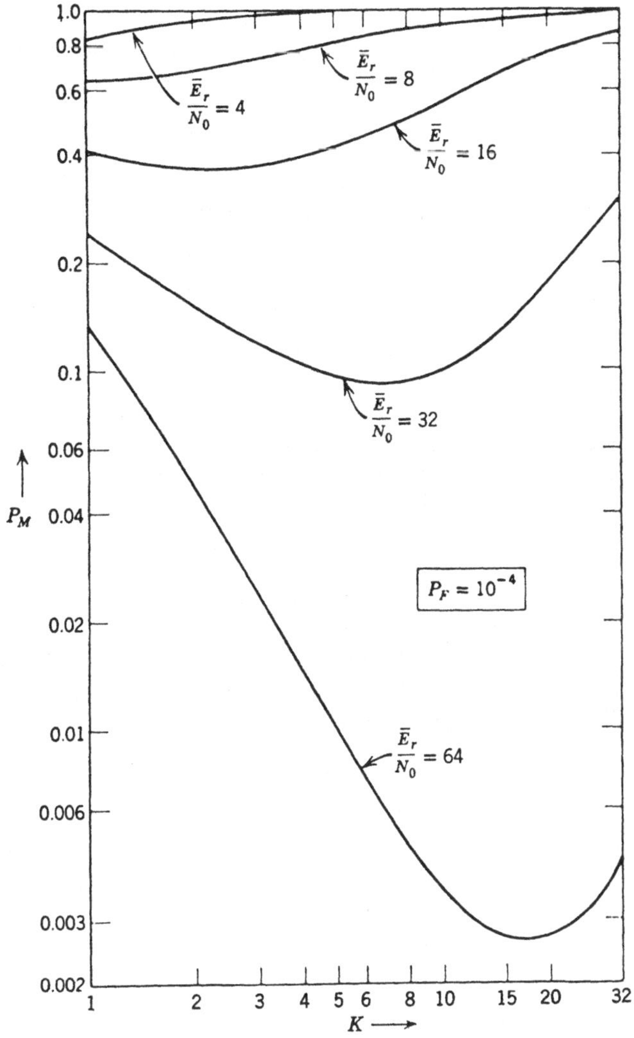


**Fig. 4.18 (Continued.)**

Each of the $K$ transmitted signals passes over a Rayleigh channel. The output is

$$r(t) = \sum_{i=1}^{K} v_i \sqrt{\frac{2}{T}} \sin [\omega_{1i}t + \phi_i] + w(t), \qquad 0 \le t \le T:H_1,$$
$$r(t) = \sum_{j=1}^{K} v_j \sqrt{\frac{2}{T}} \sin [\omega_{0j}t + \phi_j] + w(t), \qquad 0 \le t \le T:H_0. \tag{105}$$

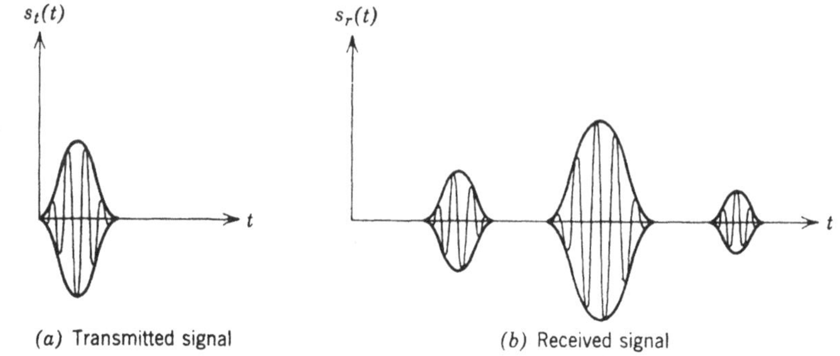


**Fig. 4.19 Ionospheric model: resolvable multipath.**

The frequencies $\omega_{1i}$ and $\omega_{0j}$ are chosen so that the outputs due to the signals are orthogonal. We shall assume that the fading in the different Rayleigh channels is statistically independent and that each channel has identical statistical characteristics. The average received signal energy in each channel is

$$E(v_i^2) = 2\sigma_b^2 \triangleq \bar{E}_{r_i}. \tag{106}$$

We see that this problem is just the binary symmetric version of the problem in Section 4.2.2. The optimum receiver structure is shown in Fig. 4.20. To evaluate the performance, we observe that this case is mathematically identical with Example 3A on pages I-130–I-132 if we let

$$N = 2K, \tag{107}$$

$$\sigma_s^2 = \sigma_b^2 = \frac{\bar{E}_{r1}}{2}, \tag{108}$$

and

$$\sigma_n^2 = \frac{N_0}{2}. \tag{109}$$

Then $\mu(\frac{1}{2})$ is given by (I-2.510) as

$$\boxed{\mu_{\text{BS,BP,SK}}(\tfrac{1}{2}) = K \ln \left\{\frac{1 + \bar{E}_{r1}/N_0}{(1 + \bar{E}_{r1}/2N_0)^2}\right\}.} \tag{110}$$

We can also obtain (110) by using Table 4.3 and (101). A bound on the Pr ($\epsilon$) follows from (I-2.473) and (110) as

$$\Pr(\epsilon) \leq \frac{1}{2}\left\{\frac{(1 + \bar{E}_{r1}/N_0)}{(1 + \bar{E}_{r1}/2N_0)^2}\right\}^K. \tag{111}$$

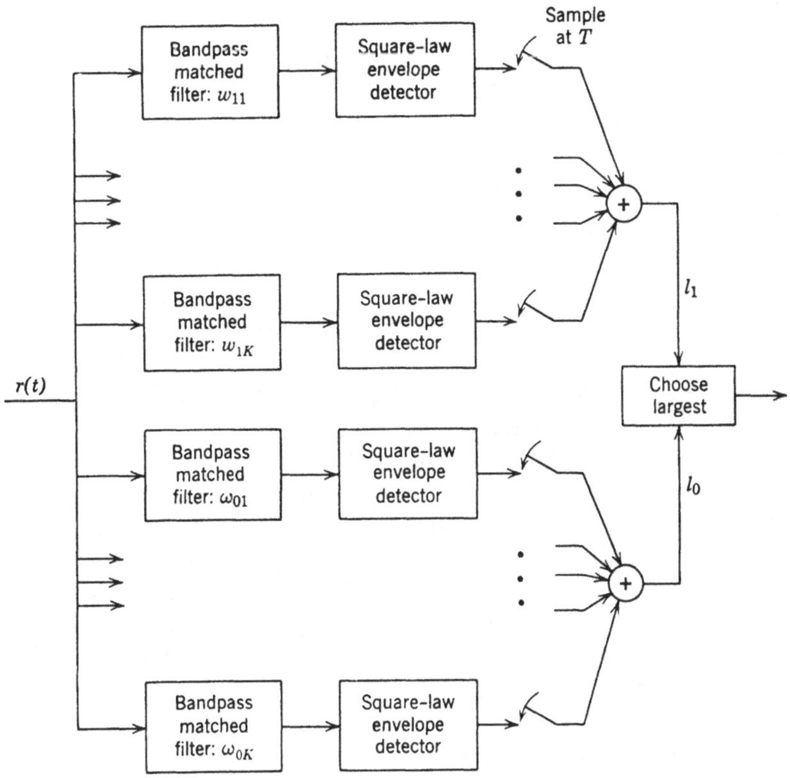


**Fig. 4.20 Frequency diversity receiver.**

An approximate error expression is given by (I-2.516) and (110) as

$$\Pr(\epsilon) \simeq \sqrt{\frac{1}{\pi K}} \frac{(1 + \bar{E}_{r1}/N_0)^K}{(\bar{E}_{r1}/N_0)(1 + \bar{E}_{r1}/2N_0)^{K/2-1}}. \tag{112}$$

Frequently the total energy available at the transmitter is fixed. We want to divide it among the various diversity branches in order to minimize the Pr ($\epsilon$). When the channel attenuations are equal, the optimum division can be calculated easily using either exact or approximate Pr ($\epsilon$) expressions. The simplest procedure is to introduce an efficiency factor for the diversity system.

$$\mu_{\text{BS,BP,SK}}(\tfrac{1}{2}) = -\frac{\bar{E}_r}{N_0}\, g_d(\lambda), \tag{113}$$

where

$$g_d(\lambda) = \frac{1}{\lambda} \ln \left\{ \frac{(1 + \lambda/2)^2}{(1 + \lambda)} \right\} \tag{114}$$

and

$$\lambda = \frac{\bar{E}_{r1}}{N_0}. \tag{115}$$

We see that $g_d(\lambda)$ is identical with $g_{BP,\infty}(\Lambda_\infty)$ in Example 4 on page 116. Thus, it is maximized by choosing

$$\lambda_{\text{OPT}} = \left(\frac{\bar{E}_{r1}}{N_0}\right)_{\text{OPT}} = 3.07, \tag{116}$$

and the Pr ($\epsilon$) is given by (78). The result in (116) says that the optimum strategy is to divide the energy so that the average energy-to-noise spectral height ratio in each Rayleigh channel is 3.07.

The comparison of these two results is interesting. In this case we optimized the performance by choosing the diversity properly. Previously we chose the signal-to-noise ratio in the signal process bandwidth properly. The relationship between the two problems is clear if we interpret both problems in terms of eigenvalues. In the case of the bandlimited spectrum, there are $4WT$ equal eigenvalues (for $WT \gg 1$) and in the diversity system there are $2K$ equal eigenvalues.

### 4.2.4 Summary: Separable Kernels

In this section we have studied the separable kernel problem. Here, the receiver output consists of a weighted sum of the squares of a finite number of statistically independent Gaussian variables. The important difference between the separable kernel case and the general Gaussian problem is that we have *finite* sums rather than *infinite* sums. Therefore, in principle at least, we can always calculate the performance exactly. As we observed in Chapter I-2, if the eigenvalues are different and $K$ is large, the procedure is tedious. If the eigenvalues are equal, the sufficient statistic has a chi-squared density (see page I-109). This leads to an exact expression for $P_F$ and $P_D$. As discussed in Section I-2.7 (page I-128), our approximate expressions based on $\mu(s)$ are accurate for moderate $K$. Thus, even in cases when an exact probability density is available, we shall normally use the approximate expressions because of their simplicity.

In the foregoing text we have considered examples in which the signal process had equal eigenvalues and the additive noise was white. In the problems in Section 4.5, we consider more general separable kernel problems.

## 4.3 LOW-ENERGY-COHERENCE (LEC) CASE†

In this section we consider the simple binary problem described in Chapter 2 (page 8). The received waveforms on the two hypotheses are

$$\begin{aligned} r(t) &= s(t) + w(t), \qquad T_i \le t \le T_f : H_1, \\ r(t) &= w(t), \qquad T_i \le t \le T_f : H_0. \end{aligned} \tag{117}$$

We assume that $w(t)$ is a white, zero-mean Gaussian process with spectral height $N_0/2$ and that $s(t)$ is a zero-mean Gaussian random process with covariance function $K_s(t, u)$. The signal covariance function can be written as a series,

$$K_s(t, u) = \sum_{i=1}^{\infty} \lambda_i^s \phi_i(t)\phi_i(u), \qquad T_i \le t, u \le T_f. \tag{118}$$

If we write $s(t)$ in a Karhunen-Loève expansion, the eigenvalue, $\lambda_i^s$, is the mean-square value of the $i$th coefficient. Physically this corresponds to the average energy along each eigenfunction. If all of the signal energy were contained in a single eigenvalue, we could write

$$s(t) = s_1\phi_1(t) \tag{119}$$

and the problem would reduce to known signal with Gaussian random amplitude that we solved in Section I-4.4. This problem is sometimes referred to as a *coherent* detection problem because all of the energy is along a single known signal.

In many physical situations we have a completely different behavior. Specifically, when we write

$$s(t) = \sum_{i=1}^{\infty} s_i\phi_i(t), \qquad T_i \le t \le T_f, \tag{120}$$

we find that the energy is distributed along a large number of coordinates and that all of the eigenvalues are small compared to the white noise level. Specifically,

$$\lambda_i^s \ll \frac{N_0}{2}, \qquad i = 1, 2, \ldots. \tag{121}$$

We refer to this case as the *low-energy-coherence* (LEC) case. In this section we study the implications of the restriction in (121) with respect

† Most of the original work in the low-energy-coherence case is due to Price [1], [2] and Middleton [3], [5], [7]. It is sometimes referred to as the "coherently undetectable" or "threshold" case. Approaching the performance through $\mu(s)$ is new, but it leads to the same results as obtained in the above references. In [9], Middleton discusses the threshold problem from a different viewpoint. Other references include [10], [11].

to the optimum receiver structure and its performance. Before we begin our discussion, several observations are worthwhile.

1. When $s(t)$ is a stationary process, we know from page I-208 that

$$\lambda_i^s \leq \lambda_{\max} \leq \max_f S_s(f). \tag{122}$$

Thus, if

$$\max_f S_s(f) \ll \frac{N_0}{2}, \tag{123}$$

the LEC condition exists.

2. It might appear that the LEC condition implies poor detection performance and is therefore uninteresting. This is not true, because the receiver output is obtained by combining a large number of components. Even though each signal eigenvalue is small, the resulting test statistic may have appreciably different probability densities on the two hypotheses.

3. We shall find that the LEC condition leads to appreciably simpler receiver configurations and performance calculations. Later we shall examine the effect of using these simpler receivers when the LEC condition is not satisfied.

We begin our discussion with the general results obtained in Section 2.1. From (2.31) we have

$$l_R = \frac{1}{N_0} \iint_{T_i}^{T_f} r(t) h_1(t, u) r(u)\, dt\, du, \tag{124}$$

and from (2.19),

$$l_R = \frac{1}{2}\left(\frac{2}{N_0}\right)^2 \sum_{i=1}^{\infty} \left(\frac{\lambda_i^s}{1 + (2/N_0)\lambda_i^s}\right) r_i^2. \tag{125}$$

To get an approximate expression, we denote the largest eigenvalue by $\lambda_{\max}^s$. If

$$\lambda_{\max}^s < \frac{N_0}{2}, \tag{126}$$

we can expand each term of the sum in (125) in a power series in $\lambda_i$,

$$l_R = \frac{1}{2}\left(\frac{2}{N_0}\right)^2 \sum_{i=1}^{\infty} \lambda_i^s \left[1 - \frac{2}{N_0}\lambda_i^s + \left(\frac{2}{N_0}\right)^2 (\lambda_i^s)^2 - \cdots\right] r_i^2. \tag{127}$$

The convergence of each expansion is guaranteed by the condition in (126). The LEC condition in (121) is more stringent that (126). When

$$\boxed{\frac{2}{N_0}\lambda_{\max}^s \ll 1, \qquad \text{[LEC condition]}} \tag{128}$$

we can approximate $l_R$ by retaining the first *two* terms in the series. The reason for retaining two terms is that they are of the order in $2\lambda_i^s/N_0$. (The reader should verify this.) The first term is

$$l_R^{(1)} = \frac{1}{2}\left(\frac{2}{N_0}\right)^2 \sum_{i=1}^{\infty} \lambda_i^s r_i^2 = \frac{1}{2}\left(\frac{2}{N_0}\right)^2 \iint_{T_i}^{T_f} r(t) K_s(t, u) r(u)\, dt\, du. \tag{129}$$

The second term is

$$l_R^{(2)} = -\frac{1}{2}\left(\frac{2}{N_0}\right)^3 \sum_{i=1}^{\infty} (\lambda_i^s)^2 r_i^2. \tag{130}$$

If we define a kernel

$$\begin{aligned} K_s^{(2)}(t, u) &= \int_{T_i}^{T_f} K_s(t, z) K_s(u, z)\, dz \\ &= \sum_{i=1}^{\infty} (\lambda_i^s)^2 \phi_i(t) \phi_i(u), \qquad T_i \le t, u \le T_f, \end{aligned} \tag{131}$$

then

$$l_R^{(2)} = -\frac{1}{2}\left(\frac{2}{N_0}\right)^3 \iint_{T_i}^{T_f} r(t) K_s^{(2)}(t, u) r(u)\, dt\, du. \tag{132}$$

Similarly, when $2\lambda_{max}/N_0 < 1$, we can expand $l_B$. From (2.33),

$$\begin{aligned} l_B &= -\frac{1}{2}\sum_{i=1}^{\infty} \ln\left(1 + \frac{2}{N_0}\lambda_i^s\right) \\ &= -\frac{1}{2}\left\{+\frac{2}{N_0}\sum_{i=1}^{\infty}\lambda_i^s - \frac{1}{2}\left(\frac{2}{N_0}\right)^2 \sum_{i=1}^{\infty}(\lambda_i^s)^2 + \frac{1}{3}\left(\frac{2}{N_0}\right)^3 \sum_{i=1}^{\infty}(\lambda_i^s)^3 + \cdots\right\}. \end{aligned} \tag{133}$$

When $2\lambda_{max}/N_0 \ll 1$, we can obtain an approximate expression by using the first *two* terms.

$$\begin{aligned} l_B &\simeq -\frac{1}{2}\left(\frac{2}{N_0}\right)\sum_{i=1}^{\infty}\lambda_i^s + \frac{1}{4}\left(\frac{2}{N_0}\right)^2 \sum_{i=1}^{\infty}(\lambda_i^s)^2 \\ &= -\frac{1}{2}\left(\frac{2}{N_0}\right)\int_{T_i}^{T_f} K_s(t, t)\, dt + \frac{1}{4}\left(\frac{2}{N_0}\right)^2 \iint_{T_i}^{T_f} K_s^2(t, u)\, dt\, du. \end{aligned} \tag{134}$$

Equations (129), (132), and (134) correspond to two parallel operations on the received data and a bias term.

We can show that as

$$\frac{2}{N_0}\lambda_{max}^s \to 0. \tag{135}$$

the ratio of the variance of $l_R^{(1)} + l_R^{(2)}$ on $H_0$ to the variance of $l_R^{(1)}$ on $H_0$ approaches zero. The same statement is true on $H_1$ because

$$\text{Var}\,[l_R^{(1)} \mid H_1] \simeq \text{Var}\,[l_R^{(1)} \mid H_0] \tag{136}$$

and

$$\text{Var}\,[l_R^{(2)} \mid H_1] \simeq \text{Var}\,[l_R^{(2)} \mid H_0]. \tag{137}$$

In this case, we may replace $l_R^{(2)}$ by its mean on $H_0$ (the means under both hypotheses are approximately equal):

$$E[l_R^{(2)} \mid H_0] = -\frac{1}{2}\left(\frac{2}{N_0}\right)^2 \iint\limits_{T_i}^{T_f} K_s^2(t, u)\, dt\, du. \tag{138}$$

Now $l_R^{(2)}$ becomes a bias term and $l_R^{(1)}$ is the only quantity that depends on $r(t)$. The resulting test is

$$l_R^{(1)} + l_B \simeq \frac{1}{2}\left(\frac{2}{N_0}\right)^2 \iint\limits_{T_i}^{T_f} r(t) K_s(t, u) r(u)\, dt\, du$$
$$- \frac{1}{2}\left(\frac{2}{N_0}\right) \int_{T_i}^{T_f} K_s(t, t)\, dt - \frac{1}{4}\left(\frac{2}{N_0}\right)^2 \iint\limits_{T_i}^{T_f} K_s^2(t, u)\, dt\, du. \tag{139}$$

Including the bias in the threshold gives the test

$$\boxed{\frac{1}{2}\left(\frac{2}{N_0}\right)^2 \iint\limits_{T_i}^{T_f} r(t) K_s(t, u) r(u)\, dt\, du \underset{H_0}{\overset{H_1}{\gtrless}} \gamma,} \tag{140}$$

where

$$\gamma = \ln \eta + \frac{1}{2}\left(\frac{2}{N_0}\right) \int_{T_i}^{T_f} K_s(t, t)\, dt + \frac{1}{4}\left(\frac{2}{N_0}\right)^2 \iint\limits_{T_i}^{T_f} K_s^2(t, u)\, dt\, du. \tag{141}$$

We refer to the receiver that performs the test in (140) as an optimum LEC receiver. Observe that it has exactly the same form as the general receiver in (124). The difference is that the kernel in the quadratic form is the signal covariance function instead of the optimum linear filter. Notice that the optimum linear filter reduces to $K_s(t, u)$ under the LEC condition. One form of the receiver is shown in Fig. 4.21. The various other realizations discussed in Section 2.1 (Figs. 2.4–2.7) can be modified for the LEC case.

When $2\lambda_{max}^s/N_0$ is less than 1 but does not satisfy (128), we can use more terms in the series of $l_R$ and $l_B$. As long as

$$\frac{2}{N_0}\lambda_{max}^s < 1, \tag{142}$$

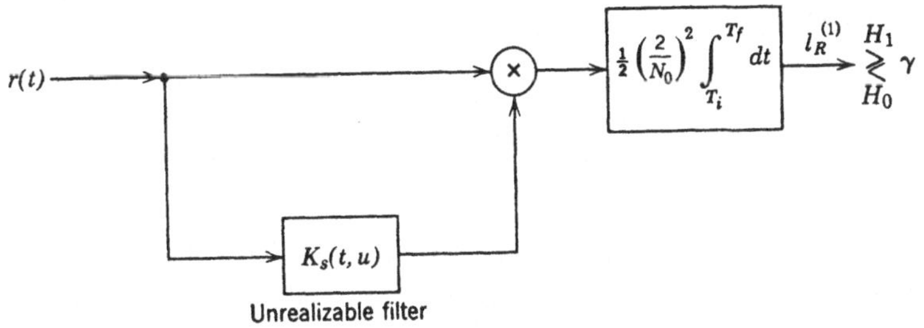


**Fig. 4.21 Optimum LEC receiver.**

we can find a series solution for the optimum detector that will converge.†

The general form follows easily,

$$l_R = -\frac{1}{2}\left(\frac{2}{N_0}\right)\int_{T_i}^{T_f} dt\, r(t) \sum_{n=1}^{\infty}\left(\frac{-2}{N_0}\right)^n \int_{T_i}^{T_f} K_s^{(n)}(t, u) r(u)\, du \tag{143}$$

and

$$l_B = \frac{1}{2}\sum_{n=1}^{\infty}\frac{1}{n}\left(-\frac{2}{N_0}\right)^n \int_{T_i}^{T_f} K_s^{(n)}(t, t)\, dt, \tag{144}$$

where

$$K_s^{(n)}(t, u) = \int_{T_i}^{T_f}\cdots\int_{T_i}^{T_f} K_s(t, z_1)K_s(z_1, z_2)\cdots K_s(z_{n-1}, u)\, dz_1 \cdots dz_{n-1}. \tag{145}$$

An interesting physical interpretation of higher-order approximations is given in Problem 4.3.2.

The final question of interest is the performance of the optimum receiver in the LEC case. We want to find a simpler expression for $\mu(s)$ by exploiting the smallness of the eigenvalues. From (2.134),

$$\mu(s) = \tfrac{1}{2}\sum_{i=1}^{\infty}\left[(1 - s)\ln\left(1 + 2\lambda_i^s/N_0\right) - \ln\left(1 + (1 - s)2\lambda_i^s/N_0\right)\right]. \tag{146}$$

Expanding the logarithms and retaining the first *two* terms, we have

$$\mu(s) \simeq \frac{1}{2}\sum_{i=1}^{\infty}\left\{(1 - s)\left[\frac{2}{N_0}\lambda_i^s - \frac{1}{2}\left(\frac{2}{N_0}\right)^2(\lambda_i^s)^2\right] - \left[(1 - s)\frac{2}{N_0}\lambda_i^s - \frac{(1 - s)^2}{2}\left(\frac{2}{N_0}\right)^2(\lambda_i^s)^2\right]\right\}. \tag{147}$$

We see that the terms linear in $\lambda_i^s$ cancel. Writing $\sum_{i=1}^{\infty}(\lambda_i^s)^2$ in closed

† This approach to finding the filter is identical with trying to solve the integral equation iteratively using a Neumann series (e.g., Middleton [5] or Helstrom [4]). This procedure is a standard technique for solving integral equations.

form, we obtain

$$\mu(s) \simeq -\frac{s(1-s)}{2}\left\{\frac{1}{2}\left(\frac{2}{N_0}\right)^2 \iint_{T_i}^{T_f} K_s^2(t, u)\, dt\, du\right\} \triangleq \mu_{\text{LEC}}(s). \tag{148}$$

The term in braces has an interesting interpretation. For the *known* signal problem, we saw in Chapter I-4 that the performance was completely determined by $d^2$, where

$$d^2 \triangleq \frac{(E[l_R^{(1)} \mid H_1] - E[l_R^{(1)} \mid H_0])^2}{\text{Var}\,[l_R^{(1)} \mid H_0]}. \tag{149}$$

Physically, this could be interpreted as the output signal-to-noise ratio. For the Gaussian signal problem discussed in this chapter, $d^2$ is no longer uniquely related to the error performance, because $l_R^{(1)}$ is *not* Gaussian. However, in the coherently undetectable case, it turns out that the term in braces in (148) is $d^2$, so that whenever our approximations are valid, the output signal-to-noise ratio leads directly to the approximate expressions for $P_F$, $P_M$, and Pr ($\epsilon$). It remains to be verified that the term in braces in (148) equals $d^2$. This result follows easily by using the fact that the expectation of four jointly Gaussian random variables can be written as sums of second moments (e.g., [8, page 168] or page I-229). (See Problem 4.3.4.)

Thus, for the LEC case,

$$\mu_{\text{LEC}}(s) = -\frac{s(1-s)}{2}\, d^2. \tag{150}$$

Substituting the expression for $\mu_{\text{LEC}}(s)$ given in (150) into (2.164) and (2.173) gives the desired error expressions as

$$P_F \simeq \text{erfc}_*\,(sd) = \text{erfc}_*\left(\frac{d}{2} + \frac{\gamma}{d}\right) \tag{151}$$

$$P_M \simeq \text{erfc}_*\,((1-s)d) = \text{erfc}_*\left(\frac{d}{2} - \frac{\gamma}{d}\right). \tag{152}$$

The ROC obtained by varying the threshold $\gamma$ is plotted in Fig. I-4.13.

The low-energy-coherence condition occurs frequently in radar astronomy and sonar problems. Price has studied the first area extensively (e.g., [6]), and we shall look at it in more detail in Chapter 11. In the sonar area the stationary process–long observation time assumption is often valid in addition to the LEC condition. The receiver and the performance are obtained by combining the results of this section with those in Section

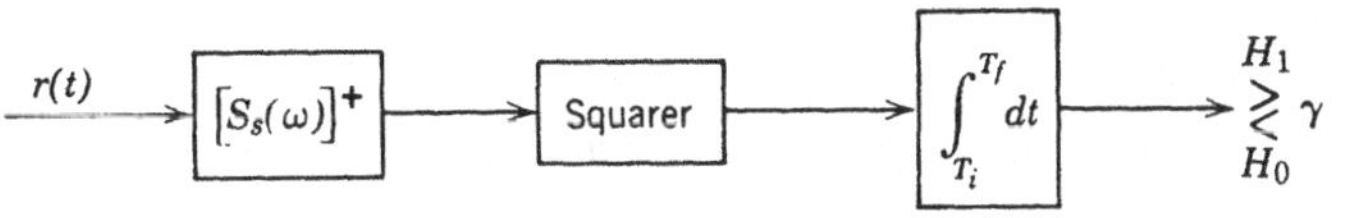


**Fig. 4.22 Optimum receiver, low-energy-coherence, stationary-process, long-observation-time case.**

4.1. A filter-squarer implementation of the resulting receiver is shown in Fig. 4.22. The value of $d^2$ is

$$d^2 = \frac{T}{2}\left(\frac{2}{N_0}\right)^2 \int_{-\infty}^{\infty} S_s^2(\omega)\,\frac{d\omega}{2\pi}. \tag{153}$$

We see that $d^2$ increases linearly with $T$, the observation time. Thus, regardless of the relative signal and noise levels, we can achieve a desired performance by observing the process over a sufficiently long observation time. We shall discuss the sonar area in more detail in *Array Processing*.

Occasionally the LEC receiver in Fig. 4.21 is used even though the LEC condition in (128) is not satisfied. In order to analyze the performance, we must treat it as a suboptimum receiver. In Chapter 5 we discuss performance analysis techniques for suboptimum receivers.

## 4.4 SUMMARY

In this chapter we have developed techniques for finding the optimum receiver and evaluating its performance for three special categories of detection problems. In Chapter 2, we had demonstrated a solution algorithm for cases in which the processes had finite-state representations.

It appears that a large portion of the physical situations that we encounter can be approximated by one of these four special cases. When this is true, we can design the optimum receiver completely and analyze its performance.

## 4.5 PROBLEMS

### P.4.1 Stationary Process, Long Observation Time (SPLOT)

Unless otherwise indicated, you should assume that the SPLOT condition is valid in all problems in this section.

SIMPLE BINARY DETECTION

**Problem 4.1.1.** Consider the model in (1). Assume that $s(t)$ is a Wiener process such that

$$K_s(t, u) = \sigma^2 \min[t, u]$$

and

$$s(0) = 0.$$

1. Find the optimum receiver.
2. Evaluate $\mu_\infty(s)$ by using (20).
3. Compare your result with that in Problem 2.1.5.

**Problem 4.1.2.** Consider the expression for the logarithm of the Fredholm determinant given in (2.75).

1. Derive the asymptotic version of (2.75).
2. Use the result in part 1 to obtain an alternative expression for $\mu_\infty(s)$.
3. Evaluate $\mu_\infty(s)$ for the model in Problem 4.1.1.

**Problem 4.1.3.** Consider the model in (1). Assume that

$$S_s(\omega) = \frac{2nP}{k}\,\frac{\sin(\pi/2n)}{1 + (\omega/k)^{2n}}.$$

Evaluate $\mu_\infty(s)$ for this case.

**Problem 4.1.4 (continuation).** In Problem 4.1.3 we derived an expression for $\mu_\infty(s)$. Fix $s$ at some value $s_0$, where

$$0 < s_0 < 1.$$

Study the behavior of $\mu_\infty(s_0)$ as a function of $n$. Consider different values of $s_0$. How does

$$\Lambda_{\text{B}} \triangleq \frac{2\pi P}{kN_0}$$

enter into the discussion?

**Problem 4.1.5 (non-zero means).** Consider the simple binary detection problem with nonzero means.

1. Derive the asymptotic version of (2.32) and (2.34).
2. Derive the asymptotic version of (2.147).

General Binary Detection

**Problem 4.1.6.** Consider the binary symmetric bandpass version of the class $A_w$ problem. Assume that the equivalent low-pass signal spectrum is

$$S_{s_{\text{L}}}(\omega) = \frac{2nP_{\text{LP}}}{k}\,\frac{\sin(\pi/2n)}{1 + (\omega/k)^{2n}},$$

where

$$P_{\text{LP}} = \frac{P}{2}.$$

Recall that

$$\mu_{\text{BS,BP},\infty}(\tfrac{1}{2}) = 4\mu_{\text{SIB,LP},\infty}(\tfrac{1}{2}).$$

1. Use the result of Problem 4.1.3 to find $\mu_{\text{BS,BP},\infty}(\frac{1}{2})$.
2. Express your answer in the form

$$\mu_{\text{BS,BP},\infty}(\tfrac{1}{2}) \triangleq -\frac{\bar{E}_r}{N_0}\cdot 4g_{\text{BP,n}}(\Lambda_{\text{B}}).$$

Plot $g_{\text{BP,n}}(\Lambda_{\text{B}})$ for various $n$. Find

$$\max_{\Lambda_{\text{B}}}\left[g_{\text{BP,n}}(\Lambda_{\text{B}})\right].$$

3. Find

$$\max_n \left\{ \max_{\Lambda_B} \left[ g_{BP,n}(\Lambda_B) \right] \right\}$$

**Problem 4.1.7.** Consider the binary symmetric bandpass version of the class $A_w$ problem.

1. Write $\mu_{BS,BP,\infty}(1/2)$ as a function of $S_{s_L}(\omega)$ (the equivalent low-pass spectrum) and $N_0/2$.

2. Constrain

$$\int_{-\infty}^{\infty} S_{s_L}(\omega) \frac{d\omega}{2\pi} = \frac{P}{2}. \tag{P.1}$$

Find the spectrum that minimizes $\mu_{BS,BP,\infty}(1/2)$ subject to the constraint in (P.1).

**Problem 4.1.8.** Consider the class $B$ problem (see Fig. 3.1). Verify that the results in Tables 4.1 and 4.2 are correct.

**Problem 4.1.9.** Consider the class $B_w$ problem in which

$$S_s(\omega) = \frac{2kP_s}{\omega^2 + k^2}$$

$$S_n(\omega) = \frac{2k_1P_n}{\omega^2 + k_1^2}$$

$$S_w(\omega) = \frac{N_0}{2}.$$

1. Find the optimum receiver.
2. Evaluate $\mu_\infty(s)$.
3. Consider the special case in which $k_1 = k$. Simplify the expressions for $\mu_\infty(s)$.

**Comment:** In the discussion of minimum Pr ($\epsilon$) tests in the text, we emphasized the case in which the hypotheses were equally likely and $\mu(s)$ was symmetric around $s = \frac{1}{2}$ (see pages 77–79). In many minimum Pr ($\epsilon$) tests the hypotheses are equally likely but $\mu(s)$ is not symmetric. We must then solve the equation

$$\dot{\mu}(s)\big|_{s=s_m} = 0 \tag{F.1}$$

for $s_m$. We then use this value of $s_m$ in (I-2.484) or (I-2.485). From the latter,

$$\Pr(\epsilon) \simeq \frac{1}{[2(2\pi\ddot{\mu}(s_m))^{\frac{1}{2}} s_m(1 - s_m)]} \exp \mu(s_m). \tag{F.2}$$

(Assumes $s_m\sqrt{\ddot{\mu}(s_m)} > 3$ and $(1 - s_m)\sqrt{\ddot{\mu}(s_m)} > 3$.) From (I-2.473),

$$\Pr(\epsilon) \le \tfrac{1}{2} \exp \mu(s_m). \tag{F.3}$$

The next several problems illustrate these ideas.

**Problem 4.1.10.** Consider the class $A_w$ problem in which

$$S_{s_1}(\omega) = \frac{2k\alpha P}{\omega^2 + k^2},$$

$$S_{s_0}(\omega) = \frac{2kP}{\omega^2 + k^2}.$$

1. Draw a block diagram of the optimum receiver. Include the necessary biases.
2. Evaluate $\mu_\infty(s)$.
3. Assume that a minimum Pr $(\epsilon)$ test is desired and that the hypotheses are equally likely. Find $s_m$ such that

$$\dot{\mu}(s)\big|_{s=s_m} = 0.$$

4. Compute the approximate Pr $(\epsilon)$ using (F.2).
5. Compute a bound on the Pr $(\epsilon)$ using (F.3).
6. Plot $\mu(s_m)$ as a function of $\alpha$.
7. Evaluate

$$\left.\frac{\partial \mu_\infty(s, \alpha)}{\partial \alpha}\right|_{\alpha=0}.$$

This result will be useful when we study parameter estimation.

**Problem 4.1.11.** Consider the class $A_w$ problem in which

$$S_{s_1}(\omega) = \begin{cases} \dfrac{\pi \alpha P}{k}, & |\omega| \le k, \\ 0, & |\omega| > k, \end{cases}$$

and

$$S_{s_0}(\omega) = \begin{cases} \dfrac{\pi P}{k}, & |\omega| \le k, \\ 0, & |\omega| > k. \end{cases}$$

Repeat Problem 4.1.10.

**Problem 4.1.12.** Consider the class $A_w$ problem in which

$$S_{s_1}(\omega) = \frac{2\beta P_1}{\omega^2 + \beta^2}$$

and

$$S_{s_0}(\omega) = \frac{2\alpha P_0}{\omega^2 + \alpha^2}.$$

1. Repeat parts 1–5 of Problem 4.1.10.
2. Evaluate the approximate Pr $(\epsilon)$ for the case in which

$$\frac{\beta P_1}{\alpha P_0} = 1$$

and

$$N_0 = 0.$$

**Problem 4.1.13.** Consider the binary symmetric class $A_w$ problem. All processes are symmetric around their respective carriers (see Section 3.4.3 and Fig. 3.9). The received waveform $r_{c_1}(t)$ is

$$\begin{aligned} r_{c_1}(t) &= s_{c_1}(t) + n_{c_1}(t) + w(t), \qquad T_i \le t \le T_f, \\ r_{c_1}(t) &= n_{c_1}(t) + w(t), \qquad T_i \le t \le T_f. \end{aligned} \tag{P.1}$$

Notice that (P.1) completely describes the problem because of the assumed symmetries. The random processes in (P.1) are statistically independent with spectra $S_s(\omega)$, $S_{n_c}(\omega)$, and $N_0/2$, respectively.

1. Derive a formula for $\mu_{BS,BP,\infty}(\frac{1}{2})$.
2. Assume that $S_s(\omega)$ and $N_0/2$ are fixed. Constrain the power in $n_{c_1}(t)$,

$$\int_{-\infty}^{\infty} S_{n_c}(\omega) \frac{d\omega}{2\pi} = \frac{P_c}{2}.$$

Choose $S_{n_c}(\omega)$ to *maximize* $\mu_{BP,BS,\infty}(\frac{1}{2})$.

3. Assume that $S_{n_c}(\omega)$ and $N_0/2$ are fixed. Constrain the power in $s_{c_1}(t)$,

$$\int_{-\infty}^{\infty} S_s(\omega) \frac{d\omega}{2\pi} = \frac{P_s}{2}.$$

Choose $S_s(\omega)$ to *minimize* $\mu_{BP,BS,\infty}(\frac{1}{2})$.

**Problem 4.1.14.** Consider the vector problem described in Problem 3.2.7. Specialize the results of this problem to the case in which the SPLOT condition is valid.

**Problem 4.1.15.** Consider the special case of Problem 4.1.14 in which

$$\mathbf{s}_1(t) = \int_{-\infty}^{\infty} \mathbf{h}(t-\tau)s(\tau)\,d\tau \tag{P.1}$$

and

$$\mathbf{s}_0(t) = \mathbf{0}. \tag{P.2}$$

The matrix filter $\mathbf{h}(\tau)$ has one input and $N$ outputs. Its transfer function is $\mathbf{H}(j\omega)$. Simplify the receiver in Problem 4.1.14.

**Problem 4.1.16.** Consider Problem 3.2.8. Specialize the results to the SPLOT case.

**Problem 4.1.17.**

1. Consider Problem 3.2.9. Specialize the results to the SPLOT case.
2. Consider the particular case described in (P.1) of Problem 4.1.15. Specialize the results of part 1.

**Problem 4.1.18.**

1. Review the results in Problem 3.5.18. Derive an expression for $\mu_\infty(s)$ for the general vector case.
2. Specialize the result in part 1 to the class $A_w$ SPLOT problem.
3. Specialize the results in part 2 to the class $B_w$ SPLOT problem.
4. Specialize the results in part 1 to the case in which the signal is described by (P.1) in Problem 4.1.15.

**Problem 4.1.19.** The received waveforms on the two hypotheses are

$$r(t) = s_1(t) + w(t), \qquad T_i \le t \le T_f : H_1,$$
$$r(t) = s_0(t) + w(t), \qquad T_i \le t \le T_f : H_0.$$

The signals $s_1(t)$ and $s_0(t)$ are stationary, zero-mean, *bandpass* Gaussian processes centered at $\omega_1$ and $\omega_0$, respectively. Their spectra are disjoint and are *not* necessarily symmetric around their carrier frequencies. The additive noise is white ($N_0/2$).

Find the optimum receiver and an expression for $\mu_{BP,\infty}(s)$. (*Hint:* Review the results of Problem 3.4.9.)

**Problem 4.1.20.** Consider the binary symmetric bandpass problem in Fig. 3.9. Assume that

$$E[r_{c_1}(t) \mid H_1] = m,$$

$$E[r_{c_0}(t) \mid H_0] = m.$$

All other means are zero.

1. Find the optimum receiver using the SPLOT assumption.
2. Evaluate $\mu_{\mathrm{BS,BS},\infty}(\frac{1}{2})$.

**Problem 4.1.21.** Consider the expression for $\mu(s)$ given in (2.208) and the expression for $\mu_\infty(s)$ given in (30).

1. Prove

$$\lim_{kT \to \infty} \mu(s) = \mu_\infty(s).$$

2. Consider the binary symmetric bandpass version of (2.208) and (30) [see Example 3, (59) and (60)]. Denote the BS, BP version of (2.208) as $\mu_{\mathrm{BS,BP}}(s, kT)$. Plot

$$f(kT) \triangleq \frac{\mu_{\mathrm{BS,BP}}(1/2, kT) - \mu_{\mathrm{BS,BP},\infty}(\frac{1}{2})}{\mu_{\mathrm{BS,BP}}(1/2, kT)}$$

as a function of $kT$ in order to study the accuracy of the SPLOT approximation.

## P.4.2 Separable Kernels

**Problem 4.2.1.** Consider the pulsed radar problem. The performance is characterized by (98)–(102). From (101),

$$\mu_{\mathrm{BP,SK}}(s) = K \ln \left[ \frac{(1 + \bar{E}_{r1}/N_0)^{1-s}}{1 + (1 - s)\bar{E}_{r1}/N_0} \right].$$

Choosing a particular value of $s$ corresponds to choosing the threshold in the LRT.

1. Fix $s = s_m$ and require

$$\mu(s_m) - s_m \dot{\mu}(s_m) = c.$$

Constrain

$$\bar{E}_r = K\bar{E}_{r1}.$$

Choose $K$ to minimize

$$F \triangleq \mu(s_m) + (1 - s_m)\dot{\mu}(s_m).$$

Explain the physical significance of this procedure.

2. Compare the results of this minimization with the results in Figs. 4.17 and 4.18.

**Problem 4.2.2.**

1. Consider the separable kernel problem in which the $a_i$ in (89) have non-zero means $a_i$. Find $\mu_D(s)$.
2. Consider the bandpass version of the model in part 1. Assume that each successive pair of $a_i$ have identical statistics. Evaluate $\mu_D(s)$ and $\mu_R(s)$.

**Problem 4.2.3.**

1. Consider the binary symmetric version of the bandpass model in Problem 4.2.2. Evaluate $\mu_{\mathrm{BS,BP,SK}}(\frac{1}{2})$.

2. Simplify the results in part 1 to the case in which all of the $a_i$ are identically distributed. Assume

$$E[a_i] = m$$

and

$$\text{Var}\,[a_i] = \sigma_s^2.$$

**Problem 4.2.4.** Consider the model in Problem 4.2.3. A physical situation in which we would encounter this model is a frequency diversity system operation over a Rician channel (see Section 4.2.2). If the energy in the transmitted signal is $E_t$, then

$$m^2 = \alpha E_t,$$

$$\sigma_s^2 = \beta E_t,$$

where $\alpha$ and $\beta$ are the strengths of the specular path and the random path, respectively.

1. Express $\mu_{\text{BS,BP,SK}}(1/2)$ in terms of $\alpha$, $\beta$, $E_t$, and $K$ (the number of paths).

2. Assume that $E_t$ is fixed. Choose $K$ to minimize $\mu_{\text{BS,BP,SK}}(\frac{1}{2})$. Explain your results intuitively and compare them with (116).

**Problem 4.2.5.** Consider the diversity system described in Section 4.2.2. If the signal eigenvalues were different, we could write the efficiency factor in (114) as

$$g_d(\boldsymbol{\lambda}) = \sum_{i=1}^{K} \frac{1}{2\lambda_i} \ln \left\{ \frac{(1 + \lambda_i/2)^2}{(1 + \lambda_i)} \right\}. \tag{P.1}$$

Assume

$$\sum_{i=1}^{K} \lambda_i = c. \tag{P.2}$$

You may choose $K$ and $\lambda_i$ subject to the restriction in (P.2). Prove that $g_d(\boldsymbol{\lambda})$ is maximized by the choice

$$\lambda_i = \begin{cases} \lambda, & i = 1, 2, \ldots, K_0, \\ 0, & i > K_0. \end{cases}$$

Find $K_0$.

**Problem 4.2.6** Consider the frequency diversity system operating over a Rayleigh channel as described in Section 4.2.

1. Generalize the model to allow for unequal path strengths, unequal energy transmitted in each channel, and unequal noise levels.

2. Consider the two-channel problem. Constrain the total transmitted power. Find the optimum division of energy to minimize $\mu_{\text{BS,BP,SK}}(1/2)$.

**Problem 4.2.7.** Consider the class $A_w$ problem in which

$$K_{s_1}(t, u) = \sum_{i=1}^{N_1} \lambda_i^{s_1} f_i(t) f_i(u), \qquad T_i \le t, u \le T_f$$

and

$$K_{s_0}(t, u) = \sum_{j=1}^{N_2} \lambda_j^{s_0} g_j(t) g_j(u), \qquad T_i \le t, u \le T_f, \tag{P.1}$$

where

$$\int_{T_i}^{T_f} f_i(t) f_k(t)\,dt = \delta_{ik}, \qquad i, k = 1, \ldots, N_1,$$

$$\int_{T_i}^{T_f} g_j(t) g_k(t)\,dt = \delta_{jk}, \qquad i, k = 1, \ldots, N_2,$$

and

$$\int_{T_i}^{T_f} f_i(t) g_j(t)\,dt = \rho_{ij}, \qquad i = 1, \ldots, N_1, \quad j = 1, \ldots, N_2. \tag{P.2}$$

1. Solve (3.33) for $h_\Delta(t, u)$.
2. Specialize part 1 to the case in which

$$\rho_{ij} = 0, \qquad i = 1, \ldots, N_1, \quad j = 1, \ldots, N_2. \tag{P.3}$$

Explain the meaning of (P.3). Give a physical situation in which (P.3) is satisfied.

3. Derive a formula for $\mu_{SK}(s)$.
4. Specialize the result in part 3 to the case in which (P.3) is satisfied.

**Problem 4.2.8.** Consider the class $B_w$ problem in which the received waveforms on the two hypotheses are

$$r(t) = s(t) + w(t), \qquad T_i \le t \le T_f : H_1,$$
$$r(t) = w(t), \qquad T_i \le t \le T_f : H_0.$$

The signal and noise processes are statistically independent, zero-mean processes with covariance functions $K_s(t, u)$ and $N_0\,\delta(t-u)/2$, respectively. The signal process $K_s(t, u)$ is separable and has $M$ equal eigenvalues,

$$K_s(t, u) = \lambda_0 \sum_{i=1}^{M} \phi_i(t)\phi_i(u), \qquad T_i \le t, u \le T_f.$$

1. Verify that the receiver in Fig. P.4.1 is optimum.

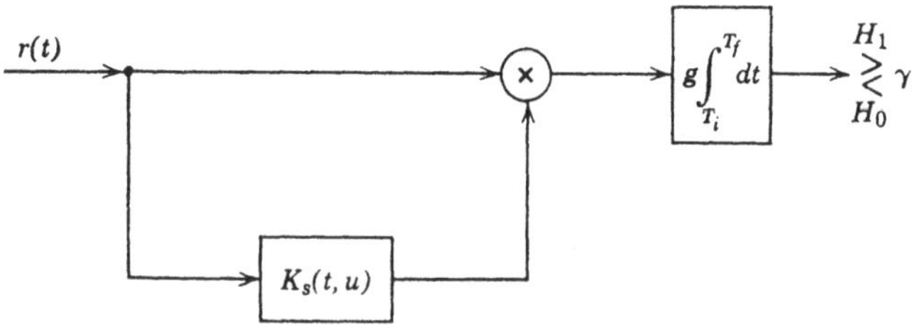


**Fig. P.4.1**

2. Compute $g$. Compare the receiver in Fig. P.4.1 with the LEC receiver of Section 4.3.

## P.4.3 Low Energy Coherence (LEC)

**Problem 4.3.1.** Consider the development in (129)–(139). Verify that the various approximations made in arriving at (139) are valid.

**Problem 4.3.2.**

1. Verify the general form in (143).

2. An easy way to remember the structure of $l_R$ in (143) is shown in Fig. P.4.2. This is an unrealizable feedback system. Verify that the output is $l_R$.

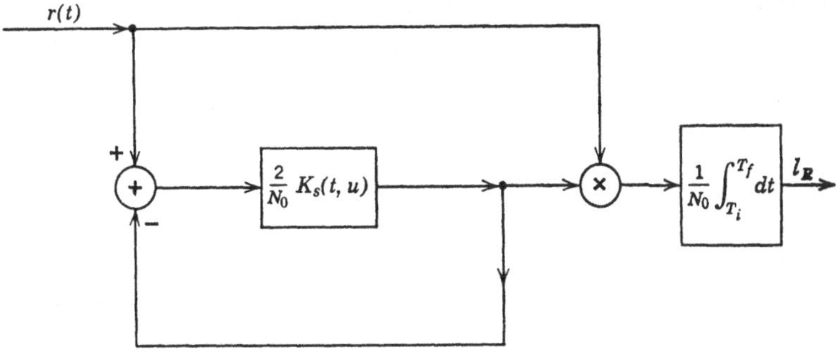


**Fig. P.4.2**

3. Why is the result in part 2 obvious? Is the receiver in Fig. P.4.2 optimum for the general case? Why is it a useful idea in the LEC case but not in the general case?

**Problem 4.3.3.** Consider the vector model in Problem 3.2.7, in which $\mathbf{s}_0(t) = \mathbf{0}$.

1. Find the optimum receiver under the LEC condition. Define precisely what the LEC condition is in the vector case.

2. Assume that both the SPLOT and LEC conditions hold. Find the optimum receiver and derive an expression for $\mu_{\infty,\text{LEC}}(s)$. Express the LEC condition in terms of the signal spectral matrix $\mathbf{S}_1(\omega)$.

**Problem 4.3.4.** Derive the result in (149).

## REFERENCES


[1] R. Price and P. E. Green, Jr., "Signal Processing in Radar Astronomy—Communication via Fluctuating Multipath Media," Massachusetts Institute of Technology, Lincoln Laboratory, Technical Report 234, October 1960.

[2] R. Price, "Detectors for Radar Astronomy," in *Radar Astronomy*, J. V. Evans and T. Hagfors Eds., New York, McGraw-Hill, 1968.

[3] D. Middleton, "On Singular and Nonsingular Optimum (Bayes) Tests for the Detection of Normal Stochastic Signals in Normal Noise," IRE Trans. IT-7, No. 2, 105–113 (April 1961).

[4] C. W. Helstrom, *Statistical Theory of Signal Detection*, Pergamon Press, New York, 1960.

[5] D. Middleton, *An Introduction to Statistical Communication Theory*, McGraw-Hill, New York, 1960.

[6] R. Price, "Output Signal-to-Noise Ratio as a Criterion in Spread-Channel Signaling," Massachusetts Institute of Technology, Lincoln Laboratory, Technical Report 388, May 13, 1965.

[7] D. Middleton, "On the Detection of Stochastic Signals in Additive Normal Noise," IRE Trans. **PGIT-3,** 86–121 (1957).

[8] W. B. Davenport and W. L. Root, *Random Signals and Noise*, McGraw-Hill, New York, 1958.

[9] D. Middleton, "Canonically Optimum Threshold Detection," IEEE Trans. Information Theory **IT-12,** No. 2, 230–243 (April 1966).

[10] J. Capon, "On the Asymptotic Efficiency of Locally Optimum Detectors," IRE Trans. Information Theory **IT-7,** 67–71 (April 1961).

[11] P. Rudnick, "Likelihood Detection of Small Signals in Stationary Noise," J. Appl. Phys. **32,** 140 (Feb. 1961).

# 5

# *Discussion: Detection of Gaussian Signals*

In Chapters 2 through 4 we studied the problem of detecting Gaussian signals in the presence of Gaussian noise. In this chapter we first discuss some related topics. We then summarize the major results of our detection theory discussion.

## 5.1 RELATED TOPICS

### 5.1.1 *M*-ary Detection: Gaussian Signals in Noise

All of our discussion in Chapters 2 through 4 dealt with the binary detection problem. In this section we discuss briefly the $M$-hypothesis problem. The general Gaussian $M$-ary problem is

$$r(t) = r_i(t), \qquad T_b \le t \le T_f : H_i, \qquad i = 1, \dots, M, \tag{1}$$

where

$$E[r_i(t) \mid H_i] = m_i(t) \tag{2}$$

and

$$E[(r_i(t) - m_i(t))(r_i(u) - m_i(u)) \mid H_i] = K_{H_i}(t, u). \tag{3}$$

Most of the ideas from the binary case carry over to the $M$-ary case with suitable modifications. As an illustration we consider a special case of the general problem.

The problem of interest is described by the following model. The received waveforms on the $M$ hypotheses are

$$r(t) = s_i(t) + n(t), \qquad T_b \le t \le T_f : H_i, \qquad i = 1, 2, \dots, M. \tag{4}$$

The additive noise $n(t)$ is a sample function from a zero-mean Gaussian process with covariance function $K_n(t, u)$. A white noise term is not necessarily present. The signal processes are sample functions from Gaussian processes and are statistically independent of the noise process. The signal processes are characterized by

$$E[s_i(t)] = m_i(t), \qquad T_b \le t \le T_f, \qquad i = 1, \dots, M \tag{5}$$

and

$$E\{[s_i(t) - m_i(t)][s_i(u) - m_i(u)]\} = K_{s_i}(t, u), \qquad T_b \le t, u \le T_f, \quad i = 1, \dots, M. \tag{6}$$

The a-priori probability of the $i$th hypothesis is $P_i$ and the criterion is minimum $\Pr(\epsilon)$. We assume that each pair of hypotheses would lead to a nonsingular binary test. The derivation of the optimum receiver is similar to the derivation for the binary case, and so we shall simply state the results. The reader can consult [1]–[3] or Problem 5.1.1 for details of the derivation.

To perform the likelihood ratio test we compute a set of $M$ sufficient statistics, which we denote by $l_i$, $i = 1, \dots, M$. The first component of the $i$th sufficient statistic is

$$l_{R_i} = \int_{T_b}^{T_f} r(t) h_i(t, u) r(u)\, dt\, du \tag{7}$$

where $h_i(t, u)$ is specified by the integral equation

$$\iint_{T_b}^{T_f} K_n(t, x) h_i(x, y)[K_n(y, u) + K_{s_i}(y, u)]\, dx\, dy = K_{s_i}(t, u), \qquad T_b \le t, u \le T_f. \tag{8}$$

The second component of the $i$th sufficient statistic is

$$l_{D_i} = \int_{T_b}^{T_f} r(t)\left[g_i(t) - \int_{T_b}^{T_f} h_i(t, u) m_i(u)\, du\right] dt, \tag{9}$$

where $g_i(t)$ is specified by the integral equation

$$\int_{T_b}^{T_f} K_n(t, u) g_i(u)\, du = m_i(t), \qquad T_b \le t \le T_f. \tag{10}$$

The bias component of the $i$th sufficient statistic is

$$l_{B_i} = -\tfrac{1}{2} \sum_{k=1}^{\infty} \ln(1 + \lambda_{ik}^*), \tag{11}$$

where the $\lambda_{ik}^*$ are the eigenvalues of the kernel,

$$K_{s_i}^*(t, u) = \iint_{T_b}^{T_f} K_n^{[-\frac{1}{2}]}(t, x) K_{s_i}(x, y) K_n^{[-\frac{1}{2}]}(u, y)\, dx\, dy. \tag{12}$$

The complete $i$th sufficient statistic is

$$l_i = l_{R_i} + l_{D_i} + l_{B_i}, \qquad i = 1, \dots, M. \tag{13}$$

The test consists of computing

$$l_i + \ln P_i, \qquad i = 1, \dots, M \tag{14}$$

and choosing the largest.

A special case of (4) that occurs frequently is

$$K_n(t, u) = \frac{N_0}{2}\,\delta(t-u), \tag{15}$$

$$m_i(t) = 0, \qquad i = 1, \dots, M. \tag{16}$$

Then $h_i(t, u)$ satisfies the equation

$$\frac{N_0}{2} h_i(t, u) + \int_{T_b}^{T_f} h_i(t, y) K_{s_i}(y, u)\, dy = K_{s_i}(t, u), \qquad T_b \le t, u \le T_f,$$
$$i = 1, \dots, M. \tag{17}$$

All of the canonical realizations in Chapter 2 are valid for this case. The bias term is

$$l_{B_i} = -\frac{1}{N_0}\int_{T_b}^{T_f} \xi_{Pi}\left(t \,\Big|\, s_i(\cdot), \frac{N_0}{2}\right) dt, \tag{18}$$

where $\xi_{Pi}(t \mid s_i(\cdot),)$ is defined as in (2.137).

The performance calculation for the general $M$-ary case is difficult. We would anticipate this because, even in the known signal case, exact $M$-ary performance calculations are usually not feasible.

An important problem in which we can get accurate bounds is that of digital communication over a Rayleigh channel using $M$-orthogonal signals. The binary version of this problem was discussed in Examples 3 and 4 of Chapter 4 (see pages 111–117). We now indicate the results for the $M$-ary problem.

The transmitted signal on the $i$th hypothesis is

$$s_t(t) = \sqrt{\frac{2E_t}{T}} \sin \omega_i t \qquad T_b \le t \le T_f : H_i \tag{19}$$

the signal passes over a fluctuating Rayleigh channel. The received waveform on the $i$th hypothesis is

$$r(t) = s_i(t) + w(t), \qquad T_b \le t \le T_f : H_i. \tag{20}$$

The $i$th signal $s_i(t)$ is a sample function of a bandpass process centered at $\omega_i$, whose spectrum is symmetric around $\omega_i$.† The signal processes

† The symmetric assumption is included to keep the notation simple. After we introduce complex notation in Chapter 9, we can handle the asymmetric case easily.

are essentially disjoint in frequency. The additive noise is a sample function from a zero-mean white Gaussian random process with spectral height $N_0/2$. The low-pass spectra of the signal processes are identical. We denote them by $S_{s_l}(\omega)$. The total received power in the signal is

$$P_r = 2\sigma_b^2 \int_{-\infty}^{\infty} S_{s_l}(\omega) \frac{d\omega}{2\pi}. \tag{21}$$

Kennedy [4] and Viterbi [5] have studied the performance for this case. Our discussion follows the latter's. Starting from a general result in [6], one can show that

$$\Pr(\epsilon) \le \exp[\rho TR] \frac{\left[D_{\mathscr{F}}\left(\frac{2}{N_0}\right)\right]^{\rho}}{\left[D_{\mathscr{F}}\left(\frac{2\rho}{N_0(1+\rho)}\right)\right]^{1+\rho}}, \qquad 0 \le \rho \le 1, \tag{22}$$

where $D_{\mathscr{F}}(z)$ is the Fredholm determinant of the low-pass process, $T$ is the length of interval, and

$$R = \ln M \tag{23}$$

is the transmission rate in nats per second. The parameter $\rho$ is used to optimize the bound. When the observation time is long, we can use (I-3.182) to obtain

$$\ln D_{\mathscr{F}}(z) = T \int_{-\infty}^{\infty} \ln(1 + zS_{s_l}(\omega)) \frac{d\omega}{2\pi}. \tag{24}$$

We now define

$$E_0(\rho) \triangleq (1+\rho) \int_{-\infty}^{\infty} \ln\left[1 + \frac{\rho}{1+\rho} \frac{2S_{s_l}(\omega)}{N_0}\right] \frac{d\omega}{2\pi} - \rho \int_{-\infty}^{\infty} \ln\left[1 + \frac{2S_{s_l}(\omega)}{N_0}\right] \frac{d\omega}{2\pi} \tag{25}$$

and

$$E(R) = \max_{0 \le \rho \le 1} [E_0(\rho) - \rho R]. \tag{26}$$

Comparing (25) and (4.21), we observe that

$$-\frac{T}{2} \frac{E_0(\rho)}{1+\rho}\bigg|_{\rho=(1-s)/s} = \mu_{\infty}(s) \tag{27}$$

and, in particular,

$$-TE_0(1) = 4\mu_{\infty}(\tfrac{1}{2}) = 2\mu_{BP,\infty}(\tfrac{1}{2}) = \mu_{BS,BP,\infty}(\tfrac{1}{2}). \tag{28}$$

Using (24)–(26), (22) reduces to

$$\Pr(\epsilon) \le e^{-TE(R)}. \tag{29}$$

The final step is to perform the maximization indicated in (26). The result is obtained as follows:

1. If

$$\dot{E}_0(\rho) \triangleq \frac{\partial E_0(\rho)}{\partial \rho} = R \tag{30}$$

has a solution for $0 \le \rho \le 1$, we denote it as $\rho_m$. Then

$$E(R) = E_0(\rho_m) - \rho_m \dot{E}_0(\rho_m), \tag{31}$$

$$R = \dot{E}_0(\rho_m), \tag{32}$$

and

$$\dot{E}_0(1) \le R \le C_{0R} = \dot{E}_0(0). \tag{33}$$

2. If (30) does not have a solution in the allowable range of $\rho$, the maximum is at $\rho = 1$, and

$$E(R) = E_0(1) - R, \tag{34}$$

$$0 < R < \dot{E}_0(1). \tag{35}$$

The results in (31) and (34) provide the exponents in the $\Pr(\epsilon)$ expression. In the problems, we include a number of examples to illustrate the application of these results.

This concludes our brief discussion of the $M$-ary problem. For a large class of processes we can find the optimum receiver, but, except for orthogonal signal processes, the performance evaluation is usually difficult.

### 5.1.2 Suboptimum Receivers

We have been able to find the optimum receiver to implement the likelihood ratio test for a large class of Gaussian signal processes. Frequently, the filters in the receiver are time-varying and may be difficult to implement. This motivates the search for suboptimum receivers, which are simpler to implement than the optimum receiver but perform almost as well as the optimum receiver. To illustrate this idea we consider a simple example.

**Example.** Consider the simple binary detection example discussed on page 104. The received waveforms on the two hypotheses are

$$\begin{aligned} r(t) &= s(t) + w(t), \qquad T_i \le t \le T_f : H_1, \\ r(t) &= w(t), \qquad T_i \le t \le T_f : H_0, \end{aligned} \tag{36}$$

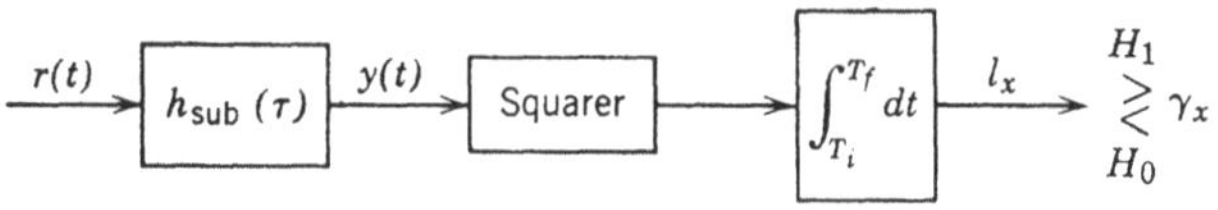


**Fig. 5.1 Suboptimum receiver.**

where $w(t)$ is a white Gaussian process with spectral height $N_0/2$ and $s(t)$ is a Gaussian process with spectrum

$$S_s(\omega) = \frac{2kP}{\omega^2 + k^2} \,. \tag{37}$$

We saw that the optimum receiver could be implemented as a cascade of a time-varying filter, a square-law device, and an integrator. The difficulty arises in implementing the time-varying filter.

A receiver that is simpler to implement is shown in Fig. 5.1. The structure is the same as the optimum receiver, but the linear filter is time-invariant,

$$h_{\text{sub}}(\tau) = e^{-\beta\tau}u_{-1}(\tau), \qquad -\infty < \tau < \infty. \tag{38}$$

We choose $\beta$ to optimize the performance. From our results in Section 4.1 (page 104) we know that if

$$\beta = k\left(1 + \frac{4P}{kN_0}\right)^{\frac{1}{2}}, \tag{39}$$

then the suboptimum receiver will be essentially optimum for long observation times. For arbitrary observation times, some other choice of $\beta$ might give better performance. Thus, the problem of interest is to choose $\beta$ to maximize the performance.

With this example as motivation, we consider the general question of suboptimum receivers. The choice of the structure for the suboptimum receiver is strongly dependent on the particular problem. Usually one takes the structure of the optimum receiver as a starting point, tries various modifications, and analyzes the resulting performance. In this section we discuss the performance of suboptimum receivers.

To motivate our development, we first recall the performance results for the optimum receiver. The optimum receiver computes $l$, the logarithm of the likelihood ratio, and compares it with a threshold. The error probabilities are

$$P_F = \Pr(\epsilon|H_0) = \int_{\gamma}^{\infty} p_{l|H_0}(L \mid H_0)\, dL \tag{40}$$

and

$$P_M = \Pr(\epsilon|H_1) = \int_{-\infty}^{\gamma} p_{l|H_1}(L \mid H_0)\, dL. \tag{41}$$

All of our performance discussion in the Gaussian signal problem has been based on $\mu(s)$, which is defined as

$$\mu(s) = \ln M_{l|H_0}(s) = \ln \int_{-\infty}^{\infty} e^{sL} p_{l|H_0}(L \mid H_0)\, dL, \tag{42}$$

the logarithm of the moment-generating function of $l$, given that $H_0$ is true. Since $l$ is the logarithm of the likelihood ratio, we can also write $\mu(s)$ in terms of $M_{l|H_1}(s)$,

$$\mu(s) = \ln M_{l|H_1}(s - 1) \tag{43}$$

(see pages I-118–I-119). Thus we can express both $P_M$ and $P_F$ in terms of $\mu(s)$.

A suboptimum receiver computes a test statistic $l_x$ and compares it with a threshold $\gamma_x$ in order to make a decision. The statistic $l_x$ is *not* equivalent to $l$ and generally is used because it is easier to compute. For suboptimum receivers, the probability densities of $l_x$ on $H_1$ and $H_0$ are not uniquely related, and so we can no longer express $P_M$ and $P_F$ in terms of a single function. This forces us to introduce two functions analogous to $\mu(s)$ and makes the performance calculations more involved.

To analyze the suboptimum receiver, we go through a development parallel to that in Sections I-2.7 and III-2.2. Because the derivation is straightforward, we merely state the results. We define

$$\mu_0(s) \triangleq \ln M_{l_x|H_0}(s) = \int_{-\infty}^{\infty} e^{sL} p_{l_x|H_0}(L \mid H_0)\, dL \tag{44}$$

and

$$\mu_1(s) \triangleq \ln M_{l_x|H_1}(s) = \int_{-\infty}^{\infty} e^{sL} p_{l_x|H_1}(L \mid H_1)\, dL. \tag{45}$$

The Chernoff bounds are

$$\Pr(\epsilon \mid H_0) \le \exp[\mu_0(s_0) - s_0\gamma], \qquad s_0 > 0, \tag{46}$$

$$\Pr(\epsilon \mid H_1) \le \exp[\mu_1(s_1) - s_1\gamma], \qquad s_1 < 0, \tag{47}$$

where

$$\dot{\mu}_0(s_0) = \gamma, \qquad s_0 > 0 \tag{48}$$

and

$$\dot{\mu}_1(s_1) = \gamma, \qquad s_1 < 0. \tag{49}$$

The equations (48) and (49) will have a unique solution if

$$E[l_x \mid H_0] \le \gamma \le E[l_x \mid H_1]. \tag{50}$$

The first-order asymptotic approximations are

$$\Pr(\epsilon \mid H_0) \simeq \operatorname{erfc}_*\left(s_0\sqrt{\ddot{\mu}_0(s_0)}\right) \exp\left[\mu_0(s_0) - s_0\dot{\mu}_0(s_0) + \frac{s_0^2}{2}\ddot{\mu}_0(s_0)\right] \tag{51}$$

and

$$\Pr(\epsilon \mid H_1) \simeq \operatorname{erfc}_*\left(-s_1\sqrt{\ddot{\mu}_1(s_1)}\right) \exp\left[\mu_1(s_1) - s_1\dot{\mu}_1(s_1) + \frac{s_1^2}{2}\ddot{\mu}_1(s_1)\right], \tag{52}$$

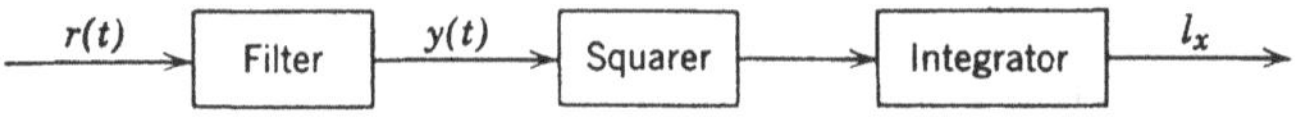


**Fig. 5.2 General filter-squarer-integrator (*FSI*) receiver.**

where $s_0$ and $s_1$ satisfy (48) and (49). Equations (51) and (52) are analogous to (179) and (188). Results similar to (181) and (189) follow easily.

The results in (44)–(52) are applicable to an arbitrary detection problem. To apply them to the general Gaussian problem, we must be able to evaluate $\mu_0(s)$ and $\mu_1(s)$ efficiently. The best technique for evaluating $\mu_0(s)$ and $\mu_1(s)$ will depend on the structure of the suboptimum receiver. We demonstrate the technique for the general filter-squarer-integrator (FSI) receiver shown in Fig. 5.2. The filter may be time-varying. For this structure the techniques that we developed in Section 2.2 (pages 35–44) are still valid. We illustrate the procedure by finding an expression for $\mu_1(s)$.

**Calculation of $\mu_1(s)$ for an FSI Receiver.** To find $\mu_1(s)$, we expand $y(t)$, the input to the squarer under $H_1$, in a Karhunen-Loève expansion. Thus

$$y_1(t) \triangleq \sum_{i=1}^{\infty} y_i \phi_{1i}(t), \qquad T_i \le t \le T_f : H_1, \tag{53}$$

where the $\phi_{1i}(t)$ are the eigenfunctions of $y(t)$ on $H_1$. The corresponding eigenvalues are $\lambda_{1i}$. We assume that the eigenvalues are ordered in magnitude so that $\lambda_{1i}$ is the largest. From (45),

$$\begin{aligned}
\mu_1(s) &= \ln \{E[e^{sl_x} \mid H_1]\} \\
&= \ln \left\{ E\left[ \exp\left( s \sum_{i=1}^{\infty} y_i^2 \right) \middle| H_1 \right] \right\} \\
&= -\tfrac{1}{2} \sum_{i=1}^{\infty} \ln (1 - 2s\lambda_{1i}), \qquad s < \frac{1}{2\lambda_{1i}}.
\end{aligned} \tag{54}$$

The expectation is a special case of Problem I-4.4.2. The sum can be written as a Fredholm determinant,†

$$\mu_1(s) = -\tfrac{1}{2} \ln D_{\mathscr{F}|H_1}(-2s), \qquad s < \frac{1}{2\lambda_{1i}}. \tag{55}$$

A similar result follows for $\mu_0(s)$,

$$\mu_0(s) = -\tfrac{1}{2} \ln D_{\mathscr{F}|H_0}(-2s), \qquad s < \frac{1}{2\lambda_{0i}}. \tag{56}$$

We now have $\mu_0(s)$ and $\mu_1(s)$ expressed in terms of Fredholm determinants. The final step is to evaluate these functions. Three cases in which we can evaluate $\ln D_{\mathscr{F}|H_i}(\cdot)$ are the following:

1. Stationary processes, long observation time.
2. Separable kernels.
3. State-representable processes.

† This result is due to Kac and Siegert [8] (e.g., [9, Chapter 9]).

The procedure for the first two cases is clear. In the third case, we can use the algorithms in section 2.2.1 or section 2.2.3 to evaluate $\mu_0(s)$ and $\mu_1(s)$. The important point to remember is that the state equation that we use to compute $\mu_1(s)$ corresponds to the system that produces $y_1(t)$ when driven by white noise. Similarly, the state equation that we use to compute $\mu_0(s)$ corresponds to the system that produces $y_0(t)$ when driven by white noise.

In this section we have developed the performance expressions needed to analyze suboptimum receivers. Because the results are straightforward modifications of our earlier results, our discussion was brief. The analysis based on these results is important in the implementation of practical receiver configurations. A number of interesting examples are developed in the problems. In Chapter 11, we encounter suboptimum receivers again and discuss them in more detail.

### 5.1.3 Adaptive Receivers

A complete discussion of adaptive receivers would take us too far afield. On the other hand, several simple observations are useful.

All of our discussion of communication systems assumed that we made a decision on each baud. This decision was independent of those made on previous bauds. If the channel process is correlated over several bauds, one should be able to exploit this correlation in order to reduce the probability of error. Since the optimum "single-baud" receiver is an estimator-correlator, a logical approach is to perform a continuous channel estimation and use this to adjust the receiver filters and gains. An easy way to perform the channel estimation is through the use of decision-directed feedback. Here we assume that all past decisions are correct in order to perform the channel estimation. As long as most of the decisions are correct, this reduces the channel estimation problem to that of a "known" signal into an unknown channel. Decision feedback schemes for simple channels have been studied by Proakis and Drouilhet [10]. More complicated systems have been studied by Glaser [11], Jakowitz, Shuey, and White [12], Scudder [13, 14], Boyd [15], and Austin [16]. Another procedure to exploit the correlation of the channel process would be to devote part of the available energy to send a known signal to measure the channel.

There has been a great deal of work done on adaptive systems. In almost all cases, the receivers are so complicated and difficult to analyze that one cannot make many useful general statements. We do feel the reader should recognize that many of these systems are logical extrapolations from the general Gaussian problem we have studied. References that deal with various types of adaptive systems include [17]–[30].

### 5.1.4 Non-Gaussian Processes

All of our results have dealt with Gaussian random processes. When the processes involved are non-Gaussian, the problems are appreciably more difficult. We shall divide our comments on these problems into four categories:

1. Processes derived from Gaussian processes.
2. Structured non-Gaussian processes.
3. Unspecified non-Gaussian processes.
4. Analysis of fixed receivers.

We shall explain the descriptions in the course of our discussion.

***Processes Derived from Gaussian Processes.*** We have emphasized cases in which the received waveform is conditionally Gaussian. A related class of problems comprises those in which $r(t)$ is a sample function of a process that can be derived from a Gaussian process. A common case is one in which either the mean-value function or the covariance function contains a random parameter set. In this case, we might have $m(t, \boldsymbol{\theta}_m)$ and $K_r(t, u:\boldsymbol{\theta}_k)$. If the probability densities of $\boldsymbol{\theta}_m$ and $\boldsymbol{\theta}_k$ are known, the parameters are integrated out in an obvious manner (conceptually, at least). Whether we can actually carry out the integration depends on how the parameters enter into the expression.

If either $\boldsymbol{\theta}_m$ or $\boldsymbol{\theta}_k$ is a nonrandom variable, we can check to see if a uniformly most powerful test exists. If it does not, a generalized likelihood ratio test may be appropriate.

***Structural Non-Gaussian Processes.*** The key to the simplicity in Gaussian problems is that we can completely characterize the process by its mean-value function and covariance function. We would expect that whenever the processes involved could be completely characterized in a reasonably simple manner, one could find the optimum receiver. An important example of such a processes is the Poisson process. References [31]–[35] discuss this problem. A second important example is Markov processes (e.g., [2-21]–[2-24]).

***Unspecified Non-Gaussian Processes.*** In this case we would like to make some general statements about the optimum receiver without restricting the process to have a particular structure. One result of this type is available in the LEC case that we studied in Section 4.3. Middleton [36], [37], derives the LEC receiver without requiring that the signal process be Gaussian. (See [39] for a different series expansion approach.) A second important result concerning unspecified Gaussian processes is given in [38]. Here, Kailath extends the realizable estimator-correlator receiver to include non-Gaussian processes.

***Analysis of Fixed Receivers.*** In this case, we consider a fixed receiver structure and analyze its performance in the presence of non-Gaussian signals and noise. Suitable examples of this type of analysis are contained in [40], [41].

These four topics illustrate some of the issues involved in the study of non-Gaussian processes. The selection was intended to be representative, not exhaustive.

### 5.1.5 Vector Gaussian Processes

We have not discussed the case in which the received signal is a vector random process. The formal extension of our results to this case is straightforward. In fact, all of the necessary equations have been developed in the problem sections of Chapters 2–4. The important issues in the vector case are the solution of the equations specifying the optimum receiver and its performance and the interpretation of the results in the context of particular physical situations. In the subsequent volume [42], we shall study the vector problem in the context of array processing in sonar and seismic systems. At that time, we shall discuss the above issues in detail.

## 5.2 SUMMARY OF DETECTION THEORY

In Chapters 2 through 5 we have studied the detection of Gaussian signals in Gaussian noise in detail. The motivation of this detailed study is to provide an adequate background for actually solving problems we encounter when modeling physical situations.

In Chapter 2 we considered the simple binary problem. The first step was to develop the likelihood ratio test. We saw that the likelihood ratio contained three components. The first was obtained by a nonlinear operation on the received waveform and arose because of the randomness in the signal. The second was obtained by a linear operation on the received waveform and was due to the deterministic part of the received signal. This component was familiar from our earlier work. The third component was the bias term, which had to be evaluated in order to conduct a Bayes test.

We next turned our attention to the problem of realizing the nonlinear operation needed to generate $l_R$. Four canonical realizations were developed:

1. The estimator-correlator receiver.
2. The filter-correlator receiver.
3. The filter-squarer receiver.
4. The optimum realizable filter receiver.

The last realization was particularly appealing when the process had finite state-variable representation. In this case we could use all of the effective state-variable procedures that we developed in Section I-6.3 actually to find the receiver.

A more difficult issue was the performance of the optimum receiver. As we might expect from our earlier work, an exact performance calculation is not feasible in many cases. By building on our earlier work on bounds and approximate expressions in Section I-2.7, we developed performance results for this problem. The key to the results was the $\mu(s)$ function defined in (2.148). We were able to express this in terms of both a realizable filtering error and the logarithm of the Fredholm determinant. We have effective computational procedures to evaluate each of these functions.

We next turned to the general binary problem in Chapter 3, where the received waveform could contain a nonwhite component on each hypothesis. The procedures were similar to the simple binary case. A key result was (3.33), whose solution was the kernel of the nonlinear part of the receiver. The modifications of the various canonical realizations were straightforward, and the performance bounds were extended. A new issue that we encountered was that of singularity. We first derived simple upper and lower bounds on the probability of error in terms of $\mu(\frac{1}{2})$. We then showed that a necessary and sufficient condition for a nonsingular test was that $\mu(\frac{1}{2})$ be finite. This condition was then expressed in terms of a square-integrability requirement on a kernel. As before, singularity was never an issue when the same white noise component was assumed to be present on both hypotheses.

In Chapter 4 we considered three special cases that led to particularly simple solutions. In Section 4.1 we looked at the stationary-process–long-time-interval case. This assumption enabled us to neglect homogeneous solutions in the integral equation specifying the kernel and allowed us to solve this equation using Fourier transform techniques. Several practical examples were considered. The separable kernel case was studied in Section 4.2. We saw that this was a suitable model for pulsed radars with slowly fluctuating targets, ionospheric communications over resolvable multipath channels, and frequency-diversity systems. The solution for this case was straightforward. Finally, in Section 4.3, we studied the low-energy-coherence case, which occurs frequently in passive sonar and radar astronomy problems. The energy in the signal process is spread over a large number of coordinates so that each eigenvalue is small when compared to the white noise level. This smallness enabled us to obtain a series solution to the integral equation. In this particular case we found that the output signal-to-noise ratio ($d^2$) is an accurate performance measure. In addition to these three special cases, we had previously developed a complete solution for the case in which the processes have a finite state representation. A large portion of the physical situations that we encounter can be satisfactorily approximated by one of these cases.

In Section 5.1 we extended our results to the $M$-ary problem. The optimum receiver is a straightforward extension of our earlier results, but the performance calculation for the general problem is difficult. A reasonably simple bound for the case of $M$-orthogonal processes was presented. In Section 5.2 we derived performance expressions for suboptimum receivers.

Our discussion of the detection problem has been lengthy and, in several instances, quite detailed. The purpose is to give the reader a thorough understanding of the techniques involved in solving actual problems. In addition to the references we have cited earlier, the reader may we wish to consult [43]–[48] for further reading in this area. In the next two chapters we consider the parameter estimation problem that was described in Chapter 1.

## 5.3 PROBLEMS

### P.5.1 Related Topics

*M*-ARY DETECTION

**Problem 5.1.1.** Consider the model described in (4)–(6). Assume that $n(t)$ contains a white noise component with spectral height $N_0/2$. Assume that

$$m_i(t) = 0, \qquad i = 1, \dots, M.$$

1. Derive (7)–(8) and (11)–(14).
2. Draw a block diagram of the optimum receiver.

**Problem 5.1.2.** Generalize the model in Problem 5.1.1 to include nonzero means and unequal costs. Derive the optimum Bayes receiver.

**Problem 5.1.3.** Consider the model in (15)–(17).

Assume that

$$K_{s_i}(t, u) = iK_s(t - u), \qquad i = 1, \dots, M$$

that

$$S_s(\omega) = \frac{2k}{\omega^2 + k^2},$$

and that the SPLOT condition is valid. The hypotheses are equally likely.

1. Draw a block diagram of the optimum receiver.
2. Consider the case in which $M = 3$. Derive a bound on the Pr ($\epsilon$).

**Problem 5.1.4.** Consider the communication system using $M$-orthogonal signals that is described in (19)–(21). On pages I-263–I-264, we derived a bound on the Pr ($\epsilon$) in an $M$-ary system in terms of the Pr ($\epsilon$) in a binary system.

1. Extend this technique to the current problem of interest.
2. Compare the bound in part 1 with the bound given by (22)–(35). For what values of $R$ is the bound in part 1 useful?

**Problem 5.1.5.** Consider the problem of detecting one of $M$-orthogonal bandpass processes in the presence of white noise. Assume that each process has $N$ eigenvalues.

1. We can immediately reduce the problem to one with $MN$ dimensions. Denote this resulting vector as $\mathbf{R}$. Compute $P_{\mathbf{r}|H_m}(\mathbf{R} \mid H_m)$.
2. In [6], Gallager derived the basic formula for a bound on the error probability,

$$\Pr(\epsilon \mid H_m) \le \int_{-\infty}^{\infty} \cdots \int d\mathbf{R}[p_{\mathbf{r}|H_m}(\mathbf{R} \mid H_m)]^{1/(1+\rho)} \times \left[\sum_{k \ne m}^{M} [p_{\mathbf{r}|H_k}(\mathbf{R} \mid H_k)]^{1/(1+\rho)}\right]^{\rho}, \qquad \rho \ge 0. \quad \text{(P.1)}$$

Use (P.1) to derive (22). (*Hint:* Use the fact that $E[x^\rho] \le (E[x])^\rho$, $0 \le \rho \le 1$.)

**Problem 5.1.6.**

1. Verify the results in (29)–(35).
2. One can show that $C_{or}$ is the capacity of the channel for this type of communication system (i.e., we require $M$-orthogonal signals and use rectangular signal envelopes). Assume

$$S_{s_l}(\omega) = \frac{2k}{\omega^2 + k^2} \qquad \text{(P.1)}$$

and

$$P_r \triangleq 2\sigma_b^2. \qquad \text{(P.2)}$$

Plot $C_{or}$ as a function of

$$\Lambda_B = \frac{\pi P_r}{kN_0}. \qquad \text{(P.3)}$$

3. Repeat for

$$S_{s_l}(\omega) = \begin{cases} \dfrac{\pi}{k}, & |\omega| \le k, \\ 0, & |\omega| > k. \end{cases} \qquad \text{(P.4)}$$

**Problem 5.1.7.** The error exponent, $E(R)$, is defined by (31) and (34).

1. Plot $E(R)$ as a function of $N_0R/P_r$ for the bandlimited message spectrum in (P.4) of Problem 5.1.6.

2. Plot $E(R)$ as a function of $N_0R/P_r$ for the one-pole message spectrum in (P.1) of Problem 5.1.6.

**Problem 5.1.8.** Assume that we want to signal at low rates so that

$$E(R) \simeq E(0).$$

1. Consider the one-pole message spectrum in (P.1) of Problem 5.1.6. Plot $E(0)/(P_r/N_0)$ as a function of $\Lambda_B$. What value of $\Lambda_B$ maximizes $E(0)/(P_r/N_0)$?
2. Repeat part 1 for the ideal bandlimited message spectrum in (P.4) of Problem 5.1.6.
3. Compare the results in parts 1 and 2 with those in (4.60) and (4.69).

**Problem 5.1.9.** Assume that we want to signal at the rate

$$R = \frac{1}{10}C_\infty = \frac{1}{10}\frac{P_r}{N_0}.$$

We want to maximize $E(R)/(P_r/N_0)$ by choosing $\Lambda_B$.

1. Carry out this maximization for the one-pole spectrum.
2. Carry out this maximization for the ideal bandlimited spectrum.
3. Compare your results with those in Problem 5.1.8.

**Problem 5.1.10.** Define

$$E_n^*(R) = \max_{\Lambda_B}\left[\frac{E(R)}{P_r/N_0}\right].$$

1. Find $E_n^*(R)$ as $R$ varies from 0 to $C_\infty$ for the ideal bandlimited spectrum.
2. Repeat part 1 for the one-pole spectrum.

**Problem 5.1.11** [5]. Assume that each signal process has a non-zero mean. Specifically,

$$E\{[s_i(t)\sqrt{2}\cos(\omega_i t)]_{\rm LP}\} = m(t),$$
$$E\{[s_i(t)\sqrt{2}\sin(\omega t_i)]_{\rm LP}\} = 0.$$

Show that the effect of the non-zero mean is to add a term to $E_0(\rho)$ in (25), which is

$$E_{0m}(\rho) = \frac{\rho}{2N_0(1+\rho)}\int_{-\infty}^{\infty}|M(j\omega)|^2\left[1+\left(\frac{\rho}{1+\rho}\right)\frac{2S_{s_l}(\omega)}{N_0}\right]^{-1}\frac{d\omega}{2\pi}.$$

**Problem 5.1.12** [7]. Consider the special case of Problem 5.1.11 in which

$$S_{s_l}(\omega) = 0.$$

1. Prove

$$\frac{E(R)}{C_\infty} = \begin{cases} \frac{1}{2} - \dfrac{R}{C_\infty}, & 0 \le \dfrac{R}{C_\infty} \le \frac{1}{4}, \\ \left(1-\sqrt{\dfrac{R}{C_\infty}}\right)^2, & \frac{1}{4} \le \dfrac{R}{C_\infty} < 1, \end{cases}$$

where

$$C_\infty = \frac{P_T}{N_0} = \frac{1}{N_0}\int_{-\infty}^{\infty} m^2(t)\,dt.$$

2. Discuss the significance of this result.

SUBOPTIMUM RECEIVERS

**Problem 5.1.13.** Consider the definitions of $\mu_0(s)$ and $\mu_1(s)$ given in (44) and (45).

1. Derive the Chernoff bounds in (46)–(49).
2. Derive the approximate error expressions in (51) and (52).

**Problem 5.1.14.** Consider the simple binary detection problem described on page 151 and the filter-squarer-integrator receiver in Fig. 5.1. The filter is time-invariant with transfer function

$$H(j\omega) = \frac{1}{j\omega + \beta}.$$

The message spectrum is given in (37).

1. Write the state equations that are needed to evaluate $\mu_1(s)$ and $\mu_0(s)$.
2. Assume that the long-time-interval approximation is valid. Find $\mu_{1\infty}(s)$ and $\mu_{0\infty}(s)$. Verify that the value of $\beta$ in (39) is optimum.

**Problem 5.1.15.**

1. Repeat part 1 of Problem 5.1.14 for the case in which $s(t)$ is a Wiener process,

$$s(0) = 0, \qquad t \geq 0,$$
$$E[s^2(t)] = \sigma^2 t, \qquad t \geq 0.$$

2. Find the optimum value of $\beta$ for long observation times.

**Problem 5.1.16.** Consider the binary symmetric communication problem whose model was given in Section 3.4.3. The quantities $r_{c_1}(t)$, $r_{s_1}(t)$, $r_{c_0}(t)$, and $r_{s_0}(t)$ were defined in Fig. 3.9. We operate on each of these waveforms as shown in Fig. 3.10. Instead of the optimum filter $h_{fr}(t, u)$, we use some arbitrary filter $h_{\text{sub}}(\tau)$ in each path. Denote the output of the top branch as $l_1$ and the output of the bottom branch as $l_0$. Define

$$l_x = l_1 - l_0.$$

The bias terms are both zero and

$$\ln \eta \triangleq \gamma_x = 0.$$

Define

$$\mu_{1_j}(s) = \ln E[e^{sl_1} \mid H_j], \qquad j = 0, 1,$$

and

$$\mu_{0_j}(s) = \ln E[e^{sl_0} \mid H_j], \qquad j = 0, 1.$$

1. Prove

$$\mu_{\text{BS},1}(s) = \mu_{11}(s) + \mu_{01}(-s),$$
$$\mu_{\text{BS},0}(s) = \mu_{00}(-s) + \mu_{10}(s) = \mu_{\text{BS},1}(-s).$$

2. Prove

$$\Pr(\epsilon) < \tfrac{1}{2}\exp(\mu_{\text{BS},1}(s_m)),$$

where

$$\dot{\mu}_{\text{BS},1}(s)\big|_{s=s_m} = 0.$$

3. Prove

$$\Pr(\epsilon) \simeq \frac{1}{2(2\pi\ddot{\mu}_{BS,1}(s_m))^{1/2} s_m(1 - s_m)} \exp \mu_{BS,1}(s_m).$$

4. Express $\mu_{BS,1}(s)$ in terms of Fredholm determinants.

**Problem 5.1.17.** Consider the binary communication system described in Problem 5.1.16. Assume that $s_1(t)$ is a sample function of a stationary process whose low-pass equivalent spectrum is $S_{\tilde{s}_1}(\omega)$ and $h(t, \tau)$ is a time-invariant filter with a rational transfer function. Assume that the SPLOT condition is valid.

1. Find an expression for $\mu_{BS,1\infty}(s)$ in terms of $S_{\tilde{s}_1}(\omega)$, $H(j\omega)$, and $N_0$.
2. Verify that $\mu_{BS,1\infty}(s)$ reduces to $\mu_{OPT,\infty}(s)$ when $H(j\omega)$ is chosen optimally.
3. Plot $\mu_{BS,1\infty}(s)$ for the case in which

$$S_{\tilde{s}_1}(\omega) = \frac{kP_r}{\omega^2 + k^2}$$

and

$$H(j\omega) = \frac{1}{j\omega + \beta}.$$

Find $s_m$.

**Problem 5.1.18 (continuation).** Consider the binary communication system discussed in Problems 5.1.16 and 5.1.17. We are interested in the case discussed in part 3 of Problem 5.1.17.

One of the problems in designing the optimum receiver is that $P_r$ may be unknown or may vary slowly. Assume that we think that

$$P_r = P_{rn}$$

and design the optimum receiver.

1. Evaluate $\mu_{BS,1\infty}(s_m)$ and $\mu_{OPT,\infty}(\frac{1}{2})$ for this receiver when

$$\Lambda_1 \triangleq \frac{2P_{rn}}{kN_0} = 100.$$

2. Now assume that

$$0.1P_{rn} \leq P_r \leq 10P_{rn}.$$

Plot $\mu_{BS,1\infty}(s_m)$. The receiver design is fixed.

3. Assume that the receiver is redesigned for each $P_r$. Compare $\mu_{OPT,\infty}(\frac{1}{2})$ with $\mu_{BS,1\infty}(s_m)$.

**Problem 5.1.19.** The LEC receiver was derived in Section 4.3 and was shown in Fig. 4.21. This receiver is sometimes used when the LEC condition is not satisfied.

1. Derive an approximate expression for the performance of this receiver.
2. Assume that $s(t)$ has a finite-dimensional state representation. Find a state equation for $\mu_1(s)$ and $\mu_0(s)$.
3. Assume that the SPLOT condition is valid. Find a simple expression for $\mu_1(s)$ and $\mu_0(s)$.

## REFERENCES


[1] R. Price, "Optimum Detection of Random Signals in Noise with Applications to Scatter-Multipath Communication. I," IRE Trans. Information Theory **IT-2,** 125–135 (Dec. 1956).

[2] G. Turin, "Communication through Noisy, Random-Multipath Channels," 1956 IRE Convention Record, Pt. 4, pp. 154–156.

[3] D. Middleton, *Introduction to Statistical Communication Theory*, McGraw-Hill, New York, 1960.

[4] R. S. Kennedy, *Fading Dispersive Communication Channels*, Wiley, New York, 1969.

[5] A. J. Viterbi, "Performance of an *M*-ary Orthogonal Communication System Using Stationary Stochastic Signals," IEEE Trans. Information Theory **IT-13,** No. 3, 414–422 (July 1967).

[6] R. G. Gallager, "A Simple Derivation of the Coding Theorem and Some Applications," IEEE Trans. Information Theory **IT-11,** 3–18 (Jan. 1965).

[7] J. M. Wozencraft and I. M. Jacobs, *Principles of Communications Engineering*, Wiley, New York, 1965.

[8] M. Kac and A. J. F. Siegert, "On the Theory of Noise in Radio-Receivers with Square-Law Detectors," J. Appl. Phys. **18,** 383–397 (April 1947).

[9] W. B. Davenport and W. L. Root, *An Introduction to the Theory of Random Signals and Noise*, McGraw-Hill, New York, 1958.

[10] J. G. Proakis and P. R. Drouilhet, "Performance of Coherent Detection Systems Using Decision-Directed Channel Measurement," IEEE Trans., PGCS, 54 (March 1964).

[11] E. M. Glaser, "Signal Detection by Adaptive Filters," IRE Trans. **IT-7,** No. 2, 87–98 (April 1961).

[12] C.V. Jakowitz, R. L. Shuey, and G. M. White, "Adaptive Waveform Recognition," TR 60-RL-2353E, General Electric Research Laboratory, Schenectady, September 1960.

[13] H. J. Scudder, "Adaptive Communication Receivers," IEEE Trans. **IT-11,** No. 2, 167–174 (April 1965).

[14] H. J. Scudder, "The Design of Some Synchronous and Asynchronous Adaptive Receivers," First IEEE Ann. Communication Conv., Boulder, Colo. 759 (June 1965).

[15] D. Boyd, "Several Adaptive Detection Systems," M.Sc. Thesis, Department of Electrical Engineering, Massachusetts Institute of Technology, July 1965.

[16] M. E. Austin, "Decision-Feedback Equalization for Digital Communication over Dispersive Channels," Sc.D. Thesis, Massachusetts Institute of Technology, May 1967.

[17] C. S. Weaver, "Adaptive Communication Filtering," IEEE Trans. Information Theory **IT-8,** S169–S178 (Sept. 1962).

[18] A. M. Breipohl and A. H. Koschmann, "A Communication System with Adaptive Decoding," Third Symposium on Discrete Adaptive Processes, 1964 National Electronics Conference, Chicago, October 19–21, pp. 72–85.

[19] J. G. Lawton, "Investigations of Adaptive Detection Techniques," Cornell Aeronautical Laboratory Report, No. RM-1744-S-2, Buffalo, N.Y., November 1964.

[20] R. W. Lucky, "Techniques for Adaptive Equalization of Digital Communication System," Bell. Syst. Tech. J. 255–286 (Feb. 1966).

[21] D. T. Magill, "Optimal Adaptive Estimation of Sampled Stochastic Processes,"

Stanford Electronics Laboratories, Tech. Rept. 6302-2, Stanford University, December 1963.

[22] K. Steiglitz and J. B. Thomas, "A Class of Adaptive Matched Digital Filters," Third Symposium on Discrete Adaptive Processes, 1964 National Electronics Conference, Chicago, October 19–21, pp. 102–115.

[23] H. L. Groginsky, L. R. Wilson, and D. Middleton, "Adaptive Detection of Statistical Signals in Noise," IEEE Trans. Information Theory **IT-12** (July 1966), pages 337–348.

[24] L. W. Nolte, "An Adaptive Realization of the Optimum Receiver for a Sporadic-Recurrent Waveform in Noise," First IEEE Ann. Communication Conv., Boulder. Colo. 593–598 (June 1965).

[25] R. Esposito, D. Middleton, and J. A. Mullen, "Advantages of Amplitude and Phase Adaptivity in the Detection of Signals Subject to Slow Rayleigh Fading," IEEE Trans, Information Theory **IT-11,** 473–482 (Oct. 1965).

[26] J. C. Hancock and P. A. Wintz, "An Adaptive Receiver Approach to the Time Synchronization Problem," IEEE Trans Commun. Syst. **COM-13,** 90–96 (March 1965).

[27] M. J. DiToro, "A New Method of High-Speed Adaptive Serial Communication through Any Time-Variable and Dispersive Transmission Medium," First IEEE Ann. Communication Conv., Boulder, Colo. 763–767 (June 1965).

[28] L. D. Davisson, "A Theory of Adaptive Filtering," IEEE Trans. Information Theory **IT-12,** 97–102 (April 1966).

[29] H. J. Scudder, "Probability of Error of Some Adaptive Pattern-Recognition Machines," IEEE Trans. Information Theory **IT-11,** 363–371 (July 1965).

[30] Y. C. Ho and R. C. Lee, "Identification of Linear Dynamic Systems," Third Symposium on Discrete Adaptive Processes, 1964 National Electronics Conference, Chicago, October 19–21, pp. 86–101.

[31] B. Reiffen and H. Sherman, "An Optimum Demodulator for Poisson Processes: Photon Sources Detectors," Proc. IEEE **51,** 1316–1320 (Oct. 1963).

[32] K. Abend, "Optimum Photon Detection," IEEE Trans. Information Theory **IT-12,** 64–65 (Jan. 1966).

[33] T. F. Curran and M. Ross, "Optimum Detection Thresholds in Optical Communication," Proc. IEEE **53,** 1770–1771 (Nov. 1965).

[34] C. W. Helstrom, "The Detection and Resolution of Optical Signals," IEEE Trans. Information Theory **IT-10,** 275–287 (Oct. 1964).

[35] I. Bar-David, "Communication under the Poisson Regime," IEEE Trans. Information Theory **IT-15,** No. 1, 31–37 (Jan. 1969).

[36] D. Middleton, "Canonically Optimum Threshold Detection," IEEE Trans. Information Theory **IT-12,** No. 2, 230–243 (April 1966).

[37] D. Middleton, *Topics in Communication Theory*, McGraw-Hill, New York, 1960.

[38] T. Kailath, "A General Likelihood-Ratio Formula for Random Signals in Gaussian Noise," IEEE Trans. Information Theory, **IT-15,** No. 3, 350–361 (May 1969).

[39] S. C. Schwartz, "A Series Technique for the Optimum Detection of Stochastic Signals in Noise," IEEE Trans. Information Theory **IT-15,** No. 3, 362–370 (May 1969).

[40] P. A. Bello, "Bounds on the Error Probability of FSK and PSK Receivers Due to Non-Gaussian Noise in Fading Channels," IEEE Trans. Information Theory **IT-12,** No. 3, 315–326 (July 1966).

[41] P. A. Bello, "Error Probabilities Due to Atmospheric Noise and Flat Fading in HF Ionospheric Communications Systems," IEEE Trans. Commun. Tech., **COM-13,** 266–279 (Sept. 1965).

[42] H. L. Van Trees, *Array Processing*, Wiley, New York, 1971.
[43] T. Kailath, "Correlation Detection of Signals Perturbed by a Random Channel," IRE Trans. Information Theory **IT-6,** 361–366 (June 1960).
[44] P. Bello, "Some Results on the Problem of Discriminating between Two Gaussian Processes," IRE Trans. Information Theory **IT-7,** No. 4, 224–233 (Oct. 1961).
[45] R. C. Davis, "The Detectability of Random Signals in the Presence of Noise," IRE Trans. Information Theory **IT-3,** 52–62, 1957.
[46] D. Slepian, "Some Comments on the Detection of Gaussian Signals in Gaussian Noise, IRE Trans. Information Theory **IT-4,** 65–68, 1958.
[47] T. T. Kadota, "Optimum Reception of Binary Sure and Gaussian Signals," Bell Syst. Tech. J. **44,** No. 8, 1621–1658 (Oct. 1965).
[48] D. Middleton and R. Esposito, "Simultaneous Optimum Detection and Estimation of Signals in Noise," IEEE Trans. Information Theory **IT-14,** No. 3, 434–445 (May 1968).

# 6

# *Estimation of the Parameters of a Random Process*

The next topic of interest is the estimation of the parameters of a Gaussian process. We study this problem in Chapters 6 and 7. Before developing a quantitative model of the problem, we discuss several physical situations in which parameter estimation problems arise.

The first example arises whenever we model a physical phenomenon using random processes. In many cases, the processes are characterized by a mean-value function, covariance function, or spectrum. We then analyze the model assuming that these functions are known. Frequently we must observe a sample function of the process and estimate the process characteristics from this observation. The measurement problems can be divided into two categories. In the first, we try to estimate an entire function, such as the power density spectrum of stationary processes. In the second, we parameterize the function and try to estimate the parameters; for example, we assume that the spectrum has the form

$$S(\omega) = \frac{P}{\omega^2 + k^2}, \tag{1}$$

and try to estimate $P$ and $k$. In many cases, this second category will fit into the parameter estimation model of this section. An adequate discussion of the first category would take us too far afield. Some of the issues are discussed in the problems. Books that discuss this problem include [1]–[3].

The second example arises in such areas as spectroscopy, radio astronomy, and passive sonar classification. The source generates a narrow-band random process whose center frequency characterizes the source. Thus, the first step in the classification problem is to estimate the center frequency of the signal process.

The third example arises in the underground nuclear blast detection problem. An important parameter in deciding whether the event was an earthquake or bomb is the depth of the source. At the station, we receive seismic waves whose angle of arrival depend on the depth of the source.

The common feature in all these examples is that in the parameters of interest are imbedded in the process characteristics. In other words, the mapping from the parameter to the signal is random. In this chapter and the next, we develop techniques for solving this type of problem.

In Chapter 6 we develop the basic results. The quantitative model of the problem is given in Section 6.1. In Section 6.2 we derive the likelihood function, the maximum likelihood equations, and the maximum a-posteriori probability equations. In Section 6.3 we develop procedures for analyzing the performance.

In our study of detection theory we saw that there were special categories of problems for which we could obtain complete solutions. In Chapter 7 we study four such special categories of problems. In Section 7.1 we consider the stationary-process, long-observation-time case. The examples in this section deal with estimating the amplitude of a known covariance function. Several issues arise that cannot be adequately resolved without developing new techniques, and so we digress and develop the needed expressions. This section is important because it illustrates how to bridge the gap between the general theory of Chapter 6 and the complete solution to an actual problem. In Sections 7.2, 7.3, and 7.4 we consider processes with a finite state representation, separable kernel processes, and low-energy-coherence problems, respectively. In Sections 7.5 and 7.6 we extend the results to include multiple parameter estimation and summarize the important results of our estimation theory discussion.

Two observations are useful before we begin our quantitative discussion.

1. The discussion is a logical extension of our parameter estimation work in Chapters I-2 and I-4. We strongly suggest that the reader review Section I-2.4 (pages 52–86), Sections I-4.2.2—I-4.2.4 (pages 271–287), and Section I-4.3.5 (pages 307–309) before beginning this section.

2. Parameter estimation problems frequently require a fair amount of calculation to get to the final result. The casual reader can skim over this detail but should be aware of the issues that are involved.

## 6.1 PARAMETER ESTIMATION MODEL

The model of the parameter estimation problem can be described easily. The received waveform $r(t)$ consists of the sum of signal waveform and a noise waveform,

$$r(t) = s(t, A) + w(t), \qquad T_i \le t \le T_f. \tag{2}$$

The waveform $s(t, A)$ is a sample function from a random process whose characteristics depend on the parameter $A$, which we want to estimate.

To emphasize the nature of the model, assume that $A$ is fixed and $w(t)$ is identically zero. Then, each time the experiment is conducted, the signal waveform $s(t, A)$ will be different because it is a sample function of a random process. By contrast, in the parameter estimation problems of Chapter I-4, the mapping from the parameter to the signal waveform is deterministic.

We assume that the signal process is a *conditionally Gaussian* process.

***Definition.*** A random process $s(t, A)$ is conditionally Gaussian if, given any value of $A$ is the allowable parameter range $\chi_a$, $s(t, A)$ is a Gaussian process.

A conditionally Gaussian process is completely characterized by a conditional mean-value function

$$E[s(t, A) \mid A] \triangleq m(t, A), \qquad T_i \le t \le T_f \tag{3}$$

and a conditional covariance function

$$E[(s(t, A) - m(t, A))(s(u, A) - m(u, A)) \mid A] \triangleq K_s(t, u : A),$$
$$T_i \le t, u \le T_f. \tag{4}$$

The noise process is a zero-mean, white Gaussian noise process with spectral height $N_0/2$ and is statistically independent of the signal process. Thus $r(t)$ is also a conditionally Gaussian process,

$$E[r(t) \mid A] = E[s(t, A) \mid A] = m(t, A), \qquad T_i \le t \le T_f, \tag{5}$$

and

$$E\{[r(t) - m(t, A)][r(u) - m(u, A)] \mid A\} \triangleq K_r(t, u : A)$$
$$= K_s(t, u : A) + \frac{N_0}{2}\,\delta(t - u), \qquad T_i \le t, u \le T_f. \tag{6}$$

Observe that any colored noise component in $r(t)$ can be included in $s(t, A)$. We assume that $m(t, A)$, $K_s(t, u: A)$, and $N_0/2$ are known.

The parameter $A$ will be modeled in two different ways. In the first, we assume that $A$ is a nonrandom parameter that lies in some range $\chi_a$, and we use maximum likelihood estimation procedures. In the second, we assume that $A$ is the value of a random variable with a known probability density $p_a(A)$. For random parameters we can use Bayes estimates with various cost functions. We shall confine our discussion to MAP estimates.

These assumptions specify our model of the parameter estimation problem. We now develop an estimation procedure.

## 6.2 ESTIMATOR STRUCTURE

Our approach to the estimation problem is analogous to the one taken in Chapters I-2 and I-4. We first find the likelihood function $\Lambda(A)$. Then, if $A$ is a nonrandom parameter and we want an ML estimate, we find the value of $A$ for which $\Lambda(A)$ is a maximum. If $A$ is the value of a random variable and we desire an MAP estimate, we construct the function

$$f(A) \triangleq \ln \Lambda(A) + \ln p_a(A), \tag{7}$$

and find that value of $A$ where it is a maximum. The only new issue is the actual construction of $\Lambda(A)$ and the processing needed to find the maximum. In this section we address these issues.

### 6.2.1 Derivation of the Likelihood Function

The derivation of the likelihood function is similar to that of the likelihood ratio in Chapter 2, and so we can proceed quickly. The first step is to find a series expansion for $r(t)$. We then find the conditional probability density of the coefficients (given $A$) and use this to find an appropriate likelihood function. The procedure is simplified if we choose the coordinate system so that the coefficients are conditionally statistically independent. This means that we must choose a coordinate system that is conditionally dependent on $A$. The coefficients are

$$r_i(A) \triangleq \int_{T_i}^{T_f} r(t)\phi_i(t:A)\,dt. \tag{8}$$

The $r_i(A)$ are Gaussian random variables whose mean and variance are functions of $A$.

$$E[r_i(A) \mid A] = E\left\{\int_{T_i}^{T_f} r(t)\phi_i(t:A)\,dt \;\Big|\; A\right\} = \int_{T_i}^{T_f} m(t, A)\phi_i(t:A)\,dt \triangleq m_i(A). \tag{9}$$

We choose the $\phi_i(t: A)$ so that

$$E[(r_i(A) - m_i(A))(r_j(A) - m_j(A)) \mid A] = \lambda_i(A)\delta_{ij}. \tag{10}$$

From our earlier work, we know that to achieve this conditional independence the $\phi_i(t\colon A)$ must be the eigenfunctions of the integral equation

$$\lambda_i(A)\phi_i(t\colon A) = \int_{T_i}^{T_f} K_s(t, u\colon A)\phi_i(u\colon A)\, du, \qquad T_i \le t \le T_f. \tag{11}$$

Because the covariance function depends on the parameter $A$, the eigenfunctions, eigenvalues, or both will depend on $A$. If $K_s(t, u\colon A)$ is positive definite, the eigenfunctions form a complete set. If $K_s(t, u\colon A)$ is only non-negative-definite, we augment the set of eigenfunctions to make it complete.

Since the resulting set is complete, we can expand the mean-value function $m(t, A)$ and the received waveform $r(t)$ in a series expansion. These series are

$$m(t, A) = \sum_{i=1}^{\infty} m_i(A)\phi_i(t\colon A), \qquad T_i \le t \le T_f \tag{12}$$

and

$$r(t) = \underset{K\to\infty}{\text{l.i.m.}} \sum_{i=1}^{K} [r_i(A) - m_i(A)]\phi_i(t, A) + m(t, A), \qquad T_i \le t \le T_f. \tag{13}$$

We denote the first $K$ coefficients by the vector $\mathbf{R}$. The probability density of $\mathbf{r}$ given the value of $A$ is

$$p_{\mathbf{r}|a}(\mathbf{R} \mid A) = \left(\prod_{i=1}^{K} \frac{1}{\sqrt{2\pi}\,(N_0/2 + \lambda_i(A))^{1/2}}\right) \exp\left[-\frac{1}{2}\sum_{i=1}^{K}\left(\frac{[R_i - m_i(A)]^2}{N_0/2 + \lambda_i(A)}\right)\right]. \tag{14}$$

Just as in the known signal case (Section I-4.2.3), it is convenient to define a likelihood function $\Lambda_K(A)$, which is obtained from $p_{\mathbf{r}|a}(\mathbf{R} \mid A)$ by dividing by some function that does not depend on $A$ (see page I-274). As before, we divide by

$$\prod_{i=1}^{K} \frac{1}{\sqrt{\pi N_0}} \exp\left(-\frac{1}{2}\frac{R_i^2}{N_0/2}\right). \tag{15}$$

Dividing (14) by (15), taking the logarithm of the result, and letting $K \to \infty$, we have

$$\boxed{\begin{aligned}\ln \Lambda(A) = {} & \frac{1}{N_0}\sum_{i=1}^{\infty}\frac{\lambda_i(A)}{\lambda_i(A) + N_0/2} R_i^2 + \sum_{i=1}^{\infty}\frac{1}{\lambda_i(A) + N_0/2} m_i(A)R_i \\ & - \frac{1}{2}\sum_{i=1}^{\infty}\ln\left(1 + \frac{2\lambda_i(A)}{N_0}\right) - \frac{1}{2}\sum_{i=1}^{\infty}\frac{m_i^2(A)}{\lambda_i(A) + N_0/2}.\end{aligned}} \tag{16}$$

Comparing (16) with the limit of (2.19) as $K \to \infty$ in our detection theory discussion, we see that there is a one-to-one correspondence. Thus, all of the closed-form expressions in the detection theory section will have obvious analogs in the estimation problem. By proceeding in a manner identical with that in Chapter 2, we can obtain four terms corresponding to those in (2.31)–(2.34).

The first term can be written as

$$\boxed{l_R(A) = \frac{1}{N_0} \iint_{T_i}^{T_f} r(t) h(t, u:A) r(u)\, dt\, du,} \tag{17}$$

where $h(t, u: A)$ satisfies the integral equation

$$\frac{N_0}{2} h(t, u:A) + \int_{T_i}^{T_f} h(t, z:A) K_s(z, u:A)\, dz = K_s(t, u:A), \qquad T_i \leq t, u \leq T_f. \tag{18}$$

We see that $h(t, u: A)$ is the optimum unrealizable filter for the problem in which we observe

$$r(t) = s(t, A) + w(t), \qquad T_i \leq t \leq T_f, \tag{19}$$

and we want to make the MMSE error estimate of $s(t, A)$ under the assumption that $A$ is known. As in the detection problem, we shall frequently use the inverse kernel $Q_r(t, u: A)$, which can be written as

$$Q_r(t, u:A) = \frac{2}{N_0} [\delta(t - u) - h(t, u:A)], \qquad T_i < t, u < T_f. \tag{20}$$

The second term in (16) can be written as

$$l_D(A) = \iint_{T_i}^{T_f} r(t) Q_r(t, u:A) m(u, A)\, dt\, du. \tag{21}$$

Recall that the subscript $D$ denotes deterministic and is used because $l_D(A)$ is analogous to the receiver output in the known signal problem. Alternatively,

$$\boxed{l_D(A) = \int_{T_i}^{T_f} r(t) g(t, A)\, dt,} \tag{22}$$

where $g(t, A)$ is defined as

$$g(t, A) = \int_{T_i}^{T_f} Q_r(t, u:A) m(u, A)\, du. \tag{23}$$

We can also specify $g(t, A)$ implicitly by the equation

$$m(t, A) = \int_{T_i}^{T_f} K_r(t, u:A)g(u, A)\,du, \qquad T_i \le t \le T_f. \tag{24}$$

The function $g(t, A)$ is familiar from the problem of estimating the parameters of a known signal in colored noise.

The remaining terms in (16) are the bias terms. The first is

$$\boxed{l_B^{[1]}(A) \triangleq -\frac{1}{2}\sum_{i=1}^{\infty} \ln\left(1 + \frac{2\lambda_i(A)}{N_0}\right) = -\frac{1}{N_0}\int_{T_i}^{T_f} \xi_P(t:A)\,dt,} \tag{25}$$

where $\xi_P(t: A)$ is the realizable mean-square filtering error for the filtering problem in (19). As in the detection case, we can also evaluate the second term in $l_B(A)$ by means of the Fredholm determinant [see (2.74)]. The second bias term is

$$\boxed{\begin{aligned} l_B^{[2]}(A) &= -\frac{1}{2}\iint_{T_i}^{T_f} m(t, A)Q_r(t, u:A)m(u, A)\,dt\,du \\ &= -\frac{1}{2}\int_{T_i}^{T_f} m(t, A)g(t, A)\,dt. \end{aligned}} \tag{26}$$

Notice that the integral in $l_B^{[2]}(A)$ is just $d^2(A)$ for the problem of detecting a known signal $m(t, A)$ in colored noise. The likelihood function is

$$\ln \Lambda(A) = l_R(A) + l_D(A) + l_B^{[1]}(A) + l_B^{[2]}(A), \tag{27}$$

where the component terms are defined in (17), (22), (25), and (26).

We can now use $\ln \Lambda(A)$ to find $\hat{a}_{map}(r(t))$ or $\hat{a}_{ml}(r(t))$. The procedure is conceptually straightforward. To find $\hat{a}_{ml}$, we construct $\ln \Lambda(A)$ as a function of $A$ and find the value of $A$ where it is a maximum. To find $\hat{a}_{map}$ we construct the function

$$f(A) \triangleq \ln \Lambda(A) + \ln p_a(A) = l_R(A) + l_D(A) + l_B(A) + \ln p_a(A) \tag{28}$$

and find the value of $A$ where it is a maximum.

Even though the procedure is well defined, the actual implementation is difficult. A receiver structure analogous to that in the PFM problem (Fig. I-4.31) of Section I-4.2.3 is usually needed.

To illustrate this, we consider the case of the maximum likelihood estimation of a parameter $A$. We assume that it lies in the interval $[A_\alpha, A_\beta]$. In addition, we assume that the mean $m(t, A)$ is zero. We divide the parameter range into intervals of length $\Delta$. The center points of these intervals are

$$A_1 = A_\alpha + \frac{\Delta}{2},$$
$$A_2 = A_\alpha + \frac{3\Delta}{2}, \tag{29}$$

and so forth. There are $M$ intervals. We then construct $\ln \Lambda(A_i)$, $i = 1, \ldots, M$, by using the parallel processing shown in Fig. 6.1. Several observations are worthwhile:

1. In general we have to solve a different integral equation to find the filter in each path. Thus the estimation problem has the same degree of complexity as an $M$-ary detection problem in the sense that we must build $M$-parallel processors.
2. The bias terms are usually functions of $A$ and cannot be neglected.
3. In analyzing the performance, we must consider both global and local errors.
4. We have to consider the effect of the grid size $\Delta$. There is a trade-off between accuracy and complexity.

Before leaving our discussion of the estimator structure, we digress briefly and derive two alternative forms for $l_R(A)$. Repeating (17),

$$l_R(A) = \frac{1}{N_0} \iint_{T_i}^{T_f} r(t)h(t, u:A)r(u)\, dt\, du. \tag{17}$$

This corresponds to Canonical Realization No. 1 in the detection problem. To obtain Canonical Realization No. 3, we define $h^{[1/2]}(t, u: A)$ implicitly,

$$h(t, u:A) = \int_{T_i}^{T_f} h^{[1/2]}(z, t:A)h^{[1/2]}(z, u:A)\, dz, \qquad T_i \le t, u \le T_f. \tag{30}$$

Then

$$l_R(A) = \frac{1}{N_0} \int_{T_i}^{T_f} dz \left[\int_{T_i}^{T_f} h^{[1/2]}(z, t:A)r(t)\, dt\right]^2. \tag{31}$$

This can be implemented by a filter-squarer-integrator for any particular $A$.

To obtain Canonical Realization No. 4, we go through an argument parallel to that on pages 19–21. The result is

$$l_R(A) = \frac{1}{N_0} \int_{T_i}^{T_f} [2r(t)\hat{s}_r(t:A) - \hat{s}_r^{\,2}(t:A)]\, dt, \tag{32}$$

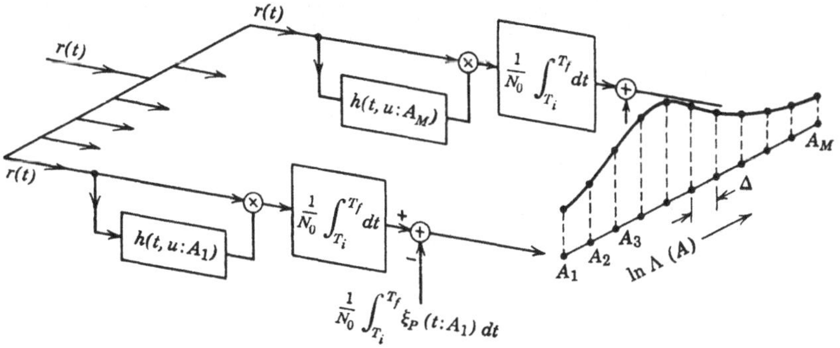


**Fig. 6.1 Generation of ln $\Lambda(A)$.**

where

$$\hat{s}_r(t:A) = \int_{T_i}^{t} h_r(t, u:A) r(u)\, du. \tag{33}$$

The filter $h_r(t, u: A)$ satisfies the equation

$$\frac{N_0}{2} h_r(t, u:A) + \int_{T_i}^{t} h_r(t, z:A) K_s(z, u:A)\, dz = K_s(t, u:A),$$

$$T_i \leq u \leq t. \tag{34}$$

For the zero-mean case, the function $\hat{s}_r(t: A)$ is the realizable MMSE estimate of $s(t: A)$, *assuming that* $A$ is given. We encounter examples of these realizations in subsequent sections. Many of the same issues that we encountered in the detection problem will also arise in the estimation problem.

Before considering some specific cases, we derive the maximum likelihood (ML) equations and the maximum a-posteriori probability (MAP) equations.

### 6.2.2 Maximum Likelihood and Maximum A-Posteriori Probability Equations

If the maximum of $\ln \Lambda(A)$ is interior to $\chi_a$ and $\ln \Lambda(A)$ has a continuous first derivative, the ML equations specify a *necessary* condition on $\hat{a}_{ml}$.

The ML equation follows easily by differentiating (27) and setting the results equal to zero. Taking the partial derivative of $\ln \Lambda(A)$ with respect to $A$, we have

$$\frac{\partial \ln \Lambda(A)}{\partial A} = \frac{\partial l_R(A)}{\partial A} + \frac{\partial l_D(A)}{\partial A} + \frac{\partial l_B(A)}{\partial A}. \tag{35}$$

To evaluate the first term, we differentiate the function in (17). The result is

$$\frac{\partial l_R(A)}{\partial A} = \frac{1}{N_0} \iint_{T_i}^{T_f} r(t) \frac{\partial h(t, u:A)}{\partial A} r(u)\, dt\, du$$

$$= -\frac{1}{2} \iint_{T_i}^{T_f} r(t) \frac{\partial Q_r(t, u:A)}{\partial A} r(u)\, dt\, du. \tag{36}$$

Notice that to find $\partial h(t, u: A)/\partial A$ we must solve (17) as a function of $A$ and then differentiate it. To evaluate the second term, we differentiate (21) and use (20) to obtain

$$\frac{\partial l_D(A)}{\partial A} = -\frac{2}{N_0} \iint_{T_i}^{T_f} r(t) \frac{\partial h(t, u:A)}{\partial A} m(u, A)\, dt\, du$$
$$+ \iint_{T_i}^{T_f} r(t) Q_r(t, u:A) \frac{\partial m(u, A)}{\partial A}\, dt\, du. \tag{37}$$

Finally, from (25) and (26),

$$\frac{\partial l_B(A)}{\partial A} = \frac{1}{N_0} \iint_{T_i}^{T_f} m(t, A) \frac{\partial h(t, u:A)}{\partial A} m(u, A)\, dt\, du$$
$$- \iint_{T_i}^{T_f} m(t, A) Q_r(t, u:A) \frac{\partial m(u, A)}{\partial A}\, dt\, du$$
$$- \frac{1}{N_0} \int_{T_i}^{T_f} \frac{\partial \xi_{Ps}(t, A)}{\partial A}\, dt. \tag{38}$$

Two alternative forms of the last term in (38) are

$$-\frac{1}{N_0} \int_{T_i}^{T_f} \frac{\partial \xi_{Ps}(t, A)}{\partial A}\, dt = \frac{1}{2} \iint_{T_i}^{T_f} K_r(t, u:A) \frac{\partial Q_r(t, u:A)}{\partial A}\, dt\, du$$
$$= -\frac{1}{2} \iint_{T_i}^{T_f} \frac{\partial K_r(t, u:A)}{\partial A} Q_r(t, u:A)\, dt\, du \tag{39}$$

(see Problem 6.2.1). Collecting terms, we have

$$\begin{aligned}\frac{\partial \ln \Lambda(A)}{\partial A} = &\frac{1}{2}\iint_{T_i}^{T_f} K_r(t, u:A)\frac{\partial Q_r(t, u:A)}{\partial A}\,dt\,du \\ &+ \iint_{T_i}^{T_f} \frac{\partial m(t, A)}{\partial A} Q_r(t, u:A)[r(u) - m(u, A)]\,dt\,du \\ &- \frac{1}{2}\iint_{T_i}^{T_f}[r(t) - m(t, A)]\frac{\partial Q_r(t, u:A)}{\partial A} \\ &\times [r(u) - m(u, A)]\,dt\,du.\end{aligned} \tag{40}$$

If we assume that the derivative exists at the maximum of $\ln \Lambda(A)$ and that the maximum is interior to the range, then a necessary condition on the maximum likelihood estimate is obtained by equating the right side of (40) to zero. To find the MAP equation, we add $(\partial \ln p_a(A)/\partial A)$ to (40) and equate the result to zero.

The likelihood equation obtained from (40) is usually difficult to solve. The reason is that even if the parameter appears linearly in the signal covariance function, it may not appear linearly in the inverse kernel. Thus, the necessary condition is somewhat less useful in the random signal case than it is in the known signal case.

## 6.3 PERFORMANCE ANALYSIS

The performance analysis is similar to that in the nonlinear estimation problems in Chapters I-2 and I-4. We divide the errors into local and global errors. The variance of the local errors can be obtained from a power series approach or by a generalized Cramér-Rao bound. The global behavior can be analyzed by an extension of the analysis on pages I-279–I-284. In this section we derive the generalized Cramér-Rao bound and discuss methods of calculating it.

### 6.3.1 A Lower Bound on the Variance

We assume that $A$ is a nonrandom variable that we want to estimate. We desire a lower bound on the variance of any unbiased estimate of $A$.

The derivation on pages I-66–I-68 extends easily to this case. The result is that for *any unbiased* estimate $\hat{a}(r(t))$ of the nonrandom variable $A$, the variance satisfies the inequality

$$\operatorname{Var}[\hat{a}(r(t)) - A] \geq -\left\{E\left[\frac{\partial^2 \ln \Lambda(A)}{\partial A^2}\right]\right\}^{-1} \tag{41}$$

with equality if and only if

$$\frac{\partial \ln \Lambda(A)}{\partial A} = \{\hat{a}(r(t)) - A\}k(A). \tag{42}$$

To evaluate the bound we differentiate (40). The result is

$$\begin{aligned}
\frac{\partial^2 \ln \Lambda(A)}{\partial A^2} &= \frac{1}{2}\iint_{T_i}^{T_f} K_r(t, u:A)\,\frac{\partial^2 Q_r(t, u:A)}{\partial A^2}\,dt\,du \\
&+ \frac{1}{2}\iint_{T_i}^{T_f} \frac{\partial K_r(t, u:A)}{\partial A}\cdot\frac{\partial Q_r(t, u:A)}{\partial A}\,dt\,du \\
&- \iint_{T_i}^{T_f} \frac{\partial m(t, A)}{\partial A}\,Q_r(t, u:A)\,\frac{\partial m(u, A)}{\partial A}\,dt\,du \\
&- \frac{1}{2}\iint_{T_i}^{T_f} [r(t) - m(t, A)]\,\frac{\partial^2 Q_r(t, u:A)}{\partial A^2}\,[r(u) - m(u, A)]\,dt\,du \\
&+ (\text{terms whose expectations are zero}).
\end{aligned} \tag{43}$$

When we take the expectation of the last integral, we find that it cancels the first term in (43). Thus any *unbiased* estimate of $A$ will have a variance satisfying the bound

$$\operatorname{Var}[\hat{a}(r(t)) - A] \geq \left\{\iint_{T_i}^{T_f} \frac{\partial m(t, A)}{\partial A}\,Q_r(t, u:A)\,\frac{\partial m(u, A)}{\partial A}\,dt\,du - \frac{1}{2}\iint_{T_i}^{T_f} \frac{\partial K_r(t, u:A)}{\partial A}\,\frac{\partial Q_r(t, u:A)}{\partial A}\,dt\,du\right\}^{-1} \tag{44}$$

For notational convenience in the subsequent discussion, we denote the first term in the braces by $J^{(1)}(A)$, the second by $J^{(2)}(A)$, and the sum by $J(A)$.

$$J^{(1)}(A) \triangleq \iint_{T_i}^{T_f} \frac{\partial m(t, A)}{\partial A} Q_r(t, u:A) \frac{\partial m(u, A)}{\partial A} \, dt \, du \tag{45}$$

and

$$J^{(2)}(A) \triangleq -\frac{1}{2} \iint_{T_i}^{T_f} dt \, du \, \frac{\partial K_r(t, u:A)}{\partial A} \frac{\partial Q_r(t, u:A)}{\partial A}. \tag{46}$$

Several observations are useful:

1. The terms in the bound depend on $A$. Thus, as we have seen before, the variance depends on the actual value of the nonrandom parameter.
2. The bound assumes that the estimate is unbiased. If the estimate is biased, a different bound must be used. (See Problem 6.3.1.)
3. The first term is familiar in the context of detection of known signals in colored noise. Specifically, it is exactly the value of $d^2$ for the simple binary detection problem in which we transmit $\partial m(t, A)/\partial A$ and the additive colored noise has a covariance function $K_r(t, u: A)$. Thus, the techniques we have developed for evaluating $d^2$ are applicable here.

We now consider efficient procedures for evaluating $J^{(2)}(A)$.

### 6.3.2 Calculation of $J^{(2)}(A)$†

The $J^{(2)}(A)$ term arises because the covariance function of the process depends on $A$. It is a term we have not encountered previously, and so we develop two convenient procedures for evaluating it. The first technique relates it to the Bhattacharyya distance (recall the discussion on pages 71–72), and the second expresses it in terms of eigenfunctions and eigenvalues.

The techniques developed in this section are applicable to arbitrary observation intervals and processes that are not necessarily stationary. In Section 7.1, we shall consider the stationary-process, long-observation-time case and develop a simple expression for $J^{(2)}(A)$.

† This section may be omitted on the first reading

***Relation to Bhattacharyya Distance.*** In this section we relate $J(A)$ to the Bhattacharyya distance. We first work with $r_K(t)$ and the vector $\mathbf{R}$ and then let $K \to \infty$ in the final answer. We define a function

$$\mu(\tfrac{1}{2}, A_1, A) \triangleq \ln \int_{-\infty}^{\infty} p_{\mathbf{r}|a}^{1/2}(\mathbf{R}|A_1) p_{\mathbf{r}|a}^{1/2}(\mathbf{R}|A)\, d\mathbf{R}. \tag{47}$$

This is simply $\mu(\frac{1}{2})$ for the general binary detection problem in which

$$p_{\mathbf{r}|H_1}(\mathbf{R} \mid H_1) = p_{\mathbf{r}|a}(\mathbf{R} \mid A_1) \tag{48}$$

and

$$p_{\mathbf{r}|H_0}(\mathbf{R} \mid H_0) = p_{\mathbf{r}|a}(\mathbf{R} \mid A). \tag{49}$$

The Bhattacharyya distance is just

$$B(A_1, A) = -\mu(\tfrac{1}{2}, A_1, A). \tag{50}$$

Using (50) and (47) leads to

$$e^{-B(A_1,A)} = \int_{-\infty}^{\infty} p_{\mathbf{r}|a}^{1/2}(\mathbf{R}|A_1) p_{\mathbf{r}|a}^{1/2}(\mathbf{R}|A)\, d\mathbf{R}. \tag{51}$$

We are interested in the case in which

$$\Delta A \triangleq A_1 - A \tag{52}$$

is small, and so we expand both sides of (51) in a series. Expanding the left side in a Taylor series in $A_1$ about the point $A_1 = A$ gives

$$e^{-B(A_1,A)} = e^{-B(A,A)} - \left[\frac{\partial B(A_1, A)}{\partial A_1} e^{-B(A_1,A)}\right]_{A_1=A} \Delta A + \frac{1}{2}\left[\left(\frac{\partial^2 B(A_1, A)}{\partial A_1^2} - \left(\frac{\partial B(A_1, A)}{\partial A_1}\right)^2\right) e^{-B(A_1,A)}\right]_{A_1=A} (\Delta A)^2 + \cdots. \tag{53}$$

From (47), it follows easily that

$$\left.\frac{\partial B(A_1, A)}{\partial A_1}\right|_{A_1=A} = 0 \tag{54}$$

(see Problem 6.3.2). Thus, (53) reduces to

$$e^{-B(A_1,A)} = 1 + \left(\left.\frac{\partial^2 B(A_1, A)}{\partial A_1^2}\right|_{A_1=A}\right)\frac{(\Delta A)^2}{2} + \cdots. \tag{55}$$

To expand the right side of (51), we use a Taylor series for the first term in the integrand and then integrate term by term. The result is

$$\int_{-\infty}^{\infty} p_{\mathbf{r}|a}^{1/2}(\mathbf{R}|A_1) p_{\mathbf{r}|a}^{1/2}(\mathbf{R}|A)\, dR \simeq 1 - \frac{(\Delta A)^2}{8}\int_{-\infty}^{\infty} \frac{([\partial p_{\mathbf{r}|a}(\mathbf{R} \mid A)]/\partial A])^2}{p_{\mathbf{r}|a}(\mathbf{R} \mid A)}\, d\mathbf{R}. \tag{56}$$

The integral on the right side of (56) can be written as

$$\int_{-\infty}^{\infty}\left(\frac{\partial \ln p_{\mathbf{r}|a}(\mathbf{R}\mid A)}{\partial A}\right)^2 p_{\mathbf{r}|a}(\mathbf{R}|A)\, d\mathbf{R} = E\left\{\frac{\partial^2 \ln p_{\mathbf{r}|a}(\mathbf{R}|A)}{\partial A^2}\right\}. \tag{57}$$

The term on the right side of (57) is just the negative of $J(A)$. Substituting (55) and (56) into (51) and equating the coefficients of $(\Delta A)^2$, we have

$$J(A) = 4\left(\frac{\partial^2 B(A_1, A)}{\partial^2 A_1}\bigg|_{A_1=A}\right). \tag{58}$$

Notice that the expression in (58) includes both $J^{(1)}(A)$ and $J^{(2)}(A)$. To calculate $J^{(2)}(A)$, we assume that the process is zero-mean and use the formula for $\mu(s)$ given in (3.60),

$$B(A_1, A) = -\frac{1}{N_0}\int_{T_i}^{T_f} dt\left\{\tfrac{1}{2}\xi_P\left(t \mid s(\cdot, A_1), \frac{N_0}{2}\right) + \tfrac{1}{2}\xi_P\left(t \mid s(\cdot, A), \frac{N_0}{2}\right) - \xi_P\left(t \mid \sqrt{\tfrac{1}{2}}\, s(\cdot, A_1) + \sqrt{\tfrac{1}{2}}\, s(\cdot, A), \frac{N_0}{2}\right)\right\}. \tag{59}$$

In the last term we have a composite process of the type discussed in (3.63). We emphasize that $s(t, A)$ and $s(t, A_1)$ are statistically independent components in this composite process. Differentiating twice and substituting into (58) gives the desired result.

$$J^{(2)}(A) = \frac{4}{N_0}\left\{\frac{\partial^2}{\partial A_1^{\,2}}\int_{T_i}^{T_f} \xi_P\left(t \mid \sqrt{\tfrac{1}{2}}\, s(\cdot, A_1) + \sqrt{\tfrac{1}{2}}\, s(\cdot, A), \frac{N_0}{2}\right) dt - \frac{1}{2}\frac{\partial^2}{\partial A_1^{\,2}}\int_{T_i}^{T_f} \xi_P\left(t \mid s(\cdot, A_1), \frac{N_0}{2}\right) dt\right\}_{A_1=A}. \tag{60}$$

It is worthwhile pointing out that in many cases it will be easier to evaluate $J^{(2)}(A)$ by using the Fredholm determinant (e.g., Section 2.2.3).

***Eigenvalue Approach.*** In this section we derive an expression for $J^{(2)}(A)$ in terms of the eigenvalues and eigenfunctions of $K_r(t, u : A)$. From (20) it is clear that we could also write $J^{(2)}(A)$ as

$$J^{(2)}(A) = \frac{1}{N_0}\iint_{T_i}^{T_f} dt\, du\, \frac{\partial K_r(t, u : A)}{\partial A}\,\frac{\partial h(t, u : A)}{\partial A}. \tag{61}$$

This expression still requires finding $h(t, u : A)$, the optimum unrealizable filter for all $t$ and $u$ in $[T_i, T_f]$. In order to express $J^{(2)}(A)$ in terms of eigenvalues and eigenfunctions, we first write $h(t, u : A)$ as the series

$$h(t, u : A) = \sum_{i=1}^{\infty}\left(\frac{\lambda_i(A)}{N_0/2 + \lambda_i(A)}\right)\phi_i(t : A)\phi_i(u : A). \tag{62}$$

Differentiating $K_r(t, u : A)\, h(t, u : A)$ and using the results in (62), we obtain

$$J^{(2)}(A) = \frac{1}{2}\sum_{i=1}^{\infty}\left(\frac{[\partial\lambda_i(A)]/\partial A}{\lambda_i(A) + N_0/2}\right)^2 + \frac{2}{N_0}\sum_{i=1}^{\infty}\left(\frac{\lambda_i^2(A) b_i(A)}{\lambda_i(A) + N_0/2}\right) - \frac{2}{N_0}\sum_{i=1}^{\infty}\sum_{j=1}^{\infty}\frac{\lambda_i(A)\lambda_j(A)}{\lambda_j(A) + N_0/2}\, a_{ij}(A), \quad (63)$$

where

$$b_i(A) \triangleq \int_{T_i}^{T_f}\left(\frac{\partial\phi_i(t, A)}{\partial A}\right)^2 dt \quad (64)$$

and

$$a_{ij}(A) \triangleq \int_{T_i}^{T_f}\frac{\partial\phi_i(t:A)}{\partial A}\,\phi_j(t:A)\, dt. \quad (65)$$

The expression in (63) is not particularly useful in the most cases. A special case of interest in which it is useful is the one in which the eigenfunctions do not depend on $A$. A common example of this case is when $A$ corresponds to the amplitude of the covariance function. Then the last two terms in (63) are zero and

$$J^{(2)}(A) = \frac{1}{2}\sum_{i=1}^{\infty}\left(\frac{[\partial\lambda_i(A)]/\partial A}{\lambda_i(A) + N_0/2}\right)^2. \quad (66)$$

The form in (66) is reasonably easy to evaluate in many problems.

A simple example illustrates the use of (66).

**Example.** The received waveform is

$$r(t) = s(t, A) + w(t), \qquad 0 \le t \le T. \quad (67)$$

The signal is a sample function of a Wiener process. It is a Gaussian process with statistics

$$E[s(t, A)] = 0, \qquad t \ge 0 \quad (68)$$

and

$$s(0, A) = 0, \quad (69)$$

$$K_s(t, u : A) = A \min(t, u), \qquad 0 \le t, u. \quad (70)$$

This process was first introduced on page I-195. The additive noise $w(t)$ is a sample function from a statistically independent white Gaussian process with spectral height $N_0/2$. We want to estimate the nonrandom parameter $A$.

In Problem 7.2.1, we shall derive the optimum receiver for this problem. In the present example, we simply evaluate the expression in (66). From page I-196, the eigenvalues are

$$\lambda_i(A) = \frac{AT^2}{(i - \frac{1}{2})^2\pi^2}, \qquad i = 1, 2, \ldots, \quad (71)$$

and the eigenfunctions do not depend on $A$. Differentiating (71) gives

$$\frac{\partial\lambda_i(A)}{\partial A} = \frac{T^2}{(i - \frac{1}{2})^2\pi^2}. \quad (72)$$

Using (71) and (72) in (66) gives

$$J(A) = J^{(2)}(A) = \frac{1}{2A^2} \sum_{i=1}^{\infty} \left( \frac{1}{1 + [N_0/2AT^2](i-1)^2\pi^2} \right)^2. \tag{73}$$

The bound on the normalized variance of any unbiased estimate is

$$\frac{\text{Var}\,[\hat{a} - A]}{A^2} \geq \frac{2}{\sum_{i=1}^{\infty} [(1 + [N_0/2AT^2](i-1)^2\pi^2)^2]^{-1}}. \tag{74}$$

The sum can be expressed in terms of polygamma functions whose values are tabulated (e.g., [4, page 265]). In Chapter 7 we shall see that for large values of $2AT^2/N_0$, the ML estimate is essentially unbiased and its variance approaches this bound. For small values of $2AT^2/N_0$, the bias is an important issue. We discuss the bias problem in detail in Section 7.1.

The final topic of interest is the performance when we estimate a random variable. In the next section we derive a lower bound on the minimum mean-square error.

### 6.3.3 Lower Bound on the Mean-Square Error

To derive the bound on the mean-square error we go through a similar procedure (e.g., page I-72). Since the derivation is straightforward, we leave it as an exercise. The result is

$$E[(\hat{a}(R) - a)^2] \geq \left\{ E_a[J^{(1)}(A)] + E_a[J^{(2)}(A)] - E_a\left[\frac{\partial^2 \ln p_a(A)}{\partial A^2}\right]\right\}^{-1}. \tag{75}$$

The expressions for $J^{(1)}(A)$ and $J^{(2)}(A)$ are given in (45) and (46). This bound holds under weak conditions analogous to those given on page I-72. Two observations are useful:

1. Since $a$ is a random variable, there is no issue of bias. The bound is on the mean-square error, *not* the variance.
2. There is an expectation over $p_a(A)$ in each term on the right side of (75). Thus the bound is not a function of the actual value of $A$. In most cases it is difficult to perform this integration over $A$.

Most of our examples in the text will deal with nonrandom variables. The extension of any particular example to the random-variable case is straightforward.

### 6.3.4 Improved Performance Bounds

Our discussion of performance has concentrated on generalizations of the Cramèr-Rao bounds. In many problems when the processes are stationary, one can show that the variance of the ML estimate approaches the bound as the observation time increases (e.g., [5]). On the other hand, as we have seen before, there are a number of problems in which the bound does not give an accurate indication of the actual performance.

One procedure for obtaining a better estimate is suggested by the structure in Fig. 6.1. We consider the problem as an $M$-ary detection problem, find the error probability, and translate this into a global estimation error. This technique was introduced for the problem of estimating deterministic signal parameters by Woodward [6] and Kotelnikov [7]. It was subsequently modified and extended [8]–[13]. We discussed the approach on pages I-278–I-284. The extension to the random signal parameter case is conceptually straightforward but usually difficult to carry out. In Problem 7.1.23, we go through the procedure for a particular estimation problem.

A second procedure for evaluating the performance is to use the Barankin bound [14]. This technique has been applied to the deterministic signal parameter problem [15]–[17]. Some progress has been made in the random signal problem by Baggeroer [18]. Once again, the basic ideas are straightforward but the actual calculations are difficult.

In Chapter 7, we study some particular estimation problems. At that point, we consider the performance question again in more detail. We may now summarize the results of this chapter.

## 6.4 SUMMARY

In this chapter we have developed the basic results needed to study the parameter estimation problem. The formal derivation of the likelihood function was a straightforward extension of our earlier detection results. The resulting likelihood function is

$$\ln \Lambda(A) = \frac{1}{N_0} \iint_{T_i}^{T_f} r(t) h(t, u : A) r(u)\, dt\, du + \int_{T_i}^{T_f} r(t) g(t, A)\, dt - \frac{1}{N_0} \int_{T_i}^{T_f} \xi_P(t : A)\, dt - \frac{1}{2} \int_{T_i}^{T_f} m(t, A) g(t, A)\, dt, \quad (76)$$

where the various functions are defined in (18), (23), and (25). To find $\hat{a}_{ml}$ we plot $\ln \Lambda(A)$ as a function of $A$ and find the point where it is a maximum.

The next step was to find the performance of the estimator. A lower bound on the variance of any unbiased estimate was given in (44).

At this point in our discussion we have derived several general results. The next, and more important, step is to see how we can use these results actually to solve a particular estimation problem. We study this question in detail in Chapter 7.

## 6.5 PROBLEMS

This problem section is brief because of the introductory nature of the chapter. Section 7.7 contains a number of interesting estimation problems.

### P.6.2 Estimator Structure

**Problem 6.2.1.** Verify the result in (39). (*Hint:* use the original definition of $Q_r(t, u: A)$ and an eigenfunction expansion of the various terms.)

**Problem 6.2.2.** Consider the vector version of the model in (2). The received waveform is

$$\mathbf{r}(t) = \mathbf{s}(t, A) + \mathbf{w}(t), \qquad T_i \leq t \leq T_f.$$

The signal process $\mathbf{s}(t, A)$ is a vector, conditionally Gaussian process with conditional mean-value function $\mathbf{m}(t, A)$ and conditional covariance function matrix $\mathbf{K_r}(t, u: A)$. The additive white Gaussian noise has a spectral matrix $(N_0/2)\mathbf{I}$.

1. Find an expression for $\ln \Lambda(A)$.
2. Find an expression for $l_R(A)$ in terms of Canonical Realizations No. 1, 3, 4, and 4S.
3. Derive the vector version of the bound in (44).

**Problem 6.2.3.** In Section 6.1, we indicated that if a colored noise component was present it could be included in $s(t, A)$. In this problem we indicate the colored noise explicitly as

$$r(t) = s_s(t, A) + n_c(t) + w(t), \qquad T_i \leq t \leq T_f.$$

The processes are zero-mean Gaussian processes with covariance functions $K_{ss}(t, u)$, $K_c(t, u)$, and $(N_0/2)\delta(t - u)$, respectively.

1. Modify (16), (17), and (25) to include the effect of the colored noise explicitly.
2. Can any of the above expressions be simplified because of the explicit inclusion of the white noise?

**Problem 6.2.4.** The model in Problem 6.2.3 is analogous to a class $B_w$ detection problem. Consider the model

$$r(t) = s(t, A) + n_c(t), \qquad T_i \leq t \leq T_f,$$

where $n_c(t)$ does not contain a white component.

1. Derive an expression for $\ln \Lambda(A)$.
2. Derive a lower bound on the variance of any unbiased estimate analogous to (44). (*Hint:* Review Section 3.5.)

**Problem 6.2.5.** Assume that

$$r(t) = s(t, A) + w(t), \qquad T_i \leq t \leq T_f,$$

with probability $p$, and that

$$r(t) = w(t), \qquad T_i \leq t \leq T_f,$$

with probability $(1 - p)$.

1. Derive an expression for $\ln \Lambda(A)$.
2. Check your answer for the degenerate cases when $p = 0$ and $p = 1$.

## P.6.3 Performance

**Problem 6.3.1.** Assume that

$$E[(\hat{a} - A)] = B(A). \tag{P.1}$$

Derive a lower bound on the variance of any estimate satisfying (P.1).

**Problem 6.3.2.** Use the definition of $B(A_1, A)$ in (47) and (50) to verify that (54) is valid.

**Problem 6.3.3.** Carry out the details of the derivation of (75).

## REFERENCES

[1] G. M. Jenkins and D. G. Watts, *Spectral Analysis and Its Applications*, Holden-Day, San Francisco, 1968.
[2] R. B. Blackman and J. W. Tukey, *The Measurement of Power Spectra from the Point of View of Communications Engineering*, Dover Press, New York, 1958.
[3] J. S. Bendat and A. G. Piersol, *Measurement and Analysis of Random Data*, Wiley, New York, 1966.
[4] M. Abramowitz and I. A. Stegun, *Handbook of Mathematical Functions*, Dover Press, New York, 1965.
[5] U. Grenander, "Stochastic Processes and Statistical Inference," Arkiv Matematik **1**, 195–277 (April 1950).
[6] P. M. Woodward, *Probability and Information Theory, with Application to Radar*, Pergamon Press, London, 1953.
[7] V. A. Kotelnikov, *The Theory of Optimum Noise Immunity*, McGraw-Hill, New York, 1959.
[8] S. Darlington, "Demodulation of Wideband, Low Power FM Signals, Bell Syst. Tech. J. **43**, No. 1, Pt. 2, 339–374 (1964).
[9] H. Akima, "Theoretical Studies on Signal-to-Noise Characteristics of an FM System," IEEE Trans. Space Electronics Tel. **SET-9**, 101–108 (1963).
[10] J. M. Wozencraft and I. M. Jacobs, *Principles of Communication Engineering*, Wiley, New York, 1965.
[11] L. A. Wainstein and V. D. Zubakov, *Extraction of Signals from Noise*, Prentice-Hall, Englewood Cliffs, N.J., 1962.
[12] J. Ziv and M. Zakai, "Some Lower Bounds on Signal Parameter Estimation," IEEE Trans. Information Theory **IT-15**, No. 3, 386–391 (May 1969).
[13] M. Zakai and J. Ziv, "On the Threshold Effect in Radar Range Estimation," IEEE Trans. Information Theory **IT-15**, No. 1, 167–170 (Jan. 1969).
[14] E. W. Barankin, "Locally Best Unbiased Estimates," Ann. Math. Statist. **20**, 477–501 (1949).
[15] P. Swerling, "Parameter Estimation for Waveforms in Additive Gaussian Noise," J. SIAM **7**, 154–166 (1959).

[16] P. Swerling, "Parameter Estimation Accuracy Formulas," IEEE Trans. Information Theory **IT-10**, 302–313 (Oct. 1964).
[17] R. J. McAulay and L. P. Seidman, "A Useful Form of the Barankin Lower Bound and Its Application to PPM Threshold Analysis," IEEE Trans. Information Theory **IT-15**, No. 2, 273–279 (March 1969).
[18] A. B. Baggeroer, "Barankin Bound on the Variance of Estimates of the Parameters of a Gaussian Random Process," Massachusetts Institute of Technology, RLE Quart. Prog. Report No. 92, 324–333, January 15, 1969.

# 7

# Special Categories of Estimation Problems

As in the detection problem, there are several categories of processes for which we can obtain a reasonably complete solution for the estimator. In this chapter we discuss four categories:

1. Stationary processes, long observation time (7.1).
2. Finite-state processes (7.2).
3. Separable kernel processes (7.3).
4. Low-energy coherence (7.4).

We exploit the similarity to the detection problem whenever possible and use the results from Chapter 4 extensively. For algebraic simplicity, we assume that $m(t, A)$ is zero throughout the chapter.

In Section 7.5, we consider some related topics. In Section 7.6, we summarize the results of our estimation theory discussion.

## 7.1 STATIONARY PROCESSES: LONG OBSERVATION TIME

The model of interest is

$$r(t) = s(t, A) + w(t), \qquad T_i \le t \le T_f. \tag{1}$$

We assume that $s(t, A)$ is a sample function from a zero-mean, stationary Gaussian random process with covariance function

$$K_s(t, u:A) \triangleq K_s(t - u:A). \tag{2}$$

The additive noise $w(t)$ is a sample function from an independent, zero-mean, white Gaussian process with spectral height $N_0/2$. Thus,

$$K_r(t, u:A) = K_s(t - u:A) + \frac{N_0}{2}\,\delta(t - u). \tag{3}$$

The power density spectrum of $r(t)$ is

$$S_r(\omega:A) = S_s(\omega:A) + \frac{N_0}{2}. \tag{4}$$

In addition, we assume that

$$T \triangleq T_f - T_i \tag{5}$$

is large enough that we can neglect transient effects at the ends of the interval. (Recall the discussion on pages 99–101.)

In this section we discuss the simplifications that result when the SPLOT condition is valid. In Section 7.1.1, we develop some general results and introduce the amplitude estimation problem. In Section 7.1.2, we study the performance of truncated estimators (we define the term at that point). In Section 7.1.3, we discuss suboptimum receivers. Finally, in Section 7.1.4, we summarize our results.

### 7.1.1 General Results

We want to find simple expressions for $l_R(A)$, $l_B(A)$, the MAP and ML equations, and the lower bound on the variance. Using (6.17), (6.18), (4), and the same procedure as on pages 100–101, we have

$$l_R(A) = \frac{1}{N_0} \iint\limits_{T_i}^{T_f} r(t) h(t - u:A) r(u)\, dt\, du, \tag{6}$$

where $h(\tau:A)$ is a time-invariant filter with the transfer function

$$\boxed{H(j\omega:A) = \frac{S_s(\omega:A)}{S_s(\omega:A) + N_0/2}.} \tag{7}$$

The filter in (7) is unrealizable and corresponds to Canonical Realization No. 1 in the detection theory problem. A simple realization can be obtained by factoring $H(j\omega:A)$.

$$\boxed{H_{fr\infty}(j\omega:A) \triangleq \left[\frac{S_s(\omega:A)}{S_s(\omega:A) + N_0/2}\right]^+.} \tag{8}$$

Then

$$l_R(A) = \frac{1}{N_0} \int_{T_i}^{T_f} dt \left[\int_{T_i}^{t} h_{fr\infty}(t - z:A) r(z)\, dz\right]^2. \tag{9}$$

This is a filter-squarer-integrator realization and is analogous to Canonical Realization No. 3. Notice that $h_{fr\infty}(\tau:A)$ is a realizable filter.

The bias term follows easily from the asymptotic mean-square-error expression. Using (4.16) and (6.25), we obtain

$$l_B(A) = -\frac{T}{2}\int_{-\infty}^{\infty} \ln\left(1 + \frac{2S_s(\omega:A)}{N_0}\right)\frac{d\omega}{2\pi}. \tag{10}$$

From (6.27) we construct

$$l(A) = l_R(A) + l_B(A) \tag{11}$$

and choose the value of $A$ where the maximum occurs. The general receiver structure, using (9) and (10), is shown in Fig. 7.1.

The ML equation is obtained by substituting (6) and (10) into (11), differentiating, and equating the result to zero for $A = \hat{a}_0$. Normally, we refer to the solution of maximum likelihood equation as $\hat{a}_{ml}$. However, the ML equation only provides a necessary condition and we must check to see that the maximum is interior to $\chi_a$ and that $\hat{a}_0$ is the absolute maximum. In several examples that we shall consider, the maximum can be at the endpoint of $\chi_a$. Therefore, we must be careful to check the conditions. The solution $A = \hat{a}_0$ can be interpreted as a candidate for $\hat{a}_{ml}$. Carrying out the indicated steps, we obtain

$$\left[-\frac{T}{2}\int_{-\infty}^{\infty}\left(\frac{\dfrac{2}{N_0}\dfrac{\partial S_s(\omega:A)}{\partial A}}{1 + \dfrac{2S_s(\omega:A)}{N_0}}\right)\frac{d\omega}{2\pi} + \frac{1}{N_0}\iint_{T_i}^{T_f} r(t)\,\frac{\partial h(t-u:A)}{\partial A}\,r(u)\,dt\,du\right]_{A=\hat{a}_0} = 0, \tag{12}$$

where

$$\frac{\partial h(\tau:A)}{\partial A} = \int_{-\infty}^{\infty}\left[\frac{(N_0/2)[\partial S_s(\omega:A)/\partial A]}{(S_s(\omega:A) + N_0/2)^2}\right]e^{j\omega\tau}\,\frac{d\omega}{2\pi}. \tag{13}$$

To find the Cramèr-Rao bound, we take the asymptotic version of (6.61).

$$\begin{aligned} J^{(2)}(A) &= \frac{1}{N_0}\iint_{T_i}^{T_f} dt\,du\,\frac{\partial K_r(t-u:A)}{\partial A}\,\frac{\partial h(t-u:A)}{\partial A} \\ &= \frac{2T}{N_0}\int_0^T\left(1 - \frac{\tau}{T}\right)\frac{\partial K_r(\tau:A)}{\partial A}\,\frac{\partial h(\tau:A)}{\partial A}\,d\tau. \end{aligned} \tag{14}$$

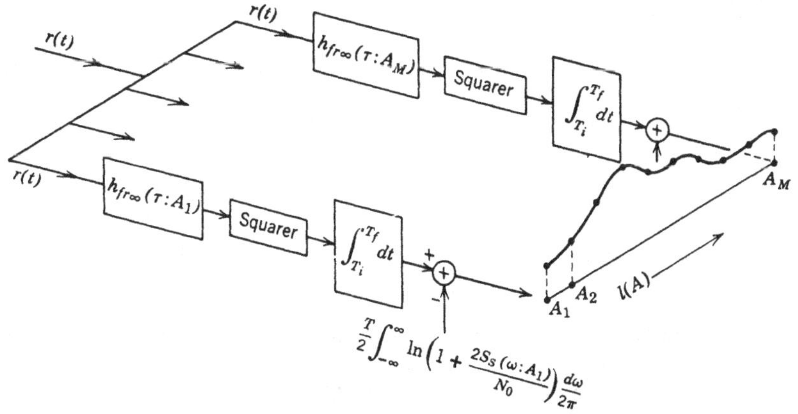


**Fig. 7.1 Generation of $l(A)$ for stationary-process, long-observation-time case.**

For large $T$, this can be written in the frequency domain as

$$J^{(2)}(A) = \frac{T}{N_0}\int_{-\infty}^{\infty} \frac{\partial S_r(\omega:A)}{\partial A}\,\frac{\partial H(j\omega:A)}{\partial A}\,\frac{d\omega}{2\pi}. \tag{15}$$

Using (13) in (15), we obtain

$$\begin{aligned} J^{(2)}(A) &= \frac{T}{2}\int_{-\infty}^{\infty} \frac{d\omega}{2\pi}\left[\frac{[\partial S_s(\omega:A)]/\partial A}{S_s(\omega:A) + N_0/2}\right]^2 \\ &= \frac{T}{2}\int_{-\infty}^{\infty} \frac{d\omega}{2\pi}\left[\frac{\partial}{\partial A}\ln\left[1 + \frac{2S_s(\omega:A)}{N_0}\right]\right]^2. \end{aligned} \tag{16}$$

From (6.44) we have

$$E\{[\hat{a} - A]^2\} \geq [J^{(2)}(A)]^{-1} \tag{17}$$

for any unbiased estimate.

To illustrate these results, we consider a series of amplitude estimation problems. The examples are important because they illustrate the difficulties that arise when we try to solve a particular problem and how we can resolve these difficulties.

**Example 1.** The first problem is that of estimating the amplitude of the spectrum of a stationary random process corrupted by additive white noise. The signal spectrum is

$$S_s(\omega:A) = AS(\omega), \tag{18}$$

where $S(\omega)$ is known. The parameter $A$ lies in the range $[A_\alpha, A_\beta]$ and is nonrandom. Substituting (18) into (7) gives

$$H(j\omega:A) = \frac{AS(\omega).}{AS(\omega) + N_0/2} \tag{19}$$

From (8) we have

$$H_{fr\infty}(j\omega:A) = \sqrt{A}\left[\frac{S(\omega)}{AS(\omega) + N_0/2}\right]^+, \tag{20}$$

and from (10),

$$l_B(A) = -\frac{T}{2}\int_{-\infty}^{\infty}\frac{d\omega}{2\pi}\left[\ln\left(1 + \frac{2AS(\omega)}{N_0}\right)\right]. \tag{21}$$

We construct

$$l(A) = \frac{1}{N_0}\iint_{T_i}^{T_f} r(t)h(t-u:A)r(u)\,dt\,du + l_B(A) \tag{22}$$

as a function of $A$ and choose that value of $A$ where it is a maximum. The resulting receiver is shown in Fig. 7.2.

To obtain the ML equation we substitute (18) into (12). The result is

$$\left\{\frac{1}{N_0}\iint_{T_i}^{T_f} r(t)\frac{\partial h(t-u:A)}{\partial A}r(u)\,dt\,du - \frac{T}{2}\int_{-\infty}^{\infty}\frac{2S(\omega)/N_0}{1 + (2AS(\omega))/N_0}\frac{d\omega}{2\pi}\right\}_{A=\hat{a}_0} = 0, \tag{23}$$

where

$$\frac{\partial H(j\omega:A)}{\partial A} = \frac{N_0S(\omega)/2}{(N_0/2 + AS(\omega))^2}. \tag{24}$$

In general, we cannot solve (23) explicitly, and so we must still implement the receiver using a set of parallel processors. If the resulting estimate is unbiased, its normalized variance is bounded by

$$\boxed{\frac{\mathrm{Var}\,[\hat{a} - A]}{A^2} \geq \left\{\frac{T}{2}\int_{-\infty}^{\infty}\left(\frac{S(\omega)}{N_0/2A^2 + S(\omega)}\right)^2\right\}^{-1}.} \tag{25}$$

We now examine the results in Example 1 in more detail for various special cases.

**Example 2.** We assume that $S(\omega)$ is strictly bandlimited:

$$S(\omega) = 0, \qquad |\omega| > 2\pi W. \tag{26}$$

We can always approximate $H_{fr\infty}(j\omega)$ arbitrarily closely and use the receiver in Fig. 7.2, but there are two limiting cases that lead to simpler realizations.

The first limiting case is when the signal spectrum is much larger than $N_0/2$. This case is sometimes referred to the high-*input* signal-to-noise ratio case:

$$AS(\omega) \gg \frac{2}{N_0}, \qquad |\omega| \leq 2\pi W. \tag{27}$$

To exploit this, we expand the terms in (24) and (23) in a power series in $(N_0/2AS(\omega)$ to obtain

$$\frac{N_0S(\omega)/2}{(N_0/2 + AS(\omega)^2} = \frac{N_0}{2A^2S(\omega)}\left\{1 - 2\left(\frac{N_0}{2AS(\omega)}\right) + \cdots\right\} \tag{28}$$

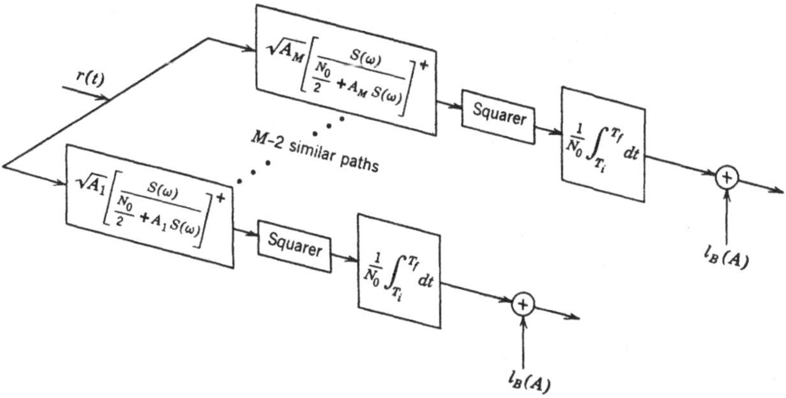


**Fig. 7.2 Generation of ln $\Lambda(A)$: Canonical Realization No. 3.**

and

$$-\frac{T}{2}\int_{-\infty}^{\infty} \frac{2S(\omega)/N_0}{1+[2AS(\omega)/N_0]}\frac{d\omega}{2\pi} = -\frac{T}{2A}\int_{-2\pi W}^{2\pi W}\left(1-\frac{N_0}{2AS(\omega)}+\cdots\right)\frac{d\omega}{2\pi}$$
$$= -\frac{WT}{A}+\frac{TN_0}{4A^2}\int_{-2\pi W}^{2\pi W}\frac{1}{2S(\omega)}\frac{d\omega}{2\pi}+\cdots. \tag{29}$$

Using (28) and (29) in (22) and neglecting powers of $1/S(\omega)$ greater than 1, we obtain

$$\hat{a}_0 = \frac{1}{2WT}\left\{\left[\iint_{T_i}^{T_f} r(t)h_\alpha(t-u)r(u)\,dt\,du\right] - \frac{TN_0}{2}\int_{-2\pi W}^{2\pi W}\frac{1}{S(\omega)}\frac{d\omega}{2\pi}\right\}, \tag{30}$$

where

$$H_\alpha(j\omega) = \begin{cases}\dfrac{1}{S(\omega)}, & |\omega| \le 2\pi W,\\ 0, & \text{elsewhere.}\end{cases} \tag{31}$$

Now we see why we were careful to denote the estimate in (30) as $\hat{a}_0$ rather than $\hat{a}_{ml}$. The reason follows by looking at the right side of (30). The first term is a random variable whose value is always non-negative. The second term is a negative bias. Therefore $\hat{a}_0$ can assume negative values. Because the parameter $A$ is the spectral amplitude, it must be non-negative. This means that the lower limit on the range of $A$, which we denoted by $A_\alpha$, must be non-negative. For algebraic simplicity we assume that $A_\alpha = 0$. Therefore, whenever $\hat{a}_0$ is negative, we choose zero for the maximum likelihood estimate,

$$\hat{a}_{ml} = \begin{cases}\hat{a}_0, & \text{if } \hat{a}_0 \ge 0,\\ 0, & \text{if } \hat{a}_0 < 0.\end{cases} \tag{32}$$

Notice that this result is consistent with our original discussion of the ML equation on page I-65. We re-emphasize that the ML equation provides a necessary condition on $\hat{a}_{ml}$ only when the maximum of the likelihood function is interior to the range of $A$. We shall find that in most cases of interest the probability that $\hat{a}_0$ will be negative is small. In the next section we discuss this issue quantitatively.

It is easy to verify that $\hat{a}_0$ is unbiased.

$$E[\hat{a}_0] = \frac{1}{2WT}\left\{E\left[\iint_{T_i}^{T_f} r(t)h_a(t-u)r(u)\,dt\,du\right] - \frac{TN_0}{2}\int_{-2\pi W}^{2\pi W} \frac{1}{S(\omega)}\frac{d\omega}{2\pi}\right\}$$

$$= \frac{1}{2WT}\left\{T\int_{-2\pi W}^{2\pi W} H_\alpha(j\omega)S_r(\omega)\frac{d\omega}{2\pi} - \frac{TN_0}{2}\int_{-2\pi W}^{2\pi W}\frac{1}{S(\omega)}\frac{d\omega}{2\pi}\right\}$$

$$= \frac{1}{2WT}\left\{T\int_{-2\pi W}^{2\pi W}\frac{AS(\omega)+[N_0/2]}{S(\omega)}\frac{d\omega}{2\pi} - \frac{TN_0}{2}\int_{-2\pi W}^{2\pi W}\frac{1}{S(\omega)}\frac{d\omega}{2\pi}\right\} = A. \tag{33}$$

Looking at (32), we see that (33) implies that $\hat{a}_{ml}$ is biased. This means that we cannot use the Cramèr-Rao bound in (17). Moreover, since it is difficult to find the bias as a function of $A$, we cannot modify the bound in an obvious manner (i.e., we cannot use the results of Problem 6.3.1). Since this issue arises in many problems, we digress and develop a technique for analyzing it.

### 7.1.2 Performance of Truncated Estimates

The reason $\hat{a}_{ml}$ is biased is that we have truncated $\hat{a}_0$ at zero. We study the effect of this truncation for the receiver shown in Fig. 7.3. This receiver is a generalization of the receiver in Example 2. We use the notation

$$\hat{a}_0 = G\{l - B\}, \tag{34}$$

where

$$l = \int_{T_i}^{T_f} r(t)h_\alpha(t-u)r(u)\,dt\,du. \tag{35}$$

Equivalently,

$$l = \int_{T_i}^{T_f} dt\left[\int_{T_i}^{T_f} h_\alpha^{[1/2]}(t-z)r(z)\,dz\right]^2, \tag{36}$$

where $h_\alpha^{[1/2]}(t-z)$ is the functional square root of $h_\alpha(t-u)$. The constants $G$ and $B$ denote "gain" and "bias," respectively.

In Example 2,

$$G = \frac{1}{2WT} \tag{37}$$

and

$$B = -\frac{TN_0}{2}\int_{-2\pi W}^{2\pi W}\frac{1}{S(\omega)}\frac{d\omega}{2\pi}. \tag{38}$$

In our initial discussion, we leave $G$ and $B$ as parameters. Later, we consider the specific values in (37) and (38). Notice that $\hat{a}_0$ will satisfy (30) only when the values in (37) and (38) are used. We shall assume that

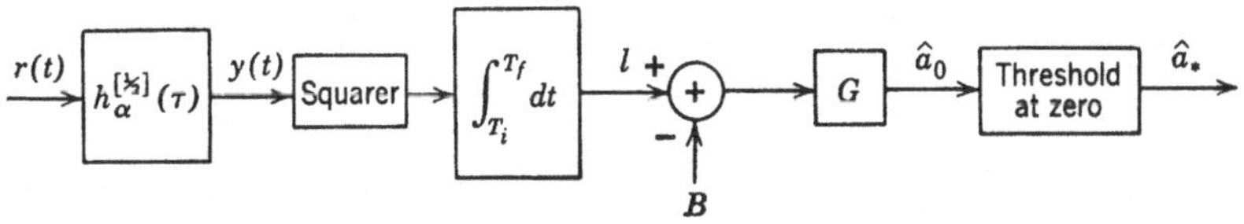


**Fig. 7.3 Filter-squarer-integrator receiver to generate $\hat{a}_*$.**

$G$ and $B$ are adjusted so that $\hat{a}_0$ is an *unbiased* estimate of $A$. We denote the truncated output as $\hat{a}_*$. In Example 2, $\hat{a}_*$ equals $\hat{a}_{ml}$, but for an arbitrary $h_\alpha(\tau)$ they will be different. Notice that we can compute the variance of $\hat{a}_0$ exactly (see Problem 7.1.1) for any $h_\alpha(\tau)$.

A typical probability density for $l$ is shown in Fig. 7.4. Notice that $A$ is a nonrandom parameter and $p_{l|a}(L \mid A)$ is our usual notation. We have shaded the region of $L$ where $\hat{a}_0$ will be truncated. If the probability that $l$ will lie in the shaded region is small, $\hat{a}_*$ will have a small bias. We would anticipate that if the probability of $l$ being in the shaded region is large, the mean-square estimation error will be large enough to make the estimation procedure unsatisfactory. We now put these statements on a quantitative basis. Specifically, we compute three quantities:

1. An upper bound on Pr $[l < B]$.
2. An upper bound on the bias of $\hat{a}_*$.
3. A lower bound on $E[(\hat{a}_* - A)^2]$.

The general expressions that we shall derive are valid for an arbitrary receiver of the form shown in Fig. 7.4 with the restriction

$$E[l] \geq B. \tag{39}$$

The general form of the results, as well as the specific answers, is important.

***Upper Bound on Pr$[l < B]$.*** Looking at Fig. 7.4, we see that the problem of interest is similar to the computation of $P_M$ for a suboptimum receiver

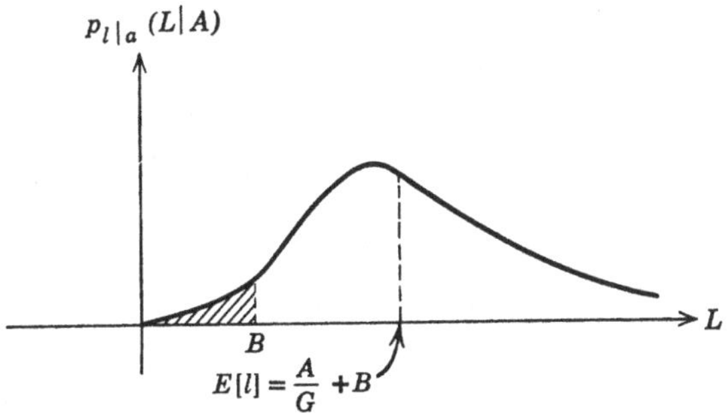


**Fig. 7.4 Typical probability density for $l$.**

that we solved in Section 5.1.2.† Using the asymptotic form of the Fredholm determinant in (5.55), we have

$$\mu(s, A) = -\frac{T}{2}\int_{-\infty}^{\infty} \ln\left(1 - 2sS_y(\omega:A)\right)\frac{d\omega}{2\pi}, \qquad s \leq 0. \tag{40}$$

The waveform $y(t)$ is the input to the squarer, and $S_y(\omega)$ is its power density spectrum. The Chernoff bound is

$$\Pr[l < B] \leq \exp[\mu(s_1, A) - s_1 B], \tag{41}$$

where $s_1$ is chosen so that

$$\left.\frac{\partial\mu(s, A)}{\partial s}\right|_{s=s_1} \triangleq \left.\dot{\mu}(s, A)\right|_{s=s_1} = B. \tag{42}$$

[Recall the result in (I-2.461) and notice the change of sign in $s$.] Since

$$B \leq E[l], \tag{43}$$

(42) will have a unique solution. Notice that this result is valid for any FSI receiver subject to the constraint on the bias in (43) [i.e., different $h_\alpha^{[1/2]}(\tau)$ could be used]. To illustrate the calculation, we consider a special case of Example 2.

**Example 2 (continuation).** We consider the receiver in Example 2. In (26) we assumed that the signal spectrum, $S(\omega)$, was strictly bandlimited to $W$ cycles per second. We now consider the special case in which $S(\omega)$ is constant over this frequency range. Thus,

$$S(\omega) = \begin{cases} \dfrac{1}{2W}, & |\omega| \leq 2\pi W, \\ 0, & \text{elsewhere.} \end{cases} \tag{44}$$

Then

$$S_y(\omega:A) = \begin{cases} (2W)\left[A\dfrac{1}{2W} + \dfrac{N_0}{2}\right] = A\left[1 + \dfrac{N_0 W}{A}\right], & |\omega| < 2\pi W, \\ 0, & \text{elsewhere,} \end{cases} \tag{45}$$

and

$$B = (2WT)(N_0 W). \tag{46}$$

Using (45) in (40) gives

$$\mu(s, A) = -WT\ln\left[1 - 2sA\left(1 + \frac{N_0 W}{A}\right)\right]. \tag{47}$$

To find $s_1$, we differentiate (47) and equate the result to $B$. The result is

$$s_1 = -\left[2\left(1 + \frac{N_0 W}{A}\right)N_0 W\right]^{-1}. \tag{48}$$

† We suggest that the reader review Section 5.1.2 and Problem 5.1.13 before reading this discussion.

Substituting (48) into (47) and (41) gives

$$\Pr[l < B] \le \left(1 + \frac{A}{N_0W}\right)^{-WT} \exp\left[\frac{WT}{(1 + N_0W/A)}\right]. \tag{49}$$

We see that the bound depends on $A/N_0W$, which is the signal-to-noise ratio in the message bandwidth, and on $WT$, which is one-half the time-bandwidth product of the signal. We have plotted the bound in Fig. 7.5 as a function of $WT$ for various values of $A/N_0W$. In most cases of interest, the probability that $\hat{a}_0$ will be negative is negligible. For example, if

$$\frac{A}{N_0W} = 10 \tag{50}$$

and

$$WT = 5, \tag{51}$$

the probability is less than $10^{-3}$ that $\hat{a}_0$ will be negative.

We used the Chernoff bound in our discussion. If it is desired, one can obtain a better approximation to the probability by using the approximate formula

$$\Pr[l < B] \simeq \frac{1}{\sqrt{2\pi s_1^2 \ddot{\mu}(s_1, A)}} \exp[\mu(s_1, A) - s_1 B] \tag{52}$$

(see Problem 5.1.13). In most cases, this additional refinement is not necessary. The next step is to bound the bias of $\hat{a}_*$.

***Upper Bound on Bias of $\hat{a}_*$.*** We can compute a bound on the bias of $\hat{a}_*$ by using techniques similar to those used to derive (41). Recall that $\hat{a}_0$ is an *unbiased* estimate of $A$ and can be written as

$$\hat{a}_0 = Gl - GB. \tag{53}$$

Therefore,

$$E[\hat{a}_0] = \int_{-\infty}^{\infty} G(L - B) p_{l|a}(L|A)\, dL = A. \tag{54}$$

Dividing the integration region into two intervals, we have

$$\int_0^B G(L - B) p_{l|a}(L|A)\, dL + \int_B^{\infty} G(L - B) p_{l|a}(L \mid A)\, dL = A. \tag{55}$$

The second integral is $E[\hat{a}_*]$. Thus, the bias of $\hat{a}_*$ is

$$\mathscr{B}(\hat{a}_*) \triangleq E[\hat{a}_*] - A = -\int_0^B G(L - B) p_{l|a}(L \mid A)\, dL. \tag{56}$$

The next step is to bound the term on the right side of (56). We develop two bounds. The first is quite simple and is adequate for most cases. The second requires a little more computation, but gives a better result.

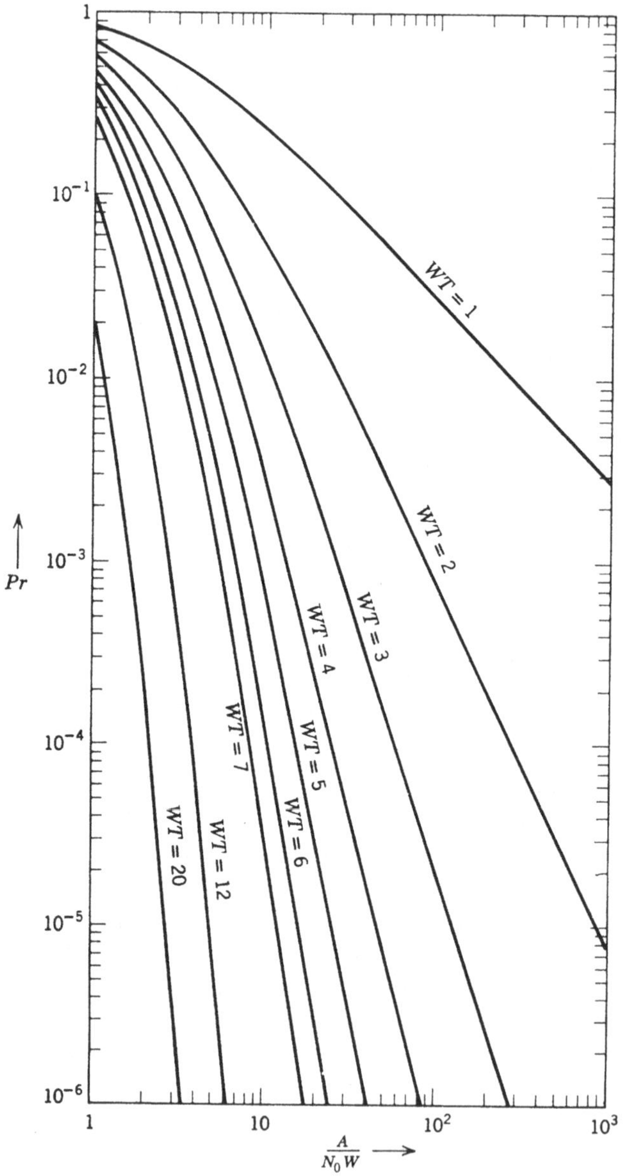


**Fig. 7.5 Bound on probability that $\hat{a}_0$ will be negative.**

***Bound No. 1.*** The first bound is

$$\begin{aligned}\mathscr{B}(\hat{a}_*) &= -\int_0^B G(L-B)p_{l|a}(L \mid A)\, dL \\ &\leq BG \int_0^B p_{l|a}(L \mid A)\, dL \\ &= BG \Pr[l \leq B] \\ &\leq BG \exp[\mu(s_1, A) - s_1 B], \end{aligned} \tag{57}$$

where $s_1$ satisfies (42). The result in (57) can be normalized to give

$$\frac{\mathscr{B}(\hat{a}_*)}{A} \leq \frac{BG}{A} \exp[\mu(s_1, A) - s_1 B]. \tag{58}$$

For the spectrum in (44), this reduces to

$$\frac{\mathscr{B}(\hat{a}_*)}{A} \leq \left(\frac{N_0 W}{A}\right)\left(1 + \frac{A}{N_0 W}\right)^{-WT} \exp\left[\frac{WT}{(1 + N_0 W)}\right]. \tag{59}$$

After deriving the second bound, we shall plot the result in (59). Notice that (59) can be plotted directly from Fig. 7.5 by multiplying each value by $N_0W/A$.

***Bound No. 2*** [2]. We can obtain a better bound by tilting the density. We define

$$p_{l_s}(L) = \exp[sL - \mu(s, A)]\, p_{l|a}(L|A), \qquad s \leq 0. \tag{60}$$

[Recall (I-2.450).] Using (60) in (57) gives

$$\begin{aligned}\mathscr{B}(\hat{a}_*) &= G \int_0^B (B - L) \exp[\mu(s, A) - sL]\, p_{l_s}(L)\, dL \\ &= G \exp[\mu(s, A) - sB] \int_0^B (B - L) \exp[s(B - L)] p_{l_s}(L)\, dL \\ &\leq G \exp[\mu(s, A) - sB] \{ \max_{0 \leq L \leq B} [(B - L) \exp[s(B - L)]] \} \\ &\quad \times \int_0^B p_{l_s}(L)\, dL \qquad s \leq 0 \end{aligned} \tag{61}$$

We now upperbound the integral by unity and let

$$Z = B - L \tag{62}$$

in the term in the braces. Thus,

$$\mathscr{B}(\hat{a}_*) \leq G \exp[\mu(s, A) - sB]\{ \max_{0 \leq Z \leq B} [Ze^{sZ}]\}, \qquad s \leq 0. \tag{63}$$

The term in the braces is maximized by

$$Z = Z_m \triangleq \min\left[B, -\frac{1}{s}\right], \qquad s \leq 0. \tag{64}$$

Using (64) in (63) gives

$$\mathscr{B}(\hat{a}_*) \leq GZ_m \exp[\mu(s, A) - s(B - Z_m)], \qquad s \leq 0. \tag{65}$$

We now minimize the bound as a function of $s$. The minimum is specified by

$$\left(\dot{\mu}(s, A) - \frac{1}{s}\right)\Bigg|_{s=s_2} = B, \qquad -\frac{1}{B} \le s_2 \le 0. \tag{66}$$

We can demonstrate that (66) has a solution in the allowable range (see Problem 7.1.2). Thus,

$$\frac{\mathscr{B}(\hat{a}_*)}{A} \le \frac{-G}{As_2} \exp\{\mu(s_2, A) - s_2 B - 1\}. \tag{67}$$

For the spectrum in (44),

$$As_2 = -\frac{C_2}{2C_1}\left[1 + \sqrt{1 + \frac{4C_1}{C_2^2}}\right], \tag{68}$$

where

$$C_1 \triangleq 4WT\frac{N_0 W}{A}\left(1 + \frac{N_0 W}{A}\right) \tag{69a}$$

and

$$C_2 \triangleq 2(1 + WT)\left(1 + \frac{N_0 W}{A}\right) - 2WT\frac{N_0 W}{A} \tag{69b}$$

(see Problem 7.1.3). In Fig. 7.6, we plot the bounds given by (59) and (67) for the case in which $WT = 5$. We see that the second bound is about an order of magnitude better than the first bound in this case, and that the bias is negligible. Similar results can be obtained for other $WT$ products. From Fig. 7.5, we see that the bias is negligible in most cases of interest. Just as on page 198, we can obtain a better approximation to the bias by using a formula similar to (52) (see Problem 7.1.4). The next step is computing a bound on the mean-square error.

***Mean-square-error Bound.*** The mean-square error using $\hat{a}_*$ is

$$\begin{aligned}\xi_* &\triangleq E[(\hat{a}_* - A)^2] \\ &= E[(\hat{a}_* - \hat{a}_0 + \hat{a}_0 - A)^2] \\ &= E[\hat{a}_* - \hat{a}_0)^2] + 2E[(\hat{a}_* - \hat{a}_0)(\hat{a}_0 - A)] + E[(\hat{a}_0 - A)^2].\end{aligned} \tag{70}$$

Observe that

$$\hat{a}_* = \hat{a}_0 \quad \text{for} \quad \hat{a}_0 \ge 0 \tag{71}$$

and

$$\hat{a}_* = 0 \quad \text{for} \quad \hat{a}_0 < 0. \tag{72}$$

Thus, (70) can be written as

$$\begin{aligned}\xi_* &= \int_0^B (\hat{a}_0(L))^2 p_{l|a}(L|A)\, dL + 2\int_0^B \hat{a}_0(L)[A - \hat{a}_0(L)]p_{l|a}(L|A)\, dL + \xi_{\hat{a}_0} \\ &= \int_0^B [2\hat{a}_0(L)A - (\hat{a}_0(L))^2]p_{l|a}(L|A)\, dL + \xi_{\hat{a}_0}.\end{aligned} \tag{73}$$

Recalling that

$$\hat{a}_0(L) = G(L - B), \tag{74}$$

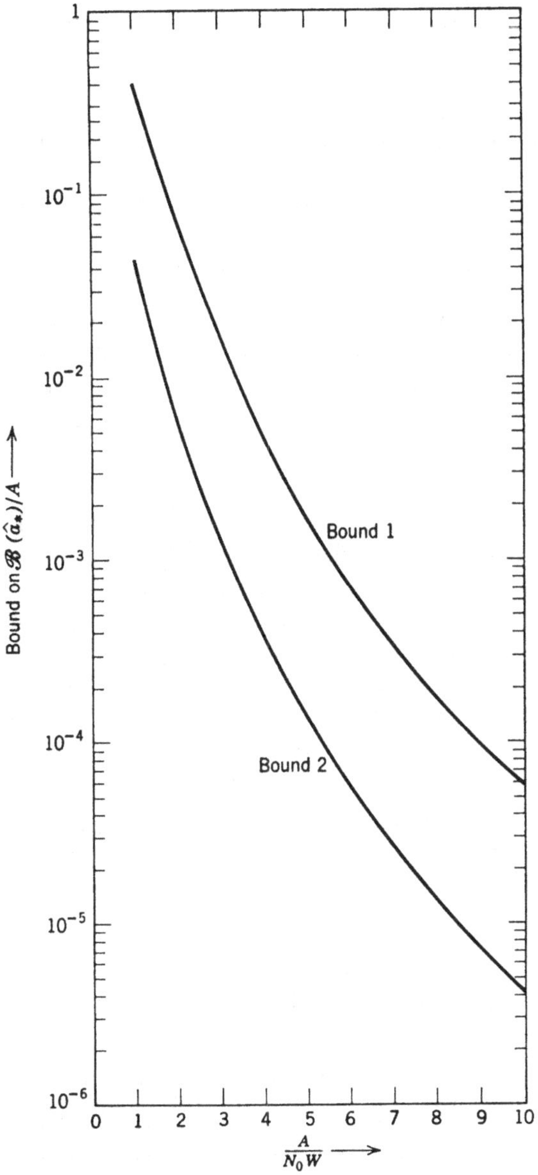


**Fig. 7.6 Comparison of bounds on normalized bias [$WT = 5$].**

we have

$$\xi_* = -\left\{\int_0^B [2AG(B-L) + G^2(B-L)^2] p_{l|a}(L|A)\, dL\right\} + \xi_{\hat{a}_0} \quad (75)$$

The term in the braces is always positive. Thus,

$$\xi_{\hat{a}_0} - \Delta\xi \le \xi_* \le \xi_{\hat{a}_0}, \quad (76)$$

where

$$\Delta\xi \triangleq \int_0^B [2AG(B-L) + G^2(B-L)^2] p_{l|a}(L|A)\, dL. \quad (77)$$

We can now proceed as on pages 199–201 to find a bound on $\Delta\xi$. A simple bound is

$$\Delta\xi \le [2AGB + G^2B^2] \Pr(l < B). \quad (78)$$

A tighter bound can be found in the same manner in which (67) was derived (see Problem 7.1.5). In many cases, the simple bound in (78) is adequate. For example, for the numerical values in (50) and (51),

$$\frac{\Delta\xi}{A^2} \le 1.25 \times 10^{-4}. \quad (79)$$

Notice that we must also find $\xi_{\hat{a}_0}$. As we indicated on page 195, we can always find $\xi_{\hat{a}_0}$ exactly (see Problem 7.1.1.).

We have developed techniques for evaluating the bias and mean-square error of a truncated estimate. A detailed discussion was included because both the results and techniques are important. In most of our subsequent discussion, we shall neglect the truncation effect and assume that the performances of $\hat{a}_*$ and $\hat{a}_0$ are essentially the same. The results of this section enable us to check that assumption in any particular problem. We now return to Example 2 and complete our discussion.

**Example 2 (continuation).** If we assume that the bias is negligible, the performance $\hat{a}_{ml}$ can be approximated by the performance of $\hat{a}_0$. The Cramèr-Rao bound on the variance of $\hat{a}_0$ is given by (18). When $2AS(\omega)/N_0$ is large, the bound is independent of the spectrum

$$\frac{\text{Var}[\hat{a}_0 - A]}{A^2} \ge \frac{2}{2WT}. \quad (80)$$

We can show that the actual variance approaches this bound as $N_0/2AS(\omega) \to 0$ Two observations are useful:

1. In the small-noise case, the *normalized* variance is independent of the *spectral height* and *spectral shape*. This is analogous to the classical estimation problem of Chapter I-2. There, we estimated the variance of a Gaussian random variable $x$. We saw that the *normalized* variance of the estimate was independent of the actual variance $\sigma_x^2$.

2. We recall from our discussion in Chapter I-3 that if a stationary process is bandlimited to $W$ and observed over an interval of length $T$, there are $N = 2WT$ significant eigenvalues. Thus, (80) can be written as

$$\frac{\text{Var}\,[\hat{a}_0 - A]}{A^2} \geq \frac{2}{N}, \tag{81}$$

which is identical with the corresponding classical estimation results.

We have used Example 2 as a vehicle for studying the performance of a biased ML estimate in detail. There were two reasons for this detailed study.

1. We encounter biased estimates frequently in estimating random process parameters. It is necessary to have a quantitative estimate of the effect of this bias. Fortunately, the effect is negligible in many problems of interest. Our bounds enable us to determine when we can neglect the bias and when we must include it.

2. The basic analytic techniques used in the study are useful in other estimation problems. Notice that we used the Chernoff bound in (41). If the probability is non-negligible and we want a better estimate of its exact value, we can use the approximate expression in (52).

We now consider another special case of the amplitude estimation problem.

**Example 3. Low-input Signal-to-Noise Ratio.** The other limiting case corresponds to a low-input signal-to-noise ratio. This case is analogous to the LEC case that we encountered in detection problems (see page 131). Assuming that

$$AS(\omega) \ll \frac{N_0}{2}, \tag{82}$$

then (22) can be expanded in a series. Carrying out the expansion and retaining the first term gives

$$\hat{a}_0 \simeq \left[T\int_{-\infty}^{\infty} S^2(\omega)\,\frac{d\omega}{2\pi}\right]^{-1}\left\{\iint_{T_i}^{T_f} r(t)h_\beta(t-u)r(u)\,dt\,du - \frac{N_0 T}{2}\int_{-\infty}^{\infty} S(\omega)\,\frac{d\omega}{2\pi}\right\}, \tag{83}$$

where

$$H_\beta(j\omega) = S(\omega). \tag{84}$$

Notice that it is not necessary to assume that $S(\omega)$ is bandlimited. The approximate *ML* estimate is

$$\hat{a}_{ml} = \begin{cases} \hat{a}_0, & \hat{a}_0 \geq 0, \\ 0, & \hat{a}_0 < 0. \end{cases} \tag{85}$$

All of the general derivations concerning bias and mean-square error are valid for this case.

For the flat bandlimited spectrum in (44), the probability that $\hat{a}_0$ will be negative is bounded by

$$\Pr[\hat{a}_0 < 0] \leq \exp\left[-WT\left(\frac{A}{WN_0}\right)\right]. \tag{86}$$

The restriction in (82) implies that

$$\frac{A}{WN_0} \ll 1. \tag{87}$$

Thus, $WT$ must be very large in order for the probability in (86) to be negligible.

The Cramèr-Rao bound on the variance of any unbiased estimate is

$$\frac{\operatorname{Var}[\hat{a} - A]}{A^2} \geq \frac{(N_0/2)^2}{(A^2T/2)\int_{-2\pi W}^{2\pi W} S^2(\omega)(d\omega/2\pi)}. \tag{88}$$

For the flat bandlimited spectrum, (88) reduces to

$$\frac{\operatorname{Var}[\hat{a} - A]}{A^2} \geq \left(\frac{WN_0}{A}\right)^2 \frac{1}{WT}. \tag{89}$$

We see that $WT$ must be large in order for this bound to be small. When this is true, we can show that the variance of $\hat{a}_0$ approaches this bound. Under these conditions the probability in (86) is negligible, so that $\hat{a}_{ml}$ equals $\hat{a}_0$ on almost all realizations of the experiment. In many cases, the probability in (86) will not be negligible, and so we use the results in (34)–(78) to evaluate the performance. This analysis is carried out in Problem 7.1.6.

In these two limiting cases of high and low signal-to-noise ratio that we studied in Examples 2 and 3, the receiver assumed the simple form shown in Fig. 7.7. In the high signal-to-noise ratio case,

$$H_C(j\omega) = \frac{1}{S(\omega)} \quad \text{[an inverse filter]}, \tag{90}$$

with

$$B = \frac{TN_0}{2}\int_{-2\pi W}^{2\pi W} \frac{1}{S(\omega)}\frac{d\omega}{2\pi} \tag{91}$$

and

$$G = \frac{1}{2WT}. \tag{92}$$

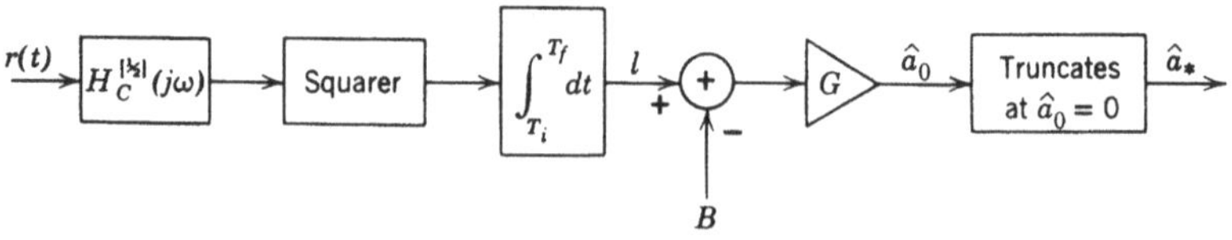


**Fig. 7.7 Amplitude estimator.**

In the low signal-to-noise ratio case,

$$H_C(j\omega) = S(\omega), \tag{93}$$

with

$$B = \frac{TN_0}{2}\int_{-2\pi W}^{2\pi W} S(\omega)\,\frac{d\omega}{2\pi} \tag{94}$$

and

$$G = \left[T\int_{-\infty}^{\infty} S^2(\omega)\,\frac{d\omega}{2\pi}\right]^{-1} \tag{95}$$

A receiver of the form in Fig. 7.7 is commonly referred to as a radiometer in the radio astronomy field [3]. It is, of course, a form of filter-squarer-integrator receiver that we have seen previously in this chapter.

The obvious advantage of the structure in Fig. 7.7 is that it generates the estimate by passing $r(t)$ through a single processing sequence. By contrast, in the general case we had to build $M$ processors, as shown in Fig. 7.1. In view of the simplicity of the filter-squarer-integrator receiver, we consider briefly a suboptimum receiver that uses the structure in Fig. 7.7 but allows us to choose $H_C(j\omega)$, $B$, and $G$.

### 7.1.3 Suboptimum Receivers

The receiver of interest is shown in Fig. 7.7. Looking at (19), we see that a logical parametric form of $H_C(j\omega)$ is

$$H_C(j\omega) = \frac{S(\omega)}{(N_0/2 + CS(\omega))^2}, \tag{96}$$

where $C$ is a constant that we shall choose. Observe that we achieve the two limiting cases of high and low signal-to-noise ratio by letting $C$ equal infinity and zero, respectively.

We choose $B$ and $G$ so that $\hat{a}_0$ will be an unbiased estimate for all values of $A$. This requires

$$B = \int_{-\infty}^{\infty} \frac{N_0 S(\omega)/2}{(N_0/2 + CS(\omega))^2}\,\frac{d\omega}{2\pi} \tag{97}$$

and

$$G = \left\{T\int_{-\infty}^{\infty} \frac{S^2(\omega)}{(N_0/2 + CS(\omega))^2}\,\frac{d\omega}{2\pi}\right\}^{-1}. \tag{98}$$

The only remaining parameter is $C$. In order to choose $C$ for the general case, we first compute the mean-square error. This is a straightforward calculation (e.g., Problem 7.1.7) whose result is

$$\xi_{\hat{a}_0} \triangleq E[\hat{a}_0 - A)^2] = \frac{GT}{2}\int_{-\infty}^{\infty} \frac{S^2(\omega)[AS(\omega) + N_0/2]^2}{(N_0/2 + CS(\omega))^4}\,\frac{d\omega}{2\pi}. \tag{99}$$

If we plot (99) as a function of $A$, we find that it is a minimum when $C = A$. This result is exactly what we would expect but does not tell us how to choose $C$, because $A$ is the unknown parameter. Frequently, we know the range of $A$. If we know that

$$A_{\alpha} \leq A \leq A_{\beta}. \tag{100}$$

then we choose a value of $C$ in $[A_{\alpha}, A_{\beta}]$ according to some criterion. Some possible criteria will be discussed in the context of an example.

Notice that we must still investigate the bias of $\hat{a}_*$ using the techniques developed above. In the regions of the most interest (i.e., good performance) the bias is negligible and we may assume $\hat{a}_* = \hat{a}_0$ on almost all experiments.

In the next example we investigate the performance of our suboptimum receiver for a particular message spectrum.

**Example 4.** In several previous examples we used an ideal bandlimited spectrum. For that spectrum, the $H_C(j\omega)$ as specified in (96) is always an ideal low-pass filter. In this example we let

$$S(\omega:A) = \frac{2kA}{\omega^2 + k^2}, \qquad -\infty < \omega < \infty. \tag{101}$$

Now $H_C(j\omega)$ will change as $C$ changes. The lower bound on the variance of any unbiased estimate is obtained by using (101) in (25). The result is (see Problem 7.1.8)

$$\frac{\text{Var}\,[\hat{a} - A]}{A^2} \geq \frac{8}{kT}\,\frac{(1 + \Lambda(A))^{3/2}}{\Lambda^2(A)}, \tag{102}$$

where $\Lambda(A)$ is the signal-to-noise ratio in the message bandwidth

$$\Lambda(A) = \frac{4A}{kN_0}. \tag{103}$$

We use the suboptimum receiver shown in Fig. 7.7. The normalized variance of $\hat{a}_0$ is obtained by substituting (101) into (99). The result is

$$\frac{\text{Var}\,[\hat{a}_0 - A]}{A^2} = \frac{1}{(kT)(\Lambda^2(A))}[1 + c_1(A)\Lambda(A)]^{3/2}[5c_2^{\,2}(A) + 2c_2^{\,2}(A) + 1], \tag{104}$$

where

$$c_1(A) = \frac{C}{A} \tag{105}$$

and

$$c_2(A) = \frac{1 + \Lambda(A)}{1 + c_1(A)\Lambda(A)}. \tag{106}$$

When $c_1(A)$ equals unity, the variance in (104) reduces to that in (102). In Fig. 7.8, we have plotted the variance using the suboptimum estimator for the case in which $kT = 100$. The value at $c_1(A) = 1$ is the lower bound on any unbiased estimate. For these parameter values we see that we could use the suboptimum receiver over a decade range

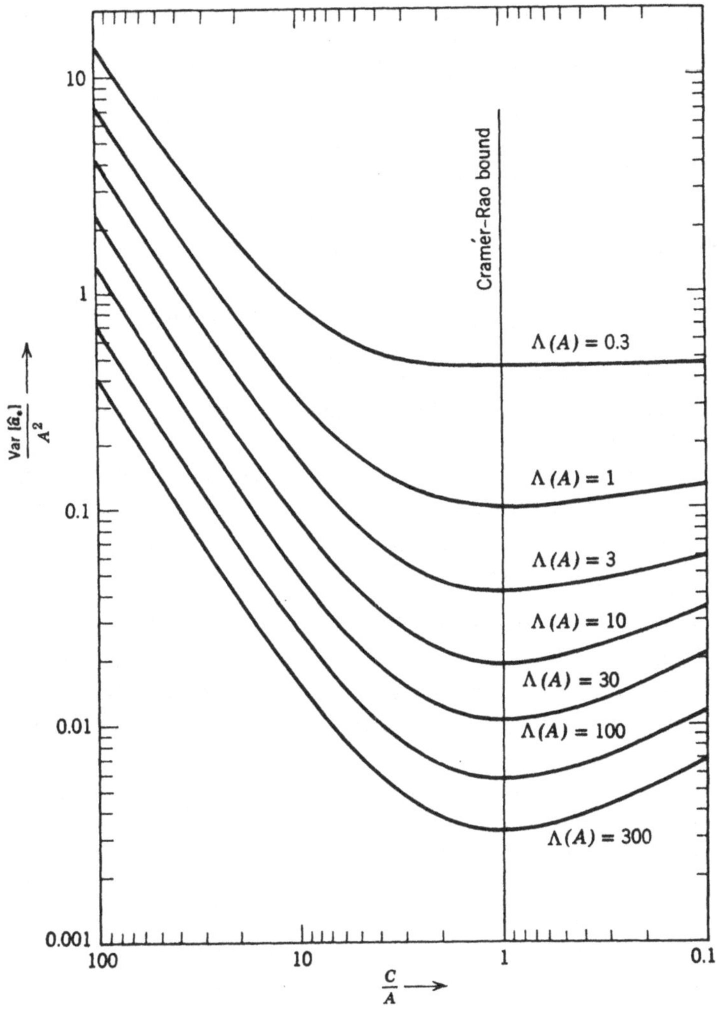


**Fig. 7.8 Performance of suboptimum receiver (from [2]).**

($A_\beta = 10A_\alpha$) at the cost of about a 50% increase in the error at the endpoints. Two observations are appropriate:

1. We must compute the bias of $\hat{a}_*$. If it is negligible, the mean-square error given in (102) can be used. If not, we evaluate (41) and (78). (See Problem 7.1.10.)
2. If the increase in variance because of the suboptimum receiver is too large, there are several possibilities. The first is to return the original receiver in Fig. 7.1. The second is to use several suboptimum receivers in parallel to cover the parameter range. The third is to estimate $A$ sequentially. We record $r(t)$ and process it once, using a suboptimum receiver to obtain an estimate that we denote as $\hat{a}_1$. We then let $C = \hat{a}_1$ in the

suboptimum receiver and reprocess $r(t)$ to obtain an estimate $\hat{a}_2$. Repeating the procedure will lead us to the desired estimate. The difficulty with the third procedure is proving when it converges to the correct estimate.

This completes our discussion of suboptimum amplitude estimators. We now summarize the results of our discussion.

### 7.1.4 Summary

In this section we have studied the stationary-process, long-observation-time case in detail. We chose this case for our detailed study because it occurs frequently in practice. When the SPLOT condition is valid, the expressions needed to generate $\ln \Lambda(A)$ and evaluate the Cramèr-Rao bound can be found easily.

Our discussion concentrated on the amplitude estimation problem because it illustrated a number of important issues. Other parameter estimation problems are discussed in Section 7.7.

The procedure in each case is similar:

1. The ML estimate is the value of $A$ that maximizes (11). In the general case, one must generate this expression as a function of $A$ and find the absolute maximum. The utility of this estimation procedure rests on being able to find a practical method of generating this function (or a good approximation).
2. In some special cases (this usually will correspond physically to either a low or high input signal-to-noise ratio), approximations can be made that lead to a unique solution for $\hat{a}_{ml}$.
3. If the estimate is unbiased, one can find a lower bound on the variance of the estimate using (16). Usually the variance of $\hat{a}_{ml}$ approaches this bound when the error is small. If the estimate is biased, we must modify our results to include the effect of the bias. If the bias can be found as a function of $A$, then the appropriate bound follows. Usually $\mathscr{B}(A)$ cannot be found exactly, and we use approximation techniques similar to those developed in Section 7.1.2.

The procedure is easy to outline, but the amount of work required to carry it out will depend on how the parameter is imbedded in the process.

In the next sections we discuss three other categories of estimation problems. Many of the issues that we have encountered here arise in the cases that we shall discuss in the next three sections. Because we have treated them carefully here, we leave many details to the reader in the ensuing discussions.

## 7.2 FINITE-STATE PROCESSES

In our discussion up to this point we have estimated a parameter of a random process, $s(t, A)$. The statistics of the process depended on $A$ through the mean-value function $m(t, A)$ and the covariance function $K_s(t, u{:}A)$. We assume that $m(t, A)$ is zero for simplicity. Instead of characterizing $s(t, A)$ by its covariance function, we can characterize it by the differential equations

$$\dot{\mathbf{x}}(t, A) = \mathbf{F}(t, A)\mathbf{x}(t, A) + \mathbf{G}(t, A)\mathbf{u}(t), \qquad t \geq T_i \tag{107}$$

and

$$s(t, A) = \mathbf{C}(t, A)\mathbf{x}(t, A), \qquad t \geq T_i, \tag{108}$$

where

$$E[\mathbf{u}(t)\mathbf{u}^T(\tau)] = \mathbf{Q}(A)\,\delta(t - \tau), \tag{109}$$

$$E[\mathbf{x}(T_i)] = \mathbf{0}, \tag{110}$$

and

$$E[\mathbf{x}(T_i)\mathbf{x}^T(T_i)] = \mathbf{P}_0(A). \tag{111}$$

Two observations are useful:

1. The parameter $A$ may appear in $\mathbf{F}(t, A)$, $\mathbf{G}(t, A)$, $\mathbf{C}(t, A)$, $\mathbf{Q}(A)$, and $\mathbf{P}_0(A)$ in the general case. Notice that there is only a single parameter. In most problems of interest only one or two of these functions will depend on $A$.
2. In the model of Section 6.1 we assumed that $s(t, A)$ was conditionally Gaussian. Thus the *linear* state equation in (107) is completely general if $s(t, A)$ is state-representable. By using the techniques described in Chapter II-7, we could study parameter estimation for Markovian non-Gaussian processes, but this is beyond the scope of our discussion.

For the zero-mean case the likelihood function is

$$l(A) = l_R(A) + l_B(A). \tag{112}$$

From (6.32),

$$\boxed{l_R(A) = \frac{2}{N_0}\int_{T_i}^{T_f} r(t)\hat{s}_r(t, A)\,dt - \frac{1}{N_0}\int_{T_i}^{T_f} \hat{s}_r^{\,2}(t, A)\,dt,} \tag{113}$$

and from (6.25),

$$l_B(A) = -\frac{1}{N_0}\int_{T_i}^{T_f} \xi_P(t{:}A)\,dt. \tag{114}$$

The function $\hat{s}_r(t, A)$ is the realizable MMSE estimate of $s(t, A)$, assuming that $A$ is known. From Chapter I-6, we know that it is specified by the differential equations

$$\dot{\hat{\mathbf{x}}}(t, A) = \mathbf{F}(t, A)\hat{\mathbf{x}}(t, A) + \boldsymbol{\xi}_P(t, A)\mathbf{C}^T(t, A)\frac{2}{N_0}[r(t) - \mathbf{C}(t, A)\hat{\mathbf{x}}(t, A)], \tag{115}$$

$$\dot{\boldsymbol{\xi}}_P(t, A) = \mathbf{F}(t, A)\boldsymbol{\xi}_P(t, A) + \boldsymbol{\xi}_P(t, A)\mathbf{F}^T(t, A) - \boldsymbol{\xi}_P(t, A)\mathbf{C}^T(t, A)\frac{2}{N_0}\mathbf{C}(t, A)\boldsymbol{\xi}_P(t, A) + \mathbf{G}(t, A)\mathbf{Q}(A)\mathbf{G}^T(t, A), \tag{116}$$

and

$$\hat{s}_r(t, A) = \mathbf{C}(t, A)\hat{\mathbf{x}}(t, A). \tag{117}$$

with appropriate initial conditions. The function $\xi_P(t:A)$ is the minimum mean-square error in estimating $s(t, A)$, assuming that $A$ is known. In almost all cases, we must generate $l(A)$ for a set of $A_i$ that span the allowable range of $A$ and choose the value at which $l(A_i)$ is the largest. This realization is shown in Fig. 7.9.

In order to bound the variance of any unbiased estimate, we use the bound in (6.44), (6.46), and (6.60). For finite-state processes, the form in (6.60) is straightforward. The expression is

$$J^{(2)}(A) = \frac{4}{N_0}\left[\frac{\partial^2}{\partial A_1^2}\left\{\int_{T_i}^{T_f} \xi_P\left(t \,\middle|\, \sqrt{\tfrac{1}{2}}\, s(\cdot, A_1) + \sqrt{\tfrac{1}{2}}\, s(\cdot, A), \frac{N_0}{2}\right) dt - \frac{1}{2}\int_{T_i}^{T_f} \xi_P\left(t \,\middle|\, s(\cdot, A_1), \frac{N_0}{2}\right) dt\right\}\right]_{A_1=A}. \tag{118}$$

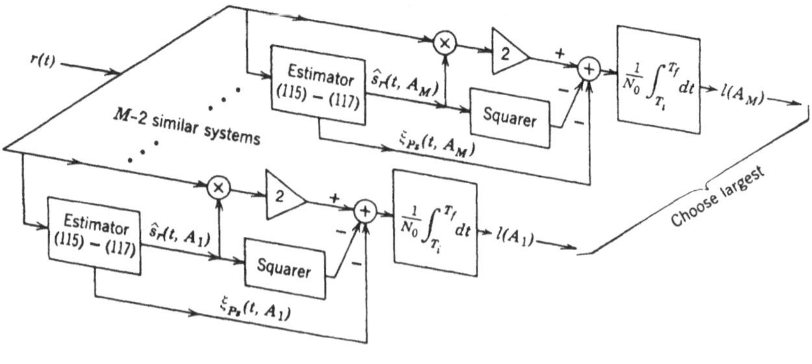


**Fig. 7.9 Likelihood function generation for finite-state processes.**

The error expression in the integrand of the second term follows from (116) as

$$\xi_P\left(t \mid s(\cdot, A_1), \frac{N_0}{2}\right) = \mathbf{C}(t, A_1)\boldsymbol{\xi}_P(t, A_1)\mathbf{C}^T(t, A_1). \qquad (119)$$

To compute the error expression in the first term, we go through a similar analysis for the composite process. Notice that the composite process is the sum of two statistically independent processes with different values of $A$. As in the analogous detection theory problem, there is an alternative form for $J^{(2)}(A)$ that is sometimes simpler to compute (see pages 179–181). The details are developed in Problem 7.2.1.

Several examples of estimating a parameter of a state-representable process are developed in the problems. We now consider separable kernel processes.

## 7.3 SEPARABLE KERNELS

In Chapter 4 we discussed several physical situations that led to detection theory problems with separable kernels. Most of these situations have a counterpart in the estimation theory context. In view of this similarity, we simply define the problems and work a simple example. The signal covariance function is $K_s(t, u:A)$. If we can write it as

$$K_s(t, u:A) = \sum_{i=1}^{K} \lambda_i(A)\phi_i(t, A)\phi_i(u, A), \qquad T_i \leq t, u \leq T_f \qquad (120)$$

for some finite $K$ and for every value of $A$ in the range $[A_\alpha, A_\beta]$, we have a separable-kernel estimation problem. Notice that both the eigenvalues and eigenfunctions may depend on $A$. To illustrate some of the implications of separability, we consider a simple amplitude estimation problem.

**Example 5.**† The received waveform is

$$r(t) = s(t, A) + w(t), \qquad T_i \leq t \leq T_f. \qquad (121)$$

The signal is zero-mean with covariance function

$$K_s(t, u:A) = AK_s(t, u), \qquad T_i \leq t, u \leq T_f. \qquad (122)$$

We assume that $K_s(t, u)$ is separable:

$$K_s(t, u) = \sum_{i=1}^{K} \lambda_i\phi_i(t)\phi_i(u), \qquad T_i \leq t, u \leq T_f. \qquad (123)$$

† This particular problem has been studied by Hofstetter [7].

The likelihood function is

$$\ln \Lambda(A) = \frac{1}{2}\sum_{i=1}^{K} \ln\left(1 + \frac{2A\lambda_i}{N_0}\right) + \frac{1}{N_0}\sum_{i=1}^{K}\left(\frac{A\lambda_i}{N_0/2 + A\lambda_i}\right) r_i^2, \tag{124}$$

where

$$r_i = \int_{T_i}^{T_f} r(t)\phi_i(t)\, dt. \tag{125}$$

To find $\hat{a}_{ml}$, we must find the value of $A$ where $\ln \Lambda(A)$ has its maximum. If $\hat{a}_{ml}$ is unbiased, its variance is bounded by

$$\frac{\text{Var}\,[(\hat{a} - A)]}{A^2} \geq 2\left[\sum_{i=1}^{K} \frac{\lambda_i^2}{(N_0/2A + \lambda_i)^2}\right]^{-1} \tag{126}$$

If the maximum is interior to the range of $A$ and $\ln \Lambda(A)$ has a continuous first derivative, then a necessary condition is obtained by differentiating (124),

$$\left\{-\frac{1}{2}\sum_{i=1}^{K} \frac{\lambda_i}{N_0/2 + \lambda_i A} + \frac{1}{2}\sum_{i=1}^{K} \frac{\lambda_i}{(N_0/2 + A\lambda_i)^2} r_i^2\right\}\Bigg|_{A=\hat{a}_0} = 0. \tag{127}$$

For arbitrary values of $N_0$, $A$, and $\lambda_i$ this condition is not too useful. There are three cases in which a simple result is obtained for the estimate:

1. The $K$ eigenvalues are all equal.
2. All of the $\lambda_i$ are much greater than $N_0/2A$.
3. All of the $\lambda_i$ are much less than $N_0/2A$.

The last two cases are analogous to the limiting cases that we discussed in Section 7.1, and so we relegate them to the problems (see Problems 7.3.1 and 7.3.2).

In the first case,

$$\lambda_i = \lambda_c, \qquad i = 1, 2, \ldots, K, \tag{128}$$

and (127) reduces to

$$\hat{a}_0 = \frac{1}{\lambda_c}\left[\frac{1}{K}\sum_{i=1}^{K} r_i^2 - \frac{N_0}{2}\right]. \tag{129}$$

Since $\hat{a}_0$ can assume negative values, we have

$$\hat{a}_{ml} = \begin{cases} \hat{a}_0, & \hat{a}_0 \geq 0, \\ 0\,, & \hat{a}_0 < 0. \end{cases} \tag{130}$$

In this particular case we can compute $p_{\hat{a}_0}(A_0)$ exactly. It is a chi-square density (page I-109) shifted by $N_0/2\lambda_c$. We can also compute the bias and variance exactly. For moderate $K$ ($K > 8$) the approximate expressions in Section 7.1.2 give an accurate answer. The various expressions are derived in the problems.

We should also observe that in the equal eigenvalue case, $\hat{a}_0$ is an *efficient* unbiased estimate of $A$. In other words, its variance satisfies (126) with an equality sign. This can be verified by computing Var $[\hat{a}_0 - A]$ directly.

This example illustrates the simplest type of separable kernel problem. In the general case we have to use the parallel processing configuration in

Fig. 6.1. Notice that *each* path will contain $K$ filter-squarers. Thus, if there are $M$ paths, the complete structure will contain $MK$ filter-squarers. In view of this complexity, we usually try to find a suboptimum receiver whose performance is close to the optimum receiver. The design of this suboptimum receiver will depend on how the parameter enters into the covariance function. Several typical cases are given in the problems.

## 7.4 LOW-ENERGY-COHERENCE CASE

In Section 4.3 of the detection theory discussion, we saw that when the largest eigenvalue was less than $N_0/2$ we could obtain an iterative solution for $h(t, u)$. We can proceed in exactly the same manner for the estimation problem. The only difference is that the largest eigenvalue may depend on $A$. From (6.16) we have

$$\begin{aligned}\ln \Lambda(A) &= l_R(A) + l_B(A)\\ &= \frac{1}{N_0}\sum_{i=1}^{\infty}\left(\frac{\lambda_i(A)}{\lambda_i(A) + N_0/2}\right)R_i^2 - \frac{1}{2}\sum_{i=1}^{\infty}\ln\left(1 + \frac{2\lambda_i(A)}{N_0}\right).\end{aligned} \tag{131}$$

Assuming that

$$\lambda_i(A) < \frac{N_0}{2} \tag{132}$$

*for all A*, we can expand each term in the sums in a convergent power series in $[2\lambda_i(A)]/N_0$. The result is

$$l_R(A) = \frac{1}{2}\left(\frac{2}{N_0}\right)^2\sum_{i=1}^{\infty}\lambda_i(A)\left[1 - \frac{2}{N_0}\lambda_i(A) + \left(\frac{2}{N_0}\right)^2(\lambda_i(A))^2\cdots\right]r_i^2 \tag{133}$$

and

$$l_B(A) = -\frac{1}{2}\left(\frac{2}{N_0}\right)\sum_{i=1}^{\infty}\left[\lambda_i(A) - \frac{1}{2}\left(\frac{2}{N_0}\right)(\lambda_i(A))^2 + \frac{1}{3}\left(\frac{2}{N_0}\right)^2(\lambda_i(A))^3\cdots\right]. \tag{134}$$

In the LEC case we have

$$\lambda_i(A) \ll \frac{N_0}{2} \tag{135}$$

for all $A$. When this inequality holds, we construct the approximate likelihood function by retaining the first term and the average value of the second term in (133) and the first two terms in (134). (See discussion on page 133.) This gives

$$\ln \Lambda(A) = \frac{1}{2}\left(\frac{2}{N_0}\right)^2 \iint_{T_i}^{T_f} r(t)K_s(t, u:A)r(u)\,dt\,du - \frac{1}{2}\left(\frac{2}{N_0}\right)\int_{T_i}^{T_f} K_s(t, t:A)\,dt - \frac{1}{4}\left(\frac{2}{N_0}\right)^2 \int_{T_i}^{T_f} K_s^2(t, u:A)\,dt\,du. \tag{136}$$

To find $\hat{a}_{ml}$ we must construct $\ln \Lambda(A)$ as a function of $A$ and choose the value of $A$ where it is a maximum.

The lower bound on the variance of any unbiased estimate is

$$\operatorname{Var}[\hat{a} - A] \geq \frac{\left(\frac{N_0}{2}\right)^2}{\frac{1}{2}\iint_{T_i}^{T_f}\left[\frac{\partial K_s(t, u:A)}{\partial A}\right]^2 dt\,du} \tag{137}$$

If $A$ is the value of a random parameter, we obtain the MAP estimate by adding $\ln p_a(A)$ to (136) and finding the maximum of the over-all function. To illustrate the simplicity caused by the LEC condition, we consider two simple examples.

**Example 6.** In this example, we want to estimate the amplitude of the correlation function of a random process,

$$K_r(t, u:A) = AK(t, u) + \frac{N_0}{2}\delta(t - u), \tag{138}$$

where $K(t, u)$ is a known covariance function. Assuming that the LEC condition is satisfied, we may use (136) to obtain

$$\ln \Lambda(A) = +\frac{1}{2}\left(\frac{2}{N_0}\right)^2 \left\{A \iint_{T_i}^{T_f} r(t)K(t, u)r(u)\,dt\,du - \frac{N_0}{2}A\int_{T_i}^{T_f} K(t, t)\,dt - \frac{A^2}{2}\iint_{T_i}^{T_f} [K(t, u)]^2\,dt\,du\right\}. \tag{139}$$

† This particular problem has been solved by Price [1] and Middleton [4].

Differentiating and equating to zero, we obtain

$$\hat{a}_0 = \frac{\displaystyle\iint_{T_i}^{T_f} r(t)K(t, u)r(u)\, dt\, du - \frac{N_0}{2}\int_{T_i}^{T_f} K(t, t)\, dt}{\displaystyle\iint_{T_i}^{T_f} [K(t, u)]^2\, dt\, du} \tag{140}$$

As before,

$$\hat{a}_{ml} = \begin{cases} \hat{a}_0, & \hat{a}_0 \geq 0, \\ 0, & \hat{a}_0 < 0. \end{cases} \tag{141}$$

We can obtain an upper bound on the bias of $\hat{a}_{ml}$ by using the techniques on pages 198–201. The lower bound on the variance on any unbiased estimate is

$$\frac{\text{Var}\,[\hat{a} - A]}{A^2} \geq \frac{2(N_0/2)^2}{A^2 \displaystyle\int_{T_i}^{T_f} [K(t, u)]^2\, dt\, du}. \tag{142}$$

We see that the right side of (142) is the reciprocal of $d^2$ in the LEC detection problem [see (4.148)]. We might expect this relation in view of the results in Chapter I-4 for amplitude estimation in the known signal case.

All our examples have considered the estimation of nonrandom parameters. We indicated that the modification to include random parameters was straightforward. In the next example we illustrate the modifications.

**Example 7.** We assume that the covariance function of the received signal is given by (138). Now we model $A$ as the value of a random variable $a$. In general the probability density is not known, and so we choose a density with several free parameters. We then vary the parameters in order to match the available experimental evidence. As we discussed on page I-142, we frequently choose a reproducing density for the a-priori density because of the resulting computational simplicity. For this example, a suitable a-priori density is the Gamma probability density,

$$p_a(A) = \begin{cases} \dfrac{\lambda^{n+1}}{n!} A^n e^{-\lambda A}, & A \geq 0, \\ 0, & A < 0, \end{cases} \tag{143}$$

where $\lambda$ is a positive constant and $n$ is a positive integer that we choose based on our a-priori information about $a$. In our subsequent discussion we assume that $n$ and $\lambda$ are known. We want to find $\hat{a}_{\text{map}}$. To do this we construct the function

$$f(A) = \ln \Lambda(A) + \ln p_a(A) \tag{144}$$

and find its maximum. Substituting (139) and (143) into (144) and collecting terms, we have

$$f(A) = -c_1 A^2 + f(r(t))A + n \ln (\lambda A) + \ln \lambda, \qquad A \geq 0, \tag{145}$$

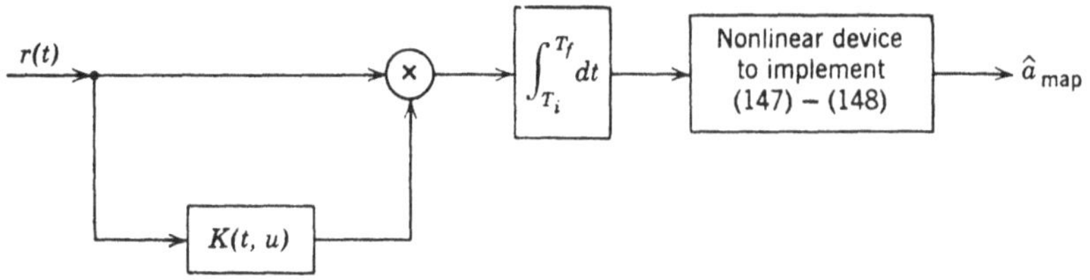


**Fig. 7.10 Realization of *MAP* amplitude estimator under *LEC* conditions.**

where

$$c_1 = \left(\frac{1}{N_0}\right)^2 \iint_{T_i}^{T_f} [K(t, u)]^2 \, dt \, du \tag{146}$$

and

$$f(r(t)) = \frac{1}{2}\left(\frac{2}{N_0}\right)^2 \left[\iint_{T_i}^{T_f} r(t)K(t, u)r(u) \, dt \, du - \frac{N_0}{2}\int_{T_i}^{T_f} K(t, t) \, dt\right] - \lambda \tag{147}$$

Differentiating $f(A)$ with respect to $A$, equating the result to zero, and solving the resulting quadratic equation, we have

$$\hat{a}_{map} = \begin{cases} \dfrac{f(r(t))}{4c_1}\left[\left(1 + \dfrac{8c_1 n}{[f(r(t))]^2\lambda}\right)^{1/2} + 1\right], & f(r(t)) \geq 0, \\[2ex] -\dfrac{f(r(t))}{4c_1}\left[\left(1 + \dfrac{8c_1 n}{[f(r(t))]^2\lambda}\right)^{1/2} - 1\right], & f(r(t)) < 0. \end{cases} \tag{148}$$

The second derivative of $f(A)$ is always negative and $f(0) = -\infty$, so that this is a unique maximum. The receiver is shown in Fig. 7.10. We see that the receiver carries out two operations in cascade. The first section computes

$$l_*(r(t)) = \iint_{T_i}^{T_f} r(t)K(t, u)r(u) \, dt \, du. \tag{149}$$

The quadratic operation is familiar from our earlier work. The second section is a nonlinear, no-memory device that implements (147) and (148). The calculation of the performance is difficult because of the nonlinear, no-memory operation, which is not a quadratic operation.

Several other examples of parameter estimation under LEC conditions are developed in the problems. This completes our discussion of special categories of estimation problems. In the next section, we discuss some related topics.

## 7.5 RELATED TOPICS

In this section we discuss two topics that are related to the parameter estimation problem we have been studying. In Section 7.5.1, we discuss the multiple-parameter estimation problem. In Section 7.5.2, we discuss generalized likelihood ratio tests.

### 7.5.1 Multiple-Parameter Estimation

As we would expect from our earlier work (e.g., Section I-4.6), the basic results for the single-parameter case can be extended easily to include the multiple-parameter case. We state the model and some of the more important results. The received waveform is

$$r(t) = s(t, \mathbf{A}) + w(t), \qquad T_i \le t \le T_f, \tag{150}$$

where $\mathbf{A}$ is an $M$-dimensional parameter vector. The signal $s(t, \mathbf{A})$ is a sample function from a Gaussian random process whose statistics depend on $\mathbf{A}$,

$$E[s(t, \mathbf{A})] = m(t, \mathbf{A}), \qquad T_i \le t \le T_f, \tag{151}$$

and

$$E[(s(t, \mathbf{A}) - m(t, \mathbf{A}))(s(u, \mathbf{A}) - m(u, \mathbf{A}))] = K_s(t, u:\mathbf{A}), \quad T_i \le t, u \le T_f. \tag{152}$$

The additive noise $w(t)$ is a sample function from a zero-mean white Gaussian process with spectral height $N_0/2$. We are interested both in the case in which $\mathbf{A}$ is a nonrandom vector and in the case in which $\mathbf{A}$ is a value of a random vector.

***Nonrandom Parameters.*** We assume that $A$ is a unknown nonrandom vector that lies in the set $\chi_a$. The likelihood function is

$$\begin{aligned} \ln \Lambda(\mathbf{A}) = {} & \frac{1}{N_0} \int_{T_i}^{T_f} dt \int_{T_i}^{T_f} du\, r(t) h(t, u:\mathbf{A}) r(u) \\ & + \int_{T_i}^{T_f} dt \int_{T_i}^{T_f} du\, r(t) Q_r(t, u:\mathbf{A}) m(u, \mathbf{A}) \\ & - \frac{1}{2} \int_{T_i}^{T_f} dt \int_{T_i}^{T_f} du\, m(t, \mathbf{A}) Q_r(t, u:\mathbf{A}) m(u, \mathbf{A}) \\ & - \frac{1}{2} \sum_{i=1}^{\infty} \ln\left(1 + \frac{2}{N_0} \lambda_i(\mathbf{A})\right), \qquad \mathbf{A} \in \chi_a. \end{aligned} \tag{153}$$

The ML estimate, $\hat{a}_{ml}$, is that value of $\mathbf{A}$ where this function is a maximum. In general, we must construct this function for some set of $\mathbf{A}_i$ that span the set $\chi_a$ and choose the value of $\mathbf{A}_i$ where $\ln \Lambda(\mathbf{A}_i)$ is the largest. If the maximum of $\ln \Lambda(\mathbf{A})$ is interior to $\chi_a$ and $\ln \Lambda(\mathbf{A})$ has a continuous first derivative, a necessary condition is given by the $M$ likelihood equations,

$$\begin{aligned} \left.\frac{\partial \ln \Lambda(\mathbf{A})}{\partial A_i}\right|_{\mathbf{A}=\hat{\mathbf{a}}_{ml}} = \Bigg\{ & \frac{1}{2} \iint_{T_i}^{T_f} K_r(t, u:\mathbf{A}) \frac{\partial Q_r(t, u:\mathbf{A})}{\partial A_i}\, dt\, du \\ & + \iint_{T_i}^{T_f} \frac{\partial m(t, \mathbf{A})}{\partial A_i} Q_r(t, u:\mathbf{A})[r(u) - m(u, \mathbf{A})]\, dt\, du \\ & - \frac{1}{2} \iint_{T_i}^{T_f} [r(t) - m(t, \mathbf{A})] \frac{\partial Q_r(t, u:\mathbf{A})}{\partial A_i} \\ & \times [r(u) - m(u, \mathbf{A})]\, dt\, du \Bigg\}\Bigg|_{\mathbf{A}=\hat{\mathbf{a}}_{ml}} = 0, \\ & i = 1, 2, \dots, M. \end{aligned} \tag{154}$$

The elements in the information matrix are

$$\begin{aligned} J_{ij}(\mathbf{A}) = \iint_{T_i}^{T_f} \Bigg[ & \frac{\partial m(t, \mathbf{A})}{\partial A_i} Q_r(t, u:\mathbf{A}) \frac{\partial m(u, \mathbf{A})}{\partial A_j} \\ & - \frac{1}{2} \frac{\partial K_r(t, u:\mathbf{A})}{\partial A_i} \frac{\partial Q_r(t, u:\mathbf{A})}{\partial A_j} \Bigg] dt\, du. \end{aligned} \tag{155}$$

The information matrix is used in exactly the same manner as in Section I-4.6. The first term is analogous to $d^2(\mathbf{A})$ and can be computed in a straightforward manner. The second term, $J_{ij}^{(2)}(\mathbf{A})$, is the one that requires some work.

We can also express $J_{ij}^{(2)}(A)$ in terms of the derivative of the Bhattacharyya distance.

$$B(\mathbf{A}_1, \mathbf{A}) = -\mu(\tfrac{1}{2}, \mathbf{A}_1, \mathbf{A}) \triangleq -\ln \int_{-\infty}^{\infty} p_{\mathbf{r}|\mathbf{a}}^{1/2}(\mathbf{R} \mid \mathbf{A}_1) p_{\mathbf{r}|\mathbf{a}}^{1/2}(\mathbf{R} \mid \mathbf{A})\, d\mathbf{R}. \tag{156}$$

Then, using a derivation similar to (6.47)–(6.60), we obtain

$$J_{ij}^{(2)}(\mathbf{A}) = 4\left(\left.\frac{\partial^2 B(\mathbf{A}_1, \mathbf{A})}{\partial A_{1i}\, \partial A_{1j}}\right|_{\mathbf{A}_1=\mathbf{A}}\right). \tag{157}$$

The expression for $B(\mathbf{A}_1, \mathbf{A})$ is an obvious modification of (6.59). This formula provides an effective procedure for computing $J_{ij}^{(2)}(\mathbf{A})$ numerically.

(Notice that the numerical calculation of the second derivative must be done carefully.)

For stationary processes and long time intervals, the second term in (155) has a simple form,

$$J_{ij}^{(2)}(\mathbf{A}) = \frac{T}{2}\int_{-\infty}^{\infty}\frac{d\omega}{2\pi}\left[\frac{\partial}{\partial A_i}\left(\ln\left[S_s(\omega, \mathbf{A}) + \frac{N_0}{2}\right]\right)\right] \times \left[\frac{\partial}{\partial A_j}\left(\ln\left[S_s(\omega, \mathbf{A}) + \frac{N_0}{2}\right]\right)\right]. \quad (158)$$

The results in (153)–(158) indicate how the single-parameter formulas are modified to study the multiple-parameter problem. Just as in the single parameter case, the realization of the estimator depends on the specific problem.

***Random Parameters.*** For the general random parameter case, the results are obtained by appropriately modifying those in the preceding section. A specific case of interest in Chapter II-8 is the case in which the parameters to be estimated are independent, zero-mean Gaussian random variables with variances $\sigma_{a_i}^2$. The MAP equations are

$$\hat{a}_i = \sigma_{a_i}^2\left\{\int_{T_i}^{T_f} dt \int_{T_i}^{T_f} du \left[\tfrac{1}{2}K_r(t, u:\mathbf{A})\frac{\partial Q_r(t, u:\mathbf{A})}{\partial A_i} + [r(u) - m(u, \mathbf{A})] \times \left[\frac{\partial m(t, \mathbf{A})}{\partial A_i} Q_r(t, u:\mathbf{A}) - \tfrac{1}{2}[r(t) - m(t, \mathbf{A})]\frac{\partial Q_r(t, u:\mathbf{A})}{\partial A_i}\right]\right]_{\mathbf{A}=\hat{\mathbf{a}}_{map}}\right\},$$

$$i = 1, 2, \ldots, M. \quad (159)$$

The terms in the information matrix are

$$J_{ij} = \frac{\delta_{ij}}{\sigma_{a_i}^2} + E_{\mathbf{a}}[J_{ij}^{(1)}(\mathbf{a}) + J_{ij}^{(2)}(\mathbf{a})], \quad (160)$$

where $J_{ij}^{(1)}(\mathbf{a})$ and $J_{ij}^{(2)}(\mathbf{a})$ are the two terms in (155). Notice that $J_{ij}$ contains an average over $p_{\mathbf{a}}(\mathbf{A})$, so that the final result does not depend on $\mathbf{A}$.

Several joint parameter estimation examples are developed in the problems.

### 7.5.2 Composite-hypothesis Tests

In some detection problems the signals contain unknown nonrandom parameters. We can write the received waveforms on the two hypotheses as

$$r(t) = s_1(t, \boldsymbol{\theta}) + w(t), \qquad T_i \le t \le T_f : H_1,$$

$$r(t) = s_0(t, \boldsymbol{\theta}) + w(t), \qquad T_i \le t \le T_f : H_0, \quad (161)$$

where $s_1(t, \boldsymbol{\theta})$ and $s_0(t, \boldsymbol{\theta})$ are conditionally Gaussian processes. This model is a generalization of the classical model in Section I-2.5. Usually a uniformly most powerful test does not exist, and so we use a generalized likelihood ratio test. This test is just a generalization of that in (I-2.305). We use the procedures of Chapters 6 and 7 to find $\hat{\boldsymbol{\theta}}_{ml}$. We then use this value as if it were correct in the likelihood ratio test of Chapter 4. Although the extension is conceptually straightforward, the actual calculations are usually quite complicated. It is difficult to make any general statements about the performance. Some typical examples are developed in the problems.

## 7.6 SUMMARY OF ESTIMATION THEORY

In Chapters 6 and 7, we have studied the problem of estimating the parameters of a Gaussian random process in the presence of additive Gaussian noise. As in earlier estimation problems, the first step was to construct the log likelihood function. For our model,

$$\ln \Lambda(A) = \frac{1}{N_0} \iint\limits_{T_i}^{T_f} r(t)h(t, u:A)r(u)\, dt\, du + \int_{T_i}^{T_f} r(t)g(t, A)\, dt - \frac{1}{N_0} \int_{T_i}^{T_f} \xi_P(t:A)\, dt - \frac{1}{2} \int_{T_i}^{T_f} m(t, A)g(t, A)\, dt. \quad (161)$$

In order to find the maximum likelihood estimate, we construct $\ln \Lambda(A)$ as a function of $A$. In practice it is usually necessary to construct a discrete approximation by computing $\ln \Lambda(A_i)$ for a set of values than span $\chi_a$.

In order to analyze the performance, we derived a lower bound of the variance of any unbiased estimate. The bound is

$$\operatorname{Var}[\hat{a} - A] \geq \left\{ \iint\limits_{T_i}^{T_f} \frac{\partial m(t, A)}{\partial A} Q_r(t, u:A) \frac{\partial m(u, A)}{\partial A}\, dt\, du - \frac{1}{2} \iint\limits_{T_i}^{T_f} \frac{\partial K_r(t, u:A)}{\partial A} \frac{\partial Q_r(t, u:A)}{\partial A}\, dt\, du \right\}^{-1} \quad (162)$$

for any unbiased estimate. In most problems we must evaluate the bound using numerical techniques. Since the estimates of the process parameters are usually not efficient, the bound in (162) may not give an accurate indication of the actual variance. In addition, the estimate may have a bias that we cannot evaluate, so that we cannot use (162) or the generalization of it derived in Problem 6.3.1. We pointed out that other bounds, such as the Barankin bound and the Kotelnikov bound, were available but did not discuss them in detail.

The discussion in Chapter 6 provided the general theory needed to study the parameter estimation problem. An equally important topic was the application of this theory to a particular problem in order actually to obtain the estimate and evaluate its performance.

In Chapter 7 we illustrate the transition from theory to practice for a particular problem. In Section 7.1 we studied the problem of estimating the mean-square value of a stationary random process. After finding expressions for the likelihood function, we considered some limiting cases in which we could generate $\hat{a}_{ml}$ easily. We encountered the issue of a truncated estimate and developed new techniques for computing the bias and mean-square error. Finally, we looked at some suboptimum estimation procedures. This section illustrated a number of the issues that one encounters and must resolve in a typical estimation problem. In Sections 7.2–7.4, we considered finite-state processes, separable kernel processes, and the low-energy-coherence problem, respectively. In all of these special cases we could solve the necessary integral equation and generate $\ln \Lambda(A)$ explicitly. It is important to emphasize that, even after we have solved the integral equation, we usually still have to construct $\ln \Lambda(A_i)$ for a set of values that span $\chi_a$. Thus, the estimation problem is appreciably more complicated than the detection problem.

We have indicated some typical estimation problems in the text and in the problem section. References dealing with various facets of parameter estimation include [5]–[15].

This chapter completes our work on detection and estimation of Gaussian processes. Having already studied the modulation theory problem in Chapter II-8, we have now completed the hierarchy of problems that we outlined in Chapter I-1. The remainder of the book deals with the application of this theory to the radar-sonar problem.

## 7.7 PROBLEMS

### P.7.1 Stationary Processes: Long Observation Time

**Problem 7.1.1.** Consider the FSI estimator in Fig. 7.3. The filter $h_\alpha^{1/2}(\tau)$ and the parameters $G$ and $B$ are arbitrary subject to the constraint

$$E[\hat{a}_0] = A. \tag{P.1}$$

1. Derive an exact expression for

$$\xi_{\hat{a}_0} \triangleq E[(\hat{a}_0 - A)^2].$$

2. Consider the condition in (26) and (27) and assume that (31), (37), and (38) are satisfied. Denote the Cramér-Rao bound in (25) as $\xi_{\mathrm{CR}}$. Prove

$$\lim_{(N_0/2AS(\omega))\to 0} [\xi_{\hat{a}_0}/A^2 - \xi_{\mathrm{CR}}] = 0.$$

**Problem 7.1.2.** The function $\mu(s, A)$ is defined in (40), and $B$ satisfies (43). Prove that (66) has a solution in the allowable range. A possible procedure is the following:

(i) Evaluate $\dot{\mu}(0, A)$ and $\dot{\mu}(-\infty, A)$.

(ii) Prove that this guarantees a solution to (66) for some $s < 0$.

(iii) Use the fact that $\ddot{\mu}(s, A) > 0$ to prove the desired result.

**Problem 7.1.3.** Assume

$$S(\omega) = \begin{cases} \dfrac{1}{2W}, & |\omega| \le 2\pi W, \\ 0, & \text{elsewhere.} \end{cases}$$

Then, from (47),

$$\mu(s, A) = -WT \ln\left[1 - 2sA\left(1 + \frac{N_0 W}{A}\right)\right].$$

1. Solve (66) for $s_2$.
2. Verify the results in (68) and (69).

**Problem 7.1.4.**

1. Modify the derivation of bounds 1 and 2 to incorporate (52).
2. Compare your results with those in Fig. 7.6.

**Problem 7.1.5** [2].

1. Consider the expression for $\xi_*$ given in (73). Use the same procedure as in (60)–(69) to obtain a bound on $\xi_*$.
2. Modify the derivation in part 1 to incorporate (52).

**Problem 7.1.6.** Recall the result in part 1 of Problem 7.1.1 and consider the receiver in Example 3.

1. Show that $\xi_{\hat{a}_0}$ approaches the Cramér-Rao bound as $WT \to \infty$.
2. Investigate the bias and the mean-square error in the ML estimate.

**Problem 7.1.7.** Recall the result in part 1 of Problem 7.1.1. Substitute (96)–(98) into this formula and verify that (99) is true.

**Problem 7.1.8.** Consider the model in Example 4. Verify that the result in (102) is true.

**Problem 7.1.9.** Consider the result in (99). Assume

$$S(\omega) = \begin{cases} \dfrac{1}{2W}, & |\omega| \le 2\pi W, \\ 0, & \text{elsewhere.} \end{cases}$$

1. Evaluate (99) to obtain $\xi_{\hat{a}_0}$ as a function of $C$.
2. Check the two limiting cases of Examples 2 and 3 as $C \to \infty$ and $C \to 0$.
3. Plot $\xi_{\hat{a}_0}$ as a function of $C$ for various values of $A/N_0W$ and $WT$.

**Problem 7.1.10.** Carry out the details of the bias calculation for the model in Example 4.

**Problem 7.1.11.** Assume that $s(t, A)$ is a Wiener process,

$$E[s^2(t)] = At,$$

and that the SPLOT assumption may be used.

1. Evaluate the Cramér-Rao bound by using (25).
2. Consider the suboptimum receiver described by (96)–(98). Evaluate $\xi_{\hat{a}_0}$ in (99). Plot $\xi_{\hat{a}_0}$ as a function of $C$.
3. Calculate a bound on Pr $[\hat{a}_0 < 0]$.

**Problem 7.1.12.** Consider the model in (1)–(5). Assume that

$$K_s(\tau{:}A) = e^{-A|\tau|}, \qquad -\infty < \tau < \infty,$$

where $A$ is a nonrandom positive parameter.

1. Draw a block diagram of a receiver to find $\hat{a}_{ml}$.
2. Evaluate the Cramér-Rao bound in (16).

**Problem 7.1.13.** Consider the model in (1)–(5). Assume that $s(t, A)$ is a bandpass process whose spectrum is

$$S_s(\omega{:}A) = S_{s,\mathrm{LP}}(-\omega - A) + S_{s,\mathrm{LP}}(\omega - A),$$

where $S_{s,\mathrm{LP}}(\omega)$ is a known low-pass spectrum.

1. Draw a block diagram of a receiver to find $\hat{a}_{ml}$.
2. Evaluate the Cramér-Rao bound in (16).
3. Specialize the result in part 2 to the case in which

$$S_{s,\mathrm{LP}}(\omega) = \frac{2kP}{\omega^2 + k^2}.$$

**Problem 7.1.14** [5]. Suppose that

$$s(t, A) = c_1[s(t) + c_2 s(t - A)],$$

where $c_1$ and $c_2$ are known constants and $s(t)$ is a sample function from a stationary random process. Evaluate the Cramér-Rao bound in (16).

**Problem 7.1.15.** In the text we assumed that $s(t, A)$ was a zero-mean process. In this problem we remove that restriction.

1. Derive the SPLOT version of (6.16)–(6.26).
2. Derive the SPLOT version of (6.45) and (6.46).

**Problem 7.1.16.** Consider the system shown in Fig. P.7.1. The input $s(t)$ is a sample function of a zero-mean Gaussian random process with spectrum $S(\omega)$. The additive noise $w(t)$ is a sample function of a white Gaussian random process with spectral height $N_0/2$. We observe $r(t)$, $T_i \le t \le T_f$, and want to find $\hat{a}_{ml}$.

1. Draw a block diagram of the optimum receiver.
2. Write an expression for the Cramér-Rao bound. Denote the variance given by this bound as $\xi_{\mathrm{CR}}$.

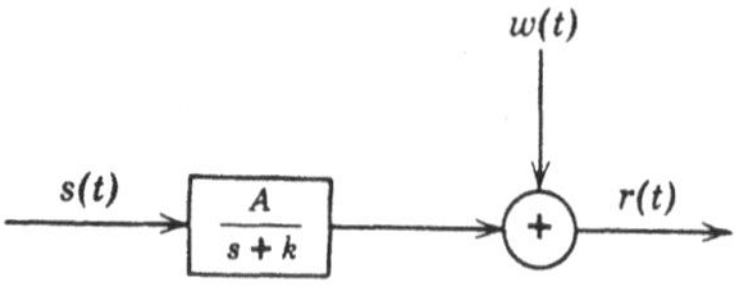


**Fig. P.7.1**

3. Constrain

$$\int_{-\infty}^{\infty} S(\omega)\frac{d\omega}{2\pi} = P.$$

Choose $S(\omega)$ to minimize $\xi_{CR}$.

**Problem 7.1.17.** The received waveform is

$$r(t) = s(t, A) + n_c(t), \qquad T_i \le t \le T_f. \tag{P.1}$$

The additive noise $n_c(t)$ is a sample function of a zero-mean, finite-power, stationary Gaussian noise process with spectrum $S_c(\omega)$. Notice that there is no white noise component.

1. Derive an expression for the Cramér-Rao bound.
2. Discuss the question of singular estimation problems. In particular, consider the case in which

$$S(\omega, A) = AS(\omega). \tag{P.2}$$

3. Assume that (P.1) and (P.2) hold and

$$\int_{-\infty}^{\infty} S_c(\omega)\frac{d\omega}{2\pi} = P_c. \tag{P.3}$$

Choose $S_c(\omega)$ to maximize the Cramér-Rao bound.

**Problem 7.1.18.** The received waveform is

$$\mathbf{r}(t) = \mathbf{s}(t, A) + \mathbf{w}(t), \qquad T_i \le t \le T_f.$$

Assume that the SPLOT conditions are valid.

1. Derive an expression for $\ln \Lambda(A)$.
2. Derive an expression for the Cramér-Rao bound.

**Problem 7.1.19.** Consider the two-element receiver shown in Fig. P.7.2. The signal $s(t)$ is a sample function of a zero-mean stationary Gaussian random process with spectrum $S(\omega)$. It propagates along a line that is $\alpha$ radians from the $y$-axis. The received signals at the two elements are

$$r_1(t) = s(t) + w_1(t), \qquad T_i \le t \le T_f,$$

$$r_2(t) = s\left(t - \frac{L \sin \alpha}{c}\right) + w_2(t), \qquad T_i \le t \le T_f,$$

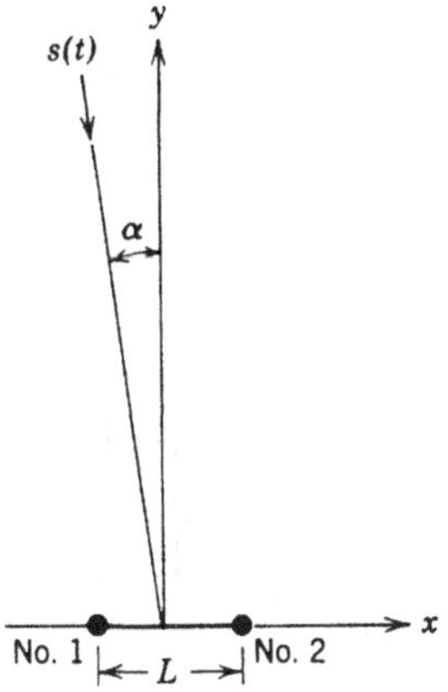


**Fig. P.7.2**

where $c$ is the velocity of propagation. The additive noises $w_1(t)$ and $w_2(t)$ are statistically independent white noises with spectral height $N_0/2$. Assume that $|\alpha| \leq \pi/8$ and that $S(\omega)$ is known.

1. Draw a block diagram of a receiver to generate $\hat{\alpha}_{ml}$.
2. Write an expression for the Cramér-Rao bound.
3. Evaluate the bound in part 2 for the case in which

$$S(\omega) = \begin{cases} \dfrac{P}{2W}, & |\omega| \leq 2\pi W, \\ 0, & \text{elsewhere.} \end{cases}$$

4. Discuss various suboptimum receiver configurations and their performance.
5. One procedure for estimating $\alpha$ is to compute

$$\varphi(\tau) = \frac{1}{T_f - T_i} \int_{T_i}^{T_f} r_1(t) r_2(t - \tau)\, dt$$

as a function of $\tau$ and find the value of $\tau$ where $\varphi(\tau)$ has its maximum. Denote this point as $\hat{\tau}_*$. Then define

$$\hat{\alpha}_* \triangleq \frac{c\hat{\tau}_*}{L}.$$

Discuss the rationale for this procedure. Compare its performance with the Cramér-Rao bound.

**Problem 7.1.20.** Consider the problem of estimating the height of the spectrum of a random process $s(t)$ at a particular frequency. A typical estimator is shown in Fig. P.7.3.

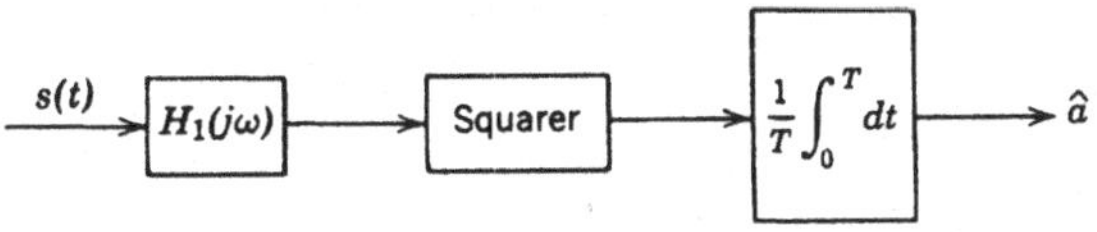


**Fig. P.7.3**

Denote the particular point in the spectrum that we want to estimate as

$$A \triangleq S(\omega_1). \tag{P.1}$$

The filter $H_1(j\omega)$ is an ideal bandpass filter centered at $\omega_1$, as shown in Fig. P.7.4.

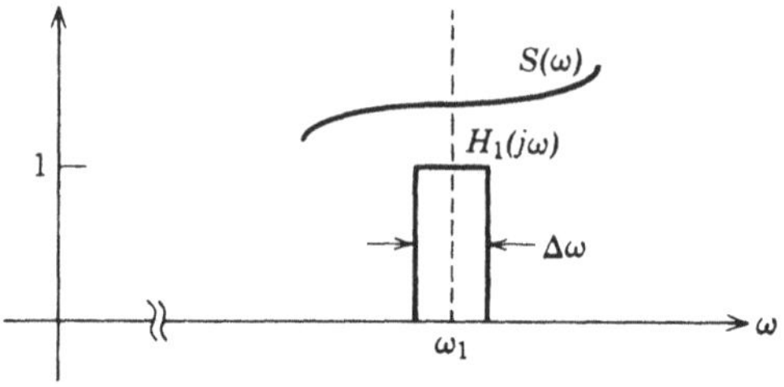


**Fig. P.7.4**

1. Compute the bias in $\hat{a}$. Under what conditions will this bias be negligible?
2. Assume that the bias is negligible. Compute the normalized variance as a function of the various parameters.
3. Demonstrate a choice of parameters such that the normalized variance goes to zero as $T \to \infty$
4. Demonstrate a choice of parameters such that the normalized variance goes to two as $T \to \infty$.

*Comment:* This problem illustrates some of the issues in power spectrum estimation. The interested reader should consult [6.1]–[6.3] for a detailed discussion.

**Problem 7.1.21.** Assume that

$$S(\omega : A) = AS(\omega).$$

Denote the expression in the Cramér-Rao bound of (25) as $\xi_{CR}$.

1. Evaluate $\xi_{CR}$ for the case

$$S(\omega) = \frac{2nP}{k} \frac{\sin(\pi/2n)}{(\omega/k)^{2n} + 1}.$$

2. Assume

$$\int_{-\infty}^{\infty} S(\omega) \frac{d\omega}{2\pi} = P.$$

Find the spectrum $S(\omega)$ that minimizes $\xi_{CR}$.

**Problem 7.1.22.** Assume that

$$S(\omega : A) = \frac{2AP}{k} \frac{\sin(\pi/2A)}{(\omega/k)^{2A} + 1},$$

where $A$ is a positive integer.

1. Find a receiver to generate $\hat{a}_{ml}$.
2. Find a lower bound on the variance of any unbiased estimate of $A$.

**Problem 7.1.23.** Consider the estimation problem described in Problem 7.1.13. Assume

$$A_\alpha < A < A_\beta,$$

$$S(\omega) = \begin{cases} \dfrac{P}{2W}, & |\omega| \le 2\pi W, \\ 0, & \text{elsewhere,} \end{cases}$$

and

$$W = \frac{A_\alpha - A_\beta}{M}.$$

1. Draw a block diagram of a receiver to generate $\hat{a}_{ml}$.
2. Is the Cramér-Rao bound valid in this problem?
3. Use the techniques on pages I-278–I-284 and the results of Section 5.1 to evaluate the receiver performance. How can you analyze the weak noise performance (local errors)?

## P.7.2 Finite-state Processes

**Problem 7.2.1.** The signal process is a Wiener process,

$$E[s^2(t)] = At.$$

We want to find $\hat{a}_{ml}$.

1. Draw a block diagram of the receiver. Specify all components (including initial conditions) completely.
2. Verify that

$$\xi_P(t, A) = \left(\frac{AN_0}{2}\right)^{1/2} \left[\frac{1 - \exp\left[-2(2A/N_0)^{1/2}t\right]}{1 + \exp\left[-2(2A/N_0)^{1/2}t\right]}\right].$$

3. Use the result in part 2 to compute $J^{(2)}(A)$ in (118).
4. Plot $J^{(2)}(A)/A^2$ as a function of $2AT^2/N_0$.

**Problem 7.2.2.** Assume

$$s(t, A) = As(t),$$

where $s(t)$ has a known finite-dimensional state representation. Consider the FSI receiver in Fig. 7.4. Assume that $h_\alpha^{[1/2]}(\tau)$ has a finite-dimensional state representation. Derive a differential equation specifying $\xi_{\hat{a}_0}$.

**Problem 7.2.3.** Consider the model in (107)–(111). Assume that $\mathbf{F}(t, A)$, $\mathbf{G}(t\, A)$, $\mathbf{Q}(A)$, and $\mathbf{P}_0(A)$ are not functions of $A$.

$$\mathbf{C}(t, A) = f(t - A)\mathbf{C},$$

where $f(t)$ is only nonzero in $[\alpha < t < \beta]$ and $\alpha - A$ and $\beta - A$ are in $[T_i, T_f]$. We want to make a maximum-likelihood estimate of $A$.

1. Draw a block diagram of the optimum receiver.
2. Write an expression for the Cramér-Rao bound.

**Problem 7.2.4.** Consider the model in (107)–(111). Assume

$$\mathbf{F}(t, A) = -k,$$

$$\mathbf{G}(t, A) = 1,$$

$$\mathbf{C}(t, A) = 1,$$

$$\mathbf{Q}(A) = 2kP,$$

and

$$P_0(A) = A.$$

We want to make a maximum-likelihood estimate of $A$.

1. Draw a block diagram of a receiver to generate $\hat{a}_{ml}$.

2. Discuss various suboptimum configurations. (*Hint:* What segment of the received waveform contains most of the information about $A$?)

3. Write an expression for the Cramér-Rao bound.

## P.7.3 Separable Kernels

**Problem 7.3.1.** Consider the model described in Example 5. Assume

$$\lambda_i \gg \frac{N_0}{2A} \quad \text{for all} \quad A \text{ in } \chi_a.$$

1. Find an expression for $\hat{a}_0$.
2. Derive an expression for Pr $[\hat{a}_0 < 0]$.
3. Derive an expression for $\xi_{\hat{a}_0}$.
4. Compare the expression in part 3 with the Cramér-Rao bound.

**Problem 7.3.2.** Repeat Problem 7.3.1 for the case in which

$$\lambda_i \ll \frac{N_0}{2A} \quad \text{for all} \quad A \text{ in } \chi_a.$$

**Problem 7.3.3.** Consider the model in (121)–(127) and assume that the equal eigenvalue condition in (128) is valid.

1. Calculate

$$\xi_{\hat{a}_0} \triangleq E[(\hat{a}_0 - A)^2].$$

2. Compute Pr $[\hat{a}_0 < 0]$ by using (41).

3. Assume

$$\sum_{i=1}^{K} \lambda_i = \bar{E}_i.$$

Choose $K$ to minimize $\xi_{a_0}$. Compare your results with those in (4.76) and (4.116).

4. Calculate $p_{\hat{a}_0}(A_0)$ exactly.

5. Evaluate Pr $[\hat{a}_0 < 0]$ using the result in part 4 and compare it with the result in part 2.

**Problem 7.3.4.** Consider the model in (120), but assume that

$$K_s(t, u:A) = \lambda_c(A) \sum_{i=1}^{K} \phi_i(t)\phi_i(u).$$

Assume that $\lambda_c^{-1}(A)$ exists.

1. Draw a block diagram of the optimum receiver to generate $\hat{a}_{ml}$.
2. Derive an expression for the Cramér-Rao bound.
3. What difficulty arises when you try to compute the performance exactly?

**Problem 7.3.5.** Consider the model in Problem 7.3.4. Let

$$\lambda_c(A) = \frac{1}{A}.$$

Assume that $A$ is the value of a random variable whose a priori density is

$$p_a(A \mid k_1, k_2) = c(A^{k_1/2-1}) \exp(-\tfrac{1}{2}Ak_1k_2), \qquad A \geq 0, \qquad k_1, k_2 > 0,$$

where $c$ is a normalization constant

1. Find $p_{a|r(t)}(A \mid r(t))$.
2. Find $\hat{a}_{ms}$.
3. Compute $E[(\hat{a}_{ms} - a)^2]$.

**Problem 7.3.6.** Consider the model in (120). Assume that

$$K_s(t, u:A) = \sum_{i=1}^{K} \lambda_i \varphi_i(t, A)\varphi_i(u, A).$$

Assume that the $\varphi_i(t, A)$ all have the same shape. The orthogonality is obtained by either time or frequency diversity (see Section 4.3 for examples).

1. Draw a block diagram of a receiver to generate $\hat{a}_{ml}$.
2. Evaluate the Cramér-Rao bound.

## P.7.4 Low-energy Coherence

**Problem 7.4.1.** Assume that both the LEC condition and the SPLOT condition are satisfied.

1. Derive the SPLOT version of (136).
2. Derive the SPLOT version of (137).

**Problem 7.4.2.** Consider the model in Example 6.

1. Evaluate (142) for the case in which

$$K(t, u) = e^{-\alpha|t-u|}. \tag{P.1}$$

2. Evaluate the SPLOT version of (142) for the covariance function in (P.1). Compare the results of parts 1 and 2.

3. Derive an expression for an upper bound on the bias. Evaluate it for the covariance function in (P.1).

**Problem 7.4.3.** Consider the model in Example 6. Assume that we use the LEC receiver in (140), even though the LEC condition may not be valid.

1. Prove that $\hat{a}_0$ is unbiased under all conditions.
2. Find an expression for $\xi_{\hat{a}_0}$.
3. Evaluate this expression for the covariance function in (P.1) of Problem 7.4.2. Compare your result with the result in part 1 of that problem.

**Problem 7.4.4.** Assume that

$$K_s(t, u:A) = Af(t)K_s(t - u)f(u)$$

and that the LEC condition is valid.

Draw a block diagram of the optimum receiver to generate $\hat{a}_{ml}$.

## P.7.5 Related Topics

**Problem 7.5.1.** Consider the estimation model in (150)–(158). Assume that

$$m(t, \mathbf{A}) = 0,$$

$$S_s(\omega, \mathbf{A}) = \frac{A_1}{\omega^2 + A_2{}^2},$$

and that the SPLOT assumption is valid:

1. Draw a block diagram of the ML estimator.
2. Evaluate $\mathbf{J}(\mathbf{A})$.
3. Compute a lower bound on the variance of any unbiased estimate of $A_1$.
4. Compute a lower bound on the variance of any unbiased estimate of $A_2$.
5. Compare the result in part 4 with that in Problem 7.1.12. What effect does the unknown amplitude have on the accuracy bounds for estimating $A_2$?

**Problem 7.5.2.** Consider the estimation model in (150)–(158). Assume that

$$m(t, \mathbf{A}) = 0$$

and

$$S_s(\omega, \mathbf{A}) = \frac{2k_1A_1}{\omega^2 + k_1{}^2} + \frac{2k_2A_2}{\omega^2 + k_2{}^2}, \qquad \text{(P.1)}$$

where $k_1$ and $k_2$ are known.

1. Draw a block of the ML estimator of $A_1$ and $A_2$.
2. Evaluate $\mathbf{J}(\mathbf{A})$.
3. Compute a lower bound on the variance of an unbiased estimate of $A_1$.
4. Compute a lower bound on the variance of an unbiased estimate of $A_2$.
5. Assume that $A_2$ is known. Compute a lower bound on the variance of any unbiased estimate of $A_1$. Compare this result with that in part 3.
6. Assume that the LEC condition is valid. Draw a block diagram of the optimum receiver.
7. Consider the behavior of the result in part 3 as $N_0 \to 0$.

**Problem 7.5.3.** Let

$$S(\omega:\mathbf{A}) = AS(\omega:\boldsymbol{\alpha}).$$

Assume that $S(\omega:\boldsymbol{\alpha})$ is bandlimited and that

$$AS(\omega:\boldsymbol{\alpha}) \gg N_0/2 \quad \text{for all } A \text{ and } \boldsymbol{\alpha}.$$

1. Assume that $\boldsymbol{\alpha}$ is fixed. Maximize $\ln \Lambda(\mathbf{A})$ over $A$ to find $\hat{a}_{ml}(\boldsymbol{\alpha})$, the ML estimate of $A$.

2. Substitute this result into $\ln \Lambda(\mathbf{A})$ to find $\ln \Lambda(\hat{a}_{ml}, \boldsymbol{\alpha})$. This function must be maximized to find $\hat{\boldsymbol{\alpha}}_{ml}$.

3. Assume that $\boldsymbol{\alpha}$ is a scalar $\alpha$. Find a lower bound on the variance of an unbiased estimate of $\alpha$. Compare this bound with the bound for the case in which $A$ is known. Under what conditions is the knowledge of $A$ unimportant?

**Problem 7.5.4.** Repeat Problem 7.5.3 for the case in which

$$AS(\omega:\boldsymbol{\alpha}) \ll N_0/2 \quad \text{for all } A \text{ and } \boldsymbol{\alpha}$$

and $S(\omega:\boldsymbol{\alpha})$ is not necessarily bandlimited.

**Problem 7.5.5.** Consider the model in Problem 7.5.2. Assume that

$$S_s(\omega, \mathbf{A}) = \frac{A_1}{\omega^2} + \frac{2k_2A_2}{\omega^2 + k_2^2}.$$

Repeat Problem 7.5.2.

**Problem 7.5.6.** Consider the model in Problem 7.5.2. Assume that

$$S_s(\omega, \mathbf{A}) = \frac{\omega^2 + A_1^2}{\omega^4 + A_2^4}.$$

Evaluate $J(\mathbf{A})$.

**Problem 7.5.7.** Derive the SPLOT versions of (153) and (154).

**Problem 7.5.8.** Consider the estimation model in (150)–(158). Assume that

$$m(t, \mathbf{A}) = A_1 m(t),$$

$$S_s(\omega, \mathbf{A}) = A_2 S(\omega).$$

1. Draw a block diagram of the receiver to generate the ML estimates of $A_1$ and $A_2$.
2. Evaluate $\mathbf{J}(\mathbf{A})$.
3. Compute a lower bound on the variance of any unbiased estimate of $A_1$.
4. Compute a lower bound on the variance of any unbiased estimate of $A_2$.

**Problem 7.5.9.** Consider the binary symmetric communication system described in Section 3.4. Assume that the spectra of the received waveform on the two hypotheses are

$$S_r(\omega) = A_1 S_1(\omega) + N_0/2 : H_1,$$

$$S_r(\omega) = A_0 S_0(\omega) + N_0/2 : H_0.$$

The hypotheses are equally likely and the criterion is minimum Pr ($\epsilon$). The parameters $A_0$ and $A_1$ are unknown nonrandom parameters. The spectra $S_1(\omega)$ and $S_0(\omega)$ are bandpass spectra that are essentially disjoint in frequency and are symmetric around their respective frequencies.

1. Derive a generalized likelihood ratio test.
2. Find an approximate expression for the Pr ($\epsilon$) of the test.

**Problem 7.5.10.** Consider the following generalization of Problem 7.5.4. The covariance function of $s(t, \mathbf{A})$ is

$$K_s(t, u:\mathbf{A}) = AK_s(t, u:\boldsymbol{\alpha}).$$

Assume that the LEC condition is valid.

1. Find $\ln \Lambda(\hat{a}_{ml}, \boldsymbol{\alpha})$. Use the vector generalization of (136) as a starting point.
2. Derive $\mathbf{J}(\boldsymbol{\alpha}; A)$, the information matrix for estimating $\boldsymbol{\alpha}$.

## REFERENCES


[1] R. Price, "Maximum-Likelihood Estimation of the Correlation Function of a Threshold Signal and Its Application to the Measurement of the Target Scattering Function in Radar Astronomy," Group Report 34-G-4, Massachusetts Institute of Technology, Lincoln Laboratory, May 1962.

[2] R. L. Adams, "Design and Performance of Some Practical Approximations to Maximum Likelihood Estimation," M.Sc. Thesis, Department of Electrical Engineering, Massachusetts Institute of Technology, February 1969.

[3] R. H. Dicke, R. Beringer, R. L. Kyhl, and A. B. Vane, "Atmospheric Absorption Measurements with a Microwave Radiometer," Phys. Rev., **70,** 340–348 (1946).

[4] D. Middleton, "Estimating the Noise Power of an Otherwise Known Noise Process," presented at IDA Summer Study, July 1963.

[5] M. J. Levin, "Power Spectrum Parameter Estimation," IEEE Trans. **IT-11,** No. 1, 100–107 (Jan. 1965).

[6] M. J. Levin, "Parameter Estimation for Deterministic and Random Signals," Group Report 34G-11, Massachusetts Institute of Technology, Lincoln Laboratory (unpublished draft).

[7] E. M. Hofstetter, "Some Results on the Stochastic Signal Parameter Problem," IEEE Trans. **IT-11,** No. 3, 422–429 (July 1965).

[8] R. O. Harger, "Maximum-Likelihood Estimation of Focus of Gaussian Signals," IEEE Trans. Information Theory **IT-13,** No. 2, 318–320 (April 1967).

[9] R. F. Pawula, "Analysis of an Estimator of the Center Frequency of a Power Spectrum," IEEE Trans. Information Theory **IT-14,** No. 5, 669–676 (Sept. 1968).

[10] H. Steinberg, P. Schultheiss, C. Wogrin, and F. Zweig, "Short-Time Frequency Measurements of Narrow-Band Random Signals by Means of a Zero-Counting Process," J. Appl. Phys. **26,** 195–201 (Feb. 1955).

[11] J. A. Mullen, "Optimal Filtering for Radar Doppler Navigators," 1962 IRE Conv. Rec., Pt. 5, pp. 40–48.

[12] D. J. Sakrison, "Efficient Recursive Estimation of the Parameters of Radar or Radio Astronomy Target," IEEE Trans. Information Theory **IT-12,** No. 1, 35–41 (Jan. 1966).

[13] N. E. Nahi and R. Gagliardi. "On the Estimation of Signal-to-Noise Ratio and Applications to Detection and Tracking Systems," USC EE Rept. 114, July 1964.

[14] N. E. Nahi and R. Gagliardi, "Use of Limiters for Estimating Signal-to-Noise Ratio," IEEE Trans. Information Theory **IT-13,** No. 1, 127–129 (Jan. 1967).

[15] S. Rauch, "Estimation of Signal-to-Noise Ratio," IEEE Trans. Information Theory **IT-15,** No. 1, 166–167 (Jan. 1969).

# 8

# *The Radar-sonar Problem*

In the second half of this book, we apply the detection and estimation theory results that we have derived to problems encountered in the analysis and design of modern radar and sonar systems. We shall confine our discussion to the signal-processing aspects of the problem. There are a number of books dealing with the practical and theoretical aspects of the design of the overall system (e.g., for radar [1]–[6], and for sonar [7]–[9].†

In this Chapter we discuss the problem qualitatively and outline the organization of the remainder of the book.

A simple model of an *active* radar system is shown in Fig. 8.1. A narrow-band signal centered around some carrier frequency $\omega_c$ is transmitted. If a target is present, the transmitted signal is reflected. The properties of the reflected signal depend on the target characteristics (e.g., shape and motion). An attenuated and possibly distorted version of the reflected signal is returned to the receiver. In the simplest case, the only source of interference is an additive Gaussian receiver noise. In more general cases, there is interference due to external noise sources or reflections from other targets. In the detection problem the receiver processes the signal to decide whether or not a target is present at a particular location. In the estimation problem the receiver processes the signal to measure some characteristics of the target, such as range and velocity.

As we pointed out in Chapter 1, there are several issues that arise in the signal processing problem.

1. The reflective characteristics of the target.
2. Transmission characteristics of the channel.
3. Characteristics of the interference.
4. Optimum and suboptimum receiver design and performance.

† We strongly suggest that readers who have no familiarity with the basic ideas of radar or sonar read at least the first chapter of one of the above references.

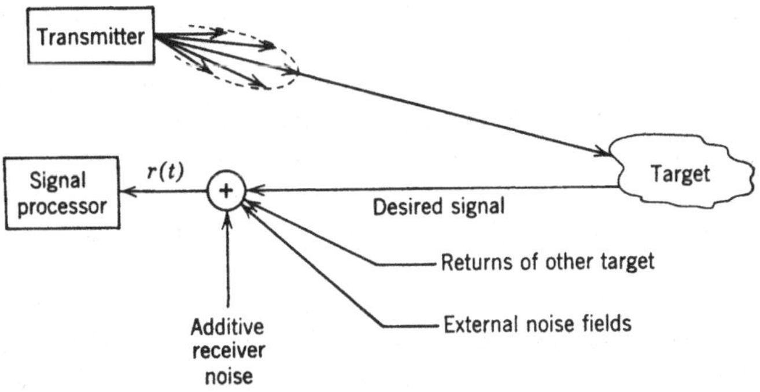


**Fig. 8.1 Model of an active radar or sonar system.**

We study these issues in detail in Chapters 9–13. Before beginning our detailed discussion, it is worthwhile outlining the hierarchy of the target models that we shall discuss.

The simplest model of the target is one that assumes that the target characteristics are fixed during the time it is illuminated by the transmitted pulse. If we further assume that its depth (measured in seconds) is negligible compared to the pulse length, we may consider it as a point reflector (with respect to the envelope of the pulse). Thus, the only effect of the target on the envelope is to attenuate and delay it. The carrier acquires a random phase shift. For this case, the attenuation and phase shift will be essentially constant over the pulse length, and we can model them as random variables. We refer to this type of target as a slowly fluctuating point target.

In Chapter 9 we discuss the problem of detecting a slowly fluctuating point target at a particular range and velocity. First, we assume that the only interference is additive white Gaussian noise and develop the optimum receiver and evaluate its performance. We then consider nonwhite Gaussian noise and find the optimum receiver and its performance. We use complex state-variable theory to obtain complete solutions for the nonwhite noise case. The final topic in the chapter is a brief discussion of signal design.

In Chapter 10 we consider the problem of *estimating* target parameters. Initially, we consider the problem of estimating the range and velocity of a single target when the interference is additive white Gaussian noise. Starting with the likelihood function, we develop the structure of the optimum receiver. We then investigate the performance of the receiver and see how the signal characteristics affect the estimation accuracy. We find that the signal enters into the analysis through a function called the *ambiguity function*; we therefore develop various properties of the ambiguity function and discuss how to design signals with desirable ambiguity functions. The final topic is the detection of a target in the presence of other interfering targets (the discrete resolution problem).

Although the material in Chapters 9 and 10 deals with the simplest type of target model it enables the reader to understand the signal processing aspects of most modern radar and sonar systems. The only background needed for these two chapters is the material in Chapter I-4.

In the remaining three chapters we study more complicated target models. Except for the reverberation discussion in Section 13.2, they assume familiarity with the material in Chapter III-2–III-4. The work in Chapters 11–13 is more advanced than that in Chapter 10 but is essential for readers doing research or development of more sophisticated signal-processing systems.

In Chapter 11, we consider a point target that fluctuates during the time the transmitted pulse is being reflected. This fluctuation causes time-selective fading, and we must model the received signal as a sample function of a random process.

In Chapter 12, we consider a slowly fluctuating point target that is distributed in range. We shall find that this type of target causes frequency-selective fading and, once again, we must model the received signal as a sample function of a random process.

In Chapter 13, we consider fluctuating, distributed targets. This model is useful in the study of reverberation in sonar systems and clutter in radar systems. It is also appropriate in radar astronomy and scatter communications problems. In the first part of the chapter we study the problems of signal and receiver design for systems operating in reverberation and clutter. This discussion will complete the resolution problem development that we begin in Chapter 10. In the second part of the chapter we study the detection of fluctuating, distributed targets and communication over fluctuating distributed channels. Finally, we study the problem of estimating the parameters of a fluctuating, distributed target.

In Chapter 14, we summarize the major results of the radar-sonar discussion. Throughout our discussion we emphasize the similarity between the radar problem and the digital communications problem. In several sections we digress from the radar-sonar development and consider specific digital communications problems.

All of our discussion in Chapters 9–13 describes the signals, systems, and processes with a complex envelope notation. In the Appendix we develop this complex representation in detail. The idea is familiar to electrical engineers in the context of phasor diagrams. Most of the results concerning signals, bandpass systems, and stationary processes will serve as a review for many readers, and our discussion merely serves to establish our notation. The material dealing with nonstationary processes, eigenfunctions, and complex state variables may be new, and so we include more examples in these parts. The purpose of the entire discussion is to develop an efficient notation for the problems of interest. The time expended in developing this notation is warranted, in view of the significant simplifications it allows in the remainder of the book. Those readers who are not familiar with the complex envelope notation should read the Appendix before starting Chapter 9.

## REFERENCES

[1] J. F. Reintjes and G. T. Coate, *Principles of Radar*, McGraw-Hill, New York, 1952.
[2] M. I. Skolnik, *Introduction to Radar Systems*, McGraw-Hill, New York, 1962.
[3] D. K. Barton, *Radar System Analysis*, Prentice-Hall, Englewood Cliffs, N.J., 1965.
[4] R. S. Berkowitz, *Modern Radar: Analysis, Evaluation, and System Design*, Wiley, New York, 1965.
[5] J. V. Di Franco and W. L. Rubin, *Radar Detection*, Prentice-Hall, Englewood Cliffs, N.J., 1968.
[6] C. E. Cook and M. Bernfeld, *Radar Signals, An Introduction to Theory and Application*, Academic Press, New York, 1967.
[7] R. J. Urick, *Principles of Underwater Sound for Engineers*, McGraw-Hill, New York, 1967.
[8] V. O. Albers, *Underwater Acoustics Handbook*, Pennsylvania State University Press, University Park, Pa., 1960.
[9] J. W. Horton, *Fundamentals of Sonar*, U.S. Naval Institute, Washington, D.C., 1957.

# 9

# *Detection of Slowly Fluctuating Point Targets*

In this chapter we discuss the problem of detecting a slowly fluctuating point target in the presence of additive noise. The first step is to develop a realistic mathematical model for the physical situations of interest. In the course of that development we shall explain the phrases "slowly fluctuating" and "point" more explicitly. Once we obtain the mathematical model, the detection problem is directly analogous to that in Sections I-4.2 and I-4.3, so that we can proceed quickly. We consider three cases:

1. Detection in white bandpass noise.
2. Detection in colored bandpass noise.
3. Detection in bandpass noise that has a finite state representation.

In all three cases, we use the complex notation that we develop in detail in the Appendix. We begin by developing a model for the target reflection process in Section 9.1. In Section 9.2, we study detection in white bandpass noise. In Section 9.3, we study detection in colored bandpass noise. In Section 9.4, we specialize the results of Section 9.3 to the case in which the bandpass noise has a finite state representation. In Section 9.5, we study the question of optimal signal design briefly.

## 9.1 MODEL OF A SLOWLY FLUCTUATING POINT TARGET

In order to develop our target model, we first assume that the radar/sonar system transmits a cosine wave continuously. Thus,

$$s_t(t) = \sqrt{2P_t}\cos\omega_c t = \sqrt{2}\,\text{Re}\,[\sqrt{P_t}\,e^{j\omega_c t}], \qquad -\infty < t < \infty. \quad (1)$$

Now assume that there is a zero-velocity target located at some range $R$ from the transmitter. We assume that the target has a physical structure that includes several reflecting surfaces. Thus the returned signal may be written as

$$s_r(t) = \sqrt{2}\,\mathrm{Re}\left\{\sqrt{P_t}\sum_{i=1}^{K} g_i \exp\,[j\omega_c(t-\tau)+\theta_i]\right\}. \tag{2}$$

The attenuation $g_i$ includes the effects of the transmitting antenna gain, the two-way path loss, the radar cross-section of the $i$th reflecting surface, and the receiving antenna aperture. The phase angle $\theta_i$ is a random phase incurred in the reflection process. The constant $\tau$ is the round-trip delay time from the target. If the velocity of propagation is $c$,

$$\tau \triangleq \frac{2R}{c}. \tag{3}$$

We want to determine the characteristics of the sum in (2). If we assume that the $\theta_i$ are statistically independent, that the $g_i$ have equal magnitudes, and that $K$ is large, we can use a central limit theorem argument to obtain

$$s_r(t) = \sqrt{2}\,\mathrm{Re}\,\{\sqrt{P_t}\,\tilde{b}\exp\,(j\omega_c(t-\tau)]\}, \tag{4}$$

where $\tilde{b}$ is a complex Gaussian random variable. The envelope, $|\tilde{b}|$, is a Rayleigh random variable whose moments are

$$E\{|\tilde{b}|\} = \sqrt{\frac{\pi}{2}}\,\sigma_b \tag{5}$$

and

$$E\{|\tilde{b}|^2\} = 2\sigma_b^2. \tag{6}$$

The value of $\sigma_b^2$ includes the antenna gains, path losses, and radar cross-section of the target. The expected value of the received power is $2P_t\sigma_b^2$. The phase of $\tilde{b}$ is uniform. In practice, $K$ does not have to very large in order for the complex Gaussian approximation to be valid. Slack [1] and Bennett [2] have studied the approximation in detail. It turns out that if $K = 6$, the envelope is essentially Rayleigh and the phase is uniform. The central limit theorem approximation is best near the mean and is less accurate on the tail of the density. Fortunately, the tail of the density corresponds to high power levels, so that it is less important that our model be exact.

We assume that the reflection process is *frequency-independent*. Thus, if we transmit

$$s_t(t) = \sqrt{2}\,\mathrm{Re}\,[\sqrt{P_t}\exp\,(j\omega_c t + j\omega t)], \tag{7}$$

we receive

$$s_r(t) = \sqrt{2}\,\mathrm{Re}\,[\sqrt{P_t}\,\tilde{b}\exp\,[j(\omega_c+\omega)(t-\tau)]]. \tag{8}$$

We also assume that the reflection process is *linear*. Thus, if we transmit

$$s_t(t) = \sqrt{2}\,\text{Re}\,[\sqrt{E_t}\tilde{f}(t)e^{j\omega_c t}]$$
$$= \sqrt{2}\,\text{Re}\left[\sqrt{E_t}\,e^{j\omega_c t}\int_{-\infty}^{\infty}\tilde{F}(j\omega)e^{j\omega t}\frac{d\omega}{2\pi}\right], \tag{9}$$

we receive

$$s_r(t) = \sqrt{2}\,\text{Re}\left[\sqrt{E_t}\,\tilde{b}\exp\,[j\omega_c(t-\tau)]\int_{-\infty}^{\infty}\tilde{F}(j\omega)\exp\,[j\omega(t-\tau)]\frac{d\omega}{2\pi}\right]$$
$$= \sqrt{2}\,\text{Re}\,[\sqrt{E_t}\,\tilde{b}\exp\,[j\omega_c(t-\tau)]\tilde{f}(t-\tau)]. \tag{10}$$

Since $\tilde{b}$ has a uniform phase, we can absorb the $e^{j\omega_c\tau}$ term in the phase. Then

$$\boxed{s_r(t) = \sqrt{2}\,\text{Re}\,[\sqrt{E_t}\,\tilde{b}\tilde{f}(t-\tau)e^{j\omega_c t}].} \tag{11}$$

The function $\tilde{f}(t)$ is the complex envelope of the transmitted signal. We assume that it is normalized:

$$\int_{-\infty}^{\infty}|\tilde{f}(t)|^2\,dt = 1. \tag{12}$$

Thus the transmitted energy is $E_t$. The expected value of the received signal energy is

$$\bar{E}_r \triangleq 2E_t\sigma_b^2. \tag{13}$$

We next consider a target with constant radial velocity $v$. The range is

$$R(t) = R_0 - vt. \tag{14}$$

The signal returned from this target is

$$s_r(t) = \sqrt{2}\,\text{Re}\,[\sqrt{E_t}\,\tilde{b}\tilde{f}(t-\tau(t))\exp\,[j\omega_c(t-\tau(t))]], \tag{15}$$

where $\tau(t)$ is the round-trip delay time. Notice that a signal received at $t$ was reflected from the target at $[t-(\tau(t)/2)]$. At that time the target range was

$$R\left(t-\frac{\tau(t)}{2}\right) = R_0 - v\left(t-\frac{\tau(t)}{2}\right). \tag{16}$$

By definition,

$$\tau(t) = \frac{2R(t-\tau(t)/2)}{c}. \tag{17}$$

Substituting (16) into (17) and solving for $\tau(t)$, we obtain

$$\tau(t) = \frac{2R_0/c}{1 + v/c} - \frac{(2v/c)t}{1 + v/c}. \tag{18}$$

For target velocities of interest,

$$\frac{v}{c} \ll 1. \tag{19}$$

Thus,

$$\tau(t) \simeq \frac{2R_0}{c} - \frac{2v}{c}t \triangleq \tau - \frac{2v}{c}t. \tag{20}$$

Substituting (20) into (15) gives

$$s_r(t) = \sqrt{2}\,\text{Re}\left[\sqrt{E_t}\,\tilde{b}\tilde{f}\left(t - \tau + \frac{2v}{c}t\right)\exp\left[j\omega_c\left(t + \frac{2v}{c}t\right)\right]\right]. \tag{21}$$

(Once again, we absorbed the $\omega_c\tau$ term in $\tilde{b}$.) We see that the target velocity has two effects:

1. A compression or stretching of the time scale of the complex envelope.
2. A shift of the carrier frequency.

In most cases we can ignore this first effect. To demonstrate this, consider the error in plotting $\tilde{f}(t)$ instead of $\tilde{f}(t - (2v/c)t)$. The maximum difference in the arguments occurs at the end of the pulse (say $T$) and equals $2vT/c$. The resulting error in amplitude is a function of the signal bandwidth. If the signal bandwidth is $W$, the signal does not change appreciably in a time equal to $W^{-1}$. Therefore, if

$$\frac{2vT}{c} \ll \frac{1}{W} \tag{22}$$

or, equivalently,

$$WT \ll \frac{c}{2v}, \tag{23}$$

we may ignore the time-scale change. For example, if the target velocity is 5000 mph, a $WT$ product of 2000 would satisfy the inequality.†

The shift in the carrier frequency is called the *Doppler shift*

$$\omega_D \triangleq \omega_c\left(\frac{2v}{c}\right). \tag{24}$$

† There are some sonar problems in which (23) is not satisfied. We shall comment on these problems in Section 10.6.

Using (24) in (21), and neglecting the time compression, we obtain

$$\boxed{s_r(t) = \sqrt{2}\,\text{Re}\,[\sqrt{E_t}\,\tilde{b}\tilde{f}(t - \tau)\exp(j\omega_c t + \omega_D t)].} \tag{25}$$

We shall use this expression for the received signal throughout our discussion of slowly fluctuating point targets. We have developed it in reasonable detail because it is important to understand the assumptions inherent in the mathematical model.

The next step is to characterize the additive noise process. We assume that there is an additive Gaussian noise $n(t)$ that has a bandpass spectrum so that we can represent it as

$$n(t) = \sqrt{2}\,\text{Re}\,[\tilde{n}(t)e^{j\omega_c t}]. \tag{26}$$

(This representation of the bandpass processes is developed in the Appendix.) Thus, the total received waveform is

$$r(t) = \sqrt{2E_t}\,\text{Re}\,\{\tilde{b}\tilde{f}(t - \tau)\exp(j\omega_c t + j\omega_D t)\} + \sqrt{2}\,\text{Re}\,\{\tilde{n}(t)\exp(j\omega_c t)\} \tag{27}$$

or, more compactly,

$$\boxed{r(t) = \sqrt{2}\,\text{Re}\,[\tilde{r}(t)e^{j\omega_c t}],} \tag{28a}$$

where

$$\boxed{\tilde{r}(t) \triangleq \tilde{b}\sqrt{E_t}\tilde{f}(t - \tau)e^{j\omega_D t} + \tilde{n}(t).} \tag{28b}$$

Up to this point we have developed a model for the return from a target at a particular point in the range-Doppler plane. We can now formulate the detection problem explicitly. We want to examine a particular value of range and Doppler and decide whether or not a target is present at that point. This is a binary hypothesis-testing problem. The received waveforms on the two hypotheses are

$$r(t) = \sqrt{2}\,\text{Re}\,\{[\tilde{b}\sqrt{E_t}\tilde{f}(t - \tau)e^{j\omega_D t} + \tilde{n}(t)]e^{j\omega_c t}\}, \qquad T_i \le t \le T_f : H_1 \tag{29a}$$

and

$$r(t) = \sqrt{2}\,\text{Re}\,\{\tilde{n}(t)e^{j\omega_c t}\}, \qquad T_i \le t \le T_f : H_0. \tag{29b}$$

Since we are considering only a particular value of $\tau$ and $\omega$, we can assume that they are zero for algebraic simplicity. The modifications for nonzero $\tau$ and $\omega$ are obvious and will be pointed out later. Setting $\tau$ and $\omega_D$ equal to zero gives

$$r(t) = \sqrt{2}\ \text{Re}\ \{[\tilde{b}\sqrt{E_t}\tilde{f}(t) + \tilde{n}(t)]e^{j\omega_c t}\}, \qquad T_i \le t \le T_f : H_1 \quad (30)$$

and

$$r(t) = \sqrt{2}\ \text{Re}\ [\tilde{n}(t)e^{j\omega_c t}], \qquad T_i \le t \le T_f : H_0. \quad (31)$$

In the next three sections, we use the model described by (30) and (31) and consider the three cases outlined on page 238.

Before we begin this development, some further comments on the model are worthwhile. All of our discussion in the text will use the Rayleigh model for the envelope $|\tilde{b}|$. In practice there are target models that cannot be adequately modeled by a Rayleigh variable, and so various other densities have been introduced. Marcum's work [8]–[10] deals with *nonfluctuating* targets. Swerling [7] uses both the Rayleigh model and a probability density assuming one large reflector and a set of small reflectors. Specifically, defining

$$z \triangleq |\tilde{b}|^2, \quad (32a)$$

the density given by the latter model is

$$p_z(Z) = \frac{Z}{\sigma_b^{\,2}} e^{-Z/\sigma_b}, \qquad Z \ge 0, \quad (32b)$$

where $\sigma_b^2$ is defined in (6). Swerling [11] also uses a chi-square density for $z$,

$$p_z(Z) = \frac{1}{(K-1)!}\frac{K}{2\sigma_b^{\,2}}\left(\frac{KZ}{2\sigma_b^{\,2}}\right)^{K-1} e^{-KZ/2\sigma_b^2}, \qquad Z \ge 0. \quad (32c)$$

The interested reader can consult the references cited above as well as [12, Chapter VI-5] and [13] for discussions of target models.

Most of our basic results are applicable to the problem of digital communication over slowly fluctuating point channels that exhibit Rayleigh fading. Other fading models can be used to accommodate different physical channels. The Rician channel [14] was introduced on page I-360. A more general fading model, the Nakagami channel [15], [16], models $|\tilde{b}|$ as

$$p_{|\tilde{b}|}(X) = \frac{2m^m X^{2m-1}}{\Gamma(m)(2\sigma_b^{\,2})^m} e^{-mX^2/2\sigma_b^2}, \qquad X \ge 0, \quad (33)$$

which is a generalization of (32c) to include noninteger $K$. Various problems using this channel model are discussed in [17]–[22].

We now proceed with our discussion of the detection of a slowly fluctuating point target.

## 9.2 WHITE BANDPASS NOISE

In this case, the complex envelopes of the received waveform on the two hypotheses are

$$\begin{aligned} \tilde{r}(t) &= \sqrt{E_t}\,\tilde{b}\tilde{f}(t) + \tilde{w}(t), \qquad & 0 \le t \le T: H_1, \\ \tilde{r}(t) &= \tilde{w}(t), & 0 \le t \le T: H_0, \end{aligned} \tag{34}$$

where $\tilde{b}$ is a zero-mean complex Gaussian random variable ($E\{|\tilde{b}|^2\} = 2\sigma_b^2$) and $\tilde{w}(t)$ is an independent zero-mean white complex Gaussian random process,

$$E[\tilde{w}(t)\tilde{w}^*(u)] = N_0\delta(t-u). \tag{35}$$

The complex envelope $\tilde{f}(t)$ has unit energy. Because the noise is white, we can make the observation interval coincident with the signal duration.

The first step is to find a sufficient statistic. Since the noise is white, we can expand using any complete orthonormal set of functions and obtain statistically independent coefficients [see (A.117)]. Just as in Section I-4.2, we can choose the signal as the first orthonormal function and the resulting coefficient will be a sufficient statistic. In the complex case we correlate $\tilde{r}(t)$ with $\tilde{f}^*(t)$ as shown in Fig. 9.1. The resulting coefficient is

$$\boxed{\tilde{r}_1 \triangleq \int_0^T \tilde{r}(t)\tilde{f}^*(t)\,dt.} \tag{36}$$

Using (34) in (36),

$$\tilde{r}_1 = \begin{cases} \sqrt{E_t}\,\tilde{b} + \tilde{w}_1 : H_1 \\ \tilde{w}_1 : H_0, \end{cases} \tag{37}$$

where $\tilde{w}_1$ is a zero-mean complex Gaussian random variable ($E\{|\tilde{w}_1|^2\} = N_0$). We can easily verify that $\tilde{r}_1$ is a sufficient statistic. The probability density of a complex Gaussian random variable is given by (A.81).

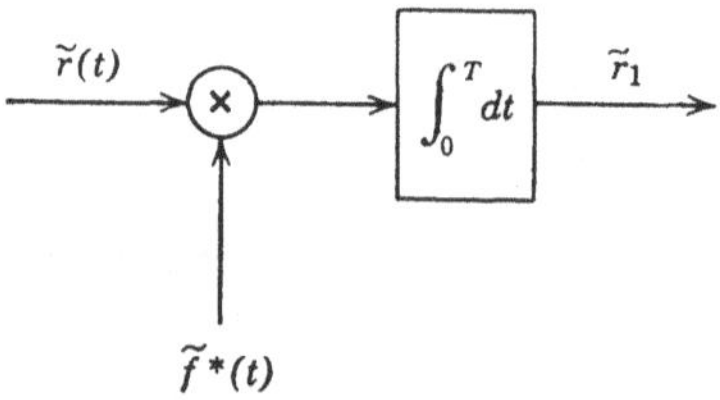


**Fig. 9.1 Generation of complex sufficient statistic.**

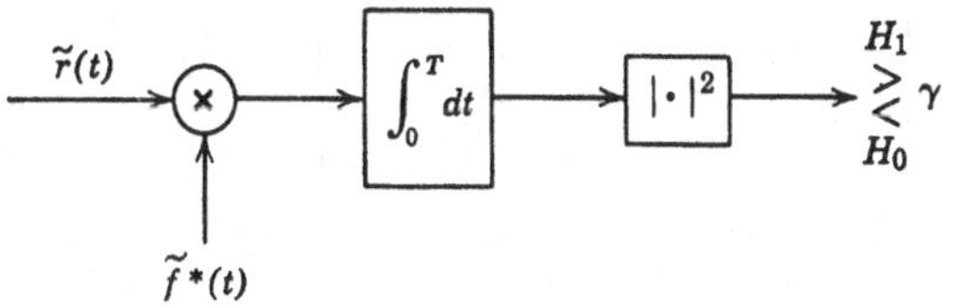


**Fig. 9.2 Correlation receiver (complex operations).**

The likelihood ratio test is

$$\Lambda(\tilde{R}_1) = \frac{p_{\tilde{r}_1|H_1}(\tilde{R}_1 \mid H_1)}{p_{\tilde{r}_1|H_0}(\tilde{R}_1 \mid H_0)}$$

$$= \frac{[\pi(2\sigma_b^2 E_t + N_0)]^{-1} \exp -[|\tilde{R}_1|^2/(2\sigma_b^2 E_t + N_0)]}{(1/\pi N_0) \exp(-|\tilde{R}_1|^2/N_0)} \underset{H_0}{\overset{H_1}{\gtrless}} \eta. \tag{38}$$

Taking the logarithm and rearranging terms, we have

$$|\tilde{R}_1|^2 \underset{H_0}{\overset{H_1}{\gtrless}} \frac{N_0(N_0 + 2\sigma_b^2 E_t)}{2\sigma_b^2 E_t} \left\{\ln \eta + \ln \left(1 + \frac{2\sigma_b^2 E_t}{N_0}\right)\right\} \triangleq \gamma. \tag{39}$$

A complex receiver using a correlation operation is shown in Fig. 9.2. A complex receiver using a matched filter is shown in Fig. 9.3. Here

$$\tilde{r}_1 = \int_0^T \tilde{r}(u)\tilde{h}(T - u)\, du, \tag{40}$$

where

$$\tilde{h}(u) = \tilde{f}^*(T - u). \tag{41}$$

The actual bandpass receiver is shown in Fig. 9.4. We see that it is a bandpass matched filter followed by a square-law envelope detector and sampler.

The calculation of the error probabilities is straightforward. We have solved this exact problem on page I-355, but we repeat the calculation here as a review. The false-alarm probability is

$$P_F = \Pr [|\tilde{r}_1|^2 > \gamma \,|H_0|]$$

$$= \int_{\sqrt{\gamma}}^{\infty} \int_0^{2\pi} \frac{1}{\pi N_0} e^{-Z^2/N_0} Z \, dZ \, d\beta, \tag{42}$$

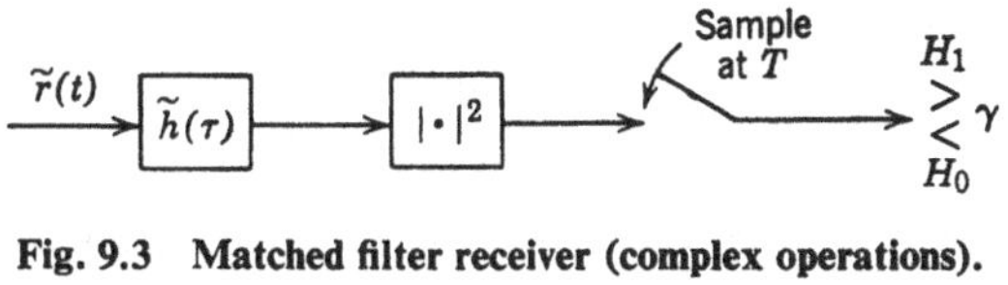


**Fig. 9.3 Matched filter receiver (complex operations).**

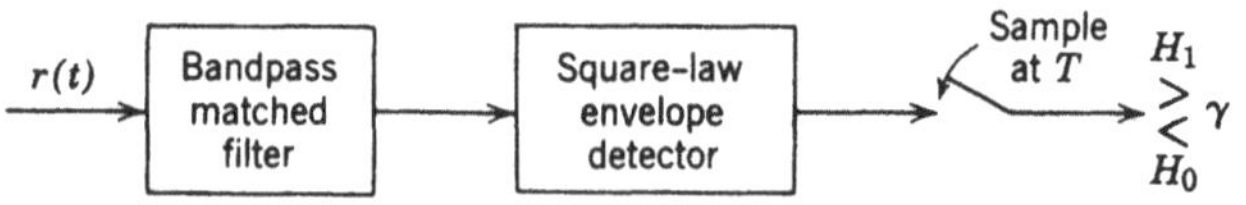


**Fig. 9.4 Optimum receiver: detection of bandpass signal in white Gaussian noise.**

where we have defined

$$\tilde{R}_1 \triangleq Ze^{j\beta}. \tag{43}$$

Thus,

$$P_F = e^{-\gamma/N_0}. \tag{44}$$

Similarly, we obtain

$$P_D = \exp\left(-\frac{\gamma}{2\sigma_b^{\,2}E_t + N_0}\right) = \exp\left(-\frac{\gamma}{\bar{E}_r + N_0}\right), \tag{45}$$

where

$$\bar{E}_r \triangleq 2\sigma_b^{\,2}E_t \tag{46}$$

is the expected value of the received signal energy. Combining (42) and (45) gives

$$\boxed{P_F = (P_D)^{\frac{N_0+\bar{E}_r}{N_0}} = (P_D)^{1+\bar{E}_r/N_0}.} \tag{47}$$

As we would expect, the performance is only a function of $\bar{E}_r/N_0$, and the signal shape $\tilde{f}(t)$ is unimportant. We also observe that the exponent of $P_D$ is the ratio of the expectation of $|\tilde{R}_1|^2$ on the two hypotheses:

$$\frac{N_0 + \bar{E}_r}{N_0} = \frac{E[|\tilde{R}_1|^2 \mid H_1]}{E[|\tilde{R}_1|^2 \mid H_0]}. \tag{48}$$

From our above development, it is clear that this result will be valid for the test in (39) whenever $\tilde{R}_1$ is a zero-mean complex Gaussian random variable on both hypotheses. It is convenient to write the result in (48) in a different form.

$$\Delta \triangleq \frac{E[|\tilde{R}_1|^2 \mid H_1]}{E[|\tilde{R}_1|^2 \mid H_0]} - 1 = \frac{E[|\tilde{R}_1|^2 \mid H_1] - E[|\tilde{R}_1|^2 \mid H_0]}{E[|\tilde{R}_1|^2 \mid H_0]}. \tag{49}$$

Now we can write

$$P_F = (P_D)^{1+\Delta}. \tag{50}$$

For the white noise case,

$$\Delta = \frac{\bar{E}_r}{N_0}. \tag{51}$$

In the next section we evaluate $\Delta$ for the nonwhite-noise case.

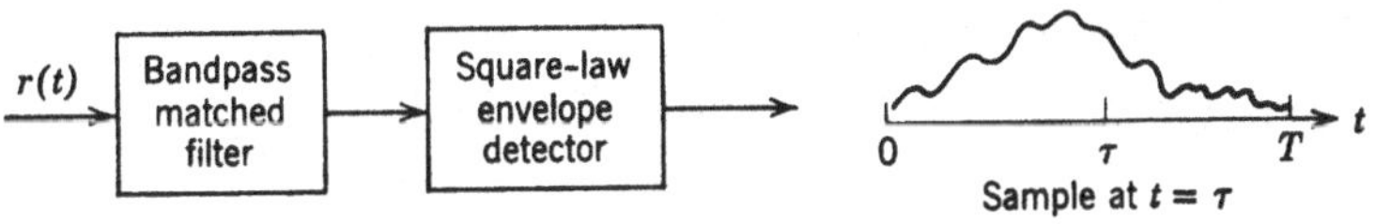


**Fig. 9.5 Optimum receiver for known Doppler shift.**

The modification to include non-zero $\tau$ and $\omega_D$ is straightforward. The desired output is

$$|\tilde{R}_1(\tau, \omega)|^2 = \left|\int_{-\infty}^{\infty} \tilde{r}(t)\tilde{f}^*(t - \tau)e^{-j\omega_D t}\, dt\right|^2. \tag{52}$$

This could be obtained by passing the received waveform through a filter whose complex impulse response is

$$\tilde{h}(u) = \tilde{f}^*(T + \tau - u)e^{j\omega_D u}\, du, \tag{53}$$

then through a square-law envelope detector, and sampling the output at

$$t = T. \tag{54}$$

Equivalently, we can use a complex impulse response

$$\tilde{h}(u) = \tilde{f}^*(-u)e^{j\omega_D u}\, du \tag{55}$$

and sample the detector output at

$$t = \tau. \tag{56}$$

The obvious advantage of this realization is that we can test all ranges with the same filter. This operation is shown in Fig. 9.5. The complex envelope of the bandpass matched filter is specified by (55). In practice, we normally sample the output waveform at the reciprocal of the signal bandwidth. To test different Doppler values, we need different filters. We discuss this issue in more detail in Chapter 10.

We now consider the case in which $\tilde{n}(t)$ is nonwhite.

## 9.3 COLORED BANDPASS NOISE

In this case, the complex envelopes on the two hypotheses are

$$\begin{aligned} \tilde{r}(t) &= \sqrt{E_t}\,\tilde{b}\tilde{f}(t) + \tilde{n}(t), \qquad && T_i \le t \le T_f : H_1, \\ \tilde{r}(t) &= \tilde{n}(t), && T_i \le t \le T_f : H_0. \end{aligned} \tag{57}$$

The additive noise $\tilde{n}(t)$ is a sample function from a zero-mean nonwhite complex Gaussian process. It contains two statistically independent components,

$$\tilde{n}(t) \triangleq \tilde{n}_c(t) + \tilde{w}(t). \tag{58}$$

The covariance of $\tilde{n}(t)$ is

$$E[\tilde{n}(t)\tilde{n}^*(u)] \triangleq \tilde{K}_{\tilde{n}}(t, u) = \tilde{K}_c(t, u) + N_0\,\delta(t - u), \qquad T_i \leq t, u \leq T_f. \tag{59}$$

Notice that the observation interval $[T_i, T_f]$ may be different from the interval over which the signal is nonzero. Any of the three approaches that we used in Section I.4.3 (pages I-287–I-301) will also work here. We use the whitening approach.† Let $\tilde{h}_{wu}(t, z)$ denote the impulse response of a complex whitening filter. When the filter input is $\tilde{n}(t)$, we denote the output as $\tilde{n}_*(t)$,

$$\tilde{n}_*(t) = \int_{T_i}^{T_f} \tilde{h}_{wu}(t, z)\tilde{n}(z)\,dz, \qquad T_i \leq t \leq T_f. \tag{60}$$

The complex impulse response $\tilde{h}_{wu}(t, z)$ is chosen so that

$$\begin{aligned} E[\tilde{n}_*(t)\tilde{n}_*^*(u)] &= E\left[\iint_{T_i}^{T_f} \tilde{h}_{wu}(t, z)\tilde{h}_{wu}^*(u, y)\tilde{n}(z)\tilde{n}^*(y)\,dz\,dy\right] \\ &= \delta(t - u), \qquad T_i \leq t, u \leq T_f. \end{aligned} \tag{61}$$

We define

$$\tilde{r}_*(t) = \int_{T_i}^{T_f} \tilde{h}_{wu}(t, z)\tilde{r}(z)\,dz, \qquad T_i \leq t \leq T_f \tag{62‡}$$

and

$$\tilde{f}_*(t) = \int_{T_i}^{T_f} \tilde{h}_{wu}(t, y)\tilde{f}(y)\,dy, \qquad T_i \leq t \leq T_f. \tag{63}$$

We may now use the results of Section 9.2 directly to form the sufficient statistic. From (36),

$$\begin{aligned} \tilde{r}_1 &= \int_{T_i}^{T_f} \tilde{r}_*(t)\tilde{f}_*^*(t)\,dt \\ &= \int_{T_i}^{T_f} dt \int_{T}^{T_f} \tilde{h}_{wu}(t, z)\tilde{r}(z)\,dz \int_{T_i}^{T_f} \tilde{h}_{wu}^*(t, y)\tilde{f}^*(y)\,dy. \end{aligned} \tag{64}$$

As before, we define an inverse kernel,

$$\tilde{Q}_{\tilde{n}}^*(z, y) \triangleq \int_{T_i}^{T_f} \tilde{h}_{wu}(t, z)\tilde{h}_{wu}^*(t, y)\,dt, \qquad T_i < z, y < T_f. \tag{65}$$

† The argument is parallel to that on pages I-290–I-297, and so we shall move quickly. We strongly suggest that the reader review the above pages before reading this section.

‡ In Section I-4.3, we associated the $\sqrt{E}$ with the whitened signal. Here it is simpler to leave it out of (63) and associate it with the multiplier $\tilde{b}$.

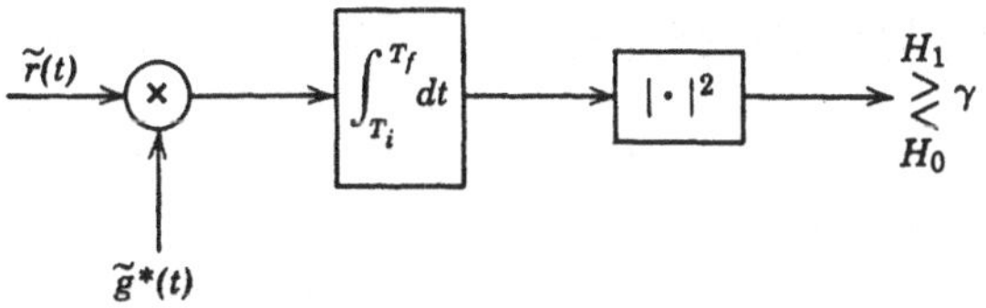


**Fig. 9.6 Optimum receiver: bandpass signal in nonwhite Gaussian noise (complex operations).**

Using (65) in (64) gives

$$\tilde{r}_1 = \iint_{T_i}^{T_f} \tilde{r}(z)\tilde{Q}_{\tilde{n}}^*(z, y)\tilde{f}^*(y)\, dz\, dy. \tag{66}$$

Defining

$$\tilde{g}(z) = \int_{T_i}^{T_f} \tilde{Q}_{\tilde{n}}(z, y)\tilde{f}(y)\, dy, \qquad T_i \le t \le T_f, \tag{67}$$

we have

$$\tilde{r}_1 = \int_{T_i}^{T_f} \tilde{r}(z)\tilde{g}^*(z)\, dz. \tag{68}$$

The optimum test is

$$\boxed{\left| \int_{T_i}^{T_f} \tilde{r}(z)\tilde{g}^*(z)\, dz \right|^2 \underset{H_0}{\overset{H_1}{\gtrless}} \gamma.} \tag{69}$$

The complex receiver is shown in Fig. 9.6, and the actual bandpass receiver is shown in Fig. 9.7.

Proceeding as in Section I-4.3.1, we obtain the following relations:

$$\int_{T_i}^{T_f} \tilde{Q}_{\tilde{n}}(t, x)\tilde{K}_{\tilde{n}}(x, u)\, dx = \delta(t - u), \qquad T_i < t, u < T_f \tag{70}$$

and

$$\tilde{Q}_{\tilde{n}}(t, u) = \frac{1}{N_0}\,[\delta(t - u) - \tilde{h}_{ou}(t, u)], \qquad T_i < t, u < T_f, \tag{71}$$

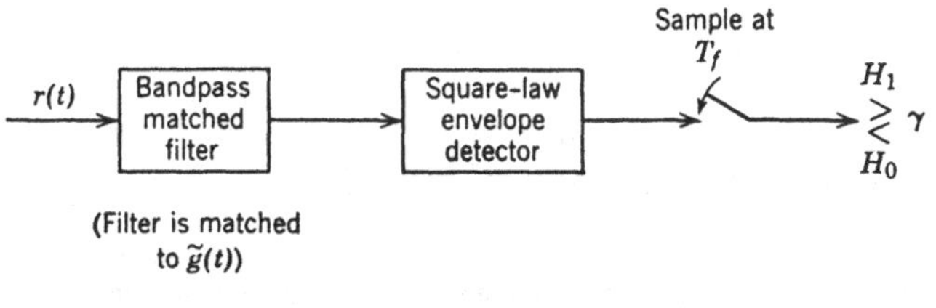


**Fig. 9.7 Optimum receiver: bandpass signal in nonwhite Gaussian noise.**

where $\tilde{h}_{ou}(t, u)$ satisfies the integral equation

$$N_0\tilde{h}_{ou}(t, u) + \int_{T_i}^{T_f} \tilde{h}_{ou}(x, u)\tilde{K}_c(t, x)\, dx = \tilde{K}_c(t, u), \qquad T_i \leq t, u \leq T_f. \quad (72)$$

The function $\tilde{h}_{ou}(t, u)$ is the impulse response of the optimum unrealizable filter for estimating $\tilde{n}_c(t)$ in the presence of white noise $\tilde{w}(t)$ of spectral height $N_0$. Using (70) in (67), we have

$$\boxed{\tilde{f}(t) = \int_{T_i}^{T_f} \tilde{K}_{\tilde{n}}(t, u)\tilde{g}(u)\, du, \qquad T_i < t < T_f} \quad (73)$$

or

$$\boxed{\tilde{f}(t) = \int_{T_i}^{T_f} \tilde{K}_c(t, u)\tilde{g}(u)\, du + N_0\tilde{g}(t), \qquad T_i \leq t \leq T_f.} \quad (74)$$

This equation is just the complex version of (I-4.169b).† In Section I-4.3.6, we discussed solution techniques for integral equations of this form. All of these techniques carry over to the complex case. A particularly simple solution is obtained when $\tilde{n}_c(t)$ is stationary and the observation interval is infinite. We can then use Fourier transforms to solve (73),

$$\tilde{G}_{\infty}(j\omega) = \frac{\tilde{F}(j\omega)}{\tilde{S}_{\tilde{n}}(\omega)}. \quad (75)$$

For finite observation intervals we can use the techniques of Section I-4.3.6. However, when the colored noise has a finite-dimensional complex state representation (see Section A.3.3), the techniques developed in the next section are computationally more efficient.

To evaluate the performance, we compute $\Delta$ using (49). The result is

$$\boxed{\Delta = \bar{E}_r \iint_{T_i}^{T_f} \tilde{f}(t)\tilde{Q}_{\tilde{n}}^*(t, u)\tilde{f}^*(u)\, dt\, du} \quad (76)$$

or

$$\boxed{\Delta = \bar{E}_r \int_{T_i}^{T_f} \tilde{f}(t)\tilde{g}^*(t)\, dt.} \quad (77)$$

Notice that $\Delta$ is a real quantity. Its functional form is identical with that of $d^2$ in the known signal case [see (I-4.198)]. The performance is obtained

† It is important for the reader to identify the similarities between the complex case and the known signal case. One of the advantages of the complex notation is that it emphasizes these similarities and helps us to exploit all of our earlier work.

from (50),

$$\boxed{P_F = (P_D)^{1+\Delta}.} \tag{78}$$

From (78) it is clear that increasing $\Delta$ always improves the performance. As we would expect, the performance of the system depends on the signal shape. We shall discuss some of the issues of signal design in Section 9.5.

## 9.4 COLORED NOISE WITH A FINITE STATE REPRESENTATION†

When the colored noise component has a finite state representation, we can derive an alternative configuration for the optimum receiver that is easy to implement. The approach is just the complex version of the derivation in the appendix in Part II. We use the same noise model as in (58).

$$\tilde{n}(t) = \tilde{n}_c(t) + \tilde{w}(t). \tag{79}$$

We assume that the colored noise can be generated by passing a complex white Gaussian noise process, $\tilde{\mathbf{u}}(t)$, through a finite-dimensional linear system. The state and observation equations are

$$\dot{\tilde{\mathbf{x}}}(t) = \tilde{\mathbf{F}}(t)\tilde{\mathbf{x}}(t) + \tilde{\mathbf{G}}(t)\tilde{\mathbf{u}}(t), \tag{80}$$

$$\tilde{n}_c(t) = \tilde{\mathbf{C}}(t)\tilde{\mathbf{x}}(t) \tag{81}$$

The initial conditions are

$$E[\tilde{\mathbf{x}}(T_i)] = \mathbf{0} \tag{82}$$

and

$$E[\tilde{\mathbf{x}}(T_i)\tilde{\mathbf{x}}^\dagger(T_i)] = \tilde{\mathbf{P}}_0. \tag{83}$$

The covariance matrix of the driving function is

$$E[\tilde{\mathbf{u}}(t)\tilde{\mathbf{u}}^\dagger(\sigma)] = \mathbf{Q}\delta(t-\sigma). \tag{84}$$

In the preceding section we showed that the optimum receiver computed the statistic

$$l_o \triangleq \left| \int_{T_i}^{T_f} \tilde{r}(z)\tilde{g}^*(z)\, dz \right|^2 \tag{85}$$

† In this section, we use the results of Section A.3.3, Problem I-4.3.4, and Problem I-6.6.5. The detailed derivations of the results are included as problems. This section can be omitted on the first reading.

and compared it with a threshold [see (69)]. The function $\tilde{g}(t)$ was specified by

$$\tilde{f}(t) = \int_{T_i}^{T_f} \tilde{K}_c(t, u)\tilde{g}(u)\, du + N_0\tilde{g}(t), \qquad T_i \le t \le T_f. \tag{86}$$

From (81) we have

$$\tilde{K}_c(t, u) = E[\tilde{n}_c(t)\tilde{n}_c^*(u)] = E[\tilde{\mathbf{C}}(t)\tilde{\mathbf{x}}(t)\tilde{\mathbf{x}}^\dagger(u)\tilde{\mathbf{C}}^\dagger(u)]$$
$$= \tilde{\mathbf{C}}(t)\tilde{\mathbf{K}}_{\tilde{\mathbf{x}}}(t, u)\tilde{\mathbf{C}}^\dagger(u). \tag{87}$$

Using (87) in (86) gives

$$\tilde{f}(t) = \tilde{\mathbf{C}}(t)\int_{T_i}^{T_f} \tilde{\mathbf{K}}_{\tilde{\mathbf{x}}}(t, u)\tilde{\mathbf{C}}^\dagger(u)\tilde{\mathbf{g}}(u)\, du + N_0\tilde{\mathbf{g}}(t), \qquad T_i \le t \le T_f. \tag{88}$$

The performance was characterized by

$$\Delta = \bar{E}_r\int_{T_i}^{T_f} \tilde{f}(t)\tilde{g}^*(t)\, dt = \bar{E}_r\int_{T_i}^{T_f} \tilde{f}^*(t)\tilde{g}(t)\, dt. \tag{89}$$

In this section we want to derive an expression for $l_0$ and $\Delta$ in terms of differential equations. These expressions will enable us to specify the receiver and its performance completely without solving an integral equation. We derive two alternative expressions. The first expression is obtained by finding a set of differential equations and associated boundary conditions that specify $\tilde{g}(t)$. The second expression is based on the realizable MMSE estimate of $\tilde{n}_c(t)$.

### 9.4.1 Differential-equation Representation of the Optimum Receiver and Its Performance: I

We define

$$\tilde{\boldsymbol{\xi}}(t) \triangleq \int_{T_i}^{T_f} \tilde{\mathbf{K}}_{\tilde{\mathbf{x}}}(t, \tau)\tilde{\mathbf{C}}^\dagger(\tau)\tilde{g}(\tau)\, d\tau, \qquad T_i \le t \le T_f. \tag{90‡}$$

Using (90) in (88) gives

$$\tilde{f}(t) = \tilde{\mathbf{C}}(t)\tilde{\boldsymbol{\xi}}(t) + N_0\tilde{g}(t), \qquad T_i \le t \le T_f \tag{91}$$

or

$$\tilde{g}(t) = \frac{1}{N_0}[\tilde{f}(t) - \tilde{\mathbf{C}}(t)\tilde{\boldsymbol{\xi}}(t)], \qquad T_i \le t \le T_f. \tag{92}$$

‡ Notice that $\tilde{\boldsymbol{\xi}}(t)$ is defined by (90). It should *not* be confused with $\tilde{\boldsymbol{\xi}}_P(t)$, the error covariance matrix.

Thus, if we can find $\tilde{\xi}(t)$, we have an explicit relation for $\tilde{g}(t)$. By modifying the derivation in the appendix of Part II, we can show that $\tilde{\xi}(t)$ is specified by the equations

$$\frac{d\tilde{\xi}(t)}{dt} = \tilde{\mathbf{F}}(t)\tilde{\xi}(t) + \tilde{\mathbf{G}}(t)\tilde{\mathbf{Q}}\tilde{\mathbf{G}}^{\dagger}(t)\tilde{\eta}(t), \tag{93}$$

$$\frac{d\tilde{\eta}(t)}{dt} = \frac{1}{N_0}\tilde{\mathbf{C}}^{\dagger}(t)\tilde{\mathbf{C}}(t)\tilde{\xi}(t) - \tilde{\mathbf{F}}^{\dagger}(t)\tilde{\eta}(t) - \frac{1}{N_0}\tilde{\mathbf{C}}^{\dagger}(t)\tilde{f}(t), \tag{94}$$

$$\tilde{\xi}(T_i) = \tilde{\mathbf{P}}_0\tilde{\eta}(T_i), \tag{95}$$

$$\tilde{\eta}(T_f) = \mathbf{0}, \tag{96}$$

and

$$\tilde{\mathbf{P}}_0 = \bar{E}_r\mathbf{I}. \tag{97}$$

This is a set of linear matrix equations that can be solved numerically.

To evaluate $\Delta$, we substitute (90) into (89) to obtain

$$\Delta = \frac{\bar{E}_r}{N_0}\left[1 - \int_{T_i}^{T_f} \tilde{f}^*(t)\tilde{\mathbf{C}}(t)\tilde{\xi}(t)\,dt\right]. \tag{98}$$

(Recall that we assume

$$\int_{T_i}^{T_f} |\tilde{f}(t)|^2\,dt = 1.) \tag{99}$$

The first term is the performance in the presence of white noise only. The second term is the degradation due to the colored noise, which we denote as

$$\boxed{\Delta_{dg} \triangleq \int_{T_i}^{T_f} \tilde{f}^*(t)\tilde{\mathbf{C}}(t)\tilde{\xi}(t)\,dt.} \tag{100}$$

Later we shall discuss how to design $\tilde{f}(t)$ to minimize $\Delta_{dg}$. Notice that $\Delta_{dg}$ is normalized and does not include the $\bar{E}_r/N_0$ multiplier.

We now develop an alternative realization based on the realizable estimate.

### 9.4.2 Differential-equation Representation of the Optimum Receiver and Its Performance: II

There are several ways to develop the desired structure. We carry out the details for two methods.

The first method is based on a whitening filter approach. In Section 9.3, we used an unrealizable whitening filter to derive $\tilde{g}(t)$. Now we use a realizable whitening filter. Let $\tilde{h}_{wr}(t, z)$ denote the impulse response of the complex realizable whitening filter. When the filter input is $\tilde{n}(t)$, the output is a sample function from a white noise process.

By extending the results of Problem I-4.3.4 to the complex case, we can show that

$$\tilde{h}_{wr}(t, z) = \left(\frac{1}{N_0}\right)^{1/2} [\delta(t - z) - \tilde{h}_o(t, \tau:t)], \tag{101}$$

where $\tilde{h}_o(t, \tau:t)$ is the linear filter whose output is the MMSE estimate of $\tilde{n}_c(t)$ when the input is $\tilde{n}_c(t) + \tilde{w}(t)$. The test statistic can be written as

$$l_o = \left|\int_{T_i}^{T_f} dt \left[\int_{T_i}^{t} \tilde{h}_{wr}(t, z)\tilde{r}(z)\, dz \int_{T_i}^{t} \tilde{h}^*_{wr}(t, y)\tilde{f}^*(y)\, dy\right]\right|^2$$

$$= \left|\int_{T_i}^{T_f} dt\, \tilde{r}_{wr}(t)\tilde{f}^*_{wr}(t)\right|^2. \tag{102}$$

The receiver is shown in Fig. 9.8. Notice that the operation inside the dashed lines does not depend on $\tilde{r}(t)$. The function $\tilde{f}_{wr}(t)$ is calculated when the receiver is designed. The operation inside the dashed lines indicates this calculation.

A state-variable implementation is obtained by specifying $\tilde{h}_o(t, \tau:t)$ in terms of differential equations. Because it can be interpreted as an optimum

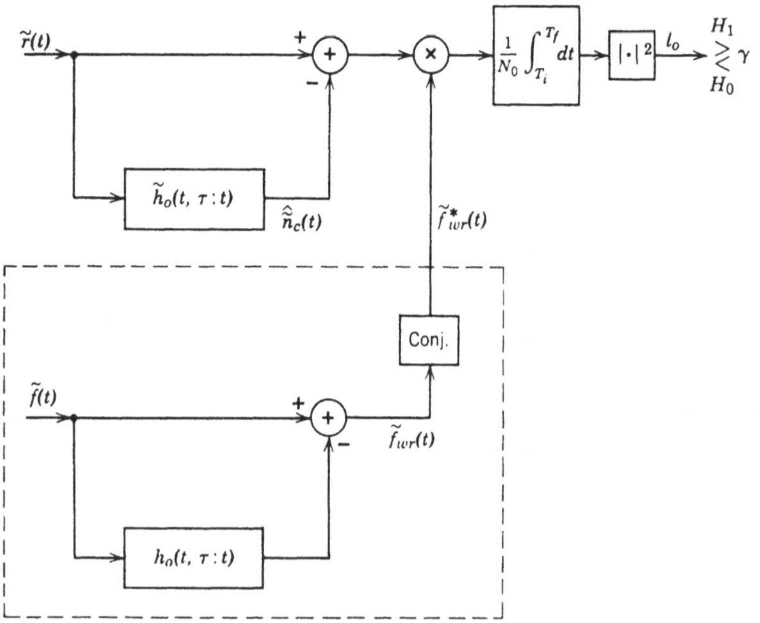


**Fig. 9.8 Optimum receiver realization using realizable whitening filters.**

estimator of $\tilde{n}_c(t)$, we can use (A.159)–(A.162). The estimator equation is

$$\frac{d\hat{\tilde{\mathbf{x}}}(t)}{dt} = \tilde{\mathbf{F}}(t)\hat{\tilde{\mathbf{x}}}(t) + \tilde{\boldsymbol{\xi}}_P(t)\tilde{\mathbf{C}}^\dagger(t)\frac{1}{N_0}[\tilde{r}(t) - \tilde{\mathbf{C}}(t)\hat{\tilde{\mathbf{x}}}(t)], \qquad T_i \le t, \quad (103)$$

and the variance equation is

$$\frac{d\tilde{\boldsymbol{\xi}}_P(t)}{dt} = \tilde{\mathbf{F}}(t)\tilde{\boldsymbol{\xi}}_P(t) + \tilde{\boldsymbol{\xi}}_P(t)\tilde{\mathbf{F}}^\dagger(t) - \tilde{\boldsymbol{\xi}}_P(t)\tilde{\mathbf{C}}^\dagger(t)\frac{1}{N_0}\tilde{\mathbf{C}}(t)\tilde{\boldsymbol{\xi}}_P(t) + \tilde{\mathbf{G}}(t)\tilde{\mathbf{Q}}\tilde{\mathbf{G}}^\dagger(t), \qquad T_i \le t, \quad (104)$$

with initial conditions

$$\hat{\tilde{\mathbf{x}}}(T_i) = E[\hat{\tilde{\mathbf{x}}}(T_i)] = \mathbf{0} \quad (105)$$

and

$$\tilde{\boldsymbol{\xi}}_P(T_i) = E[\hat{\tilde{\mathbf{x}}}(T_i)\hat{\tilde{\mathbf{x}}}^\dagger(T_i)]. \quad (106)$$

The estimate of $\tilde{n}_c(t)$ is

$$\hat{\tilde{n}}_{cr}(t) = \tilde{\mathbf{C}}(t)\tilde{\mathbf{x}}(t). \quad (107)$$

Notice that this is the MMSE realizable estimate, assuming that $H_0$ is true. Using (103), (107), and Fig. 9.8, we obtain the receiver shown in Fig. 9.9.

The performance expression follows easily. The output of the whitening filter in the bottom path is $\tilde{f}_{wr}(t)$. From (76),

$$\Delta = \frac{\bar{E}_r}{N_0}\int_{T_i}^{T_f} |\tilde{f}_{wr}(t)|^2\,dt. \quad (108)$$

From Fig. 9.8 or 9.9 we can write

$$\tilde{f}_{wr}(t) = \{\tilde{f}(t) - \tilde{f}_r(t)\}, \quad (109)$$

where $\tilde{f}_r(t)$ is the output of the optimum realizable filter when its input is $\tilde{f}(t)$. Using (109) in (108), we have

$$\Delta = \frac{\bar{E}_r}{N_0}\left\{1 - \int_{T_i}^{T_f}\{2\tilde{f}^*(t)\tilde{f}_r(t) - |\tilde{f}_r(t)|^2\}\,dt\right\}. \quad (110)$$

From (100) we see that

$$\boxed{\Delta_{dg} = \int_{T_i}^{T_f}\{2\tilde{f}^*(t)\tilde{f}_r(t) - |\tilde{f}_r(t)|^2\}\,dt.} \quad (111)$$

We can also derive the optimum receiver directly from (85) and (92). Because this technique can also be used for other problems, we carry out the details.

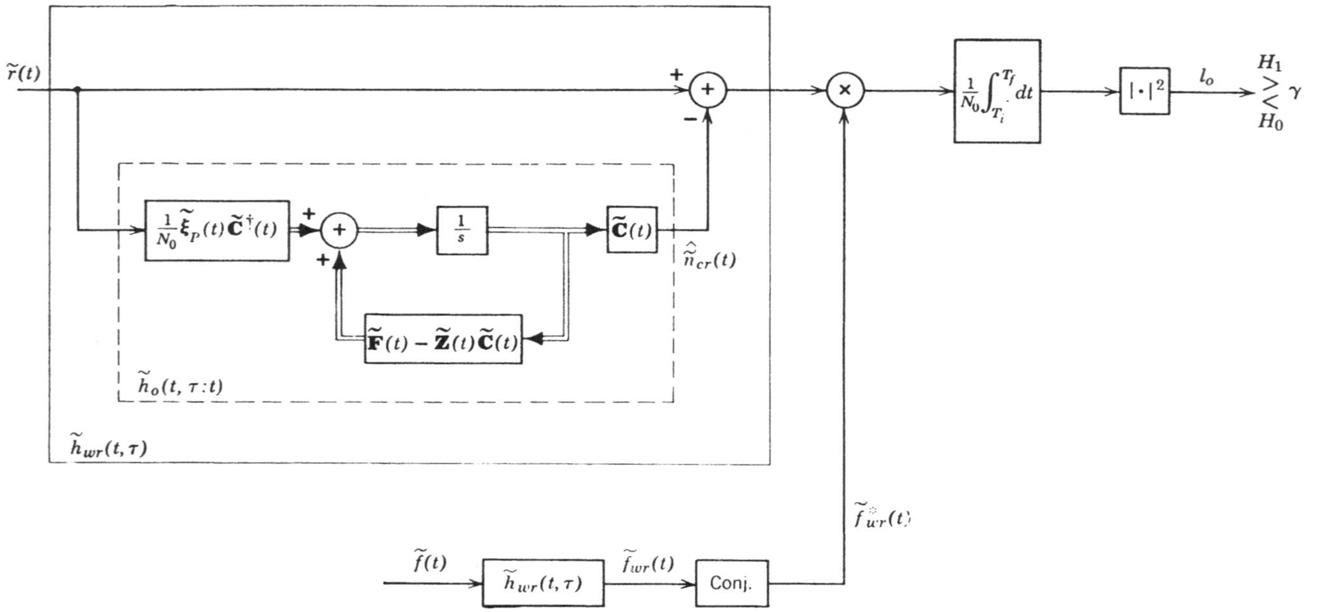

**Fig. 9.9** State-variable realization of the optimum receiver for detecting a slowly fluctuating point target in nonwhite noise.

**An Alternative Derivation.**† This method is based on the relationship

$$\tilde{\boldsymbol{\xi}}(t) = \tilde{\boldsymbol{\Sigma}}(t)\tilde{\boldsymbol{\eta}}(t) + \tilde{\boldsymbol{\xi}}_r(t), \qquad T_i \le t \le T_f, \tag{112}$$

where $\tilde{\boldsymbol{\Sigma}}(t)$ and $\tilde{\boldsymbol{\xi}}_r(t)$ are matrices that we now specify. Differentiating (112) and using (93)–(96), we find that $\tilde{\boldsymbol{\Sigma}}(t)$ must satisfy

$$\frac{d\tilde{\boldsymbol{\Sigma}}(t)}{dt} = \tilde{\mathbf{F}}(t)\tilde{\boldsymbol{\Sigma}}(t) + \tilde{\boldsymbol{\Sigma}}(t)\tilde{\mathbf{F}}^{\dagger}(t) - \frac{1}{N_0}\,\tilde{\boldsymbol{\Sigma}}(t)\tilde{\mathbf{C}}^{\dagger}(t)\tilde{\mathbf{C}}(t)\tilde{\boldsymbol{\Sigma}}(t) + \tilde{\mathbf{G}}(t)\mathbf{Q}\tilde{\mathbf{G}}^{\dagger}(t), \tag{113}$$

with

$$\tilde{\boldsymbol{\Sigma}}(T_i) = \tilde{\mathbf{P}}_0, \tag{114}$$

which is familiar as the variance equation (104). [Thus, $\tilde{\boldsymbol{\Sigma}}(t) = \tilde{\boldsymbol{\xi}}_P(t)$.] The function $\tilde{\boldsymbol{\xi}}_r(t)$ must satisfy

$$\frac{d\tilde{\boldsymbol{\xi}}_r(t)}{dt} = \tilde{\mathbf{F}}(t)\tilde{\boldsymbol{\xi}}_r(t) + \frac{1}{N_0}\,\tilde{\boldsymbol{\Sigma}}(t)\tilde{\mathbf{C}}^{\dagger}(t)[\tilde{f}(t) - \tilde{\mathbf{C}}(t)\tilde{\boldsymbol{\xi}}_r(t)], \tag{115}$$

with

$$\tilde{\boldsymbol{\xi}}_r(T_i) = \mathbf{0}. \tag{116}$$

This has the same structure as the estimator equation, except that $\tilde{r}(t)$ is replaced by $\tilde{f}(t)$.

In order to carry out the next step, we introduce a notation for $\tilde{\boldsymbol{\xi}}(t)$ and $\tilde{\boldsymbol{\eta}}(t)$ to indicate the endpoint of the interval. We write $\tilde{\boldsymbol{\xi}}(t, T_f)$ and $\tilde{\boldsymbol{\eta}}(t, T_f)$. These functions satisfy (93)–(96) over the interval $T_i \le t \le T_f$.

The test statistic is

$$\begin{aligned} l_o &= \left| \int_{T_i}^{T_f} \tilde{r}(t)\tilde{g}^*(t)\,dt \right|^2 \\ &= \left| \frac{1}{N_0} \int_{T_i}^{T_f} \tilde{r}(\tau)[\tilde{f}(\tau) - \tilde{\mathbf{C}}(\tau)\tilde{\boldsymbol{\Sigma}}(\tau)\tilde{\boldsymbol{\eta}}(\tau, T_f) - \tilde{\mathbf{C}}(\tau)\tilde{\boldsymbol{\xi}}_r(\tau)]^*\,d\tau \right|^2. \end{aligned} \tag{117}$$

To obtain the desired result, we use the familiar technique of differentiation and integration.

$$l_o = \left| \frac{1}{N_0} \int_{T_i}^{T_f} \left[ \frac{d}{dt} \left\{ \int_{T_i}^{t} \tilde{r}(\tau)[\tilde{f}(\tau) - \tilde{\mathbf{C}}(\tau)\tilde{\boldsymbol{\Sigma}}(\tau)\tilde{\boldsymbol{\eta}}(\tau, t) - \tilde{\mathbf{C}}(\tau)\tilde{\boldsymbol{\xi}}_r(\tau)]^*\,d\tau \right\} \right] dt \right|^2. \tag{118}$$

Differentiating the terms in braces gives

$$\frac{d}{dt}\{\cdot\} = \tilde{r}(t)[\tilde{f}(t) - \tilde{\mathbf{C}}(t)\tilde{\boldsymbol{\xi}}_r(t)]^* + \int_0^t \tilde{r}(\tau)\left[ -\tilde{\mathbf{C}}(\tau)\tilde{\boldsymbol{\Sigma}}(\tau)\,\frac{\partial \tilde{\boldsymbol{\eta}}(\tau, t)}{\partial t} \right]^* d\tau. \tag{119}$$

We can show (see Problem 9.4.5) that the second term reduces to

$$\int_0^t \tilde{r}(\tau)\left[ -\tilde{\mathbf{C}}(\tau)\tilde{\boldsymbol{\Sigma}}(\tau)\,\frac{\partial \tilde{\boldsymbol{\eta}}(\tau, t)}{dt} \right]^* d\tau = [\tilde{f}(t) - \tilde{\mathbf{C}}(t)\tilde{\boldsymbol{\xi}}_r(t)]^*[-\tilde{\mathbf{C}}(t)\hat{\tilde{\mathbf{x}}}(t)], \tag{120}$$

† This alternative derivation can be omitted on the first reading.

where $\hat{\mathbf{x}}(t)$ is the state vector of the optimum realizable linear filter when its input is $\tilde{r}(t)$. Using (120) in (119) and the result in (118), we obtain

$$l_o = \left| \frac{1}{N_0} \int_{T_i}^{T_f} [\tilde{r}(t) - \tilde{\mathbf{C}}(t)\hat{\mathbf{x}}(t)][\tilde{f}(t) - \tilde{\mathbf{C}}(t)\tilde{\boldsymbol{\xi}}_r(t)]^* \, dt \right|^2. \tag{121}$$

The receiver specified by (121) is identical with the receiver shown in Fig. 9.9.

In this section we have developed two state-variable realizations for the optimum receiver to detect a bandpass signal in colored noise. The performance degradation was also expressed in terms of a differential equation. These results are important because they enable us to specify completely the optimum receiver and its performance for a large class of colored noise processes. They also express the problem in a format in which we can study the question of optimal signal design. We discuss this problem briefly in the next section.

## 9.5 OPTIMAL SIGNAL DESIGN

The performance in the presence of colored noise is given by (77). This can be rewritten as

$$\begin{aligned}
\Delta &= \bar{E}_r \int_{T_i}^{T_f} \tilde{f}(t)\tilde{g}^*(t) \, dt \\
&= \bar{E}_r \int_{T_i}^{T_f} \tilde{f}(t)\left[\int_{T_i}^{T_f} \tilde{Q}_{\tilde{n}}^*(t, u)\tilde{f}^*(u) \, du\right] dt \\
&= \bar{E}_r \int_{T_i}^{T_f} \tilde{f}(t)\left[\frac{1}{N_0}\tilde{f}^*(t) - \frac{1}{N_0}\int_{T_i}^{T_f} \tilde{h}_{ou}^*(t, u)\tilde{f}^*(u) \, du\right] dt \\
&= \frac{\bar{E}_r}{N_0}\left[1 - \int_{T_i}^{T_f}\int_{T_i}^{T_f} \tilde{f}(t)\tilde{h}_{ou}^*(t, u)\tilde{f}^*(u) \, dt \, du\right].
\end{aligned} \tag{122}$$

In the last equality, we used (99). The integral in the second term is just $\Delta_{dg}$, which was defined originally in (100). [An alternative expression for $\Delta_{dg}$ is given in (111).]

We want to choose $\tilde{f}(t)$ to minimize $\Delta_{dg}$. In order to obtain a meaningful problem, we must constrain both the energy and bandwidth of $\tilde{f}(t)$ (see discussion on page I-302). We impose the following constraints. The energy constraint is

$$\int_{T_i}^{T_f} |\tilde{f}(t)|^2 \, dt = 1. \tag{123}$$

The mean-square bandwidth constraint is

$$\int_{T_i}^{T_f} \left| \frac{d\tilde{f}(t)}{dt} \right|^2 dt = B^2. \tag{124}$$

In addition, we require

$$\tilde{f}(T_i) = \tilde{f}(T_f) = 0 \tag{125}$$

to avoid discontinuities at the endpoints.

The function that we want to minimize is

$$J = \iint_{T_i}^{T_f} \tilde{f}(t)\tilde{h}_{ou}^*(t, u)\tilde{f}^*(u)\, dt\, du + \lambda_E\left[\int_{T_i}^{T_f} |\tilde{f}(t)|^2\, dt - 1\right] + \lambda_B\left[\int_{T_i}^{T_f} |\dot{\tilde{f}}(t)|^2\, dt - B^2\right], \tag{126}$$

where $\lambda_E$ and $\lambda_B$ are Lagrange multipliers. To carry out the minimization, we let

$$\tilde{f}(t) = \tilde{f}_o(t) + \varepsilon\tilde{f}_\varepsilon(t) \tag{127}$$

and require that

$$\left.\frac{dJ}{d\varepsilon}\right|_{\varepsilon=0} = 0 \tag{128}$$

for all $\tilde{f}_\varepsilon(t)$ satisfying (123)–(125). Substituting (127) into (126) and carrying out the indicated steps, we obtain

$$\operatorname{Re}\left\{\iint_{T_i}^{T_f} \tilde{f}_\varepsilon(t)\tilde{h}_{ou}^*(t, u)\tilde{f}_o^*(u)\, du\, dt + \lambda_E\int_{T_i}^{T_f} \tilde{f}(t)\tilde{f}_o^*(t)\, dt + \lambda_B\int_{T_i}^{T_f} \dot{\tilde{f}}_\varepsilon(t)\dot{\tilde{f}}(t)\, dt\right\} = 0. \tag{129}$$

Integrating the last term by parts, using (125), and collecting terms, we have

$$\operatorname{Re}\left\{\int_{T_i}^{T_f} \tilde{f}_\varepsilon(t)\, dt\left[\int_{T_i}^{T_f} \tilde{h}_{ou}^*(t, u)\tilde{f}_o^*(u)\, du + \lambda_E\tilde{f}_o^*(t) - \lambda_B\ddot{\tilde{f}}_o^*(t)\right]\right\} = 0. \tag{130}$$

Since $\tilde{f}_\varepsilon(t)$ is arbitrary, the term in the brackets must be identically zero. From (92) and (122), we observe that

$$\int_{T_i}^{T_f} \tilde{h}_{ou}^*(t, u)\tilde{f}_o^*(u)\, du = [\tilde{\mathbf{C}}(t)\tilde{\boldsymbol{\xi}}(t)]^* \qquad \text{when } \varepsilon = 0. \tag{131}$$

We define

$$\tilde{p}_f(t) = -\lambda_B \dot{\tilde{f}}_o(t). \tag{132}$$

We now have the following set of differential equations that specify $\tilde{f}_o(t)$:

$$\dot{\tilde{p}}_f(t) = -\lambda_E \tilde{f}_o(t) - \tilde{\mathbf{C}}(t)\tilde{\boldsymbol{\xi}}(t), \tag{133}$$

$$\dot{\tilde{f}}_o(t) = -\frac{1}{\lambda_B}\tilde{p}_f(t), \tag{134}$$

$$\dot{\tilde{\boldsymbol{\xi}}}(t) = \tilde{\mathbf{F}}(t)\tilde{\boldsymbol{\xi}}(t) + \tilde{\mathbf{G}}(t)\mathbf{Q}\tilde{\mathbf{G}}^\dagger(t)\tilde{\boldsymbol{\eta}}(t), \tag{135}$$

$$\dot{\tilde{\boldsymbol{\eta}}}(t) = \frac{1}{N_0}\tilde{\mathbf{C}}^\dagger(t)\tilde{\mathbf{C}}(t)\tilde{\boldsymbol{\xi}}(t) - \tilde{\mathbf{F}}^\dagger(t)\tilde{\boldsymbol{\eta}}(t) - \frac{1}{N_0}\tilde{\mathbf{C}}^\dagger(t)\tilde{f}_o(t), \tag{136}$$

with boundary conditions

$$\tilde{f}_0(T_i) = \tilde{f}_0(T_f) = 0, \tag{137}$$

$$\tilde{\boldsymbol{\xi}}(T_i) = \tilde{\mathbf{P}}_0\tilde{\boldsymbol{\eta}}(T_i), \tag{138}$$

$$\tilde{\boldsymbol{\eta}}(T_f) = \mathbf{0}. \tag{139}$$

If the process state vector is $n$-dimensional, we have $2n + 2$ linear equations. We must solve these as a function of $\lambda_E$ and $\lambda_B$ and then evaluate $\lambda_E$ and $\lambda_B$ by using the constraint equations (123) and (124). Since (128) is only a necessary condition, we get several solutions that satisfy (133)–(139) and (123)–(125). Therefore, we must choose the solution that gives the absolute minimum. Baggeroer [3], [4] originally derived (133)–(139) using Pontryagin's principle, and carried out the solution for some typical real-valued processes. The interested reader should consult these two references for further details.

Frequently we want to impose hard constraints on the signal instead of the quadratic constraints in (123) and (124). For example, we can require

$$|\tilde{f}(t)| < A, \qquad T_i \le t \le T_f. \tag{140}$$

In this case we can use Pontryagin's principle (cf. [5] or [6]) to find the equations specifying the optimal signal.

The purpose of this brief discussion is to demonstrate how the state-variable formulation can be used to study optimal signal design. Other signal design problems will be encountered as we proceed through the text.

## 9.6 SUMMARY AND RELATED ISSUES

In this chapter we have discussed the problem of detecting the return from a slowly fluctuating point target in additive noise. The derivations were all straightforward extensions of our earlier work. Several important results should be emphasized:

1. When the additive noise is white, the optimum receiver is as shown in Fig. 9.4. The received waveform is passed through a bandpass matched filter and a square-law envelope detector. The output of the envelope detector is sampled and compared with a threshold. The performance is a monotonic function of $\bar{E}_r/N_0$,

$$P_F = (P_D)^{1+\bar{E}_r/N_0}. \tag{141}$$

2. When the additive noise is nonwhite, the optimum receiver is as shown in Fig. 9.7. The only difference is in the impulse response of the matched filter. The performance is a function of $\Delta$,

$$\Delta = \bar{E}_r \int\!\!\int_{T_i}^{T_f} \tilde{f}(t)\tilde{Q}_{\tilde{n}}^*(t, u)\tilde{f}^*(u)\, dt\, du. \tag{142}$$

Specific nonwhite noises will be studied later.

3. When the colored noise has a finite-dimensional state representation, the optimum receiver implementation is as shown in Fig. 9.9. The advantage of this implementation is that it avoids solving an integral equation.

There are several related issues that should be mentioned. In many radar/sonar systems it is necessary to illuminate the target with a number of pulses in order to achieve satisfactory performance. A typical transmitted sequence is shown in Fig. 9.10. Once again we assume that the Rayleigh reflection model developed in Section 9.1 is valid. We must now specify how the returns from successive pulses are related. There are three cases of interest.

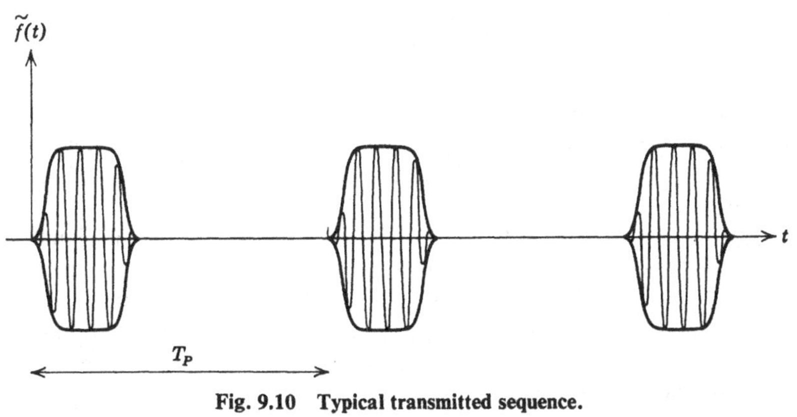


**Fig. 9.10 Typical transmitted sequence.**

In the first case, the target does not fluctuate during the time the *entire* sequence illuminates it. In this case we can write the received signal from a zero-velocity target as

$$s_r(t) = \sqrt{2}\,\text{Re}\left\{\tilde{b}\sum_{i=1}^{N}\tilde{f}_s(t - iT_p - \tau)e^{j\omega_c t}\right\}. \tag{143}$$

Notice that there is a single complex multiplier, $\tilde{b}$. This model might be appropriate for a radar with a high pulse repetition rate and a target where small movements do not affect the return appreciably. Comparing (25) and (143), we see that this reduces to the problem that we just solved if we define

$$\tilde{f}(t) \triangleq \sum_{i=1}^{N}\tilde{f}_s(t - iT_p). \tag{144}$$

In a white noise environment, the optimum receiver has a bandpass filter matched to the subpulse. The sampled outputs are added *before* envelope detection. The performance is determined by

$$\Delta = \frac{\bar{E}_r}{N_0} = \frac{N(2\sigma^2 E_i)}{N_0}, \tag{145}$$

where $E_i$ is the transmitted energy in each subpulse.

In the second case we assume that $|\tilde{b}|$ has the same value on all pulses, but we model the phase of each pulse as a statistically independent, uniformly distributed random variable. This model might be appropriate in the same target environment as case 1 when the radar does not have pulse-to-pulse coherence. The optimum receiver for the problem is derived in Problem 9.6.1.

In the third case, the target fluctuates enough so that the returns from successive subpulses are *statistically independent*. Then

$$s_r(t) = \sqrt{2}\,\text{Re}\left\{\sum_{i=1}^{N}\tilde{b}_i\tilde{f}_s(t - iT_p - \tau)e^{j\omega_c t}\right\}. \tag{146}$$

The $\tilde{b}_i$ are zero-mean, statistically independent, complex Gaussian random variables with identical statistics. This model is appropriate when small changes in the target orientation give rise to significant changes in the reflected signal.

This model corresponds to the separable-kernel Gaussian signal-in-noise problem that we discussed in Section 4.2. The optimum receiver passes the received waveform through a bandpass filter matched to the subpulse and a square-law envelope detector. The detector output is sampled every $T_p$ seconds, and the samples are summed. The sum is compared with a threshold in order to make a decision.

The performance is evaluated just as in Section 4.2 (see Problem 9.6.2). The performance for this particular model has been investigated extensively by Swerling (Case II in [7]).

A second related issue is that of digital communication over a slowly fluctuating Rayleigh channel using a binary or $M$-ary signaling scheme. Here the complex envelope of the received signal is

$$\tilde{r}(t) = \sqrt{E_k}\,\tilde{b}\tilde{f}_k(t) + \tilde{n}(t), \qquad T_i \leq t \leq T_f : H_k, \qquad k = 1, \ldots, M. \tag{147}$$

The optimum receiver follows easily (see Problem 9.6.7). We shall return to the performance in a later chapter.

This completes our initial discussion of the detection problem. In the next chapter we consider the parameter estimation problem. Later we consider some further topics in detection.

## 9.7 PROBLEMS

### P.9.2 Detection in White Noise

#### Suboptimum Receivers

**Problem 9.2.1.** The optimum receiver in the presence of white noise is specified by (36) and (39). Consider the suboptimum receiver that computes

$$\tilde{l}_v = \int_{T_i}^{T_f} \tilde{r}(t)\tilde{v}^*(t)\,dt \tag{P.1}$$

and compares $|\tilde{l}_v|^2$ with a threshold $\gamma$. The function $\tilde{v}(t)$ is arbitrary.

1. Verify that the performance of this receiver is completely characterized by letting $\Delta = \Delta_v$ in (50), where

$$\Delta_v \triangleq \frac{E[|\tilde{l}_v|^2 \mid H_1] - E[|\tilde{l}_v|^2 \mid H_0]}{E[|\tilde{l}_v|^2 \mid H_0]}. \tag{P.2}$$

2. Calculate $\Delta_v$ for the input

$$\tilde{r}(t) = \tilde{f}(t) + \tilde{w}(t), \qquad T_i \leq t \leq T_f.$$

3. The results in parts 1 and 2 give an expression for $\Delta_v$ as a functional of $\tilde{v}(t)$. Find the function $\tilde{v}(t)$ that minimizes $\Delta_v$. [This is the structured approach to the optimum receiver of (36) and (39).]

**Problem 9.2.2.** Assume that

$$\tilde{f}(t) = \begin{cases} \sqrt{\dfrac{1}{T}}, & 0 \leq t \leq T, \\ 0, & \text{elsewhere.} \end{cases}$$

The complex envelope of the received waveform is

$$\tilde{r}(t) = \sqrt{E}\tilde{f}(t) + \tilde{w}(t), \qquad -\infty < t < \infty.$$

1. Plot the output of the matched filter as a function of time.

2. Assume that instead of using a matched filter, we use a bandpass filter centered at $\omega_c$, whose complex envelope's transfer function is

$$\tilde{H}(f) = \begin{cases} 1, & |f| \le \dfrac{W}{2} \\ 0, & |f| > \dfrac{W}{2}. \end{cases}$$

Denote the output of this filter due to the signal as $\tilde{f}_0(t)$, and the output due to noise as $\tilde{w}_0(t)$. Define

$$\Delta_c = \frac{\max\limits_t |\tilde{f}_0(t)|^2}{E[|\tilde{w}_0(t)|^2]}.$$

Verify that this quantity corresponds to $\Delta_v$ as defined in (P.2) of Problem 9.2.1. Plot

$$\Lambda_{cn} = \frac{\Delta_c}{\bar{E}_r/N_0}$$

as a function of $WT$. What is the optimum value of $WT$? What is $\Delta_{cn}$ in decibels at this optimum value? Is $\Delta_{cn}$ sensitive to the value of $WT$?

**Problem 9.2.3.** The complex envelope of the transmitted signal is

$$\tilde{f}(t) = a \sum_{i=1}^{N} \tilde{u}(t - iT_p),$$

where

$$\tilde{u}(t) = \begin{cases} \dfrac{1}{\sqrt{T_s}}, & 0 \le t \le T_s, \\ 0, & \text{elsewhere}, \end{cases}$$

and

$$T_p \gg T_s.$$

1. Plot the Fourier transform of $\tilde{f}(t)$.

2. The matched filter for $\tilde{f}(t)$ is sometimes referred to as a "comb filter." Consider the filter response

$$\tilde{H}\{f\} = \sum_{i=-M}^{M} \tilde{Y}\{f - iW_p\},$$

where

$$\tilde{Y}\{f\} = \begin{cases} 1, & |f| < \dfrac{W_s}{2}, \\ 0, & \text{elsewhere}. \end{cases}$$

and

$$W_p \gg W_s.$$

Assume that

$$W_p = \frac{1}{T_p},$$

$$W_s = \frac{2}{NT_p},$$

and that $M$ is the smallest integer greater than $T_p/T_s$.

(i) Sketch $\tilde{H}\{f\}$.

(ii) Find the degradation in $\Delta$ due to this suboptimum filter.

(iii) Why might one use $\tilde{H}\{f\}$ instead of the optimum filter?

COMMUNICATION SYSTEMS

**Problem 9.2.4.** An on-off signaling scheme operates over two frequency-diversity channels. The received waveforms on the two hypotheses are

$$r(t) = \sqrt{E_t}\ \text{Re}\ \{\tilde{b}_1\tilde{f}(t)e^{j\omega_1 t} + \tilde{b}_2\tilde{f}(t)e^{j\omega_2 t}\} + n(t), \qquad 0 \le t \le T: H_1,$$

$$r(t) = n(t), \qquad 0 \le t \le T: H_0,$$

The multipliers $\tilde{b}_1$ and $\tilde{b}_2$ are statistically independent, zero-mean complex Gaussian random variables

$$E[\tilde{b}_1\tilde{b}_1^*] = E[\tilde{b}_2\tilde{b}_2^*] = 2\sigma_b^2.$$

The frequencies $\omega_1$ and $\omega_2$ are such that the signals are essentially disjoint. The *total* energy transmitted is $E_t$. (There is $E_t/2$ in each channel.) The additive noise $n(t)$ is a sample function from a zero-mean, white Gaussian process with spectral height $N_0/2$.

1. Find the optimum receiver.
2. Find $P_D$ and $P_F$ as a function of the threshold.
3. Assume a minimum probability-of-error criterion and equal a priori probabilities. Find the threshold setting and the resulting Pr ($\epsilon$).

**Problem 9.2.5.** Consider the model in Problem 9.2.4. Assume that the two channels have unequal strengths and that we use unequal energies in the two channels. Thus,

$$r(t) = \sqrt{2}\ \text{Re}\ \{\sqrt{E_1}\ \tilde{b}_1\tilde{f}(t)e^{j\omega_1 t} + \sqrt{E_2}\ \tilde{b}_2\tilde{f}(t)e^{j\omega_2 t}\} + n(t), \qquad 0 \le t \le T: H_1,$$

where

$$E_1 + E_2 = E_t. \tag{P.1}$$

The received waveform on $H_0$ is the same as in Problem 9.2.4. The mean-square values of the channel variables are

$$E[\tilde{b}_1\tilde{b}_1^*] = 2\sigma_1^2$$

and

$$E[\tilde{b}_2\tilde{b}_2^*] = 2\sigma_2^2.$$

1. Find the optimum receiver.
2. Find $P_D$ and $P_F$ as functions of the threshold.
3. Assume a minimum probability-of-error criterion. Find the threshold setting and the resulting Pr ($\epsilon$).

**Problem 9.2.6.** Consider the model in Problem 9.2.5. Now assume that the channel gains are correlated.

$$\tilde{b} \triangleq \begin{bmatrix} \tilde{b}_1 \\ \tilde{b}_2 \end{bmatrix}$$

and

$$E[\tilde{b}\tilde{b}^{\dagger}] = \tilde{\Lambda}_{\tilde{b}}.$$

Repeat parts 1–3 of Problem 9.2.5.

**Problem 9.2.7.** In an on-off signaling system a signal is transmitted over $N$ Rayleigh channels when $H_1$ is true. The received waveforms on the two hypotheses are

$$r(t) = \sqrt{\frac{2E_t}{N}} \operatorname{Re}\left\{\sum_{i=1}^{N} \tilde{b}_i \tilde{f}(t) e^{j\omega_i t}\right\} + n(t), \qquad 0 \le t \le T: H_1,$$

$$r(t) = n(t), \qquad 0 \le t \le T: H_0,$$

The channel multipliers are statistically independent, zero-mean, complex Gaussian random variables

$$E[\tilde{b}_i \tilde{b}_j^*] = 2\sigma_b^2 \delta_{ij}.$$

The frequencies are such that the signal components are disjoint. The total energy transmitted is $E_t$. The additive noise $n(t)$ is a zero-mean Gaussian process with spectral height $N_0/2$.

1. Find the optimum receiver.
2. Find $\mu(s)$.
3. Assume that the criterion has a minimum probability of error. Find an approximate Pr ($\epsilon$). (*Hint*: Review Section I-2.7.)

**Problem 9.2.8.** Consider the model in Problem 9.2.7. Assume that the transmitted energy in the $i$th channel is $E_i$, where

$$\sum_{i=1}^{N} E_i = E_t.$$

Assume that the channel multipliers are correlated:

$$E[\tilde{\mathbf{b}}\tilde{\mathbf{b}}^{\dagger}] = \tilde{\mathbf{\Lambda}}_{\tilde{\mathbf{b}}}.$$

1. Find the optimum receiver.
2. Find $\mu(s)$.

## Alternative Target Models

**Problem 9.2.9.** Consider the target model given in (32*a*) and (32*b*). Assume that the phase is a uniform random variable.

1. Derive the optimum receiver.
2. Calculate $P_D$ and $P_F$.
3. Assume that we require the same $P_F$ in this system and the system corresponding to the Rayleigh model. Find an expression for the ratio of the values of $P_D$ in the two systems.

**Problem 9.2.10.** Consider the target model in (32*c*). Repeat parts 1 and 2 of Problem 9.2.9.

## P.9.3 Detection in Colored Noise

**Problem 9.3.1.** Consider the receiver specified in (P.1) of Problem 9.2.1. The inputs on the two hypotheses are specified by (57)–(59).

1. Verify that the results in part 1 of Problem 9.2.1 are still valid.
2. Calculate $\Delta_v$ for the model in (57)–(59).
3. Find the function $\tilde{v}(t)$ that minimizes $\Delta_v$.

**Problem 9.3.2.** Consider the model in (57)–(59). Assume that

$$\tilde{f}(t) = \begin{cases} \dfrac{1}{\sqrt{T_s}}, & 0 \le t \le T_s, \\ 0, & \text{elsewhere,} \end{cases}$$

and that

$$\tilde{S}_c(\omega) = \frac{2kP_c}{\omega^2 + k^2}, \qquad -\infty < \omega < \infty.$$

The observation interval is infinite.

1. Find $g_{\infty}(\tau)$.
2. Evaluate $\Delta_0$ as a function of $E_t$, $k$, $T_s$, $P_c$, and $N_0$.
3. What value of $T_s$ maximizes $\Delta_0$? Explain this result intuitively.

**Problem 9.3.3.** Assume that

$$\tilde{n}_c(t) = n_1(t) - jn_2(t), \qquad -\infty < t < \infty.$$

The function $n_1(t)$ is generated by passing $u_1(t)$ through the filter

$$H_1(j\omega) = \frac{\sqrt{2k}}{j\omega + k},$$

and the function $n_2(t)$ is generated by passing $u_2(t)$ through an identical filter. The inputs $u_1(t)$ and $u_2(t)$ are sample functions of real, white Gaussian processes with unity spectral height and

$$E[u_1(t_1)u_2(t_2)] = \alpha\delta(t_1 - t_2 - \Delta).$$

1. Find $\tilde{S}_{\tilde{n}_c}(\omega)$.

2. Consider the model in (57)–(59) and assume that the observation interval is infinite. Find an expression for a realizable whitening filter whose inverse is also realizable.

**Problem 9.3.4.** Assume that

$$\tilde{n}_c(t) = \sum_{i=1}^{N} \tilde{a}_i \tilde{k}_i(t),$$

where the $\tilde{k}_i(t)$ are known functions with unit energy and the $\tilde{a}_i$ are statistically independent, complex Gaussian random variables with

$$E[|\tilde{a}_i|^2] = 2\sigma_i^2.$$

The observation interval is infinite. The model in (57)-(59) is assumed.

1. Find $\tilde{g}(t)$. Introduce suitable matrix notation to keep the problem simple.
2. Consider the special case in which

$$\tilde{k}_i(t) = \tilde{f}(t - \tau_i)e^{j\omega_i t},$$

where $\tau_i$ and $\omega_i$ are known constants. Draw a block diagram of the optimum receiver.

**Problem 9.3.5.** Assume that

$$\tilde{f}(t) = \sum_{i=1}^{N} \tilde{f}_i \tilde{u}(t - iT_s), \tag{P.1}$$

where

$$\tilde{u}(t) = \begin{cases} \dfrac{1}{\sqrt{T_s}}, & 0 \le t \le T_s, \\ 0, & \text{elsewhere}, \end{cases} \tag{P.2}$$

and

$$\sum_{i=1}^{N} |\tilde{f}_i|^2 = 1. \tag{P.3}$$

Assume that we use the receiver in Problem 9.2.1 and that

$$\tilde{v}(t) = \sum_{i=1}^{N} \tilde{v}_i \tilde{u}(t - iT_s), \tag{P.4}$$

where

$$\sum_{i=1}^{N} |\tilde{v}_i|^2 = 1. \tag{P.5}$$

The model in (57)–(59) is valid and the observation is infinite. Define a filter-weighting vector as

$$\mathbf{v} = \begin{bmatrix} \tilde{v}_1 \\ \tilde{v}_2 \\ \cdot \\ \cdot \\ \cdot \\ \tilde{v}_N \end{bmatrix}. \tag{P.6}$$

1. Find an expression for $\Delta_v$ in terms of $\tilde{\mathbf{f}}$, $\tilde{\mathbf{v}}$, $E_t$, $N_0$, and $\tilde{K}_c(t, u)$. Introduce suitable matrices.
2. Choose $\tilde{\mathbf{v}}$ subject to the constraint in (P.5) in order to maximize $\Delta_v$.

**Problem 9.3.6.** The complex envelopes on the two hypotheses are

$$\begin{aligned} r(t) &= \sqrt{E_t}\,\tilde{b}\tilde{f}(t) + \tilde{n}_c(t) + \tilde{w}(t), \qquad -\infty < t < \infty : H_1, \\ r(t) &= \tilde{n}_c(t) + \tilde{w}(t), \qquad -\infty < t < \infty : H_0. \end{aligned} \tag{P.1}$$

The signal has unit energy

$$\int_{-\infty}^{\infty} |\tilde{f}(t)|^2\, dt = 1. \tag{P.2}$$

The colored noise is a sample function of a zero-mean complex Gaussian process with spectrum $S_c(\omega)$, where

$$\int_{-\infty}^{\infty} \tilde{S}_c(\omega)\frac{d\omega}{2\pi} = 2\sigma_c^2. \tag{P.3}$$

The white noise is a zero-mean complex Gaussian process with spectral height $N_0$. The multiplier $\tilde{b}$ is a zero-mean complex Gaussian random variable,

$$E[\tilde{b}\tilde{b}^*] = 2\sigma_b^2. \tag{P.4}$$

The various random processes and random variables are all statistically independent.

1. Find the colored noise spectrum $\tilde{S}_c(\omega)$ that satisfies the constraint in (P.3) and minimizes $\Delta$ as defined in (76). Observe that $\Delta$ can also be written as

$$\Delta = \bar{E}_r \int_{-\infty}^{\infty} \frac{|\tilde{F}(j\omega)|^2}{S_{\tilde{n}}(\omega)}\frac{d\omega}{2\pi}.$$

(*Hint:* Recall the technique in Chapter II-5. Denote the minimum $\Delta$ as $\Delta_m$.)

2. Evaluate $\Delta_m$ for the signal

$$\tilde{f}(t) = \begin{cases} \alpha e^{-\alpha t}, & t \geq 0, \\ 0, & t < 0. \end{cases}$$

3. Evaluate $\Delta_m$ for the signal

$$\tilde{F}\{f\} = \begin{cases} \dfrac{1}{\sqrt{W}}, & |f| \leq W \\ 0, & |f| > W. \end{cases}$$

**Problem 9.3.7.** Consider the same model as Problem 9.3.6. We want to design the optimum signal subject to an energy and bandwidth constraint. Assume that $\tilde{S}_c(\omega)$ is symmetric around zero and that we require

$$\int_{-\infty}^{\infty} \omega\tilde{F}(j\omega) = 0, \tag{P.1}$$

$$\int_{-\infty}^{\infty} \omega^2\,|\tilde{F}(j\omega)|^2 \leq \Omega_B. \tag{P.2}$$

1. Verify that $\Delta$ depends only on the signal shape through

$$S_{\tilde{f}}(\omega) \triangleq |\tilde{F}(j\omega)|^2.$$

2. Find the $S_{\tilde{f}}(\omega)$ subject to the constraints in (P.1) and (P.2) of this problem and in (P.2) of Problem 9.3.6, such that $\Delta$ is maximized.

3. Is your answer to part 2 intuitively correct?

4. What is the effect of removing the symmetry requirement on $\tilde{S}_c(\omega)$ and the requirement on $\tilde{F}(j\omega)$ in (P.1)? Discuss the implications in the context of some particular spectra.

**Problem 9.3.8.** Consider the model in Problem 9.3.5. Assume that the complex envelope of the desired signal is

$$\tilde{f}_d(t) = \tilde{f}(t)e^{j\omega_d t}$$

and that $\tilde{\mathbf{v}}$ is chosen to maximize $\Delta_v$ for this desired signal. Assume that $\tilde{K}_c(t, u)$ is a stationary process whose spectrum is

$$\tilde{S}_c(\omega) \triangleq \frac{E_t P_c}{N_0} |\tilde{F}(j\omega)|^2.$$

1. Assume that $N = 2$. Find $\tilde{v}_1$ and $\tilde{v}_2$.
2. Assume that

$$\tilde{f}_i = 1$$

and $N = 3$. Find $\tilde{v}_1$, $\tilde{v}_2$, and $\tilde{v}_3$.

## P.9.4 Finite-state Noise Processes

**Problem 9.4.1.** Consider the model in (79)–(92). Derive the results in (93)–(96). *Hint:* Read Sections A.4–A.6 of the Appendix to Part II.

**Problem 9.4.2.** Assume that $\tilde{n}_c(t)$ has the state representation in (A.137)–(A.140). Write out (93)–(96) in detail.

**Problem 9.4.3.** Assume that $\tilde{n}_c(t)$ has the state representation in (A.148)–(A.153). Write out (93)–(96) in detail.

**Problem 9.4.4.** Consider the model in Section 9.4.2. Assume that $\tilde{n}_c(t)$ is a complex Gaussian process whose real and imaginary parts are statistically independent Wiener processes. Find the necessary functions for the receiver in Fig. 9.9.

**Problem 9.4.5.** Verify the result in (120).

**Problem 9.4.6.** Consider the model in Section 9.4.2. Assume that

$$\tilde{n}_c(t) = \sum_{i=1}^{N} \tilde{b}_i(t)\tilde{f}(t - \tau_i)e^{j\omega_i t},$$

where the $\tilde{b}_i(t)$ are statistically independent, complex Gaussian processes with the state representation in (A.137)–(A.140). Draw a block diagram of the optimum receiver in Fig. 9.9. Write out the necessary equations in detail.

## P.9.5 Optimum Signal Design

**Problem 9.5.1.** Consider the optimum signal design problem in Section 9.5. Assume that

$$\tilde{S}_c(\omega) = \frac{2\alpha}{\omega^2 + \alpha^2}, \qquad -\infty < \omega < \infty.$$

Write the equations specifying the optimum signal in detail.

**Problem 9.5.2.** The optimal signal-design problem is appreciably simpler if we constrain the form of the signal and receiver. Assume that $\tilde{f}(t)$ is characterized by (P.1–P.3) in Problem 9.3.5, and that we require

$$\tilde{v}(t) = \tilde{f}(t)$$

[see (P.4)–(P.5) in Problem 9.3.5].

1. Express $\Delta_v$ in terms of the $\tilde{f}_i$, $E_t$, $N_0$, and $\tilde{K}_c(t, u)$.
2. Maximize $\Delta_v$ by choosing the $\tilde{f}_i$ optimally.

3. Assume that

$$\tilde{K}_c(t, u) = e^{-k|t-u|}$$

and $N = 2$. Solve the equations in part 2 to find the optimum value of $\tilde{f}_1$ and $\tilde{f}_2$.

4. Consider the covariance in part 3 and assume that $N = 3$. Find the optimum values of $\tilde{f}_1$, $\tilde{f}_2$, and $\tilde{f}_3$.

**Problem 9.5.3.** Consider the generalization of Problem 9.5.2, in which we let

$$\tilde{f}_i = \tilde{a}_i \, e^{j\omega_i t},$$

where the $\tilde{a}_i$ are complex numbers such that

$$\sum_{i=1}^{N} |\tilde{a}_i|^2 = 1$$

and the $\omega_i$ may take on values

$$|\omega_i| \leq 2\pi W.$$

The remainder of the model in Problem 9.5.2 is still valid.

1. Express $\Delta_v$ in terms of $\tilde{\mathbf{a}}$, $\omega_i$, $E_t$, $N_0$, and $K_c(t, u)$.
2. Explain how the $\omega_i$ should be chosen in order to maximize $\Delta_v$.
3. Carry out the procedure in part 2 for the covariance function in part 3 of Problem 9.5.2 for $N = 2$. Is your result intuitively obvious?

**Problem 9.5.4.** Consider the models in Problems 9.3.5 and 9.5.2. Assume that $\tilde{\mathbf{v}}$ is chosen to maximize $\Delta_v$. Call the maximum value $\Delta_{v_o}$.

1. Express $\Delta_{v_o}$ as a function of $\tilde{\mathbf{f}}$, $E_t$, $N_0$, and $\tilde{K}_c(t, u)$.
2. Find that value of $\tilde{\mathbf{f}}$ that maximizes $\Delta_{v_0}$.
3. Consider the special case in part 3 of Problem 9.5.2. Find the optimum $\tilde{\mathbf{f}}$ and compare it with the optimum $\tilde{\mathbf{f}}$ in part 3 of Problem 9.5.2.
4. Repeat part 3 for $N = 3$.

## P.9.6 Related Issues

### Multiple Observations

**Problem 9.6.1.** The complex envelopes on the received waveforms on the two hypotheses are

$$\tilde{r}(t) = \sqrt{\frac{E_t}{N}} \sum_{i=1}^{N} |\tilde{b}|\, \tilde{u}(t - iT_p) e^{j\theta_i} + \tilde{w}(t), \qquad -\infty < t < \infty,$$

$$\tilde{r}(t) = \tilde{w}(t), \qquad -\infty < t < \infty,$$

where

$$\tilde{u}(t) = \begin{cases} \dfrac{1}{\sqrt{T_s}}, & 0 \leq t \leq T_s, \\ 0, & \text{elsewhere.} \end{cases}$$

The multiplier $|\tilde{b}|$ is a Rayleigh random variable with mean-square value $2\sigma_b^2$. The $\theta_i$ are statistically independent, uniform random variables.

1. Find the optimum receiver.
2. Evaluate $P_F$.
3. Set up the expressions to evaluate $P_D$. Extensive performance results for this model are given by Swerling [7], [10].

**Problem 9.6.2.** Consider the target model in (146). Review the discussion in Section 4.2.2.

1. Draw a block diagram of the optimum receiver.
2. Review the performance results in Section 4.2.2. Observe that fixing $s$ in the $\mu_{\text{BP,SK}}(s)$ expression fixes the threshold and, therefore, $P_F$. Fix $s$ and assume that

$$K\bar{E}_{r_1} \triangleq \bar{E}_r$$

is fixed. Find the value of $K$ that minimizes $\mu_{\text{BS,SK}}(s)$ as a function of $s$. Discuss the implications of this result in the context of an actual radar system.

**Problem 9.6.3.** Consider the model in Problem 9.6.1. Define

$$z \; |\tilde{b}|^2$$

and assume that $z$ has the probability density given in (32b).

1. Derive the optimum receiver.
2. Evaluate $P_F$.
3. Set up the expressions to evaluate $P_D$. Results for this model are given in [7] and [10] (Case III in those references).

**Problem 9.6.4.** Consider the model in (146). Write

$$\tilde{b}_i = |\tilde{b}_i| \, e^{j\theta_i}.$$

Assume that the $\theta_i$ are statistically independent random variables with a uniform probability density, Assume that each

$$z_i \triangleq |\tilde{b}_i|^2$$

has the probability density in (32*b*) and the $z_i$ are statistically independent.

1. Derive the optimum receiver.
2. Evaluate $P_F$.
3. Set up the expressions to evaluate $P_D$. See [7] and [10] for performance results (Case IV in those references). Chapter 11 of [23] has extensive performance results based on Swerling's work.

### Communication Systems

**Problem 9.6.5.** The complex received waveforms on the two hypotheses in a binary communication system are

$$\tilde{r}(t) = \sqrt{E_t}\,\tilde{b}\tilde{f}(t)e^{j\omega_\Delta t} + \tilde{w}(t), \qquad 0 \le t \le T: H_1,$$

$$\tilde{r}(t) = \sqrt{E_t}\,\tilde{b}\tilde{f}(t) + \tilde{w}(t), \qquad 0 \le t \le T: H_0,$$

where $\omega_\Delta$ is large enough for the two signal components to be orthogonal. The hypotheses are equally likely, and the criterion is minimum probability of error.

1. Draw a block diagram of the optimum receiver.
2. Calculate the probability of error.

**Problem 9.6.6.** The complex received waveforms on the two hypotheses in a binary communication system are

$$\tilde{r}(t) = \sqrt{E_t}\,\tilde{b}\tilde{f}_1(t) + \tilde{w}(t), \qquad 0 \le t \le T: H_1,$$
$$\tilde{r}(t) = \sqrt{E_t}\,\tilde{b}\tilde{f}_0(t) + \tilde{w}(t), \qquad 0 \le t \le T: H_0,$$

where

$$\int_0^T \tilde{f}_0(t)\tilde{f}_1^*(t)\,dt = \tilde{\rho}_{01}.$$

1. Draw a block diagram of the optimum receiver.
2. Calculate the probability of error.

**Problem 9.6.7.** Consider the model in (147) and assume that

$$\int_{T_i}^{T_f} \tilde{f}_k(t)\tilde{f}_m^*(t)\,dt = \delta_{km}.$$

The hypotheses are equally likely, and the criterion is minimum probability of error.

1. Draw a block diagram of the optimum receiver.
2. Use the union bound on pages I-263–I-264 to approximate Pr ($\epsilon$).

(*Comment:* The reader who is interested in other communications problems should look at Sections P.4.4 and P.4.5 in Part I (Pages I-394–I-416). Most of those problems could also be included at this point.)

## REFERENCES


[1] M. Slack, "Probability Densities of Sinusoidal Oscillations Combined in Random Phase," J. IEE **93**, Pt. III, 76–86 (1946).

[2] W. R. Bennett, "Distribution of the Sum of Randomly Phased Components," Quart. J. Appl. Math. **5**, 385–393 (Jan. 1948).

[3] A. B. Baggeroer, "State Variables, the Fredholm Theory, and Optimal Communications," Sc.D. Thesis, Department of Electrical Engineering, Massachusetts Institute of Technology, January 1968.

[4] A. B. Baggeroer, *State Variables and Communication Theory*, Massachusetts Institute of Technology Press, Cambridge, Mass., 1970.

[5] M. Athans and P. L. Falb, *Optimal Control*, McGraw-Hill, New York, 1966.

[6] L. S. Pontryagin, V. Boltyanskii, R. Gamkrelidze, and E. Mishchenko, *The Mathematical Theory of Optimal Processes*, Interscience Publishers, New York, 1962.

[7] P. Swerling, "Probability of Detection for Fluctuating Targets," RAND Report RM-1217, March 1954.

[8] J. I. Marcum, "A Statistical Theory of Target Detection by Pulsed Radar," RAND Report RM-754, December 1947.

[9] J. I. Marcum, "A Statistical Theory of Target Detection by Pulsed Radar, Mathematical Appendix," RAND Report RM-753, July 1948.

[10] J. I. Marcum and P. Swerling, "Studies of Target Detection by Pulsed Radar," IRE Trans. Information Theory **IT-6,** No. 2, (April 1960) (This is a reprint of [7]–[9].)

[11] P. Swerling, "Detection of Fluctuating Pulsed Signals in the Presence of Noise," IRE Trans. Information Theory **IT-3,** 175–178 (Sept. 1957).

[12] R. S. Berkowitz (Ed.), *Modern Radar, Analysis, Evaluation and System Design,* Wiley, New York, 1965.

[13] T. S. Edrington, "The Amplitude Statistic of Aircraft Radar Echoes," IEEE Trans. Military Electronics **MIL-9,** No. 1, 10–16 (Jan. 1965).

[14] M. Nakagami, "Statistical Characteristics of Short-Wave Fading," J. Inst. Elec. Commun. Engrs. Japan **239** (Feb. 1943).

[15] M. Nakagami, in *Statistical Methods in Radio Wave Propagation* (W. C. Hoffman, Ed.), Pergamon Press, New York, 1960.

[16] S. O. Rice, "Mathematical Analysis of Random Noise," Bell Syst. Tech. J. **23,** 283–332 (1944); **24,** 46–156 (1945).

[17] B. B. Barrow, "Error Probabilities for Data Transmission over Fading Radio Paths," SHAPE ADTC, The Hague, Rept. TM-26, 1962.

[18] M. Nesenberg, "Binary Error Probability Due to an Adaptable Fading Model," IEEE Trans. Commun. Syst. **CS-12,** 64–73 (March 1964).

[19] N. Hveding, "Comparison of Digital Modulation and Detection Techniques for a Low-Power Transportable Troposcatter System," Proc. First IEEE Annual Commun. Conv., Boulder, Colo. 691–694 (June 1965).

[20] R. Esposito and J. A. Mullen, "Amplitude Distributions for Fading Signals: Nakagami and Edgeworth Expansions," Fall URSI-IEEE Meeting, 1965.

[21] R. Esposito, "Error Probabilities for the Nakagami Channel," IEEE Trans. Information Theory **IT-13,** 145–148 (Jan. 1961).

[22] M. Nakagami, S. Wada, and S. Fujimura, "Some Considerations on Random Phase Problems from the Standpoint of Fading," J. Inst. Elec. Commun. Engrs. Japan, (Nov. 1953).

[23] J. V. Di Franco and W. L. Rubin, *Radar Detection,* Prentice-Hall, Englewood Cliffs, N.J., 1968.

# 10

# *Parameter Estimation: Slowly Fluctuating Point Targets*

At the beginning of Chapter 9, we developed a model for the return from a slowly fluctuating point target that was located at a particular range and was moving at a particular velocity. The received signal in the absence of noise was

$$s(t) = \sqrt{2}\,\mathrm{Re}\,[\sqrt{E_t}\,\tilde{b}\tilde{f}(t - \tau)e^{j\omega_D t}]. \tag{1}$$

In the detection problem we assumed that $\tau$ and $\omega_D$ were known, and made a decision on the presence or absence of a target. We now consider the problem in which $\tau$ and $\omega_D$ are *unknown*, *nonrandom* parameters that we want to estimate.

Since the chapter is long, we briefly describe its organization. In Section 10.1 we derive the optimum receiver and discuss signal design qualitatively. In Section 10.2 we analyze the performance of the optimum receiver. We find that a function called the ambiguity function plays a central role in the performance discussion. In Section 10.3 we develop a number of properties of this function, which serve as a foundation for the signal design problem. In Section 10.4 we investigate the performance of coded pulse sequences. In Section 10.5 we consider the situation in which there are interfering targets in addition to the desired target whose parameters we want to estimate. Finally, in Section 10.6, we summarize our results and discuss several related topics.

## 10.1 RECEIVER DERIVATION AND SIGNAL DESIGN

The target reflection model was discussed in Section 9.1, and the received signal in the absence of noise is given in (1). We assume that the additive noise is white bandpass Gaussian noise with spectral height

$N_0/2$. We shall assume that the observation interval is infinite. For notational simplicity we drop the subscript $D$ from the frequency shift. Thus, the complex envelope of the received waveform is

$$\tilde{r}(t) = \tilde{b}\sqrt{E_t}\tilde{f}(t - \tau)e^{j\omega t} + \tilde{w}(t), \qquad -\infty < t < \infty. \tag{2}$$

The multiplier, $\tilde{b}$, is a zero-mean complex Gaussian random variable,

$$E[|\tilde{b}|^2] = 2\sigma_b^2. \tag{3}$$

The complex signal envelope is normalized as in (A.15), so that $E_t$ is the transmitted energy. The average received signal energy is

$$\bar{E}_r = 2\sigma_b^2 E_t. \tag{4}$$

The complex white noise has a covariance function

$$\tilde{K}_{\tilde{w}}(t, u) = N_0\, \delta(t - u), \qquad -\infty < t, u < \infty. \tag{5}$$

The parameters $\tau$ and $\omega$ are unknown nonrandom parameters whose values we shall estimate.

The first step is to find the likelihood function. Recalling from Chapter I-4 the one-to-one correspondence between the likelihood function and the likelihood ratio, we may use (9.36), (9.38), and (9.39) to obtain the answer directly. The result is

$$\ln \Lambda_1(\tau, \omega) = \frac{1}{N_0} \frac{\bar{E}_r}{N_0 + \bar{E}_r} \{|\tilde{L}(\tau, \omega)|^2\}, \tag{6}$$

where

$$\tilde{L}(\tau, \omega) = \int_{-\infty}^{\infty} \tilde{r}(t)\tilde{f}^*(t - \tau)e^{-j\omega t}\, dt. \tag{7}$$

The coefficient in (6) is of importance only when we compute the Cramér-Rao bound, and we can suppress it in most of our discussion. Then we want to compute

$$\ln \Lambda(\tau, \omega) = |\tilde{L}(\tau, \omega)|^2 \tag{8}$$

as a function of $\tau$ and $\omega$. The values of $\tau$ and $\omega$ where this function has its maximum are $\hat{\tau}_{ml}$ and $\hat{\omega}_{ml}$. Because we are considering only maximum likelihood estimates, we eliminate the subscript in subsequent expressions.

We now must generate $\ln \Lambda(\tau, \omega)$ for the values of $\tau$ and $\omega$ in the region of interest. For any particular $\omega$, say $\omega_1$, we can generate $\ln \Lambda(\tau, \omega_1)$ as a function of time by using a bandpass matched filter and square-law envelope detector (recall Fig. 9.5). For different values of $\omega$ we must use different filters. By choosing a set of $\omega_i$ that span the frequency range of interest, we can obtain a discrete approximation to $\ln \Lambda(\tau, \omega)$. For the moment we shall not worry about how fine the frequency grid must be in order to obtain a satisfactory approximation. The processing system is a bank of matched filters and square-law envelope detectors as shown in Fig. 10.1. We now want to investigate the properties of the output of the processor. For simplicity, we view it as a continuous function of $\tau$ and $\omega$.

Let us assume that the actual delay and Doppler shift are $\tau_a$ and $\omega_a$, respectively. (Recall that $\tau$ and $\omega$ are the variables in the likelihood function.) Then we may write

$$\tilde{L}(\tau, \omega) = \int_{-\infty}^{\infty} \tilde{r}(t)\tilde{f}^*(t - \tau)e^{-j\omega t}\, dt$$

$$= \int_{-\infty}^{\infty} [\sqrt{E_t}\, \tilde{b}\tilde{f}(t - \tau_a)e^{j\omega_a t} + \tilde{w}(t)][\tilde{f}^*(t - \tau)e^{-j\omega t}]\, dt, \qquad (9)$$

or

$$\tilde{L}(\tau, \omega) = \sqrt{E_t}\, \tilde{b} \int_{-\infty}^{\infty} \tilde{f}(t - \tau_a)\tilde{f}^*(t - \tau)e^{j(\omega_a - \omega)t}\, dt$$

$$+ \int_{-\infty}^{\infty} \tilde{w}(t)\tilde{f}^*(t - \tau)e^{-j\omega t}\, dt. \qquad (10)$$

To simplify this expression we define

$$\tau' = \tau - \tau_a, \qquad (11)$$

$$\omega' = \omega - \omega_a, \qquad (12)$$

and

$$\tilde{n}(\tau, \omega) = \int_{-\infty}^{\infty} \tilde{w}(t)\tilde{f}^*(t - \tau)e^{-j\omega t}\, dt. \qquad (13)$$

The effect of (11) and (12) is to shift the origin to the point in the $\tau$, $\omega$ plane where the target is located. This is shown in Fig. 10.2. Using (11)–(13) in (10), we have

$$\ln \Lambda(\tau, \omega) = E_t\, |\tilde{b}|^2 \left\{\left| \int_{-\infty}^{\infty} \tilde{f}(t - \tau)\tilde{f}^*(t - \tau + \tau')e^{j\omega' t}\, dt \right|^2\right\}$$

$$+ 2\, \mathrm{Re} \left\{\sqrt{E_t}\, \tilde{b} \left(\int_{-\infty}^{\infty} \tilde{f}^*(t - \tau)\tilde{f}(t - \tau + \tau')e^{-j\omega' t}\, dt\right) \tilde{n}^*(\tau, \omega)\right.$$

$$+ |\tilde{n}(\tau, \omega)|^2. \qquad (14)$$

The first term in (14) is due entirely to the signal and is the only term that would be present in the absence of noise. By making the substitution

$$z = t - \tau + \frac{\tau'}{2}, \qquad (15)$$

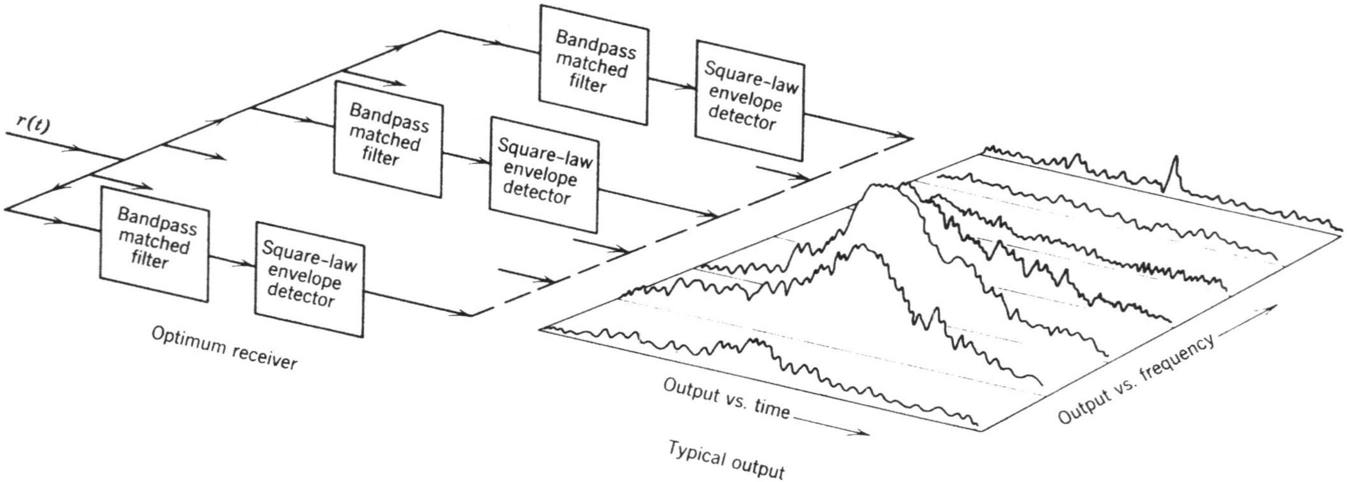


**Fig. 10.1 Receiver to generate ln $\Lambda$ ($\tau$, $\omega$).**

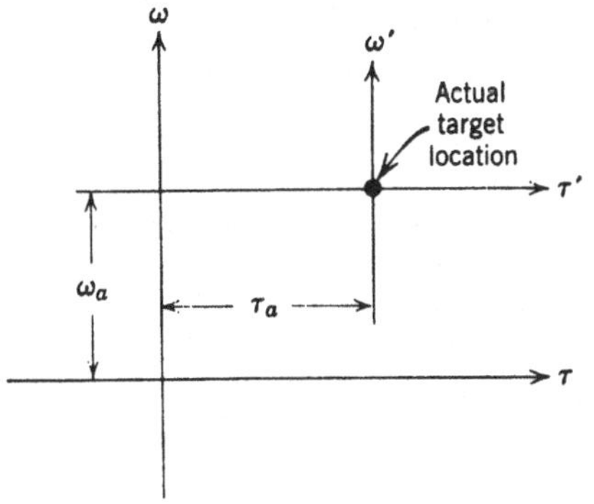


**Fig. 10.2 Coordinate systems in the τ, ω plane.**

we see that it is not a function of $\tau$ and $\omega$. We denote the term in braces as $\theta(\tau', \omega')$,

$$\theta(\tau', \omega') \triangleq \left| \int_{-\infty}^{\infty} \tilde{f}\left(z - \frac{\tau'}{2}\right)\tilde{f}^*\left(z + \frac{\tau'}{2}\right) e^{j\omega' z}\, dz \right|^2. \tag{16}$$

It corresponds to a scaled (by $E_t\, |\tilde{b}|^2$) version of the output of the receiver in the absence of noise.

We define the function inside the magnitude signs as the *time-frequency autocorrelation function* of $\tilde{f}(t)$ and denote it by $\phi(\tau', \omega')$,†

$$\boxed{\phi(\tau', \omega') \triangleq \int_{-\infty}^{\infty} \tilde{f}\left(t - \frac{\tau'}{2}\right)\tilde{f}^*\left(t + \frac{\tau'}{2}\right) e^{j\omega' t}\, dt.} \tag{17}$$

It is a measure of the degree of similarity between a complex envelope and a replica of it that is shifted in time and frequency. Clearly,

$$\boxed{\theta(\tau', \omega') = |\tilde{\phi}(\tau', \omega')|^2.} \tag{18}$$

The function $\theta(\tau', \omega')$ was introduced originally by Ville [1] and is referred to as the *ambiguity function*. Later we shall see why this is an appropriate name. It is sometimes referred to as Woodward's ambiguity function because of his pioneering work with it [8], [60].

Because $\tilde{f}(t)$ is normalized it follows that

$$\phi(0, 0) = 1. \tag{19}$$

From the Schwarz inequality,

$$|\phi(\tau', \omega')| \leq \phi(0, 0) = 1 \tag{20}$$

† There is a certain degree of choice in defining the time-frequency autocorrelation function, and various definitions are used in the literature.

and

$$\theta(\tau', \omega') \leq \theta(0, 0) = 1. \tag{21}$$

Thus, the output of the receiver is a surface in the $\tau$, $\omega$ plane that contains three components. The first is $\theta(\tau', \omega')$, which is a positive function whose maximum value is at that point in the plane where the target is located. The second and third terms are due to the additive noise. In a moment, we shall consider the effect of these two terms, but first we look at $\theta(\tau', \omega')$ in more detail.

To get some feeling for the behavior of $\theta(\tau, \omega)$ and $\phi(\tau, \omega)$ for some typical signals, we consider several examples.

**Example 1. Single Rectangular Pulse.** Let $\tilde{f}(t)$ be a real rectangular pulse,

$$\tilde{f}(t) = \begin{cases} \dfrac{1}{\sqrt{T}}, & -\dfrac{T}{2} < t < \dfrac{T}{2}, \\ 0, & \text{elsewhere}. \end{cases} \tag{22}$$

Then

$$\phi(\tau, \omega) = \begin{cases} \dfrac{1}{T} \displaystyle\int_{-\frac{1}{2}(T-|\tau|)}^{\frac{1}{2}(T-|\tau|)} e^{j\omega t}\, dt = \left(1 - \dfrac{|\tau|}{T}\right)\left(\dfrac{\sin\,[(\omega T/2)(1-(|\tau|/T))]}{(\omega T/2)(1-(|\tau|/T))}\right), & \tau \leq T, \\ 0, & \text{elsewhere}, \end{cases} \tag{23}$$

and

$$\theta(\tau, \omega) = \begin{cases} \left(1 - \dfrac{|\tau|}{T}\right)^2 \left(\dfrac{\sin\,[(\omega T/2)(1-(|\tau|/T))]}{(\omega T/2)(1-(|\tau|/T))}\right)^2, & |\tau| \leq T, \\ 0, & \text{elsewhere}. \end{cases} \tag{24}$$

The magnitude of the time-frequency autocorrelation function is shown in Fig. 10.3. (Actually we show some cuts through the surface along constant $\tau$ and constant $\omega$ lines.) Notice that the function is symmetric about both axes.

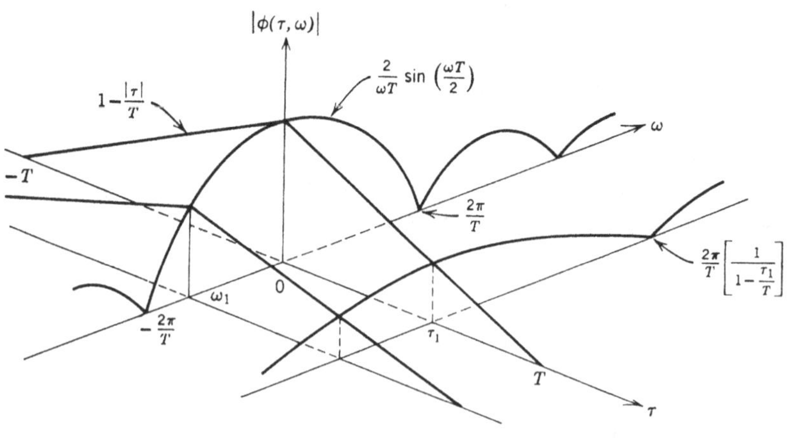


**Fig. 10.3 Magnitude of the time-frequency correlation function for a rectangular pulse.**

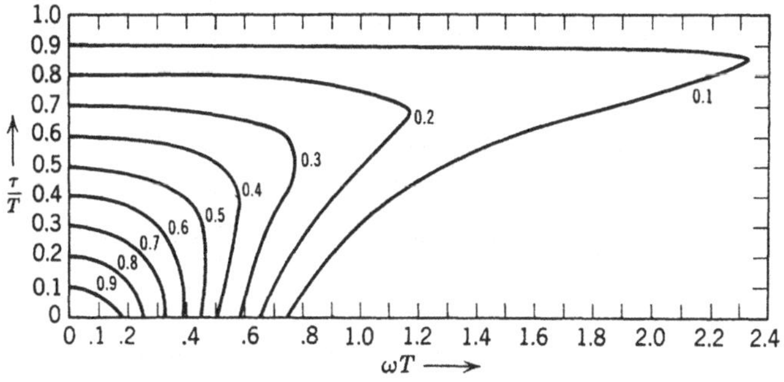


**Fig. 10.4 Equal-height contours of ambiguity function of rectangular pulse.**

A convenient method of representing the ambiguity function is shown in Fig. 10.4. The curves are equal-height contours of $\theta(\tau, \omega)$. Notice that the $\tau$-axis and $\omega$-axis are scaled by factors of $T^{-1}$ and $T$, respectively. Notice also that the ambiguity function has a single peak whose width along the $\tau$-axis is *directly* proportional to $T$ and whose width along the $\omega$-axis is *inversely* proportional to $T$.

Before considering a second example it is worthwhile discussing *qualitatively* how the other two terms in (14) affect the estimate of $\tau$ and $\omega$ in a typical realization of the experiment. In order to see this, we first consider a vertical cut along the $\tau$-axis of $\ln \Lambda(\tau, \omega)$ as shown in Fig. 10.5*a*. From (14) we see that the function consists of $E_t\,|\tilde{b}|^2\,\theta(\tau, 0)$ plus the contributions due to noise indicated by the second and third terms. In Fig. 10.5*b*, we show a top view of $\ln \Lambda(\tau, \omega)$. The shaded surface is the $E_t\,|\tilde{b}|^2\,\theta(\tau, \omega)$ from Fig. 10.4. The contour lines are the equal-height loci of $\ln \Lambda(\tau, \omega)$. The values of $\tau$ and $\omega$ where the surface has its maximum are $\hat{\tau}_{ml}$ and $\hat{\omega}_{ml}$. We see that in the absence of noise we always choose the correct values. There will be an error if the noise contributions at some $\tau' \neq 0$ and $\omega' \neq 0$ are large enough to move the peak of the total function away from the origin. Therefore, in order to minimize the errors, we should try to find an $\tilde{f}(t)$ whose ambiguity function is one at the origin and zero elsewhere. An *ideal* $\theta(\tau, \omega)$ function might be the one shown in Fig. 10.6*a*. We expect that it will be difficult to find an $\tilde{f}(t)$ that has such a discontinuous ambiguity function. However, a close approximation such as is shown in Fig. 10.6*b* *might* be practical.

Thus, it appears that we want to choose $\tilde{f}(t)$ so that $\theta(\tau, \omega)$ is a narrow spike. From (24) or Fig. 10.3, it is clear that, with a rectangular pulse, we can make the peak arbitrarily narrow in either direction (but *not* both) by varying $T$.

Since the rectangular pulse does not lead to the ambiguity function in Fig. 10.6*b*, we shall try some other signals.

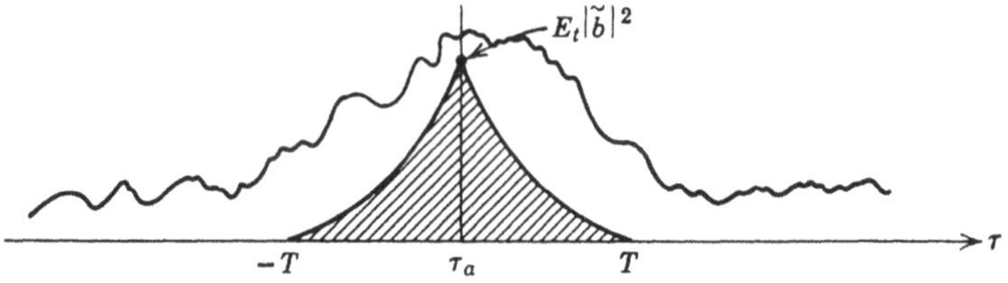


(a) Cut of ln Λ (0, ω)

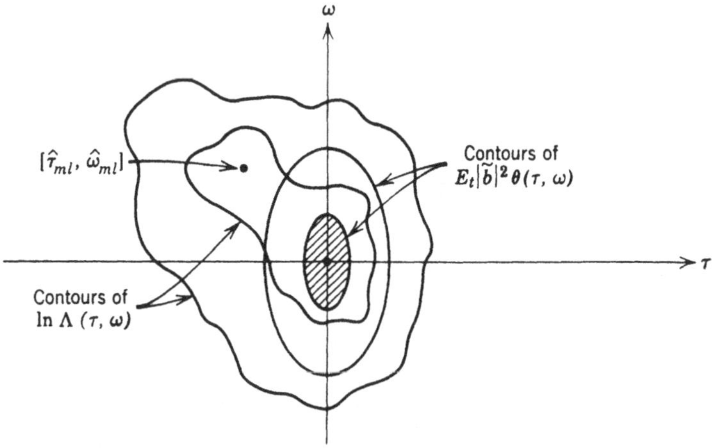


(b) Contours of typical output

**Fig. 10.5 Output of receiver on a particular experiment.**

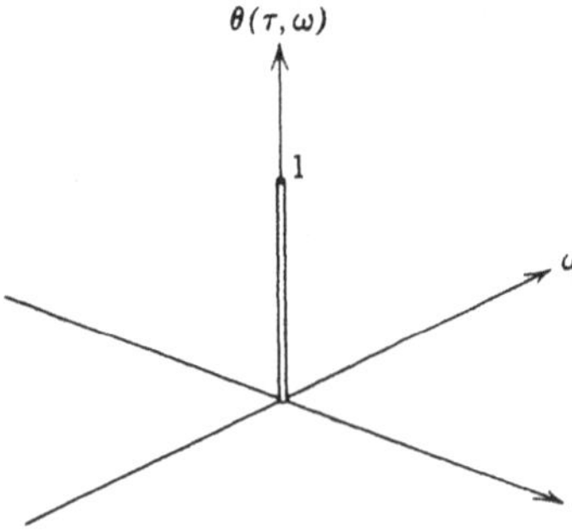


(a) Ideal ambiguity function

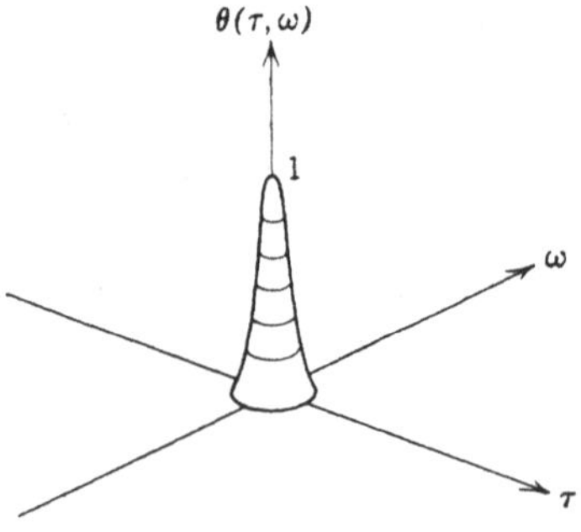


(b) Possible approximation to ideal ambiguity function

**Fig. 10.6 Desirable ambiguity functions.**

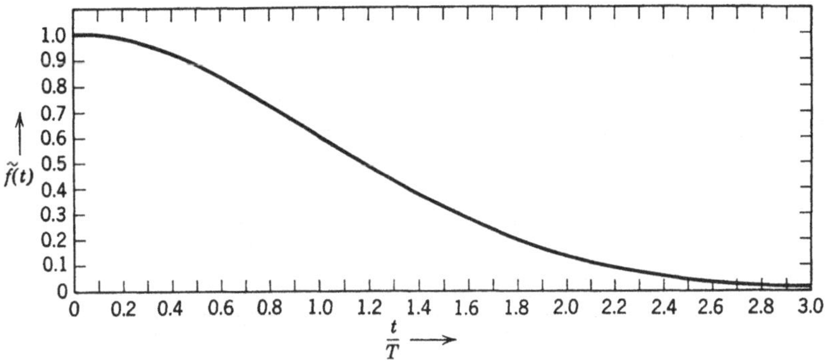


**Fig. 10.7 Gaussian pulse.**

**Example 2. Simple Gaussian Pulse.** A pulse that frequently serves as a useful analytic idealization is the Gaussian pulse of Fig. 10.7.

$$\tilde{f}(t) = \left(\frac{1}{\pi T^2}\right)^{1/4} \exp\left(-\frac{t^2}{2T^2}\right), \qquad -\infty < t < \infty. \tag{25}$$

The effective duration is proportional to $T$. The time-frequency autocorrelation function is

$$\phi(\tau, \omega) = \int_{-\infty}^{\infty} \left(\frac{1}{\pi T^2}\right)^{1/4} \exp\left[-\frac{(t-\tau/2)^2}{2T^2} - \frac{(t+\tau/2)^2}{2T^2} + j\omega t\right]. \tag{26}$$

Completing the square and integrating, we obtain

$$\phi(\tau, \omega) = \exp\left[-\tfrac{1}{4}\left(\frac{\tau^2}{T^2} + T^2\omega^2\right)\right]. \tag{27}$$

The ambiguity function is

$$\theta(\tau, \omega) = \exp\left[-\tfrac{1}{2}\left(\frac{\tau^2}{T^2} + T^2\omega^2\right)\right]. \tag{28}$$

The equal-height contours of $\theta(\tau, \omega)$ are ellipses, as shown in Fig. 10.8. Just as in Example 1, a single parameter, the pulse duration, controls both the range and Doppler accuracy.

These two examples suggest that if we are going to improve our range and Doppler estimates *simultaneously*, we must try a more complicated signal. Apparently, we need a signal that contains several parameters which we can vary to optimize the performance. We shall consider two broad classes of signals.

***Coded Pulse Sequences.*** This class of signals is constructed by operations on single subpulse, $\tilde{u}(t)$. A commonly used subpulse is the rectangular

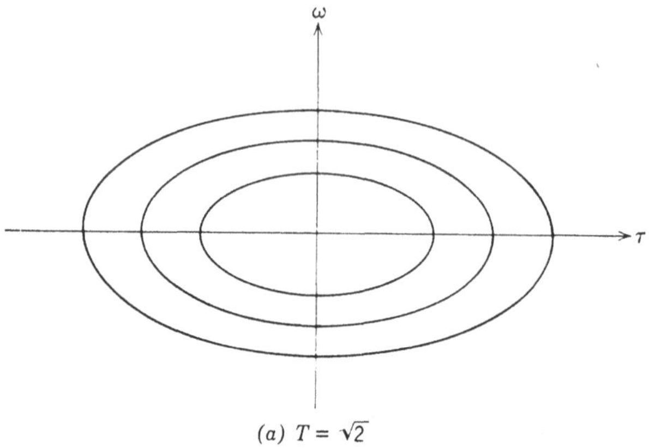


**Fig. 10.8 Equal-height contours for the ambiguity function of a Gaussian pulse with $T = \sqrt{2}$.**

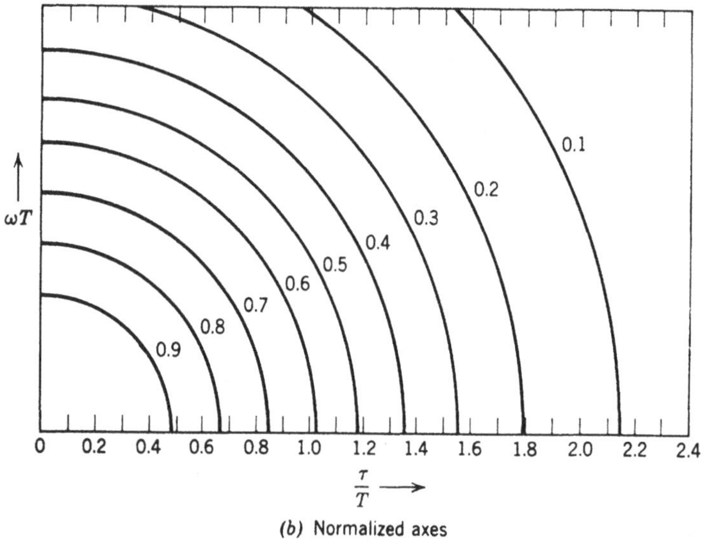


**Fig. 10.8 Equal-height contours for ambiguity function of Gaussian pulse (normalized axes).**

pulse in Example 1,

$$\tilde{u}(t) = \begin{cases} \dfrac{1}{\sqrt{T}}, & -\dfrac{T}{2} \le t \le \dfrac{T}{2}, \\ 0, & \text{elsewhere.} \end{cases} \tag{29}$$

The subpulses are delayed, amplitude-weighted, frequency-shifted, phase-shifted, and then summed. Thus,

$$\tilde{f}(t) = c \sum_{n=1}^{N} \tilde{a}_n \tilde{u}(t - nT) \exp[j(\omega_n t + \theta_n)]. \tag{30}$$

The constant $c$ normalizes $\tilde{f}(t)$. We discuss a simple example of this class of signal in Example 3. In Section 10.4, we study the class in detail.

***Modulated Analog Waveforms.*** This class is obtained by modulating the signal in amplitude and/or frequency to achieve the desired properties. A simple example of this class of signals is given in Examples 4 and 5.

We now derive the ambiguity function for several useful signals. These examples give us some feeling for the general properties that we might expect.

**Example 3. Pulse Train with Constant Repetition Rate.** Consider the sequence of rectangular pulses shown in Fig. 10.9. It is characterized by the pulse duration $T$, the interpulse spacing $T_p$, and the total number of pulses $(2n + 1)$. This sequence is frequently used in radar and sonar systems for the following reasons:

1. It is easy to generate.
2. The optimum receiver is easy to implement.
3. The parameters can be varied to match different operating conditions.

We assume that $T \ll T_p$. The interpulse spacing is not necessarily a multiple of $T$. The duration of the entire sequence is $T_d$,

$$T_d \triangleq 2nT_p + T. \tag{31}$$

Denoting the pulse as $\tilde{u}(t)$ [see (29)], we can write the complex envelope of the transmitted signal as

$$\tilde{f}(t) = \frac{1}{(2n+1)T^{1/2}} \sum_{k=-n}^{k=n} \tilde{u}(t - kT_p). \tag{32}$$

Notice that our model assumes that the target does not fluctuate in the $T_d$ seconds during which the signal illuminates it.

We now derive $\phi(\tau, \omega)$ and $\theta(\tau, \omega)$. First, we consider values of $|\tau| < T$. Using (32) in (17) gives

$$\phi(\tau, \omega) = \left(\frac{1}{(2n+1)T}\right) \sum_{k=-n}^{k=n} \int_{kT_p - \frac{1}{2}(T-|\tau|)}^{kT_p + \frac{1}{2}(T-|\tau|)} \tilde{u}\left((t - kT_p - \frac{\tau}{2}\right) \tilde{u}^*\left(t - kT_p + \frac{\tau}{2}\right) \times e^{j\omega t}\, dt \qquad |\tau| \le |T|. \tag{33}$$

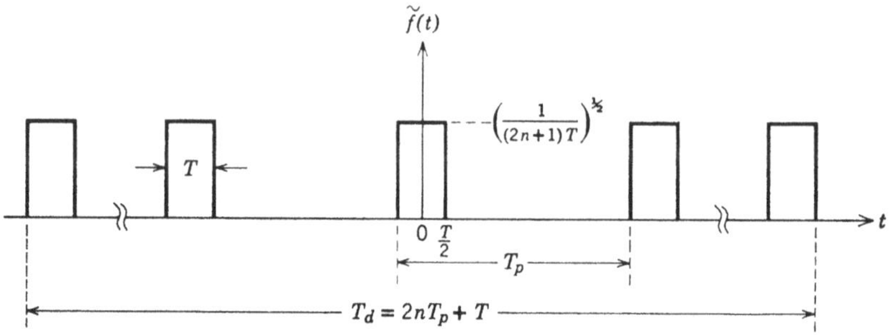


Fig. 10.9 Sequence of pulses.

Letting

$$z = t - kT_p, \tag{34}$$

$$\phi(\tau, \omega) = \left(\frac{1}{(2n+1)T}\right)\sum_{k=-n}^{k=n} e^{j\omega kT_p}\left\{\int_{-T/2+|\tau|/2}^{T/2-|\tau|/2} \tilde{u}\left(z - \frac{\tau}{2}\right)\tilde{u}^*\left(z + \frac{\tau}{2}\right)e^{j\omega z}\,dz\right\}. \tag{35}$$

The term in the braces is $\phi_{\tilde{u}}(\tau, \omega)$. The sum is a finite geometric series. Thus, (35) reduces to

$$\phi(\tau, \omega) = \frac{1}{(2n+1)}\left\{\frac{\sin\left[\omega(n+\frac{1}{2})T_p\right]}{\sin\left[\omega T_p/2\right]}\right\}\phi_{\tilde{u}}(\tau, \omega), \qquad |\tau| \le T. \tag{36}$$

We see that the subpulse characteristics only enter into the last term. The bracketed term is a function of $\omega$ only and is determined by $T_p$, the pulse repetition rate, and $n$, the number of pulses. The bracketed term is shown in Fig. 10.10*a*. We see that the first zero is at

$$\omega = \frac{2\pi}{(2n+1)T_p} \simeq \frac{2\pi}{T_d}, \tag{37a}$$

and the subsidiary peaks occur at

$$\omega = \frac{2\pi}{T_p}. \tag{37b}$$

In Fig. 10.10*b* we show $\phi_{\tilde{u}}(0, \omega)$ for a rectangular pulse. The two plots indicate the effect of the parameters $T$, $T_p$, and $T_d$. Recalling that

$$T_d > T_p \gg T, \tag{37c}$$

we see that the shape of $\phi(0, \omega)$ is controlled by the term in Fig. 10.10*a*. Thus, the width of the main peak decreases as the over-all duration $T_d$ increases. Subsidary peaks occur at intervals of $1/T_p$ on the frequency axis. When $\omega = 0$, the bracketed term equals $(2n + 1)$, so that

$$\phi(\tau, 0) = \phi_{\tilde{u}}(\tau, 0), \qquad |\tau| \le T. \tag{38}$$

Next we consider values of $\tau > T$. There is no overlap until $\tau = T_p - T$. At this point, the situation is similar to that at $\tau = -T$, except that there is one less pulse

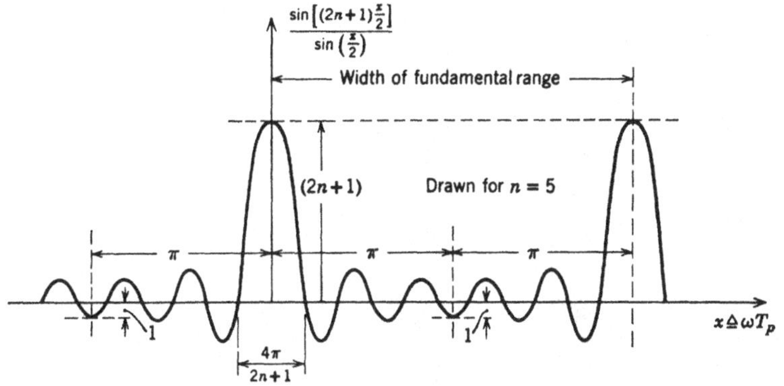


**Fig. 10.10a Bracketed term in (36). (After [45].)**

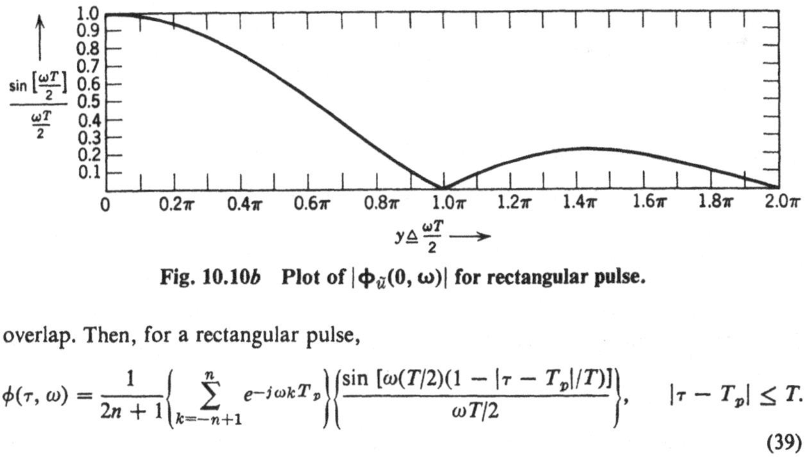


**Fig. 10.10b Plot of $|\phi_{\tilde{u}}(0, \omega)|$ for rectangular pulse.**

overlap. Then, for a rectangular pulse,

$$\phi(\tau, \omega) = \frac{1}{2n+1}\left\{\sum_{k=-n+1}^{n} e^{-j\omega k T_p}\right\}\left\{\frac{\sin\,[\omega(T/2)(1 - |\tau - T_p|/T)]}{\omega T/2}\right\}, \qquad |\tau - T_p| \le T. \tag{39}$$

On the $\tau$-axis we have the same expression as in (38) except for a scale factor and a shift,

$$\phi(\tau, 0) = \left(\frac{2n}{2n+1}\right)\phi_{\tilde{u}}(\tau - T_p, 0), \qquad |\tau - T_p| < T. \tag{40}$$

A similar result follows for larger $\tau$. Every $T_p$ seconds there is a peak, but the magnitude is reduced. A different representation of the ambiguity function $\theta(\tau, \omega)$ is shown in Fig. 10.11. This type of plot was introduced by Siebert [9]. The dark shaded areas indicate regions where the height of $\theta(\tau, \omega)$ is significant (usually, the border corresponds to $\theta(\tau, \omega) = \frac{1}{2}$). In the light shaded areas $\theta(\tau, \omega)$ is small, but nonzero. In the unshaded areas $\theta(\tau, \omega)$ is zero.

Several new features of the signal design problem are indicated by Example 3:

1. We can decrease the width of the major peak in the frequency (Doppler) direction by increasing $T_d$ (or $n$).

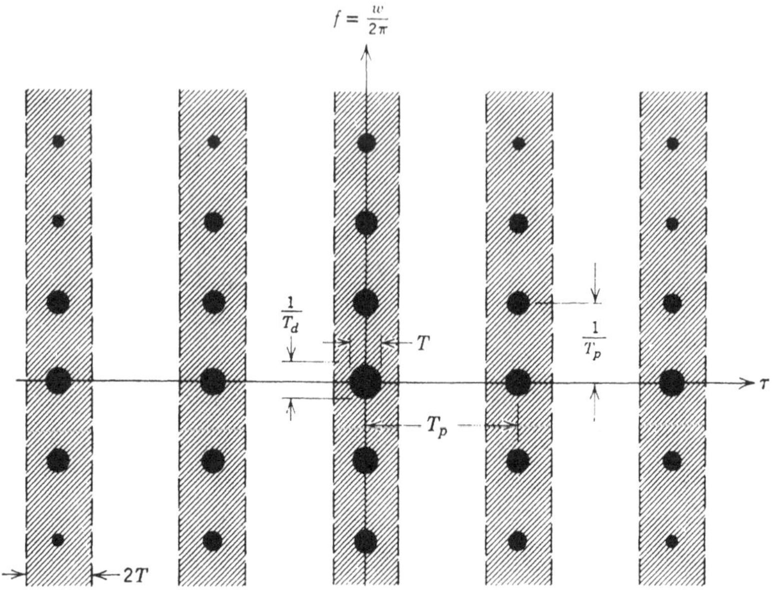


**Fig. 10.11 An approximate contour plot of $\theta(\tau, \omega)$ for pulse train [from [9] and [9.12]].**

2. We can decrease the width of the major peak in the time (range) direction by decreasing $T$. (This corresponds to an increased bandwidth.) Thus, by allowing more parameters in our signal design, we can obtain an ambiguity function whose major peak is narrow in both the range and Doppler direction.

3. This particular signal accomplishes this at the cost of including subsidiary peaks. It is easy to see the effects of these subsidiary peaks. A small noise contribution can cause the *total* value at a subsidiary peak to exceed the value at the correct peak. The importance of these subsidiary peaks depends on our a-priori knowledge of the area in the $\tau,\omega$ plane in which the target may be located. Two cases are shown in Fig. 10.12. In the first case, the set of subsidiary peaks lies outside the area of interest for all possible $\tau$, $\omega$. Thus, they will not cause any trouble. In the second case, they are inside the area of interest, and even in the presence of weak noise we may choose the wrong peak.

This discussion illustrates two of the issues that we encounter in a performance discussion. The first is local accuracy (i.e., given that we are on the correct peak, how small will the error be?). The second is global accuracy (i.e., how often will there be large errors?). This is, of course, the same phenomenon that we encountered in the PFM and PPM problems of Chapter I-4 and in the angle-modulation problems of Chapter II-2.

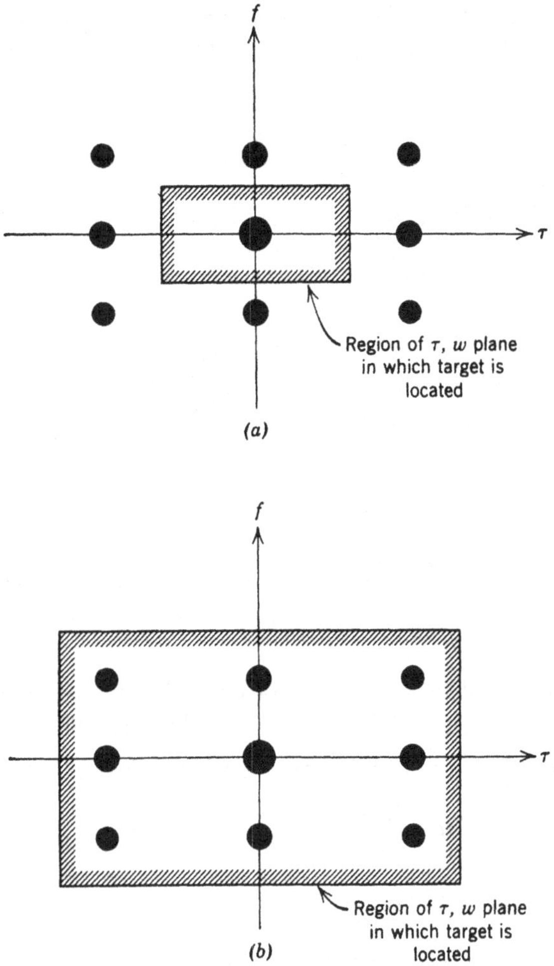


**Fig. 10.12 Regions in which target can be located.**

Before studying these two issues quantitatively, it is interesting to look at the ambiguity function for several other signals.

We next consider an example of a *modulated analog waveform.* All of our signals up to this point were obtained by amplitude-modulating a constant carrier. In order to introduce more freedom into the signal design, we now consider the possibility of frequency-modulating the carrier. Specifically, we consider a linear frequency sweep, i.e.,

$$\phi_{\tilde{f}}(t) = bt^2. \tag{41}$$

[Recall that $\phi_{\tilde{f}}(t)$ is the phase of $\tilde{f}(t)$.]

Instead of computing the ambiguity function for a particular pulse directly, we use an interesting property for arbitrary $\tilde{f}(t)$.

**Property 1.** If

$$\tilde{f}_1(t) \sim \phi(\tau, \omega) \sim \theta(\tau, \omega), \tag{42a}$$†

then

$$\tilde{f}_2(t) \triangleq \tilde{f}_1(t)\, e^{jbt^2} \sim \phi(\tau, \omega - 2b\tau) \sim \theta(\tau, \omega - 2b\tau). \tag{42b}$$

This result follows directly from the definitions in (17) and (18).

$$\begin{aligned}
\phi_2(\tau, \omega) &= \int_{-\infty}^{\infty} \tilde{f}_2\left(t - \frac{\tau}{2}\right) \tilde{f}_2^*\left(t + \frac{\tau}{2}\right) e^{j\omega t}\, dt \\
&= \int_{-\infty}^{\infty} \tilde{f}_1\left(t - \frac{\tau}{2}\right) \tilde{f}_1^*\left(t + \frac{\tau}{2}\right) \\
&\quad \times \exp\left[jb\left(t - \frac{\tau}{2}\right)^2 - jb\left(t + \frac{\tau}{2}\right)^2 + j\omega t\right] dt \\
&= \int_{-\infty}^{\infty} \tilde{f}_1\left(t - \frac{\tau}{2}\right) \tilde{f}_1^*\left(t + \frac{\tau}{2}\right) \exp\left[jt[\omega - 2b\tau]\right] dt \\
&= \phi_1(\tau, \omega - 2b\tau).
\end{aligned} \tag{43}$$

Thus, a linear frequency sweep *shears* the ambiguity diagram parallel to the $\omega$-axis. We now apply this property to the Gaussian pulse in Example 2.

**Example 4. Gaussian Pulse with Linear Frequency Modulation.** Now

$$\tilde{f}(t) = \left(\frac{1}{\pi T^2}\right)^{1/4} \exp\left[-\left(\frac{1}{2T^2} - jb\right)t^2\right]. \tag{44a}$$

Then, from (28) and (42b), we obtain

$$\theta(\tau, \omega) = \exp\left[-\frac{1}{2}\left(\frac{\tau^2}{T^2} + T^2(\omega - 2b\tau)^2\right)\right]. \tag{44b}$$

The equal-height contour lines are the ellipses

$$\frac{1}{2}\left[T^2\omega^2 - 4bT^2\omega\tau + \left(4b^2T^2 + \frac{1}{T^2}\right)\tau^2\right] = c^2. \tag{45}$$

For convenience in plotting, we introduce $\overline{\omega^2}$, $\overline{\omega t}$, and $\overline{t^2}$, which are defined in the Appendix. For the signal in (43),

$$\overline{t^2} = \frac{T^2}{2}, \tag{46}$$

$$\overline{\omega t} = bT^2, \tag{47}$$

† The symbol $\sim$ means "corresponds to."

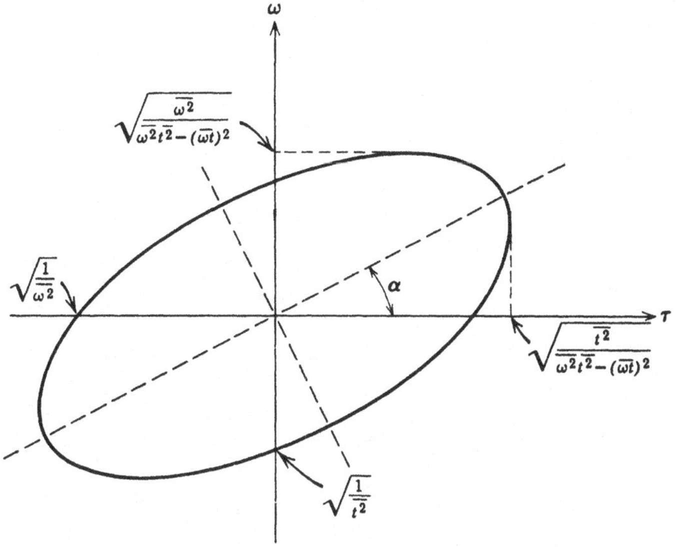


**Fig. 10.13 Contour of $\theta(\tau, \omega)$ for Gaussian pulse with linear *FM*.**

and

$$\overline{\omega^2} = \frac{1}{2T^2} + 2b^2T^2. \tag{48}$$

Then (45) reduces to

$$\overline{t^2}\omega^2 - 2\overline{\omega t}\omega\tau + \overline{\omega^2}\tau^2 = c^2. \tag{49}$$

In Fig. 10.13 we have plotted (49) for the case when $c = 1$. The major axis is at an angle $\alpha$, defined by

$$\alpha = \tfrac{1}{2}\tan^{-1}\left(\frac{4b}{1-(1/4T^4+b^2)/\pi^2}\right) = \tfrac{1}{2}\tan^{-1}\left(\frac{2\overline{\omega t}}{\overline{t^2}-\overline{\omega^2}/(2\pi)^2}\right), \qquad |\alpha| < \pi. \tag{50}$$

Along the $\tau$-axis,

$$\theta(0, \tau) = \exp\,[-\overline{\omega^2}\tau^2] = \exp\left[-\left(\frac{1}{2T^2} + 2b^2T^2\right)\tau^2\right]. \tag{51a}$$

Similarly,

$$\theta(\omega, 0) = \exp\,[-\overline{t^2}\omega^2] = \exp\left[-\frac{T^2\omega^2}{2}\right]. \tag{51b}$$

We see that the width on the $\tau$-axis is inversely proportional to the root-mean-square signal bandwidth and the width on the $\omega$-axis is inversely proportional to the root-mean-square signal duration. Thus, by increasing both $b$ and $T$, we decrease the width on both the $\tau$- and $\omega$-axes simultaneously. Therefore, we can accurately measure the range of a target with known velocity, or we can accurately measure the velocity of a target with known range. However, if both parameters are unknown there is an ambiguous region in the $\tau$, $\omega$ plane. For positive values of $b$, the ambiguous region lies in the first and third quadrants, as shown in Fig 10.13. Whether or not this ambiguity is important depends

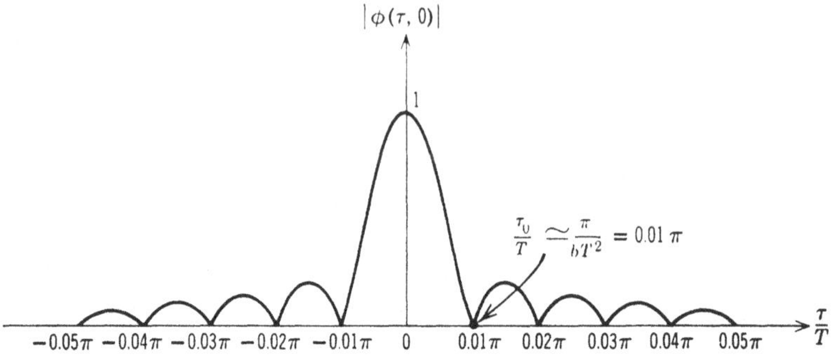


**Fig. 10.14. Plot of $|\phi(\tau, 0)|$: rectangular pulse with linear *FM* ($bT^2 = 100$).**

on the physical situation (i.e., can targets occur along this line in the $\tau$, $\omega$ plane?). One way to resolve the ambiguity is to transmit a second pulse with the opposite frequency sweep.

A similar result follows for the rectangular pulse with a linear frequency modulation.

**Example 5. Rectangular Pulse, Linear Frequency Modulation.**

$$\tilde{f}(t) = \begin{cases} \dfrac{1}{\sqrt{T}}\, e^{jbt^2}, & -\dfrac{T}{2} \le t \le \dfrac{T}{2}, \\ 0, & \text{elsewhere.} \end{cases} \tag{52}$$

Using (23) and (42b), we have

$$|\phi(\tau, \omega)| = \begin{cases} \left(1 - \dfrac{|\tau|}{T}\right) \left| \dfrac{\sin\left(((\omega - 2b\tau)/2)(T - |\tau|)\right)}{((\omega - 2b\tau)/2)(T - |\tau|)} \right|, & \tau \le T, \\ 0, & \text{elsewhere.} \end{cases} \tag{53}$$

Along the $\tau$-axis,

$$|\phi(\tau, 0)| = \begin{cases} \left(1 - \dfrac{|\tau|}{T}\right) \dfrac{\sin\left[b\tau(T - |\tau|)\right]}{b\tau(T - |\tau|)}, & |\tau| < T, \\ 0, & \text{elsewhere.} \end{cases} \tag{54}$$

Along the $\omega$-axis,

$$|\phi(0, \omega)| = \left| \frac{\sin(\omega T/2)}{\omega T/2} \right|. \tag{55}$$

In Fig. 10.14, we have plotted $|\phi(\tau, 0)|$ for the case when $bT^2 = 100$. We see that the first zero is near the point

$$\tau_0 = \frac{T}{2}\left[1 - \left(1 - \frac{4\pi}{bT^2}\right)^{\frac{1}{2}}\right] \simeq \frac{\pi}{bT} = \frac{1}{W_0}, \tag{56}$$

where

$$2\pi W_0 \triangleq 2bT \tag{57}$$

is the range of the frequency sweep. Thus, as we would expect from our discussion of the Gaussian pulse, the range estimation accuracy for a known velocity target is proportional to the signal bandwidth. Once again there is a region of ambiguity in the first and third quadrants for positive $b$.

The performance of the receiver for Example 5 has an interesting interpretation. The input is the "long" pulse shown in Fig. 10.15*a*. Its instantaneous frequency increases with time, as shown in Fig. 10.15*b*. Now, the transfer function of the matched filter has a phase characteristic that is quadratic with respect to frequency. The delay of the envelope of a bandpass signal through any filter is proportional to the derivative of the

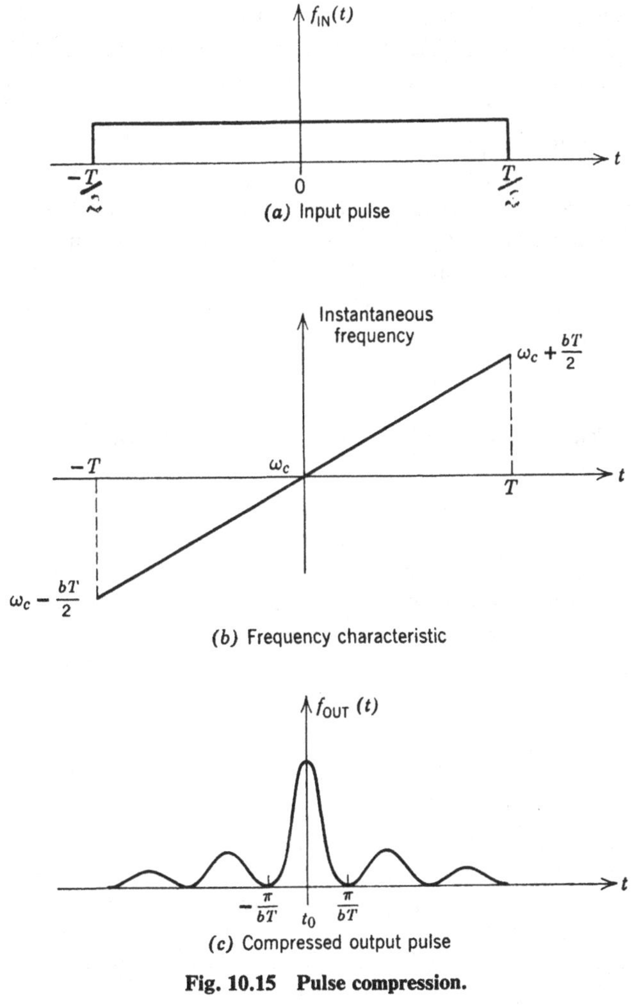


**Fig. 10.15 Pulse compression.**

phase characteristic of the filter with respect to frequency (e.g., [2]. For the linear *FM* pulse, the derivative of the phase of the matched filter decreases linearly with increasing frequency. Thus, the low-frequency components, which occur at the beginning, are delayed more than the high-frequency-components at the end of the pulse. The result is the "short" pulse shown in Fig. 10.15*c*. The effect of the receiver is to compress the long pulse at the input to the receiver into a short pulse at the output of the processor, with an accompanying increase in range measurement accuracy. This type of system is commonly referred to as a "*pulse-compression*" *radar*. Its obvious advantage is that if the system is peak-power-limited, one can increase the *transmitted energy* by transmitting a longer pulse without losing range accuracy. The idea of pulse compression through the use of frequency modulation was derived independently in the United States (Dicke [3] and Darlington [4]) and in Germany (Huttman [5] and Cauer [6]). An interesting discussion is given by Cook [7].

This series of examples illustrates the fundamental role that the ambiguity function plays in the range-Doppler estimation problem. We now return to the general case and derive some quantitative performance results. In Section 10.2, we derive expressions for the estimation accuracies in terms of the ambiguity function. In Section 10.3, we develop some general properties of the ambiguity function. Then, in Section 10.4, we return to signal design problems.

## 10.2 PERFORMANCE OF THE OPTIMUM ESTIMATOR

In this section, we discuss the accuracy of our estimates of $\tau$ and $\omega$. We first consider the case in which the energy-to-noise ratio is high and the errors are small. We refer to this as the *local accuracy* problem.

The accuracy problem for range measurement was studied by Woodward [60]. The accuracy problem for range and velocity measurement has been studied by Manasse [76] and by Kelly, Reed, and Root [77].

### 10.2.1 Local Accuracy

We approach the local accuracy problem in two steps. First, we derive the Cramér-Rao bound on the accuracy of any unbiased estimates. We then argue that the errors using maximum likelihood estimates approach these bounds under certain conditions. We discuss these conditions in detail in Section 10.2.2.

The derivation of the Cramér-Rao bound is a straightforward application of the techniques in Sections I-4.2.3 and I-4.6. We recall that the first step was to derive an *information matrix* **J** whose elements are

$$J_{ij} = -E\left[\frac{\partial^2 \ln \Lambda(\mathbf{A})}{\partial A_i\, \partial A_j}\right] \tag{58}$$

(see page I-372). In this case the parameters of interest, $\tau$ and $\omega$, are nonrandom, so that the expectation is over $r(t)$ [or $n(t)$]. Here the information matrix is two-dimensional:

$$\mathbf{J} = \begin{bmatrix} J_{11} & J_{12} \\ J_{21} & J_{22} \end{bmatrix}. \tag{59}$$

We identify the subscript 1 with $\tau$ and the subscript 2 with $\omega$. From (6), (58), and (59),

$$J_{11} = -E\left[\frac{\partial^2 \ln \Lambda_1(\tau, \omega)}{\partial \tau^2}\right], \tag{60}$$

$$J_{22} = -E\left[\frac{\partial^2 \ln \Lambda_1(\tau, \omega)}{\partial \omega^2}\right], \tag{61}$$

$$J_{12} = J_{21} = -E\left[\frac{\partial^2 \ln \Lambda_1(\tau, \omega)}{\partial \tau\, \partial \omega}\right]. \tag{62}$$

The evaluation of these three quantities is a straightforward manipulation. We shall state the results first and then carry out the derivation. The elements of the information matrix are

$$J_{11} = C[\overline{\omega^2} - (\bar{\omega})^2] = C\sigma_\omega^2, \tag{63}$$

$$\text{and} \quad J_{12} = C[\overline{\omega t} - \bar{\omega}\bar{t}] = C\rho_{\omega t}, \tag{64}$$

$$J_{22} = C[\overline{t^2} - (\bar{t})^2] = C\sigma_t^2, \tag{65}$$

where

$$C \triangleq \frac{2\bar{E}_r}{N_0}\left(\frac{\bar{E}_r}{\bar{E}_r + N_0}\right) \tag{66}$$

and

$$\overline{\omega^2} = \int_{-\infty}^{\infty} \omega^2 |\tilde{F}(j\omega)|^2 \frac{d\omega}{2\pi}, \tag{67}$$

$$\overline{\omega t} \triangleq \operatorname{Im} \int_{-\infty}^{\infty} u \tilde{f}(u) \frac{\partial \tilde{f}^*(u)}{\partial u}\, du, \tag{68}$$

$$\overline{t^2} = \int_{-\infty}^{\infty} u^2 |\tilde{f}(u)|^2\, du. \tag{69}$$

We assume that the quantities in (67)–(69) are finite.

We now carry out the derivation for a typical term and then return to discuss the implications.

**Derivation of the terms in *J*.** We consider $J_{11}$ first. From (6),

$$\frac{\partial \ln \Lambda_1(\tau, \omega)}{\partial \tau} = C'\left[\tilde{L}(\tau, \omega)\frac{\partial \tilde{L}^*(\tau, \omega)}{\partial \tau} + \frac{\partial \tilde{L}(\tau, \omega)}{\partial \tau}\tilde{L}^*(\tau, \omega)\right]$$

$$= 2C' \operatorname{Re}\left[\tilde{L}(\tau, \omega)\frac{\partial \tilde{L}^*(\tau, \omega)}{\partial \tau}\right], \tag{70}$$

where

$$C' \triangleq \frac{1}{N_0}\frac{\bar{E}_r}{N_0 + \bar{E}_r}. \tag{71}$$

Differentiating again, we obtain

$$\frac{\partial^2 \ln \Lambda_1(\tau, \omega)}{\partial \tau^2} = 2C' \operatorname{Re}\left[\frac{\partial \tilde{L}(\tau, \omega)}{\partial \tau}\cdot\frac{\partial \tilde{L}^*(\tau, \omega)}{\partial \tau} + \tilde{L}(\tau, \omega)\frac{\partial^2 \tilde{L}^*(\tau, \omega)}{\partial \tau^2}\right]. \tag{72}$$

Similarly,

$$\frac{\partial^2 \ln \Lambda_1(\tau, \omega)}{\partial \tau\, \partial \omega} = 2C' \operatorname{Re}\left[\frac{\partial \tilde{L}(\tau, \omega)}{\partial \omega}\cdot\frac{\partial \tilde{L}^*(\tau, \omega)}{\partial \tau} + \tilde{L}(\tau, \omega)\frac{\partial^2 \tilde{L}^*(\tau, \omega)}{\partial \tau\, \partial \omega}\right] \tag{73}$$

and

$$\frac{\partial^2 \ln \Lambda_1(\tau, \omega)}{\partial \omega^2} = 2C' \operatorname{Re}\left[\frac{\partial \tilde{L}(\tau, \omega)}{\partial \omega}\frac{\partial \tilde{L}^*(\tau, \omega)}{\partial \omega} + \tilde{L}(\tau, \omega)\frac{\partial^2 L^*(\tau, \omega)}{\partial \omega^2}\right]. \tag{74}$$

Now recall from (7) that

$$\tilde{L}(\tau, \omega) = \int_{-\infty}^{\infty} \tilde{r}(t)\tilde{f}^*(t - \tau)\, e^{-j\omega t}\, dt. \tag{75}$$

Differentiating (75) twice with respect to $\tau$ and using the results in (70) and (72), we have

$$J_{11} = -E\left\{\frac{\partial^2 \ln \Lambda_1(\tau, \omega)}{\partial \tau^2}\right\}$$

$$= -2C'\left\{\iint_{-\infty}^{\infty} \frac{\partial \tilde{f}^*(t - \tau)}{\partial \tau}\frac{\partial \tilde{f}(u - \tau)}{\partial \tau} e^{jw(t-u)} E[\tilde{r}(t)\tilde{r}^*(u)]\, dt\, du\right.$$

$$\left. + \iint_{-\infty}^{\infty} \tilde{f}^*(t - \tau)\frac{\partial^2 \tilde{f}(u - \tau)}{\partial \tau^2} e^{jw(t-u)} E[\tilde{r}(t)\tilde{r}^*(u)]\, dt\, du\right\}. \tag{76}$$

The correlation function of $\tilde{r}(t)$ is

$$E[\tilde{r}(t)\tilde{r}^*(u)] = 2\sigma_b{}^2 E_t \tilde{f}(t - \tau)\tilde{f}^*(u - \tau)e^{jw(t-u)} + N_0\, \delta(t - u). \tag{77}$$

Substituting (77) into (76), we obtain

$$J_{11} = -2C'\left\{\bar{E}_r\left|\int_{-\infty}^{\infty}\frac{\partial\tilde{f}(t-\tau)}{\partial t}\tilde{f}^*(t-\tau)\,dt\right|^2 + N_0\int_{-\infty}^{\infty}\left|\frac{\partial\tilde{f}(t-\tau)}{\partial t}\right|^2 dt\right.$$
$$+\operatorname{Re}\left[\bar{E}_r\int_{-\infty}^{\infty}|\tilde{f}(t-\tau)|^2\,dt\int_{-\infty}^{\infty}\frac{\partial^2\tilde{f}^*(u-\tau)}{\partial\tau^2}\tilde{f}(u-\tau)\,du\right]$$
$$\left.+\operatorname{Re}\left[N_0\int_{-\infty}^{\infty}\tilde{f}(t-\tau)\frac{\partial^2\tilde{f}^*(t-\tau)}{\partial\tau^2}\,dt\right]\right\}. \tag{78}$$

(Recall that $\bar{E}_r \triangleq 2\sigma_b{}^2E_t$.) We now simplify this expression by demonstrating that the first term is $(\bar{\omega})^2$ and that the sum of the second and fourth terms is zero. To do this, we first observe that

$$\int_{-\infty}^{\infty}|\tilde{f}(t-\tau)|^2\,dt = 1 \tag{79}$$

for all $\tau$. In other words, the energy does not depend on the delay. Differentiating both sides of (79) with respect to $\tau$, we have

$$\operatorname{Re}\left\{\int_{-\infty}^{\infty}\left[\frac{\partial\tilde{f}(t-\tau)}{\partial\tau}\tilde{f}^*(t-\tau)\right]dt\right\} = 0. \tag{80}$$

Differentiating again gives

$$\operatorname{Re}\left\{\int_{-\infty}^{\infty}\left(\frac{\partial^2\tilde{f}(t-\tau)}{\partial\tau^2}\tilde{f}^*(t-\tau) + \frac{\partial\tilde{f}(t-\tau)}{\partial\tau}\frac{\partial\tilde{f}^*(t-\tau)}{\partial\tau}\right)dt\right\} = 0. \tag{81}$$

Thus,

$$\operatorname{Re}\left[\int_{-\infty}^{\infty}\frac{\partial^2\tilde{f}^*(t-\tau)}{\partial\tau^2}\tilde{f}(t-\tau)\,dt\right] = -\int_{-\infty}^{\infty}\left|\frac{\partial\tilde{f}(t-\tau)}{\partial\tau}\right|^2 dt \tag{82}$$

Thus, the second term in (78) cancels the fourth term in (78).

Comparing the first term in (78) and the definition of $\bar{\omega}$ in the Appendix (A.16), and using Parseval's theorem, we see that the first term is $\bar{E}_r(\bar{\omega})^2$.

To simplify the third term, we use (79) and then observe that

$$\int_{-\infty}^{\infty}\left|\frac{\partial\tilde{f}(t-\tau)}{\partial\tau}\right|^2 dt = \int_{-\infty}^{\infty}\omega^2|\tilde{F}(\omega)|^2\frac{d\omega}{2\pi} = \overline{\omega^2}. \tag{83}$$

Using the above results in (78) gives

$$J_{11} = 2C'\bar{E}_r[\overline{\omega^2} - (\bar{\omega})^2], \tag{84}$$

which is (63). As pointed out in the Appendix, we usually choose the carrier so that

$$\bar{\omega} = 0. \tag{85}$$

The derivation of $J_{12}$ and $J_{22}$ is similar. (See Problem 10.2.1.)

The information matrix is specified by (63)–(66) as

$$\mathbf{J} = \frac{2\bar{E}_r}{N_0}\left(\frac{\bar{E}_r}{\bar{E}_r + N_0}\right)\begin{bmatrix}\overline{\omega^2} & \overline{\omega t}\\ \overline{\omega t} & \overline{t^2}\end{bmatrix}. \tag{86}$$

The information matrix is useful in two ways. If we denote the error covariance matrix for some pair of unbiased estimates as $\mathbf{\Lambda}_\varepsilon$,

$$\mathbf{J} - \mathbf{\Lambda}_\varepsilon^{-1} \tag{87}$$

is non-negative definite.

We now interpret this statement in relation to the maximum-likelihood-estimation procedure. When the joint probability density of the errors using *ML* estimation is Gaussian, this result has a simple interpretation. Denote the errors by the vector

$$\mathbf{a}_\varepsilon \triangleq \begin{bmatrix} \hat{\tau}_{ml} - \tau \\ \hat{\omega}_{ml} - \omega \end{bmatrix} \triangleq \begin{bmatrix} \tau_\varepsilon \\ \omega_\varepsilon \end{bmatrix}. \tag{88}$$

If $\mathbf{a}_\varepsilon$ has a Gaussian density, then

$$p_{\mathbf{a}\varepsilon}(\mathbf{A}_\varepsilon) = \frac{1}{2\pi\,|\mathbf{\Lambda}_\varepsilon|^{1/2}} \exp\left(-\frac{\mathbf{A}_\varepsilon{}^T \mathbf{\Lambda}_\varepsilon^{-1} \mathbf{A}_\varepsilon}{2}\right). \tag{89}$$

The equal-height contours are ellipses that are given by the equation

$$\mathbf{A}_\varepsilon \mathbf{\Lambda}_\varepsilon^{-1} \mathbf{A}_\varepsilon = k_i^2, \qquad i = 1, 2, \ldots \tag{90}$$

and are shown in Fig. 10.16. The result in (87) says that if we construct the bound ellipses,

$$\mathbf{A}_\varepsilon \mathbf{J} \mathbf{A}_\varepsilon = k_i^2, \tag{91}$$

they will lie wholly inside the actual ellipses. Since the probability of lying in the region outside an ellipse is $e^{-k^2}$ (see page I-77), we can bound the actual probabilities.

In general, the errors do not have a Gaussian density. However, we shall show that under certain conditions, the *ML* estimates are unbiased and the probability density of the errors approaches a joint Gaussian density. (As we would expect, the Gaussian approximation is best near the mean of the density.)

The second way in which the information matrix is useful is to bound the variance of the individual errors. The variance of *any* unbiased estimate is bounded by the diagonal elements in $\mathbf{J}^{-1}$. Thus

$$\boxed{\operatorname{Var}[\hat{\tau} - \tau] \geq \left[\frac{2\bar{E}_r}{N_0}\left(\frac{\bar{E}_r}{\bar{E}_r + N_0}\right)\right]^{-1} \left(\frac{\overline{t^2}}{\overline{\omega^2}\,\overline{t^2} - (\overline{\omega t})^2}\right)} \tag{92}$$

and

$$\boxed{\operatorname{Var}[\hat{\omega} - \omega] \geq \left[\frac{2\bar{E}_r}{N_0}\left(\frac{\bar{E}_r}{\bar{E}_r + N_0}\right)\right]^{-1} \left[\frac{\overline{\omega^2}}{\overline{\omega^2}\,\overline{t^2} - (\overline{\omega t})^2}\right].} \tag{93}$$

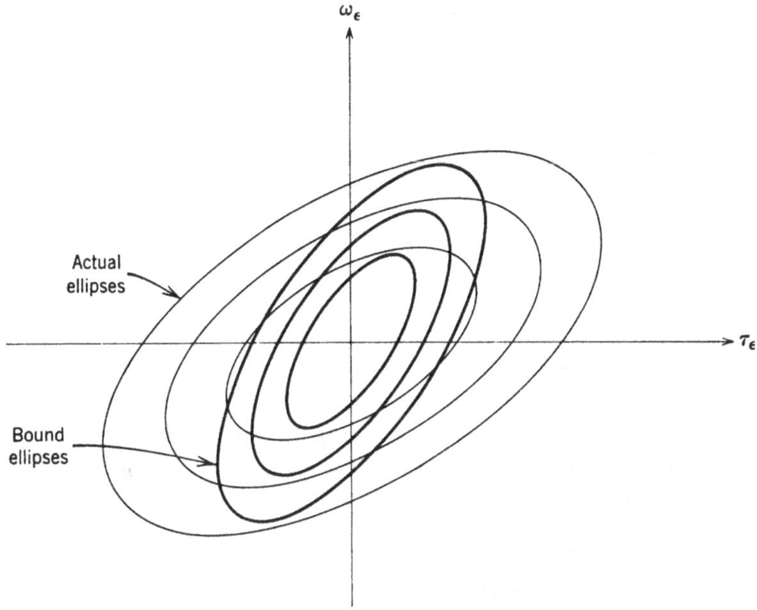


**Fig. 10.16 Error ellipses.**

Looking at (68), we see that a sufficient condition for the bounds on the estimation errors to be uncoupled is that the complex envelope be real. In this case,

$$\operatorname{Var}[\hat{\tau} - \tau] \geq \left[\frac{2\bar{E}_r}{N_0}\left(\frac{\bar{E}_r}{\bar{E}_r + N_0}\right)\right]^{-1}\left[\frac{1}{\overline{\omega^2}}\right] \tag{94}$$

and

$$\operatorname{Var}[\hat{\omega} - \omega] \geq \left[\frac{2\bar{E}_r}{N_0}\left(\frac{\bar{E}_r}{\bar{E}_r + N_0}\right)\right]^{-1}\left[\frac{1}{\overline{t^2}}\right] \tag{95}$$

for all *unbiased* estimates with $\overline{\omega t} = 0$. [Notice that (94) and (95) are bounds even with $\overline{\omega t} \neq 0$, but they are not as tight as (92) and (93).]

The first terms in (92)–(95) are functions of the ratio of the average received signal energy to the white noise level. The second terms indicate the effect of the signal shape. Looking at (94), we see that the bound on the delay estimation accuracy is determined by effective bandwidth. This is logical because, as we increase the signal bandwidth, we can design a signal with a faster rise time. From (95), we see that the bound on the Doppler estimation accuracy is determined by the effective pulse length.

Recalling the definition of the elements in the information matrix (60)–(62) and that of the ambiguity function, we would expect that the elements of **J** could be expressed directly in terms of the ambiguity function.

**Property 2.**† The elements in the matrix in (86) can be expressed as

$$\overline{\omega^2} - (\bar{\omega})^2 = -\frac{1}{2}\frac{\partial^2\theta(\tau, \omega)}{\partial\tau^2}\bigg|_{\omega,\tau=0}, \tag{96}$$

$$\overline{\omega t} - \bar{\omega}\bar{t} = -\frac{1}{2}\frac{\partial^2\theta(\tau, \omega)}{\partial\tau\,\partial\omega}\bigg|_{\omega,\tau=0}, \tag{97}$$

$$\overline{t^2} - (\bar{t})^2 = -\frac{1}{2}\frac{\partial^2\theta(\tau, \omega)}{\partial\omega^2}\bigg|_{\omega,\tau=0}. \tag{98}$$

These results follow from (17), (18) and (67)–(69) (see Problem 10.2.2).

Thus the information matrix can be expressed in terms of the behavior of the ambiguity function at the origin. Property 2 substantiates our intuitive observation on page 281 regarding desirable ambiguity functions.

The final step in our local accuracy discussion is to investigate when the actual estimation error variances approach the bounds given in (92) and (93). To motivate our discussion, we recall some results from our earlier work.

In our discussion of classical estimation theory on page I-71, we quoted some asymptotic properties of maximum likelihood estimates. They can be restated in the context of the present problem. Assume that we have $N$ independent observations of the target. In other words, we receive

$$\tilde{r}_i(t) = \tilde{b}_i\sqrt{E_t}\tilde{f}(t-\tau)e^{j\omega t} + \tilde{w}_i(t), \qquad -\infty < t < \infty, \quad i = 1, 2, \ldots, N, \tag{99}$$

where the $\tilde{b}_i$ and $\tilde{w}_i(t)$ are characterized as in (3)–(5) and are statistically-independent. Physically this could be obtained by transmitting pulses at different times (the time-shift is suppressed in $\tilde{f}(t-\tau)$). Then, as $N \to \infty$,

1. The solution to the likelihood equation,

$$\frac{\partial \ln \Lambda_N(\tau, \omega)}{\partial\tau}\bigg|_{\substack{\tau=\hat{\tau}_{ml}\\ \omega=\hat{\omega}_{ml}}} = 0, \tag{100a}$$

$$\frac{\partial \ln \Lambda_N(\tau, \omega)}{\partial\omega}\bigg|_{\substack{\tau=\hat{\tau}_{ml}\\ \omega=\hat{\omega}_{ml}}} = 0, \tag{100b}$$

where

$$\ln \Lambda_N(\tau, \omega) = \sum_{i=1}^{N} \ln \Lambda_i(\tau, \omega), \tag{100c}$$

$$\ln \Lambda_i(\tau, \omega) = \frac{1}{N_0}\frac{\bar{E}_r}{N_0 + \bar{E}_r}\{|\tilde{L}_i(\tau, \omega)|^2\}, \tag{100d}$$

† We shall derive a number of properties of the ambiguity function in this chapter, and so we use a common numbering system for ease in reference.

and

$$\tilde{L}_i(\tau, \omega) \triangleq \int_{-\infty}^{\infty} \tilde{r}_i(t)\tilde{f}^*(t-\tau)e^{-j\omega t}\,dt, \tag{100e}$$

converges in probability to the correct value $\tau_a$, $\omega_a$ as $N \to \infty$. Thus the $ML$ estimates are consistent.

2. The $ML$ estimates are efficient; that is

$$\lim_{N\to\infty} \frac{\operatorname{Var}[\hat{\tau}_{ml} - \tau_a]}{\dfrac{1}{N}\left[\left(\dfrac{N_0}{2\bar{E}_r}\right)\left(\dfrac{\bar{E}_r + N_0}{\bar{E}_r}\right)\left(\dfrac{\overline{t^2}}{\overline{\omega^2 t^2} - \overline{(\omega t)}^2}\right)\right.} = 1, \tag{101}$$

and a similar relation for $\operatorname{Var}[\hat{\omega}_{ml} - \omega_a]$.

3. The $ML$ estimates are asymptotically jointly Gaussian with covariance matrix $\mathbf{J}^{-1}$.

These results relate to error behavior as the number of observations increase.

In Chapter I-4 (pages I-273 to I-287), we saw that the error variances approached the Cramér-Rao bound for large $E/N_0$. This is a different type of "asymptotic" behavior (asymptotic as $E/N_0 \to \infty$, not as $N \to \infty$.) In the present problem we would like to demonstrate that, using only one pulse, the error variance approaches the Cramér-Rao bound as $\bar{E}_r/N_0$ approaches infinity. Unfortunately, this does not seem to be true (see Problem 10.2.3). Thus the $ML$ estimates are asymptotically efficient in the classical sense ($N \to \infty$) instead of in the high $E/N_0$ sense we encountered in Chapter 4.

There are two other issues concerning asymptotic behavior that should be mentioned:

1. Suppose that we use a fixed number of pulses, $N$ (where $N > 1$) and let $\bar{E}_r/N_0$ increase. Do the error variances approach the bound? We have not been able to resolve this question to our satisfaction.

2. An alternative model that is sometimes used is

$$\tilde{r}(t) = \tilde{b}E_t\tilde{f}(t-\tau)e^{j\omega t} + \tilde{w}(t), \qquad -\infty < t < \infty, \tag{102a}$$

where $|\tilde{b}|$ is either a known amplitude or an unknown *nonrandom* amplitude. The local accuracy results [(58) to (69)] are valid for this model if we let

$$C \triangleq \frac{2E_t|\tilde{b}|^2}{N_0}, \tag{102b}$$

instead of using the value in (66). In this case we can show that the actual error variances approach the bounds as $C \to \infty$ (e.g., [77]).

All of our discussion in this section assumes that the errors are small. The next important question is the behavior of the errors when they are not small. We refer to this as the global accuracy (or ambiguity) problem.

### 10.2.2 Global Accuracy (or Ambiguity)

In this section we study the performance of the system when the errors are not necessarily small.

We can perform an approximate performance analysis by using the same technique as as in the *FM* threshold analysis on pages I-278–I-286.† The basic idea is straightforward. We assume that the region of the $\tau$, $\omega$ plane that we must consider is a rectangle with dimensions $\Omega_*$ and $T_*$. We divide this region into rectangular cells, as shown in Fig. 10.17. The dimensions of the cell are proportional to the dimensions of the central peak of the signal ambiguity function. We shall use a grid with dimensions

$$\Delta_\tau = \frac{1}{\sigma_\omega} \tag{103a}$$

and

$$\Delta_\omega = \frac{1}{\sigma_t}, \tag{103b}$$

where

$$\sigma_t^2 = \overline{t^2} - (\bar{t})^2 \tag{103c}$$

and

$$\sigma_\omega^2 = \overline{\omega^2} - (\bar{\omega})^2. \tag{103d}$$

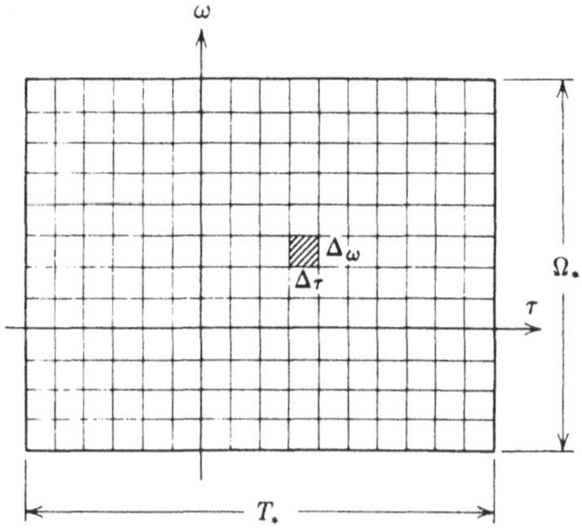


**Fig. 10.17 Region in τ, ω plane in which targets may be present.**

† We suggest that the reader review these pages, because our analysis in this section follows it closely.

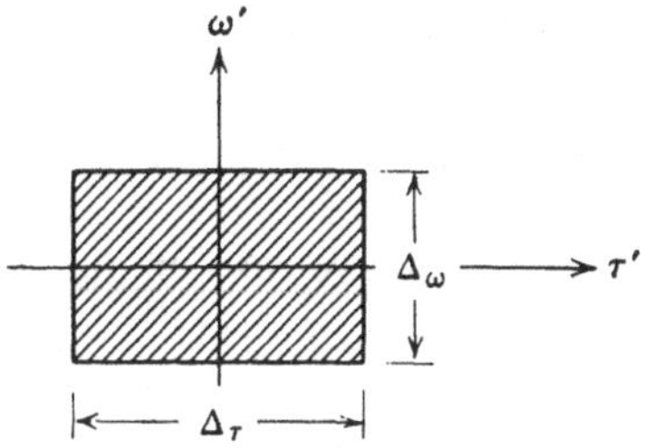


**Fig. 10.18 Range-Doppler cell.**

The cell is shown in Fig. 10.18. Notice that if $\overline{\omega t}$ is significant (e.g., in a linear *FM* signal) a parallelogram cell would be logical. We assume $\overline{\omega t}$ equals zero for simplicity.

We process the received signal in two steps. We first decide which of the cells the signal is in. We next perform a localization within the chosen cell. Thus, there are two kinds of errors: decision errors, because of picking the wrong cell, and local errors within the cell. The local errors were discussed in the last section. We now analyze the decision errors.

To analyze the errors we assume that the signal lies at the center of one of the cells. We denote the center point of the $i$th cell as $(\tau_i, \omega_i)$. We assume that the a-priori probability that a signal will lie in any cell is equal. Thus we have an $M$-hypothesis problem where

$$M = \Omega_* \sigma_t T_* \sigma_\omega. \tag{104a}$$

The LRT consists of computing $|\tilde{L}_i|^2$,

$$|\tilde{L}_i|^2 = \left| \int_{-\infty}^{\infty} \tilde{r}(t) \tilde{f}^*(t - \tau_i) e^{-j\omega_i t}\, dt \right|^2, \tag{104b}$$

and choosing the largest. To analyze the performance, we must consider two cases.

Case 1. The signal ambiguity function has a central peak and no subsidiary peaks. The signal output in all incorrect cells is negligible.

Case 2. The signal ambiguity function has subsidiary peaks whose amplitudes are non-negligible.

We now analyze the first of these cases. The analysis for the second case is outlined briefly. The first case corresponds to transmitting one of $M$-orthogonal signals over a Rayleigh channel. The Pr $(\epsilon)$ for this case was derived in Problem I-4.4.24 (see also [I-80]).

For our purposes the approximate expression derived in Problem

I-4.4.25 (see also [I-90]) is most useful. The result is

$$\Pr(\epsilon) \simeq \frac{N_0}{\bar{E}_r}\left(\ln M - \frac{1}{2M} + 0.577\right). \tag{105}$$

As in Example 2 on page I-282, we can compute a mean-square error, given an interval error

$$E[\tau_\epsilon^{\ 2} \mid \text{interval error}] \leq 2\sigma_{T_*}^{\ 2} = \frac{T_*^2}{6}, \tag{106a}$$

$$E[\omega_\epsilon^{\ 2} \mid \text{interval error}] \leq 2\sigma_{\Omega_*}^{\ 2} = \frac{\Omega_*^2}{6}. \tag{106b}$$

In (106a) we have ignored the set of decision errors that cause no range error (i.e., choosing a cell at the correct range, but the wrong Doppler).

We now restrict our attention to the range estimation error. We can combine the various results to obtain an overall variance.

$$E(\tau_e^{\ 2}) = E(\tau_e^{\ 2} \mid \text{no decision error}) \Pr(\text{no decision error}) + E(\tau_e^{\ 2} \mid \text{decision error}) \Pr(\text{decision error}). \tag{107a}$$

The only term that we have not evaluated is $E(\tau_e^{\ 2} \mid \text{no decision error})$. We can obtain a good approximation to this term, but it is too complicated for our present purposes. It is adequate to observe that the first term is nonnegative, so that the normalized error can be bounded by using only the second term. The result is

$$\frac{E(\tau_e^{\ 2})}{\text{Var}[T_*]} = \frac{E(\tau_e^{\ 2})}{T_*^2/12} \geq \frac{12}{T_*^2} E(\tau_e^{\ 2} \mid \text{decision error}) \Pr(\text{decision error})$$
$$= \frac{2N_0}{\bar{E}_r}\left(\ln[\Omega_*\sigma_t T_*\sigma_\omega] - \frac{1}{2\Omega_*\sigma_t T_*\sigma_\omega} + 0.577\right). \tag{107b}$$

From Section 10.2.1, we know that we also can bound the variance by using the Cramér-Rao bound. For large $\bar{E}_r/N_0$, the normalized Cramér-Rao bound is approximately

$$\frac{E(\tau_e^{\ 2})}{\text{Var}[T_*]} \geq \frac{N_0}{2\bar{E}_r}\frac{1}{(T_*\sigma_\omega)^2}. \tag{107c}$$

Comparing (107b) and (107c), we see that the right sides of both expressions have the same dependence on $\bar{E}_r/N_0$. However, for the parameter values of interest, the right side of (107b) is always appreciably larger than the right side of (107c). Thus, we conclude that, for a single transmitted pulse and a Rayleigh target model, the probability of a decision error dominates the error behavior, and we never achieve the variance indicated by the Cramér-Rao bound. The reason for this behavior is that in the

Rayleigh model we encounter targets with small $|\tilde{b}|^2$ regardless of how large $\bar{E}_r/N_0$ is. The large estimation errors associated with these small amplitude targets keep the average error from approaching the bound.

In Section 10.2.1, we indicated that the ML estimates were asymptotically efficient as the number of observations (that is, the number of pulses transmitted) approached infinity. We can now discuss the behavior as a function of $N$. For $N$ pulses the Pr $(\epsilon)$ is approximately

$$\Pr(\epsilon) \simeq (\Omega_* \sigma_t T_* \sigma_\omega) \sqrt{\frac{1}{\pi N}}\, \frac{(1 + \bar{E}_r/N_0)^N}{(\bar{E}_r/N_0)(1 + \bar{E}_r/N_0)^{2N-1}} \tag{108a}$$

[use (I-2.516) and (I-4.64)]. If we assume that $E(\tau_e^2 \mid \text{no decision error})$ can be approximated by the right side of (92) then

$$\frac{E(\tau_e^2)}{\operatorname{Var}[T_*]} \simeq \frac{1}{(T_* \sigma_\omega)^2} \frac{6}{N\bar{E}_r/N_0} \frac{1 + N\bar{E}_r/N_0}{N\bar{E}_r/N_0} (1 - \Pr(\epsilon)) + 2 \Pr(\epsilon). \tag{108b}$$

In Fig. 10.19, we have plotted the reciprocal of the normalized mean-square error as a function of $N$ for various $\bar{E}_r/N_0$, $\Omega_* \sigma_t$, and $T_* \sigma_\omega$ values. As we would expect, there is a definite threshold effect. Below threshold the variance increases as $(\sigma_\omega T_*)(\sigma_t \Omega_*)$ is increased. Increasing $(\sigma_\omega T_*)(\sigma_t \Omega_*)$ also moves the threshold point to a larger value of $N$. Notice that even when $\bar{E}_r/N_0$ equals 10, it takes about 10 pulses to get above threshold.

In our discussion at the end of Section 10.2.1 we mentioned an alternative target model in which $|\tilde{b}|$ was modeled as a fixed quantity. It is interesting to discuss the global accuracy for this model. By a power series approach one can show that

$$E(\tau_e^2 \mid \text{no decision error}) = \frac{N_0}{2E_r} \frac{1}{\sigma_\omega^2}, \tag{109a}$$

where

$$E_r = E_t\, |\tilde{b}|^2, \tag{109b}$$

(see for example [77]). Using Problem 4.4.7 and (I-4.64),

$$\Pr(\epsilon) \leq \frac{\Omega_* \sigma_t T_* \sigma_\omega}{2} \exp\left(-\frac{E_r}{2N_0}\right). \tag{110}$$

Using (107a), (109), and (110), we obtain an expression for the normalized mean-square error. The reciprocal of the normalized mean-square error is plotted in Fig. 10.20 as a function of $E_r/2N_0$ for various values of $\Omega_* \sigma_t$ and $T_* \sigma_\omega$. Once again, we observe the threshold effect. Notice that if $(\sigma_\omega T_*)(\sigma_t \Omega_*)$ equals $10^4$, then we need an $E_r/N_0$ of about 40 to get above threshold.

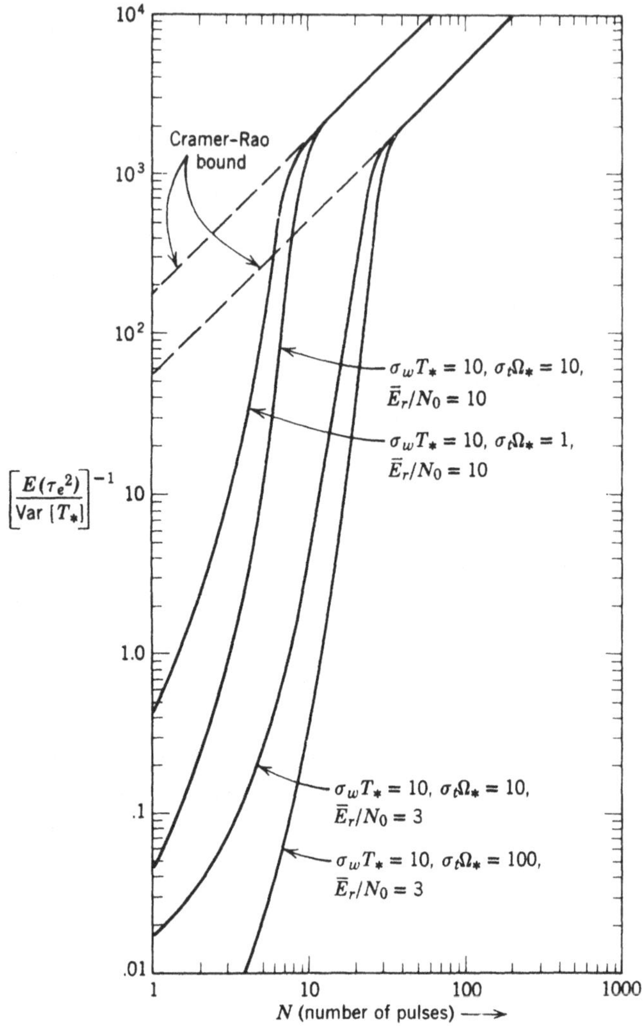


**Fig. 10.19 Reciprocal of normalized mean-square error versus number of pulses: Rayleigh target model.**

We should point out that in both Figs. 10.19 and 10.20, the location of the threshold is a function of the grid size that we selected. An analysis of the effect of grid size on the local and global errors could be carried out, but we have not done so.

All of our discussion up to this point has considered a signal whose ambiguity function had no subsidiary peaks. If the ambiguity function has subsidiary peaks, then the decision problem corresponds to the $M$-ary decision problem with nonorthogonal signals. For a particular signal

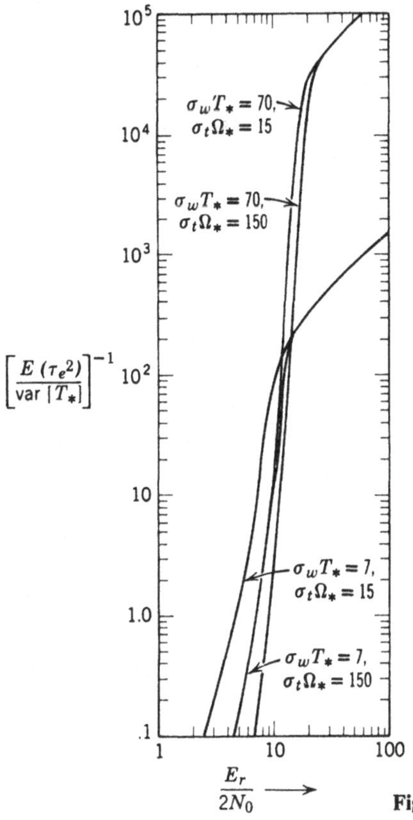


**Fig. 10.20 Reciprocal of normalized mean-square error versus $E_r/2N_0$: nonfluctuating target model.**

ambiguity function an approximate answer can be obtained, but the exact numerical results add little insight into the general case.

We should point out that there are other techniques available for studying the global accuracy problem. The Barankin bound discussed on pages I-71, I-147, and I-286 is quite useful (see for example [78]–[81]). Other references dealing with the global accuracy problem include [82]–[83].

### 10.2.3 Summary

In this section we have studied the performance of the optimum receiver. We have found that the local accuracy depends on the shape of the ambiguity function, $\theta(\tau, \omega)$, near the origin. We also studied the global accuracy (or ambiguity) problem. Here the performance depends on the behavior of $\theta(\tau, \omega)$ in the entire $\tau$, $\omega$ plane.

Thus, we have established that in both the accuracy and ambiguity issues the functions $\phi(\tau, \omega)$ and $\theta(\tau, \omega)$ play a fundamental role. In the next section, we derive some useful properties of these functions.

## 10.3 PROPERTIES OF TIME-FREQUENCY AUTOCORRELATION FUNCTIONS AND AMBIGUITY FUNCTIONS

The autocorrelation and ambiguity functions were first introduced by Ville [1]. Their properties have been studied in detail by Woodward [8], Siebert [9], [10], Lerner [11], and Price [12].

The first property that we shall derive in this section concerns the volume under the ambiguity function. One implication of this property is that the ideal ambiguity function of Fig. 10.6 cannot exist.

**Property 3 (Volume Invariance).** The total volume under the ambiguity function is invariant to the choice of signal. Specifically,

$$\boxed{\int_{-\infty}^{\infty}\int_{-\infty}^{\infty} \theta(\tau, \omega)\, d\tau \frac{d\omega}{2\pi} = 1.} \tag{111}$$

*Proof.* The proof follows directly from the definitions in (17) and (18). We have

$$\iint_{-\infty}^{\infty} \theta(\tau, \omega)\, d\tau \frac{d\omega}{2\pi} = \int_{-\infty}^{\infty}\int_{-\infty}^{\infty}\int_{-\infty}^{\infty}\int_{-\infty}^{\infty} d\tau \frac{d\omega}{2\pi}\, dt\, du\, \tilde{f}\left(t - \frac{\tau}{2}\right)$$

$$\tilde{f}^*\left(t + \frac{\tau}{2}\right) e^{j\omega t} \tilde{f}^*\left(u - \frac{\tau}{2}\right) \tilde{f}\left(u + \frac{\tau}{2}\right) e^{-j\omega u}. \tag{112}$$

Integrating with respect to $\omega$ gives $2\pi\delta(t - u)$. Then integrating with respect to $u$ changes $u$ to $t$. This gives

$$\iint_{-\infty}^{\infty} \theta(\tau, \omega)\, d\tau \frac{d\omega}{2\pi} = \int_{-\infty}^{\infty}\int_{-\infty}^{\infty} d\tau\, dt \left|\tilde{f}\left(t - \frac{\tau}{2}\right)\right|^2 \left|\tilde{f}\left(t + \frac{\tau}{2}\right)\right|^2. \tag{113}$$

Let $z = t - (\tau/2)$. Then

$$\iint_{-\infty}^{\infty} \theta(\tau, \omega)\, d\tau \frac{d\omega}{2\pi} = \int_{-\infty}^{\infty} dz\, |\tilde{f}(z)|^2 \int_{-\infty}^{\infty} d\tau\, |\tilde{f}(\tau + z)|^2. \tag{114}$$

The inner integral equals unity for all $z$, since the energy is invariant to a time shift. The remaining integral equals unity, which is the desired result.

The implication of this result, which is frequently called the *radar uncertainty principle*, is clear. If we change the signal in order to narrow the main peak and improve the accuracy, we then must check to see where the displaced volume reappears in the $\tau$, $\omega$ plane and check the effect on system performance. The radar uncertainty principle is probably the most important property of the ambiguity function. There are a number of other properties that are less fundamental but are useful in signal analysis and design.

The first group of properties deals principally with the time-frequency autocorrelation function (most of these were indicated in [10]). The proofs are all straightforward and many are left as exercises.

**Property 4 (Symmetry).**

$$\phi(\tau, \omega) = \phi^*(-\tau, -\omega) \tag{115}$$

and

$$\theta(\tau, \omega) = \theta(-\tau, -\omega). \tag{116}$$

**Property 5 (Alternative Representations).** An alternative representation of the time-frequency autocorrelation function is

$$\phi(\tau, \omega) = \frac{1}{2\pi}\int_{-\infty}^{\infty} \tilde{F}\left(j\alpha - \frac{j\omega}{2}\right)\tilde{F}^*\left(j\alpha + \frac{j\omega}{2}\right)e^{-j\alpha\tau}\,d\alpha. \tag{117}$$

At this point it is convenient to introduce a time-frequency autocorrelation function whose second argument is in cycles per second. We denote it by

$$\boxed{\phi\{\tau, f\} = \int_{-\infty}^{\infty} \tilde{f}\left(t - \frac{\tau}{2}\right)\tilde{f}^*\left(t + \frac{\tau}{2}\right)e^{j2\pi f t}\,dt.} \tag{118}$$

Similarly,

$$\theta\{\tau, f\} = |\phi\{\tau, f\}|^2. \tag{119}$$

The braces $\{\cdot\}$ indicate the definitions in (118) and (119), while the parentheses $(\cdot)$ indicate the original definition.†

**Property 6 (Duality).** The result in Property 5 points out an interesting duality that we shall exploit in detail later. Consider two signals, $\tilde{g}_1(t)$ and $\tilde{g}_2(t)$, such that

$$\tilde{g}_2\{f\} \triangleq \int_{-\infty}^{\infty} \tilde{g}_1(t)e^{-j2\pi f t}\,dt, \tag{120}$$

† We apologize for this diabolical notation. In most cases, the definition being used is obvious from the context and one can be careless. In duality problems one must be careful, because the definition cannot be inferred from the argument.

that is, $\tilde{g}_2\{\cdot\}$ is the Fourier transform of $\tilde{g}_1(\cdot)$. The result in (117) implies that

$$\phi_2\{f, -\tau\} = \phi_1\{\tau, f\}. \tag{121}$$

Thus the effect of transmitting the Fourier transform of a signal is to rotate the time-frequency diagram 90° in a *clockwise* direction.

Similarly,

$$\theta_2\{f, -\tau\} = \theta_1\{\tau, f\}. \tag{122}$$

**Property 7 (Scaling).** If

$$\tilde{f}(t) \sim \phi(\tau, \omega), \tag{123}$$

then

$$\sqrt{\alpha}\,\tilde{f}(\alpha t) \sim \phi\left(\alpha\tau, \frac{\omega}{\alpha}\right), \qquad \alpha > 0. \tag{124}$$

The ambiguity function is scaled in an identical manner.

**Property 8.** If

$$\tilde{F}(j\omega) \sim \phi(\tau, \omega), \tag{125}$$

then

$$\tilde{F}(j\omega)e^{j\alpha\omega} \sim \phi(\tau + 2\alpha\omega, \omega). \tag{126}$$

This is the frequency domain dual of Property 1 (page 290). The ambiguity function is changed in an identical manner.

**Property 9 (Rotation).** A generalization of the duality relation in Property 6 is the rotation property. Assume that

$$\tilde{f}_1(t) \sim \phi_1(\tau, \omega). \tag{127}$$

If we desire a new time-frequency function that is obtained by rotating the given $\phi_1(\cdot, \cdot)$, that is,

$$\phi_2(\tau, \omega) = \phi_1(\omega \sin\alpha + \tau\cos\alpha, \omega\cos\alpha - \tau\sin\alpha), \qquad 0 < \alpha < \frac{\pi}{2}, \tag{128}$$

we can obtain this by transmitting

$$\tilde{f}_2(t) = \frac{1}{\sqrt{\cos\alpha}} \exp\left(j\frac{t^2\tan\alpha}{2}\right) \int_{-\infty}^{\infty} \tilde{F}(j\omega) \exp\left(j\left(\frac{\omega^2\tan\alpha}{2} + \frac{\omega t}{\cos\alpha}\right)\right) \frac{d\omega}{2\pi}. \tag{129}$$

The ambiguity function is also rotated by $\alpha$ radians.

**Property 10.** A question of interest is: Given some function of two variables, $\phi\{\tau, f\}$, is it a time-frequency correlation function of some signal? We can answer this by taking the inverse transform of (118). Since

$$\phi\{\tau, f\} = \int_{-\infty}^{\infty} \tilde{f}\left(t - \frac{\tau}{2}\right)\tilde{f}^*\left(t + \frac{\tau}{2}\right)e^{j2\pi f t}\,dt, \tag{130}$$

$$\int_{-\infty}^{\infty} \phi\{\tau, f\}e^{-j2\pi f t}\,df = \tilde{f}\left(t - \frac{\tau}{2}\right)\tilde{f}^*\left(t + \frac{\tau}{2}\right). \tag{131}$$

Thus, if the transform of $\phi\{\tau, f\}$ can be written in the form shown in (131), it is a legitimate time-frequency correlation function and $\tilde{f}(t)$ is the corresponding signal. By a change of variables ($x = t - \tau/2$ and $y = t + \tau/2$), the relation in (131) can be rewritten as

$$\tilde{f}(x)\tilde{f}^*(y) = \int_{-\infty}^{\infty} \phi\{y - x, f\} \exp\left(-j2\pi f\left(\frac{x + y}{2}\right)\right) df. \tag{132}$$

By duality (Property 6), this can be written as

$$\tilde{F}\{x\}\tilde{F}^*\{y\} = \int_{-\infty}^{\infty} \phi\{f, x - y\} \exp\left(-j2\pi f\left(\frac{x + y}{2}\right)\right) df. \tag{133}$$

The relations in (132) and (133) enable us to determine the signal directly. Notice that the signal is unique except for a constant phase angle. Thus,

$$\tilde{f}_a(x) \triangleq \tilde{f}(x)e^{j\alpha} \tag{134}$$

also satisfies (132).

A similar relation has not been derived for the ambiguity function. Thus, if we are given a function $\theta\{\tau, f\}$, we do not have a test that is both necessary and sufficient for $\theta\{\cdot, \cdot\}$ to be an ambiguity function. In addition, we do not have any direct procedure for finding an $\tilde{f}(t)$ that will produce a desired ambiguity function.

**Property 11 (Multiplication).** If

$$\tilde{f}_1(t) \sim \phi_1\{\tau, f\} \tag{135}$$

and

$$\tilde{f}_2(t) \sim \phi_2\{\tau, f\}, \tag{136}$$

then

$$\tilde{f}_1(t)\tilde{f}_2(t) \sim \int_{-\infty}^{\infty} \phi_1\{\tau, x\}\phi_2\{\tau, f - x\}\,dx \tag{137}$$

(i.e., convolution with respect to the frequency-variable), and

$$\tilde{F}_1\{f\}\tilde{F}_2\{f\} \sim \int_{-\infty}^{\infty} \phi_1\{x, f\}\phi_2\{\tau - x, f\}\,dx \tag{138}$$

(i.e., convolution with respect to the time variable).

**Property 12 (Axis-Intercept Functions).** The time-frequency correlation function evaluated at $\omega = 0$ is just the time-correlation function of the complex envelope,

$$\phi(\tau, 0) = \int_{-\infty}^{\infty} \tilde{f}\left(t - \frac{\tau}{2}\right)\tilde{f}^*\left(t + \frac{\tau}{2}\right) dt. \tag{139}$$

The time-frequency correlation evaluated at $\tau = 0$ has two interesting interpretations. It is the Fourier transform of the squared magnitude of the complex envelope,

$$\phi(0, \omega) = \int_{-\infty}^{\infty} |\tilde{f}(t)|^2 \, e^{j\omega t} \, dt. \tag{140}$$

From (117), it is the correlation function of the Fourier transform of the complex envelope,

$$\phi\{0, f\} = \int_{-\infty}^{\infty} \tilde{F}\left\{\left(x + \frac{f}{2}\right)\right\}\tilde{F}^*\left\{\left(x - \frac{f}{2}\right)\right\} df. \tag{141}$$

The final property of interest applies only to the ambiguity function.

**Property 13 (Self-Transform).** An ambiguity function is its own two-dimensional Fourier transform,

$$\iint_{-\infty}^{\infty} \theta\{\tau, f\} \exp\left[j2\pi(vf - u\tau)\right] d\tau \, df = \theta\{v, u\}. \tag{142}$$

Observe the sign convention in the definition of the double transform [minus on the time (first) variable and plus on the frequency (second) variable]. This choice is arbitrary and is made to agree with current radar/sonar literature. It is worth noting that the converse statement is not true; the self-transform property does not guarantee that a particular function is an ambiguity function.

In this section we have derived a number of useful properties of the time-frequency autocorrelation function and the ambiguity function. Several other properties are derived in the problems. In addition, the properties are applied to some typical examples.

Notice that we have not been able to find a necessary and sufficient condition for a function to be an ambiguity function. Even if we know (or assume) that some two-variable function is an ambiguity function, we do not have an algorithm for finding the corresponding complex envelope. Thus, we can not simply choose a desired ambiguity function and then solve for the required signal. An alternative approach to the signal design problem is to look at certain classes of waveforms, develop the resulting ambiguity function, and then choose the best waveform in the class. This is the approach that we shall use.

In Section 10.2, we examined modulated analog waveforms. We now look at a class of waveforms that we call coded pulse sequences.

## 10.4 CODED PULSE SEQUENCES

In this section we study complex envelopes consisting of a sequence of pulses that are amplitude-, phase-, and frequency-modulated. Each pulse in the sequence can be expressed as a delayed version of an elementary signal, $\tilde{u}(t)$, where

$$\tilde{u}(t) \triangleq \begin{cases} \dfrac{1}{\sqrt{T_s}}, & 0 \le t \le T_s, \\ 0, & \text{elsewhere.} \end{cases} \tag{143}$$

We denote the delayed version as $\tilde{u}_n(t)$,

$$\tilde{u}_n(t) \triangleq \tilde{u}(t - nT_s). \tag{144}$$

The complex envelope of interest is

$$\tilde{f}(t) = c \sum_{n=1}^{N} a_n \tilde{u}_n(t) \exp(j\omega_n t + j\theta_n). \tag{145}$$

We see that $a_n$ is a constant amplitude modulation on the $n$th pulse, $\omega_n$ is a constant frequency modulation of the $n$th pulse, and $\theta_n$ is the phase modulation on the $n$th pulse. The constant $c$ is used to normalize the envelope. The signal in (145) has $3N$ parameters that can be adjusted. We shall investigate the effect of various parameters.

Our discussion is divided into three parts:

1. A brief investigation of on-off sequences.
2. A development of a class of signals whose ambiguity functions are similar to the ideal ambiguity function of Fig. 10.6.
3. A brief commentary on other classes of signals that may be useful for particular applications.

### 10.4.1 On-off Sequences

The simplest example of an on-off sequence is the periodic pulse sequence in Example 3. Clearly, it can be written in the form of (145). To illustrate this, we assume that we have a periodic pulse sequence with interpulse spacing $T_p = 10T_s$ and a total of 10 pulses, as shown in Fig. 10.21. In the

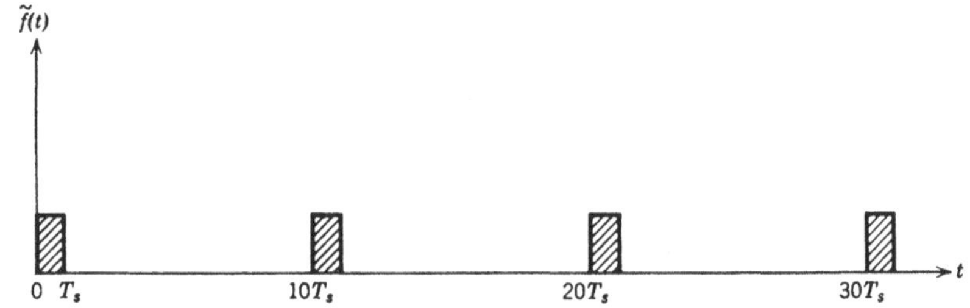


**Fig. 10.21 Periodic pulse train.**

notation of (145),

$$a_1 = a_{11} = a_{21} = \cdots a_{91} = 1, \tag{146}$$

and all other $a_n = 0$. Both $\omega_n$ and $\theta_n = 0$ for all $n$.

The disadvantage of the periodic sequence was the presence of large subsidiary peaks in the ambiguity function. Since the peaks are caused by the periodic structure, we can try to eliminate them by using a nonuniform pulse-repetition rate. One way to construct such a sequence would be to have 100 possible pulse positions and insert 10 pulses randomly. This type of procedure has been investigated in detail (e.g., Rihaczek [13], Cook and Bernfeld [14, pages 232–240], Kaiteris and Rubin [15], and Resnick [16]). It can be shown (the easiest way is experimentally) that staggering the PRF causes a significant reduction in the sidelobe level. (A sidelobe is a subsidiary peak in the $\tau$, $\omega$ plane.) The interested reader can consult the above references for a detailed discussion.

### 10.4.2 Constant Power, Amplitude-modulated Waveforms

In this section, we consider the special case of (145) in which the waveforms can be written as

$$\tilde{f}(t) = c\sum_{i=1}^{N} a_n\tilde{u}_n(t), \tag{147}$$

where

$$a_n = \pm 1. \tag{148}$$

To motivate the use of this class of waveforms, let us recall the properties that an "ideal" ambiguity function should have:

1. The central peak should be narrow along the $\tau$-axis. The minimum width of the central peak is governed by the signal bandwidth $W$. Here the bandwidth is the reciprocal of the length of the elemental pulse, $T_s$. Outside the region of the central peak, the ambiguity function should be

reasonably flat. From Property 12,

$$\phi(\tau, 0) = \int_{-\infty}^{\infty} \tilde{f}\left(t - \frac{\tau}{2}\right)\tilde{f}^*\left(t + \frac{\tau}{2}\right) dt. \tag{149}$$

Thus, we want a signal whose correlation function has the behavior shown in Fig. 10.22.

2. The central peak should be narrow along the $\omega$-axis. From Property 12,

$$\phi(0, \omega) = \int_{-\infty}^{\infty} |\tilde{f}(t)|^2 e^{j\omega t}\, dt. \tag{150}$$

By making $|\tilde{f}(t)|$ constant over the entire signal sequence, we make $\phi(0, \omega)$ a narrow function of $\omega$. This suggests choosing

$$a_n = \pm 1, \qquad n = 1, 2, \ldots, N. \tag{151}$$

Then

$$|\tilde{f}(t)| = \frac{1}{\sqrt{N}}, \qquad 0 \le t < NT_s = T, \tag{152}$$

and the width on the $f$-axis is approximately $2/T$.

3. The ambiguity function should be reasonably flat except for the central peak. This requirement is harder to interpret in terms of a requirement on the signal. Therefore, we design signals using the first two requirements and check their behavior in the $\tau$, $\omega$ plane to see if it is satisfactory.

4. The volume-invariance property indicates that if the ambiguity function is approximately flat away from the origin, its height must be such that the total volume integrates to unity. To compute this height, we observe that the total length of the ambiguity function is $2T$ if the duration of the complex envelope is $T$ (recall $T = NT_s$). The ambiguity function does not have a finite width on the $f$-axis for a finite duration signal. However, we can approximate it by a width of $2W$ cycles per second, where $W$ is the effective signal bandwidth. (In this case, $W = T_s^{-1}$.) With these approximations we have the desired ambiguity function shown in

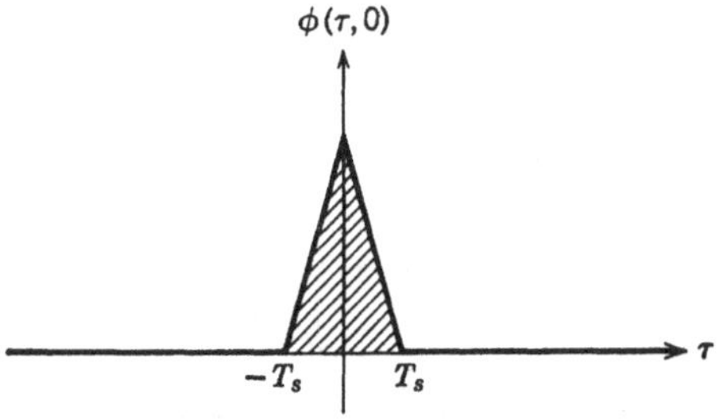


**Fig. 10.22 A desirable signal correlation function.**

Fig. 10.23. We can approximate the volume of the central peak by $1/TW$. Thus, the height of the flat region must satisfy the equation

$$h \cdot 4WT + \frac{1}{WT} \simeq 1 \tag{153}$$

or

$$h \simeq \frac{1}{4WT}\left(1 - \frac{1}{WT}\right). \tag{154}$$

For large $TW$ products,

$$h \simeq \frac{1}{4WT}. \tag{155}$$

This result gives us a general indication of the type of behavior we may be able to obtain. From the radar uncertainty principle, we know that this is the lowest *uniform* height we can obtain. Depressing certain areas further would require peaks in other areas.

With these four observations as background, we now try to find a waveform that leads to the ambiguity function in Fig. 10.23. In the absence of any obvious design procedure, a logical approach is to use the intuition we have gained from the few examples we have studied and the properties we have derived.

***Barker Codes.*** A plausible first approach is to let $N$ equal a small number and investigate all possible sequences of $a_n$. For example, if

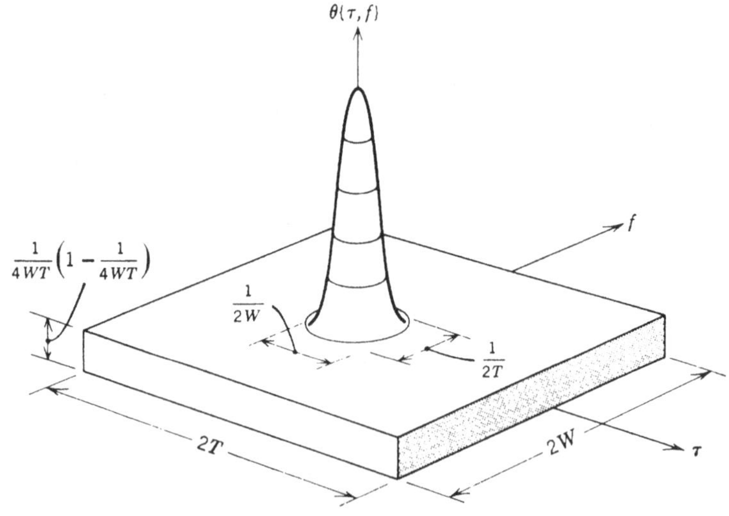


**Fig. 10.23 A desirable ambiguity function.**

$N = 3$, there are $2^3$ arrangements. We can indicate the sequence by the amplitudes. Thus,

$$+ + - \tag{156}$$

denotes

$$\tilde{f}(t) = \frac{1}{\sqrt{3}}[u_1(t) + u_2(t) - u_3(t)]. \tag{157}$$

We can compute the correlation function $\phi(\tau, 0)$ easily. Since we are shifting rectangular pulses, we just compute

$$\phi(nT_s, 0), \qquad n = 1, 2, 3 \tag{158}$$

and connect these values with straight lines. The resulting correlation function is shown in Fig. 10.24. We see that the correlation function has the property that

$$|\phi(nT_s, 0)| \leq \frac{1}{N}, \qquad n \neq 0. \tag{159}$$

Notice that the complement of this sequence, $- - +$, the reverse, $- + +$, and the complement of the reverse, $+ - -$, all have the same property. We can verify that none of the other sequences of length 3 has this property. Barker [17] developed sequences that satisfy the condition in (159) for various $N \leq 13$. These sequences, which are referred to as Barker codes, are tabulated in Table 10.1. Unfortunately, Barker codes with lengths greater than 13 have not been found. It can be proved that no odd sequences greater than 13 exist and no even sequences with $N$ between

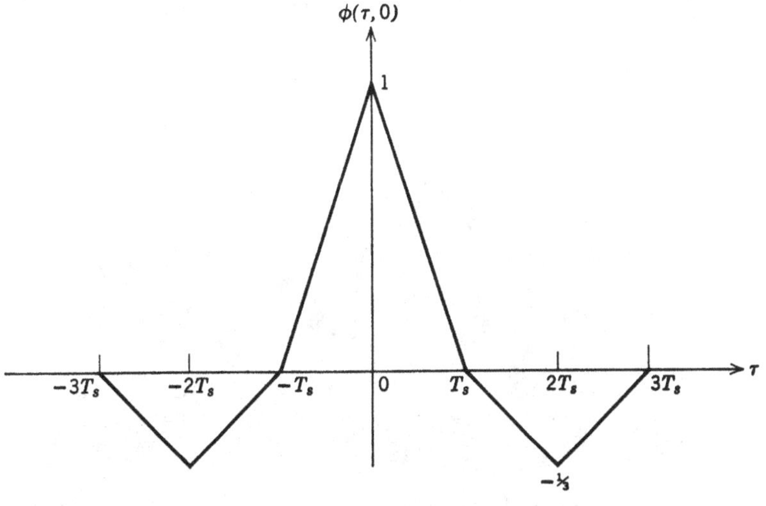


**Fig. 10.24 Correlation function of three-element Barker code.**

**Table 10.1 Barker Sequences**

| $N$ | Sequences |
|---|---|
| 2 | + +, − + |
| 3 | + + − |
| 4 | + + − +, + + + − |
| 5 | + + + − + |
| 7 | + + + − − + − |
| 11 | + + + − − − + − − + − |
| 13 | + + + + + − − + + − + − + |

4 and 6084 have been found [18]. The magnitude of the time-frequency correlation function for a Barker code of length 13 is shown in Fig. 10.25. We see that there are two ridges of non-negligible height.

***Shift-Register Sequences.*** A second approach to signal design is suggested by an example that is usually encountered in random-process courses. Consider the experiment of flipping a fair coin. If the outcome is a head, we let $a_1 = 1$. If the outcome is a tail, we let $a_1 = -1$. The $n$th toss determines the value of $a_n$. The result is a sample function, the familiar Bernoulli process. As $N \to \infty$, we have the property that

$$\phi(nT_s, 0) = 0, \qquad n \neq 0. \tag{160}$$

Thus, for large $N$, we would expect that the waveform would have a satisfactory correlation function. One possible disadvantage of this procedure is the storage requirement. From our results in Section 10.1, we know that the receiver must have available a replica of the signal in order to construct the matched filter. Thus, if $N = 1000$, we would need to store a sequence of 1000 amplitudes.

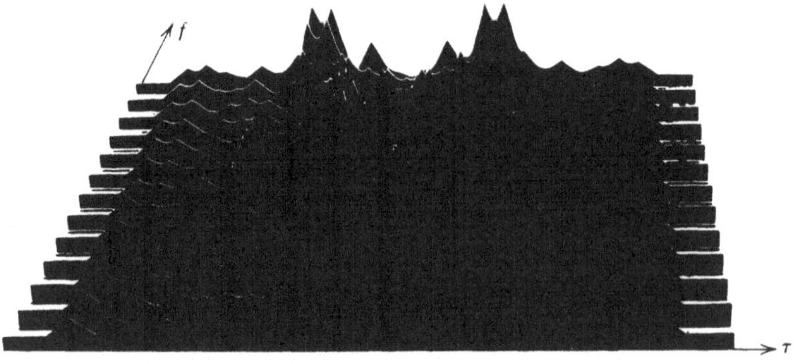


**Fig. 10.25 Magnitude of time-frequency correlation function: Barker code of length 13 (From [19.])**

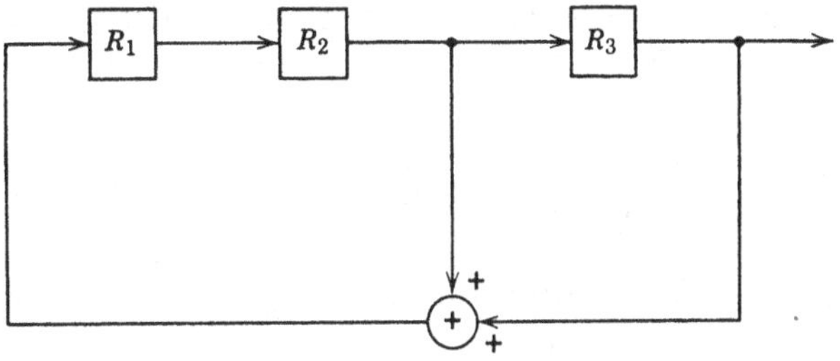


**Fig. 10.26 Feedback shift register.**

Fortunately, there exists a class of deterministic sequences that have many of the characteristics of Bernoulli sequences and can be generated easily. The device used to generate the sequence is called a *feedback shift register*. A typical three-stage configuration is shown in Fig. 10.26. Each stage has a binary digit stored in it. Every $T_s$ seconds a clock pulse excites the system. Two operations then take place:

1. The contents shift one stage to the right. The digit in the last stage becomes the system output.
2. The output of various stages are combined using mod 2 addition. The output of these mod 2 additions is the new content of the first stage. In the system shown, the contents of stages 2 and 3 are added to form the input.
3. Since all of the operations are linear, we refer to this as a *linear shift register*.

The operation of this shift register is shown in detail in Table 10.2. We see that after the seventh clock pulse the contents are identical with

**Table 10.2**

| | Contents | Output sequence |
|---|---|---|
| Initial | 1 1 1 | |
| 1 | 0 1 1 | 1 |
| 2 | 0 0 1 | 1 1 |
| 3 | 1 0 0 | 1 1 1 |
| 4 | 0 1 ~~1~~ 0 | 0 1 1 1 |
| 5 | 1 0 1 | 0 0 1 1 1 |
| 6 | 1 1 0 | 1 0 0 1 1 1 |
| 7 | 1 1 1 | 0 1 0 0 1 1 1 |
| 8 | ↓ | ↓ |
| 9 | | |
| 10 | Repeats | Repeats |

the initial contents. Since the output is determined completely from the contents, the sequence will repeat itself. The period of this sequence is

$$L \triangleq 2^N - 1 = 7. \tag{161}$$

It is clear that we cannot obtain a sequence with a longer period, because there are only $2^N = 8$ possible states, and we must exclude the state 000. (If the shift register is in the 000 state, it continues to generate zeros.) Notice that we have chosen a particular feedback connection to obtain the period of 7. Other feedback connections can result in a shorter period.

To obtain the desired waveform, we map

$$\begin{aligned} 1 &\to +1, \\ 0 &\to -1. \end{aligned} \tag{162}$$

If we assume that the periodic sequence is *infinite* in extent, the correlation function can be computed easily. For convenience we normalize the energy per period to unity instead of normalizing the total energy. The correlation function of the resulting waveform is shown in Fig. 10.27. We see that

$$\phi_p(nT_s, 0) = \begin{cases} 1, & n = 0, L, 2L, \ldots \\ -\frac{1}{7}, & n \neq kL, \quad k = 0, 1, \ldots . \end{cases} \tag{163}$$

All our comments up to this point pertain to the three-stage shift register in Fig. 10.26. In the general case we have an $N$-stage shift register.

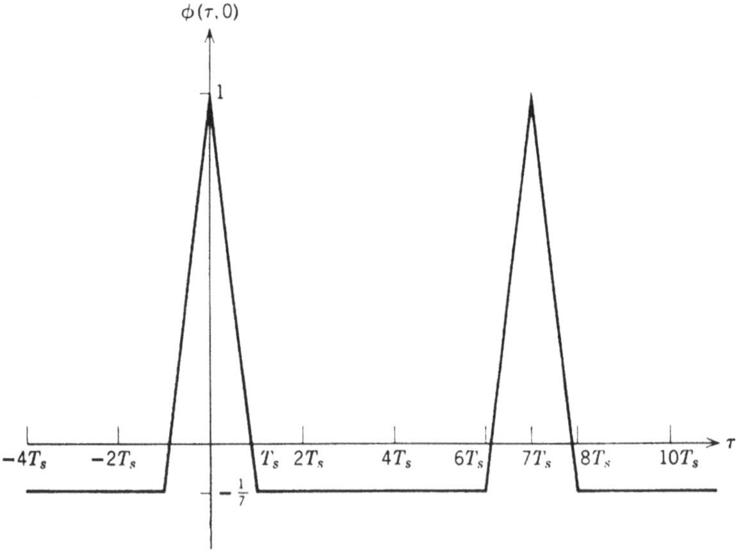


**Fig. 10.27 Correlation function of a periodic pseudorandom sequence.**

Since there are $2^N$ states, we cannot generate a sequence whose period is greater than $2^N - 1$. However, it is not clear that there exists a feedback configuration that will generate a sequence of this period. To prove the existence of such a sequence and find the configuration that produces one requires mathematical background that we have not developed. The properties of shift-register sequences have been studied extensively by Golomb, [20], Huffman [21], Zierler [22], Peterson [23], and Elspas [24]. Very readable tutorial discussions are given by Ristenblatt [25], [26] and Golomb [27, Chapters 1 and 2]. The result of interest to us is that for all $N$ there exists at least one feedback connection such that the output sequence will have a period

$$L = 2^N - 1. \tag{164}$$

These sequences are called *maximal-length shift-register* sequences. Notice that the length is an exponential function of the number of stages. A list of connections for $N \leq 31$ is given by Peterson [23]. A partial list for $N \leq 6$ is given in Problems 10.4.5 and 10.4.6. These sequences are also referred to as *pseudo-random* (*PR*) *sequences*. The "random" comes from the fact that they have many of the characteristics of a Bernoulli sequence, specifically the following:

1. In a Bernoulli sequence, the number of ones and zeros is approximately equal. In a PR sequence, the number of ones per period is one more than the number of zeros per period.
2. A run of a length $n$ means that we have $n$ consecutive outputs of the same type. In a Bernoulli sequence, approximately half the runs are of length 1, one-fourth of length 2, one-eighth of length 3, and so forth. The *PR* sequences have the same run characteristics.
3. The autocorrelation functions are similar.

The "pseudo" comes from the fact that the sequences are perfectly deterministic. The correlation function shown in Fig. 10.27 assumes that the sequence is periodic. This assumption would be valid in a continuous-wave (CW) radar. Applications of this type are discussed in [28] and [29]. Continuous PR sequences are also used extensively in digital communications. In many radar systems we transmit one period of the sequence. Since the above properties assumed a periodic waveform, we must evaluate the behavior for the truncated waveform. For small $N$, the correlation function can be computed directly (e.g., Problem 10.4.3). We can show that for large $N$ the sidelobe levels on the $\tau$-axis approaches $\sqrt{N}$ (or $\sqrt{WT}$). The time-frequency correlation function can be obtained experimentally.

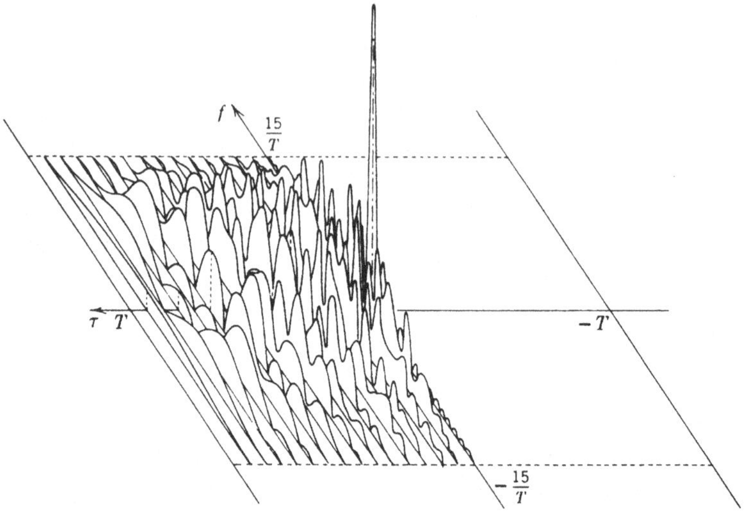


**Fig. 10.28** $|\phi\{\tau, f\}|$ **for pseudorandom sequence of length** $N = 15$ **(From [35]).**

A plot for $N = 15$ is shown in Fig. 10.28. In many applications, the detailed structure of $\phi\{\tau, f\}$ is not critical. Therefore, in most of our discussion we shall use the approximate function shown in Fig. 10.29. This function has the characteristics hypothesized on page 316. Thus, it appears that the shift-register sequences provide a good solution to the combined ambiguity and accuracy problem.

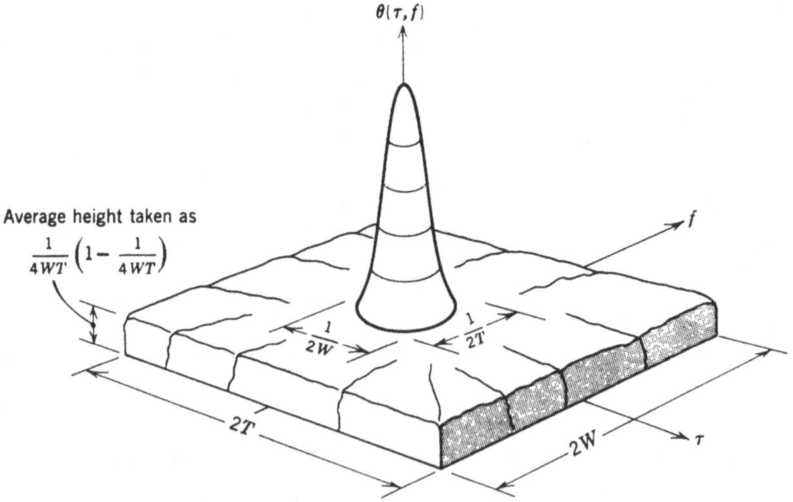


**Fig. 10.29 Approximation to the ambiguity function of a pseudorandom sequence.**

### 10.4.3 Other Coded Sequences

Before leaving our discussion of coded pulse sequences, we should point out that there are several alternatives that we have not discussed. The interested reader can consult [14, Chapter 8] and [30] for a tutorial discussion, or the original papers by Huffman [31], Golomb and Scholtz [32], Heimiller [33], and Frank [34].

We should emphasize that pulse sequences are frequently used in practice because they are relatively easy to generate, the optimum receiver is relatively easy to implement, and they offer a great deal of flexibility. Readers who are specializing in radar signal design should spend much more time on the study of these waveforms than our general development has permitted.

Up to this point we have assumed that only a single target is present. In many cases, additional targets interfere with our observation of the desired target. This problem is commonly called the resolution problem. We discuss it in the next section.

## 10.5 RESOLUTION

The resolution problem in radar or sonar is the problem of detecting or estimating the parameters of a desired target in the presence of other targets or objects that act as reflectors. These reflectors may be part of the environment (e.g., other airplanes, missiles, ships, rain) or may be deliberately placed by an enemy to cause confusion (e.g., decoys, electronic countermeasures, or chaff). It is convenient to divide the resolution

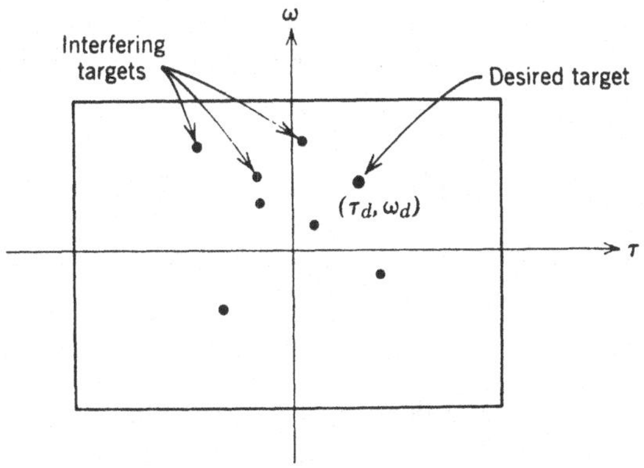


**Fig. 10.30 Target geometry for discrete resolution problem.**

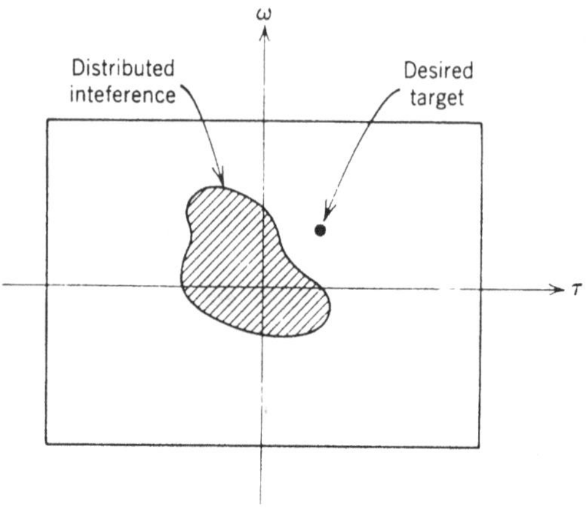


**Fig. 10.31 Geometry for continuous resolution problem.**

problem into two categories:

1. Resolution in a discrete environment.
2. Resolution in a continuous (or dense) environment.

We now discuss our model for these two categories.

In Figure 10.30, we show the region of the $\tau$, $\omega$ plane that we must investigate. The desired target is at coordinates $\tau_d$, $\omega_d$. A set of $K$ interfering targets are at various points in the $\tau$, $\omega$ plane. The desired and interfering targets are assumed to be slowly fluctuating Rayleigh targets. In general, the strengths of the various targets may be unequal. (Unfortunately, we occasionally encounter the problem in which the interfering target strengths are appreciably larger than the desired target strength.) We shall give a detailed model for the discrete resolution problem in the next section.

In the continuous resolution problem, the interference is modeled as a continuum of small reflectors distributed over some area in the $\tau$, $\omega$ plane, as shown in Fig. 10.31. This model is appropriate to the reverberation problem in sonar and the clutter problem in radar. We whall discuss it in more detail in Chapter 13 as part of our discussion of distributed targets.

We now consider the discrete resolution problem in detail.

### 10.5.1 Resolution In a Discrete Environment: Model

In this section, we consider a typical resolution problem. The particular example that we shall discuss is a *detection* problem, but similar results can be obtained for estimation problems.

We want to decide whether or not a target is present at a particular point $(\tau_d, \omega_d)$ in the $\tau$, $\omega$ plane. For algebraic simplicity, we let

$$\tau_d = 0 \tag{165}$$

and

$$\omega_d = 0. \tag{166}$$

There are two sources of interference:

1. Bandpass white noise with spectral height $N_0/2$.
2. A set of $K$ interfering targets located at $(\tau_i, \omega_i)$, $i = 1, 2, \ldots, K$. We model these targets as slowly fluctuating point targets whose location and average strength are known. The fact that exactly $K$ interfering targets are present is also assumed to be known.

The transmitted signal is

$$s_t(t) = \sqrt{2E_t}\,\text{Re}\,[\tilde{f}(t)e^{j\omega_c t}], \tag{167}$$

where $\tilde{f}(t)$ is the normalized complex envelope.

The complex envelope of the received signal on $H_0$ is

$$\tilde{r}(t) = \sqrt{E_t}\left[\sum_{i=1}^{K} \tilde{b}_i \tilde{f}(t-\tau_i)e^{j\omega_i t}\right] + \tilde{w}(t), \qquad -\infty < t < \infty : H_0. \tag{168}$$

When the desired target is present, the complex envelope is

$$\tilde{r}(t) = \sqrt{E_t}\left[\tilde{b}_d \tilde{f}(t) + \sum_{i=1}^{K} \tilde{b}_i \tilde{f}(t-\tau_i)e^{j\omega_i t}\right] + \tilde{w}(t), \qquad -\infty < t < \infty : H_1. \tag{169}$$

The multipliers $\tilde{b}_d$ and $\tilde{b}_i$ are zero-mean complex Gaussian variables that are statistically independent with unequal variances:

$$E[\tilde{b}_d \tilde{b}_d^*] = 2\sigma_d^2, \tag{170}$$

$$E[\tilde{b}_i \tilde{b}_j^*] = 2\sigma_i^2\,\delta_{ij}, \qquad i, j = 1, 2, \ldots, K, \tag{171}$$

and

$$E[\tilde{b}_d \tilde{b}_d] = E[\tilde{b}_i \tilde{b}_i] = E[\tilde{b}_d \tilde{b}_i^*] = E[\tilde{b}_d \tilde{b}_i] = 0, \qquad i = 1, \ldots, K. \tag{172}$$

There are several issues of interest with respect to this model. The first two concern the receiver design, and the next two concern the signal design.

1. We might assume that the receiver is designed *without* knowledge of the interfering signals. The resulting receiver will be the bandpass matched filter that we derived previously. We can then compute the effect of the

interfering targets on the receiver performance. We refer to this as the *conventional* receiver problem.

2. We can design the receiver using the assumed statistical properties of the interference. We shall see that this is a special case of the problem of detection in colored bandpass noise that we discussed in Section 9.3. We refer to this as the optimum receiver problem.

3. We can require the receiver to be a matched filter (as in part 1) and then choose $\tilde{f}(t)$ to minimize the interference effects.

4. We can use the optimum receiver (as in part 2) and choose the signal to minimize the interference effects.

We now discuss these four issues.

### 10.5.2 Conventional Receivers

The performance of the conventional receiver can be obtained in a straightforward manner. If we use a bandpass filter matched to $\tilde{f}(t)$, the optimum receiver performs the test

$$|\tilde{l}_{wo}|^2 \triangleq \left|\int_{-\infty}^{\infty} \tilde{r}(t)\tilde{f}^*(t)\,dt\right|^2 \underset{H_0}{\overset{H_1}{\gtrless}} \gamma \tag{173}$$

[see (9.36) and (9.39).] We use the subscript *wo* to indicate that the test would be *optimum* if the noise were *white*. Now, since $\tilde{l}_{wo}$ is a complex Gaussian variable under both hypotheses, the performance is completely determined by

$$\Delta_{wo} \triangleq \frac{E[|\tilde{l}_{wo}|^2 \mid H_1] - E[|\tilde{l}_{wo}|^2 \mid H_0]}{E[|\tilde{l}_{wo}|^2 \mid H_0]} \tag{174}$$

[see (9.49).] To evaluate the two expectations in (174), we substitute (168) and (169) into the definition in (173). The denominator in (174) is

$$E[|\tilde{l}_{wo}|^2|H_0] = E\left\{\int_{-\infty}^{\infty}\left[\sqrt{E_t}\sum_{i=1}^{K}\tilde{b}_i\tilde{f}(t-\tau_i)e^{j\omega_i t} + \tilde{w}(t)\right]\tilde{f}^*(t)\,dt \right.$$
$$\left. \times \int_{-\infty}^{\infty}\left[\sqrt{E_t}\sum_{k=1}^{K}\tilde{b}_k^*\tilde{f}^*(u-\tau_k)e^{-j\omega_k u} + \tilde{w}^*(u)\right]\tilde{f}(u)\,du\right\}. \tag{175}$$

Using the independence of the $\tilde{b}_i$ and the definition in (18), this reduces to

$$E[|\tilde{l}_{wo}|^2 \mid H_0] = \sum_{i=1}^{K} \bar{E}_{r_i}\theta(\tau_i,\omega_i) + N_0, \tag{176}$$

where

$$\bar{E}_{r_i} \triangleq 2E_t\sigma_i^2 \tag{177}$$

is the average energy received from the $i$th interfering target. Similarly,

$$E[|\tilde{l}_{wo}|^2 \mid H_1] - E[|\tilde{l}_{wo}|^2 \mid H_0] = \bar{E}_{r_d}, \tag{178}$$

where

$$\bar{E}_{r_d} \triangleq 2E_t\sigma_d^2 \tag{179}$$

is the average received energy from the desired target. Then

$$\boxed{\Delta_{wo} = \frac{\dfrac{\bar{E}_{r_d}}{N_0}}{1 + \sum_{i=1}^{K} (\bar{E}_{r_i}/N_0)\theta(\tau_i, w_i)}.} \tag{180}$$

The numerator corresponds to the performance when only white noise is present. The second term in the denominator represents the degradation in the performance due to the interfering targets. We see that the performance using the conventional matched filter is completely characterized by the average strength of the return from the interfering targets and the value of the ambiguity function at their delay and Doppler location.

Conceptually, at least, this result provides the answer to the third issue. We design a signal whose ambiguity function equals zero at the $K$ points in the $\tau$, $\omega$ plane where the interfering signals lie. Even if we could carry out the design, several practical difficulties remain with the solution:

1. The resulting waveform will undoubtedly be complicated.
2. Each time the environment changes, the transmitted signal will have to change.
3. The performance may be sensitive to the detailed assumptions of the model (i.e., the values of $\tau_i$ and $\omega_i$).

On the other hand, there are a number of physical situations in which our solution gives a great deal of insight into how to design good signals. A simple example illustrates the application of the above results.

**Example.** Consider the multiple-target environment shown in Fig. 10.32. We are interested in detecting zero-velocity targets. The interfering targets are moving at a velocity such that there is a *minimum* Doppler shift of $\omega_0$. We want to design a signal such that

$$\theta(\tau, \omega) = 0, \qquad |\omega| > \omega_0. \tag{181}$$

We could accomplish this exactly by transmitting

$$\tilde{f}(t) = \sqrt{\frac{4\pi}{\omega_0}}\,\frac{\sin \omega_0 t}{t}. \tag{182}$$

This result can be verified by looking at the ambiguity function of the rectangular pulse in Example 1 (page 280) and using the duality result in Property 6 (pages 309–310).

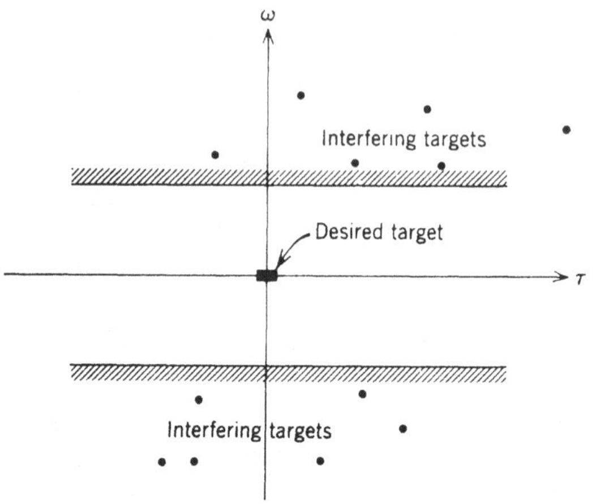


**Fig. 10.32 Multiple-target geometry.**

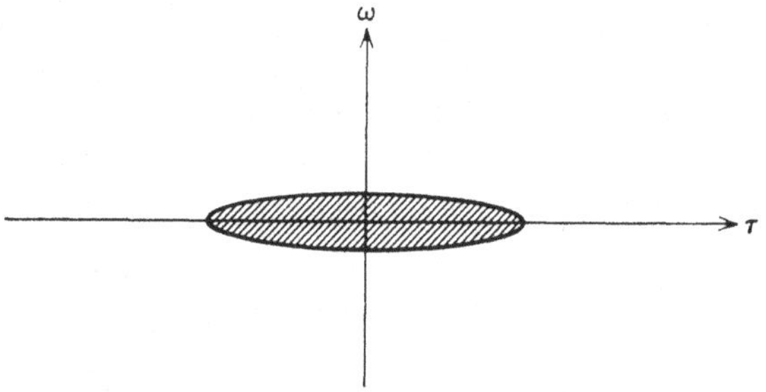


**Fig. 10.33 Contour of ambiguity function of rectangular pulse.**

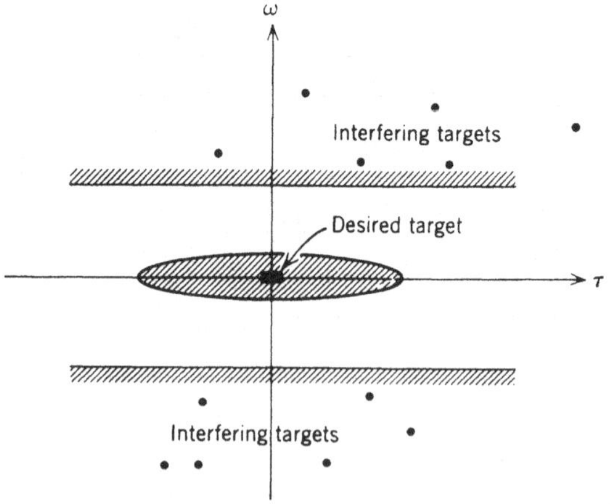


**Fig. 10.34 Signal-interference relation.**

On the other hand, we can make

$$\theta(\tau, \omega) \simeq 0, \qquad |\omega| > \omega_0 \tag{183}$$

by transmitting

$$\tilde{f}(t) = \begin{cases} \dfrac{1}{\sqrt{T}}, & 0 \le t \le T, \\ 0, & \text{elsewhere}, \qquad T \gg \dfrac{2\pi}{\omega_0}. \end{cases} \tag{184}$$

This solution is more practical. A typical contour of the resulting ambiguity function is given in Fig. 10.33, and is shown superimposed on the interference plot in Fig. 10.34. The reason for this simple solution is that the interference and desired targets have some separation in the delay-Doppler plane. If we had ignored the resolution problem, we might have used some other signal, such as a short pulse or a PR waveform. In the target environment shown in Fig. 10.32, the interfering targets can cause an appreciable degradation in performance.

The result of this example suggests the conclusion that we shall probably reach with respect to signal design. No single signal is optimum from the standpoints of accuracy, ambiguity, and resolution under all operating conditions. The choice of a suitable signal will depend on the anticipated target environment.

Now we turn to the second issue. Assuming that we know the statistics of the interference, how do we design the optimum receiver?

### 10.5.3 Optimum Receiver: Discrete Resolution Problem

Looking at (168) and (169), we see that the sum of the returns from the interfering targets can be viewed as a sample function from complex Gaussian noise processes. If we denote the first term in (168) as $\tilde{n}_c(t)$, then

$$\tilde{n}_c(t) = \sqrt{E_t}\left[\sum_{i=1}^{K} \tilde{b}_i \tilde{f}(t - \tau_i) e^{j\omega_i t}\right], \tag{185}$$

and we have the familiar problem of detection in nonwhite complex Gaussian noise (see Section 9.3). The covariance function of $\tilde{n}_c(t)$ is

$$\tilde{K}_c(t, u) = \sum_{i=1}^{K} E_{r_i} \tilde{f}(t - \tau_i) \tilde{f}^*(u - \tau_i) e^{j\omega_i(t-u)}, \qquad -\infty < t, u < \infty. \tag{186}$$

We have used an infinite observation for algebraic simplicity. Usually $\tilde{f}(t)$ has a finite duration, so that $\tilde{K}_c(t, u)$ will be zero outside some region in the $(t, u)$ plane. From (9.69), the optimum receiver performs the operation

$$\left| \int_{T_i}^{T_f} \tilde{r}(t) \tilde{g}^*(t)\, dt \right|^2 \gtrless \gamma, \tag{187}$$

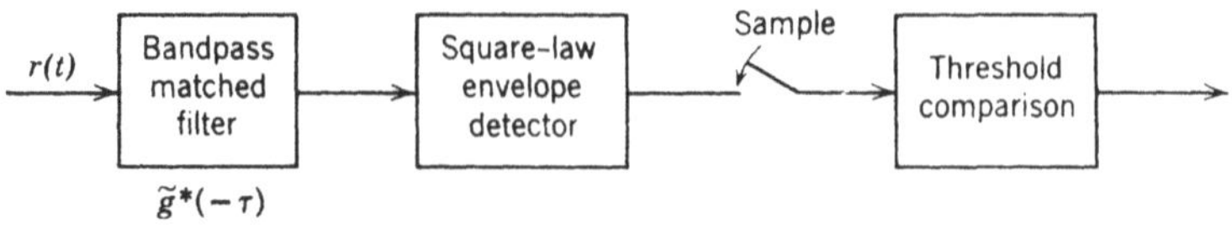


**Fig. 10.35 Optimum receiver.**

where $g(t)$ satisfies (9.74). To find $\tilde{f}(t)$, we substitute (188) into (9.74). The result is

$$\tilde{f}(t) = \int_{-\infty}^{\infty} \left[\sum_{i=1}^{K} E_{r_i} \tilde{f}(t - \tau_i)\tilde{f}^*(u - \tau_i)e^{j\omega_i(t-u)}\right]\tilde{g}(u)\, du + N_0\tilde{g}(t),$$

$$-\infty < t < \infty. \quad (188)$$

We see that this is an integral equation with a separable kernel (see pages I-322–I-325). It can be rewritten as

$$\tilde{f}(t) = \sum_{i=1}^{K} E_{r_i} \tilde{f}(t - \tau_i)e^{j\omega_i t}\left(\int_{-\infty}^{\infty} \tilde{f}^*(u - \tau_i)e^{-j\omega_i u}\tilde{g}(u)\, du\right) + N_0\tilde{g}(t),$$

$$-\infty < t < \infty. \quad (189)$$

The solution to (189) is

$$\tilde{g}(t) = \tilde{g}_d\tilde{f}(t) + \sum_{i=1}^{K} \tilde{g}_i\tilde{f}(t - \tau_i)e^{j\omega_i t}, \qquad -\infty < t < \infty, \qquad (190)$$

where $\tilde{g}_d$ and $\tilde{g}_i$, $i = 1, \ldots, K$, are constants that we must find. The optimum receiver is shown in Fig. 10.35. The calculation of the constants is a straightforward but tedious exercise in matrix manipulation. Since this type of manipulation arises in other situations, we shall carry out the details. The results are given in (201) and (202).

**Calculation of Filter Coefficients.** We first define four matrices. The coefficient matrix $\tilde{\mathbf{b}}$ is

$$\tilde{\mathbf{b}} \triangleq \begin{bmatrix} \tilde{b}_1 \\ \tilde{b}_2 \\ \cdot \\ \cdot \\ \cdot \\ \tilde{b}_K \end{bmatrix}. \qquad (191)$$

The interference matrix $\tilde{\mathbf{f}}_I(t)$ is

$$\tilde{\mathbf{f}}_I(t) \triangleq \begin{bmatrix} \tilde{f}(t-\tau_1)e^{j\omega_1 t} \\ \tilde{f}(t-\tau_2)e^{j\omega_2 t} \\ \cdot \\ \cdot \\ \cdot \\ \tilde{f}(t-\tau_K)e^{j\omega_K t} \end{bmatrix}. \tag{192}$$

In addition, we define

$$\tilde{\mathbf{\Lambda}} \triangleq E_t[E[\tilde{\mathbf{b}}\tilde{\mathbf{b}}^\dagger]] = 2E_t \begin{bmatrix} \sigma_1^2 & & & & 0 \\ & \sigma_2^2 & & & \\ & & \cdot & & \\ & & & \cdot & \\ & & & & \cdot \\ 0 & & & & \sigma_K^2 \end{bmatrix}$$

$$= \begin{bmatrix} \bar{E}_{\tau_1} & & & & 0 \\ & \bar{E}_{\tau_2} & & & \\ & & \cdot & & \\ & & & \cdot & \\ & & & & \cdot \\ 0 & & & & \bar{E}_{\tau_K} \end{bmatrix} \tag{193}$$

and

$$\tilde{\boldsymbol{\rho}} = \int_{T_i}^{T_f} \tilde{\mathbf{f}}_I(t)\tilde{\mathbf{f}}_I^\dagger(t)\,dt. \tag{194}$$

Looking at (192) and (194), we see that all the elements in $\tilde{\boldsymbol{\rho}}$ can be written in terms of time-frequency correlation functions of the signal. The covariance function of the colored noise is

$$\tilde{K}_c(t, u) = \tilde{\mathbf{f}}_I^T(t)\tilde{\mathbf{\Lambda}}\tilde{\mathbf{f}}_I^*(u). \tag{195}$$

Rewriting (190) in matrix notation gives

$$\begin{aligned} \tilde{g}(t) &= \tilde{g}_d\tilde{f}(t) + \tilde{\mathbf{g}}^T\tilde{\mathbf{f}}_I(t) \\ &= \tilde{g}_d\tilde{f}(t) + \tilde{\mathbf{f}}_I^T(t)\tilde{\mathbf{g}}, \end{aligned} \tag{196}$$

where

$$\tilde{\mathbf{g}} \triangleq \begin{bmatrix} \tilde{g}_1 \\ \cdot \\ \cdot \\ \cdot \\ \tilde{g}_K \end{bmatrix}. \tag{197}$$

Substituting (195) and (196) into (188), we have

$$\tilde{f}(t) = \int_{-\infty}^{\infty} \{\tilde{\mathbf{f}}_I^T(t)\tilde{\mathbf{\Lambda}}\tilde{\mathbf{f}}_I^*(u) + N_0\delta(t-u)\}\{\tilde{g}_d\tilde{f}(u) + \tilde{\mathbf{f}}_I^T(u)\tilde{\mathbf{g}}\}\,du. \tag{198}$$

This reduces to

$$\tilde{f}(t) = \tilde{g}_d \tilde{\mathbf{f}}_I^T(t)\mathbf{\Lambda}\tilde{\boldsymbol{\rho}}_d + N_0\tilde{g}_d\tilde{f}(t) + \tilde{\mathbf{f}}_I^T(t)\tilde{\mathbf{\Lambda}}\tilde{\boldsymbol{\rho}}^*\tilde{\mathbf{g}} + N_0\tilde{\mathbf{f}}_I^T(t)\tilde{\mathbf{g}}, \tag{199}$$

where

$$\tilde{\boldsymbol{\rho}}_d = \int_{-\infty}^{\infty} \tilde{\mathbf{f}}_I^*(u)\tilde{f}(u)\,du. \tag{200}$$

Solving (199), we have

$$\tilde{g}_d = \frac{1}{N_0} \tag{201}$$

and

$$\tilde{\mathbf{g}} = -\frac{1}{N_0^2}\left[\mathbf{I} + \frac{1}{N_0}\tilde{\mathbf{\Lambda}}\tilde{\boldsymbol{\rho}}^*\right]^{-1}\tilde{\mathbf{\Lambda}}\tilde{\boldsymbol{\rho}}_d. \tag{202}$$

This completely specifies the optimum receiver.

Using (196), (201), and (202) in (9.77), we find that the performance is determined by

$$\Delta_o = \frac{\bar{E}_r}{N_0}\left\{1 - \frac{1}{N_0}\tilde{\boldsymbol{\rho}}_d^{\dagger}\left[\mathbf{I} + \frac{1}{N_0}\tilde{\mathbf{\Lambda}}\tilde{\boldsymbol{\rho}}^*\right]^{-1}\tilde{\mathbf{\Lambda}}\tilde{\boldsymbol{\rho}}_d\right\}. \tag{203}$$

To illustrate these results, we consider a simple example.

**Example. Single Interfering Target.** In this particular case, the complex envelope of the return from the desired signal is $\sqrt{E_t}\,\tilde{b}_d\tilde{f}(t)$ and the complex envelope of the return from the single interfering target is

$$\tilde{f}_1(t) = \sqrt{E_t}\,\tilde{b}_1\tilde{f}(t - \tau_1)e^{j\omega_1 t}. \tag{204}$$

Thus, $\tilde{f}_I(t)$, $\tilde{\mathbf{g}}$, $\mathbf{\Lambda}$, $\boldsymbol{\rho}$, and $\boldsymbol{\rho}_d$ are scalars. Using (202), we obtain

$$\tilde{g}_1 = -\frac{1}{N_0}\left(\frac{2\sigma_1^2}{N_0 + 2\sigma_1^2}\right)\tilde{\rho}_d. \tag{205}$$

observing that

$$\tilde{\rho}_d = \int_{-\infty}^{\infty} \tilde{\mathbf{f}}_I^*(u)\tilde{f}(u)\,du = \int_{-\infty}^{\infty} \tilde{f}(u)\tilde{f}^*(u - \tau_1)e^{j\omega_1 u}\,du \tag{206}$$

and

$$\bar{E}_1 = 2E_t\sigma_1^2, \tag{207}$$

we can write the performance expression in (203) as

$$\Delta_o = \frac{\bar{E}_r}{N_0}\left\{1 - \frac{\bar{E}_1/N_0}{1 + \bar{E}_1/N_0}\,\theta(\tau_1, \omega_1)\right\}. \tag{208}$$

The ratio of $\Delta_o$ to $\bar{E}_r/N_0$ is plotted in Fig. 10.36. This indicates the degradation in performance due to the interfering target.

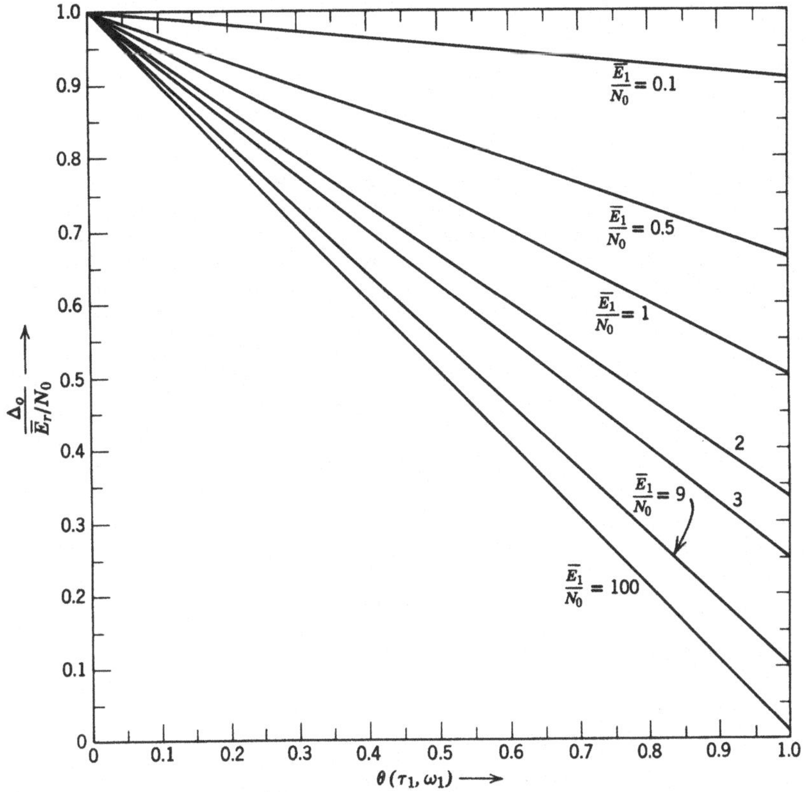


**Fig. 10.36 Performance of optimum receiver with single interfering target.**

The improvement obtained by using optimum filtering instead of conventional filtering can be found by comparing (208) and (180). The ratio of $\Delta_o$ to $\Delta_{wo}$ (the performance using a conventional matched filter) is

$$R \triangleq \frac{\Delta_o}{\Delta_{wo}} = \frac{\dfrac{\bar{E}_r}{N_0}\left(1 - \dfrac{\bar{E}_1/N_0}{1 + [\bar{E}_1/N_0]}\,\theta(\tau_1, \omega_1)\right)}{\dfrac{\bar{E}_r}{N_0}\left(1 + \dfrac{\bar{E}_1}{N_0}\,\theta(\tau_1, \omega_1)\right)^{-1}}. \tag{209}$$

This reduces to

$$R = 1 + \frac{(\bar{E}_1/N_0)^2}{(1 + \bar{E}_1/N_0)}\,\theta(\tau_1, \omega_1)[1 - \theta(\tau_1, \omega_1)]. \tag{210}$$

The ratio is plotted in Fig. 10.37 for various values of $\bar{E}_1/N_0$ and $\theta(\tau_1, \omega_1)$. We see that the function is symmetric about $\theta(\tau_1, \omega_1) = 0.5$. The behavior at the endpoints can be explained as follows.

1. As $\theta(\tau_1, \omega_1) \to 1$, the interference becomes highly correlated with the signal. This means that

$$\tilde{f}_1(t) = \tilde{f}(t - \tau_1)e^{j\omega_1 t} \simeq \tilde{f}(t), \tag{211}$$

so that

$$\tilde{g}(t) = \frac{1}{N_0}\tilde{f}(t) + \tilde{g}_1\tilde{f}_1(t) \simeq \left(\frac{1}{N_0} + \tilde{g}_1\right)\tilde{f}(t). \tag{212}$$

Thus, the optimum and conventional receivers only differ by a gain. Notice that the performance of both receivers become worse as $\theta(\tau_1, \omega_1)$ approaches unity.

2. As $\theta(\tau_1, \omega_1) \to 0$, the interference becomes essentially uncorrelated with the signal, so that the optimum and conventional receivers are the same. Thus, if we have complete freedom to choose the signal, we design it to make $\theta(\tau_1, \omega_1)$ small and the conventional matched filter will be essentially optimum.

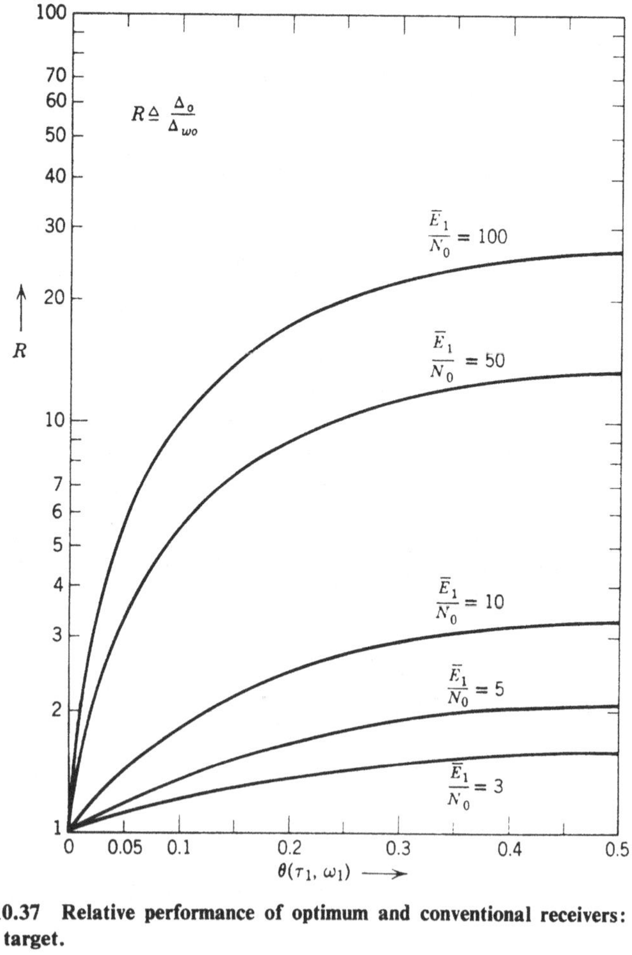


**Fig. 10.37 Relative performance of optimum and conventional receivers: single interfering target.**

The conclusions of this simple example carry over to the general problem. If possible, we would like to make $\theta(\tau_i, \omega_i)$ zero for all other interfering targets. If we can accomplish this, the optimum and conventional receiver will be the same. When $\theta(\tau_i, \omega_i)$ and $E_i/N_0$ are large for some values of $i$, appreciable improvement can be obtained by using the optimum receiver. Notice that our design of the optimum receiver assumes that we know the range and velocity of the interfering targets. In many physical situations this is not a realistic assumption. In these cases we would first have to estimate the parameters of the interfering targets and use these estimates to design the optimum receiver. This procedure is complex, but would be feasible when the interference environment is constant over several target encounters.

This completes our discussion of the discrete resolution problem, and we may now summarize our results.

### 10.5.4 Summary of Resolution Results

In our study of the discrete resolution problem, we have found that there are two important issues. The first issue is the effect of signal design. Signals that may be very good from the standpoint of accuracy and ambiguity may be very poor in a particular interference environment. Thus, one must match the signal to the expected environment whenever possible.

The second issue is the effect of optimum receiver design. Simple examples indicate that this is fruitful only when the correlation between the interference and the desired signal is moderate. For either small or large correlations, the improvement over conventional matched filtering becomes small. If one can design the signal so that the target return is uncorrelated with the interference, the optimum receiver reduces to the conventional matched filter. In cases when this is not practical, the optimum receiver should be used to improve performances.

We have chosen a particular model of the resolution problem in order to illustrate some of the important issues. Various modifications in the model can be made to accommodate particular physical situations. Two typical problems are the following:

1. The location and number of the interfering targets are not known. We design a receiver that estimates the environment and uses this estimate to detect the target.

2. The targets are known to be located in a certain region (say $\Omega_I$) of the $\tau$, $\omega$ plane. The signal is fixed. We design a receiver that reduces the subsidiary peaks (sidelobes) in $\Omega_I$ without reducing the value at the correct

target location too much. This problem is usually referred to as the "mismatched filter" or "sidelobe reduction" problem.

A great deal of research has been done on the resolution problem, because it is the critical issue in many radar/sonar systems. There are a number of good references that the interested reader can consult (e.g., [47]–[59]). There is also a great deal of literature on the intersymbol interference problem in digital communication which is of interest (e.g., [61]–[64]).

We have confined our discussion in this section to the discrete resolution problem. After we have developed a model for singly-spread and doubly-spread targets, we shall return to the continuous resolution problem.†

## 10.6 SUMMARY AND RELATED TOPICS

We first summarize the results that we have obtained, and then discuss some related topics.

### 10.6.1 Summary

In this section we have studied the problem of estimating the range and velocity of a slowly fluctuating point target. The model of interest is characterized by several features:

1. The signals and random processes of interest are bandlimited around some carrier frequency. This property enabled us to represent the signals and processes either by two real low-pass waveforms or by one complex waveform. Choosing the latter, we reformulated our earlier results in terms of complex quantities. Other than factors of 2 and conjugates in various places, the resulting equations and structures are familiar.
2. The effect of the slowly fluctuating target is to multiply the signal by a complex Gaussian variable. Physically, this corresponds to a random amplitude and phase being introduced into the reflected signal. By assuming that the signal is narrow-band, we can model the effect of the target velocity as a Doppler shift. Thus, the received signal is

$$s(t) = \sqrt{2E_t}\,\text{Re}\,[\tilde{b}\tilde{f}(t - \tau)e^{j(\omega_c + \omega)t}]. \tag{213}$$

Using this model, we looked at the problem of estimating range and velocity. The likelihood function led us directly to the optimum receiver.

† The reader who is only interested in the resolution can read pages 459–482 at this point.

In evaluating the performance of the receiver, we encountered the ambiguity function of the signal. Three separate problems were found to be important:

*Accuracy.* If we can be certain that the error is small (i.e., we are looking at the correct region of the $\tau$, $\omega$ plane), the shape of the ambiguity function near the origin completely determines the accuracy. The quantitative accuracy results are obtained by use of the Cramér-Rao inequality.

*Ambiguity.* The volume-invariance property of the ambiguity function shows that as the volume in the central peak is reduced to improve accuracy, the function has to increase somewhere else in the $\tau$, $\omega$ plane. Periodic pulse trains, linear *FM* signals, and pseudo-random sequences were investigated from the standpoint of accuracy and ambiguity.

*Resolution.* The possible presence of additional interfering targets gives rise to the discrete resolution problem. The principal result of our discussion is the conclusion that the signal should, if possible, be matched to the environment. If we can make the value of the ambiguity function, $\theta(\tau, \omega)$, small at those points in the $\tau$, $\omega$ plane where interfering targets are expected, a conventional matched filter receiver is close to optimum.

We now mention some related topics.

### 10.6.2 Related Topics

***Generalized Parameter Sets.*** We have emphasized the problems of range and Doppler estimation. In many systems, there are other parameters of interest. Typical quantities might be azimuth angle or elevation angle. Because the extension of the results to an arbitrary parameter set is straightforward, we can merely state the results.

We assume that the received signal is $r(t)$, where

$$r(t) = \sqrt{2}\,\mathrm{Re}\,[\{\tilde{b}\sqrt{E_t}\tilde{f}(t, \mathbf{A}) + \tilde{w}(t)\}e^{j\omega_c t}], \qquad -\infty < t < \infty. \quad (214a)$$

Here $\mathbf{A}$ is a nonrandom vector parameter that we want to estimate, and $\tilde{w}(t)$ is a complex white Gaussian process. We also assume

$$\int_{-\infty}^{\infty} |\tilde{f}(t, \mathbf{A})|^2\, dt = 1 \qquad \text{for all } \mathbf{A} \in \chi_{\mathbf{a}}. \quad (214b)$$

The complex function generated by the optimum receiver is

$$\tilde{l}(\mathbf{A}) = \int_{-\infty}^{\infty} \tilde{r}(t)\tilde{f}^*(t, \mathbf{A})\, dt, \quad (215)$$

and the log likelihood function is

$$\ln \Lambda(\mathbf{A}) = \frac{1}{N_0} \frac{\bar{E}_r}{\bar{E}_r + N_0} |\tilde{l}(\mathbf{A})|^2 \tag{216}$$

[by analogy with (6)]. The function in (216) is calculated as a function of $M$ parameters; $A_1, A_2, \ldots, A_M$. The value of $\mathbf{A}$ where $\ln \Lambda(\mathbf{A})$ has its maximum is $\hat{\mathbf{a}}_{ml}$. Just as on page 277, we investigate the characteristics of $\ln \Lambda(\mathbf{A})$ by assuming that the actual signal is $f(t, \mathbf{A}_a)$. This procedure leads us to a generalized correlation function,

$$\phi(\mathbf{A}, \mathbf{A}_a) \triangleq \int_{-\infty}^{\infty} \tilde{f}(t, \mathbf{A}_a)\tilde{f}^*(t, \mathbf{A})\, dt, \tag{217}$$

and a generalized ambiguity function,

$$\theta(\mathbf{A}, \mathbf{A}_a) \triangleq |\phi(\mathbf{A}, \mathbf{A}_a)|^2. \tag{218}$$

We should also observe that the specific properties derived in Section 10.3 apply only to the time-frequency functions. The problems of accuracy, ambiguity, and resolution in a general parameter space can all be studied in terms of this generalized ambiguity function.

The accuracy formulas follow easily. Specifically, one can show that the elements in the information matrix are

$$J_{ij} = -\frac{\bar{E}_r}{N_0}\left(\frac{\bar{E}_r}{\bar{E}_r + N_0}\right)\left\{\frac{\partial^2}{\partial A_i\, \partial A_j}\theta(\mathbf{A}, \mathbf{A}_a)\right\}_{\mathbf{A}=\mathbf{A}_a} \tag{219}$$

Some interesting examples to illustrate these relations are contained in the problems and in *Array Processing* (see [36] also).

***Mismatched Filters.*** There are several cases in which the filters in the receivers are not matched to the signal. One example is estimation in the presence of colored noise. Here the optimum filter is the solution to an integral equation whose kernel is the noise covariance function (e.g., pages 247–251, 329–334). A second example arises when we deliberately mismatch the filter to reduce the sidelobes. The local accuracy performance is no longer optimum, but it may still be satisfactory.

If the filter is matched to $\tilde{g}^*(t)$, the receiver output is

$$\left|\int_{-\infty}^{\infty} \tilde{r}(t)\tilde{g}^*(t)\, dt\right|^2. \tag{220}$$

By analogy with (17) and (18), we define a time-frequency cross-correlation function,

$$\phi_{fg}(\tau, \omega) \triangleq \int_{-\infty}^{\infty} \tilde{f}\left(t - \frac{\tau}{2}\right)\tilde{g}^*\left(t + \frac{\tau}{2}\right)e^{j\omega t}\, dt, \tag{221}$$

and a cross-ambiguity function,

$$\theta_{fg}(\tau, \omega) \triangleq |\phi_{fg}(\tau, \omega)|^2. \tag{222}$$

The properties of these functions have been studied by Stutt [37], [44] and Root [38]. (See also Problems 10.6.2–10.6.5 and [68], [74], and [75].)

***Detection of a Target with Unknown Parameters.*** A problem of frequent interest is the detection of targets whose range and velocity are unknown. The model is

$$\tilde{r}(t) = \sqrt{2E_t}\tilde{f}(t - \tau)e^{j\omega t} + \tilde{w}(t), \qquad -\infty < t < \infty : H_1, \tag{223}$$

$$\tilde{r}(t) = \tilde{w}(t), \qquad -\infty < t < \infty : H_0, \tag{224}$$

where $\tau$ and $\omega$ are unknown nonrandom parameters. Two approaches come to mind.

The first approach is to use a generalized likelihood ratio test (see Section I-2.5). The test is

$$\max_{\tau,\omega}\left\{\left|\int_{-\infty}^{\infty} \tilde{r}(t)\tilde{f}^*(t - \tau)e^{-j\omega t}\,dt\right|^2\right\} \underset{H_0}{\overset{H_1}{\gtrless}} \gamma. \tag{225}$$

The threshold $\gamma$ is adjusted to give the desired $P_F$. The performance of this test is discussed in detail by Helstrom [39].

The second approach is to divide the region of the $\tau$, $\omega$ plane where targets are expected into $M$ rectangular segments (see discussion on pages 302–303 on how to choose the segments). We denote the $\tau$, $\omega$ coordinates at the center of the $i$th cell as $\tau_i$, $\omega_i$. We then consider the binary hypothesis problem

$$\tilde{r}(t) = \sqrt{2E_t}\tilde{f}(t - \tau_i)e^{j\omega_i t} + \tilde{w}(t), \qquad -\infty < t < \infty,$$

$$\text{with probability} \quad p_i = \frac{1}{M}, \qquad i = 1, 2, \ldots, M : H_1 \tag{226}$$

and

$$\tilde{r}(t) = \tilde{w}(t), \qquad -\infty < t < \infty : H_0. \tag{227}$$

This is just the problem of detecting one of $M$ signals that we discussed on page I-405. As we would expect, the performances of the two systems are similar.

***Large Time-Bandwidth Signals.*** In some sonar systems the bandwidth of the signal is large enough that the condition

$$WT \ll \frac{c}{2v} \tag{228}$$

given in (9.23) is not a valid assumption. In these cases we cannot model the time compression as a Doppler shift. Several references [69]–[73] treat this problem, and the interested reader should consult them.

This completes our discussion of slowly fluctuating targets and channels. We now consider the next level of target in our hierarchy.

## 10.7 PROBLEMS

### P.10.1 Receiver Derivation and Signal Design

**Problem 10.1.1.** The random variable $\tilde{n}(\tau, \omega)$ is defined in (13). Prove that the probability density of $\tilde{n}(\tau, \omega)$ is not a function of $\tau$ and $\omega$.

**Problem 10.1.2.** Consider the Gaussian pulse with linear FM in (44a).

1. Verify the results in (46)–(48) directly from the definitions in the Appendix.
2. Verify the results in (46)–(48) by using (96)–(98).

**Problem 10.1.3.** Let

$$\tilde{f}(t) = c \sin^2\left(\frac{2\pi t}{T}\right), \qquad 0 \le t \le T,$$

where $c$ is a normalizing constant. Find $\theta(\tau, \omega)$.

**Problem 10.1.4.** Let

$$\tilde{f}(t) = \tilde{f}_1(t) + \tilde{f}_2(t),$$

where

$$\tilde{f}_1(t) = \begin{cases} \dfrac{1}{\sqrt{T_1}}, & 0 \le t \le T_1, \\ 0, & \text{elsewhere,} \end{cases}$$

and

$$\tilde{f}_2(t) = \begin{cases} \dfrac{1}{\sqrt{T_2}}\, e^{-j\omega_2 t}, & T_p \le t \le T_p + T_2, \\ 0, & \text{elsewhere.} \end{cases}$$

Assume that

$$T_p > T_1.$$

1. Find $\theta(\tau, \omega)$.
2. Plot for the case

$$T_1 \ll T_2,$$

$$\frac{\omega_2}{2\pi} \gg \frac{1}{T_1}.$$

## P.10.2 Performance Analysis

**Problem 10.2.1.** Derive the expressions for $J_{12}$ and $J_{22}$ that are given in (64) and (65).

**Problem 10.2.2.** Derive the expressions in (96)–(98).

**Problem 10.2.3.** Consider the expression in (6). Expand $\tilde{L}(\tau, \omega)$ in a power series in $\tau$ and $\omega$ around the point

$$\tau = \hat{\tau}_{ml},$$
$$\omega = \hat{\omega}_{ml}.$$

Assuming that the errors are small, find an expression for their variance.

## P.10.3 Properties of $\phi(\tau, \omega)$ and $\theta(\tau, \omega)$

**Problem 10.3.1.** In Property 5 we derived an alternative representation of $\phi(\tau, \omega)$. Prove that another alternative representation is

$$\phi(\tau, \omega) = \frac{e^{-j\omega\tau/2}}{2\pi} \iint_{-\infty}^{\infty} \tilde{f}\left(\beta - \frac{\tau}{2}\right) \tilde{F}^*(j\omega - j\alpha) e^{j\alpha\beta + j\alpha\tau/2}\, d\alpha\, d\beta.$$

**Problem 10.3.2.** The transmitted waveform, $\tilde{f}(t)$, has a Fourier transform,

$$\tilde{F}\{f\} = \begin{cases} \dfrac{1}{\sqrt{W}}, & |f| \le W, \\ 0, & |f| > W. \end{cases}$$

Find $\phi\{\tau, f\}$.

**Problem 10.3.3.** The transmitted waveform, $\tilde{f}(t)$, has a Fourier transform which can be written as

$$\tilde{F}\{f\} = c \sum_{k=-n}^{k=n} \tilde{U}\{f - kW_s\},$$

where

$$\tilde{U}\{f\} = \begin{cases} \dfrac{1}{\sqrt{W}}, & |f| \le W, \\ 0, & |f| > W, \end{cases}$$

$$W_s \gg W,$$

and $c$ is a normalizing factor.

1. Find $\phi\{\tau, f\}$.
2. How would you synthesize $\tilde{f}(t)$?

**Problem 10.3.4. Partial Volume Invariances.** Prove

$$\int_{-\infty}^{\infty} |\tilde{F}\{x\}|^2\, |\tilde{F}\{x + f\}|^2\, dx = \int_{-\infty}^{\infty} \theta\{\tau, f\}\, df.$$

Notice that this is a partial volume invariance property. The total volume in a strip of width $\Delta\tau$ at some value of $\tau$ cannot be altered by phase modulation of the waveform.

**Problem 10.3.5.**

1. Prove

$$\int_{-\infty}^{\infty} |\tilde{f}(t)|^2 \, |\tilde{f}(t+\tau)|^2 \, dt = \int_{-\infty}^{\infty} \theta\{\tau, f\} \, d\tau \tag{P.1}$$

directly from the definition.

2. Prove the relationship in (P.1) by inspection by using the result of Problem 10.3.4 and the duality principle.

*Note:* This is another partial volume invariance property.

**Problem 10.3.6.**

1. Expand the ambiguity function, $\theta\{\tau, f\}$, in a Taylor series around the origin.
2. Express the coefficients of the quadratic terms as functions of $\overline{t^2}$, $\overline{\omega t}$, and $\overline{\omega^2}$.

**Problem 10.3.7** [40]. Derive the following generalization of Property 9. Assume

$$\begin{vmatrix} c_{11} & c_{12} \\ c_{21} & c_{22} \end{vmatrix} = \pm 1$$

and

$$\phi_1\{\tau, f\} \sim \tilde{f}_1(t).$$

If

$$\tilde{f}_2(t) = |c_{11}|^{1/2} \int_{-\infty}^{\infty} df \exp\left[j2\pi c_{11} f\left(t - \frac{c_{12}}{2} f\right)\right]$$

$$\times \int_{-\infty}^{\infty} \tilde{f}_1(z) \exp\left[-j2\pi z\left(f - \frac{c_{21}}{2c_{11}} z\right)\right] dz,$$

then

$$|\phi_2\{\tau, f\}| = |\phi_1\{c_{11}\tau + c_{12}f, -c_{21}\tau - c_{22}f\}|.$$

**Problem 10.3.8** [12]. Derive the following "volume" invariance property:

$$\iint_{-\infty}^{\infty} \theta\{\tau, f\}^p \, d\tau \, df \le \frac{1}{p} \, \theta\{\tau, f\},$$

where $p$ is an integer greater than or equal to 1.

**Problem 10.3.9** [41]–[43]. Assume that we expand $\tilde{f}(t)$ using a CON set,

$$\tilde{f}(t) = \sum_{i=1}^{\infty} f_i \tilde{u}_i(t), \qquad -\infty < t < \infty.$$

Let

$$\phi_{ik}\{\tau, f\} \triangleq \int_{-\infty}^{\infty} \tilde{u}_i\left(t - \frac{\tau}{2}\right) \tilde{u}_k^*\left(t + \frac{\tau}{2}\right) e^{j2\pi f t} \, dt$$

denote the time-frequency cross-correlation function of $\tilde{u}_i(t)$ and $\tilde{u}_k(t)$.

1. Prove

$$\iint_{-\infty}^{\infty} \phi_{ik}\{\tau, f\}\phi_{mn}^*\{\tau, f\}\, d\tau\, df = \delta_{kn}\, \delta_{im}.$$

2. Compare this result with Property 3.
3. Prove

$$\phi_{\tilde{f}}\{\tau, f\} = \sum_{i=1}^{\infty} \sum_{k=1}^{\infty} f_i f_k^* \phi_{ik}\{\tau, f\}$$

and

$$c_{ik} \triangleq f_i f_k^* = \iint_{-\infty}^{\infty} \phi_{\tilde{f}}\{\tau, f\}\phi_{ik}^*\{\tau, f\}\, d\tau\, df.$$

**Problem 10.3.10.** The Hermite waveforms are

$$\tilde{f}_n(t) = \frac{2^{1/4}}{\sqrt{n!}} e^{-\pi t^2} H_n(2\sqrt{\pi} t), \qquad -\infty < t < \infty, \quad n = 1, 2, \ldots, \tag{P.1}$$

where $H_n(t)$ is the $n$th-order Hermite polynomial,

$$H_n(t) = (-1)^n e^{t^2/2} \frac{d^n}{dt^n} e^{-t^2/2}, \qquad -\infty < t < \infty. \tag{P.2}$$

1. Find $\tilde{F}\{f\}$.
2. Prove that

$$\phi_n\{\tau, f\} = \exp\left[-\frac{\pi}{2}(\tau^2 + f^2)\right] L_n[\pi(\tau^2 + \phi^2)], \tag{P.3}$$

where $L_n(x)$ is the $n$th-order Laguerre polynomial

$$L_n(x) = \begin{cases} \dfrac{1}{n!} e^x \dfrac{d^n}{dx^n}(x^n e^{-x}), & x \geq 0, \\ 0, & x < 0. \end{cases}$$

3. Verify that your answer reduces to Fig. 10.8 for $n = 1$.

*Comment:* Plots of these waveforms are given in [46].

4. Notice that the time-frequency autocorrelation function is rotationally symmetric.
   a. Use this fact to derive $\tilde{F}\{f\}$ by inspection (except for a phase factor).
   b. What does Property 9 imply with respect to the Hermite waveforms?

**Problem 10.3.11** [14]. Let

$$\tilde{f}(t) = \begin{cases} \sqrt{\dfrac{1}{T}} \exp\left[j\,\Delta\theta \sin \dfrac{2\pi t}{T}\right], & |t| < \dfrac{T}{2}, \\ 0, & \text{elsewhere.} \end{cases}$$

1. Find $\tilde{F}\{f\}$.
2. Find $|\phi\{\tau, f\}|$.

**Problem 10.3.12** [14]. Let

$$\tilde{f}(t) = \left(\frac{2k^2}{\pi}\right)^{1/4} e^{-(k^2 - jct)t^2}, \qquad -\infty < t < \infty.$$

This is a pulse with a Gaussian envelope and parabolic frequency modulation.
Find $|\phi\{\tau, f\}|$.

**Problem 10.3.13.** A waveform that is useful for analytic purposes is obtained from the $\tilde{f}(t)$ in (32) by letting

$$T \to 0$$

and

$$n \to \infty.$$

while holding $nT$ constant.
We denote the resulting signal as $i_\delta(t)$.

1. Plot $|\phi\{\tau, f\}|$ for this limiting case. Discuss suitable normalizations.
2. Define

$$\tilde{f}(t) = \int_{-\infty}^{\infty} \tilde{h}(\tau) i_\delta(t - \tau)\, d\tau.$$

Express $\phi_{\tilde{f}}\{\tau, f\}$ in terms of $\phi_{i_\delta}\{\tau, f\}$.

**Problem 10.3.14.** Consider the problem in which we transmit two disjoint pulses, $\tilde{f}_1(t)$ and $\tilde{f}_2(t)$. The complex envelope of received waveform on $H_1$ is

$$\tilde{r}(t) = \sqrt{E_1}\, \tilde{b}_1 \tilde{f}_1(t - \tau) e^{j\omega t} + \sqrt{E_2}\, \tilde{b}_2 \tilde{f}_2(t - \tau) e^{j\omega t} + \tilde{w}(t), \quad -\infty < t < \infty : H_1$$

The multipliers $\tilde{b}_1$ and $\tilde{b}_2$ are statistically independent, zero-mean complex Gaussian random variables,

$$E[|\tilde{b}_i|^2] = 2\sigma_b^2.$$

On $H_0$, only $\tilde{w}(t)$ is present,

1. Find the likelihood function and the optimum receiver.
2. How is the signal component at the output of the optimum receiver related to $\phi_1\{\tau, f\}$ and $\phi_2\{\tau, f\}$?

**Problem 10.3.15.** Consider a special case of Problem 10.3.14 in which $\tilde{f}_1(t)$ is a short rectangular pulse and $\tilde{f}_2(t)$ is a long rectangular pulse.

1. Sketch the signal component at the output of the optimum receiver.
2. Discuss other receiver realizations that might improve the global accuracy. For example, consider a receiver whose signal output consists of $\theta_1\{\tau, f\}$ times $\theta_2\{\tau, f\}$.

## P.10.4 Coded Pulse Sequences

**Problem 10.4.1.** Consider the periodic pulse sequence in Fig. 10.21. Assume that

$$\omega_n = (n - 1)\omega_\Delta,$$

where

$$\frac{\omega_\Delta}{2\pi} = \frac{1}{T_s}.$$

1. Find $|\phi(\tau, \omega)|$.
2. Plot the result in part 1.

**Problem 10.4.2.** Consider the signal in (145). Assume that

$$
\begin{aligned}
a_1 &= 1, & & \\
a_n &= 1, & &\text{with probability } \tfrac{1}{3}, \text{ for } n = 2, \ldots, 7, \\
a_n &= 0, & &\text{with probability } \tfrac{2}{3}, \text{ for } n = 2, \ldots, 7, \\
a_8 &= 1. & &
\end{aligned}
$$

1. Find $E\{|\phi(\tau, \omega)|\}$.
2. Discuss the sidelobe behavior in comparison with a periodic pulse train.
3. How would you use these results to design a practical signal?

**Problem 10.4.3.** Consider the three-stage shift register in Fig. 10.26. Compute $\phi\{\tau, \omega\}$ for various initial states of the shift register. Assume that the output consists of one period.

**Problem 10.4.4.** Consider the shift register system is Fig. P.10.1.

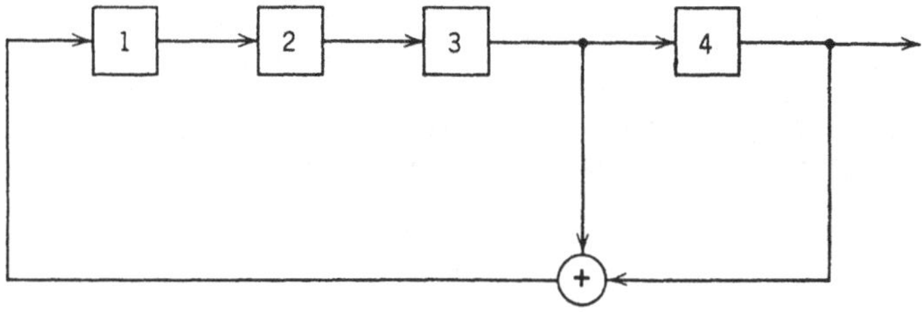


**Fig. P.10.1**

1. Verify that this feedback connection gives a maximum length sequence. We use the notation [4, 3] to indicate that the outputs of stages 3 and 4 are fedback.

2. Does the shift register with connections [4, 2] generate a maximum length sequence?

**Problem 10.4.5.**

1. Consider a shift register with connections [5, 3]. Verify that it generates a maximal length sequence.

2. Does the shift register with connections [5, 4, 3, 2] generate a maximal length sequence?

3. Does the shift register with connections [5, 4, 3, 1] generate a maximal length sequence?

**Problem 10.4.6.** Verify that shift registers with connections [6, 5], [6, 5, 4, 1], and [6, 5, 3, 2] generate maximal length sequences.

## P.10.5 Resolution

**Problem 10.5.1.** Consider the following three hypothesis problem.

$$\tilde{r}(t) = \sqrt{E_1}\,\tilde{b}_1\tilde{f}(t-\tau_1)e^{j\omega_1 t} + \sqrt{E_2}\,\tilde{b}_2\tilde{f}(t-\tau_2)e^{j\omega_2 t} + \tilde{w}(t), \qquad -\infty < t < \infty : H_2,$$

$$\tilde{r}(t) = \sqrt{E_1}\,\tilde{b}_1\tilde{f}(t=\tau_1)e^{j\omega_1 t} + \tilde{w}(t), \qquad -\infty < t < \infty : H_1,$$

$$\tilde{r}(t) = \tilde{w}(t), \qquad -\infty < t < \infty : H_0.$$

The multipliers $\tilde{b}_1$ and $\tilde{b}_2$ are zero-mean complex Gaussian random variables with mean-square values $2\sigma_1^2$ and $2\sigma_2^2$. The parameters $\tau_1$, $\tau_2$, $\omega_1$, and $\omega_2$ are assumed known. The additive noise is a complex white Gaussian process with spectral height $N_0$.

Find the optimum Bayes test. Leave the costs as parameters.

**Problem 10.5.2.** Consider the same model as in Problem 10.5.1. We design a test using the MAP estimates of $|\tilde{b}_1|$ and $|\tilde{b}_2|$.

1. Find $\widehat{|\tilde{b}_1|}$, given that $H_1$ is true. Find $\widehat{|\tilde{b}_1|}$ and $\widehat{|\tilde{b}_2|}$, given that $H_2$ is true.
2. Design a test based on the above estimates. Compare it with the Bayes test in Problem 10.5.1.
3. Define $P_F$ as

$$P_F = \Pr\,[\text{say } H_1 \text{ or } H_2 \mid H_0].$$

Find $P_{D_1}$ and $P_{D_2}$.

**Problem 10.5.3.** Assume that a rectangle in the $\tau, \omega$ plane with dimension $T_* \times \Omega_*$ is of interest. We use a grid so that there are $M$ cells of interest (see discussion on pages 302–303). In each cell there is at most one target. We want to estimate the number of targets that are present and the cells which they occupy.

Discuss various procedures for implementing this test. Consider both performance and complexity.

**Problem 10.5.4.** An alternative way to approach the optimum receiver problem in Section 10.5.3 is to find the eigenfunctions and eigenvalues of the interfering noise process. From (195),

$$\tilde{K}_c(t, u) = \tilde{\mathbf{f}}_I^T(t)\tilde{\mathbf{\Lambda}}\tilde{\mathbf{f}}_I^*(u), \qquad -\infty < t, u < \infty.$$

We want to write

$$\tilde{K}_c(t, u) = \sum_{i=1}^{\infty} \tilde{\lambda}_i\tilde{\phi}_i(t)\tilde{\phi}_i^*(u), \qquad -\infty < t, u < \infty.$$

1. Find $\tilde{\lambda}_i$ and $\tilde{\phi}_i(t)$. How many eigenvalues are nonzero?
2. Use the result in part 1 to find the optimum receiver.
3. Find $\Delta_o$.

**Problem 10.5.5.** We want to communicate over the resolvable multipath channel shown in Fig. P.10.2. To investigate the channel structure, we transmit a *known* sounding signal with complex envelope $\tilde{f}(t)$. The complex envelope of the received waveform is

$$\tilde{r}(t) = \sum_{i=1}^{3} \sqrt{2}\,\tilde{b}_i\tilde{f}(t-\tau_i) + \tilde{w}(t),$$

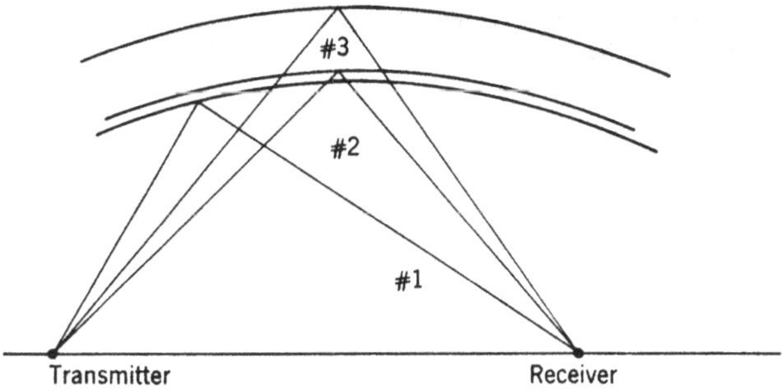


**Fig. P.10.2**

where the $\tilde{b}_i$ are independent zero-mean complex Gaussian variables with variances $2\sigma_i^2$ and $\tilde{w}(t)$ is a zero-mean complex Gaussian process with spectral height $N_0$. The signal outputs of the three channels are disjoint in time. The $\tau_i$ are modeled as independent, uniformly distributed, random variables, $U[-T, T]$, where $T$ is large.

Derive the optimum receiver for estimating $\hat{\tau}_{i,\text{map}}$.

**Problem 10.5.6.** Consider the following detection problem:

$$\tilde{r}(t) = \sqrt{2E_t}\,\tilde{b}\tilde{f}(t) + \tilde{w}(t), \qquad T_i \le t \le T_f : H_1,$$

$$\tilde{r}(t) = \tilde{w}(t), \qquad T_i \le t \le T_f : H_0.$$

The deterministic signal $\tilde{f}(t)$ has unit energy and is zero outside the interval $(0, T)$. [This interval is included in $(T_i, T_f)$.] The multiplier $\tilde{b}$ is a zero-mean complex Gaussian random variable,

$$E\{|\tilde{b}|^2\} = 2\sigma_b^2.$$

The additive noise, $\tilde{w}(t)$, is a complex zero-mean Gaussian process with spectral height $N_0$.

1. Find the optimum detector. Compute $\Delta_o$, $P_D$, and $P_F$.
2. Now assume that the received signal on $H_1$ is actually

$$\tilde{r}(t) = \sqrt{2E_t}\,\tilde{b}\tilde{f}(t - \tau)e^{j\omega t} + \tilde{w}(t), \qquad T_i \le t \le T_f : H_1,$$

where $\tau$ and $\omega$ are small. We process $\tilde{r}(t)$ using the detector in part 1. Compute the change in $\Delta$ as a function of $\tau$ and $\omega$. Express your answer in terms of $\theta(\tau, \omega)$, the ambiguity function of $\tilde{f}(t)$.

3. Design several signals that are insensitive to small changes in $\tau$ and $\omega$. Explain your design procedure.

**Problem 10.5.7.** Consider the resolution problem described in (173)–(180). A conventional receiver is used. The degradation in performance is given by

$$\Delta_I = \sum_{i=1}^{K} \frac{\bar{E}_{r_i}}{N_0}\,\theta(\tau_i, \omega_i).$$

Assume that the interfering targets all have zero Doppler relative to the desired target and that $\bar{E}_{r_i} = \bar{E}_I$. Then

$$\Delta_I = \frac{\bar{E}_I}{N_0} \sum_{i=1}^{K} \theta(\tau_i, 0).$$

Now assume that the targets are uniformly spaced on the $\tau$-axis such that there is a target in each successive $\Delta\tau$ interval. Defining

$$\bar{E}_I = A\,\Delta\tau$$

and letting $K \to \infty$ and $\Delta\tau \to 0$, we obtain the integral

$$\Delta_I = \frac{A}{N_0} \int_{-\infty}^{\infty} \theta(\tau, 0)\, d\tau = \frac{A}{N_0} \Delta_R.$$

*Comment:* The resolution constant $\Delta_R$ was first introduced by Woodward [60].

Prove that

$$\Delta_R = \int_{-\infty}^{\infty} |F\{f\}|^4\, df.$$

Notice that the signal $\tilde{f}(t)$ is always normalized.

**Problem 10.5.8.** Extend the ideas of Problem 10.5.7 to find a Doppler resolution constant

$$\Delta_D \triangleq \int_{-\infty}^{\infty} \theta\{0, f\}\, df.$$

Prove that

$$\Delta_D = \int_{-\infty}^{\infty} |\tilde{f}(t)|^4\, dt.$$

**Problem 10.5.9.** Compute $\Delta_R$, $\Delta_D$, and $\Delta_R\Delta_D$ for the following:

1. A simple Gaussian pulse (25).
2. A rectangular pulse.

**Problem 10.5.10** [39, Chapter X]. Consider the following resolution problem. The received waveforms on the four hypotheses are

$$\begin{aligned}
r(t) &= w(t), & 0 \le t \le T: H_0,\\
r(t) &= Af(t) + w(t), & 0 \le t \le T: H_1,\\
r(t) &= Bg(t) + w(t), & 0 \le t \le T: H_2,\\
r(t) &= Af(t) + Bg(t) + w(t), & 0 \le t \le T: H_3.
\end{aligned}$$

The multipliers $A$ and $B$ are unknown nonrandom variables. The signals $f(t)$ and $g(t)$ are known waveforms with unit energy, and

$$\int_0^T f(t)g(t)\, dt = \rho.$$

The additive noise $w(t)$ is a sample function of a white Gaussian random process with spectral height $N_0/2$. (Notice that the waveforms are not bandpass.) We want to derive a generalized likelihood ratio test (see Section I-2.5).

1. Assume that $H_3$ is true. Prove that

$$\hat{a}_{ml} = \frac{1}{1-\rho^2}\int_0^T [f(t) - \rho g(t)]r(t)\,dt.$$

2. Find $\hat{b}_{ml}$.

3. Calculate $E[\hat{a}_{ml}]$, $E[\hat{b}_{ml}]$, Var $[\hat{a}_{ml}]$, Var $[\hat{b}_{ml}]$, and Cov $[\hat{a}_{ml}, \hat{b}_{ml}]$ under the four hypotheses.

4. Assume that we use the test

$$\hat{a}_{ml} \underset{H_0 \text{ or } H_2}{\overset{H_1 \text{ or } H_3}{\gtrless}} \gamma_a,$$

$$\hat{b}_{ml} \underset{H_0 \text{ or } H_1}{\overset{H_2 \text{ or } H_3}{\gtrless}} \gamma_b.$$

Calculate $P_{Fa}$, $P_{Da}$, $P_{Fb}$, and $P_{Db}$. Verify that both $P_{Da}$ and $P_{Db}$ decrease monotonically as $\rho$ increases.

**Problem 10.5.11** [39, Chapter X]. Consider the bandpass version of the model in Problem 10.5.10. Assume that the complex envelopes on the four hypotheses are

$$\begin{aligned}
\tilde{r}(t) &= \tilde{w}(t), & 0 \le t \le T: H_0,\\
\tilde{r}(t) &= Ae^{j\varphi_a}\tilde{f}(t) + \tilde{w}(t), & 0 \le t \le T: H_1,\\
\tilde{r}(t) &= Be^{j\varphi_b}\tilde{g}(t) + \tilde{w}(t), & 0 \le t \le T: H_2,\\
\tilde{r}(t) &= Ae^{j\varphi_a}\tilde{f}(t) + Be^{j\varphi_b}\tilde{g}(t) + \tilde{w}(t), & 0 \le t \le T: H_3.
\end{aligned}$$

The multipliers $A$ and $B$ are unknown nonrandom variables. The phases $\varphi_a$ and $\varphi_b$ are statistically independent, uniformly distributed random variables on $(0, 2\pi)$. The complex envelopes $\tilde{f}(t)$ and $\tilde{g}(t)$ are known unit energy signals with

$$\int_{-\infty}^{\infty} \tilde{f}(t)\tilde{g}^*(t)\,dt = \tilde{\rho}_{fg}.$$

The additive noise $\tilde{w}(t)$ is a sample function of a complex white Gaussian process with spectral height $N_0$.

1. Find $\hat{a}_{ml}$ and $\hat{b}_{ml}$ under the assumption that $H_3$ is true.

2. Calculate $E[\hat{a}_{ml}]$, $E[\hat{b}_{ml}]$, Var $[\hat{a}_{ml}]$, Var $[\hat{b}_{ml}]$, and Cov $[\hat{a}_{ml}, \hat{b}_{ml}]$ under the four hypotheses.

3. The test in part 4 of Problem 10.5.10 is used. Calculate the performance.

**Problem 10.5.12** [39, Chapter X]. The model in Problem 10.5.11 can be extended to include an unknown arrival time. The complex envelope of the received waveform on $H_3$ is

$$\tilde{r}(t) = Ae^{j\varphi_a}\tilde{f}(t-\tau) + Be^{j\varphi_b}\tilde{g}(t-\tau) + \tilde{w}(t), \quad -\infty < t < \infty: H_3.$$

The other hypotheses are modified accordingly.

1. Find a receiver that generates $\hat{a}_{ml}$, $\hat{b}_{ml}$, and $\hat{\tau}_{ml}$.

2. Find the corresponding likelihood ratio test.

**Problem 10.5.13.** Consider the following model of the resolution problem. The complex envelopes of the received waveforms on the two hypotheses are

$$\tilde{r}(t) = \tilde{b}_d \tilde{f}(t) + \sum_{i=1}^{N} B_i e^{j\varphi_i} \tilde{f}(t - \tau_i) e^{j\omega_i t} + \tilde{w}(t); \qquad -\infty < t < \infty : H_1,$$

$$\tilde{r}(t) = \sum_{i=1}^{N} B_i e^{j\varphi_i} \tilde{f}(t - \tau_i) e^{j\omega_i t} + \tilde{w}(t), \qquad -\infty < t < \infty : H_0.$$

The model is the same as that in Section 10.5.3 (page 329) except that the $B_i$ are assumed to be *unknown, nonrandom* variables. The $\varphi_i$ are statistically independent, uniformly distributed random variables. The $\tau_i$ and $\omega_i$ are assumed known.

1. Find the generalized likelihood ratio test for this model. (*Hint:* Review Section I-2.5.)
2. Evaluate $P_F$ and $P_D$.

*Comment:* This problem is similar to that solved in [59].

## P.10.6 Summary and Related Topics

**Problem 10.6.1. Generalized Likelihood Ratio Tests.** Consider the following composite hypothesis problem

$$\tilde{r}(t) = \sqrt{2E_t}\, \tilde{b} \tilde{f}(t - \tau) e^{j\omega t} + \tilde{w}(t), \qquad -\infty < t < \infty : H_1,$$

$$\tilde{r}(t) = \tilde{w}(t), \qquad -\infty < t < \infty : H_0.$$

The multiplier $\tilde{b}$ is a zero-mean complex Gaussian variable,

$$E\{|\tilde{b}|^2\} = 2\sigma_b^2,$$

and $\tilde{w}(t)$ is a complex zero-mean Gaussian white noise process with spectral height $N_0$. The quantities $\tau$ and $\omega$ are *unknown nonrandom* variables whose ranges are known:

$$\tau_0 < \tau \leq \tau_1,$$

$$\omega_0 < \omega \leq \omega_1.$$

Find the generalized likelihood ratio test and draw a block diagram of the optimum receiver.

**Problem 10.6.2.** The time-frequency cross-correlation function is defined by (221) as

$$\phi_{fg}\{\tau, f\} = \int_{-\infty}^{\infty} \tilde{f}\left(t - \frac{\tau}{2}\right) \tilde{g}^*\left(t + \frac{\tau}{2}\right) e^{j2\pi f t}\, dt.$$

In addition,

$$\theta_{fg}\{\tau, f\} \triangleq |\phi_{fg}\{\tau, f\}|^2.$$

Assume that

$$\int_{-\infty}^{\infty} |\tilde{f}(t)|^2\, dt = \int_{-\infty}^{\infty} |\tilde{g}(t)|^2\, dt = 1.$$

1. Prove

$$\iint_{-\infty}^{\infty} \theta_{fg}\{\tau, f\}\, d\tau\, df = 1.$$

2. Does $\theta_{fg}\{0, 0\}$ always equal unity? Justify your answer.

3. Is the equality

$$\theta_{fg}\{\tau, f\} \leq \theta_{fg}\{0, 0\}$$

true? Justify your answer.

**Problem 10.6.3** [44]. As in (221), we define the time-frequency cross-correlation function as

$$\phi_{fg}\{\tau, f\} = \int_{-\infty}^{\infty} \tilde{f}\left(t - \frac{\tau}{2}\right)\tilde{g}^*\left(t + \frac{\tau}{2}\right)e^{j2\pi ft}\, dt.$$

Verify the following properties.

1.
$$\phi_{fg}\{\tau, f\} = \int_{-\infty}^{\infty} \tilde{G}^*\left(x + \frac{f}{2}\right)\tilde{F}\left(x - \frac{f}{2}\right)e^{j2\pi f\tau}\, dx.$$

2.
$$\phi_{fg}\{\tau, f\} = \int_{-\infty}^{\infty} \tilde{f}(x)\tilde{G}^*\{y\} \exp\left[j2\pi\left(\frac{\tau f}{2} - xy - fx + \tau y\right)\right] dx\, dy.$$

3.
$$\phi_{fg}\{\tau, f\} = \phi_{gf}\{-\tau, -f\}.$$

4.
$$\theta_{fg}\{\tau, f\} = \theta_{gf}\{-\tau, -f\}.$$

**Problem 10.6.4** [44]. Prove

$$\iint_{-\infty}^{\infty} \phi_{12}\{\tau, f\}\phi_{34}^*\{\tau, f\}e^{j2\pi[fx - \tau y]}\, d\tau\, df = \phi_{13}\{x, y\}\phi_{24}^*\{x, y\}.$$

**Problem 10.6.5** [44]. Prove

$$\iint_{-\infty}^{\infty} \theta_{fg}\{\tau, f\}e^{j2\pi(fx - \tau y)}\, d\tau\, df = \phi_f(x, y)\phi_g^*(x, y).$$

**Problem 10.6.6.** Consider the following estimation problem. The complex envelope of the received signal is

$$\tilde{r}(t) = \sqrt{E_t}\,\tilde{f}(t - \tau)e^{j\omega t} + \tilde{n}_c(t) + \tilde{w}(t), \qquad -\infty < t < \infty.$$

The colored Gaussian noise has a spectrum

$$S_{\tilde{n}_c}(\omega) = \frac{2\alpha P_c}{\omega^2 + \alpha^2}.$$

The complex white noise has spectral height $N_0$.

1. Find the optimum filter for estimating $\tau$ and $\omega$. Express it as a set of paths consisting of a cascade of a realizable whitening filter, a matched filter, and a square-law envelope detector.

2. Write an expression for the appropriate ambiguity function at the output of the square-law envelope detector. Denote this function as $\theta_{fw}(\tau, \omega)$.

3. Denote the impulse response of the whitening filter as $\tilde{h}_{wr}(u)$. Express the ambiguity function $\theta_{fw}(\tau, \omega)$ in terms of $\phi_f(\tau, \omega)$, $\theta_f(\tau, \omega)$, and $\tilde{h}_{wr}(u)$. (Recall Property 11 on page 311.)

**Problem 10.6.7 (continuation).**

1. Derive an expression for the elements in the information matrix **J** [see (63)–(65)]. These formulas give us a bound on the accuracy in estimating $\tau$ and $\omega$.

2. Is your answer a function of the actual value of $\tau$ or the actual value of $\omega$? Is your result intuitively logical?

**Problem 10.6.8.** Consider the special case of Problem 10.6.6 in which

$$N_0 = 0$$

and

$$\tilde{f}(t) = \begin{cases} c\sin^2\left(\dfrac{2\pi}{T}t\right), & 0 \le t \le T, \\ 0, & \text{elsewhere.} \end{cases}$$

Evaluate $\theta_{fw}(\tau, \omega)$.

**Problem 10.6.9.** Generalize the results of Problem 10.6.6 so that they include a colored noise with an arbitrary rational spectrum.

**Problem 10.6.10.** Consider the model in (214*a*), but do not impose the constraint in (214*b*).

1. Find the log likelihood function.
2. Evaluate the Cramér-Rao bound.

**Problem 10.6.11.** In this problem we consider the simultaneous estimation of range, velocity, and acceleration.

Assume that the complex envelope of the received waveform is

$$\tilde{r}(t) = \sqrt{E_t}\,\tilde{b}\sqrt{1 - \dot{\tau}(t)}\,\tilde{f}(t - \tau(t))e^{-j2\omega f_c \tau(t)} + \tilde{w}(t), \quad -\infty < t < \infty,$$

where

$$\tau(t) \triangleq \tau + \frac{v}{f_c}t + \frac{\alpha}{2f_c}t^2.$$

The parameter $\tau$, $v$, and $\alpha$ are unknown nonrandom parameters.

1. Find $\hat{\tau}_{ml}$, $\hat{v}_{ml}$, and $\hat{\alpha}_{ml}$.
2. Compute the Cramér-Rao bound.

See [65] or [66] for a complete discussion of this problem.

## REFERENCES


[1] J. Ville, "Theorie et application de la notion de signal analytique," Cables et Transmission **2,** No. 1, 61–74 (1948).

[2] S. J. Mason and H. J. Zimmerman, *Electronic Circuits, Signals, and Systems*, Wiley, New York, 1960, p. 367.

[3] R. H. Dicke, U.S. Patent, 2,624,876, September 14, 1945.

[4] S. Darlington, U.S. Patent 2,678,997, December 31, 1949.

[5] E. Huttman, German Patent 768,068, March 22, 1940.

[6] W. A. Cauer, German Federal Republic Patent 892,774.

[7] C. E. Cooke, "Pulse Compression—Key to More Efficient Radar Transmission," Proc. IRE **48,** 310–316 (1960).

[8] P. M. Woodward and I. L. Davies, "A Theory of Radar Information," Phil. Mag. **41,** 1001–1017 (1950).

[9] W. M. Siebert, "A Radar Detection Philosophy," IRE Trans. Information Theory **IT-2,** 204–221 (Sept. 1956).

[10] W. M. Siebert, "Studies of Woodward's Uncertainty Function," Research Laboratory of Electronics, Massachusetts Institute of Technology, Cambridge, Mass., Quarterly Progress Report, April 1958, pp. 90–94.

[11] R. M. Lerner, "Signals with Uniform Ambiguity Functions," 1958 IRE Trans. Natl. Conv. Rec., Pt. 4, pp. 27–36.

[12] R. Price and E. M. Hofstetter, "Bounds on the Volume and Height Distributions of the Ambiguity Function," IEEE Trans. Information Theory **IT-11,** 207–214 (1965).

[13] A. W. Rihaczek, "Radar Resolution Properties of Pulse Trains," Proc. IEEE **52,** 153–164 (1964).

[14] C. E. Cook and M. Bernfeld, *Radar Signals: An Introduction to Theory and Application*, Academic Press, New York, 1967.

[15] C. Kaiteris and W. L. Rubin, "Pulse Trains with Low Residue Ambiguity Surfaces That Minimize Overlapping Target Echo Suppression in Limiting Receivers," Proc. IEEE **54,** 438–439 (1966).

[16] J. B. Resnick, "High Resolution Waveforms Suitable for a Multiple Target Environment," M.Sc. Thesis, Massachusetts Institute of Technology, June 1962.

[17] R. H. Barker, "Group Sychronization of Binary Digital Systems," in *Communication Theory* (W. Jackson, Ed.), Academic Press, New York, 1953.

[18] R. Turyn, "On Barker Codes of Even Length," Proc. IRE **51,** 1256 (1963).

[19] E. N. Fowle, "The Design of Radar Signals," Mitre Corp. Report SR-98, November 1, 1963, Bedford, Mass.

[20] S. W. Golomb, "Sequence with Randomness Properties," Glenn L. Martin Co., Internal Report, Baltimore, June 1955.

[21] D. A. Huffman, "The Synthesis of Linear Sequential Coding Networks," in *Information Theory* (C. Cherry, Ed.), Academic Press, New York, 1956.

[22] N. Zierler, "Several Binary-Sequence Generators," Technical Report No. 95, Massachusetts Institute of Technology, Lincoln Laboratory, September 1955.

[23] W. W. Peterson, *Error Correcting Codes*, Massachusetts Institute of Technology Press, Cambridge, Mass., 1961.

[24] B. Elspas, "The Theory of Autonomous Linear Sequential Networks," IRE Trans. **CT-6,** 45–60 (March 1959).

[25] M. P. Ristenblatt, "Pseudo-Random Binary Coded Waveforms," Chapter 4 in [9-4].

[26] T. G. Birdsall and M. P. Ristenblatt, "Introduction to Linear Shift-Register Generated Sequences," Cooley Electronics Laboratory, Technical Report No. 90, University of Michigan Research Institute, October 1958.

[27] S. Golomb (Ed.), *Digital Communications*, Prentice-Hall, Englewood Cliffs, N.J., 1964.

[28] S. E. Craig, W. Fishbein, and O. E. Rittenbach, "Continuous-Wave Radar with High Range Resolution and Unambiguous Velocity Determination," IRE Trans. PGMIL 153–161 (April 1962).

[29] W. Fishbein and O. E. Rittenbach, "Correlation Radar Using Pseudo-Random Modulation," 1961 IRE Int. Conv. Rec. Pt. 5, pp. 259–277.

[30] R. S. Berkowitz, *Modern Radar: Analysis, Evaluation and System Design*, Wiley, New York, 1965.

[31] D. Huffman, "The Generation of Impulse-Equivalent Pulse Trains," IRE Trans. Information Theory **IT-8,** S10–S16 (Sept. 1962).

[32] S. W. Golomb and R. A. Scholtz, "Generalized Barker Sequences," IEEE Trans. Information Theory **IT-11,** 533–537 (1965).

[33] R. C. Heimiller, "Phase Shift Codes with Good Periodic Correlation Properties," IRE Trans. **IT-7,** 254–257 (1961).

[34] R. L. Frank, "Polyphase Codes with Good Nonperiodic Correlation Properties, IEEE Trans. **IT-9,** 43–45 (1963).

[35] T. Sakamoto, Y. Taki, H. Miyakawa, H. Kobayashi, and T. Kanda, "Coded Pulse Radar System," J. Faculty Eng. Univ. Tokyo **27,** 119–181 (1964).

[36] H. Urkowitz, C. A. Hauer, and J. F. Koval, "Generalized Resolution in Radar Systems, Proc. IRE **50,** No. 10, 2093–2105 (1962).

[37] C. A. Stutt, "A Note on Invariant Relations for Ambiguity and Distance Functions," IRE Trans. **IT-5,** 164–167 (Dec. 1959).

[38] W. L. Root, Paper presented at IDA Summer Study, 1963.

[39] C. W. Helstron, *Statistical Theory of Signal Detection*, Pergamon Press, New York, 1960.

[40] F. B. Reis, "A Linear Transformation of the Ambiguity Function Plane," IRE Trans. Information Theory **IT-8,** 59 (1962).

[41] C. H. Wilcox, "A Mathematical Analysis of the Waveform Design Problem," General Electric Co., Schenectady, N.Y., Report No. R57EMH54, October 1957.

[42] C. H. Wilcox, "The Synthesis Problem for Radar Ambiguity Functions," Mathematics Research Center, University of Wisconsin, MRC Technical Summary Report No. 157, April 1960.

[43] C. A. Stutt, "The Application of Time/Frequency Correlation Functions to the Continuous Waveform Encoding of Message Symbols," WESCON, Sect. 9-1, 1961.

[44] C. A. Stutt, "Some Results on Real Part/Imaginary Part and Magnitude/Phase Relations in Ambiguity Functions," IEEE Trans. Information Theory **IT-10,** No. 4, 321–327 (Oct. 1964).

[45] E. A. Guillemin, *Mathematics of Circuit Analysis*, Technology Press, Cambridge, Mass., and Wiley, New York, 1949.

[46] J. Klauder, "The Design of Radar Signals Having Both High Range Resolution and High Velocity Resolution," Bell Syst. Tech. J. **39,** 809–819 (1960).

[47] C. W. Helstrom, "The Resolution of Signals in White Gaussian Noise," Proc. IRE **43,** 1111–1118 (Sept. 1955).

[48] W. L. Root, "Radar Resolution of Closely Spaced Targets," IRE Trans. PGMIL **MIL-6,** No. 2 (April 1962).

[49] B. Elspas, "A Radar System Based on Statistical Estimation and Resolution Considerations," Stanford Electronics Laboratories, Stanford University, Technical Report No. 361-1, August 1, 1955.

[50] J. Allan, M. S. Thesis, Department of Electrical Engineering, Massachusetts Institute of Technology,

[51] P. Swerling, "The Resolvability of Point Sources," Proc. Symp. Decision Theory, Rome Air Development Center, Rome, N.Y., Technical Report No. 60-70A, April 1960.

[52] N. J. Nilsson, "On the Optimum Range Resolution of Radar Signals in Noise," IRE Trans. Information Theory **IT-7,** 245–253 (Oct. 1961).

[53] I. Selin, "Estimation of the Relative Delay of Two Similar Signals of Unknown Phases in White Gaussian Noise," IEEE Trans. Information Theory **IT-10,** No. 3, 189–191 (1964).

[54] E. Brookner, "Optimum Clutter Rejection," IEEE Trans. Information Theory **IT-11,** No. 4, 597–598 (Oct. 1965).

[55] H. Urkowitz, "Filters for the Detection of Small Radar Signals in Noise," J. Appl. Phys. **24,** 1024–1031 (Aug. 1953).

[56] E. L. Key, E. N. Fowle, and R. D. Haggerty, "A Method of Designing Signals of Large Time-Bandwidth Product," 1961 IRE Int. Conv. Rec. Pt. 4, pp. 146–154.

[57] D. A. George, "Matched Filters for Interfering Signals," IEEE Trans. Information Theory **IT-11,** No. 1 (Jan. 1965).

[58] F. C. Schweppe and D. L. Gray, "Radar Signal Design Subject to Simultaneous Peak and Average Power Constraints," IEEE Trans. Information Theory **IT-12,** No. 1, 13–26 (Jan. 1966).

[59] M. G. Lichtenstein and T. Y. Young, "The Resolution of Closely Spaced Signals," IEEE Trans. Information Theory **IT-14,** No. 2, 288–293 (March 1968).

[60] P. M. Woodward, *Probability and Information Theory, with Application to Radar,* Pergamon Press, New York, 1953.

[61] R. W. Lucky, "Techniques for Adaptive Equalization of Digital Communication Systems," Bell Syst. Tech. J. 255–286 (Feb. 1966).

[62] M. J. DiToro, "A New Method of High-Speed Adaptive Serial Communication through Any Time-Variable and Dispersive Transmission Medium," First IEEE Ann. Commun. Conv., Boulder, Colo., Conv. Rec., p. 763.

[63] M R. Aaron and D. W. Tufts, "Intersymbol Interference and Error Probability," IEEE Trans. **IT-12,** No. 1 (Jan. 1966).

[64] I. Gerst and J. Diamond, "The Elimination of Intersymbol Interference by Input Signal Shaping," Proc. IRE **49,** 1195–1203 (1961).

[65] P. A. Bello, "Joint Estimation of Delay, Doppler, and Doppler Rate," IRE Trans. Information Theory **IT-6,** No. 3, 330–341 (June 1960).

[66] E. J. Kelly, "The Radar Measurement of Range, Velocity, and Acceleration," IRE Trans. Mil. Elec. **MIL-5,** 51–57 (April 1961).

[67] F. C. Schweppe, "Radar Frequency Modulations for Accelerating Targets under a Bandwidth Constraint," IEEE Trans. Mil. Elec. **MIL-9,** 25–32 (Jan. 1965).

[68] R. deBuda, "An Extension of Green's Condition to Cross-Ambiguity Functions," IEEE Trans. Information Theory **IT-13,** No. 1, 75–82 (Jan. 1967).

[69] E. J. Kelly and R. P. Wishner, "Matched-Filter Theory for High-Velocity Targets," IEEE Trans. Mil. Elec. **MIL-9,** 56–69 (Jan. 1965).

[70] R. L. Gassner and G. R. Cooper, "Note on a Generalized Ambiguity Function," IEEE Trans. Information Theory **IT-13,** 126 (Jan. 1967).

[71] D. A. Swick, "Wideband Ambiguity Function of Pseudo-Random Sequences: An Open Problem," IEEE Trans. Information Theory **IT-14,** 602–603 (1968).

[72] J. M. Speiser, "Wide-Band Ambiguity Functions," IEEE Trans. Information Theory **IT-13,** 122–123 (Jan. 1967).

[73] R. J. Purdy and G. R. Cooper," A Note on the Volume of Generalized Ambiguity Functions," IEEE Trans. Information Theory **IT-14,** No. 1, 153–154 (Jan. 1968).

[74] E. L. Titlebaum, "A Generalization of a Two-Dimensional Fourier Transform Property for Ambiguity Functions," IEEE Trans. Information Theory **IT-12,** No. 1, 80–81 (Jan. 1966).

[75] C. W. Helstrom, "An Expansion of a Signal in Gaussian Elementary Signals," IEEE Trans. Information Theory **IT-12,** No. 1, 81–82 (Jan. 1966).

[76] R. Manasse, "Range and Velocity Accuracy from Radar Measurements," Massachusetts Institute of Technology, Lincoln Laboratory, Group 312-26, February 1955.

[77] E. J. Kelly, I. S. Reed, and W. L. Root, "The Detection of Radar Echoes in Noise. I and II," J. SIAM **8**, 309-341, 481-507 (1960).

[78] P. Swerling, "Parameter Estimation Accuracy Formulas," IEEE Trans. Infor. Thy., **IT-10**, No. 1, 302-313 (Oct. 1964).

[79] R. J. McAulay and E. M. Hofstetter, "Barankin Bounds on Parameter Estimation Accuracy Applied to Communications and Radar Problems," Massachusetts Inst. of Tech. Lincoln Laboratory Technical Note 1969-20, March 13, 1969.

[80] R. J. McAulay and L. P. Seidman, "A Useful Form of the Barankin Lower Bound and its Application to PPM Threshold Analysis," IEEE Trans. on Info. Thy., **IT-15**, 273-279 (March 1969).

[81] A. B. Baggeroer, "The Barankin Bound on the Estimation of Parameters Imbedded in the Covariance of a Gaussian Process," Proc. M. J. Kelly Conf. on Comm., Univ. of Missouri at Rolla, Oct. 6-8, 1970.

[82] M. Zakai and J. Ziv, "On the Threshold Effect in Radar Range Estimation," IEEE Trans. on Info. Thy., **IT-15**, 167-170 (Jan. 1969).

[83] J. Ziv and M. Zakai, "Some Lower Bounds on Signal Parameter Estimation," IEEE Trans. on Info. Thy., **IT-15**, 386-392 (May 1969).

# 11

# *Doppler-Spread Targets and Channels*

In Chapters 9 and 10 we confined our attention to slowly fluctuating point targets. They were characterized by a "perfect" reflection of the envelope of the incident signal. The returned signal differed from the transmitted signal in four ways:

1. Random amplitude.
2. Random phase angle.
3. Doppler shift.
4. Delay.

The amplitude and phase were due to the reflective characteristics of the target and could be modeled as random variables. The Doppler shift and delay were due to the velocity and range of the target and were modeled as unknown nonrandom variables.

In this chapter we consider point targets that cannot be modeled as slowly fluctuating targets. We begin our development with a qualitative discussion of the target model.

A simple example is shown in Fig. 11.1. The geometry could represent the reflective structure of an airplane, a satellite, or a submarine. The direction of signal propagation is along the $x$-axis. The target orientation changes as a function of time. Three positions are shown in Fig. 11.1*a–c*. As the orientation changes, the reflective characteristics change.

Now assume that we illuminate the target with a long pulse whose complex envelope is shown in Fig. 11.2*a*. A typical returned signal envelope is shown in Fig. 11.2*b*. We see that the effect of the changing orientation of the target is a time-varying attenuation of the envelope, which is usually referred to as *time-selective fading*.

Notice that if we transmit a short pulse as shown in Fig. 11.2*c*, the received signal envelope is undistorted (Fig. 11.2*d*) and the target can be modeled as a slowly fluctuating target. Later we shall see that all of our

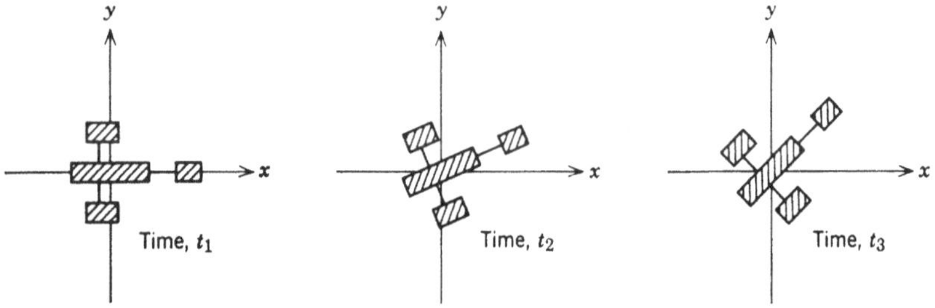


(a) (b) (c)

**Fig. 11.1 Target orientations.**

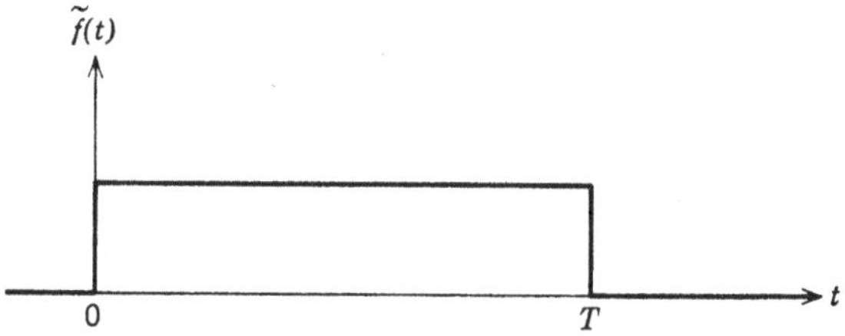


(a) Complex envelope of transmitted signal

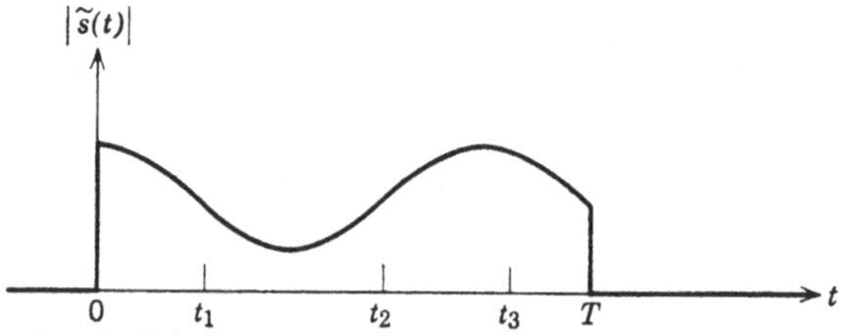


(b) Typical returned envelope (shifted time origin)

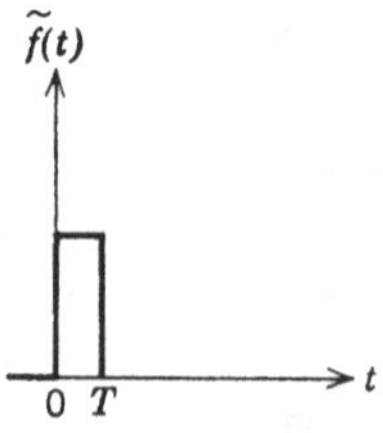


(c) Short pulse

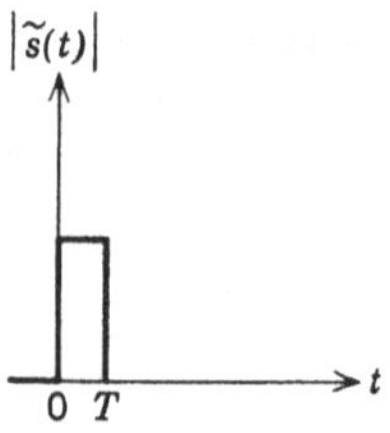


(d) Return from short pulse

**Fig. 11.2 Signals illustrating time-selective fading.**

results in Chapters 9 and 10 can be viewed as limiting cases of the results from the more general model in this chapter.

The energy spectrum of the long transmitted pulse is shown in Fig. 11.3*a*. Since the time-varying attenuation is an amplitude modulation, the spectrum of the returned signal is spread in frequency, as shown in Fig. 11.3*b*. The amount of spreading depends on the rate at which the target's reflective characteristics are changing. We refer to this type of target as a *frequency-spread* or *Doppler-spread* target. (Notice that frequency spreading and time-selective fading are just two different ways of describing the same phenomenon.)

Our simple example dealt with a radar problem. We have exactly the same mathematical problem when we communicate over a channel whose reflective characteristics change during the signaling interval. We refer to such channels as *Doppler-spread channels*, and most of our basic results will be applicable to both the radar/sonar and communications problems.

At this point we have an intuitive understanding of how a fluctuating target causes Doppler spreading. In Section 11.1 we develop a mathematical model for a fluctuating target. In Section 11.2 we derive the optimum receiver to detect a Doppler-spread target and evaluate its performance. In Section 11.3 we study the problem of digital communication systems

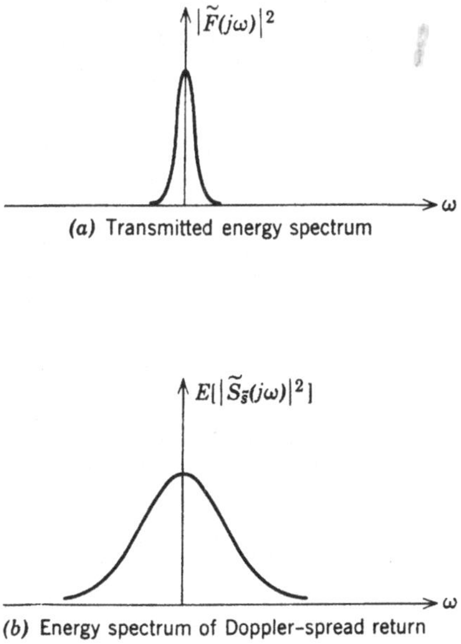


(*b*) Energy spectrum of Doppler-spread return

**Fig. 11.3 Energy spectra of transmitted and returned signals.**

operating over Doppler-spread channels. In Section 11.4 we consider the problem of estimating the parameters of a Doppler-spread target. Finally, in Section 11.5, we summarize our results.

## 11.1 MODEL FOR DOPPLER-SPREAD TARGET (OR CHANNEL)

The model for the point target with arbitrary fluctuations is a straightforward generalization of the slow-fluctuation model.† Initially we shall discuss the model in the context of an active radar or sonar system. If a sinusoidal signal

$$\sqrt{2}\cos\omega_c t = \sqrt{2}\,\mathrm{Re}\,[e^{j\omega_c t}] \tag{1}$$

is transmitted, the return from a target located at a point $\lambda$ (measured in units of round-trip travel time) is

$$b(t) = \sqrt{2}\left[b_c\left(t-\frac{\lambda}{2}\right)\cos\left[\omega_c(t-\lambda)\right] + b_s\left(t-\frac{\lambda}{2}\right)\sin\left[\omega_c(t-\lambda)\right]\right]. \tag{2}$$

The $\lambda/2$ arises because the signal arriving at the receiver at time $t$ left the transmitter at $t-\lambda$ and was reflected from the target at $t-\lambda/2$. We assume that $b_c(t)$ and $b_s(t)$ are sample functions from low-pass, zero-mean, stationary, Gaussian random processes and that $b(t)$ is a *stationary* bandpass process.

Defining

$$\tilde{b}_D(t) \triangleq b_c(t) - jb_s(t), \tag{3}$$

we have

$$b(t) = \sqrt{2}\,\mathrm{Re}\left[\tilde{b}_D\left(t-\frac{\lambda}{2}\right)e^{j\omega_c(t-\lambda)}\right], \tag{4}$$

where $\tilde{b}_D(t)$ is a sample function from a complex Gaussian process. (The subscript $D$ denotes Doppler.) We assume that $\tilde{b}_D(t)$ varies slowly compared to the carrier frequency $\omega_c$. Because $\tilde{b}_D(t)$ has a uniform phase at any time, this assumption allows us to write (4) as

$$b(t) = \sqrt{2}\,\mathrm{Re}\left[\tilde{b}_D\left(t-\frac{\lambda}{2}\right)e^{j\omega_c t}\right]. \tag{5}$$

† This model has been used by a number of researchers (e.g., Price and Green [1] and Bello [2]).

The random process $\tilde{b}_D(t)$ is completely characterized by its complex covariance function,

$$E[\tilde{b}_D(t)\tilde{b}_D^*(u)] \triangleq \tilde{K}_D(t-u) = \tilde{K}_D(\tau). \tag{6}$$

From our development in the Appendix,

$$E[\tilde{b}_D(t)\tilde{b}_D(u)] = 0 \qquad \text{for all } t \text{ and } u. \tag{7}$$

In all of our discussion we assume that $\tilde{K}_D(\tau)$ is known. In Section 11.4 we discuss the problem of measuring the parameters of $\tilde{K}_D(\tau)$.

Notice that if we assume

$$K_D(\tau) = K_D(0) \qquad \text{for all } \tau, \tag{8}$$

we would have the slowly fluctuating model of Chapter 9 (see page 242). To be consistent with that model, we assume

$$\tilde{K}_D(0) = 2\sigma_b^2. \tag{9}$$

Because the target reflection process is assumed to be stationary, we can equally well characterize it by its spectrum:†

$$\tilde{S}_D\{f\} = \int_{-\infty}^{\infty} \tilde{K}_D(\tau)e^{-j2\pi f\tau}\,d\tau. \tag{10}$$

We refer to $\tilde{S}_D\{f\}$ as the Doppler scattering function. From (A.56) we know that $\tilde{S}_D\{f\}$ is a real function and that the spectrum of the actual bandpass signal is

$$S_D\{f\} = \tfrac{1}{2}\tilde{S}_D\{f-f_c\} + \tfrac{1}{2}\tilde{S}_D\{-f-f_c\}. \tag{11}$$

Some typical spectra are shown in Fig. 11.4. We assume that the transmitted signal is a long pulse with a rectangular envelope. It has a narrow energy spectrum, as shown in Fig. 11.4*a*. In Fig. 11.4*b*, we show the energy spectrum of the returned signal when the target is fluctuating and has a zero average velocity. In Fig. 11.4*c*, we show the spectrum corresponding to a target that has a nonzero average velocity but is not

† In most of the discussion in the next three chapters it is convenient to use $f$ as an argument in the spectrum and Fourier transform. The braces { } around the argument imply the $f$ notation. The $f$ notation is used throughout Chapters 11–13, so that the reader does not need to watch the $\{\cdot\}$. Notice that for deterministic signals,

$$\tilde{F}\{f\} \triangleq \int_{-\infty}^{\infty} \tilde{f}(t)e^{-j2\pi ft}\,dt.$$

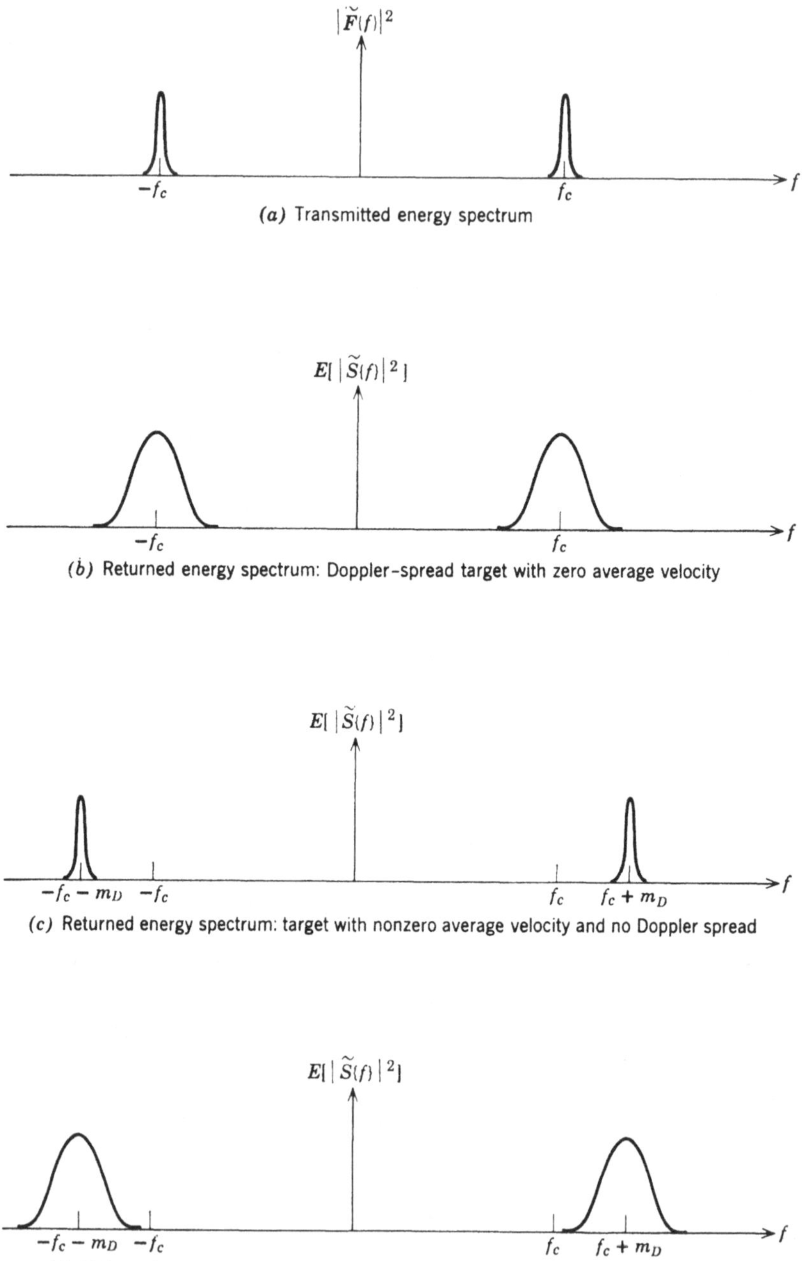


*(a)* Transmitted energy spectrum

*(b)* Returned energy spectrum: Doppler-spread target with zero average velocity

*(c)* Returned energy spectrum: target with nonzero average velocity and no Doppler spread

*(d)* Returned energy spectrum: Doppler-spread target with nonzero average velocity

**Fig. 11.4 Typical energy spectra.**

fluctuating. This is the model of Chapter 9. In Fig. 11.4*d*, we show the spectrum for a fluctuating target with a nonzero average velocity.

We introduce two quantities to describe the gross behavior of the target. The first is the *mean Doppler shift*, which is defined as

$$\boxed{m_D \triangleq \frac{1}{2\sigma_b^2} \int_{-\infty}^{\infty} f \tilde{S}_D\{f\}\, df.} \tag{12}$$

We next define

$$\overline{f_D^2} \triangleq \frac{1}{2\sigma_b^2} \int_{-\infty}^{\infty} f^2 S_D\{f\}\, df. \tag{13}$$

Combining (12) and (13) gives a quantity that we refer to as a *mean-square Doppler spread*,

$$\boxed{\sigma_D^2 \triangleq \overline{f_D^2} - m_D^2 = \frac{1}{2\sigma_b^2} \int_{-\infty}^{\infty} f^2 S_D\{f\}\, df - m_D^2.} \tag{14}$$

We see that $m_D$ and $\sigma_D^2$ are identical with the mean and variance of a random variable.

Our discussion up to this point has a sinusoidal transmitted signal. However, because we assume that the reflection process is linear and frequency-independent, (2) characterizes the target behavior. Therefore, if we assume that the transmitted waveform is a known narrow-band signal,

$$f(t) = \sqrt{2}\, \text{Re}\, [\sqrt{E_t} \tilde{f}(t) e^{j\omega_c t}], \qquad -\infty < t < \infty, \tag{15}$$

the returned signal in the absence of noise is

$$s(t) = \sqrt{2}\, \text{Re} \left[ \sqrt{E_t} \tilde{f}(t - \lambda) \tilde{b}\left(t - \frac{\lambda}{2}\right) e^{j\omega_c t} \right]. \tag{16}$$

The complex envelope is

$$\boxed{\tilde{s}(t) \triangleq \sqrt{E_t} \tilde{f}(t - \lambda) \tilde{b}\left(t - \frac{\lambda}{2}\right),} \tag{17}$$

and the actual signal can be written as

$$s(t) = \sqrt{2}\, \text{Re}\, [\sqrt{E_t}\, \tilde{s}(t) e^{j\omega_c t}]. \tag{18}$$

The complex covariance function of the signal process is

$$\tilde{K}_{\tilde{s}}(t, u) = E[\tilde{s}(t) \tilde{s}^*(u)], \tag{19}$$

or

$$\boxed{\tilde{K}_{\tilde{s}}(t, u) = E_t \tilde{f}(t - \lambda) \tilde{K}_D(t - u) \tilde{f}^*(u - \lambda).} \tag{20}$$

Now (20) completely specifies the characteristics of the received signal.

The total received waveform is $s(t)$ plus an additive noise. Thus,

$$r(t) = \sqrt{2}\,\text{Re}\,[\tilde{s}(t)e^{j\omega_c t}] + \sqrt{2}\,\text{Re}\,[\tilde{w}(t)e^{j\omega_c t}], \qquad T_i \le t \le T_f, \quad (21)$$

or

$$r(t) = \sqrt{2}\,\text{Re}\,[\tilde{r}(t)\, e^{j\omega_c t}], \quad (22)$$

where

$$\tilde{r}(t) = \tilde{s}(t) + \tilde{w}(t). \quad (23)$$

The complete model is shown in Fig. 11.5.

We assume that the additive noise is a sample function from a zero-mean, stationary Gaussian process that is statistically independent of the reflection process and has a flat spectrum of height $N_0/2$ over a band wide compared to the signals of interest. Then

$$E[\tilde{w}(t)\tilde{w}^*(u)] = N_0\,\delta(t-u), \quad (24)$$

and the covariance function of $\tilde{r}(t)$ is

$$\boxed{\begin{aligned}\tilde{K}_{\tilde{r}}(t,u) = E_t\tilde{f}(t-\lambda)\tilde{K}_D(t-u)\tilde{f}^*(u-\lambda)\\ + N_0\,\delta(t-u), \qquad T_i \le t, u \le T_f.\end{aligned}} \quad (25)$$

The covariance function in (25) completely characterizes the received waveform that we have available for processing.

Whenever the reflection process $\tilde{b}_D(t)$ has a rational spectrum, we may also characterize it using complex state variables.† The state equation is

$$\dot{\tilde{\mathbf{x}}}(t) = \tilde{\mathbf{F}}\tilde{\mathbf{x}}(t) + \tilde{\mathbf{G}}\tilde{u}(t), \qquad t \ge T_i, \quad (26)$$

where

$$E[\tilde{u}(t)\tilde{u}(\tau)] = \tilde{Q}\,\delta(t-\tau) \quad (27)$$

and

$$E[\tilde{\mathbf{x}}(T_i)\tilde{\mathbf{x}}^\dagger(T_i)] = \tilde{\mathbf{P}}_0. \quad (28)$$

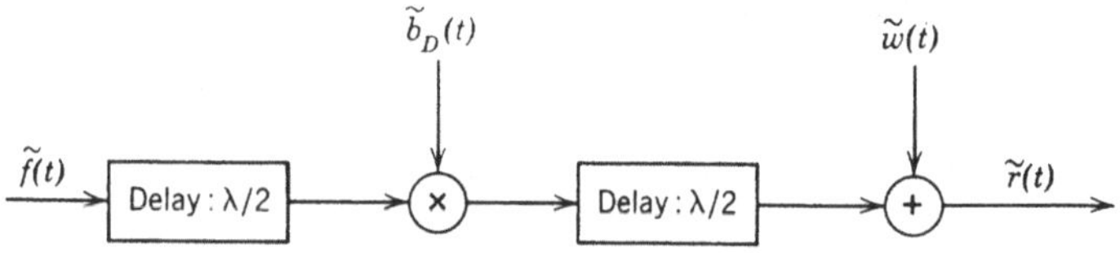


**Fig. 11.5 Model for Doppler-spread target problem.**

† Complex state variables are discussed in Section A.3.3.

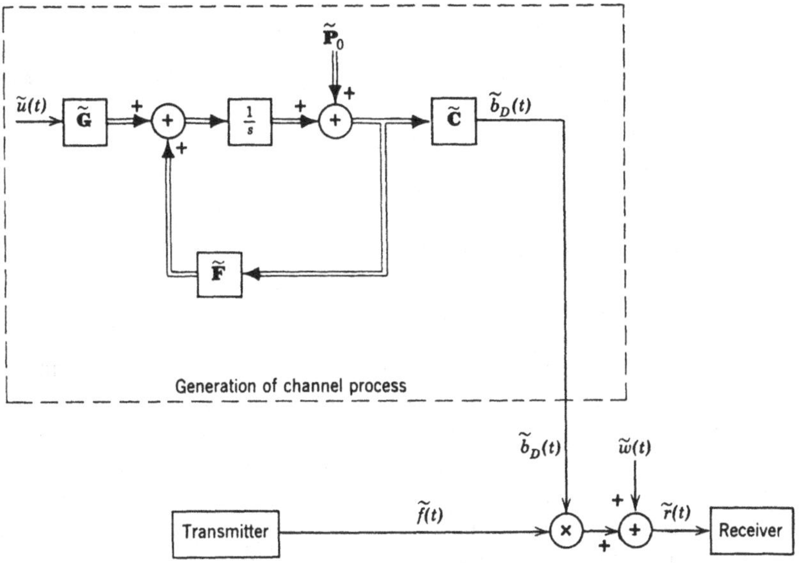


**Fig. 11.6 State-variable model for Doppler-spread target (or channel).**

The process $\tilde{b}_D(t)$ is obtained by the relation

$$\tilde{b}_D(t) = \tilde{\mathbf{C}}\tilde{\mathbf{x}}(t), \qquad t > T_i. \tag{29}$$

We shall find this representation useful in many problems of interest. This model is shown in Fig. 11.6 for the special case in which $\lambda = 0$.

This completes our formulation of the model for a Doppler-spread target. All of our work in the text deals with this model. There are two simple generalizations of the model that we should mention:

1. Let $\tilde{b}_D(t)$ be a non-zero-mean process. This corresponds to a fluctuating Rician channel.
2. Let $\tilde{b}_D(t)$ be a nonstationary complex Gaussian process.

Both these generalizations can be included in a straightforward manner and are discussed in the problems. There are targets and channels that do not fit the Rayleigh or Rician model (recall the discussion on page 243). The reader should consult the references cited earlier for a discussion of these models. We now turn our attention to the optimum detection problem.

## 11.2 DETECTION OF DOPPLER-SPREAD TARGETS

In this section we consider the problem of detecting a Doppler-spread target. The complex envelope of the received waveform on the two hypotheses is

$$\tilde{r}(t) = \sqrt{E_t}\tilde{f}(t-\lambda)\tilde{b}_D\left(t-\frac{\lambda}{2}\right) + \tilde{w}(t), \qquad T_i \le t \le T_f : H_1, \tag{30}$$

and

$$\tilde{r}(t) = \tilde{w}(t), \qquad T_i \le t \le T_f : H_0. \tag{31}$$

The signal process is a sample function from a zero-mean complex Gaussian process whose covariance function is

$$\tilde{K}_{\tilde{s}}(t,u) = E_t\tilde{f}(t-\lambda)\tilde{K}_D(t-u)\tilde{f}^*(u-\lambda), \qquad T_i \le t, u \le T_f. \tag{32}$$

The additive noise $\tilde{w}(t)$ is a sample function of a statistically independent, zero-mean complex white Gaussian process with spectral height $N_0$. The range parameter $\lambda$ is known.

We see that this problem is just the complex version of the Gaussian signal in Gaussian noise problem that we discussed in detail in Chapter 2.† Because of this strong similarity, we state many of our results without proof. The four issues of interest are:

1. The likelihood ratio test.
2. The canonical receiver realizations to implement the likelihood ratio test.
3. The performance of the optimum receiver.
4. The classes of spectra for which complete solutions can be obtained.

We discuss all of these issues briefly.

### 11.2.1 Likelihood Ratio Test

The likelihood ratio test can be derived by using series expansion as in (A.116), or by starting with (2.31) and exploiting the bandpass character of the processes (e.g., Problems 11.2.1 and 11.2.2, respectively). The result is

$$\boxed{l = \frac{1}{N_0}\iint_{T_i}^{T_f} \tilde{r}^*(t)\tilde{h}(t,u)\tilde{r}(u)\,dt\,du \underset{H_0}{\overset{H_1}{\gtrless}} \gamma,} \tag{33}$$

† As we pointed out in Chapter 2, the problem of detecting Gaussian signals in Gaussian noise has been studied extensively. References that deal with problems similar to that of current interest include Price [8]–[10], Kailath [11], [12], Turin [13], [14], and Bello [15]. The fundamental Gaussian signal detection problem is discussed by Middleton [16]–[18]. Book references include Helstrom [19, Chapter 11] and Middleton [20, Part 4].

where $\tilde{h}(t, u)$ satisfies the integral equation

$$N_0\tilde{h}(t, u) + \int_{T_i}^{T_f} \tilde{h}(t, z)\tilde{K}_{\tilde{s}}(z, u)\, dz = \tilde{K}_{\tilde{s}}(t, u), \qquad T_i \leq t, u \leq T_f \tag{34}$$

and

$$\tilde{K}_{\tilde{s}}(t, u) = E_t\tilde{f}(t - \lambda)\tilde{K}_D(t - u)\tilde{f}^*(u - \lambda), \qquad T_i \leq t, u \leq T_f. \tag{35}$$

The threshold $\gamma$ is determined by the costs and a-priori probabilities in a Bayes test and by the desired $P_F$ in a Neyman-Pearson test. In the next section we discuss various receiver realizations to generate $l$.

### 11.2.2 Canonical Receiver Realizations

The four realizations of interest were developed for real processes in Chapter 2 (see pages 15–32). The extension to the complex case is straightforward. We indicate the resulting structures for reference.

***Estimator-correlator Receiver.*** Realization No. 1 is shown in complex notation in Fig. 11.7*a*. The filter $\tilde{h}(t, u)$ is the optimum unrealizable filter for estimating $\tilde{s}(t)$ and satisfies (34). The actual bandpass realization is shown in Fig. 11.7*b*. Notice that the integrator eliminates the high-frequency component of the multiplier output.

***Filter-squarer-integrator (FSI) Receiver.*** To obtain this realization, we factor $\tilde{h}(t, u)$ as

$$\int_{T_i}^{T_f} \tilde{g}^*(z, t)\tilde{g}(z, u)\, dz = \tilde{h}(t, u), \qquad T_i \leq t, u \leq T_f. \tag{36}$$

Then

$$l = \frac{1}{N_0}\int_{T_i}^{T_f} dz \left| \int_{T_i}^{T_f} \tilde{g}(z, t)\tilde{r}(t)\, dt \right|^2. \tag{37}$$

The complex operations are indicated in Fig. 11.8*a*, and the actual receiver is shown in Fig. 11.8*b*.

***Optimum Realizable Filter Receiver.*** For this realization, we rewrite the LRT as

$$l = \frac{1}{N_0}\int_{T_i}^{T_f} \{2\,\mathrm{Re}\,[\tilde{r}^*(t)\hat{\tilde{s}}_r(t)] - |\hat{\tilde{s}}_r(t)|^2\}\, dt \underset{H_0}{\overset{H_1}{\gtrless}} \gamma, \tag{38}$$

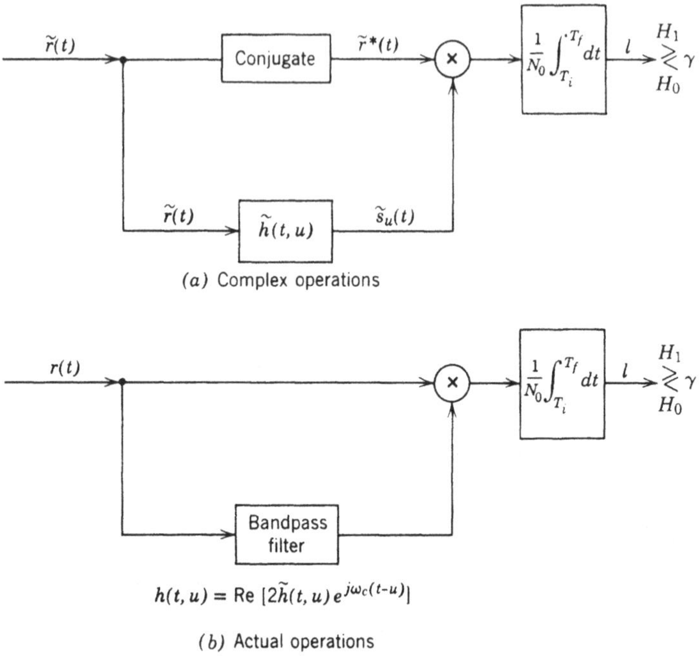


**Fig. 11.7 Estimator-correlator receiver (Canonical Realization No. 1).**

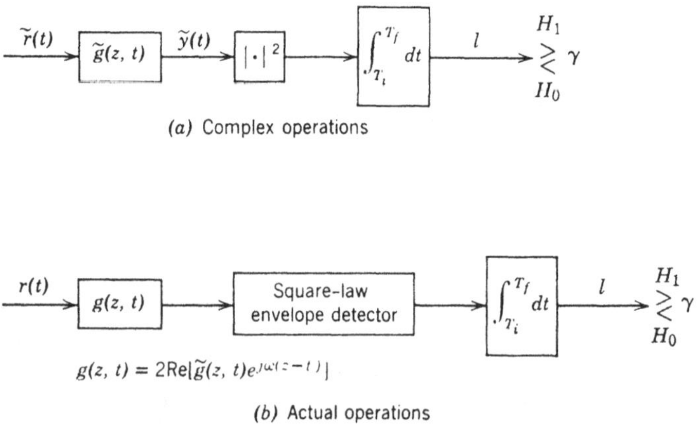


**Fig. 11.8 Filter-squarer-integrator receiver (Canonical Realization No. 3).**

where $\hat{\tilde{s}}_r(t)$ is the realizable MMSE estimate of $\tilde{s}(t)$ when $H_1$ is true (e.g., Problem 11.2.3). It is obtained by passing $\tilde{r}(t)$ through a filter $\tilde{h}_{or}(t, u)$, whose impulse response is specified by

$$N_0\tilde{h}_{or}(t, u) + \int_{T_i}^{t} \tilde{h}_{or}(t, z)\tilde{K}_{\tilde{s}}(z, u)\, dz = \tilde{K}_{\tilde{s}}(t, u), \qquad T_i \le u \le t \tag{39}$$

and

$$\hat{\tilde{s}}_r(t) \triangleq \int_{T_i}^{t} \tilde{h}_{or}(t, u)\tilde{r}(u)\, du. \tag{40}$$

The complex receiver is shown in Fig. 11.9.

***State-Variable Realization.***† When $\tilde{b}_D(t)$ has a finite-dimensional complex state representation, it is usually more convenient to obtain $\hat{\tilde{s}}_r(t)$ through the use of state-variable techniques. Recall that we are considering a point target and $\lambda$ is assumed known. Therefore, for algebraic simplicity, we can let $\lambda = 0$ with no loss of generality.

If we denote the state vector of $\tilde{b}_D(t)$ as $\tilde{\mathbf{x}}(t)$, then

$$\tilde{b}_D(t) = \tilde{\mathbf{C}}\tilde{\mathbf{x}}(t) \tag{41}$$

and

$$\tilde{s}(t) = \tilde{f}(t)\tilde{\mathbf{C}}\tilde{\mathbf{x}}(t) \triangleq \tilde{\mathbf{C}}_s(t)\tilde{\mathbf{x}}(t). \tag{42}$$

The state vector $\tilde{\mathbf{x}}(t)$ satisfies the differential equation

$$\dot{\tilde{\mathbf{x}}}(t) = \tilde{\mathbf{F}}(t)\tilde{\mathbf{x}}(t) + \tilde{\mathbf{G}}(t)\tilde{u}(t), \qquad t > T_i, \tag{43}$$

where

$$E[\tilde{u}(t)\tilde{u}^*(\sigma)] = \tilde{Q}\,\delta(t - \sigma) \tag{44}$$

and

$$E[\tilde{\mathbf{x}}(T_i)] = \mathbf{0}, \tag{45}$$

$$E[\tilde{\mathbf{x}}(T_i)\tilde{\mathbf{x}}^\dagger(T_i)] = \tilde{\mathbf{P}}_0 \tag{46}$$

(see page 590).

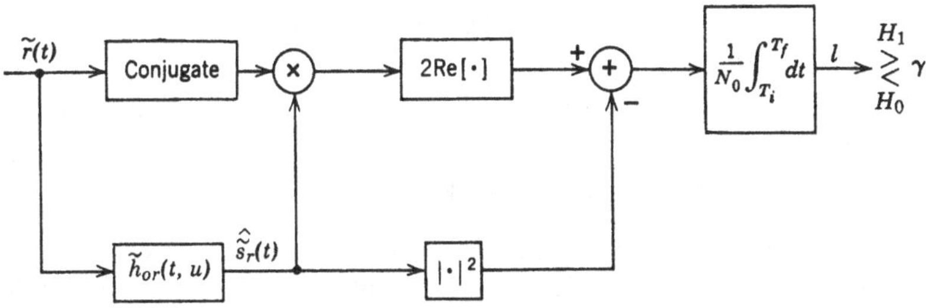


**Fig. 11.9 Optimum receiver: optimum realizable filter realization (Canonical Realization No. 4).**

† This section assumes familiarity with Section A.3.3.

The optimum estimate is specified as a solution of the differential equations

$$\dot{\hat{\mathbf{x}}}(t) = \tilde{\mathbf{F}}(t)\hat{\tilde{\mathbf{x}}}(t) + \tilde{\boldsymbol{\xi}}_P(t)\tilde{\mathbf{C}}_s^{\dagger}(t)\frac{1}{N_0}[\tilde{r}(t) - \tilde{\mathbf{C}}_s(t)\hat{\tilde{\mathbf{x}}}(t)], \quad t > T_i \tag{47a}$$

and

$$\hat{\tilde{s}}_r(t) = \tilde{f}(t)\tilde{\mathbf{C}}\hat{\tilde{\mathbf{x}}}(t) = \tilde{\mathbf{C}}_s(t)\hat{\tilde{\mathbf{x}}}(t). \tag{47b}$$

The variance equation is

$$\dot{\tilde{\boldsymbol{\xi}}}_P(t) = \tilde{\mathbf{F}}(t)\tilde{\boldsymbol{\xi}}_P(t) + \tilde{\boldsymbol{\xi}}_P(t)\tilde{\mathbf{F}}^{\dagger}(t) - \tilde{\boldsymbol{\xi}}_P(t)\tilde{\mathbf{C}}_s^{\dagger}(t)\frac{1}{N_0}\tilde{\mathbf{C}}_s(t)\tilde{\boldsymbol{\xi}}_P(t) + \tilde{\mathbf{G}}(t)\tilde{\mathbf{Q}}\tilde{\mathbf{G}}(t),$$
$$t > T_i, \tag{48a}$$

and

$$\xi_P(t, \tilde{s}(t), N_0) \triangleq E[|\tilde{s}(t) - \hat{\tilde{s}}_r(t)|^2] = \tilde{\mathbf{C}}_s(t)\tilde{\boldsymbol{\xi}}_P(t)\tilde{\mathbf{C}}_s^{\dagger}(t). \tag{48b}$$

Notice that the covariance matrix $\tilde{\boldsymbol{\xi}}_P(t)$ is a Hermitian matrix. Substituting (42) into (48$a$) gives

$$\dot{\tilde{\boldsymbol{\xi}}}_P(t) = \tilde{\mathbf{F}}(t)\tilde{\boldsymbol{\xi}}_P(t) + \tilde{\boldsymbol{\xi}}_P(t)\tilde{\mathbf{F}}^{\dagger}(t) - \tilde{\boldsymbol{\xi}}_P(t)\tilde{\mathbf{C}}^{\dagger}\left[\frac{|\tilde{f}(t)|^2}{N_0}\right]\tilde{\mathbf{C}}\tilde{\boldsymbol{\xi}}_P(t) + \tilde{\mathbf{G}}(t)\tilde{Q}\tilde{\mathbf{G}}^{\dagger}(t),$$
$$t > T_i. \tag{49}$$

We see that the mean-square error is only affected by the *envelope* of the transmitted signal. When we discuss performance we shall find that it can be expressed in terms of the mean-square error. Thus, performance is not affected by phase-modulating the signal. If the target has a mean Doppler shift, this is mathematically equivalent to a phase modulation of the signal. Thus, a mean Doppler shift does not affect the performance.

It is important to observe that, even though the reflection process is stationary, the returned signal process will be nonstationary unless $\tilde{f}(t)$ is a real pulse with constant height. This nonstationarity makes the state-variable realization quite important, because we can actually find the necessary functions to implement the optimum receiver.

This completes our initial discussion of receiver configurations. We now consider the performance of the receiver.

### 11.2.3 Performance of the Optimum Receiver

To evaluate the performance, we follow the same procedure as in Chapter 2 (pages 32–42). The key function is $\tilde{\mu}(s)$. First assume that there are $K$ complex observables, which we denote by the vector $\tilde{\mathbf{r}}$. Then

$$\tilde{\mu}_K(s) \triangleq \ln \int_{\infty}^{\infty} [p_{\tilde{\mathbf{r}}|H_1}(\tilde{\mathbf{R}}|H_1)]^s [p_{\tilde{\mathbf{r}}|H_0}(\tilde{\mathbf{R}}|H_0)]^{1-s}\, d\tilde{\mathbf{R}}. \tag{50}$$

Using (A.116) and (A.117), we have

$$p_{\tilde{\mathbf{r}}|H_1}(\tilde{\mathbf{R}}|H_1) = \prod_{i=1}^{K} \frac{1}{\pi(\tilde{\lambda}_i + N_0)} \exp\left(-\frac{|\tilde{R}_i|^2}{(\tilde{\lambda}_i + N_0)}\right), \tag{51}$$

where $\tilde{\lambda}_i$ is the eigenvalue of the signal process $\tilde{s}(t)$, and

$$p_{\tilde{\mathbf{r}}|H_0}(\tilde{\mathbf{R}}|H_0) = \prod_{i=1}^{K} \frac{1}{\pi N_0} \exp\left(-\frac{|\tilde{R}_i|^2}{N_0}\right). \tag{52}$$

Substituting (51) and (52) into (50), evaluating the integral (or comparing with 2.131), and letting $K \to \infty$, we obtain

$$\boxed{\tilde{\mu}(s) = \sum_{i=1}^{\infty} \left[(1-s)\ln\left(1 + \frac{\tilde{\lambda}_i}{N_0}\right) - \ln\left(1 + (1-s)\frac{\tilde{\lambda}_i}{N_0}\right)\right], \quad 0 \le s \le 1.} \tag{53}$$

Notice that $\tilde{\mu}(s)$ is a real function and is identical with (2.134), except for factors of 2. This can be expressed in a closed form in several ways. As in (2.138), it is the integral of two mean-square realizable filtering errors.

$$\boxed{\tilde{\mu}(s) = \frac{1-s}{N_0}\int_{T_i}^{T_f} dt\left(\xi_P(t, \tilde{s}(t), N_0) - \xi_P\left(t, \tilde{s}(t), \frac{N_0}{1-s}\right)\right].} \tag{54}$$

It can also be expressed by a formula like that in (2.195) if the signal process has a finite-dimensional state equation. For complex state-variable processes the appropriate formulas are

$$\ln \tilde{D}_{\mathscr{F}}(z) = \sum_{i=1}^{\infty} \ln(1 + z\tilde{\lambda}_i) = \ln \det \tilde{\mathbf{\Gamma}}_2(T_f) + \operatorname{Re}\int_{T_i}^{T_f} \operatorname{Tr}[\tilde{\mathbf{F}}(t)]\, dt, \tag{55}$$

where $\tilde{\mathbf{\Gamma}}_2(t)$ is specified by

$$\frac{d}{dt}\begin{bmatrix} \tilde{\mathbf{\Gamma}}_1(t) \\ \tilde{\mathbf{\Gamma}}_2(t) \end{bmatrix} = \begin{bmatrix} \tilde{\mathbf{F}}(t) & \tilde{\mathbf{G}}(t)\tilde{Q}\tilde{\mathbf{G}}^{\dagger}(t) \\ z\tilde{\mathbf{C}}^{\dagger}(t)\tilde{\mathbf{C}}(t) & -\tilde{\mathbf{F}}^{\dagger}(t) \end{bmatrix} \begin{bmatrix} \tilde{\mathbf{\Gamma}}_1(t) \\ \tilde{\mathbf{\Gamma}}_2(t) \end{bmatrix} \tag{56}$$

with initial conditions

$$\tilde{\mathbf{\Gamma}}_1(T_i) = \tilde{\mathbf{P}}_0, \tag{57}$$

$$\tilde{\mathbf{\Gamma}}_2(T_i) = \mathbf{I}. \tag{58}$$

Notice that $\tilde{D}_{\mathscr{F}}(z)$ is a real function.

To evaluate the performance, we use (53) in (2.166) and (2.174) to obtain

$$P_F \simeq \left[\sqrt{2\pi s^2 \ddot{\tilde{\mu}}(s)}\right]^{-1} e^{\tilde{\mu}(s) - s\dot{\tilde{\mu}}(s)}, \qquad 0 \le s \le 1 \tag{59}$$

and

$$P_M \simeq \left[\sqrt{2\pi(1-s)^2\ddot{\tilde{\mu}}(s)}\right]^{-1} e^{\tilde{\mu}(s)+(1-s)\dot{\tilde{\mu}}(s)} \qquad 0 \leq s \leq 1. \tag{60}$$

As a final topic we discuss the classes of returned signal processes for which complete solutions for the optimum receiver and its performance can be obtained.

### 11.2.4 Classes of Processes

There are four cases in which complete results can be obtained.

**Case 1. Reflection Process with Finite State Representation.** In this case $\tilde{b}_D(t)$ can be described by differential equations, as in (41)–(46). Because we have limited ourselves to stationary processes, this is equivalent to the requirement that the spectrum $\tilde{S}_D\{f\}$ be rational. In this case, (38) and (47)–(49) apply directly, and the receiver can be realized with a feedback structure. The performance follows easily by using (54) in (59) and (60).

**Case 2. Stationary Signal Process: Long Observation Time.** This case is the bandpass analog of the problem discussed in Section 4.1. Physically it could arise in several ways. Two of particular interest are the following:

1. The complex envelope of the transmitted signal is a real rectangular pulse whose length is appreciably longer than the correlation time of the reflection process.
2. In the passive detection problem the signal is generated by the target, and if this process is stationary, the received envelope is a stationary process.

For this case, we can use asymptotic formulas and obtain much simpler expressions. We solve (34) using transforms. The result is

$$\tilde{H}\{f\} = \frac{\tilde{S}_{\tilde{s}}\{f\}}{\tilde{S}_{\tilde{s}}\{f\} + N_0}. \tag{61}$$

A common realization in this case is the filter-squarer realization [see (36) and (37)]. We can find a solution to (36) that is a realizable filter:

$$\tilde{G}\{f\} = \left[\frac{\tilde{S}_{\tilde{s}}\{f\}}{\tilde{S}_{\tilde{s}}\{f\} + N_0}\right]^{+}. \tag{62}$$

Recall that the superscript "+" denotes the term containing the left-half-plane poles and zeros.

**Case 3. Separable Kernels.** In this case, the reflection process has a finite number of eigenvalues (say $K$). Looking at (20), we see that this means that the received signal process must also have $K$ eigenvalues. In this case the problem is mathematically identical with a reflection from $K$ slowly fluctuating targets.

**Case 4. Low-Energy-Coherence Case.** In this case, the largest eigenvalue is much smaller than the white noise level. Then, as on pages 131–137, we can obtain a series solution to the integral equation specifying $\tilde{h}(t, u)$. By an identical argument, the likelihood ratio test becomes

$$l = \frac{1}{N_0^2} \iint_{T_i}^{T_f} \tilde{r}^*(t)\tilde{K}_{\tilde{s}}(t, u)\tilde{r}(u)\, du \underset{H_0}{\overset{H_1}{\gtrless}} \gamma. \tag{63a}$$

Using (35),

$$\boxed{l = \frac{E_t}{N_0^2} \iint_{T_i}^{T_f} \tilde{r}^*(t)\tilde{f}(t-\lambda)\tilde{K}_D(t-u)\tilde{f}^*(u-\lambda)\tilde{r}(u)\, dt\, du.} \tag{63b}$$

We can write $\tilde{K}_D(t-u)$ in a factored form as

$$\tilde{K}_D(t-u) = \int_{T_i}^{T_f} \tilde{k}^*(z, t)\tilde{k}(z, u)\, dz, \qquad T_i \le t, u \le T_f. \tag{64a}$$

Using (64*a*) in (63*b*) gives

$$\boxed{l = \frac{E_t}{N_0^2} \int_{T_i}^{T_f} dz \left| \int_{T_i}^{T_f} k(z, u)\tilde{f}^*(u-\lambda)\tilde{r}(u)\, du \right|^2.} \tag{64b}$$

This realization is shown in Fig. 11.10. (The reader should verify that the receiver in Fig. 11.10*b* generates the desired output. The bandpass filter at $\omega_\Delta$ is assumed to be ideal.)

For an arbitrary time interval, the factoring indicated in (64*a*) may be difficult to carry out. However, in many cases of interest the time interval is large and we can obtain an approximate solution to (64*a*) by using Fourier transforms.

$$\tilde{K}_\infty\{f\} = [\tilde{S}_D\{f\}]^+. \tag{64c}$$

In this case we obtain the receiver structure shown in Fig. 11.11. Notice that we do not require $\tilde{f}(t)$ to be a constant, and so the result is more general than the SPLOT condition in Case 2.

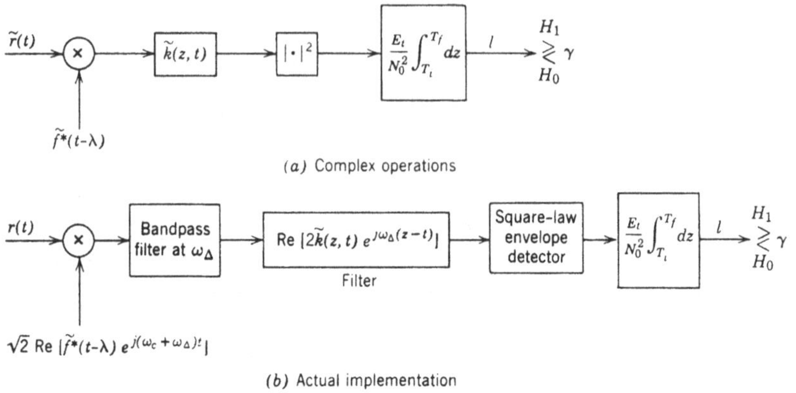


**Fig. 11.10 Optimum LEC receiver.**

The performance in the LEC case is obtained from (53). Expanding the logarithm in a series and neglecting the higher-order terms, we have

$$\tilde{\mu}(s) = -\frac{s(1-s)}{2N_0^2}\sum_{i=1}^{\infty}(\tilde{\lambda}_i)^2$$

$$= -\frac{s(1-s)}{2N_0^2}\iint_{-\infty}^{\infty}|\tilde{K}_{\tilde{s}}(t, u)|^2\, dt\, du. \tag{65}$$

Using (20) in (65) gives

$$\tilde{\mu}(s) = -\frac{s(1-s)E_t^2}{2N_0^2}\iint_{-\infty}^{\infty}|\tilde{f}(t-\lambda)|^2\,|\tilde{K}_D(t-u)|^2\,|\tilde{f}(u-\lambda)|^2\, dt\, du, \tag{66a}$$

which may also be written as

$$\boxed{\tilde{\mu}(s) = -\frac{s(1-s)E_t^2}{2N_0^2}\iint_{-\infty}^{\infty}\tilde{S}_D\{f_1\}\theta\{0, f_1 - f_2\}\tilde{S}_D\{f_2\}\, df_1\, df_2.} \tag{66b}$$

Thus the performance can be obtained by performing a double integration. We use $\tilde{\mu}(s)$ in (59) and (60) to find $P_F$ and $P_M$.

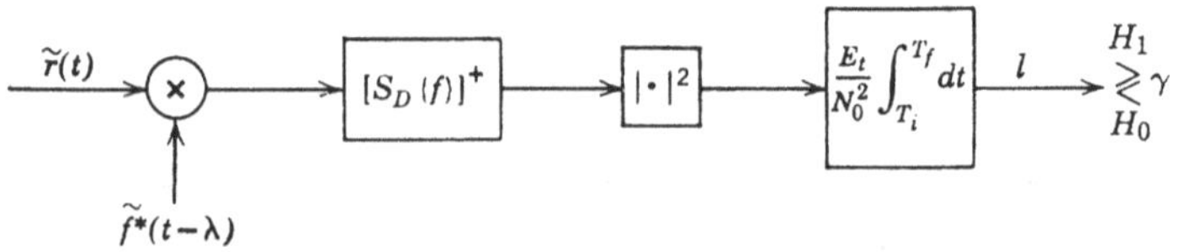


**Fig. 11.11 Optimum LEC receiver: long observation time.**

### 11.2.5 Summary

In this section we have discussed the detection of Doppler-spread targets. The four important issues were the derivation of the likelihood ratio test, canonical realizations of the optimum receiver, performance of the optimum receiver, and the classes of spectra that permit complete solutions. All the results are the complex versions of the results in Chapters 2–4.

Throughout our discussion we have tried to emphasize the similarities between the current problem and our earlier work. The reader should realize that these similarities arise because we have introduced complex notation. It is difficult to solve the bandpass problem without complex notation unless the quadrature components of the signal process are statistically independent (recall Problem 3.4.9). Using (20) in (A.67), we see that

$$\text{Im}\,[\tilde{K}_s(t, u)] = \text{Im}\,[\tilde{f}(t)\tilde{K}_D(t - u)\tilde{f}^*(u)] = 0 \tag{67}$$

must be satisfied in order for the quadrature components to be statistically independent. The restriction in (67) would severely limit the class of targets and signals that we could study [e.g., a linear FM signal would violate (67)].

A second observation is that we are almost always dealing with non-stationary signal processes in the current problem. This means that the complex state-variable approach will prove most effective in many problems.

The problem that we have studied is a simple binary problem. All of the results can be extended to the bandpass version of the general binary problem of Chapter 3. Some of these extensions are carried out in the problems at the end of this chapter.

We next consider the problem of digital communication over Doppler-spread channels. This application illustrates the use of the formulas in this section in the context of an important problem.

## 11.3 COMMUNICATION OVER DOPPLER-SPREAD CHANNELS

In this section we consider the problem of digital communication over a Doppler-spread channel. In the first three subsections, we consider binary systems. In Section 11.3.1 we derive the optimum receiver for a binary system and evaluate its performance. In Section 11.3.2 we derive a bound on the performance of any binary system, and in Section 11.3.3 we study suboptimum receivers. In Section 11.3.4 we consider $M$-ary systems, and in Section 11.3.5 we summarize our results.

### 11.3.1 Binary Communications Systems: Optimum Receiver and Performance

We consider a binary system in which the transmitted signals on the two hypotheses are

$$\begin{aligned}&\sqrt{2E_t}\,\mathrm{Re}\,[\tilde{f}(t)e^{j\omega_0 t}]:H_0\\&\sqrt{2E_t}\,\mathrm{Re}\,[\tilde{f}(t)e^{j\omega_1 t}]:H_1.\end{aligned} \tag{68}$$

We assume that $\omega_1 - \omega_0$ is large enough so that the output signal processes on the two hypotheses are in disjoint frequency bands. The received waveforms are

$$r(t) = \begin{cases}\sqrt{2E_t}\,\mathrm{Re}\,[\tilde{b}(t)\tilde{f}(t)e^{j\omega_0 t}] + w(t), & T_i \le t \le T_f:H_0,\\ \sqrt{2E_t}\,\mathrm{Re}\,[\tilde{b}(t)\tilde{f}(t)e^{j\omega_1 t}] + w(t), & T_i \le t \le T_f.:H_1.\end{cases} \tag{69}$$

The hypotheses are equally likely, and the criterion is minimum probability of error. The optimum receiver consists of two parallel branches centered at $\omega_1$ and $\omega_0$. The first branch computes

$$l_1 = \frac{1}{N_0}\iint_{T_i}^{T_f} \tilde{r}^*(t)\tilde{h}(t, u)\tilde{r}(u)\,dt\,du, \tag{70}$$

where the complex envelopes are referenced to $\omega_1$. The second branch computes

$$l_0 = \frac{1}{N_0}\iint_{T_i}^{T_f} \tilde{r}^*(t)\tilde{h}(t, u)\tilde{r}(u)\,dt\,du, \tag{71}$$

where the complex envelopes are referenced to $\omega_0$. In both cases $\tilde{h}(t, u)$ is specified by

$$N_0\tilde{h}(t, u) + \int_{T_i}^{T_f} \tilde{h}(t, z)\tilde{K}_{\tilde{s}}(z, u)\,dz = \tilde{K}_{\tilde{s}}(t, u), \qquad T_i \le t, u \le T_f \tag{72}$$

where

$$\tilde{K}_{\tilde{s}}(t, u) = E_t\tilde{f}(t - \lambda)\tilde{K}_D(t - u)\tilde{f}^*(u - \lambda), \qquad T_i \le t, u \le T_f. \tag{73}$$

The receiver performs the test

$$l_1 \underset{H_0}{\overset{H_1}{\gtrless}} l_0. \tag{74}$$

The receiver configuration is shown in Fig. 11.12. Notice that each branch is just the simple binary receiver of Fig. 11.7. This simple structure arises because the signal processes on the two hypotheses are in disjoint frequency bands.

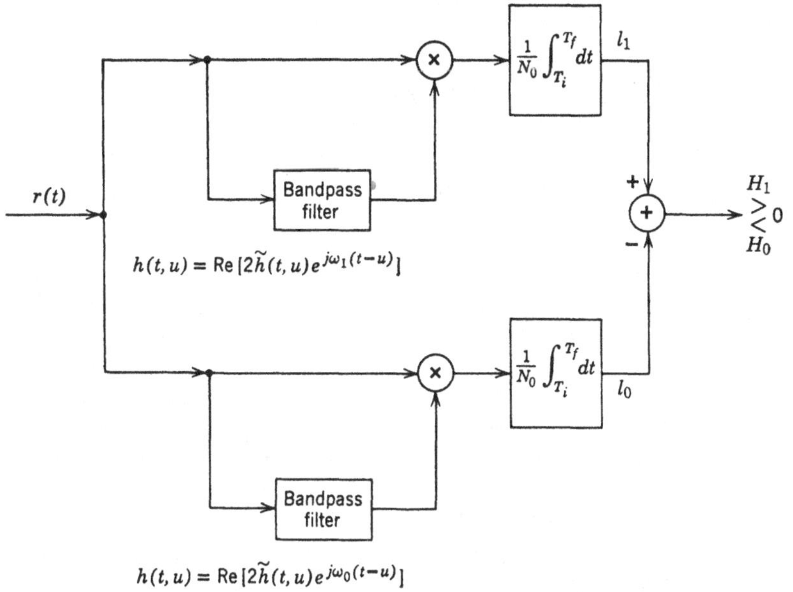


**Fig. 11.12 Optimum receiver: binary FSK system operating over a Doppler-spread channel (Canonical Realization No. 1).**

An alternative configuration is obtained by factoring $\tilde{h}(t, u)$ as indicated in (36). This configuration is shown in Fig. 11.13.

Because this is a binary symmetric bandpass problem,† we may use the bounds on the probability of error that we derived in (3.111).

$$\frac{e^{\tilde{\mu}_{BS}(1/2)}}{2[1 + ((\pi/8)\ddot{\tilde{\mu}}_{BS}(\frac{1}{2}))^{1/2}]} \le \Pr(\epsilon) \le \frac{e^{\tilde{\mu}_{BS}(1/2)}}{2[1 + ((\frac{1}{8})\ddot{\tilde{\mu}}_{BS}(\frac{1}{2}))^{1/2}]} \le \frac{e^{\tilde{\mu}_{BS}(1/2)}}{2}, \tag{75}$$

where

$$\begin{aligned} \tilde{\mu}_{BS}(s) &= \tilde{\mu}_{SIB}(s) + \tilde{\mu}_{SIB}(1 - s) \\ &= \sum_{i=1}^{\infty}\left[\ln\left(1 + \frac{\tilde{\lambda}_i}{N_0}\right) - \ln\left(1 + \frac{s\tilde{\lambda}_i}{N_0}\right) - \ln\left(1 + \frac{(1-s)\tilde{\lambda}_i}{N_0}\right)\right] \end{aligned} \tag{76}$$

and

$$\tilde{\mu}_{BS}(\tfrac{1}{2}) = \sum_{i=1}^{\infty}\left[\ln\left(1 + \frac{\tilde{\lambda}_i}{N_0}\right) - 2\ln\left(1 + \frac{\tilde{\lambda}_i}{2N_0}\right)\right]. \tag{77}$$

The $\tilde{\lambda}_i$ are the eigenvalues of the output signal process $\tilde{s}(t)$, whose covariance is given by (73). We can also write $\tilde{\mu}_{BS}(s)$ in various closed-form expressions such as (54) and (55).

† Notice that "binary symmetric" refers to the hypotheses. The processes are not necessarily symmetric about their respective carriers.

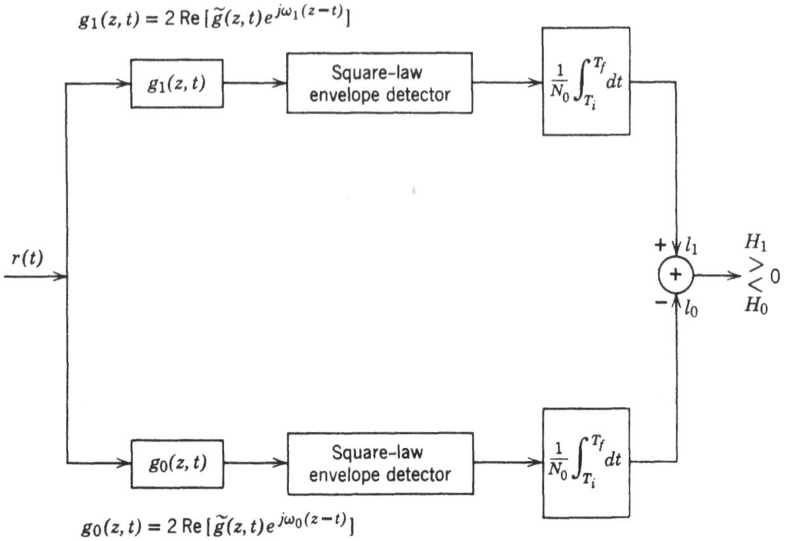


**Fig. 11.13 Optimum receiver for binary communication over a Doppler-spread channel: filter squarer-integrator realization.**

There are three questions of interest with respect to the binary communication problem:

1. What is the performance of the optimum system when the signal $\sqrt{E_t}\tilde{f}(t)$, the channel covariance function $\tilde{K}_D(\tau)$, and the noise level $N_0$ are fixed?
2. If we use a suboptimum receiver, how does its performance compare with that of the optimum receiver for a particular $\tilde{f}(t)$, $E_t$, $\tilde{K}_D(\tau)$, and $N_0$?
3. If the channel covariance function $\tilde{K}_D(\tau)$, the noise level $N_0$, and the transmitted energy $E_t$ are fixed, how can we choose $\tilde{f}(t)$ to minimize the probability of error?

We can answer the first question for a large class of problems by evaluating $\tilde{\mu}_{BS}(s)$ and using (75). Specific solution techniques were discussed on pages 35–44. We can answer the second question by using the techniques of Section 5.1.2. We shall discuss this question in Section 11.3.3. We now consider the third question.

### 11.3.2 Performance Bounds for Optimized Binary Systems

We assume that the channel covariance function $\tilde{K}_D(\tau)$, the noise level $N_0$, and the transmitted energy $E_t$ are fixed. We would like to choose $\tilde{f}(t)$ to minimize the probability of error. In practice it is much simpler to minimize $\tilde{\mu}_{BS}(\frac{1}{2})$. This minimizes the exponent in the bound in (75). Our procedure consists of two steps:

1. We consider the covariance function of the output signal process $\tilde{K}_{\tilde{s}}(t, u)$ and its associated eigenvalues $\tilde{\lambda}_i$. We find the set of $\tilde{\lambda}_i$ that will minimize $\tilde{\mu}_{BS}(\frac{1}{2})$. In this step we do not consider whether a transmitted signal exists that would generate the optimum set of $\tilde{\lambda}_i$ through the relation in (73). The result of this step is a bound on the performance of any binary system.
2. We discuss how to choose $\tilde{f}(t)$ to obtain performance that is close to the bound derived in the first step.

We first observe that a constraint on the input energy implies a constraint on the expected value of the output energy. From (73), the expected value of the total received signal energy is

$$E[|\tilde{s}(t)|^2] = \int_{T_i}^{T_f} \tilde{K}_{\tilde{s}}(t, t)\, dt = 2E_t\sigma_b^2 = \bar{E}_r. \tag{78a}$$

(Recall that

$$\int_{T_i}^{T_f} |\tilde{f}(t)|^2\, dt = 1.) \tag{78b}$$

Notice that this constraint is independent of the signal shape. In terms of eigenvalues of the output process, the constraint is

$$\sum_{i=1}^{\infty} \tilde{\lambda}_i = \bar{E}_r. \tag{79}$$

We now choose the $\tilde{\lambda}_i$, subject to the constraint in (79), to minimize $\tilde{\mu}_{BS}(\frac{1}{2})$. Notice that it is not clear that we can find an $\tilde{f}(t)$ that can generate a particular set of $\tilde{\lambda}_i$.

We first define normalized eigenvalues,

$$\tilde{\lambda}_{in} \triangleq \frac{\tilde{\lambda}_i}{\bar{E}_r}. \tag{80}$$

We rewrite (77) as†

$$\tilde{\mu}_{BS}(\tfrac{1}{2}) = \frac{-\bar{E}_r}{N_0}\left\{\left(\sum_{i=1}^{\infty} \tilde{\lambda}_{in}\tilde{g}(\tilde{\lambda}_{in})\right)\right\}, \tag{81}$$

where

$$\tilde{g}(x) \triangleq \frac{2}{\gamma x}\left[-\tfrac{1}{2}\ln(1+\gamma x) + \ln\left(1+\frac{\gamma x}{2}\right)\right] \tag{82}$$

and

$$\gamma \triangleq \frac{\bar{E}_r}{N_0}. \tag{83}$$

† This derivation is due to Kennedy [3].

We refer to the term in braces in (81) as the *efficiency factor*. (Recall the discussion on page 118.) The function $\tilde{g}(x)$ is plotted in Fig. 11.14. We see that $\tilde{g}(x)$ is a positive function whose unique maximum occurs at $x = \hat{x}$, which is the solution to

$$\frac{\gamma^2\hat{x}/4}{(1+\gamma\hat{x})(1+\gamma\hat{x}/2)} = \frac{1}{\hat{x}}\left[-\tfrac{1}{2}\ln(1+\gamma\hat{x}) + \ln\left(1+\frac{\gamma\hat{x}}{2}\right)\right]. \tag{84}$$

The solution is $\gamma\hat{x} = 3.07$. For all positive $\gamma$,

$$\hat{x} = \frac{3.07}{\gamma}. \tag{85}$$

We can use the result in (84) and (85) to bound $\tilde{\mu}_{BS}(\frac{1}{2})$. From (81),

$$-\tilde{\mu}_{BS}(\tfrac{1}{2}) = \frac{\bar{E}_r}{N_0}\sum_{i=1}^{\infty}\tilde{\lambda}_{in}\tilde{g}(\tilde{\lambda}_{in}) \le \frac{\bar{E}_r}{N_0}\sum_{i=1}^{\infty}\lambda_{in}\tilde{g}(\hat{x}). \tag{86}$$

Using (79) and (80), (86) reduces to

$$-\tilde{\mu}_{BS}(\tfrac{1}{2}) \le \frac{\bar{E}_r}{N_0}\,\tilde{g}(\hat{x}). \tag{87}$$

Thus we have a bound on how negative we can make $\tilde{\mu}_{BS}(\frac{1}{2})$. We can achieve this bound exactly by letting

$$\tilde{\lambda}_{in} = \begin{cases} \hat{x} = \dfrac{3.07}{\gamma}, & i = 1, 2, \dots, D_o, \\ 0, & i > D_o, \end{cases} \tag{88†}$$

where

$$D_o = \frac{\gamma}{3.07} = \frac{\bar{E}_r/N_0}{3.07}. \tag{89}$$

This result says that we should choose the first $D_o$ normalized eigenvalues to be equal to $3.07/\gamma$ and choose the others to be zero. Using (88) in (82) and the result in (87) gives

$$\tilde{\mu}_{BS}(\tfrac{1}{2}) \ge -0.1488\left(\frac{\bar{E}_r}{N_0}\right). \tag{90}$$

Substituting (90) into (75) gives an upper bound of the probability of error as

$$\boxed{\Pr(\epsilon) \le \tfrac{1}{2}\exp\left(-0.1488\,\frac{\bar{E}_r}{N_0}\right).} \tag{91}$$

† This result assumes that $\bar{E}_r/N_0$ is a integer multiplier of 3.07. If this is not true, (86) is still a bound and the actual performance is slightly worse.

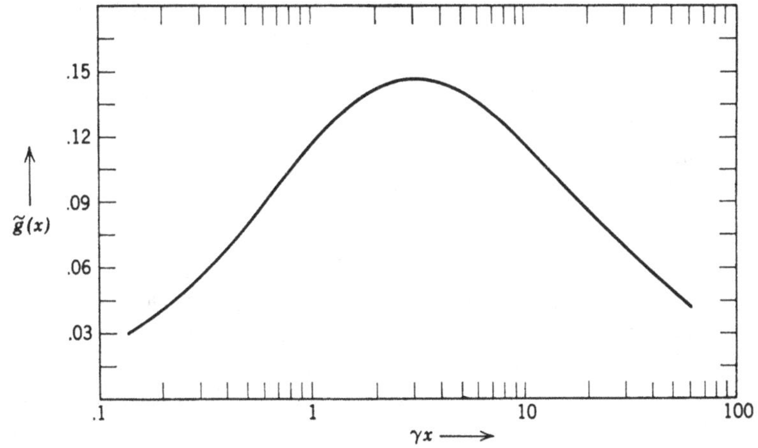


**Fig. 11.14 Plot of $\tilde{g}(x)$ versus $\gamma x$ (from[3]).**

We have encountered this type of result in some of earlier examples. In Section 4.2.3, we studied a frequency diversity system in which the received energies in all channels were required to be equal. We found that we could minimize $\mu_{BS,BP,SK}(\frac{1}{2})$ by dividing the available energy among the channels so that

$$\frac{\bar{E}_{r1}}{N_0} = 3.07 \tag{92}$$

(see 4.116). In that case we achieved the optimum performance by an *explicit diversity* system.

In the present case there is only one channel. In order to achieve the optimum performance, we must transmit a signal so that the covariance function

$$\tilde{K}_{\tilde{s}}(t, u) = \tilde{f}(t)\tilde{K}_D(t - u)\tilde{f}^*(u) \tag{93}$$

has $D_o$ equal eigenvalues. We can think of this as an *implicit diversity* system.

The result in (91) is quite important, because it gives us a performance bound that is independent of the shape of the Doppler scattering functions. It provides a standard against which we can compare any particular signaling scheme. Once again we should emphasize that there is no guarantee that we can achieve this bound for all channel-scattering functions.

The next step in our development is to consider some specific Doppler scattering functions and see whether we can design a signal that achieves the bound in (90) with equality. We first review two examples that we studied in Section 4.1.2.

**Example 1.** This example is identical with Example 4 on page 116. The channel-scattering function is

$$\tilde{S}_D\{f\} = \begin{cases} \dfrac{\sigma_b^{\,2}}{B}, & |f| \le B, \\ 0, & |f| > B. \end{cases} \tag{94}$$

The transmitted signal is

$$\tilde{f}(t) = \begin{cases} \dfrac{1}{\sqrt{T}}, & 0 \le t \le T, \\ 0, & \text{elsewhere}. \end{cases} \tag{95}$$

We assume that $BT$ is large enough that we may use the SPLOT formulas. We constrain $\bar{E}_r$ and choose $T$ to minimize $\tilde{\mu}_{BS,\infty}(\frac{1}{2})$. From (4.76), the optimum value is $T_o$, which is specified by

$$\frac{\bar{E}_r/N_0}{2BT_o} = 3.07. \tag{96}$$

Then

$$[\tilde{\mu}_{\text{BS},\infty}(\tfrac{1}{2})]_{\text{opt}} = -0.1488\left(\frac{\bar{E}_r}{N_0}\right), \tag{97}$$

and we achieve the bound with equality. The result in (96) assumes the SPLOT condition If we require

$$BT_o \ge 5 \tag{98}$$

to assure the validity of the SPLOT assumption, the result in (96) requires that

$$\frac{\bar{E}_r}{N_0} \ge 30.7. \tag{99}$$

When (99) is satisfied, the optimum binary system for a channel whose scattering function is given by (94) is one that transmits a rectangular pulse of duration $T_o$. [Notice that the condition in (99) is conservative.]

**Example 2.** This example is identical with Example 3 on page 111. The channel-scattering function is

$$\tilde{S}_D\{f\} = \frac{4k\sigma_b^{\,2}}{(2\pi f)^2 + k^2}, \tag{100}$$

and the transmitted signal is given in (95). Previously, we used the SPLOT assumption to evaluate $\tilde{\mu}_{\text{BS}}(s)$. In this example we use complex state variables. The channel state equations are

$$\tilde{F}(t) = -k, \tag{101}$$

$$\tilde{G}(t) = \tilde{C}(t) = 1, \tag{102}$$

$$\tilde{Q} = 4k\sigma_b^{\,2}, \tag{103}$$

$$\tilde{\mathbf{P}}_0 = 2\sigma_b^{\,2}. \tag{104}$$

We evaluate $\tilde{\mu}_{\text{BS}}(\frac{1}{2})$ by using (48*a*), (48*b*), (54), and (76) and then minimize over $T$. The result is shown in Fig. 11.15. We see that if $\bar{E}_r/N_0$ is small, we can transmit a very short pulse such that

$$kT_o \simeq 0. \tag{105}$$

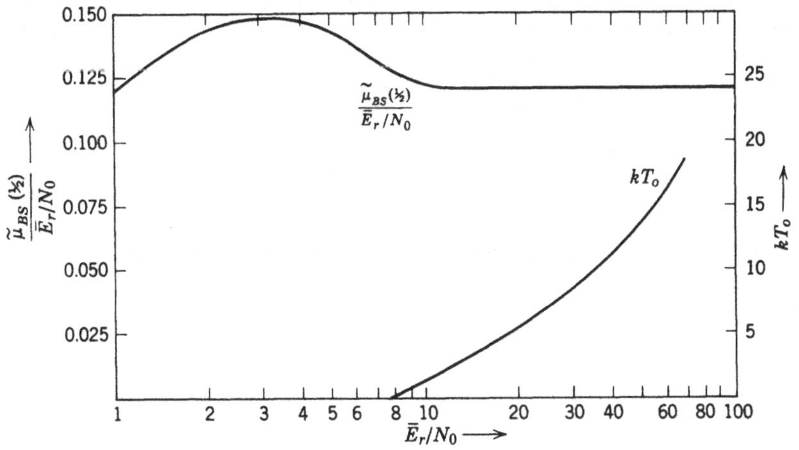


**Fig. 11.15 Optimum error exponent and $kT_o$ product as a function of $\bar{E}_r/N_0$ (first-order fading, rectangular pulse (from [4]).**

A short pulse causes a single eigenvalue at the channel output, because the channel does not fluctuate during the signaling interval. (This is the model in Chapter 9.) When

$$\frac{\bar{E}_r}{N_0} = 3.07, \tag{106}$$

the condition in (89) is satisfied and the bound in (90) is achieved. As long as

$$\frac{\bar{E}_r}{N_0} \leq 7.8, \tag{107}$$

a single eigenvalue is still optimum, but the system only achieves the bound in (90) when (106) is satisfied. As the available $\bar{E}_r/N_0$ increases, the optimum $kT$ product increases. For

$$\frac{\bar{E}_r}{N_0} > 13, \tag{108}$$

the results coincide with the SPLOT results of Example 3 on page 111:

$$kT_o \simeq \frac{\bar{E}_r/N_0}{3.44} \tag{109}$$

and

$$\tilde{\mu}_{BS}(\tfrac{1}{2}) = -0.118\frac{\bar{E}_r}{N_0}. \tag{110}$$

The result in (110) indicates that a rectangular pulse cannot generate the equal eigenvalue distribution required to satisfy the bound.

In the next example we consider a more complicated signal in an effort to reach the bound in (90) for a channel whose scattering function is given in (100).

**Example 3.** The channel-scattering function is given in (100). To motivate our signal choice, we recall that a short pulse generates a single eigenvalue. By transmitting a sequence of pulses whose time separation is much greater than the channel correlation time, we can obtain the desired number of equal eigenvalues at the output.

The signal of interest is shown in Fig. 11.16. It consists of a train of rectangular pulses with width $T_s$ and interpulse spacing $T_p$. The number of pulses is

$$n \triangleq D_o = \frac{\bar{E}_r/N_0}{3.07}. \tag{111}$$

The height of each pulse is chosen so that the *average received energy per pulse* is

$$\bar{E}_{ri} = 3.07N_0, \qquad i = 1, 2, \ldots, D_o. \tag{112}$$

We can write

$$\tilde{f}(t) = \sum_{i=1}^{D_o} c\ \tilde{u}(t - iT_p), \tag{113a}$$

where

$$\tilde{u}(t) = \begin{cases} \dfrac{1}{\sqrt{T_s}}, & 0 \leq t \leq T, \\ 0, & \text{elsewhere,} \end{cases} \tag{113b}$$

and $c$ normalizes $\tilde{f}(t)$ to have unit energy. The covariance function of the output signal process is

$$\tilde{K}_s(t, \tau) = \sum_{i=1}^{D_o} \sum_{k=1}^{D_o} c^2 \tilde{u}(t - iT_p)\tilde{K}_D(t - \tau)\tilde{u}^*(\tau - kT_p). \tag{114}$$

We can evaluate the performance for any particular $T_s$ and $T_p$. The case of current interest is obtained by letting

$$T_s \to 0 \tag{115}$$

and

$$T_p \to \infty. \tag{116}$$

In this limit the covariance function becomes the separable function

$$\tilde{K}_{\tilde{s}}(t, \tau) = \sum_{i=1}^{D_o} c^2 \tilde{K}_D(0)\tilde{u}(t - iT_p)\tilde{u}^*(\tau - kT_p). \tag{117}$$

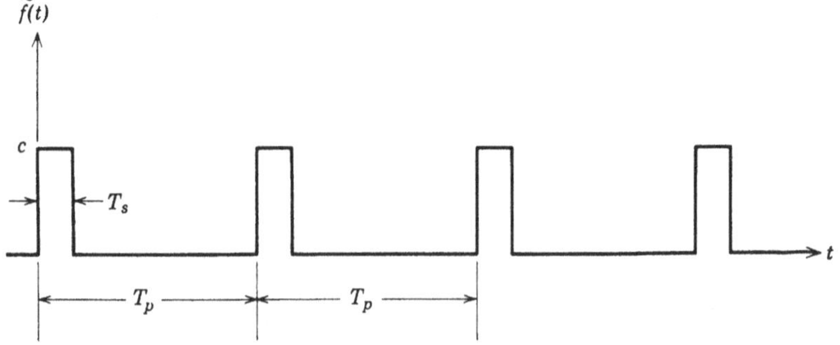


**Fig. 11.16 Transmitted signal in Example 3.**

We now have $D_0$ equal eigenvalues whose magnitudes satisfy (88). Therefore the performance satisfies the upper bound in (90).

The limiting case is not practical, but we can frequently obtain a good approximation to it. We need to make $T_s$ appreciably shorter than the correlation time of $\tilde{b}_D(t)$. This will make the amplitude of each returned pulse approximately constant. We need to make $T_p$ appreciably longer than the correlation time of $\tilde{b}_D(t)$, so that the amplitude of different pulses will be statistically independent. The result approximates an optimum-diversity system. Notice that the optimum receiver reduces to two branches like that in Fig. 4.16 in the limiting case. There are no integral equations to solve.

There may be constraints that make it impossible to use this solution:

1. If there is a peak power limitation, we may not be able to get enough energy in each pulse.
2. If there is a bandwidth limitation, we may not be able to make $T_s$ short enough to get a constant amplitude on each received pulse.
3. If there is a time restriction on the signaling interval, we may not be able to make $T_p$ long enough to get statistically independent amplitudes.

These issues are investigated in Problems 11.3.6 and 11.3.7. If any of the above constraints makes it impossible to achieve the bound with this type of signal, we can return to the signal design problem and try a different strategy.

Before leaving this example, we should point out that a digital system using the signal in Fig. 11.16 would probably work in a time-division multiplex mode (see Section II-9.11) and interleave signals from other message sources in the space between pulses.

We should also observe that the result does not depend on the detailed shape of $\tilde{S}_D\{f\}$.

In this section we have studied the problem of digital communication over a channel that exhibits time-selective fading by using binary orthogonal signals. The basic receiver derivation and performance analysis were straightforward extensions of the results in Section 11.2.

The first important result of the section was the bound in (90). For any scattering function,

$$\tilde{\mu}_{BS}(\tfrac{1}{2}) \geq -0.1488\left(\frac{\bar{E}_r}{N_0}\right). \tag{118}$$

In order to achieve this bound, the transmitted signal must generate a certain number of equal eigenvalues in the output signal process.

The second result of interest was the demonstration that we could essentially achieve the bound for various channel-scattering functions by the use of simple signals.

There are two topics remaining to complete our digital communication discussion. In Section 11.3.3, we study the design of suboptimum receivers. In Section 11.3.4, we discuss $M$-ary communication briefly.

### 11.3.3 Suboptimum Receivers

For a large number of physical situations we can find the optimum receiver and evaluate its performance. Frequently, the optimum receiver is complicated to implement, and we wish to study suboptimum receiver designs. In this section we develop two logical suboptimum receiver configurations and analyze their performance.

To obtain the first configuration, we consider a typical sample function of $\tilde{b}_D(t)$ as shown in Fig. 11.17*a*. For discussion purposes we assume that $\tilde{b}_D(t)$ is bandlimited to $\pm B/2$ cps. We could approximate $\tilde{b}_D(t)$ by the piecewise constant function shown in Fig. 11.17*b*. In this approximation we have used segments equal to the reciprocal of the bandwidth. A more general approximation is shown in Fig. 11.17*c*. Here we have left the

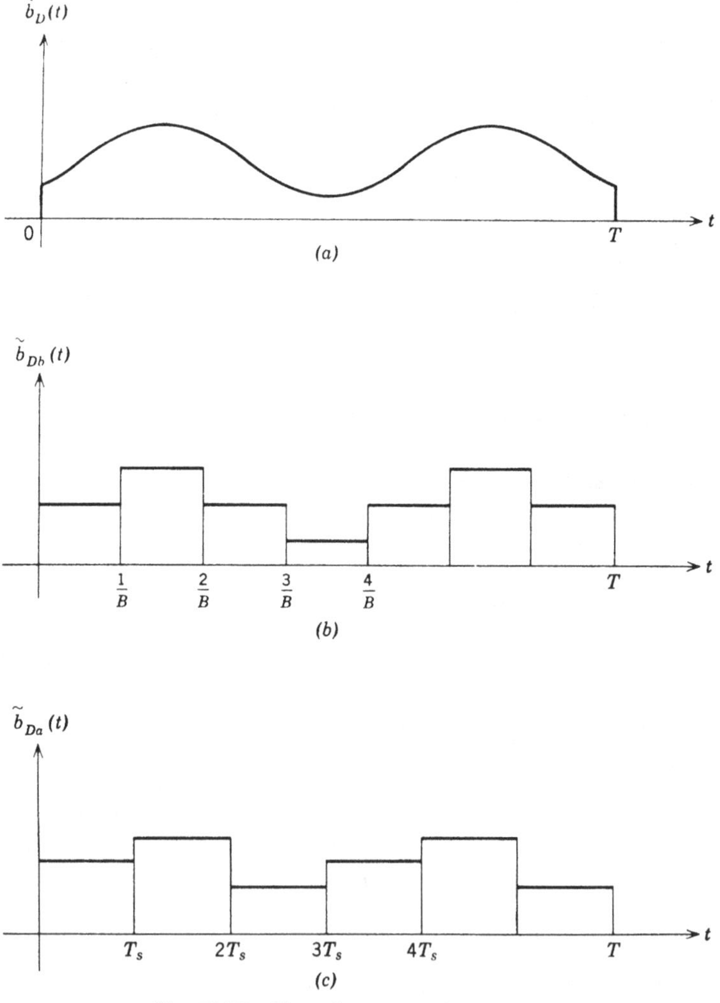


**Fig. 11.17 Channel process and approximations.**

length of the subintervals as a parameter. We would expect $T_s$ to be less than the reciprocal of the fading bandwidth in order for the approximation to be valid.

In order to design the first suboptimum receiver, we assume that the function in Fig. 11.17*c* is exact *and* that the values in each subinterval are statistically independent. Notice that the two assumptions are somewhat contradictory. As $T_s$ decreases, the approximation is more exact, but the values are more statistically dependent. As $T_s$ increases, the opposite behavior occurs. The fact that the assumptions are not valid is the reason why the resulting receiver is suboptimum.

We write

$$\tilde{b}_{Da}(t) = \sum_{i=1}^{N} \tilde{b}_i \tilde{u}(t - iT_s), \tag{119}$$

where $\tilde{u}(t)$ is the unit pulse defined in (113*b*). Using (119) in (20) gives

$$\tilde{K}_{\tilde{s}}(t, v) = \sum_{i=1}^{N} \frac{2\sigma_b^2 E_t}{N} \tilde{f}(t)\tilde{u}(t - iT_s)\tilde{f}^*(v)\tilde{u}^*(v - iT_s). \tag{120}$$

The covariance function in (120) is separable, and so the receiver structure is quite simple.

The branch of the resulting receiver that generates $l_1$ is shown in Fig. 11.18. A similar branch generates $l_0$.† The different weightings arise because

$$E_i = \int_{(i-1)T_s}^{iT_s} |\tilde{f}(t)|^2 \, dt \tag{121}$$

is usually a function of $i$, so that the eigenvalues are unequal. From Problem 11.2.1,

$$g_i = \frac{2\sigma_b^2 E_i}{2\sigma_b^2 E_i + N_0}. \tag{122}$$

The receiver in Fig. 11.18 is easy to understand but is more complicated than necessary. Each path is gating out a $T_s$ segment of $\tilde{r}(t)$ and operating on it. Thus we need only one path if we include a gating operation. This version is shown in Fig. 11.19. A particularly simple version of the receiver arises when $\tilde{f}(t)$ is constant over the entire interval. Then the weightings are unnecessary and we have the configuration in Fig. 11.20. We have replaced the correlation operation with a matched filter to emphasize the interchangeability.

† In Figs. 11.18 to 11.22, we use complex notation to show one branch of various receivers. The complex envelopes in the indicated branch are referenced to $\omega_1$ so that the output is $l_1$. As discussed at the beginning of Section 11.3.1, we compute $l_0$ by using the same complex operations referenced to $\omega_0$.

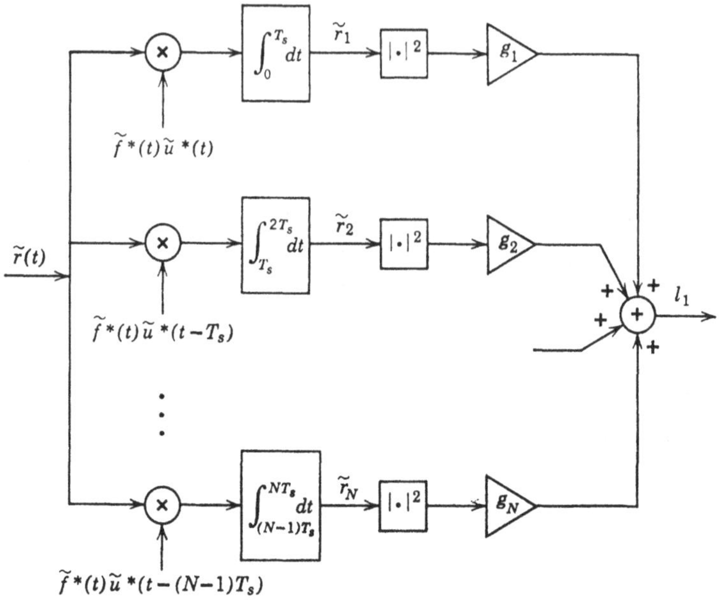


**Fig. 11.18 Suboptimum receiver No. 1 (one branch).**

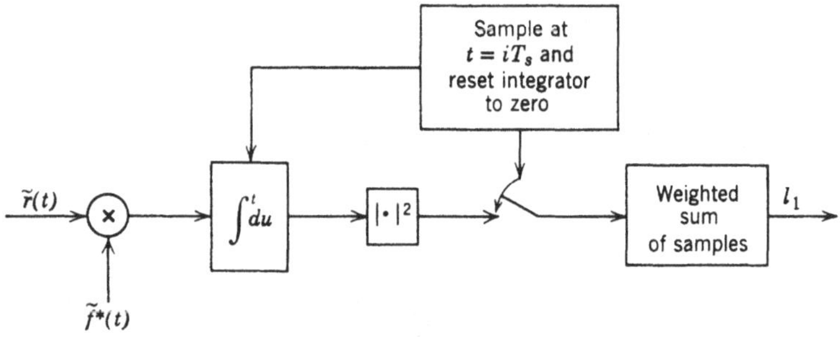


**Fig. 11.19 Alternative version of suboptimum receiver No. 1 (one branch).**

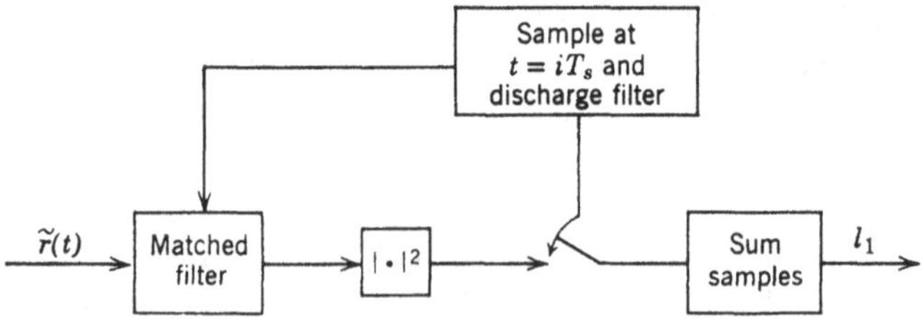


**Fig. 11.20 Suboptimum receiver No. 1 for constant $\tilde{f}(t)$ (one branch).**

This completes our development of our first suboptimum receiver structure. We refer to it as a GFSS (gate-filter-square-sum) receiver. Before analyzing its performance, we develop a second suboptimum receiver configuration.

In our development of the second suboptimum receiver, we restrict our attention to channel processes with finite-dimensional state representations. The second suboptimum receiver configuration is suggested by the optimum receiver that we obtain when both the LEC condition and the long observation time assumption are valid. This receiver is shown in Fig. 11.21. (This is the receiver of Fig. 11.11 redrawn in state-variable notation with $\lambda = 0$.) Notice that the state-variable portion corresponds exactly to the system used to generate $\tilde{b}_D(t)$.

We retain the basic structure in Fig. 11.21. To obtain more design flexibility, we do not require the filter matrices to be capable of generating $\tilde{b}_D(t)$, but we do require them to be time-invariant. The resulting receiver is shown in Fig. 11.22. (This type of receiver was suggested in [4].) The receiver equations are

$$\dot{\tilde{\mathbf{x}}}_r(t) = \tilde{\mathbf{F}}_r\tilde{\mathbf{x}}_r(t) + \tilde{\mathbf{G}}_r\tilde{f}^*(t)\tilde{r}(t), \tag{123}$$

$$\tilde{l}_i(t) = |\tilde{\mathbf{C}}_r\tilde{\mathbf{x}}_r(t)|^2, \tag{124}$$

$$E[\tilde{\mathbf{x}}_r(T_i)\tilde{\mathbf{x}}_r^\dagger(T_i)] = \tilde{\mathbf{P}}_r, \tag{125}$$

and

$$\tilde{\mathbf{x}}_r(T_i) = \mathbf{0}. \tag{126}$$

This specifies the structure of the second suboptimum receiver [we refer to it as an FSI (filter-squarer-integrator) receiver]. We must specify $\tilde{\mathbf{F}}_r$, $\tilde{\mathbf{G}}_r$, $\tilde{\mathbf{C}}_r$, and $\tilde{\mathbf{P}}_r$ to maximize its performance.

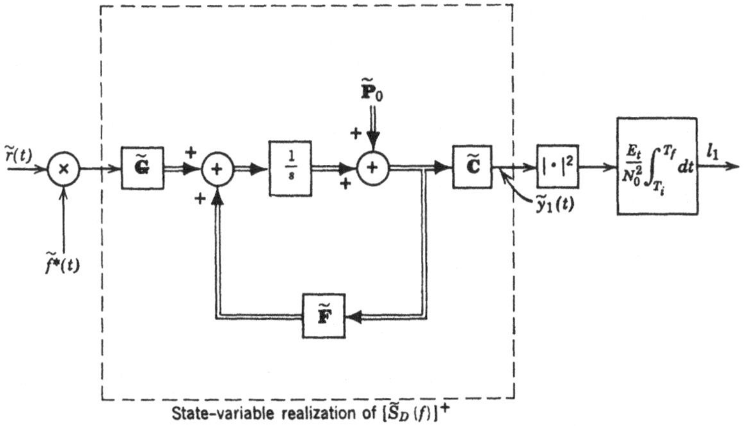


**Fig. 11.21 Optimum LEC receiver: long observation time (one branch).**

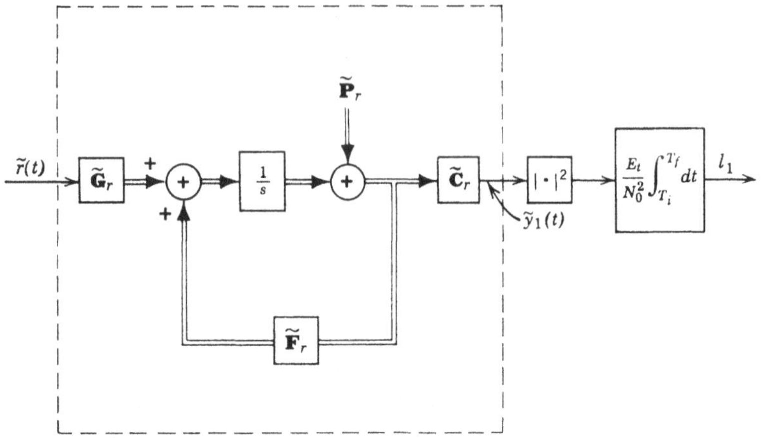


**Fig. 11.22 Suboptimum receiver No. 2 (one branch).**

We now proceed as follows.

1. We derive a bound on the performance of the suboptimum receivers. This is a straightforward extension of the bound in Section 5.1.2 to include complex processes. This bound is valid for both receiver configurations.
2. We develop expressions for the quantities in the bound for the two receivers.
3. We optimize each receiver and compare its performance with that of the optimum receiver.

All the steps are straightforward, but complicated. Many readers will prefer to look at the results in Figs. 11.24 and 11.25 and the accompanying conclusions.

***Performance Bounds for Suboptimum Receivers.*** Because we have a binary symmetric system with orthogonal signals, we need to modify the results of Problem 5.1.16. (These bounds were originally derived in [5].) The result is

$$\Pr(\epsilon) < \tfrac{1}{2}e^{\tilde{\mu}_{BS}(s)}, \tag{127}$$

where

$$\tilde{\mu}_{BS}(s) \triangleq \tilde{\mu}_{11}(s) + \tilde{\mu}_{01}(-s) \tag{128}$$

and

$$\tilde{\mu}_{11}(s) \triangleq \ln E[e^{sl_1} \mid H_1], \tag{129}$$

$$\tilde{\mu}_{01}(s) \triangleq \ln E[e^{sl_0} \mid H_1]. \tag{130}$$

The next step is to evaluate $\tilde{\mu}_{BS}(s)$ for the two receivers.

***Evaluation of*** $\tilde{\mu}_{BS}(s)$ ***for Suboptimum Receiver No. 1.*** In this case, $l_1$ and $l_0$ are finite quadratic forms and the evaluation is straightforward (see Problem 11.3.9; the original result was given in [4]).

$$\tilde{\mu}_{11}(s) = -\ln \det (\mathbf{I} - s\tilde{\mathbf{W}}[\tilde{\mathbf{\Lambda}}_s + \tilde{\mathbf{\Lambda}}_n]) \tag{131}$$

and

$$\tilde{\mu}_{01}(s) = -\ln \det (\mathbf{I} - s\tilde{\mathbf{W}}\tilde{\mathbf{\Lambda}}_n), \tag{132}$$

where

$$\tilde{\mathbf{W}} = \begin{bmatrix} g_1 & & & & 0 \\ & g_2 & & & \\ & & g_3 & & \\ & & & \ddots & \\ 0 & & & & g_N \end{bmatrix}, \tag{133}$$

$$\tilde{\mathbf{\Lambda}}_n = N_0 \begin{bmatrix} E_1 & & & 0 \\ & E_2 & & \\ & & \ddots & \\ 0 & & & E_N \end{bmatrix}, \tag{134}$$

and

$$\tilde{\mathbf{\Lambda}}_{s,ij} \triangleq \int_{(i-1)T_s}^{iT_s} \int_{(j-1)T_s}^{jT_s} |\tilde{f}(t)|^2 \tilde{K}_D(t-u)\, |\tilde{f}(u)|^2 \, dt\, du. \tag{135}$$

Notice that we include the statistical dependence between the various subintervals in the performance analysis. Using (131)–(135) in (127) gives the performance bound for any particular system.

***Evaluation*** **of** $\tilde{\mu}_{BS}(s)$ ***for Suboptimum Receiver No. 2.*** In this case, we can write $\tilde{\mu}_{11}(s)$ and $\tilde{\mu}_{01}(s)$ as Fredholm determinants,

$$\begin{aligned} \tilde{\mu}_{11}(s) &= -\sum_{i=1}^{\infty} \ln (1 - s\tilde{\lambda}_{11,i}) \\ &= \tilde{D}_{\mathscr{F}_{11}}(-s), \qquad s < 0. \end{aligned} \tag{136}$$

Here the $\tilde{\lambda}_{11,i}$ are the ordered eigenvalues of $\tilde{y}_1(t)$ when $H_1$ is true. Notice that $\tilde{y}_i(t)$ is the input to the squarer in the $i$th branch. Similarly,

$$\begin{aligned} \tilde{\mu}_{01}(s) &= -\sum_{i=1}^{\infty} \ln (1 + s\tilde{\lambda}_{01,i}) \\ &= \tilde{D}_{\mathscr{F}_{01}}(s), \qquad -\frac{1}{\tilde{\lambda}_{01,i}} < s \le 0, \end{aligned} \tag{137}$$

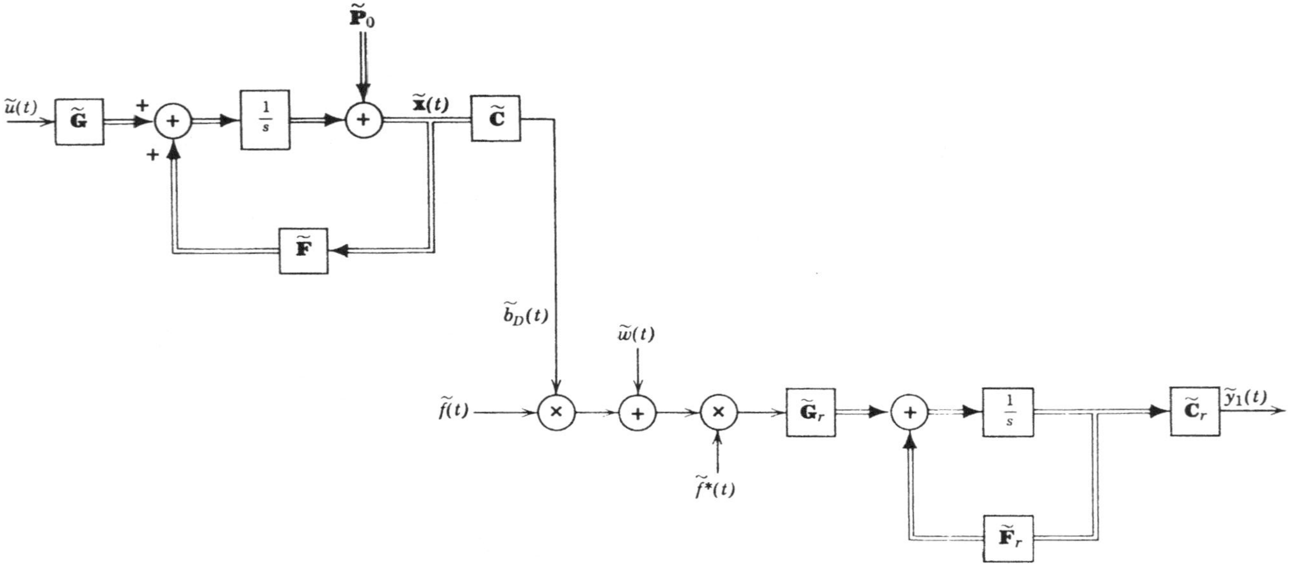


**Fig. 11.23 Generation of $\tilde{y}_1(t)$.**

where the $\lambda_{01,i}$ are the ordered eigenvalues of $\tilde{y}_0(t)$ when $H_1$ is true. We now illustrate how to evaluate $\tilde{D}_{\mathscr{F}_{11}}(s)$ using state-variable techniques.

***Evaluation of*** $D_{\mathscr{F}_{11}}(-s)$. To do this we must write $\tilde{y}_1(t)$ given $H_1$ as the output of a linear dynamic sytesm excited by a white noise process. We assume the relations

$$\dot{\tilde{\mathbf{x}}}_c(t) = \tilde{\mathbf{F}}_c(t)\tilde{\mathbf{x}}_c(t) + \tilde{\mathbf{G}}_c(t)\tilde{\mathbf{u}}_c(t), \qquad t \geq T_i, \tag{138}$$

$$\tilde{y}_1(t) = \tilde{\mathbf{C}}_c(t)\tilde{\mathbf{x}}_c(t), \tag{139}$$

$$E[\tilde{\mathbf{u}}_c(t)\tilde{\mathbf{u}}_c^{\dagger}(\tau)] = \tilde{\mathbf{Q}}_c, \tag{140}$$

$$E[\tilde{\mathbf{x}}_c(T_i)\tilde{\mathbf{x}}_c^{\dagger}(T_i)] = \tilde{\mathbf{P}}_c. \tag{141}$$

We must specify $\tilde{\mathbf{F}}_c(t)$, $\tilde{\mathbf{G}}_c(t)$, $\tilde{\mathbf{Q}}_c$, $\tilde{\mathbf{C}}_c(t)$, and $\tilde{\mathbf{P}}_c$.

On $H_1$, $\tilde{y}_i(t)$ is generated as shown in Fig. 11.23. We must express this system in the form of (127)–(130). We do this by adjoining the state vectors $\tilde{\mathbf{x}}(t)$ and $\tilde{\mathbf{x}}_r(t)$ to obtain

$$\tilde{\mathbf{x}}_c(t) = \left[\begin{array}{c} \tilde{\mathbf{x}}(t) \\ \hline \tilde{\mathbf{x}}_r(t) \end{array}\right]. \tag{142}$$

The resulting system matrices are

$$\tilde{\mathbf{F}}_c(t) = \left[\begin{array}{c|c} \tilde{\mathbf{F}} & \mathbf{0} \\ \hline \tilde{f}^*(t)\tilde{\mathbf{G}}_r\tilde{\mathbf{C}}_r & \tilde{\mathbf{F}}_r \end{array}\right], \tag{143}$$

$$\tilde{\mathbf{G}}_c(t) = \left[\begin{array}{c|c} \tilde{\mathbf{G}} & \mathbf{0} \\ \hline \mathbf{0} & \tilde{\mathbf{G}}_r \end{array}\right], \tag{144}$$

$$\tilde{\mathbf{C}}_c(t) = [\mathbf{0} \;\vdots\; \tilde{\mathbf{C}}_r], \tag{145}$$

$$\tilde{\mathbf{Q}}_c = \left[\begin{array}{c|c} \tilde{Q} & 0 \\ \hline 0 & N_0 \end{array}\right], \tag{146}$$

and

$$\tilde{\mathbf{P}}_c = \left[\begin{array}{c|c} \tilde{\mathbf{P}} & \mathbf{0} \\ \hline \mathbf{0} & \tilde{\mathbf{P}}_r \end{array}\right]. \tag{147}$$

Once we have represented $\tilde{y}_1(t)$ in this manner, we know that

$$\tilde{D}_{\mathscr{F}_{11}}(-s) = \ln \det \tilde{\mathbf{\Gamma}}_2(T_f) + \operatorname{Re} \int_{T_i}^{T_f} \operatorname{Tr}\, [\tilde{\mathbf{F}}_c(t)]\, dt, \tag{148}$$

where

$$\frac{d}{dt}\left[\begin{array}{c} \tilde{\mathbf{\Gamma}}_1(t) \\ \hline \tilde{\mathbf{\Gamma}}_2(t) \end{array}\right] = \left[\begin{array}{c|c} \tilde{\mathbf{F}}_c(t) & \tilde{\mathbf{G}}_c(t)\tilde{\mathbf{Q}}_c\tilde{\mathbf{G}}_c^{\dagger}(t) \\ \hline -s\tilde{\mathbf{C}}_c^{\dagger}(t)\tilde{\mathbf{C}}_c(t) & -\tilde{\mathbf{F}}_c^{\dagger}(t) \end{array}\right]\left[\begin{array}{c} \tilde{\mathbf{\Gamma}}_1(t) \\ \hline \tilde{\mathbf{\Gamma}}_2(t) \end{array}\right], \tag{149}$$

and

$$\tilde{\mathbf{\Gamma}}_1(T_i) = \tilde{\mathbf{P}}_c, \tag{150}$$

$$\tilde{\mathbf{\Gamma}}_2(T_i) = \mathbf{I} \tag{151}$$

(see pages 42–44).

The results in (143)–(151) completely specify the first Fredholm determinant. We can carry out the actual evaluation numerically. The second Fredholm determinant can be calculated in a similar manner. Thus, we have formulated the problem so that we can investigate any set of filter matrices.

**Example** [4]. We consider a first-order Butterworth fading spectrum and a transmitted signal with a constant envelope. The scattering function is

$$\tilde{S}_D\{f\} = \frac{4k\sigma_b^{\,2}}{(2\pi f)^2 + k^2} \tag{152}$$

and

$$\tilde{f}(t) = \sqrt{\frac{1}{T}} \qquad 0 \le t \le T. \tag{153}$$

The average received energy in the signal component is

$$\bar{E}_r = 2\sigma_b^{\,2} E_t. \tag{154}$$

To evaluate the performance of Receiver No. 1, we calculate $\tilde{\mu}_{BS}(s)$ by using (128) and (131)–(135). We then minimize over $s$ to obtain the tightest bound in (127). Finally we minimize over $T_s$, the subinterval length, to obtain the best suboptimum receiver. The result is a function

$$\min_{T_s} \left[\min_{s} (\tilde{\mu}_{BS}(s))\right], \tag{155}$$

which is a measure of performance for Receiver No. 1.

In Receiver No. 2 we use a first-order filter. Thus,

$$\dot{\tilde{y}}(t) = -k_r \tilde{y}(t) + \tilde{f}^*(t) r(t). \tag{156}$$

We also assume that

$$\tilde{\mathbf{P}}_r = \mathbf{0} \tag{157}$$

for simplicity.

We evaluate $\tilde{\mu}_{BS}(s)$ as a function of $k_rT$. For each value of $k_rT$ we find

$$\min_{s} [\tilde{\mu}_{BS}(s)] \tag{158}$$

to use in the exponent of (127). We then choose the value of $k_rT$ that minimizes (158). The resulting value of

$$\min_{k_rT} \min_{s} [\tilde{\mu}_{BS}(s)] \tag{159}$$

is a measure of performance of Receiver No. 2. In Figs. 11.24 and 11.25, we have plotted the quantities in (155) and (159) for the cases in which $\bar{E}_r/N_0$ equals 5 and 20, respectively. We also show $\tilde{\mu}_{BS}(\frac{1}{2})$ for the optimum receiver. The horizontal axis is $kT$, and the number in parentheses on the Receiver No. 1 curve is $T/T_s$, the number of subintervals used. In both cases, the performance of Receiver No. 1 approaches that of

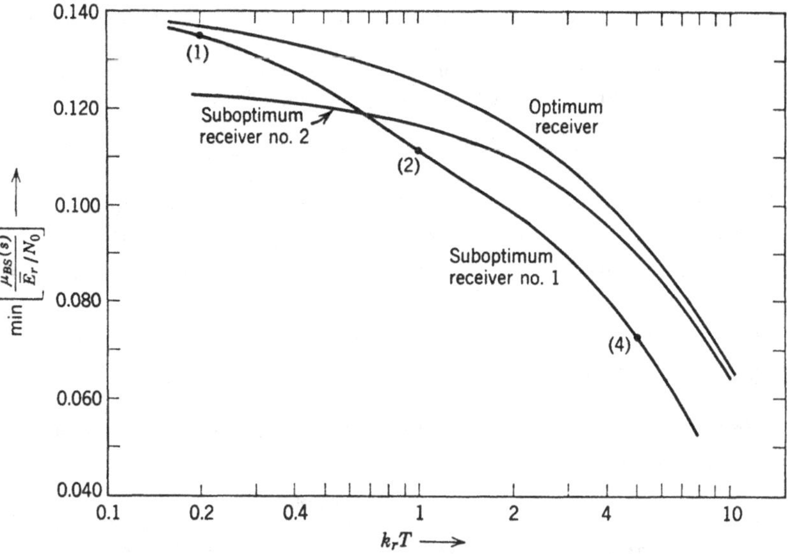


**Fig. 11.24 Normalized error-bound exponents for optimum and suboptimum receivers: Doppler-spread channel with first-order fading, $\bar{E}_r/N_0 = 5$. (from [4].)**

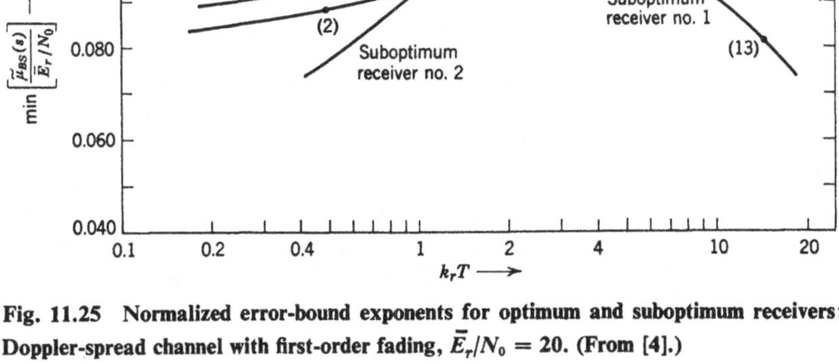

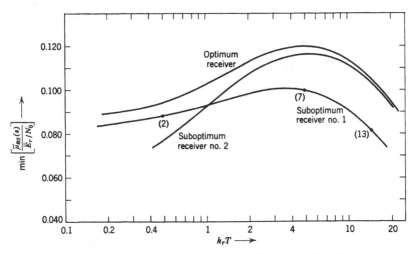

**Fig. 11.25 Normalized error-bound exponents for optimum and suboptimum receivers: Doppler-spread channel with first-order fading, $\bar{E}_r/N_0 = 20$. (From [4].)**

the optimum receiver as $kT$ approaches zero, and the performance of Receiver No. 2 approaches that of the optimum receiver as $kT$ becomes large. This behavior is just what we would expect. We also see that one of the receivers is within .01 of the optimum over the entire range of $kT$. Thus, for this particular example, the simplicity afforded by the suboptimum receivers is probably worth the slight decrease in performance.

We should comment that the above example is not adequate to verify that these suboptimum receivers will be satisfactory in all cases. A more severe test of the suboptimum receiver structure would require a non-constant signal. Our problem formulation allows us to carry this analysis out for any desired $\tilde{f}(t)$. Other references that deal with suboptimum receiver analysis include [21] and [22].

### 11.3.4 *M*-ary Systems

We consider an $M$-ary system in which the transmitted signal on the $i$th hypothesis is

$$s_{ti}(t) = \sqrt{2E_t}\,\text{Re}\,[\tilde{f}(t)e^{j\omega_i t}]:H_i. \tag{160}$$

We assume that the $\omega_i$ are chosen so that the output signal processes on the different hypotheses are in disjoint frequency bands. The received waveform on the $i$th hypothesis is

$$r(t) = \sqrt{2E_t}\,\text{Re}\,[\tilde{b}(t)\tilde{f}(t)e^{j\omega_i t}] + w(t), \qquad 0 \le t \le T:H_i. \tag{161}$$

The hypotheses are equally likely, and the criterion is minimum probability of error.

The optimum receiver is an obvious generalization of the binary receiver in Figs. 11.12 and 11.13. To calculate the performance, we extend (5.22) to include nonstationary complex processes. The result is

$$\Pr(\epsilon) < e^{\rho TR}\frac{[\tilde{D}_{\mathscr{F}}(1/N_0)]^{\rho}}{[\tilde{D}_{\mathscr{F}}(\rho/N_0(1+\rho))]^{1+\rho}}, \qquad 0 \le \rho \le 1, \tag{162}$$

where

$$\tilde{D}_{\mathscr{F}}(z) = \prod_{i=1}^{\infty}(1 + z\tilde{\lambda}_i), \tag{163}$$

and the $\tilde{\lambda}_i$ are the eigenvalues of (20). We then minimize over $\rho$ as in (5.29–5.35) to obtain $E(R)$. The next step is to find the distribution of eigenvalues that minimizes $E(R)$. Kennedy has carried out this minimization, and the result is given in [3]. Once again, the minimum is obtained by using a certain number of equal eigenvalues. The optimum number depends on $\bar{E}_r/N_0$ and the rate $R$. The final step is to try to find signals to give the appropriate eigenvalue distribution. The techniques of the binary case carry over directly to this problem.

This completes our discussion of the $M$-orthogonal signal problem. The interested reader should consult [3] for a complete discussion.

### 11.3.5 Summary: Communication over Doppler-spread Channels

In this section we have studied the problem of digital communication over Doppler-spread channels. There are several significant results that should be re-emphasized:

1. The optimum receiver can be realized exactly when the channel process has a state-variable representation.
2. Tight bounds on the probability of error are given by (75). These can be evaluated for any desired $\tilde{f}(t)$ if $\tilde{b}_D(t)$ has a state-variable representation.
3. There exists an upper bound on the probability of error for *any* $\tilde{f}(t)$ that does not depend on $\tilde{S}_D\{f\}$. For any binary system,

$$\Pr(\epsilon) \leq \tfrac{1}{2} \exp\left(-0.1488 \frac{\bar{E}_r}{N_0}\right). \tag{164}$$

4. In many cases we can choose signals that give performance close to the bound in (164).
5. Two suboptimum receiver configurations were developed that are much simpler to implement than the optimum receiver. In many cases they will perform almost as well as the optimum receiver.
6. The basic results can be extended to include systems using $M$-orthogonal signals.

This completes our discussion of digital communication.

The reader may wonder why we have included a detailed discussion of digital communication in the middle of a radar/sonar chapter. One obvious reason is that it is an important problem and this is the first place where we possess the necessary background to discuss it. This reason neglects an important point. The binary symmetric problem is one degree easier to analyze than the radar-detection problem, because the symmetry makes $\tilde{\mu}(\frac{1}{2})$ the important quantity. In the radar problem we must work with $\tilde{\mu}(s)$, $0 \leq s \leq 1$, until a specific threshold (or $P_F$) is chosen. This means that all the signal-design ideas and optimum eigenvalue distributions are harder to develop. Now that we have developed them for the symmetric communications case, we could extend them to the asymmetric problem. The quantitative results are different, but the basic concepts are the same.

## 11.4 PARAMETER ESTIMATION: DOPPLER-SPREAD TARGETS

The model for the estimation problem is a straightforward modification of the detection model. Once again,

$$\tilde{r}(t) = \sqrt{E_t}\tilde{f}(t - \lambda)\tilde{b}_D\left(t - \frac{\lambda}{2}\right) + \tilde{w}(t), \qquad T_i \leq t \leq T_f. \tag{165}$$

There are two cases of the parameter-estimation problem that we shall consider. In the first case, the only unknown parameters are the range to the target and its mean Doppler shift, $m_D$. We assume that the scattering function of $\tilde{b}_D(t)$ is completely known except for its mean. The covariance function of the signal returned from the target is

$$\tilde{K}_{\tilde{s}}(t, u:\lambda, m_D) = E_t\tilde{f}(t - \lambda)e^{j2\pi m_D t}\tilde{K}_{D_0}(t - u)e^{-j2\pi m_D u}\tilde{f}^*(u - \lambda), \tag{166}$$

where $\tilde{K}_{D_0}(t - u)$ is the covariance function of $\tilde{b}_D(t)$ with its mean Doppler removed. In other words,

$$\tilde{K}_D(t - u) \triangleq e^{j2\pi m_D t}\tilde{K}_{D_0}(t - u)e^{-j2\pi m_D u}. \tag{167}$$

We observe $\tilde{r}(t)$ and want to estimate $\lambda$ and $m_D$. Notice that the parameters of interest can be separated out of the covariance function.

In the second case, the covariance function of $\tilde{b}_D(t)$ depends on a parameter (either scalar or vector) that we want to estimate. Thus

$$\tilde{K}_{\tilde{s}}(t, u:\lambda, \mathbf{A}) = E_t\tilde{f}(t - \lambda)\tilde{K}_D(t - u:\mathbf{A})\tilde{f}^*(u - \lambda). \tag{168}$$

A typical parameter of interest might be the amplitude, or the root-mean-square Doppler spread. In this case the parameters cannot necessarily be separated out of the covariance function. Notice that the first case is included in the second case.

Most of the necessary results for both cases can be obtained by suitably combining the results in Chapters 6, 7, and 10. To illustrate some of the ideas involved, we consider the problem outlined in (166) and (167).

We assume that the target is a point target at range $R$, which corresponds to a round-trip travel time of $\lambda$. It is moving at a constant velocity corresponding to a Doppler shift of $m_D$ cps. In addition, it has a Doppler spread characterized by the scattering function $\tilde{S}_{D_0}\{f\}$, where

$$\tilde{S}_{D_0}\{f_1\} \triangleq \tilde{S}_D\{f_1 - m_D\}. \tag{169}$$

The complex envelope of the received signal is

$$\tilde{r}(t) = \sqrt{E_t}\tilde{f}(t - \lambda)e^{j\omega t}\tilde{b}_{D_0}\left(t - \frac{\lambda}{2}\right) + \tilde{w}(t), \qquad -\infty < t < \infty. \tag{170}$$

We assume an infinite observation interval for simplicity.

The covariance function of the returned signal process is given in (166). The likelihood function is given by

$$l(\lambda, m_D) = \frac{1}{N_0} \iint_{-\infty}^{\infty} \tilde{r}^*(t)\tilde{h}_{ou}(t, u:\lambda, m_D)\tilde{r}(u)\, dt\, du, \tag{171}$$

where $\tilde{h}_{ou}(t, u:\lambda, m_D)$ is specified by

$$N_0\tilde{h}_{ou}(t, u:\lambda, m_D) + \int_{-\infty}^{\infty} \tilde{h}_{ou}(t, z:\lambda, m_D)\tilde{K}_{\tilde{s}}(z, u:\lambda, m_D)\, dz$$
$$= \tilde{K}_{\tilde{s}}(t, u:\lambda, m_D), \qquad -\infty < t, u < \infty. \tag{172}$$

(Notice that the bias term is not a function of $\lambda$ or $m_D$, and so it has been omitted.) In order to construct the likelihood function, we must solve (172) for a set of $\lambda_i$ and $m_{D_j}$ that span the region of the range-Doppler plane in which targets may be located. Notice that, unlike the slowly fluctuating case in Chapter 10, we must normally use a discrete approximation in both range and Doppler. (There are other realizations for generating $l(\lambda, m_D)$ that may be easier to evaluate, but (171) is adequate for discussion purposes.) The maximum likelihood estimates are obtained by finding the point in the $\lambda$, $m_D$ plane where $l(\lambda, m_D)$ has its maximum.

To analyze the performance, we introduce a spread ambiguity function. As before, the ambiguity function corresponds to the output of the receiver when the additive noise $\tilde{w}(t)$ is absent. In this case the signal is a sample function of a random process, so that the output changes each time the experiment is conducted. A useful characterization is the expectation of this output. The input in the absence of noise is

$$\tilde{r}(t) = \sqrt{E_t}\tilde{f}(t - \lambda_a)e^{j2\pi m_{Da}t}\tilde{b}_{D_0}\left(t - \frac{\lambda_a}{2}\right). \tag{173}$$

We substitute (173) into (171) and take the expectation over $\tilde{b}_{D_0}(t)$. The result is

$$\begin{aligned}\theta_{\Omega_D}\{\lambda_a, \lambda: m_{Da}, m\} &= \frac{1}{N_0} \iint_{-\infty}^{\infty} \tilde{h}_{ou}(t, u:\lambda, m_D)\tilde{K}_{\tilde{s}}^*(t, u:\lambda_a, m_{Da})\, dt\, du \\ &= \frac{E_t}{N_0} \iint_{-\infty}^{\infty} \tilde{h}_{ou}(t, u:\lambda, m_D)\tilde{f}^*(t - \lambda_a) \\ &\quad \times e^{-j2\pi m_{Da}(t-u)}\tilde{f}(u - \lambda_a)\tilde{K}_{\tilde{D}_0}^*(t - u)\, dt\, du,\end{aligned} \tag{174}$$

which we define to be the *Doppler-spread ambiguity function*. Notice that it is a function of four variables, $\lambda_a$, $\lambda$, $m_{D_a}$, and $m_D$. This function provides a basis for studying the accuracy, ambiguity, and resolution problems when the target is Doppler-spread. The local accuracy problem can be studied by means of Cramér-Rao bounds. The elements in the **J** matrix are of the form

$$J_{\lambda\lambda}(\lambda_a, m_{Da}) = c\,\frac{\partial^2\theta_{\Omega_D}\{\lambda_a, \lambda : m_{Da}, m\}}{\partial\lambda\,\partial\lambda_a}\Bigg|_{\substack{\lambda=\lambda_a\\ m=m_{Da}}} \tag{175}$$

(see Problem 11.4.7). The other elements have a similar form.

We do not discuss the ambiguity and resolution issues in the text. Several properties of the Doppler-spread ambiguity function are developed in the problems. Notice that $\theta_{\Omega_D}\{\lambda_a, \lambda : m_{D_a}, m\}$ may be written in several other forms that may be easier to evaluate.

In general, the spread ambiguity function is difficult to use. When the LEC condition is valid,

$$\begin{aligned} h_{ou}(t, u : \lambda, m) &\simeq \frac{1}{N_0} K_{\tilde{s}}(t, u : \lambda, m) \\ &= \frac{E_t}{N_0} \tilde{f}(t-\lambda)e^{j2\pi mt}\tilde{K}_{D_0}(t-u)e^{-j2\pi mu}\tilde{f}^*(u-\lambda). \end{aligned} \tag{176}$$

Using (176) in (174) gives

$$\begin{aligned} &\theta_{\Omega_D,\mathrm{LEC}}\{\lambda_a, \lambda : m_a, m\} \\ &\quad = \frac{E_t^2}{N_0^2}\iint_{-\infty}^{\infty} \tilde{f}(t-\lambda)\tilde{f}^*(t-\lambda_a)e^{j2\pi(m-m_a)t}|\tilde{K}_{D_0}(t-u)|^2 \\ &\qquad \times e^{-j2\pi(m-m_a)u}\tilde{f}^*(u-\lambda)\tilde{f}(u-\lambda_a)\,dt\,du. \end{aligned} \tag{177}$$

(We suppressed the $D$ subscript on $m$ for notational simplicity.) This can be reduced to the two-variable function

$$\boxed{\begin{aligned} \theta_{\Omega_D,\mathrm{LEC}}\{\lambda_e, m_e\} &= \frac{E_t^2}{N_0^2}\iint_{-\infty}^{\infty} \tilde{f}\left(t-\frac{\lambda_e}{2}\right)\tilde{f}^*\left(t+\frac{\lambda_e}{2}\right)e^{j2\pi m_e t}\,|\tilde{K}_{D_0}(t-u)|^2 \\ &\quad \times e^{-j2\pi m_e u}\tilde{f}^*\left(u-\frac{\lambda_e}{2}\right)\tilde{f}\left(u+\frac{\lambda_e}{2}\right)dt\,du. \end{aligned}} \tag{178}$$

Some of the properties of $\theta_{\Omega_D,\mathrm{LEC}}\{\cdot,\cdot\}$ are developed in the problems.

A final comment concerning ambiguity functions is worthwhile. In the general parameter estimation problem, the likelihood function is

$$l(\mathbf{A}) = \frac{1}{N_0} \iint_{-\infty}^{\infty} \tilde{r}^*(t)\tilde{h}_{ou}(t, u:\mathbf{A})\tilde{r}(u)\, dt\, du + l_B(\mathbf{A}), \qquad \mathbf{A} \in \psi_{\mathbf{a}}, \quad (179)$$

where $\tilde{h}_{ou}(t, u:\mathbf{A})$ satisfies

$$N_0\tilde{h}_{ou}(t, u:\mathbf{A}) + \int_{-\infty}^{\infty} \tilde{h}_{ou}(t, z:\mathbf{A})\tilde{K}_{\tilde{s}}(z, u:\mathbf{A})\, dz = \tilde{K}_{\tilde{s}}(t, u:\mathbf{A}),$$

$$-\infty < t, u < \infty, \qquad \mathbf{A} \in \psi_{\mathbf{a}}, \quad (180)$$

and $l_B(\mathbf{A})$ is the bias. For this problem we define the *generalized spread ambiguity function* as

$$\theta_{\Omega}(\mathbf{A}_a, \mathbf{A}) = \frac{1}{N_0} \iint_{-\infty}^{\infty} \tilde{h}_{ou}(t, u:\mathbf{A})\tilde{K}_{\tilde{s}}(t, u:\mathbf{A}_a)\, dt\, du, \qquad \mathbf{A}_a, \mathbf{A} \in \psi_{\mathbf{a}}. \quad (181)$$

We shall encounter this function in Chapters 12 and 13.

This completes our discussion of the estimation problem. Our discussion has been brief because most of the basic concepts can be obtained by modifying the results in Chapters 6 and 7 in a manner suggested by our work in Chapter 10.

## 11.5 SUMMARY: DOPPLER-SPREAD TARGETS AND CHANNELS

In this chapter we have studied detection and parameter estimation in situations in which the target (or channel) caused the transmitted signal to be spread in frequency. We modeled the complex envelope of the received signal as a sample function of a complex Gaussian random process whose covariance is

$$\tilde{K}_{\tilde{s}}(t, u) = E_t \tilde{f}(t - \lambda)\tilde{K}_D(t - u)\tilde{f}^*(u - \lambda). \quad (179)$$

The covariance function $\tilde{K}_D(t - u)$ completely characterized the target (or channel) reflection process. We saw that whenever the transmitted pulse was longer than the reciprocal of the reflection process, the target (or channel) caused time-selective fading. We then studied three problems.

In Section 11.2, we formulated the optimum detection problem and gave the formulas that specify the optimum receiver and its performance. This problem was just the bandpass version of the Gaussian signal in noise problem that we had solved in Chapters 2–5. By exploiting our complex representation, all of the results carried over easily. We observed that whenever the reflection process could be modeled as a complex finite-state process, we could find a complete solution for the optimum receiver and obtain a good approximation to the performance. This technique is particularly important in this problem, because the reflected signal process is usually nonstationary. Another special case that is important is the LEC case. Here the optimum receiver and its performance can be evaluated easily. The results for the LEC condition also suggest suboptimum receivers for other situations.

In Section 11.3, we studied binary communication over Doppler-spread channels. The first important result was a bound on the probability of error that was independent of the channel-scattering function. We then demonstrated how to design signals that approached this bound. Techniques for designing and analyzing suboptimum receivers were developed. In the particular example studied, the performance of the suboptimum receivers was close to that of the optimum receiver. The extension of the results to $M$-ary systems was discussed briefly.

The final topic was the parameter-estimation problem. In Section 11.4, we formulated the problem and indicated some of the basic results. We defined a new function, the spread-ambiguity function, which could be used to study the issues of accuracy, ambiguity, and resolution. A number of questions regarding estimation are discussed in the problems. We study parameter estimation in more detail in Section 13.4.

We now turn to the other type of singly-spread target discussed in Chapter 8. This is the case in which the transmitted signal is spread in range.

## 11.6 PROBLEMS

### P.11.2 Detection of Doppler-spread Targets

**Problem 11.2.1.** We want to derive the result in (33) Define

$$\tilde{r}_i = \int_{T_i}^{T_f} \tilde{r}(t)\tilde{\varphi}_i^*(t)\,dt, \tag{P.1}$$

where $\tilde{\varphi}_i(t)$ is the $i$th eigenfunction of $\tilde{K}_{\tilde{s}}(t, u)$. Observe from (A.116) that

$$p_{\tilde{r}_i|H_1}(\tilde{R}_i \mid H_1) = \frac{1}{\pi(\tilde{\lambda}_i + N_0)} \exp\left[-\frac{|\tilde{R}_i|^2}{\tilde{\lambda}_i + N_0}\right], \qquad -\infty < \tilde{R}_i < \infty. \tag{P.2}$$

Using (P.1) and (P.2) as a starting point, derive (33).

**Problem 11.2.2.** Derive (33) directly from (2.31) by using bandpass characteristics developed in the Appendix.

**Problem 11.2.3.** Derive the result in (38) in two ways:

1. Use (33) and (34) as a starting point.
2. Use (2.86) as a starting point.

**Problem 11.2.4.** Consider the detection problem specified below.

$$\tilde{r}(t) = \sqrt{E_t}\,\tilde{f}(t)\tilde{b}_D(t) + \tilde{w}(t), \qquad T_i \le t \le T_f : H_1,$$

$$\tilde{r}(t) = \tilde{w}(t), \qquad T_i \le t \le T_f : H_0.$$

The Doppler scattering function is

$$\tilde{S}_D\{f\} = \frac{4k\sigma_b^{\,2}}{(2\pi f)^2 + k^2}.$$

The complex white noise has spectral height $N_0$.

1. Draw a block diagram of the optimum receiver. Write out explicitly the differential equations specifying the system.
2. Write out the equations that specify $\tilde{\mu}(s)$. Indicate how you would use $\tilde{\mu}(s)$ to plot the receiver operating characteristic.

**Problem 11.2.5.** Consider the same model as in Problem 11.2.4. Assume that

$$\tilde{f}(t) = \begin{cases} \sqrt{\dfrac{1}{T}}, & T_i \le t \le T_f, \\ 0, & \text{elsewhere,} \end{cases}$$

where

$$T = T_f - T_i,$$

and that $T$ is large enough that the asymptotic formulas are valid.

1. Draw the filter-squarer realization of the optimum receiver. Specify the transfer function of the filter.
2. Draw the optimum realizable filter realization of the optimum receiver. Specify the transfer function of the filter.
3. Compute $\tilde{\mu}_\infty(s)$.

**Problem 11.2.6.** Consider the same model as in Problem 11.2.4. Assume that $\tilde{f}(t)$ is a piecewise constant signal,

$$\tilde{f}(t) = \begin{cases} c\displaystyle\sum_{i=1}^{k} \tilde{f}_i \tilde{u}(t - iT_s), & 0 \le t \le T, \\ 0, & \text{elsewhere,} \end{cases}$$

where

$$\tilde{u}(t) = \begin{cases} \dfrac{1}{\sqrt{T_s}}, & 0 \le t \le T_s, \\ 0, & \text{elsewhere,} \end{cases}$$

and

$$T_s = \frac{T}{K}.$$

The $\tilde{f}_i$ are complex weighting coefficients, and $c$ is a constant chosen so that $\tilde{f}(t)$ has unit energy.

1. Draw a block diagram of the optimum receiver. Write out explicitly the differential equations specifying the system.
2. Write out the equations specifying $\tilde{\mu}(s)$.

**Problem 11.2.7.** Repeat Problem 11.2.4 for the Doppler scattering function

$$\tilde{S}_D\{f\} = \frac{4k\sigma_b^2}{[2\pi(f - m_D)]^2 + k^2}, \qquad -\infty < f < \infty.$$

**Problem 11.2.8.** Repeat Problem 11.2.4 for the case in which the target reflection process is characterized by the spectrum in (A.148).

**Problem 11.2.9.** Consider the following detection problem:

$$\tilde{r}(t) = \sqrt{E_t}\tilde{f}(t)\tilde{b}_D(t) + \tilde{n}_c(t) + \tilde{w}(t), \qquad 0 \le t \le T: H_1,$$
$$\tilde{r}(t) = \tilde{n}_c(t) + \tilde{w}(t), \qquad 0 \le t \le T: H_0.$$

The colored noise is a zero-mean complex Gaussian process with covariance function $\tilde{K}_c(t, u)$. It is statistically independent of both $\tilde{b}_D(t)$ and $\tilde{w}(t)$.

1. Derive the equations specifying the optimum receiver.
2. Derive a formula for $\tilde{\mu}(s)$.

**Problem 11.2.10.** Consider the model in Problem 11.2.9. Assume that $\tilde{n}_c(t)$ has a complex finite state representation.

1. Write out the differential equations specifying the optimum receiver.
2. Write out the differential equations specifying $\tilde{\mu}(s)$.

**Problem 11.2.11.** Consider the following detection problem.

$$\tilde{r}(t) = \sqrt{E_t}\{\tilde{f}(t - \lambda_1)e^{j\omega_1 t}\tilde{b}_{D1}(t) + \tilde{f}(t - \lambda_2)e^{j\omega_2 t}\tilde{b}_{D2}(t) + \tilde{w}(t)\}, \quad T_i \le t \le T_f: H_1,$$
$$= \sqrt{E_t}\tilde{f}(t - \lambda_2)e^{j\omega_2 t}\tilde{b}_{D2}(t) + \tilde{w}(t), \qquad T_i \le t \le T_f: H_0.$$

The quantities $\lambda_1$, $\lambda_2$, $\omega_1$, and $\omega_2$ are known. The two reflection processes are statistically independent, zero-mean complex Gaussian processes with covariance functions $\tilde{K}_{D1}(\tau)$ and $\tilde{K}_{D2}(\tau)$. Both processes have finite state representations.

1. Find the optimum receiver.
2. Find an expression for $\tilde{\mu}(s)$.

**Problem 11.2.12.** Consider the model in Problem 11.2.11. Assume that $\tilde{b}_{D2}(t)$ is a random variable instead of a random process.

$$\tilde{b}_{D2}(t) = \tilde{b}_{D2}.$$

1. Find the optimum receiver.
2. Find an expression for $\tilde{\mu}(s)$.

**Problem 11.2.13.** Consider the model in Problem 11.2.11. Assume that $\tilde{b}_{D1}(t)$ is a random variable instead of a random process.

$$\tilde{b}_{D1}(t) = \tilde{b}_{D1}.$$

Assume that $\tilde{b}_{D2}(t)$ has a finite state representation.

1. Find the optimum receiver. Specify both a correlator realization and a realizable filter realization.
2. Recall that

$$P_F = (P_D)^{1+\Delta}$$

for this type of model (see page 251). Find an integral expression for $\Delta$. Find the set of differential equations that specify $\Delta$.

3. Assume that

$$\tilde{S}_{D2}\{f\} = \frac{2kP_2}{(2\pi f)^2 + k^2}.$$

Write out the differential equations specifying the optimum receiver and $\Delta$.

**Problem 11.2.14.** Consider the model in Problem 11.2.13.

$$\tilde{r}(t) = \sqrt{E_t}\,\tilde{b}_{D1}\tilde{f}(t-\lambda_1)e^{j\omega_1 t} + \sqrt{E_t}\,\tilde{b}_{D2}(t)\tilde{f}(t-\lambda_2)e^{j\omega_2 t} + \tilde{w}(t), \quad T_i \le t \le T_f : H_1,$$

$$\tilde{r}(t) = \sqrt{E_t}\,\tilde{b}_{D2}(t)\tilde{f}(t-\lambda_2)e^{j\omega_2 t} + \tilde{w}(t), \quad T_i \le t \le T_f : H_0.$$

We want to design the optimum signal subject to an energy and bandwidth constraint.

$$\int_{T_i}^{T_f} |\tilde{f}(t)|^2\,dt = 1,$$

$$\int_{T_i}^{T_f} f^2 |\tilde{F}\{f\}|^2\,dt = B^2.$$

1. Assume that we use an optimum receiver. Find the differential equations that specify the optimum *signal* (see Section 9.5).
2. Assume that we use a conventional receiver (see Section 10.5). Find the differential equations that specify the optimum *signal.*
3. What is the fundamental difference between the equations in parts 1 and 2 and the equations in Section 9.5 (9.133)–(9.139)?

**Problem 11.2.15.** Consider the following detection problem:

$$\tilde{r}(t) = \sqrt{E_t}\,\tilde{b}_D\tilde{f}(t) + \sqrt{E_t}\left\{\sum_{i=1}^{K}\tilde{b}_{Di}(t)\tilde{f}(t-\lambda_i)e^{j\omega_i t}\right\} + \tilde{w}(t), \quad T_0 \le t \le T_f : H_1,$$

$$\tilde{r}(t) = \sqrt{E_t}\left\{\sum_{i=1}^{K}\tilde{b}_{Di}(t)\tilde{f}(t-\lambda_i)e^{j\omega_i t}\right\} + \tilde{w}(t), \quad T_0 \le t \le T_f : H_0.$$

The $\tilde{b}_{Di}(t)$ are statistically independent, zero-mean complex Gaussian processes with covariance functions $\tilde{K}_{Di}(\tau)$. The $\lambda_i$ and $\omega_i$ are known. The target reflection $\tilde{b}_D$ is a complex zero-mean Gaussian variable with mean-square value $2\sigma_b^2$.

1. Find the optimum receiver and an expression for $\Delta$.
2. Assume that a conventional receiver is used (see Section 10.5). Find an expression for $\Delta_{wo}$. Write this expression in terms of $\theta\{\tau, f\}$ and $S_{Di}\{f\}$.

**Problem 11.2.16.** Consider the multiple hypothesis detection problem:

$$\tilde{r}(t) = \tilde{w}(t), \quad T_i \le t \le T_f : H_0,$$

$$\tilde{r}(t) = \sqrt{E_t}\,\tilde{b}_{D1}\tilde{f}(t) + \tilde{w}(t), \quad T_i \le t \le T_f : H_1,$$

$$\tilde{r}(t) = \sqrt{E_t}\,\tilde{b}_{D2}(t)\tilde{f}(t) + \tilde{w}(t), \quad T_i \le t \le T_f : H_2.$$

We see that the three hypotheses correspond to noise only, noise plus a point-non-fluctuating target, and noise plus a fluctuating target.

Assume the following cost matrix:

$$\mathbf{C} = \begin{bmatrix} 0 & C_M & C_M \\ C_F & 0 & C_X \\ C_F & C_X & 0 \end{bmatrix}.$$

1. Find the optimum Bayes receiver.
2. Consider the special case when $C_X = 0$. Draw the optimum receiver. Find an expression for $\tilde{\mu}(s)$.
3. Assume that the following criteria are used:
a. $P_F \triangleq \{\Pr [\text{say } H_1 \text{ or } H_2 \mid H_0 \text{ is true}]\}$.
b. $P_D \triangleq \{\Pr [\text{say } H_1 \text{ or } H_2 \mid H_1 \text{ or } H_2 \text{ is true}]\}$.
c. Maximize $P_D$ subject to constraint that $P_F \leq \alpha$.
d. If the receiver says that a target is present, we want to classify it further. Define

$$P_{F_2} \triangleq \{\Pr [\text{say } H_2 \mid H_1 \text{ is true, target decision positive}]\}$$

and

$$P_{D_2} \triangleq \{\Pr [\text{say } H_2 \mid H_2 \text{ is true, target decision positive}]\}.$$

Maximize $P_{D_2}$ subject to constraint $P_{F_2} \leq \alpha_2$.

Explain how the over-all receiver operates. Can you write this in terms of a Bayes test?

**Problem 11.2.17.** Consider the detection problem in (30) and (31). Assume that

$$E[\tilde{b}_D(t)] = \tilde{m},$$

where $\tilde{m}$ is itself a complex Gaussian random variable with mean-square value $2\sigma^2$. The rest of the model remains the same. Find the optimum receiver.

## P.11.3 Digital Communication over Doppler-Spread Channels

**Problem 11.3.1.** Consider the binary FSK system described in Section 11.3.1. Assume that

$$\tilde{S}_D\{f\} = \frac{4k\sigma_b^2}{(2\pi f)^2 + k^2}.$$

1. Write out the differential equations specifying the receiver in detail.
2. Write out the differential equations specifying $\tilde{\mu}_{BS}(\frac{1}{2})$.

**Problem 11.3.2.** The performance of a binary FSK system operating over a Doppler-spread channel is given by

$$\tilde{\mu}_{BS}(\tfrac{1}{2}) = \sum_{i=1}^{\infty} \left[ \ln\left(1 + \frac{\tilde{\lambda}_i}{N_0}\right) - 2\ln\left(1 + \frac{\tilde{\lambda}_i}{2N_0}\right) \right]. \qquad \text{(P.1)}$$

For constant transmitted signals and large time-bandwidth products, we can use the SPLOT formulas.

1. Write the SPLOT formula corresponding to (P.1).
2. Evaluate $\tilde{\mu}_{BS,\infty}(\frac{1}{2})$ for

$$\tilde{S}_D\{f\} = \frac{4n\sigma_b^2}{k} \frac{\sin(\pi/2n)}{(2\pi f/k)^{2n} + 1}.$$

The transmitted signal has energy $E_t$ and duration $T$.

3. Find the optimum value of $kT$. Show that if the optimum value of $kT$ is used, $\tilde{\mu}_{BS,\infty}(\frac{1}{2})$ will decrease monotonically with $n$.

**Problem 11.3.3.** Consider the binary FSK system described in Section 11.3.1. Assume that

$$\tilde{f}(t) = \left(\frac{1}{\pi T^2}\right)^{1/4} e^{-t^2/2T^2}, \qquad -\infty < t < \infty$$

and

$$\tilde{S}_D\{f\} = \frac{2\sigma_b^{\,2}}{\sqrt{\pi}\, B_c} e^{-f^2 B_c^{\,2}}, \qquad -\infty < f < \infty.$$

The observation interval is infinite.

1. Find the output eigenvalues. (*Hint:* Use Mehler's expansion [e.g., [6] or [7].)
2. Evaluate $\tilde{\mu}_{BS,\infty}(\frac{1}{2})$.

**Problem 11.3.4.** Consider a binary communication system operating under LEC conditions.

1. Show that $\tilde{\mu}_{BS}(\frac{1}{2})$ can be expressed in terms of $\Delta$ [see (9.49)].
2. Use the results of part 1 in (75) to find a bound on the probability of error.
3. Find an expression for $\Delta$ in terms of $\tilde{f}(t)$ and $S_D\{f\}$.

**Problem 11.3.5.** Consider a $K$-channel frequency-diversity system using orthogonal FSK in each channel. The received waveform in the $i$th channel is

$$r(t) = \begin{cases} \sqrt{\dfrac{2E_t}{K}}\, \text{Re}\,[\tilde{b}_i(t)\tilde{f}(t)e^{j\omega_{1i}t}] + w(t), & T_0 \le t \le T_f : H_1, \\ \sqrt{\dfrac{2E_t}{K}}\, \text{Re}\,[\tilde{b}_i(t)\tilde{f}(t)e^{j\omega_{0i}t}] + w(t), & T_0 \le t \le T_f : H_0, \qquad i = 1, 2, \dots, K. \end{cases}$$

The channel fading processes are statistically independent and have identical scattering functions. Assume that the SPLOT condition is valid.

1. Evaluate $\tilde{\mu}_{BS}(\frac{1}{2})$.
2. Assume that

$$\tilde{S}_D\{f\} = \frac{4k\sigma_b^{\,2}}{(2\pi f)^2 + k^2}\cdot$$

The single-channel system with this scattering function was discussed in Example 2 on page 382. How would you use the additional freedom of a frequency-diversity system to improve the performance over that of the system in Example 2?

**Problem 11.3.6.** Consider the model in Example 3 on page 384. We want to investigate the probability of error as a function of $T_s$. One of the two branches of the receiver is

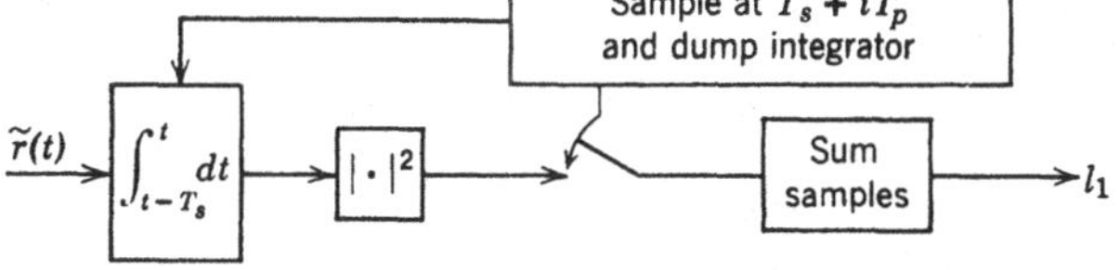


**Fig. P.11.1**

shown in Fig. P.11.1. This branch is referenced to $f_1$; the other branch, to $f_0$. Assume that $T_p$ is large enough that the outputs due to each pulse are statistically independent.

1. Find an expression for Pr ($\epsilon$) as a function of $\bar{E}_r$, $N_0$, $k$, and $T_s$. Assume that $D_o$ pulses are used. [*Hint:* Recall the results in (I-2.434) and (I-2.516).]
2. Plot

$$\frac{\ln \Pr(\epsilon)}{-0.1488\bar{E}_r/N_0}$$

as a function of $kT_s$.

**Problem 11.3.7.** Consider the model in Example 3 on page 384. We want to investigate the probability of error as a function of $T_s$ and $T_p$. The receiver in Problem 11.3.6 is used. Derive an expression for $\tilde{\mu}_{BS}(\frac{1}{2})$.

**Problem 11.3.8.** Consider the piecewise constant channel model in Fig. 11.17 and assume that $\tilde{f}(t)$ is a rectangular pulse. We generate a set of random variables $\tilde{r}_i$ as shown in Fig. 11.18. However, instead of using a weighted sum of their squared magnitudes, we operate on them in an optimum manner.

1. Find the optimum test based on the observed vector $\tilde{\mathbf{r}}$.
2. Find an expression for $\tilde{\mu}_{BS}(s)$ for this test.
3. Prove that the receiver in part 1 approaches the optimum receiver of Section 11.3.1 as $T_s$ approaches zero.

**Problem 11.3.9.** The definitions of $\tilde{\mu}_{11}(s)$ and $\tilde{\mu}_{01}(s)$ are given in (129) and (130).

1. Verify that the results in (131) and (132) are correct.
2. Verify the result in (136).

**Problem 11.3.10.** Consider the $M$-ary problem described in Section 11.3.4. Draw a block diagram of the optimum receiver.

**Problem 11.3.11.** Consider a binary communication system operating over a discrete multipath channel. The complex envelopes of the received waveforms are

$$\tilde{r}(t) = \sqrt{E_t}\left\{\sum_{i=1}^{K} \tilde{b}_{Di}(t)\tilde{f}(t-\lambda_i)\right\} + \tilde{w}(t), \qquad T_i \le t \le T_f : H_1,$$

where the complex representation is with reference to $\omega_1$, and

$$\tilde{r}(t) = \sqrt{E_t}\left\{\sum_{i=1}^{K} \tilde{b}_{Di}(t)\tilde{f}(t-\lambda_i)\right\} + \tilde{w}(t), \qquad T_i \le t \le T_f : H_0,$$

where the complex representation is with reference to $\omega_0$. The $\lambda_i$ are known and the $\tilde{f}_{Di}(t)$ are statistically independent, zero-mean complex Gaussian random processes with rational spectra. The signal components on the two hypotheses are in disjoint frequency bands.

1. Find the optimum receiver.
2. How is the receiver simplified if $\tilde{f}(t)$ and $\lambda_i$ are such that the path outputs are disjoint in time (resolvable multipath)?

**Problem 11.3.12.** Consider the detection problem described in (30)–(32). Assume that we use a gated correlator-squarer-summer-receiver of the type shown in Fig. 11.18.

1. Modify the results of Chapter 5 to obtain formulas that can be used to evaluate suboptimum bandpass receivers.
2. Use the results of part 1 to obtain performance expressions for the above receiver.

## P.11.4 Parameter Estimation

**Problem 11.4.1.** Consider the estimation problem described in (168)–(175). Assume that the LEC condition is valid.

1. Verify that the result in (178) is correct.
2. Evaluate $\theta_{\Omega_D,\text{LEC}}\{0,0\}$.
3. Prove

$$\theta_{\Omega_D,\text{LEC}}\{\lambda, m\} \leq \theta_{\Omega_D,\text{LEC}}\{0,0\}.$$

4. Is there a volume invariance relation for $\theta_{\Omega_D,\text{LEC}}\{\lambda, m\}$?

**Problem 11.4.2.** Assume

$$\tilde{f}(t) = \left(\frac{1}{\pi T^2}\right)^{1/4} e^{-t^2/2T^2}, \qquad -\infty < t < \infty$$

and

$$\tilde{S}_{D_0}\{f\} = \frac{2\sigma_b^{\,2}}{\sqrt{2\pi}\,\sigma_D} e^{-f^2/2\sigma_D^2}, \qquad -\infty < f < \infty.$$

Evaluate $\theta_{\Omega_D,\text{LEC}}\{\lambda, m\}$.

**Problem 11.4.3.** Consider the LEC estimation problem discussed in Problem 11.4.1.

1. Derive an expression for the elements of the **J** matrix in terms of $\theta_{\Omega_D,\text{LEC}}\{\tau, m\}$.
2. Evaluate the **J** matrix for the signal and scattering function in Problem 11.4.2.

**Problem 11.4.4.** Go through the list of properties in Section 10.3 and see which ones can be generalized to the spread-ambiguity function, $\theta_{\Omega_D,\text{LEC}}\{\tau, m\}$.

**Problem 11.4.5.** Assume that we are trying to detect a Doppler-spread target in the presence of white noise and have designed the optimum LEC receiver.

1. In addition to the desired target, there is a second Doppler-spread target with an identical scattering function. Evaluate the effect of the second target in terms of $\theta_{\Omega_D,\text{LEC}}\{\lambda, m\}$. (Notice that the receiver is not changed from the original design.)
2. Extend the result to $K$ interfering targets with identical scattering functions.
3. What modifications must be made if the Doppler scattering functions are not identical? (This is the *spread cross-ambiguity function.*)
4. In part 3, we encountered a spread cross-ambiguity function. A more general definition is

$$\psi_{\Omega}\{\lambda, f\} \triangleq \frac{1}{N_0}\iint_{-\infty}^{\infty} \tilde{g}\left(t - \frac{\lambda}{2}\right)\tilde{f}^*\left(t - \frac{\lambda}{2}\right) e^{j2\pi f t}\tilde{K}_g(t-u)\tilde{K}^*_{D_0}(t-u)e^{-j2\pi f t}$$
$$\times\, \tilde{g}^*\left(u - \frac{\lambda}{2}\right)\tilde{f}\left(u + \frac{\lambda}{2}\right) dt\, du. \tag{P.1}$$

Where would this function be encountered? How is it related to the ordinary cross-ambiguity function $\theta_{fg}\{\lambda, f\}$?

**Problem 11.4.6.** Consider the degenerate case of Problem 11.4.5, in which we are trying to detect a slowly fluctuating point target in the presence of white noise and have designed the optimum receiver.

What effect will the presence of a set of a set of Doppler-spread targets have on the performance of the above receiver.?

**Problem 11.4.7.** Derive the term in (175) and the other elements in the **J** matrix.

**Problem 11.4.8.** Consider the problem of estimating the amplitude of a scattering function. Thus,

$$\tilde{K}_D(\tau:A) = A\tilde{K}_D(\tau), \tag{P.1}$$

and $\tilde{K}_D(\tau)$ is assumed to be known. The complex envelope of the transmitted signal is $\sqrt{E_t}\tilde{f}(t)$. The complex envelope of the returned waveform is

$$\tilde{r}(t) = \sqrt{E_t}\,\tilde{b}_D(t, A)\tilde{f}(t) + \tilde{w}(t), \qquad T_i \le t \le T_f,$$

where $\tilde{b}_D(t, A)$ is a complex Gaussian random process whose covariance is given in (P.1). Assume that the LEC condition is valid.

1. Find a receiver to generate $\hat{a}_{ml}$.
2. Is $\hat{a}_{ml}$ unbiased?
3. Assume that the bias of $\hat{a}_{ml}$ is negligible. (How could you check this?) Calculate

$$E[(\hat{a}_{ml} - A)^2].$$

4. Calculate a bound on the normalized variance of any unbiased estimate of $A$.
5. Express the bound in part 4 in terms of $\theta\{\tau, f\}$ and $\tilde{S}_D\{f\}$.
6. Assume that

$$\tilde{f}(t) = \left(\frac{1}{\pi T^2}\right)^{1/4} e^{-t^2/2T^2}$$

and

$$\tilde{S}_D\{f\} = \frac{1}{\sqrt{2\pi}B} e^{-f^2/2B^2}$$

Evaluate the bound in part 4 for this case. Discuss the behavior as a function of $BT$. Would this behavior be the same if the LEC condition were not satisfied?

7. Express the largest eigenvalue in terms of $A$, $B$, and $T$.

**Problem 11.4.9.** The complex envelope of the received waveform is

$$\tilde{r}(t) = \sqrt{E_t}\tilde{f}(t)[e^{j\omega_1 t} + e^{j\omega_0 t}]\tilde{b}_D(t) + \tilde{w}(t), \qquad -\infty < t < \infty.$$

We want to estimate the quantity $\omega_\Delta = \omega_1 - \omega_0$. The process $\tilde{b}_D(t)$ is a zero-mean complex Gaussian process whose bandwidth is much less than $\omega_\Delta$.

1. Find a receiver to generate the maximum likelihood estimate of $\omega_\Delta$.
2. Find an expression for the Cramér-Rao bound.

**Problem 11.4.10.** Assume that

$$\tilde{S}_D\{f:A\} = \tilde{S}_{D_1}\left\{\frac{f}{A}\right\},$$

where $\tilde{S}_{D_1}\{\cdot\}$ is known. We want to estimate $A$, the scale of the frequency axis. Assume that

$$\tilde{r}(t) = \sqrt{E_t}\tilde{b}_D(t, A)\tilde{f}(t) + \tilde{w}(t), \qquad -\infty < t < \infty,$$

and that the LEC condition is valid.

1. Draw a block diagram of a receiver to generate $\hat{a}_{ml}$.
2. Evaluate the Cramér-Rao bound.

**Problem 11.4.11.** Assume that the target consists of two reflectors at different ranges. The complex envelope of the returned waveform is

$$\tilde{r}(t) = \sqrt{E_t} \sum_{i=1}^{2} \tilde{b}_i \tilde{f}(t - \lambda_i) + \tilde{w}(t), \qquad -\infty < t < \infty,$$

where the $\tilde{b}_i$ are statistically independent complex Gaussian random variables ($E|\tilde{b}_i|^2 = 2\sigma_i^2$). We want to estimate the mean range, which we define as

$$\lambda_r = \tfrac{1}{2}(\lambda_1 + \lambda_2).$$

1. Draw the block diagram of a receiver to generate $\hat{\lambda}_{r,ml}$.
2. Does

$$\hat{\lambda}_{r,ml} = \tfrac{1}{2}(\hat{\lambda}_{1,ml} + \hat{\lambda}_{2,ml})?$$

3. Evaluate the Cramér-Rao bound of the variance of the estimate.

**Problem 11.4.12.** Consider the problem of estimating the range and mean Doppler when the amplitude of the scattering function is unknown. Thus,

$$\tilde{K}_{\tilde{s}}(t, u:\text{A}) = AE_t \tilde{f}(t - \lambda)e^{j2\pi mt}\tilde{K}_{D_0}(t - u)e^{-j2\pi mu}\tilde{f}^*(u - \lambda).$$

Assume that the LEC condition is valid and that the bias on $\hat{a}_{ml}$ can be ignored.

1. Find $l(\hat{a}_{ml}, \lambda, m)$.
2. Draw a block diagram of the optimum receiver to generate $\hat{\lambda}_{ml}$, $\hat{m}_{ml}$.
3. Evaluate **J**. Does the fact that $A$ is unknown increase the bounds on the variances of $\hat{\lambda}_{ml}$ and $\hat{m}_{ml}$?

## REFERENCES

[1] R. Price and P. E. Green, "Signal Processing in Radar Astronomy—Communication via Fluctuating Multipath Media," Technical Report No. 234, Lincoln Laboratory, Massachusetts Institute of Technology, 1960.

[2] P. A. Bello, "Characterization of Randomly Time-Variant Linear Channels," IRE Trans. Commun. Syst. **CS-11** (Dec. 1963).

[3] R. S. Kennedy, *Fading Dispersive Communications Channels*, Wiley, New York, 1969.

[4] R. R. Kurth, "Distributed-Parameter State-Variable Techniques Applied to Communication over Dispersive Channels," Sc.D. Thesis, Department of Electrical Engineering, Massachusetts Institute of Technology, June 1969.

[5] L. D. Collins and R. R. Kurth, "Asymptotic Approximations to the Error Probability for Square-Law Detection of Gaussian Signals," Massachusetts Institute of Technology. Research Laboratory of Electronics, Quarterly Progress Report No. 90, July 15, 1968, 191–200.

[6] J. L. Brown, "On the Expansion of the Bivariate Gaussian Probability Density Using the Results of Nonlinear Theory, IEEE Trans. Information Theory **IT-14,** No. 1, 158–159 (Jan. 1968).

[7] N. Wiener, *The Fourier Integral and Certain of Its Applications*, Cambridge University Press, London, 1933.

[8] R. Price, "Statistical Theory Applied to Communication through Multipath Disturbances," Massachusetts Institute of Technology Research Laboratory of Electronics, Tech. Rept. 266, September 3, 1953.

[9] R. Price, "Detection of Signals Perturbed by Scatter and Noise," IRE Trans. **PGIT-4,** 163–170 (Sept. 1954).

[10] R. Price, "Optimum Detection of Random Signals in Noise, with Application to Scatter-Multipath Communication. I," IRE Trans. **PGIT-6,** 125–135 (Dec. 1956).

[11] T. Kailath, "Correlation Detection of Signals Perturbed by a Random Channel," IRE Trans. Information Theory **IT-6-** 361–366 (June 1960).

[12] T. Kailath, "Optimum Receivers for Randomly Varying Channels," Proc. Fourth London Symp. Information Theory, Butterworths, London, 1960.

[13] G. Turin, "Communication through Noisy, Random-Multipath Channels," 1956 IRE Convention Record, Pt. 4, pp. 154–156.

[14] G. Turin, "Error Probabilities for Binary Symmetric Ideal Reception through Nonselective Slow Fading and Noise," Proc. IRE **46,** 1603–1619 (Sept. 1958).

[15] P. Bello, "Some Results on the Problem of Discriminating between Two Gaussian Processes," IRE Trans. Information Theory **IT-7,** No. 4, 224–233 (Oct. 1961).

[16] D. Middleton, "On the Detection of Stochastic Signals in Additive Normal Noise. I," IRE Trans. Information Theory **IT-3,** 86–121 (June 1957).

[17] D. Middleton, "On the Detection of Stochastic Signals in Additive Normal Noise. II," IRE Trans. Information Theory **IT-6,** 349–360 (June 1960).

[18] D. Middleton, "On Singular and Nonsingular Optimum (Bayes) Tests for the Detection of Normal Stochastic Signals in Normal Noise," IRE Trans. Information Theory **IT-7,** 105–113 (April 1961).

[19] C. W. Helstrom, *Statistical Theory of Signal Detection*, Pergamon Press, New York, 1960.

[20] D. Middleton, *Introduction to Statistical Communication Theory*, McGraw-Hill, New York, 1960.

[21] P. A. Bello and B. D. Nelin, "The Influence of Fading Spectrum on the Binary Error Probabilities of Incoherent and Differentially Coherent Matched Filter Receivers," IRE Trans. Commun. Syst. **CS-10,** 160–168 (June 1962).

[22] P. A. Bello and B. D. Nelin, "Predetection Diversity Combining with Selectively-Fading Channels," IRE Trans. Commun. Syst. **CS-10** (March 1962).

# 12

# *Range-Spread Targets and Channels*

In Chapters 9 and 10, we studied slowly fluctuating point targets. In Chapter 11, we studied point targets that could fluctuate at an arbitrary rate. In this chapter, we consider slowly fluctuating targets that are spread in range.

A typical case is shown in Fig. 12.1. We transmit a short pulse as shown in Fig. 12.1*a*. The target configuration is shown in Fig. 12.1*b*. The surface is rough, so that energy is reflected in the direction of the receiver. The target has length $L$ (measured in seconds of travel time). To characterize the reflected signal, we divide the target in $\Delta\lambda$ increments. The return from each increment is a superposition of a number of reflections, and so we can characterize it as a complex Gaussian random variable. Thus the return from the first increment is

$$\sqrt{E_t}\,\tilde{b}(\lambda_0)\tilde{f}(t-\lambda_0)\,\Delta\lambda, \tag{1}$$

the return from the second increment is

$$\sqrt{E_t}\,\tilde{b}(\lambda_1)\tilde{f}(t-\lambda_1)\,\Delta\lambda, \tag{2}$$

and so forth. The total return is

$$\tilde{s}(t) = \sqrt{E_t}\sum_{i=0}^{N}\tilde{b}(\lambda_i)\tilde{f}(t-\lambda_i)\,\Delta\lambda. \tag{3}$$

We see that it consists of delayed versions of the signal, which are weighted with complex Gaussian variables and summed. A typical returned signal is shown in Fig. 12.1*c*. We see that the signal is spread out in time (or range), and so we refer to this type of target as a *range-spread target*. Other adjectives commonly used are *delay-spread* and *dispersive*.

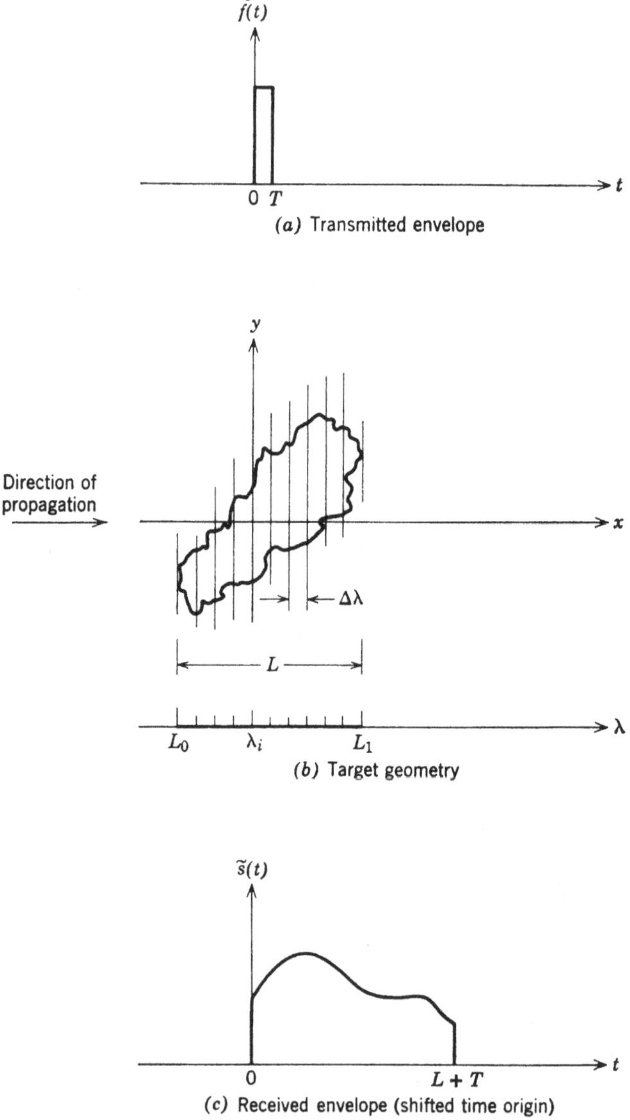


**Fig. 12.1 Range-spread model.**

In this chapter we study detection and parameter estimation for range-spread targets. In Section 12.1 we develop a quantitative model for range-spread targets and channels and show how this type of target causes *frequency-selective fading*. In Section 12.2, we discuss optimum receiver configurations briefly. In Section 12.3, we develop the concept of time-frequency duality. This development enables us to translate all range-spread channels into equivalent Doppler-spread channels. We can then use all of the results in Chapter 11 directly. We also discuss a number of applications in Section 12.3. Finally, in Section 12.4, we summarize our results.

## 12.1 MODEL AND INTUITIVE DISCUSSION

We begin our model development with the relation in (3). The increments are useful for explanatory purposes, but the reflections actually occur from a continuous range of $\lambda$. s A$\Delta\lambda \rightarrow 0$, the sum in (3) becomes the integral

$$\tilde{s}(t) = \sqrt{E_t}\int_{L_0}^{L_1} \tilde{f}(t-\lambda)\tilde{b}_R(\lambda)\,d\lambda. \tag{4}$$

Now $\tilde{b}_R(\lambda)$ is a sample function from a zero-mean complex Gaussian process whose independent variable is the spatial variable $\lambda$. Notice that $\tilde{b}_R(\lambda)$ is *not* a function of time. We see that a range-spread target behaves exactly as a linear time-invariant filter with a random complex impulse response $\tilde{b}_R(\lambda)$. To characterize $\tilde{b}_R(\lambda)$ completely, we need the two complex covariance functions

$$\tilde{K}_{\tilde{b}_R}(\lambda, \lambda_1) = E[\tilde{b}_R(\lambda)\tilde{b}_R^*(\lambda_1)] \tag{5}$$

and

$$E[\tilde{b}_R(\lambda)\tilde{b}_R(\lambda_1)] = 0, \qquad \text{for all} \quad \lambda, \lambda_1, \tag{6}$$

where the result in (6) is a restriction we impose.

We shall assume that the returns from different ranges are statistically independent. To justify this assumption, we return to the incremental model in Fig. 12.1. The value of $\tilde{b}_R(\lambda_i)$ will be determined by the relative phases and strengths of the component reflections in the $i$th interval. Assuming that the surface is rough compared to the carrier wavelength, the values of $\tilde{b}_R(\lambda_i)$ in different intervals will not be related. In the continuous model this implies that

$$\tilde{K}_{\tilde{b}_R}(\lambda, \lambda_1) = \delta(\lambda - \lambda_1)E\{|\tilde{b}_R(\lambda)|^2\}. \tag{7}$$

Notice that the relation in (7) is an idealization analogous to white noise in the time domain. The reflected signal is given by the convolution in (4).

As long as the correlation distance of $\tilde{b}_R(\lambda)$ is much shorter than the reciprocal of the bandwidth of $\tilde{f}(t)$, then (7) will be a good approximation.

Physically, the expectation in (7) is related to the expected value of energy returned (or scattered) from an incremental element located at $\lambda$. We define

$$\tilde{S}_R(\lambda) \triangleq E\{|\tilde{b}(\lambda)|^2\}, \qquad -\infty < \lambda < \infty \tag{8}$$

and refer to it as the *range-scattering function*. For convenience, we shall always define $\tilde{S}_R(\lambda)$ for an infinite range. The finite target length will be incorporated in the functional definition.

The covariance of the received signal in the absence of additive noise is

$$\begin{aligned} \tilde{K}_{\tilde{s}}(t, u) &= E[\tilde{s}_r(t)\tilde{s}_r^*(u)] \\ &= E\left\{E_t \int_{\Omega_L} d\lambda\, \tilde{f}(t-\lambda)\tilde{b}_R(\lambda) \int_{\Omega_L} d\lambda_1\, \tilde{f}^*(u-\lambda_1)\tilde{b}_R^*(\lambda_1)\right\}. \end{aligned} \tag{9}$$

Using (7) and (8) in (9) gives

$$\boxed{\tilde{K}_{\tilde{s}}(t, u) = E_t \int_{-\infty}^{\infty} \tilde{f}(t-\lambda)\tilde{S}_R(\lambda)\tilde{f}^*(u-\lambda)\, d\lambda.} \tag{10}$$

The relation in (10) completely characterizes the signal returned from a range-spread target.

Notice that the total received energy is

$$\bar{E}_r = \int_{-\infty}^{\infty} \tilde{K}_{\tilde{s}}(t, t)\, dt = E_t \int_{-\infty}^{\infty} \tilde{S}_R(\lambda)\, d\lambda \int_{-\infty}^{\infty} |\tilde{f}(t-\lambda)|^2\, dt = E_t \int_{-\infty}^{\infty} \tilde{S}_R(\lambda)\, d\lambda. \tag{11}$$

We see that

$$\tilde{S}_R(\lambda)\, d\lambda = \frac{\text{expected value of the energy returned from } (\lambda, \lambda + d\lambda)}{E_t}. \tag{12}$$

The result in (12) is a quantitative statement of the idea expressed in (8). In order to be consistent with the point target model, we assume that

$$\int_{-\infty}^{\infty} \tilde{S}_R(\lambda)\, d\lambda = 2\sigma_b^{\,2}. \tag{13}$$

This completes our specification of the reflection model. Before beginning our optimum receiver development, it is useful to spend some time on an intuitive discussion. In Chapter 11 we saw that a Doppler-spread target causes time-selective fading. Now we want to demonstrate that a range-spread target causes frequency-selective fading.

The Fourier transform of $\tilde{s}(t)$ is a well-defined quantity when the target length is finite. Thus,

$$\tilde{S}\{f\} \triangleq \int_{-\infty}^{\infty} \tilde{s}(t)e^{-j2\pi f_1 t}\,dt$$
$$= \int_{-\infty}^{\infty} e^{-j2\pi f_1 t}\,dt \int_{-\infty}^{\infty} \tilde{f}(t-\lambda)\tilde{b}_R(\lambda)\,d\lambda. \tag{14}$$

Notice that $\tilde{S}\{f\}$ is a sample function of a complex Gaussian process.

We want to compute the cross-correlation between $\tilde{S}\{f\}$ at two different frequencies.

$$E[\tilde{S}\{f_1\}\tilde{S}^*\{f_2\}] = E\bigg\{\int_{-\infty}^{\infty} e^{-j2\pi f_1 t_1}\,dt_1 \int_{-\infty}^{\infty} \tilde{f}(t_1-\lambda_1)\tilde{b}_R(\lambda_1)\,d\lambda_1$$
$$\times \int_{-\infty}^{\infty} e^{j2\pi f_2 t_2}\,dt_2 \int_{-\infty}^{\infty} \tilde{f}^*(t_2-\lambda_2)\tilde{b}_R^*(\lambda_2)\,d\lambda_2\bigg\}. \tag{15}$$

Bringing the expectation inside the integrals, using (7) and (8), we obtain

$$E[\tilde{S}\{f_1\}\tilde{S}^*\{f_2\}] = \tilde{F}\{f_1\}\tilde{F}^*\{f_2\}\int_{-\infty}^{\infty} e^{-j2\pi\lambda(f_1-f_2)}\tilde{S}_R(\lambda)\,d\lambda, \tag{16}$$

where $\tilde{F}\{f_1\}$ is the Fourier transform of $\tilde{f}(t)$. To interpret (16), we define

$$\boxed{\tilde{K}_R\{v\} \triangleq \int_{-\infty}^{\infty} e^{-j2\pi\lambda v}\tilde{S}_R(\lambda)\,d\lambda.} \tag{17}$$

Using (17) in (16), we obtain

$$E[\tilde{S}\{f_1\}\tilde{S}^*\{f_2\}] = \tilde{F}\{f_1\}\tilde{F}^*\{f_2\}\tilde{K}_R\{f_1-f_2\}, \tag{18}$$

or

$$\tilde{K}_R\{f_1-f_2\} = \frac{E[\tilde{S}\{f_1\}\tilde{S}^*\{f_2\}]}{\tilde{F}\{f_1\}\tilde{F}^*\{f_2\}}. \tag{19}$$

The function $\tilde{K}_R\{v\}$ is called the *two-frequency correlation function*. It measures the correlation between the fading at different frequencies. Notice that it is the Fourier transform of the range-scattering function. Therefore,

$$\boxed{\tilde{S}_R(\lambda) = \int_{-\infty}^{\infty} e^{j2\pi\lambda v}\tilde{K}_R\{v\}\,dv,} \tag{20}$$

and we can use either $\tilde{K}_R\{v\}$ or $\tilde{S}_R(\lambda)$ to characterize the target.

To illustrate the implication of the result in (18), we consider the scattering function in Fig. 12.2*a*,

$$\tilde{S}_R(\lambda) = \begin{cases} \dfrac{2\sigma_b^2}{L}, & -\dfrac{L}{2} \le \lambda \le \dfrac{L}{2}, \\ 0, & \text{elsewhere.} \end{cases} \tag{21}$$

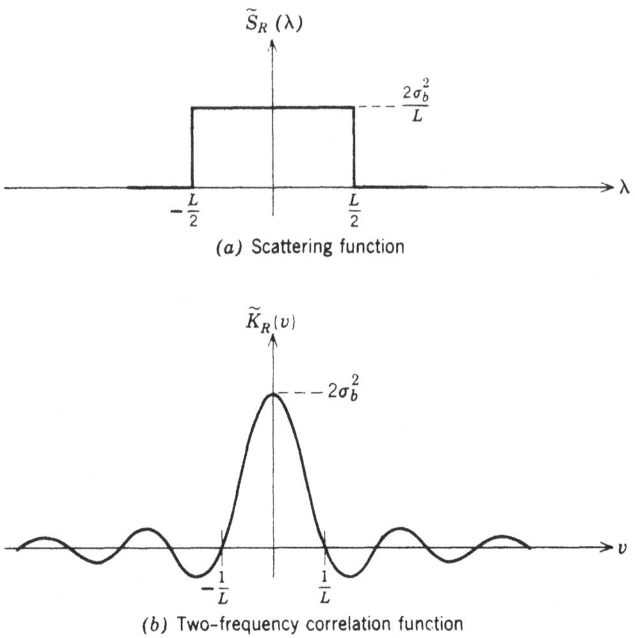


**Fig. 12.2 Functions for a uniform range-spread target.**

Thus,

$$\tilde{K}_R\{v\} = 2\sigma_b^{\,2}\,\frac{\sin \pi L v}{\pi L v}, \qquad -\infty < v < \infty, \tag{22}$$

as shown in Fig. 12.2*b*. We see that frequency components separated by more than $1/L$ cps will be essentially uncorrelated (and statistically independent, because they are jointly Gaussian).

Now assume that we transmit a signal whose Fourier transform is

$$\tilde{F}\{f\} = \begin{cases} \dfrac{1}{\sqrt{W}}, & -\dfrac{W}{2} \le f \le \dfrac{W}{2}, \\ 0, & \text{elsewhere.} \end{cases} \tag{23}$$

In Fig. 12.3*a*, we show the case in which

$$W \gg \frac{1}{L}. \tag{24}$$

In Fig. 12.3*b*, we show the transform of a typical sample function of $\tilde{s}(t)$. The amplitudes at frequencies separated by more than $1/L$ cps are essentially statistically independent, and so we refer to this behavior as frequency-selective fading.

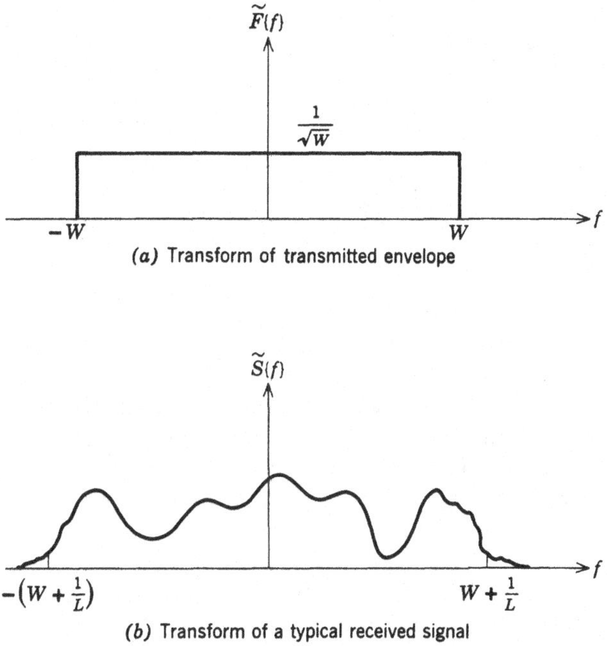


**Fig. 12.3 Functions to illustrate frequency-selective fading.**

The function in Fig. 12.3*b* is very similar to that in Fig. 11.2*b*, except that the axis is frequency instead of time. We shall exploit this similarity (or duality) in detail in Section 12.3.

Notice that if the signal bandwidth is such that

$$W \ll \frac{1}{L}, \tag{25}$$

the returned signal will be undistorted. This is, of course, the slowly fluctuating point target model of Chapters 9 and 10. The relation in (25) tells us when we can model the target as a point target.

We now have a quantitative model for range-spread targets and an intuitive understanding of how they affect the transmitted signal. The next step is to consider the problem of optimum receiver design.

## 12.2 DETECTION OF RANGE-SPREAD TARGETS

In this section we consider the binary detection problem, in which the complex envelopes of the received waveforms on the two hypotheses are

$$\tilde{r}(t) = \tilde{s}(t) + \tilde{w}(t), \qquad -\infty < t < \infty : H_1, \tag{26}$$

and

$$\tilde{r}(t) = \tilde{w}(t), \qquad -\infty < t < \infty : H_0. \tag{27}$$

The signal is a sample function from a zero-mean complex Gaussian process,

$$\tilde{s}(t) = \sqrt{E_t} \int_{-\infty}^{\infty} \tilde{f}(t - \lambda)\tilde{b}_R(\lambda)\, d\lambda, \tag{28}$$

whose covariance function is

$$\tilde{K}_{\tilde{s}}(t, u) = E_t \int_{-\infty}^{\infty} \tilde{f}(t - \lambda)\tilde{S}_R(\lambda)\tilde{f}^*(u - \lambda)\, d\lambda. \tag{29}$$

The additive noise, $\tilde{w}(t)$, is a sample function from a statistically independent, zero-mean, complex white Gaussian process with spectral height $N_0$. We have assumed an infinite observation interval for simplicity.

The expression for the optimum test follows directly from (11.33) as

$$l = \frac{1}{N_0} \iint_{-\infty}^{\infty} \tilde{r}^*(t)\tilde{h}(t, u)\tilde{r}(u)\, dt\, du \underset{H_0}{\overset{H_1}{\gtrless}} \gamma, \tag{30}$$

where $h(t, u)$ satisfies the equation

$$N_0\tilde{h}(t, u) + \int_{-\infty}^{\infty} \tilde{h}(t, z)\tilde{K}_{\tilde{s}}(z, u)\, dz = \tilde{K}_{\tilde{s}}(t, u), \qquad -\infty < t < \infty. \tag{31}$$

The difficulty arises in solving (31). There are two cases in which the solution is straightforward, the separable kernel case and the low-energy-coherence case. The separable kernel analysis is obvious, and so we relegate it to the problems. The LEC condition leads to an interesting receiver configuration, however, and so we discuss it briefly.

When the LEC condition is valid, the solution to (31) may be written as

$$\tilde{h}(t, u) = \frac{1}{N_0} \tilde{K}_{\tilde{s}}(t, u). \tag{32}$$

Using (29) in (32) and the result in (30) gives

$$l = \frac{E_t}{N_0^2} \iiint_{-\infty}^{\infty} \tilde{r}^*(t)\tilde{f}(t - \lambda)\tilde{S}_R(\lambda)\tilde{f}^*(u - \lambda)\tilde{r}(u)\, dt\, du\, d\lambda \underset{H_1}{\overset{H_1}{\gtrless}} \gamma. \tag{33}$$

This can be rewritten as

$$l_1 = \int_{-\infty}^{\infty} \tilde{S}_R(\lambda)\, |\tilde{x}(\lambda)|^2\, d\lambda \underset{H_0}{\overset{H_1}{\gtrless}} \gamma_1, \tag{34a}$$

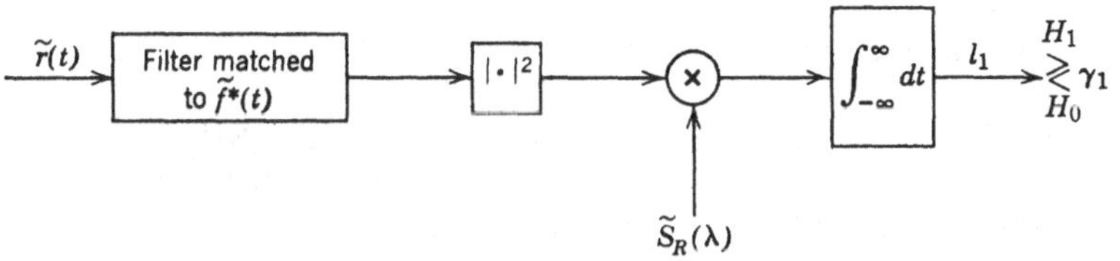


**Fig. 12.4 Two-filter radiometer: optimum detector under the LEC condition.**

where

$$\tilde{x}(\lambda) \triangleq \int_{-\infty}^{\infty} \tilde{r}(u)\tilde{f}^*(u - \lambda)\, du, \tag{34b}$$

and we have absorbed the constant in the threshold. The operation in (34*a*) may be realized as shown in Fig. 12.4. This receiver is called a two-filter radiometer and is due to Price [1].

When the LEC condition is valid, the performance is specified by (11.65). Using (10) in (11.65) and integrating gives

$$\boxed{\tilde{\mu}(s) = -\frac{s(1-s)E_t^2}{N_0^2}\iint_{-\infty}^{\infty} \tilde{S}_R(\lambda_1)\theta\{\lambda_1 - \lambda_2, 0\}\tilde{S}_R(\lambda_2)\, d\lambda_1\, d\lambda_2,} \tag{35}$$

which is the desired result.

When the LEC condition is not valid, it is difficult to solve (31) directly. In the next section we develop a procedure for solving it by transforming it into an equivalent Doppler-spread target problem.

## 12.3 TIME-FREQUENCY DUALITY

The utility of the duality concept is well known in classical network theory. Bello [2] has developed the concept of time-frequency duality in a more general framework and applied it to communications problems. In Section 12.3.1, we develop the basic duality concepts. In Section 12.3.2, we consider the dual relations in range-spread and Doppler-spread channels. In Section 12.3.3, we apply the results to specific cases.

One comment is worthwhile before beginning our development. We shall develop a number of properties and formal relationships. These are useful in solving specific problems. The reader who only learns these properties and applies them blindly will miss what we think is a major benefit of duality. This benefit is the guidance it offers in thinking about a particular problem. Often by just thinking about the duality, one can solve a problem directly without going through the formal manipulations.

### 12.3.1 Basic Duality Concepts

Our discussion consists of a series of definitions and properties with some examples interspersed to illustrate the ideas. Notice that throughout the discussion all functions are defined over infinite intervals.

**Definition 1.** Consider the two complex time functions $\tilde{y}_1(t)$ and $\tilde{y}_2(t)$. If

$$\tilde{y}_2(f) = \mathscr{F}[\tilde{y}_1(t)] \triangleq \tilde{Y}_1\{f\} \triangleq \int_{-\infty}^{\infty} \tilde{y}_1(t)e^{-j2\pi ft}\,dt, \tag{36}$$

then $\tilde{y}_2(t)$ is the *dual* of $\tilde{y}_1(t)$. If

$$\tilde{y}_2(t) = \mathscr{F}^{-1}[\tilde{y}_1(f)] \triangleq \int_{-\infty}^{\infty} \tilde{y}_1(f)e^{j2\pi ft}\,df, \tag{37}$$

then $\tilde{y}_2(t)$ is the *inverse dual* of $\tilde{y}_1(t)$.

**Example 1.** Let

$$\tilde{y}_1(t) = \begin{cases} 1, & -T \le t \le T, \\ 0, & \text{elsewhere.} \end{cases} \tag{38}$$

The dual of $\tilde{y}_1(t)$ is

$$\tilde{y}_2(t) = \frac{\sin 2\pi Tt}{\pi t}, \qquad -\infty < t < \infty. \tag{39}$$

**Definition 2. Dual Processes.** The complex Gaussian process $\tilde{y}_2(t)$ is the *statistical dual* of the complex Gaussian process $\tilde{y}_1(t)$ if

$$\begin{aligned}\tilde{K}_{\tilde{y}_2}\{f_1, f_2\} &\triangleq E[\tilde{y}_2(f_1)\tilde{y}_2^*(f_2)] \\ &= \mathscr{F}[\tilde{K}_{\tilde{y}_1}(t_1, t_2)] \\ &\triangleq \iint_{-\infty}^{\infty} \exp[-j2\pi f_1t_1 + j2\pi f_2t_2]\tilde{K}_{\tilde{y}_1}(t_1, t_2)\,dt_1\,dt_2.\end{aligned} \tag{40}$$

Note the sign convention in the direct Fourier transform. The complex Gaussian process $\tilde{y}_2(t)$ is the *statistical inverse dual* of the complex Gaussian process $\tilde{y}_1(t)$ if

$$\begin{aligned}\tilde{K}_{\tilde{y}_2}(t_1, t_2) &= \mathscr{F}^{-1}[\tilde{K}_{\tilde{y}_1}\{f_1, f_2\}] \\ &\triangleq \iint_{-\infty}^{\infty} \exp[+j2\pi f_1t_1 - j2\pi f_2t_2]\tilde{K}_{\tilde{y}_1}\{f_1, f_2\}\,df_1\,df_2.\end{aligned} \tag{41}$$

**Property 2** [3]. Assume that $\tilde{y}_2(t)$ is the statistical dual of $\tilde{y}_1(t)$, which is a *nonstationary* process whose expected energy is finite. We expand both processes over the infinite interval. The eigenvalues of $\tilde{y}_2(t)$ are identical with those of $\tilde{y}_1(t)$, and the eigenfunctions of $\tilde{y}_2(t)$ are Fourier transforms of the eigenfunctions of $\tilde{y}_1(t)$. This property follows by direct substitution.

**Example 2.** Let

$$\tilde{K}_{\tilde{y}_1}(t_1, t_2) = \sum_{i=1}^{\infty} \tilde{\lambda}_i \tilde{\varphi}_i(t_1)\tilde{\varphi}_i^*(t_2), \qquad -\infty < t_1, t_2 < \infty. \tag{42}$$

The expansion of the dual process is

$$\tilde{K}_{\tilde{y}_2}\{f_1, f_2\} = \sum_{i=1}^{\infty} \tilde{\lambda}_i \tilde{\Phi}_i\{f_1\}\tilde{\Phi}_i^*\{f_2\}, \qquad -\infty < f_1, f_2 < \infty. \tag{43}$$

At this point the reader should see why we are interested in dual processes. The performance of detection and estimation systems depends on eigenvalues, *not* eigenfunctions. Thus, systems in which the various processes are dual will perform in an identical manner.

**Property 3.** White complex Gaussian noise is a statistically self-dual process.

**Property 4.** If $\tilde{y}_2(t)$ is the dual of $\tilde{y}_1(t)$, where $\tilde{y}_1(t)$ is any sample function from a zero-mean random process, then $\tilde{K}_{\tilde{y}_2}\{f_1, f_2\}$ is the double Fourier transform of $K_{\tilde{y}_1}(t_1, t_2)$.

**Definition 3.** Consider the two deterministic functionals

$$\tilde{z}_1(t) = g_1(\tilde{y}_1(\cdot), t) \tag{44}$$

and

$$\tilde{z}_2(t) = g_2(\tilde{y}_2(\cdot), t). \tag{45}$$

Assume that $\tilde{y}_2(t)$ is the dual of $\tilde{y}_1(t)$. If this always implies that $\tilde{z}_2(t)$ is the dual of $\tilde{z}_1(t)$, then $g_2(\cdot, \cdot)$ is the *dual operation* of $g_1(\cdot, \cdot)$.

To illustrate this idea, we consider a simple example of a dual operation.

**Example 3.** Let $g_1(\cdot, \cdot)$ correspond to a delay line with a delay of $a$ seconds. Thus,

$$\tilde{z}_1(t) = \tilde{y}_1(t - a). \tag{46}$$

The dual operation is the frequency translation

$$\tilde{z}_2(t) = \tilde{y}_2(t)e^{-j2\pi at}. \tag{47}$$

To verify this, we observe that

$$\tilde{Z}_1\{f\} = \tilde{Y}_1\{f\}e^{-j2\pi fa}. \tag{48}$$

**Property 5.** The operations listed in Table 12.1 are duals (see Problems 12.3.4–12.3.10).

**Table 12.1**

| Operation | Dual operation |
|---|---|
| Delay line | Frequency translation |
| Time-varying gain | Filter |
| Gate | Low-pass or bandpass filter |
| Adder | Adder |
| Convolution | Multiplier |
| Aperiodic correlator | Square-law envelope detector |

Thus, if

$$\tilde{y}_2(t) = \tilde{Y}_1\{t\}, \tag{49}$$

then

$$\tilde{z}_2(t) = \tilde{Z}_1\{t\}, \tag{50}$$

which is the required result.

**Property 6.** Assume that the input to $g_1(\tilde{y}_1(\cdot), t)$ is a sample function of a complex Gaussian random process and that the input to $g_2(\tilde{y}_2(\cdot), t)$ is a sample function of a dual process. If $g_1(\cdot, \cdot)$ is the dual operation of $g_1(\cdot, \cdot)$, then $\tilde{z}_2(t)$ is the dual process of $\tilde{z}_1(t)$.

This completes our introductory discussion. We now turn to the specific problem of interest.

### 12.3.2 Dual Targets and Channels

In this section we introduce the idea of a dual target or channel and demonstrate that a nonfluctuating dispersive target is the dual of a fluctuating point target.

To motivate the definition, we recall the relations for the Doppler-spread and range-spread targets. The reflected signal from a Doppler-spread target at zero range is a signal

$$\tilde{s}_D(t) = \sqrt{E_t}\, \tilde{b}_D(t)\tilde{f}_D(t), \qquad -\infty < t < \infty, \tag{51}$$

whose covariance function is

$$\tilde{K}_{\tilde{s}_D}(t, u) = E_t \tilde{f}_D(t)\tilde{K}_D(t-u)\tilde{f}_D^*(u), \qquad -\infty < t, u < \infty. \tag{52}$$

The reflected signal from a range-spread target is a signal

$$\tilde{s}_R(t) = \sqrt{E_t}\int_{-\infty}^{\infty} \tilde{f}_R(t-\lambda)\tilde{b}_R(\lambda)\, d\lambda, \qquad -\infty < t < \infty, \tag{53}$$

whose covariance function is

$$\tilde{K}_{\tilde{s}_R}(t, u) = E_t \int_{-\infty}^{\infty} \tilde{f}_R(t - \lambda)\tilde{S}_R(\lambda)\tilde{f}_R^*(u - \lambda)\, d\lambda, \qquad -\infty < t, u < \infty. \tag{54}$$

We may now define dual targets and channels.

**Definition 4.** Let $\tilde{f}_1(t)$ denote the transmitted signal in system 1, and $\tilde{z}_1(t)$ the returned signal. Let $\tilde{f}_2(t)$ denote the transmitted signal in system 2, and $\tilde{z}_2(t)$ the returned signal.

If the condition that $\tilde{f}_2(t)$ is the dual of $\tilde{f}_1(t)$ implies that $\tilde{z}_2(t)$ is the statistical dual of $\tilde{z}_1(t)$, system 2 is the *dual system* of system 1. (Notice that "systems" have randomness in them whereas the "operations" in Definition 3 were deterministic.)

We now apply this definition to the targets of interest.

**Property 7.** If

$$\tilde{K}_D(\tau) = \tilde{K}_R\{\tau\} \tag{55}$$

or, equivalently,

$$\tilde{S}_D\{\lambda\} = \tilde{S}_R(-\lambda), \tag{56}$$

then the Doppler-spread target (or channel) is a *dual system* with respect to the range-spread target (or channel).

*Proof.* We must prove that

$$\tilde{K}_{\tilde{s}_D}\{f_1, f_2\} = \mathscr{F}[\tilde{K}_{\tilde{s}_R}(t_1, t_2)]. \tag{57}$$

Now

$$\mathscr{F}[\tilde{K}_{\tilde{s}_R}(t_1, t_2) \triangleq \iint_{-\infty}^{\infty} e^{-j2\pi[f_1t_1 - f_2t_2]}\tilde{K}_{\tilde{s}_R}(t_1, t_2)\, dt_1 dt_2$$

$$= \iiint_{-\infty}^{\infty} e^{-j2\pi[f_1t_1 - f_2t_2]}\tilde{f}_R(t_1 - \lambda)\tilde{S}_R(\lambda)\tilde{f}_R^*(t_2 - \lambda)\, dt_1\, dt_2\, d\lambda$$

$$= \int_{-\infty}^{\infty} \tilde{S}_R(\lambda)e^{-j2\pi\lambda[f_1 - f_2]}\, d\lambda \int_{-\infty}^{\infty} \tilde{f}_R(t_1 - \lambda)e^{-j2\pi f_1(t_1 - \lambda)}\, dt_1$$

$$\times \int_{-\infty}^{\infty} \tilde{f}_R(t_2 - \lambda)e^{j2\pi f_2(t_2 - \lambda)}\, dt_2. \tag{58}$$

Using (17), this reduces to

$$\mathscr{F}[\tilde{K}_{\tilde{s}_R}(t_1, t_2)] = \tilde{K}_R\{f_1 - f_2\}\tilde{F}_R\{f_1\}\tilde{F}_R^*\{f_2\}. \tag{59}$$

If

$$\tilde{f}_D(\cdot) = \tilde{F}_R\{\cdot\} \tag{60}$$

and

$$\tilde{K}_D(\cdot) = \tilde{K}_R\{\cdot\}, \tag{61}$$

then

$$\tilde{K}_{\tilde{s}_D}\{f_1, f_2\} = \tilde{F}_R\{f_1\}\tilde{K}_R\{f_1 - f_2\}\tilde{F}_R^*\{f_2\}, \tag{62}$$

which is the desired result.

This result is of fundamental importance, because it implies that we can work with the most tractable target model. We formalize this idea with the following definition.

**Definition 5. Dual Detection Problems.** The received waveforms on the two hypotheses in system $A$ are

$$\tilde{r}_{A1}(t), \qquad -\infty < t < \infty : H_1 \tag{63}$$

and

$$\tilde{r}_{A0}(t), \qquad -\infty < t < \infty : H_0. \tag{64}$$

The received waveforms on the two hypotheses in system $B$ are

$$\tilde{r}_{B1}(t), \qquad -\infty < t < \infty : H_1 \tag{65}$$

and

$$\tilde{r}_{B0}(t), \qquad -\infty < t < \infty : H_0. \tag{66}$$

All waveforms are sample functions of complex Gaussian processes.

If $\tilde{r}_{B1}(t)$ is the dual process to $\tilde{r}_{A1}(t)$ and $\tilde{r}_{B0}(t)$ is the dual process to $\tilde{r}_{A0}(t)$, problem $B$ is the dual detection problem of problem $A$.

The following properties are straightforward to verify.

**Property 8.** If the a-priori probabilities and costs are the same in both systems, the Bayes risks in equal detection problems are identical.

**Property 9.** We can always realize the optimum receiver for system $A$ as shown in Fig. 12.5*a*. We can always realize the optimum receiver for system $B$ as shown in Fig. 12.5*b*.

Property 9 means that being able to construct the optimum receiver for either one of the two dual systems is adequate. Techniques for implementing the Fourier transformer are discussed in numerous references (e.g., [4–7]). There is some approximation involved in this operation, but we shall ignore it in our discussion. In Section 12.3.3 we shall discuss direct implementations by using dual operations.

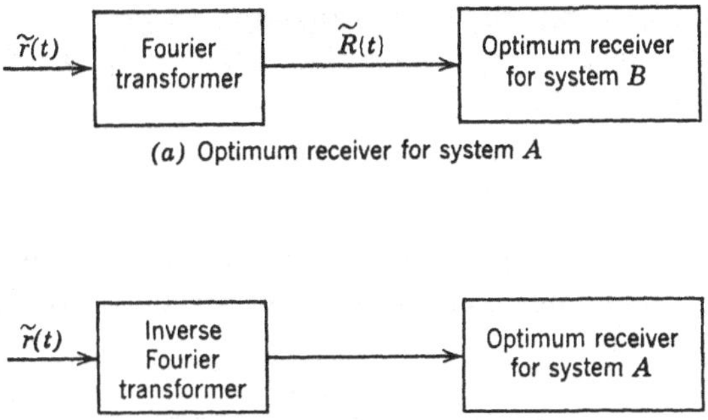


(*a*) Optimum receiver for system $A$

(*b*) Optimum receiver for system $B$

**Fig. 12.5 Optimum receivers for dual detection problems.**

**Property 10.** Consider the problem of detecting a range-spread target. The received signals on the two hypotheses are

$$\tilde{r}(t) = \tilde{s}_R(t) + \tilde{w}(t), \qquad -\infty < t < \infty : H_1 \tag{67}$$

and

$$\tilde{r}(t) = \tilde{w}(t), \qquad -\infty < t < \infty : H_0. \tag{68}$$

Consider the problem of detecting a Doppler-spread target. The received signals on the two hypotheses are

$$\tilde{r}(t) = \tilde{s}_D(t) + \tilde{w}(t), \qquad -\infty < t < \infty : H_1 \tag{69}$$

and

$$\tilde{r}(t) = \tilde{w}(t), \qquad -\infty < t < \infty : H_0. \tag{70}$$

In both cases $\tilde{w}(t)$ is a sample function from a complex Gaussian white noise process with spectral height $N_0$.

If the Doppler-spread target is a dual system to the range-spread target, the second detection problem is the dual of the first detection problem.

This property follows by using Properties 3 and 7 in Definition 5. It is important because it enables us to apply all of the results in Chapter 11 to the range-spread problem.

The result in Definition 5 concerned binary detection. The extension to $M$-ary problems and estimation problems is straightforward.

At this point in our discussion we have a number of general results available. In the next section we consider some specific cases.

### 12.3.3 Applications

In this section we apply the results of our duality theory discussion to some specific cases.

**Case 1. Dual of a Finite-State Doppler-spread Target.** The spectrum of the reflection process for a finite-state Doppler-spread target is the rational function

$$\tilde{S}_D\{f\} = \frac{a_n f^{2n-2} + \cdots + a_0}{b_n f^{2n} + \cdots + b_0}. \tag{71}$$

Notice that it is a real, non-negative, not necessarily even function of frequency. To obtain dual systems, the range-scattering function must be the rational function of $\lambda$,

$$\tilde{S}_R(\lambda) = \tilde{S}_D\{-\lambda\} = \frac{a_n(-\lambda)^{2n-2} + \cdots + a_0}{b_n(-\lambda)^{2n} + \cdots + b_0}. \tag{72}$$

If the transmitted signal for the dispersive target is $\tilde{f}\{t\}$, the Doppler-spread target system which is its dual will transmit $\tilde{F}\{t\}$. The optimum receiver is shown in Fig. 12.6.

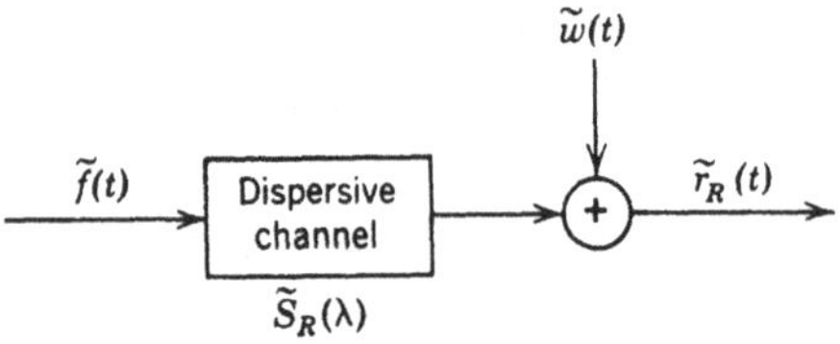


*(a)* Actual channel

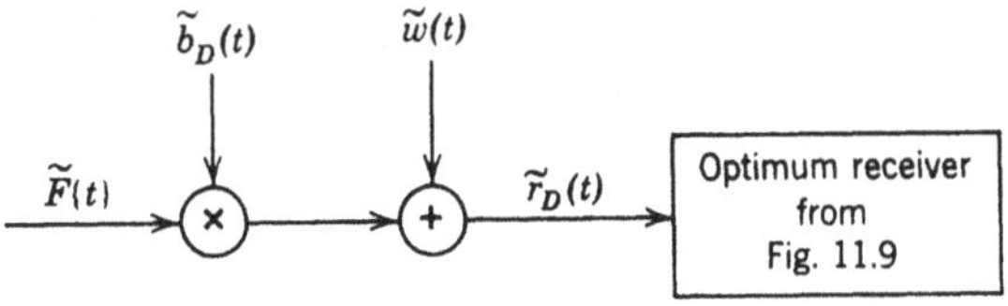


*(b)* Dual system and optimum receiver

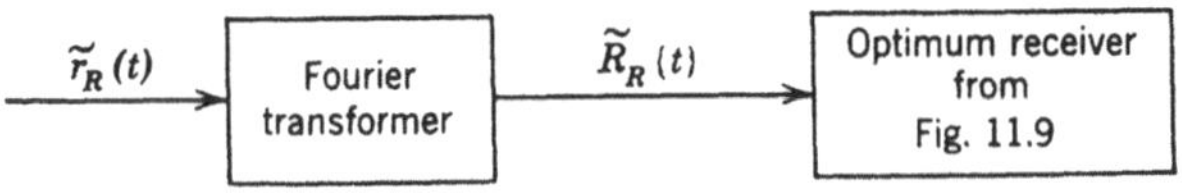


*(c)* Optimum receiver for dispersive channel

**Fig. 12.6 Finite-state range-spread target.**

To illustrate this case, we consider an example.

**Example 1.** Consider the range-spread target detection problem in which

$$\tilde{S}_R\{\lambda\} = \frac{4k\sigma_b^2}{(2\pi\lambda)^2 + k^2}, \qquad -\infty < \lambda < \infty \tag{73}$$

and

$$\tilde{f}_R\{t\} = \begin{cases} \dfrac{1}{\sqrt{T}}, & -\dfrac{T}{2} \le t \le \dfrac{T}{2}, \\ 0, & \text{elsewhere.} \end{cases} \tag{74}$$

The dual of this is the Doppler-spread problem in which

$$\tilde{S}_D\{f\} = \frac{4k\sigma_b^2}{(2\pi f)^2 + k^2}, \qquad -\infty < f < \infty \tag{75}$$

and

$$\tilde{f}_D\{t\} = \sqrt{T}\,\frac{\sin \pi T t}{\pi T t}, \qquad -\infty < t < \infty. \tag{76}$$

Combining the results in Fig. 12.6 and (11.38)–(11.49) (see also Prob. 11.2.4) gives the optimum receiver in Fig. 12.7. The performance is obtained from the result in Section 11.2.3.

We should observe that the dual of a finite-state Doppler-spread target is a range-spread target that is infinite in extent. This is never true in practice, but frequently we obtain an adequate approximation to $\tilde{S}_R(\lambda)$ with a rational function.

**Case 2. SPLOT Condition.** In the Doppler-spread case we obtained simple results when

$$\tilde{f}_D(t) = \begin{cases} \dfrac{1}{\sqrt{T}}, & -\dfrac{T}{2} \le t \le \dfrac{T}{2} \\ 0, & \text{elsewhere} \end{cases} \tag{77}$$

and $T$ was large compared to the correlation time of $\tilde{b}_D(t)$ as measured by the covariance function $\tilde{K}_D(\tau)$. The dual of this case arises when

$$\tilde{F}_R\{f\} = \begin{cases} \dfrac{1}{\sqrt{W}}, & -\dfrac{W}{2} \le f \le \dfrac{W}{2} \\ 0, & \text{elsewhere} \end{cases} \tag{78}$$

and $W$ is large compared to the two-frequency correlation distance as measured by $K_R\{v\}$.

A filter-squarer-integrator receiver for the Doppler-spread case is shown in Fig. 12.8. The gating operation is added to take into account the finite

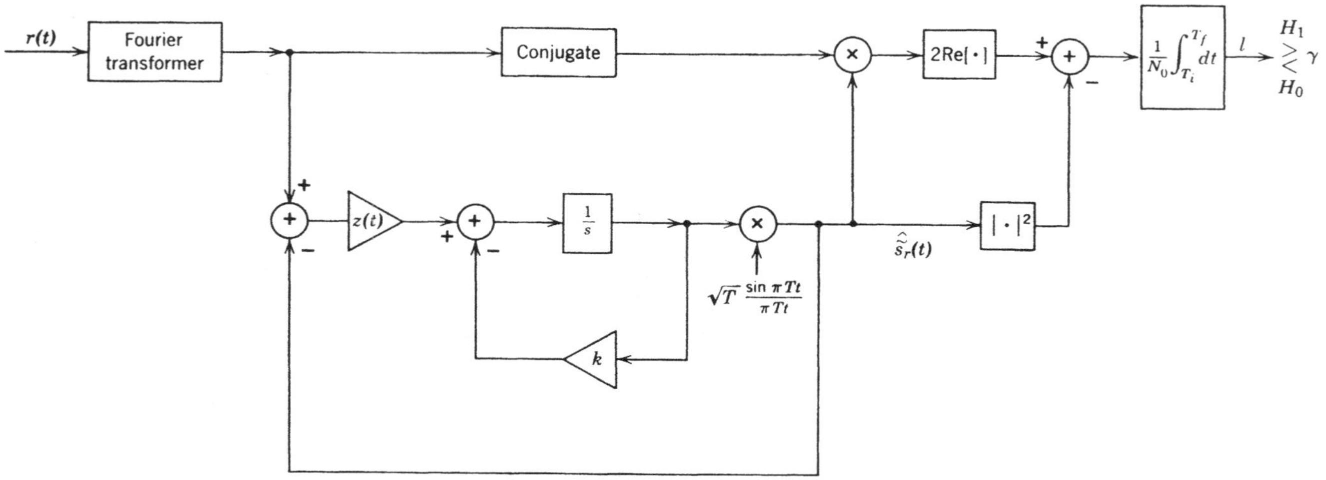


**Fig. 12.7 Optimum receiver: finite-state dispersive channel.**

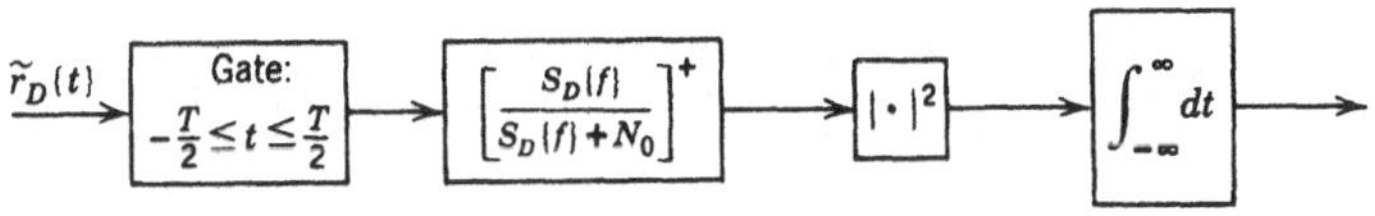


**Fig. 12.8 Optimum receiver for Doppler-spread target: SPLOT condition.**

observation time. We could implement the optimum receiver by using a Fourier transformer cascaded with the system in Fig. 12.8. In this particular problem it is easier just to implement the inverse dual of Fig. 12.8 Using the properties in Table 12-1, we obtain the system in Fig. 12.9. (We reversed the two zero-memory operations to avoid factoring the spectrum.) Notice that the transmitted signal is

$$\tilde{f}(t) = \sqrt{W} \frac{\sin \pi W t}{\pi W t}, \qquad -\infty < t < \infty. \tag{79}$$

This pulse will never be used exactly. However, if the transmitted pulse has a transform that is relatively flat over a frequency band, the receiver in Fig. 12.9 should be close to optimum.

**Case 3. LEC Condition.** When the LEC condition is valid, we can solve the problem directly for either the range-spread or Doppler-spread target. In Fig. 12.10 we show the two receivers. It is easy to verify that they are duals.

**Case 4. Resolvable Multipath.** The resolvable multipath problem corresponds to a scattering function,

$$\tilde{S}_R\{\lambda\} = \sum_{i=1}^{K} \tilde{b}_i \delta\{\lambda - \lambda_i\}, \tag{80}$$

where the $\lambda_i$ are sufficiently separated so that the output due to each path may be identified. This is the dual of the Doppler channel with the scattering function

$$\tilde{S}_D\{f\} = \sum_{i=1}^{K} \tilde{b}_i \delta\{f - f_i\}. \tag{81}$$

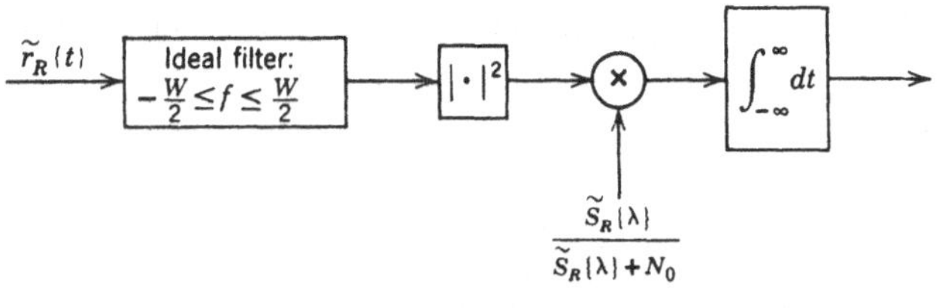


**Fig. 12.9 Optimum receiver for range-spread target: SPLOT condition.**

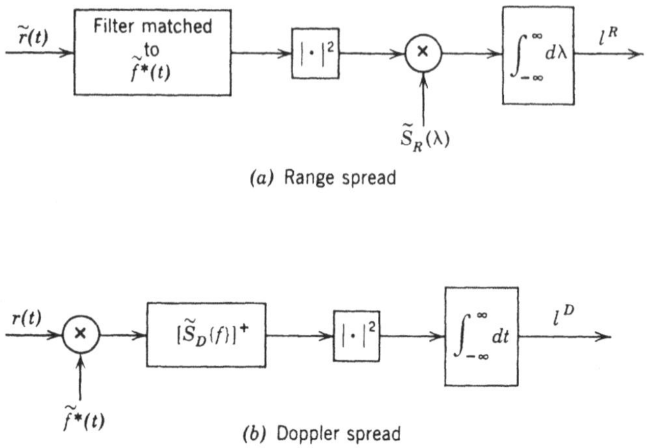


**Fig. 12.10 Optimum receiver: LEC condition.**

Notice that (81) does not describe a frequency-diversity system. It corresponds to a set of slowly fluctuating point targets moving at different velocities.

**Case 5. Optimum Binary Communication.** The model for a binary communication system operating over a range-spread channel is analogous to that in Section 11.3. The transmitted signals are given by (11.68). The receiver consists of two simple binary receivers referenced to $\omega_1$ and $\omega_0$. The actual implementation will depend on the physical situation, but it will correspond to one of the structures developed in this chapter.

The point of current interest is the performance. The derivation in Section 11.3.1 did not rely on the channel characteristics. Thus, the bound in (11.91),

$$\Pr(\epsilon) \le \tfrac{1}{2} \exp\left[-0.1488\,\frac{\bar{E}_r}{N_0}\right], \tag{82}$$

is also valid for range-spread channels. We now consider two examples of signal design to show how we can approach the bound.

**Example 2.** Let

$$\tilde{S}_R(\lambda) = \begin{cases} \dfrac{\sigma_b^2}{L}, & |\lambda| \le L, \\ 0, & |\lambda| > L. \end{cases} \tag{83}$$

This is the dual of the channel in (11.94). From the results of that example, we know that if $\bar{E}_r/N_0$ is large we can achieve the bound by transmitting

$$\tilde{f}(t) = \sqrt{W}\frac{\sin \pi Wt}{\pi Wt}, \qquad -\infty < t < \infty, \tag{84}$$

with $W$ chosen properly. From (11.96) the optimum value of $W$ is

$$W_o = \frac{\bar{E}_r/N_0.}{3.07(2L)} \tag{85}$$

Notice that results assume

$$W_o L \gg 1. \tag{86}$$

The signal in (84) is not practical. However, any signal whose transform is reasonably constant over $[-W_o, W_o]$ should approach the performance in (82).

**Example 3.** Let

$$\tilde{S}_R\{\lambda\} = \frac{4k\sigma_b^2}{(2\pi\lambda)^2 + k^2}, \qquad -\infty < \lambda < \infty. \tag{87}$$

This is the dual of the channel in Examples 2 and 3 on pages 382 and 384. The dual of the signal in Fig. 11.16 is

$$\tilde{F}\{f\} = \sum_{i=1}^{D_o} a\tilde{Y}(f - iW_p), \tag{88}$$

where

$$\tilde{Y}\{f\} \triangleq \begin{cases} \dfrac{1}{\sqrt{W_s}}, & 0 \le f \le W_s, \\ 0, & \text{elsewhere,} \end{cases} \tag{89}$$

and $D_o$ satisfies (11.111).

If

$$W_s \ll \frac{2\pi}{k} \tag{90}$$

and

$$W_p \gg \frac{2\pi}{k}, \tag{91}$$

then we approach the bound in (82).

The signal in (88) corresponds to transmitting $D_o$ frequency-shifted pulses simultaneously. An identical result can be obtained by transmitting them sequentially (see Problem 12.3.14). The shape in (89) is used to get an exact dual. Clearly, the shape is unimportant as long as (90) is satisfied.

These results deal with binary communication. The results in Section 11.3.4 on $M$-ary systems carry over to range-spread channels in a similar manner.

**Case 6. Suboptimum Receiver No. 1.** In Section 11.3.3 we developed a suboptimum receiver for the Doppler-spread channel by using a piecewise

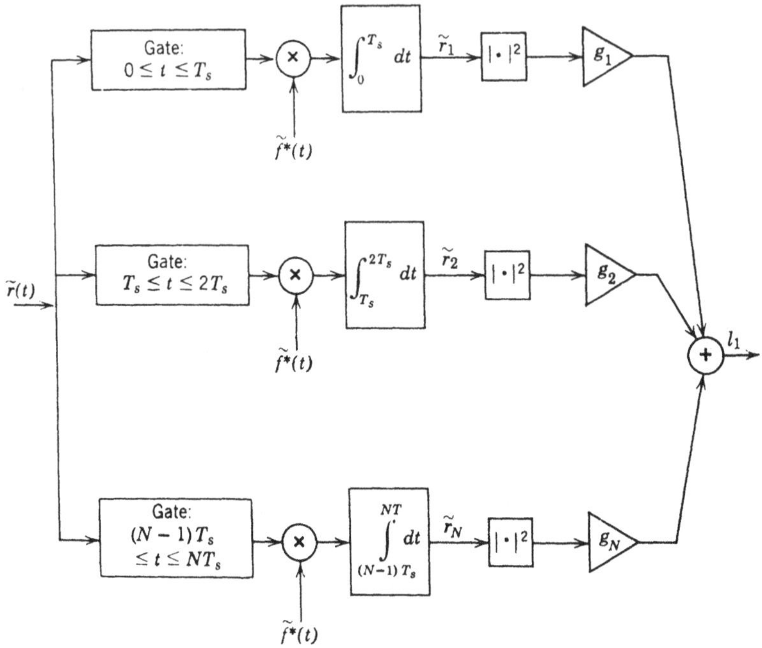


**Fig. 12.11 Suboptimum receiver No. 1 for Doppler-spread channel.**

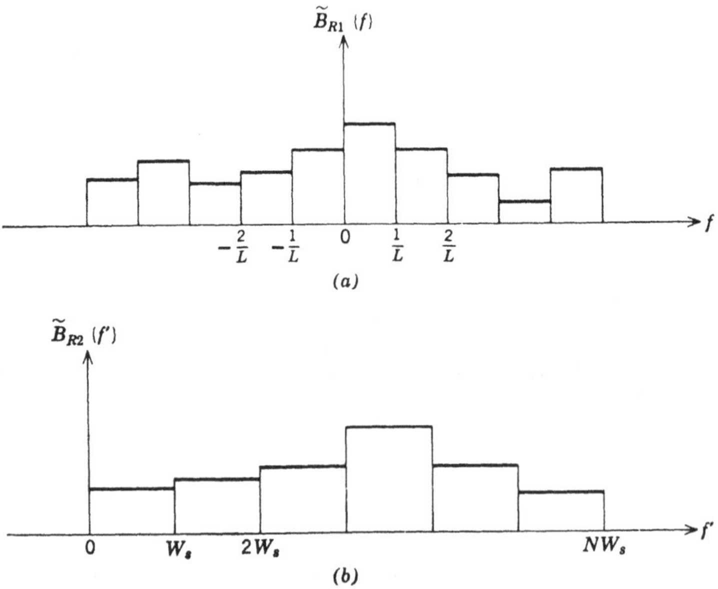


**Fig. 12.12 Piecewise constant approximations to the transform of $\tilde{b}_R(t)$.**

constant approximation to the channel fading process. We have repeated Fig. 11.18 (redrawn slightly) in Fig. 12.11. In Fig. 12.12*a* and *b*, we show two types of piecewise approximations to the transform of the channel fading process. In the first approximation, we use segments of $L^{-1}$ and have a total of

$$D_R \triangleq WL \tag{92}$$

segments. In the second approximation, we let the segment length equal $W_s$ and regard it as a design parameter. We also shift the origin for notational simplicity. The resulting receiver is shown in Fig. 12.13. This is the dual of the receiver in Fig. 12.11. The performance can be analyzed in exactly the same manner as in Section 11.3.3.

**Case 7. Suboptimum Receiver No. 2.** The dual of the suboptimum FSI receiver in Fig. 11.20 is the two-filter radiometer in Fig. 12.14. The multiplier $\tilde{G}(\lambda)$ is a function that we choose to optimize the performance. In the LEC case

$$\tilde{G}(\lambda) = \tilde{S}_R(\lambda) \tag{93}$$

(see Case 3), while in the SPLOT case

$$\tilde{G}(\lambda) = \frac{\tilde{S}_R(\lambda)}{\tilde{S}_R(\lambda) + N_0}. \tag{94}$$

This type of receiver is analyzed in [8].

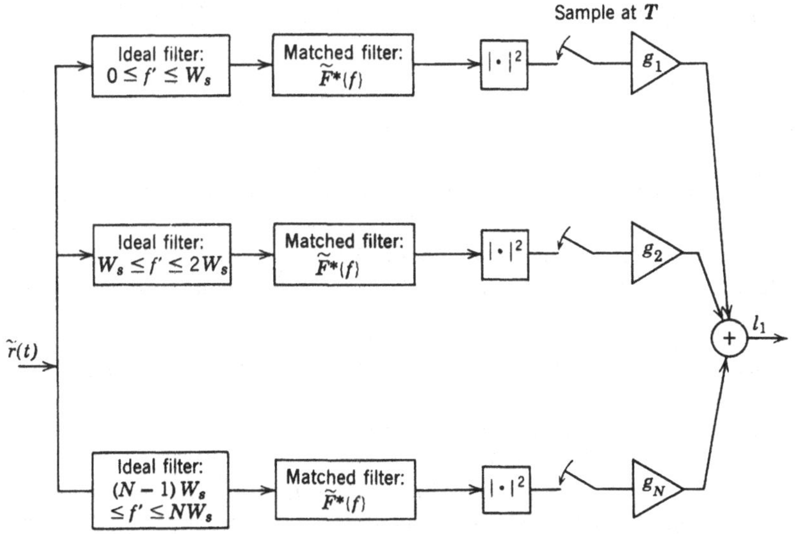


**Fig. 12.13 Suboptimum receiver No. 1 for range-spread channel.**

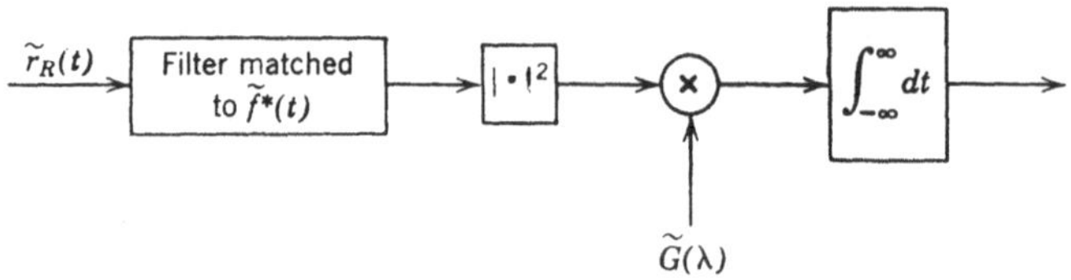


**Fig. 12.14 Suboptimum two-filter radiometer.**

**Case 8. Dual Estimation Problems.** In Section 11.4, we introduced the problem of estimating the range and mean Doppler of a fluctuating point target. The dual problem is that of estimating the Doppler and mean range of a nonfluctuating range-spread target.

We assume that the target is a nonfluctuating range-spread target whose mean range is

$$m_R \triangleq \frac{1}{2\sigma_b^2}\int_{-\infty}^{\infty} \lambda \tilde{S}_R(\lambda)\, d\lambda. \tag{95}$$

It is moving at a constant velocity corresponding to a Doppler shift $f$. The complex envelope of the returned waveform is

$$\tilde{r}(t) = \sqrt{E_t}\, e^{j2\pi f t}\int_{-\infty}^{\infty} \tilde{f}(t-\lambda)\tilde{b}_R(\lambda)\, d\lambda + \tilde{w}(t), \qquad -\infty < t < \infty. \tag{96}$$

The covariance function of the first term is

$$E[\tilde{s}(t)\tilde{s}^*(u)] = E_t e^{j2\pi f[t-u]}\int_{-\infty}^{\infty} \tilde{f}(t-\lambda_1-m_R)\tilde{S}_{R_0}(\lambda_1)\tilde{f}^*(u-\lambda_1-m_R)\, d\lambda_1,$$
$$-\infty < t, u < \infty, \tag{97}$$

where

$$\tilde{S}_{R_0}(\lambda) \triangleq \tilde{S}_R(\lambda - m_R). \tag{98}$$

This problem is the dual of that in Section 11.4. The various results of interest are developed in the problems.

This completes our discussion of the applications of time-frequency duality. Our interesting examples are developed in the problems. Before leaving the subject, several observations are important:

1. The discussion assumes infinite limits, so that there is an approximation involved.
2. If the system is implemented with a Fourier transformer, there is an approximation.
3. The concept as a guide to thinking about problems is as useful as the formal manipulations. The result of the manipulations should always be checked to see whether they operate as intended.

If one remembers these points, duality theory provides a powerful tool for solving and understanding problems. We now summarize the results of the chapter.

## 12.4 SUMMARY: RANGE-SPREAD TARGETS

In this chapter we have considered range-spread targets. The returned signal is modeled as a sample function of a zero-mean complex Gaussian process that is described by the relation

$$\tilde{s}(t) = \sqrt{E_t}\int_{-\infty}^{\infty} \tilde{b}_R(\lambda)\tilde{f}(t-\lambda)\,d\lambda. \tag{99}$$

The covariance function of the returned signal is

$$\tilde{K}_{\tilde{s}}(t, u) = E_t\int_{-\infty}^{\infty} \tilde{f}(t-\lambda)\tilde{S}_R(\lambda)\tilde{f}^*(u-\lambda)\,d\lambda. \tag{100}$$

We observed that a range-spread target caused frequency-selective fading.

The detection problem was once again that of detecting a sample function of a Gaussian random process in the presence of additive noise. The

**Table 12.2 Singly spread target results**

| | Doppler-spread target | Range-spread target |
|---|---|---|
| Reflected signal | $\tilde{s}(t) = \sqrt{E_t}\,\tilde{b}_D(t)\tilde{f}(t-\lambda)$ | $\tilde{s}(t) = \sqrt{E_t}\int_{-\infty}^{\infty} \tilde{f}(t-\lambda)\tilde{b}_R(\lambda)\,d\lambda$ |
| Covariance function | $E_t\tilde{f}(t-\lambda)\tilde{K}_D(t-u)\tilde{f}^*(u-\lambda)$ | $E_t\int_{-\infty}^{\infty} \tilde{f}(t-\lambda)\tilde{S}_R(\lambda)\tilde{f}^*(u-\lambda)\,d\lambda$ |
| Scattering functions | $\tilde{S}_D\{f\}$ | $\tilde{S}_R(\lambda)$ |
| Correlation functions | $\tilde{K}_D(\tau)$ | $\tilde{K}_R\{v\}$ Two-frequency correlation function |
| Type of fading | Time-selective | Frequency-selective |
| Approximate diversity, $WT \approx 1$ | $(B+W)T \simeq BT+1$ | $(L+T)W \simeq WL+1$ |
| Condition for flat fading | $T \ll \frac{1}{B}$ | $W \ll \frac{1}{L}$ |

structure of the optimum receiver followed easily, but the integral equation of interest was difficult to solve. For the LEC case and the separable kernel case we could obtain a complete solution.

In order to obtain solutions for the general case, we introduced the concept of time-frequency duality. This duality theory enabled us to apply all of our results for Doppler-spread targets to range-spread targets. In retrospect, we can think of Doppler-spread and range-spread targets as examples of *single-spread targets* (either time or frequency, but not both). In Chapter 13, we shall encounter other examples of singly-spread targets. In Table 12.2, we have collected some of the pertinent results for singly-spread targets.

Our discussion has concentrated on the detection problem in the presence of white noise. Other interesting topics, such as parameter estimation, detection in the presence of colored noise, and the resolution problem, are developed in the problems of Section 12.5. We now turn our attention to targets and channels that are spread in both range and Doppler.

## 12.5 PROBLEMS

### P.12.2 Detection of Range-Spread Targets

**Problem 12.2.1.** Consider the covariance function in (10). Prove that $\tilde{K}_{\tilde{s}}(t, u)$ can be written as

$$\tilde{K}_{\tilde{s}}(t, u) = \iint_{-\infty}^{\infty} df_1\, df_2\, e^{j2\pi f_1 t} \tilde{F}\{f_1\} \tilde{K}_R\{f_1 - f_2\} \tilde{F}^*\{f_2\} e^{-j2\pi f_2 u}. \tag{P.1}$$

**Problem 12.2.2.** Assume that

$$\tilde{S}_R(\lambda) = \sum_{k=1}^{N} 2\sigma_k^2\, \delta(\lambda - \lambda_k).$$

1. Find $\tilde{K}_{\tilde{s}}(t, u)$.
2. Find the optimum receiver. Specify all components completely.

**Problem 12.2.3.** Assume that $\tilde{f}(t)$ is bandlimited to $\pm W/2$ cps. We approximate $\tilde{S}_R(\lambda)$ as

$$\tilde{S}_R(\lambda) = \sum_{k=1}^{N} \frac{1}{W} \tilde{S}_R\left(\frac{k}{W}\right) \delta\left(\lambda - \frac{k}{W}\right), \tag{P.1}$$

where

$$N = LW,$$

which is assumed to be an integer.

1. Draw a block diagram of the optimum receiver.
2. Justify the approximation in (P.1) in the following way:

a. Use finite limits on the expression in (P.1) in Problem 12.2.1.
b. Expand $K_R\{\cdot\}$ using Mercer's theorem.
c. Use the asymptotic properties of the eigenfunctions that were derived in Section I-3.4.6 (page I-206).

**Problem 12.2.4.** Assume that

$$\tilde{S}_R(\lambda) = \frac{2\sigma_b^2}{\sqrt{2\pi}\,\sigma_R} e^{-\lambda^2/2\sigma_R^2}, \qquad -\infty < \lambda < \infty,$$

and that

$$\tilde{f}(t) = \left(\frac{1}{\pi T^2}\right)^{1/4} e^{-t^2/2T^2} \qquad -\infty < t < \infty.$$

The LEC condition is valid.

1. Evaluate $\tilde{\mu}(s)$.
2. What choice of $T$ minimizes $\tilde{\mu}(s)$? Explain your result intuitively.

**Problem 12.2.5.**

1. Prove that the expression in (35) can also be written as

$$\tilde{\mu}(s) = -\frac{s(1-s)E_t^2}{N_0} \int_{-\infty}^{\infty} \theta\{x, 0\}\beta_R(x)\,dx,$$

where

$$\beta_R(x) = \int_{-\infty}^{\infty} \tilde{S}_R(\lambda_2 + x)\tilde{S}_R(\lambda_2)\,d\lambda_2.$$

2. Express $\beta_R(x)$ in terms of $\tilde{K}_R\{v\}$.
3. Combine parts 1 and 2 to obtain another expression for $\tilde{\mu}(s)$.

**Problem 12.2.6.** Assume that

$$\tilde{S}_R(\lambda) = \begin{cases} \dfrac{2\sigma_b^2}{L}, & |\lambda| \le \dfrac{L}{2}, \\ 0, & \text{elsewhere} \end{cases}$$

and

$$\tilde{f}(t) = \begin{cases} \dfrac{1}{\sqrt{T}}, & |t| \le \dfrac{T}{2} \\ 0, & \text{elsewhere.} \end{cases}$$

The LEC condition is valid. Evaluate $\tilde{\mu}(s)$.

## P. 12.3 Time-Frequency Duality

**Problem 12.3.1.** The signal $\tilde{f}(t)$ is

$$\tilde{f}(t) = a \sum_{i=1}^{K} \tilde{u}(t - iT_p),$$

where $\tilde{u}(t)$ is defined in (10.29). Find the dual signal.

**Problem 12.3.2.** The signal $\tilde{f}(t)$ is

$$\tilde{f}(t) = \left(\frac{1}{\pi T^2}\right)^{1/4} e^{-t^2/2T^2}, \qquad -\infty < t < \infty.$$

Find the dual signal.

**Problem 12.3.3.** Find the duals of the Barker codes in Table 10.1.

**Problem 12.3.4. Time-Varying Gain.** Let

$$\tilde{z}_1(t) = \tilde{a}(t)\tilde{y}_1(t),$$

where $\tilde{a}(t)$ is a known function. Find the dual operation.

**Problem 12.3.5. Filter.** Let

$$\tilde{z}_1(t) = \int_{-\infty}^{\infty} \tilde{h}(t-\tau)\tilde{y}_1(\tau)\,d\tau,$$

where $\tilde{h}(\cdot)$ is a known function. Find the dual operation.

**Problem 12.3.6. Gate.** Let

$$\tilde{z}_1(t) = \begin{cases} \tilde{y}_1(t), & T_1 \le t \le T_2, \\ 0, & \text{elsewhere.} \end{cases}$$

Find the dual operation.

**Problem 12.3.7. Ideal Filters.** Let

$$\tilde{z}_1(t) = \int_{-\infty}^{\infty} \tilde{h}(t-\tau)\tilde{y}_1(\tau)\,d\tau,$$

where

$$\tilde{H}\{f\} = \begin{cases} 1, & F_0 \le f \le F_1, \\ 0, & \text{elsewhere.} \end{cases}$$

Find the dual operation.

**Problem 12.3.8 Aperiodic Cross-Correlator.** Let

$$\tilde{z}_1(t) = \int_{-\infty}^{\infty} \tilde{g}_1^*(t+\tau)\tilde{g}_2(\tau)\,d\tau.$$

Find the dual operation.

**Problem 12.3.9.** Let

$$\tilde{z}_1(t) = \int_{-\infty}^{\infty} \tilde{g}_1^*(t+\tau)\tilde{g}_1(\tau)\,d\tau.$$

Find the dual operation.

**Problem 12.3.10.**

1. Let

$$\tilde{z}_1(t) = \int_{t-T}^{t} \tilde{y}_1(u)\,du.$$

Find the dual operation.

2. Let

$$\tilde{z}_1(t) = \int_{0}^{t} \tilde{y}_1(u)\,du.$$

Find the dual operation.

**Problem 12.3.11.** Consider the detection problem specified in (67) and (68).

$$\tilde{S}_R(\lambda) = \frac{2\sqrt{2}\,P/k}{(2\pi f/k)^4 + 1}, \qquad -\infty < \lambda < \infty$$

and

$$\tilde{f}(t) = c \sin^2\left(\frac{2\pi t}{T}\right), \qquad 0 \leq t \leq T.$$

Draw a block diagram of the optimum receiver.

**Problem 12.3.12.** Consider Case 2 and the signal in (78). Derive the receiver in Fig. 12.9 directly from (29)–(31) without using duality theory.

**Problem 12.3.13.** Consider the two systems in Fig. 12.10. Verify that the receiver in Fig. 12.10*b* is the dual of the receiver in Fig. 12.10*a*.

**Problem 12.3.14.** Consider the example on page 433. Assume that we transmit

$$\tilde{f}(t) = \sum_{k=1}^{D_0} a\tilde{u}(t - kT_s)e^{j2\pi kW_p},$$

where $\tilde{u}(t)$ satisfies (11.113b).

1. Describe $\tilde{f}(t)$.
2. Verify that this signal achieves the same performance as that in (88) when the parameters are chosen properly.

**Problem 12.3.15.** Assume that $L = 200\ \mu$sec in (83). The available signal power-to-noise level ratio at the channel output is

$$\frac{P_R}{N_0} = 10^5.$$

The required Pr ($\epsilon$) is $10^{-4}$. We use a binary FSK system.

1. What is the maximum rate in bits per second that we can communicate over this channel with a binary system satisfying the above constraints?
2. Design a signaling scheme to achieve the rate in part 1.

**Problem 12.3.16.** Consider Case 6. Derive all of the expressions needed to analyze the performance of suboptimum receiver No. 1.

**Problem 12.3.17.** Consider Case 7. Derive all of the expressions needed to analyze the performance of suboptimum receiver No. 2.

**Problem 12.3.18.** In Case 8 (95)–(98), we formulated the problem of estimating the Doppler shift and mean range of a nonfluctuating range-spread target.

1. Starting with the general definition in (11.181), show that the *range-spread ambiguity function* is

$$\theta_{\Omega_R}\{m_{Ra}, m_R{:}f_a, f\} = \frac{E_t}{N_0}\iiint_{-\infty}^{\infty} h_{ou}(t, u{:}m_R, f)e^{-j2\pi f_a(t-u)}\tilde{f}^*(t - \lambda_1 - m_{Ra})\tilde{S}^*_{R_0}(\lambda_1)$$
$$\times \tilde{f}(u - \lambda_1 - m_{Ra})\, d\lambda_1\, dt\, du.$$

2. When the LEC condition is valid, the expression can be simplified. Show that one expression is

$$\theta_{\Omega_R,\text{LEC}}\{m_R'{:}f'\} = \frac{E_t^2}{N_0^2}\int_{-\infty}^{\infty} dx\ \theta\{x + m_R, f'\}\int_{-\infty}^{\infty} \tilde{S}_{R_0}(x + \lambda)\tilde{S}^*_{R_0}(\lambda)\, d\lambda,$$

where

$$m'_R \triangleq m_R - m_{Ra}$$

and

$$f' \triangleq f - f_a.$$

3. Express $\theta_{\Omega_R,\mathrm{LEC}}\{\cdot,\cdot\}$ in several different ways.

**Problem 12.3.19.** Prove that $\theta_{\Omega_D,\mathrm{LEC}}\{\lambda, m_D\}$, is the dual of $\theta_{\Omega_D,\mathrm{LEC}}\{m_R, f\}$. Specifically, if

$$\tilde{f}_D(t) = \tilde{F}_R\{t\}$$

and

$$\tilde{S}_D\{\lambda\} = \tilde{S}_R(-\lambda),$$

then

$$\theta_{\Omega_D,\mathrm{LEC}}\{y, x\} = \theta_{\Omega_R,\mathrm{LEC}}\{x, y\}.$$

**Problem 12.3.20.** Derive the elements in the information matrix **J** in terms of $\theta_{\Omega_R,\mathrm{LEC}}\{\cdot,\cdot\}$ and its derivatives.

**Problem 12.3.21.** Assume that

$$\tilde{f}(t) = \left(\frac{1}{\pi T^2}\right)^{1/4} e^{-t^2/2T^2}, \qquad -\infty < t < \infty$$

and

$$\tilde{S}_{R_0}\{\lambda\} = \frac{2\sigma_b^{\,2}}{\sqrt{2\pi}\,\sigma_R} e^{-\lambda^2/2\sigma_R^2}, \qquad -\infty < \lambda < \infty.$$

1. Evaluate $\theta_{\Omega_R,\mathrm{LEC}}\{m_R, f\}$.
2. Calculate the **J** matrix.

**Problem 12.3.22.** Consider the problem of estimating the amplitude of a scattering function. Thus,

$$\tilde{S}_R(\lambda:A) = A\tilde{S}_R(\lambda)$$

and $\tilde{S}_R(\lambda)$ is assumed known. Assume that the LEC condition is valid.

1. Find a receiver to generate $\hat{a}_{ml}$.
2. Is $\hat{a}_{ml}$ unbiased?
3. Assume that the bias on $\hat{a}_{ml}$ is negligible. Calculate

$$E[(\hat{a}_{ml} - A)^2].$$

4. Calculate a bound on the normalized variance of any unbiased estimate of $A$. Compare this bound with the result in part 3.
5. Compare the results of this problem with those in Problem 11.4.8.

**Problem 12.3.23.** Assume that

$$\tilde{S}_R\{\lambda:A\} = \tilde{S}_{R_1}\left\{\frac{\lambda}{A}\right\},$$

where $\tilde{S}_{R_1}\{\cdot\}$ is known. We want to estimate $A$, the scale of the range axis. Assume that the LEC condition is valid.

1. Draw a block diagram of a receiver to generate $\hat{a}_{ml}$.
2. Evaluate the Cramér-Rao bound.

**Problem 12.3.24.** Assume that

$$\tilde{S}_R(\lambda) = 2\sigma_1^2\,\delta(\lambda - \lambda_1) + 2\sigma_2^2\,\delta(\lambda - \lambda_2).$$

We want to estimate $\lambda_1$ and $\lambda_2$.

1. Find a receiver to generate $\hat{\lambda}_{1,ml}$ and $\hat{\lambda}_{2,ml}$.
2. Evaluate the Cramér-Rao bound.
3. How does this problem relate to the discrete resolution problem of Section 10.5?

**Problem 12.3.25.** Assume that we design the optimum receiver to detect a slowly fluctuating point target located at $\tau = 0, f = 0$, in the presence of white noise. We want to calculate the effect of various types of interfering targets. Recall from (9.49) that $\Delta$ characterizes the performance. Calculate the decrease in $\Delta$ due to the following:

1. A slowly fluctuating point target located at $(\tau, f)$.
2. A range-spread target with scattering function $\tilde{S}_R(\lambda)$ and Doppler shift of $f$ cps.
3. A Doppler-spread target with scattering function $\tilde{S}_D\{f\}$ and range $\lambda$.
4. Interpret the above results in terms of the ambiguity function. Discuss how you would design signals to minimize the interference.

**Problem 12.3.26.** Assume that we design the optimum LEC receiver to detect a range-spread target in the presence of white noise. We want to calculate the effect of various types of interfering targets. For simplicity, assume that the desired target has zero velocity and zero mean range. Calculate the degradation due to the following:

1. A slowly fluctuating point target at $(\tau, f)$.
2. A range-spread target with scattering function $\tilde{S}_R(\lambda)$ and Doppler shift of $f$ cps.
3. A Doppler-spread target with scattering function $\tilde{S}_D\{f\}$ and range $\lambda$.
4. Can the results in parts 1, 2, and 3 be superimposed to give a general result?

## REFERENCES


[1] R. Price and P. E. Green, "Signal Processing in Radar Astronomy—Communication via Fluctuating Multipath Media," Massachusetts Institute of Technology, Lincoln Laboratory, TR 234, October 1960.

[2] P. A. Bello, "Time-Frequency Duality," IEEE Trans. Information Theory **IT-10,** No. 1, 18–33 (Jan. 1964).

[3] R. S. Kennedy, *Fading Dispersive Communication Channels*, Wiley, New York, 1969.

[4] J. W. Cooley and J. W. Tukey, "An algorithm for the machine computation of complex Fourier series," Math. Comput. Vol. 19, 297–301 (April 1965).

[5] G. Bergland, "A guided tour of the fast Fourier transform," IEEE Spectrum, **6,** 41–52 (July 1969).

[6] B. Gold and C. Rader, *Digital Processing of Signals*, McGraw-Hill, New York, 1969.

[7] J. W. Cooley, P. A. W. Lewis, and P. D. Welch, "Historical notes on the fast Fourier transforms," IEEE Trans. Audio and Electroacoustics, **AU-15,** 76–79 (June 1967).

[8] R. R. Kurth, "Distributed-Parameter State-Variable Techniques Applied to Communication over Dispersive Channels," Sc.D. Thesis, Department of Electrical Engineering, Massachusetts Institute of Technology, June 1969.

# 13

# *Doubly-Spread Targets and Channels*

In this chapter, we generalize our model to include targets that are doubly-spread. There are several areas in which this type of target (or channel) is encountered.

A simple example of the first area arises in sonar systems. When an acoustic pulse is transmitted into the ocean, energy is returned from a large number of objects distributed throughout the ocean. If a submarine or ship is present, it also reflects the pulse. The latter reflection provides the information bearing signal and the scattered returns from other objects constitute the interference. The reflectors that cause the interference have various velocities and reflective cross-sections. In a reasonable model we assign a random amplitude and phase to each reflection. The location of the scatterers can be modeled as a spatial Poisson process whose mean-value function governs the average density. The velocity of the scatterers can be modeled by assigning a range-dependent probability density for the velocity of each scatterer. If we use this approach and assume a large number of scatterers, the result is a reverberation return that is a sample function from a Gaussian random process. The Poisson model has a great deal of physical appeal. It is developed quantitatively in [1]–[4]. In the next section we obtain the same Gaussian result in a less physical, but computationally simpler, manner. The first distinguishing feature of this type of problem is that the *spread* target corresponds to an *unwanted* return that we want to eliminate. The second feature is that the target is "soft" (i.e., its physical structure is not fixed).

A second area in which doubly-spread targets occur is one in which the details of the target are the quantities of interest. This area is encountered in mapping radars (e.g., ground mapping from airplanes or satellites; moon or planet mapping from the ground). Here we try to measure the detailed structure of the scattered return. We shall find that the return from the different segments of the target acts as a source of interference in the measurement problem. A second feature of this area is that the target is "hard" (i.e., its physical structure is fixed).

A third area of interest arises when we try to communicate over dispersive fluctuating channels (e.g., ionospheric and tropospheric channels, underwater acoustic channels, the orbiting dipole channel, local battlefield communication using chaff clouds). Here the return from the scatterers corresponds to the wanted signal, and the interference is some additive noise process. In addition, we see that the channel is "soft" (i.e., its physical structure changes).

The fourth area is radar astronomy. Here we want to measure the range and velocity of a rotating target that has an appreciable depth. Once again, the return from the spread target contains the desired information and the interference is an additive noise process. In this case, the target is "hard" and has a surface that is rough compared to the carrier wavelength.

We see that the areas of interest can be divided into two groups. In the first, the return from the spread target (or channel) is a source of interference. In the second, the return from the spread target (or channel) contains the desired information. In this chapter we study representative problems from the two groups.

In Section 13.1, we develop a simple model for a doubly-spread target and discuss its effect on the transmitted signal. We then use this model to study a sequence of problems.

In Section 13.2, we study the continuous resolution problem in active radar and sonar systems. This discussion is an extension of the discrete resolution problem in Section 10.5. The desired target is a slowly fluctuating point target. The interference is modeled as a continuum of reflectors, using the doubly-spread target model of Section 13.1. We then study receiver design and signal design for this type of environment. This problem, which is referred to as the reverberation problem in the sonar field and the clutter problem in the radar field, completes the discussion that we began in Chapter 10.

In Section 13.3, we study the detection of the return from a doubly-spread target in the presence of noise. This problem is just one of detecting a complex Gaussian process in complex white noise, which we first encountered in Section 11.2. However, the covariance function of the signal process is quite complicated, and it is appreciably harder to solve the problem.

In Section 13.4, we study the parameter-estimation problem briefly. Specifically, we consider the problem of estimating the amplitude of a doubly-spread target and the problem of estimating the mean range and mean Doppler of a doubly-spread target.

The major sections of the Chapter, 13.2 and 13.3, are almost independent. The reader who is interested only in communications can proceed directly from Section 13.1 to 13.3. Section 13.4 can also be read after Section 13.1. The results of the chapter are summarized in Section 13.5.

## 13.1 MODEL FOR A DOUBLY-SPREAD TARGET

Our model for a fluctuating range-spread target is a simple combination of the models in Chapters 11 and 12.

### 13.1.1 Basic Model

To illustrate the ideas involved, we consider the rotating sphere shown in Fig. 13.1. The surface of the sphere is rough compared to the wavelength of the carrier. We transmit a signal whose complex envelope is $\tilde{f}(t)$ and examine the reflected signal from the range interval $(\lambda, \lambda + d\lambda)$. The signal is the superposition of the number of reflections with random phases and can be modeled as a Rayleigh random variable. Since the orientation and composition of the reflectors that contribute to the returned signal change as functions of time (see Fig. 13.1*b* and *c*), we must model the reflection as a random process. Thus,

$$\tilde{s}(t, \lambda) = \sqrt{E_t}\tilde{f}(t - \lambda)\tilde{b}\left(t - \frac{\lambda}{2}, \lambda\right) d\lambda, \tag{1}$$

where $\tilde{b}(t, \lambda)$ is a complex Gaussian process whose independent variables are *both* time and space. The return from the entire target is a superposition of the returns from the incremental intervals. The complex envelope,

$$\tilde{s}(t) = \int_{-\infty}^{\infty} \sqrt{E_t}\tilde{f}(t - \lambda)\tilde{b}\left(t - \frac{\lambda}{2}, \lambda\right) d\lambda, \tag{2}$$

is a sample function from a zero-mean complex Gaussian process. It can be characterized by the covariance function

$$\begin{aligned}
\tilde{K}_{\tilde{s}}(t, u) &\triangleq E[\tilde{s}(t)\tilde{s}^*(u)] \\
&= E\left\{\iint_{-\infty}^{\infty} E_t\tilde{f}(t - \lambda)\tilde{f}^*(u - \lambda_1)\tilde{b}\left(t - \frac{\lambda}{2}, \lambda\right)\tilde{b}^*\left(u - \frac{\lambda_1}{2}, \lambda_1\right) d\lambda\, d\lambda_1\right\} \\
&= E_t\iint_{-\infty}^{\infty}\tilde{f}(t - \lambda)\left\{E\left[\tilde{b}\left(t - \frac{\lambda}{2}, \lambda\right)\tilde{b}^*\left(u - \frac{\lambda_1}{2}, \lambda_1\right)\right]\right\} \\
&\quad\times \tilde{f}^*(u - \lambda_1)\, d\lambda\, d\lambda_1.
\end{aligned} \tag{3}$$

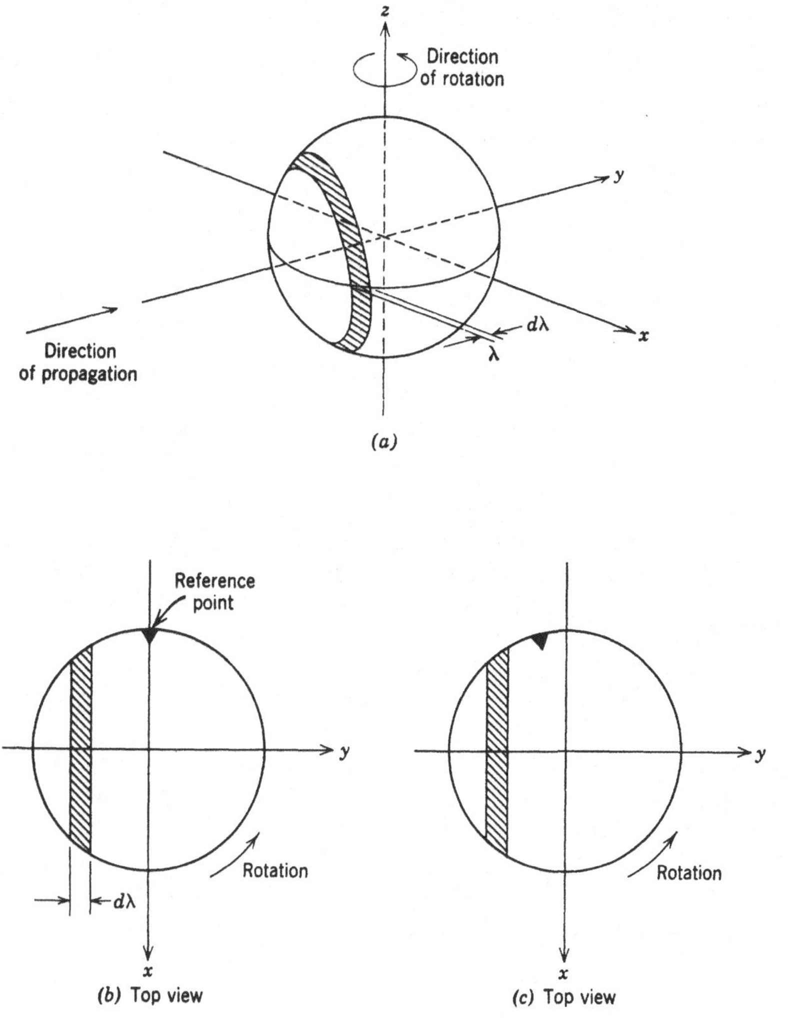


**Fig. 13.1 Rough rotating sphere.**

The statistical characteristics of the target specify the term in braces. We make two assumptions:

1. The returns from different intervals are statistically independent. (This corresponds to the model in Chapter 12.)
2. The return from each interval is a sample function of a *stationary*, zero-mean complex Gaussian random process. (This corresponds to the model in Chapter 11.)

Using these two assumptions, we can write

$$E[\tilde{b}(t, \lambda)\tilde{b}^*(u, \lambda_1)] = \tilde{K}_{DR}(t - u, \lambda)\delta(\lambda - \lambda_1). \tag{4}$$

The function $K_{DR}(\tau, \lambda)$ is a two-variable function that depends on the reflective properties of the target. Using (4) in (3) gives

$$\boxed{\tilde{K}_{\tilde{s}}(t, u) = E_t \int_{-\infty}^{\infty} \tilde{f}(t - \lambda)\tilde{K}_{DR}(t - u, \lambda)\tilde{f}^*(u - \lambda)\, d\lambda} \tag{5}$$

as a complete characterization of the returned signal process.

Just as in the singly-spread case, it is convenient to introduce a *scattering function*, which is defined as

$$\boxed{\tilde{S}_{DR}\{f, \lambda\} = \int_{-\infty}^{\infty} e^{-j2\pi f\tau}\tilde{K}_{DR}(\tau, \lambda)\, d\tau.} \tag{6}$$

Physically, $\tilde{S}_{DR}\{f, \lambda\}$ represents the spectrum of the process $\tilde{b}(t, \lambda)$. It is a *real, non-negative* function of $f$ and $\lambda$. The scattering function of a rough rotating sphere is shown in Fig. 13.2. (This result is derived in [5].) Two other scattering functions that we shall use as models are shown in Figs 13.3 and 13.4. In Fig. 13.3,

$$\tilde{S}_{DR}\{f, \lambda\} = \frac{2ck(\lambda)}{(2\pi f)^2 + k^2(\lambda)}, \qquad -\infty < f < \infty, \qquad 0 < \lambda < L. \tag{7}$$

At each value of $\lambda$, the spectrum is a first-order Butterworth process, but the pole location is a function of $\lambda$. This is an approximate model for some communications channels. In Fig. 13.4,

$$\tilde{S}_{DR}\{f, \lambda\} = \frac{\sigma_b^2}{\pi\sigma_R\sigma_D}\exp\left[-\frac{f^2}{2\sigma_D^2} - \frac{\lambda^2}{2\sigma_R^2}\right], \qquad -\infty < f, \lambda < \infty. \tag{8}$$

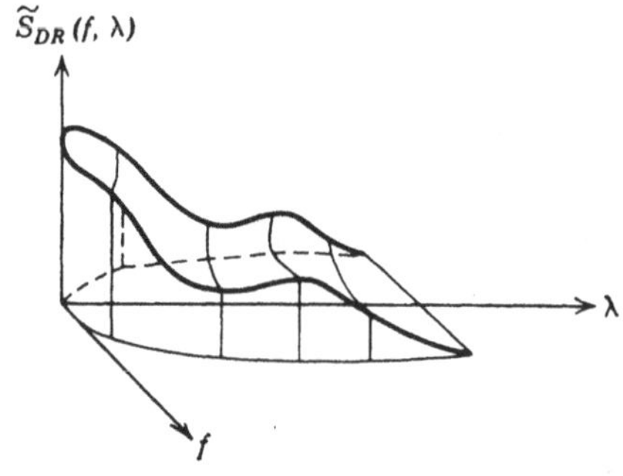


**Fig. 13.2 Scattering function of a rough sphere [from [5]].**

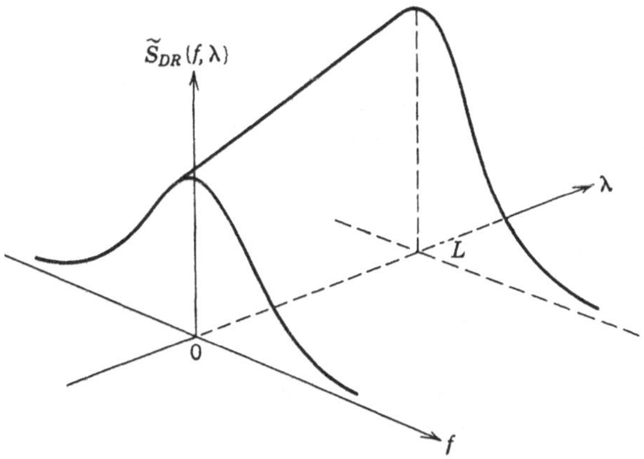


Fig. 13.3 A scattering function example.

The doubly Gaussian scattering function never occurs exactly, but is a useful approximation in many situations. Other examples of scattering functions are given on pages 35–39 of [37].

We can also write (5) using the scattering function as

$$\tilde{K}_{\tilde{s}}(t, u) = E_t \iint_{-\infty}^{\infty} \tilde{f}(t - \lambda)\tilde{S}_{DR}\{f, \lambda\}\tilde{f}^*(u - \lambda)e^{j2\pi f(t-u)}\, df\, d\lambda. \tag{9}$$

There are several properties of the model that will be useful, and we include them at this point.

**Property 1. Received Energy.** The *average* value of the received energy is

$$E[E_r] \triangleq \bar{E}_r = \int_{-\infty}^{\infty} \tilde{K}_{\tilde{s}}(t, t)\, dt. \tag{10}$$

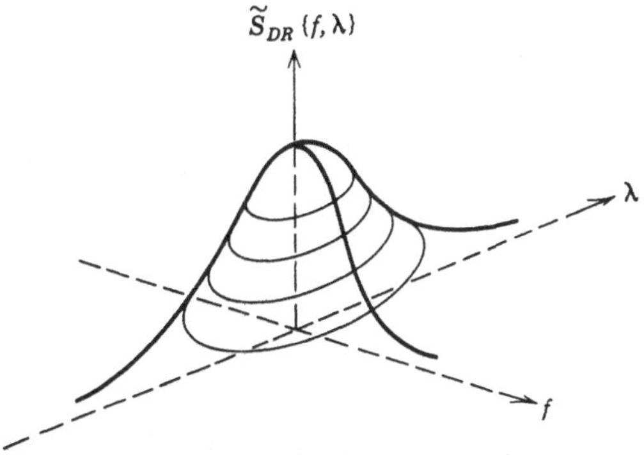


Fig. 13.4 A doubly-Gaussian scattering function (drawn for $\sigma_D < \sigma_R$).

Using (9) gives

$$\bar{E}_r = E_t \int_{-\infty}^{\infty} df \int_{-\infty}^{\infty} d\lambda \int_{-\infty}^{\infty} dt\, |\tilde{f}(t-\lambda)|^2\, \tilde{S}_{DR}\{f, \lambda\}. \tag{11}$$

The integral with respect to $t$ equals 1 for all $\lambda$. Thus,

$$\bar{E}_r = E_t \iint_{-\infty}^{\infty} \tilde{S}_{DR}\{f, \lambda\}\, df\, d\lambda. \tag{12}$$

To be consistent with our earlier models, we assume that

$$\iint_{-\infty}^{\infty} \tilde{S}_{DR}\{f, \lambda\}\, df\, d\lambda = 2\sigma_b^2. \tag{13}$$

We see that the double integral of the scattering function is the ratio of the expected value of the received energy to the transmitted energy. Notice that the received energy is not a function of the signal shape.

**Property 2.** When a scattering function is concentrated in one region of the $f, \lambda$ plane, we can characterize it grossly in terms of its moments. The mean delay is

$$m_R \triangleq \frac{1}{2\sigma_b^2} \int_{-\infty}^{\infty} d\lambda\, \lambda \int_{-\infty}^{\infty} df\, \tilde{S}_{DR}\{f, \lambda\}. \tag{14}$$

The mean-square delay spread is

$$\sigma_R^2 \triangleq \frac{1}{2\sigma_b^2} \int_{-\infty}^{\infty} d\lambda\, \lambda^2 \int_{-\infty}^{\infty} df\, \tilde{S}_{DR}\{f, \lambda\} - m_R^2. \tag{15}$$

The mean Doppler shift is

$$m_D \triangleq \frac{1}{2\sigma_b^2} \int_{-\infty}^{\infty} df\, f \int_{-\infty}^{\infty} d\lambda\, \tilde{S}_{DR}\{f, \lambda\}. \tag{16}$$

The mean-square Doppler spread is

$$\sigma_D^2 \triangleq \frac{1}{2\sigma_b^2} \int_{-\infty}^{\infty} df\, f^2 \int_{-\infty}^{\infty} d\lambda\, \tilde{S}_{DR}\{f, \lambda\} - m_D^2. \tag{17}$$

The skewness is measured by

$$\rho_{DR} = \frac{\overline{f\lambda} - m_R m_D}{\sigma_D \sigma_R}, \tag{18}$$

where

$$\overline{f\lambda} \triangleq \frac{1}{2\sigma_b^2} \int_{-\infty}^{\infty} df\, f \int_{-\infty}^{\infty} d\lambda\, \lambda \tilde{S}_{DR}\{f, \lambda\}. \tag{19}$$

In contrast with these mean-square measures, we frequently encounter scattering functions that are strictly bandlimited to $B$ cps and/or are strictly limited in length to $L$ seconds. In these cases the absolute measure is usually more useful.

**Property 3. Alternative Characterizations.** We saw that two functions that characterized the target were $\tilde{K}_{DR}(\tau, \lambda)$ and $\tilde{S}_{DR}\{f, \lambda\}$. Two other functions obtained by Fourier-transforming with respect to $\lambda$ are

$$\tilde{P}_{DR}\{f, v\} \triangleq \int_{-\infty}^{\infty} d\lambda\, e^{-j2\pi v\lambda} \tilde{S}_{DR}\{f, \lambda\} \tag{20}$$

and

$$\tilde{R}_{DR}\{\tau, v\} \triangleq \int_{-\infty}^{\infty} d\lambda\, e^{-j2\pi v\lambda} \tilde{K}_{DR}(\tau, \lambda). \tag{21}$$

Notice the sign convention in the Fourier transform. Transforming from $t$ to $f$ and from $\lambda$ to $v$, we use the minus sign in the exponent. (Remember that both $v$ and $f$ are frequency variables.) These functions are summarized in Fig. 13.5. The significance of the various variables should be emphasized. We can consider the $f$, $\tau$ pair and the $\lambda$, $v$ pair separately.

1. A "short" target is narrow on the $\lambda$-axis and therefore is wide on the $v$-axis.
2. A point target is an impulse on the $\lambda$-axis and therefore is a constant for all values of $v$.

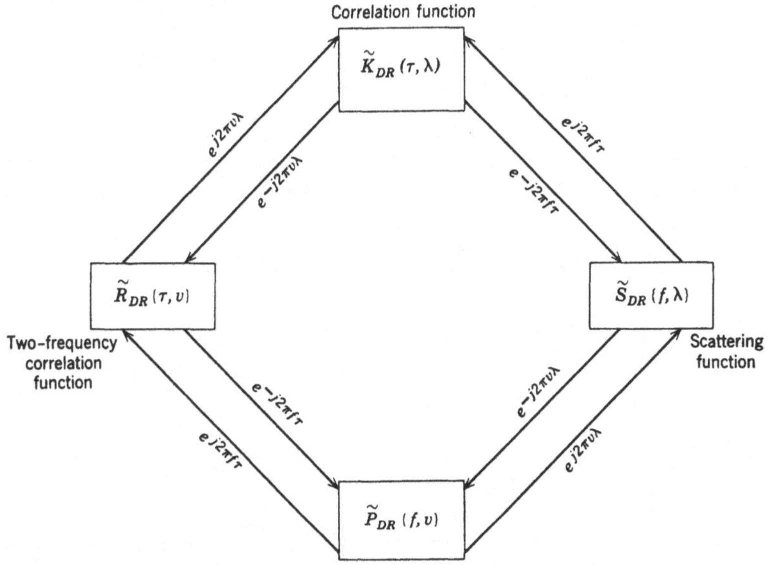


**Fig. 13.5 Target and channel characterizations.**

3. An infinite-depth target is constant along the $\lambda$-axis and therefore is an impulse on the $v$-axis.
4. A slowly fluctuating target is narrow on the $f$-axis and therefore is wide on the $\tau$-axis.
5. A fixed target is an impulse on the $f$-axis and therefore is a constant on the $\tau$-axis.
6. A rapidly fluctuating target is wide on the $f$-axis and impulsive on the $\tau$-axis.

The advantage of having several characterizations is that one is frequently easier to use than the others. We shall encounter several examples at various points in the chapter. Notice that, except for a $2\sigma_b^2$ factor, the scattering function, $\tilde{S}_{DR}\{f, \lambda\}$, has all of the properties of a joint probability density. Thus, $\tilde{R}_{DR}\{\tau, v\}$ is analogous to a joint characteristic function.

**Property 4. Degenerate Targets.** Both singly-spread target models can be viewed as limiting cases of a doubly-spread target. Repeating (5), we have

$$\tilde{K}_{\tilde{s}}(t, u) = E_t \int_{-\infty}^{\infty} \tilde{f}(t - \lambda)\tilde{K}_{DR}(t - u, \lambda)\tilde{f}^*(u - \lambda)\, d\lambda. \tag{22}$$

To obtain the point target model, we assume that the length of the target along the $\lambda$-axis is much less than the reciprocal of the signal bandwidth; that is,

$$L \ll \frac{1}{W}. \qquad (23)\dagger$$

Then we can treat $\tilde{f}(t - \lambda)$ as a constant function of $\lambda$ over the range where $\tilde{K}_{DR}(t - u, \lambda)$ is nonzero. Consequently we can approximate $\tilde{K}_{DR}(t - u, \lambda)$ as

$$\tilde{K}_{DR}(t - u, \lambda) \simeq \tilde{K}_D(t - u)\delta(\lambda - \bar{\lambda}). \tag{24}$$

Using (24) in (22) gives

$$\tilde{K}_{\tilde{s}}(t, u) = E_t \tilde{f}(t - \bar{\lambda})\tilde{K}_D(t - u)\tilde{f}^*(u - \bar{\lambda}), \tag{25}$$

which is (11.20). If (23) is satisfied, we have a fluctuating point target whose fading is *not* frequency-selective.

† In (23), (27), and (30) we use $B$, $L$, $W$, and $T$ in an intuitive manner. For a particular signal and target, the statements can be made more precise.

To obtain the nonfluctuating model, we start with the expression in (9),

$$\tilde{K}_{\tilde{s}}(t, u) = E_t \iint_{-\infty}^{\infty} \tilde{f}(t - \lambda)\tilde{S}_{DR}\{f, \lambda\}\tilde{f}^*(u - \lambda)e^{j2\pi f(t-u)}\, df\, d\lambda. \tag{26}$$

If

$$B \ll \frac{1}{T}, \tag{27}$$

then we can use the approximation

$$\tilde{S}_{DR}\{f, \lambda\} \simeq \tilde{S}_R\{\lambda\}\delta\{f - \bar{f}\}. \tag{28}$$

Using (28) in (26), we obtain

$$\tilde{K}_{\tilde{s}}(t, u) = E_t \int_{-\infty}^{\infty} \tilde{f}(t - \lambda)e^{j2\pi \bar{f} t}\tilde{S}_R\{\lambda\}\tilde{f}^*(u - \lambda)e^{-j2\pi \bar{f} u}\, d\lambda, \tag{29}$$

which corresponds to the return from a nonfluctuating range-spread target moving at a constant velocity. If (27) is satisfied, the fading is *not* time-selective.

In order to have an undistorted returned signal (i.e., fading that is flat in both time and frequency), both (23) and (27) must be satisfied. Combining them, we have

$$BL \ll \frac{1}{WT}. \tag{30}$$

Because

$$WT > 1 \tag{31}$$

for all signals, the condition in (30) can only be satisfied for $BL < 1$.

We refer to targets (or channels) for which

$$BL < 1 \tag{32}$$

as *underspread* targets (or channels). If

$$BL > 1, \tag{33}$$

we say that the target (or channel) is *overspread*. We shall look at further implications of the $BL$ product as we proceed through the chapter. Our present discussion shows that only underspread targets that satisfy (30) will ever degenerate into slowly fluctuating point targets for certain signals.

It is worthwhile recalling that the Doppler spread depends on both the target velocity spread and the carrier frequency (see (9.24)). Thus the $BL$ product for a particular target will depend on the carrier frequency.

Up to this point we have characterized the reflection process $\tilde{b}(t, \lambda)$ in terms of its covariance function or spectrum. It is frequently convenient to use a differential-equation model of the channel process. We develop this model in the next section.

### 13.1.2 Differential-Equation Model for a Doubly-Spread Target (or Channel)†

In our model for a doubly-spread target we have assumed that th returns from different range elements are statistically independent. The covariance function of the reflection process is

$$E[\tilde{b}(t, \lambda)\tilde{b}^*(u, \lambda')] = \tilde{K}_{DR}(t - u, \lambda)\delta(\lambda - \lambda'). \tag{34}$$

In many cases of interest the Fourier transform of $\tilde{K}_{DR}(t - u)$ is a rational function of $f$. In these cases we can derive a state-variable model for the doubly-spread channel. Notice that (34) implies that there is no relationship between the target processes at different values of $\lambda$. Thus we can treat the target process at any point (say $\lambda_1$) as a random process with a single independent variable $t$. Then the state representations developed in Section I–6.3.3 are immediately applicable. The first new feature is that the state equations will contain $\lambda$ as a parameter. Other new features will be encountered as we proceed through the development.

Since the output of the channel is given by

$$\tilde{s}(t) = \int_{-\infty}^{\infty} \sqrt{E_t}\tilde{f}(t - \lambda)\tilde{b}\left(t - \frac{\lambda}{2}, \lambda\right) d\lambda, \tag{35}$$

it is convenient to define a new process

$$\tilde{b}_x(t, \lambda) \triangleq \tilde{b}\left(t - \frac{\lambda}{2}, \lambda\right). \tag{36}$$

Notice that $\tilde{b}_x(t, \lambda)$ is a zero-mean complex Gaussian process whose covariance function is

$$\begin{aligned} E[\tilde{b}_x(t, \lambda)\tilde{b}_x^*(u, \lambda')] &= E\left[\tilde{b}\left(t - \frac{\lambda}{2}, \lambda\right)\tilde{b}^*\left(u - \frac{\lambda'}{2}, \lambda'\right)\right] \\ &= \tilde{K}_{DR}(t - u, \lambda)\,\delta(\lambda - \lambda'). \end{aligned} \tag{37}$$

We have modeled the channel reflection process as a random process that depends on both time and space. We now want to characterize the time dependence by a state-variable representation. The spatial dependence is included by making the state representation a function of the

† This channel model was developed by R. Kurth in his doctoral thesis [7]. It is used primarily in sections 13.2.2 and 13.3. One can defer reading this section until that point.

spatial variable $\lambda$. The representation that we need to describe a doubly-spread channel incorporates the spatial dependence in a straightforward manner.

We denote the state vector of $\tilde{b}_x(t, \lambda)$ as $\tilde{\mathbf{x}}(t, \lambda)$. The state equation is

$$\boxed{\frac{\partial \tilde{\mathbf{x}}(t, \lambda)}{\partial t} = \tilde{\mathbf{F}}(\lambda)\tilde{\mathbf{x}}(t, \lambda) + \tilde{\mathbf{G}}(\lambda)\tilde{u}(t, \lambda), \qquad t \geq T_i,} \tag{38}$$

where

$$\boxed{E[\tilde{u}(t, \lambda)\tilde{u}^*(\tau, \lambda')] = \tilde{Q}(\lambda)\,\delta(t - \tau)\,\delta(\lambda - \lambda').} \tag{39}$$

The initial covariance of the state vector is

$$\boxed{E[\tilde{\mathbf{x}}(T_i, \lambda)\tilde{\mathbf{x}}^\dagger(T_i, \lambda')] = \tilde{\mathbf{P}}_0(\lambda)\,\delta(\lambda - \lambda').} \tag{40}$$

The channel process is

$$\tilde{b}_x(t, \lambda) = \tilde{\mathbf{C}}(\lambda)\tilde{\mathbf{x}}(t, \lambda). \tag{41}$$

Notice that there is no coupling between the different values of $\lambda$ in the description. The state equation is written as a partial differential equation, but it is actually just an ordinary differential equation containing $\lambda$ as a parameter. Because of this parametric dependence, we can write a covariance equation easily. We define

$$\begin{aligned}\tilde{\mathbf{K}}_{\tilde{\mathbf{x}}}(t, t':\lambda, \lambda') &\triangleq E[\tilde{\mathbf{x}}(t, \lambda)\tilde{\mathbf{x}}^\dagger(t', \lambda')] \\ &= \tilde{\mathbf{K}}_{\tilde{\mathbf{x}}}(t, t':\lambda)\,\delta(\lambda - \lambda').\end{aligned} \tag{42}$$

As before, $\tilde{\mathbf{K}}_{\tilde{\mathbf{x}}}(t, t':\lambda)$ can be related to $\tilde{\mathbf{K}}_{\tilde{\mathbf{x}}}(t, t:\lambda)$ by the relation

$$\tilde{\mathbf{K}}_{\tilde{\mathbf{x}}}(t, t':\lambda) = \begin{cases}\tilde{\boldsymbol{\theta}}(t - t':\lambda)\tilde{\mathbf{K}}_{\tilde{\mathbf{x}}}(t', t:\lambda), & t \geq t', \\ \tilde{\mathbf{K}}_{\tilde{\mathbf{x}}}(t, t)\tilde{\boldsymbol{\theta}}^\dagger(t' - t:\lambda), & t \leq t',\end{cases} \tag{43}$$

where $\tilde{\boldsymbol{\theta}}(t:\lambda)$ is the transition matrix and is the solution to

$$\frac{\partial \tilde{\boldsymbol{\theta}}(t:\lambda)}{\partial t} = \tilde{\mathbf{F}}(\lambda)\tilde{\boldsymbol{\theta}}(t:\lambda), \tag{44}$$

with the initial condition

$$\tilde{\boldsymbol{\theta}}(0, \lambda) = \mathbf{I}. \tag{45}$$

Notice that the transition matrix has a single time argument, because $\tilde{\mathbf{F}}(\lambda)$ and $\tilde{\mathbf{G}}(\lambda)$ are not functions of time.

Since we have assumed that the channel process is stationary, $\tilde{\mathbf{K}}_{\tilde{\mathbf{x}}}(t, t:\lambda)$ is not a function of time. Thus we can write

$$\tilde{\mathbf{K}}_{\tilde{\mathbf{x}}}(\lambda) \triangleq \tilde{\mathbf{K}}_{\tilde{\mathbf{x}}}(t, t:\lambda), \qquad t \geq T_i. \tag{46}$$

The matrix $\tilde{\mathbf{K}}_{\tilde{\mathbf{x}}}(\lambda)$ is just the solution to

$$\mathbf{0} = \tilde{\mathbf{F}}(\lambda)\tilde{\mathbf{K}}_{\tilde{\mathbf{x}}}(\lambda) + \tilde{\mathbf{K}}_{\tilde{\mathbf{x}}}(\lambda)\tilde{\mathbf{F}}^{\dagger}(\lambda) + \tilde{\mathbf{G}}(\lambda)\tilde{\mathbf{Q}}(\lambda)\tilde{\mathbf{G}}^{\dagger}(\lambda) \tag{47}$$

[see (I-6.333*a*)]. Notice that the stationarity assumption requires

$$\tilde{\mathbf{P}}_0(\lambda) = \tilde{\mathbf{K}}_{\tilde{\mathbf{x}}}(\lambda). \tag{48}$$

The channel covariance function is obtained by using (41) and (42) in (37) to obtain

$$\boxed{\tilde{K}_{DR}(\tau, \lambda) = \tilde{\mathbf{C}}(\lambda)\tilde{\mathbf{K}}_{\tilde{\mathbf{x}}}(\tau, \lambda)\tilde{\mathbf{C}}^{\dagger}(\lambda).} \tag{49}$$

Once again we emphasize that all the results are ordinary state-variable results with a parametric dependence on $\lambda$. To complete our model, we must describe the observed signal process. Using (36) in (35) gives

$$\tilde{s}(t) = \int_{-\infty}^{\infty} \sqrt{E_t}\, \tilde{f}(t - \lambda)\tilde{b}_x(t, \lambda)\, d\lambda. \tag{50}$$

Using (41) in (50), we have

$$\boxed{\tilde{s}(t) = \int_{-\infty}^{\infty} \sqrt{E_t}\, \tilde{f}(t - \lambda)\tilde{\mathbf{C}}(\lambda)\tilde{\mathbf{x}}(t, \lambda)\, d\lambda.} \tag{51}$$

We see that (51) contains an integration over the spatial variable $\lambda$. This spatial functional is the new feature of the problem and will require extension of our earlier state-variable theory. Notice that it is a *linear* functional and is analogous to the modulation matrix in Section 6.6.3. It is sometimes convenient to rewrite (51) as

$$\boxed{\tilde{s}(t) = \tilde{s}(t : \tilde{\mathbf{x}}(t, \lambda)).} \tag{52}$$

This completes our differential equation model of the doubly-spread channel. To illustrate the techniques involved, we consider an example.

**Example** [7]. We consider a complex first-order state equation

$$\frac{\partial \tilde{x}(t, \lambda)}{\partial t} = -\tilde{k}(\lambda)\tilde{x}(t, \lambda) + \tilde{u}(t, \lambda), \tag{53}$$

and

$$\tilde{b}_x(t, \lambda) = c\tilde{x}(t, \lambda). \tag{54}$$

These equations correspond to (38)–(41) with

$$\tilde{\mathbf{F}}(\lambda) = -\tilde{k}(\lambda) = -k_r(\lambda) - jk_i(\lambda), \tag{55}$$

$$\tilde{\mathbf{G}}(\lambda) = 1, \tag{56}$$

and

$$\tilde{\mathbf{C}}(\lambda) = c. \tag{57}$$

We assume

$$\tilde{Q}(\lambda) \geq 0 \tag{58}$$

and

$$k_r(\lambda) > 0. \tag{59}$$

From (44) and (45), we have

$$\tilde{\theta}(\tau, \lambda) = \exp\,[-k_r(\lambda)\,|\tau| - jk_i(\lambda)\tau], \tag{60}$$

and from (47),

$$\tilde{\mathbf{K}}_{\tilde{\mathbf{x}}}(\lambda) = \frac{\tilde{Q}(\lambda)}{2k_r(\lambda)}. \tag{61}$$

Using (60) and (61) in (43) and the result in (49) gives the channel covariance function as

$$\tilde{K}_{DR}(\tau, \lambda) = \frac{c^2\tilde{Q}(\lambda)}{2k_r(\lambda)} \exp\,[-k_r(\lambda)\,|\tau| - jk_i(\lambda)\tau]. \tag{62}$$

Transforming gives the channel-scattering function as

$$\tilde{S}_{DR}\{f, \lambda\} = \frac{c^2\tilde{Q}(\lambda)}{(2\pi f + k_i(\lambda))^2 + k_r^2(\lambda)}. \tag{63}$$

Notice that

$$\iint_{-\infty}^{\infty} S_{DR}\{f, \lambda\}\, df\, d\lambda = c^2 \int_{-\infty}^{\infty} \frac{\tilde{Q}(\lambda)}{2k_r(\lambda)}\, d\lambda \triangleq 2\sigma_b^2. \tag{64}$$

The scattering function in (63), considered as a function of the frequency at any value of $\lambda$, is a one-pole spectrum centered at $f = -k_i(\lambda)/2\pi$ with a peak value $c^2\tilde{Q}(\lambda)/k_r^2(\lambda)$ and 3-db points $\pm k_r(\lambda)/2\pi$ about the center frequency.

In Fig. 13.6 we show the scattering function for the case when

$$\tilde{Q}(\lambda) = \begin{cases} 1 - \cos\left(\dfrac{2\pi\lambda}{L}\right), & 0 \leq \lambda \leq L, \\ 0, & \text{elsewhere}, \end{cases} \tag{65}$$

and

$$\tilde{k}(\lambda) = \begin{cases} k\left(1 - \frac{1}{2}\sin\left(\dfrac{\pi\lambda}{L}\right)\right), & 0 \leq \lambda \leq L, \\ 0, & \text{elsewhere}. \end{cases} \tag{66}$$

Except for the constraints of (58), (59), and (64), $\tilde{Q}(\lambda)$ and $k(\lambda)$ are arbitrary. This permits considerable flexibility in the choise of $\tilde{S}_{DR}\{f, \lambda\}$, even for this first-order model. For instance, if $k_i(\lambda)$ is proportional to $\lambda$, then $\tilde{S}_{DR}\{f, \lambda\}$ is sheared in the $\lambda, f$ plane. We can choose $\tilde{Q}(\lambda)$ to give a multimodal (in $\lambda$) scattering function. In Fig. 13.7 we show a scattering function that exhibits both the multimodal behavior and the

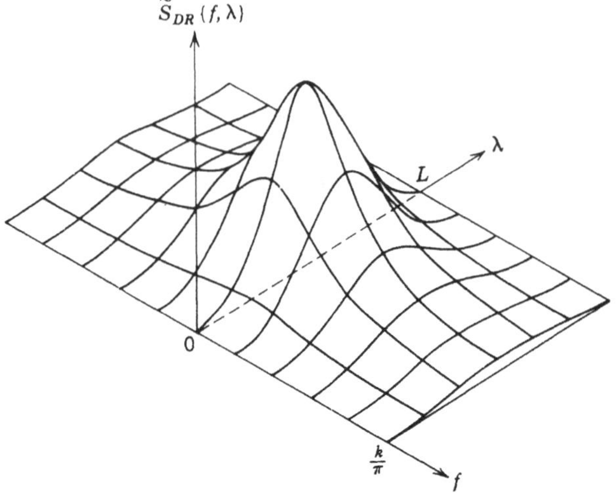


**Fig. 13.6 Scattering function specified by (65) and (66) (from [7]).**

shearing effect. Here

$$\tilde{Q}(\lambda) = \begin{cases} 1 - \cos\left(\dfrac{2\pi\lambda}{L}\right), & 0 \le \lambda \le \dfrac{L}{4}, \quad \dfrac{3L}{4} \le \lambda \le L, \\ 2 + \cos\left(\dfrac{\pi\lambda}{L}\right), & \dfrac{L}{4} \le \lambda \le \dfrac{3L}{4}, \\ 0, & \text{elsewhere,} \end{cases} \tag{67}$$

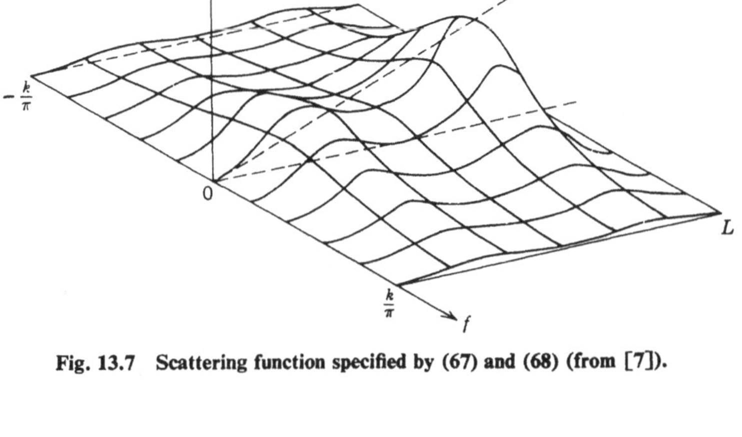


**Fig. 13.7 Scattering function specified by (67) and (68) (from [7]).**

and

$$\tilde{k}(\lambda) = k\left(1 - \frac{\lambda}{2L}\right) - j\left(\frac{3k\lambda}{5\pi L}\right). \tag{68}$$

This example illustrates the flexibility available with a first-order model. By using a higher-order system, we can describe any scattering function that is a rational function in $f$ for each value of $\lambda$. To obtain multimodal behavior in $f$ requires at least a second-order state model.

Just as in our previous work, the main advantage of the state-variable formulation is that it enables us to express the optimum receiver and its performance in a form such that we can actually compute an explicit answer. We shall look at specific examples of this model in Sections 13.2 and 13.3.

### 13.1.3 Model Summary

In this section we have developed a model for a doubly-spread target. The target return is characterized by either a scattering function or a distributed state-variable model.

In the next three sections we discuss various situations in which doubly-spread targets are the central issue. As we pointed out earlier, because the sections deal with different physical problems they can almost be read independently. (There are a few cross-references, but these can be understood out of context.)

## 13.2 DETECTION IN THE PRESENCE OF REVERBERATION OR CLUTTER (RESOLUTION IN A DENSE ENVIRONMENT)

In this section we consider the problem of detecting the return from a slowly fluctuating point target in the presence of distributed interference. The problem is an extension of our discrete resolution discussion in Section 10.5.

This type of problem is often encountered in active sonar systems. The complex envelope of the transmitted signal is $\sqrt{E_t}\tilde{f}(t)$. The target of interest is a slowly fluctuating point target that is located at a known delay $\tau_d$ and known Doppler $\omega_d$. As the transmitted signal travels through the ocean, it encounters various inhomogeneities and numerous objects that cause reflections. A possible target environment is shown in Fig. 13.8. The return of the distributed interference is referred to as reverberation in the sonar case and as clutter in the radar case.

These reflections can be modeled as a spatial Poisson random process. In [1] we have developed the model in detail (see [2], [3] also). When there

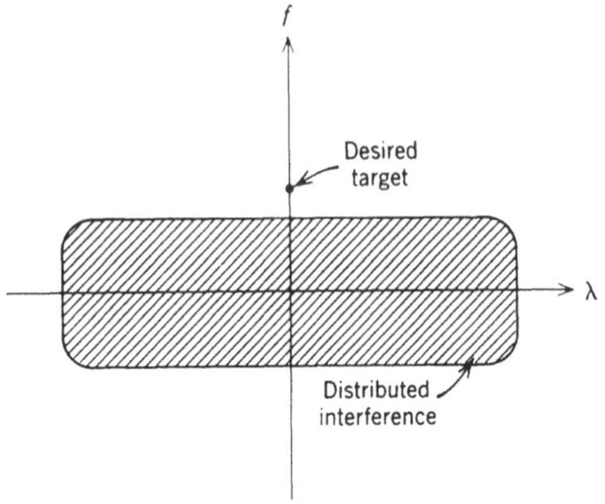


**Fig. 13.8 Target environment in (λ, f) plane.**

are a large number of reflectors we are led back to the complex Gaussian model of Section 13.1. There has been a great deal of work done on reverberation models, which the interested reader may consult (e.g., [8]–[18]). It appears that in many situations our spatial Poisson model is an adequate description of the environment.

Denoting the complex envelope of the reverberation return as $\tilde{n}_r(t)$, we have

$$\tilde{n}_r(t) = \sqrt{E_t}\int_{-\infty}^{\infty} \tilde{f}(t-\lambda)\tilde{b}\left(t-\frac{\lambda}{2}, \lambda\right) d\lambda. \tag{69}$$

This is a zero-mean complex Gaussian process with the covariance function

$$\tilde{K}_{\tilde{n}_r}(t, u) = E_t\int_{-\infty}^{\infty} \tilde{f}(t-\lambda)\tilde{K}_{DR}\{t-u, \lambda\}\tilde{f}^*(u-\lambda)\, d\lambda. \tag{70}$$

[These are just (2) and (5) repeated.] Alternatively, we can write (70) as

$$\tilde{K}_{\tilde{n}_r}(t, u) = E_t\iint_{-\infty}^{\infty} \tilde{f}(t-\lambda)\tilde{S}_{DR}\{f, \lambda\}\tilde{f}^*(u-\lambda)e^{j2\pi f(t-u)}\, df\, d\lambda. \tag{71}$$

The function $\tilde{S}_{DR}\{f, \lambda\}$ is the scattering function of the reverberation and characterizes its distribution in range and Doppler. In addition to the reverberation return there is an additive, statistically independent complex white noise process $\tilde{w}(t)$. Thus, we have the following hypothesis testing problem:

$$\begin{aligned} \tilde{r}(t) &= \tilde{b}\tilde{f}(t-\tau_d)e^{j\omega_d t} + \tilde{n}_r(t) + \tilde{w}(t), \qquad -\infty < t < \infty: H_1,\\ \tilde{r}(t) &= \tilde{n}_r(t) + \tilde{w}(t), \qquad -\infty < t < \infty: H_0. \end{aligned} \tag{72}$$

This is just a detection in colored noise problem that we encountered previously in Section 9.2. The only new feature is the dependence of the colored noise covariance function on the transmitted signal. Notice that this problem is just the continuous version of the discrete resolution problem discussed in Section 10.5. Just as in that case, we can consider two types of receivers:

1. The *conventional receiver*, which is designed assuming that only white noise is present.
2. The *optimum receiver*, whose design is based on the assumed statistical knowledge of the reverberation.

In Section 13.2.1 we study the performance of the conventional receiver. In this case we try to eliminate the reverberation by choosing the signal properly. In Section 13.2.2 we study the optimum receiver problem. In this case we try to eliminate the reverberation by both signal design and receiver design.

### 13.2.1 Conventional Receiver

We consider the problem of detecting a target at some known delay $\tau_d$ and Doppler $\omega_d$. If there were no reverberation, the results in Section 9.2 would be directly applicable. From (9.36), we compute

$$\tilde{l} \triangleq \int_{-\infty}^{\infty} \tilde{r}(t)\tilde{f}^*(t - \tau_d)e^{-j\omega_d t}\,dt. \tag{73}$$

The test consists of comparing $|\tilde{l}|^2$ with a threshold,

$$|\tilde{l}|^2 \underset{H_0}{\overset{H_1}{\gtrless}} \gamma. \tag{74}$$

When reverberation is present, this receiver is not optimum. It is frequently used for several reasons:

1. It is simpler than the optimum receiver.
2. The scattering function may not be known, and so we cannot design the optimum receiver.
3. Our analysis will demonstrate that it works almost as well as the optimum receiver in many situations.

It is straightforward to calculate the effect of the reverberation on the performance of the conventional receiver. The output $\tilde{l}$ is still a complex Gaussian random variable, so that $\Delta$ as defined in (9.49) is a complete performance measure. The definition of $\Delta$ is

$$\Delta_{wo} \triangleq \frac{E\{|\tilde{l}|^2 \mid H_1\} - E\{|\tilde{l}|^2 \mid H_0\}}{E\{|\tilde{l}|^2 \mid H_0\}}. \tag{75}$$

As in Section 10.5, we use the subscript *wo* to denote that the receiver would be *optimum* in the presence of *white* noise only. Using (70), (72), and (73), we have

$$\begin{aligned} E\{|\tilde{l}|^2 \mid H_0\} &= E\left\{\int_{-\infty}^{\infty} [\tilde{n}_r(t) + \tilde{w}(t)]\tilde{f}^*(t - \tau_d)e^{-j\omega_d t}\, dt \right. \\ &\quad \left. \times \int_{-\infty}^{\infty} [\tilde{n}_r^*(u) + \tilde{w}^*(u)]\tilde{f}(u - \tau_d)e^{j\omega_d u}\, du\right\} \\ &= \int_{-\infty}^{\infty} dt \int_{-\infty}^{\infty} du\, \tilde{f}^*(t - \tau_d)\tilde{f}(u - \tau_d)e^{-j\omega_d(t-u)} \\ &\quad \times \left[E_t \int_{-\infty}^{\infty} \tilde{f}(t - \lambda)\tilde{K}_{DR}(t - u, \lambda)\tilde{f}^*(u - \lambda)\, d\lambda + N_0\, \delta(t - u)\right]. \end{aligned} \tag{76}$$

Now recall that

$$\tilde{K}_{DR}(t - u, \lambda) = \int_{-\infty}^{\infty} \tilde{S}_{DR}\{f, \lambda\}e^{j2\pi f(t-u)}\, df. \tag{77}$$

Using (77) in (76) and rearranging terms, we obtain

$$\begin{aligned} E\{|\tilde{l}|^2 \mid H_0\} = N_0 + E_t &\int_{-\infty}^{\infty} df \int_{-\infty}^{\infty} d\lambda\, \tilde{S}_{DR}\{f, \lambda\} \\ &\times \left[\int_{-\infty}^{\infty} dt\, \tilde{f}(t - \lambda)\tilde{f}^*(t - \tau_d)e^{j2\pi(f-f_d)t}\right] \\ &\times \left[\int_{-\infty}^{\infty} du\, \tilde{f}^*(u - \lambda)\tilde{f}(u - \tau_d)e^{-j2\pi(f-f_d)u}\right]. \end{aligned} \tag{78}$$

The product of the quantities in the two brackets is just the signal ambiguity function, $\theta\{\tau_d - \lambda, f - f_d\}$. Thus,

$$E\{|\tilde{l}|^2 \mid H_0\} = N_0 + E_t \iint_{-\infty}^{\infty} df\, d\lambda\, \tilde{S}_{DR}\{f, \lambda\}\theta\{\tau_d - \lambda, f - f_d\}. \tag{79}$$

Thus, the effect of the reverberation is obtained by convolving the *reverberation scattering function* with the *signal ambiguity function*, $\theta\{\tau, -f\}$.†
Similarly,

$$E[|\tilde{l}|^2 \mid H_1] = \bar{E}_r + E[|\tilde{l}|^2 \mid H_0], \tag{80}$$

† This result was first obtained by Steward and Westerfeld [19], [20].

so that

$$\Delta_{wo} = \frac{\bar{E}_r/N_0}{1 + E_t/N_0 \displaystyle\iint_{-\infty}^{\infty} df\, d\lambda\, \tilde{S}_{DR}\{f, \lambda\}\theta\{\tau_d - \lambda, f - f_d\}}. \tag{81}$$

The second term in the denominator represents the degradation due to the reverberation,

$$\rho_r \triangleq \frac{E_t}{N_0} \iint_{-\infty}^{\infty} df\, d\lambda\, \tilde{S}_{DR}\{f, \lambda\}\theta\{\tau_d - \lambda, f - f_d\}. \tag{82}$$

Using Property 4 given in (10.116), we can write $\rho_r$ as

$$\rho_r = \frac{E_t}{N_0} \iint_{-\infty}^{\infty} df\, d\lambda\, \tilde{S}_{DR}\{f, \lambda\}\theta\{\lambda - \tau_d, f_d - f\}. \tag{83}$$

We see that $\rho_r$ increases as we increase the transmitted energy. This means that we cannot combat reverberation by simply increasing the transmitted energy. This result is not surprising, because the reverberation is caused by a reflection of the transmitted signal.

The result in (83) has the simple graphical interpretation shown in Fig. 13.9. We assume that transmitted signal has a Gaussian envelope and a linear frequency characteristic (see Example 4 on page 290)

$$\tilde{f}(t) = \left(\frac{1}{\pi T^2}\right)^{1/4} \exp\left[-\left(\frac{1}{2T^2} - jb\right)t^2\right]. \tag{84}$$

The equal-height contours of the ambiguity function are shown in Fig. 13.9*a*. The equal-height contours of the reverberation scattering function are shown in Fig. 13.9*b*. To evaluate $\rho_r$, we center $\theta\{\tau, -f\}$ at the desired target location $(\tau_d, f_d)$ as shown in Fig. 13.9*c*. The value of $\rho_r$ will be determined by the amount of overlap of the two functions. For this particular target we can decrease the overlap in one of two ways:

1. Let $b = 0$ and make $T$ large.
2. Let $T$ be small and $b$ any value.

These observations can be verified qualitatively by sketching the resulting functions. We shall verify them quantitatively later in the section.

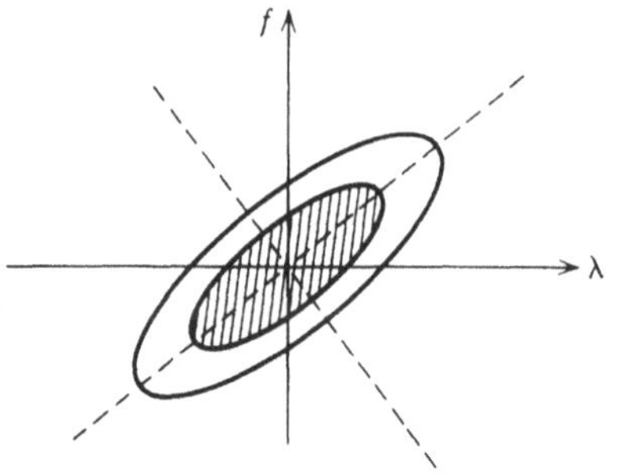


(a) Equal-height contours of $\theta(\tau, f)$

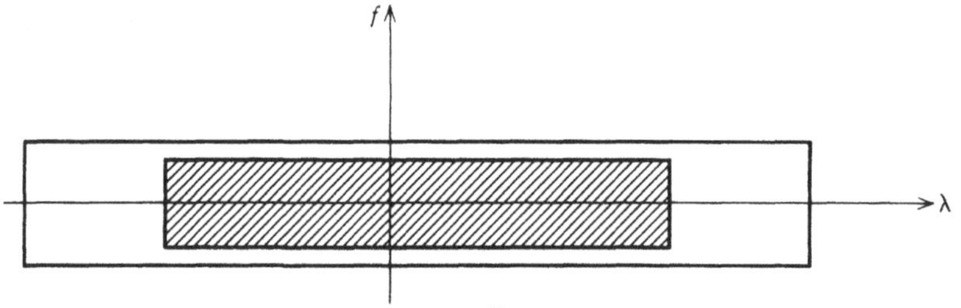


(b) Equal-height contours of $\tilde{S}_{DR}(f, \lambda)$

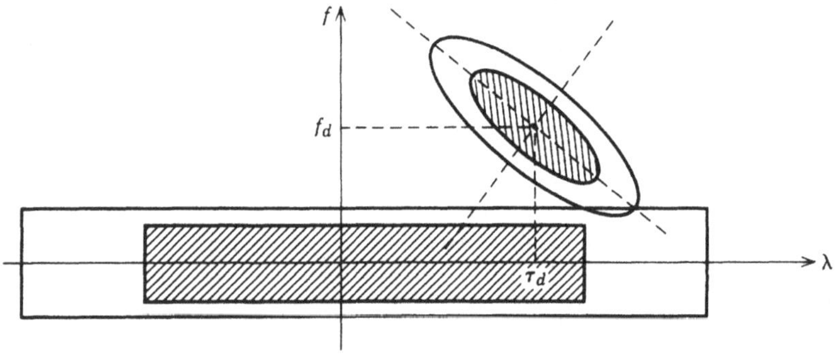


(c) Superposition

**Fig. 13.9 Graphical interpretation of $\rho_r$ integral.**

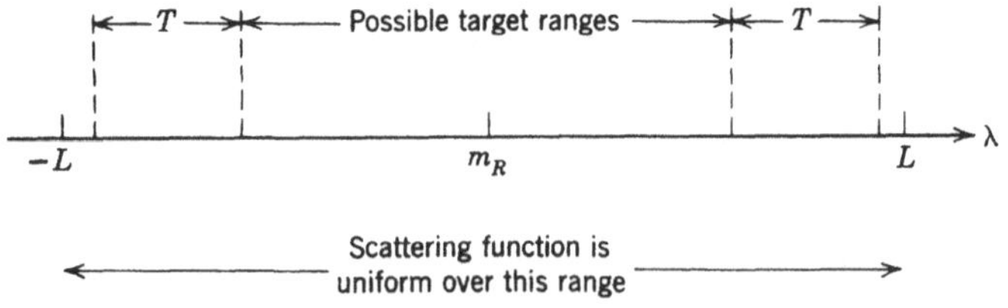


**Fig. 13.10 Geometry for range-invariant approximation to the scattering function.**

For future reference we observe that we also write $\rho_r$ in terms of the two-frequency correlation function as

$$\rho_r = \frac{E_t}{N_0} \iint_{-\infty}^{\infty} d\tau\, dv\, \tilde{R}_{DR}\{\tau, v\}\theta\{\tau, v\} \exp\left[-j2\pi(\tau f_d - v\tau_d)\right]. \qquad (85)$$

The results in (83) and (85) are valid for an arbitrary scattering function. A special case of interest is when the scattering function is essentially infinite in length and has a uniform Doppler profile.

***Range-Invariant Scattering Functions.*** When the scattering function has a uniform Doppler profile and extends beyond the range of a possible target by a distance greater than $T$, we can treat it as if it were infinite in extent. The geometry is shown in Fig. 13.10. We write

$$\tilde{S}_{DR}\{f, \lambda\} \triangleq \tilde{S}_{Du}\{f\}, \qquad -\infty < \lambda < \infty. \qquad (86)$$

Notice that both sides of (86) have units $\text{cps}^{-1} \times \text{meter}^{-1}$. Thus,

$$E_t \int_{-\infty}^{\infty} \tilde{S}_{Du}\{f\}\, df \qquad (87)$$

is the average received energy per meter of reverberation.

We can calculate $\rho_r$ by specializing the result in (83). We can also calculate $\rho_r$ directly from the original model. We carry out the second procedure because it has more intuitive appeal.

When the reverberation is infinite in extent, the reverberation return is a sample function of a stationary process. Using (86) in (71), we have

$$\tilde{K}_{\tilde{n}_r}(t, u) = E_t \tilde{K}_{Du}(t - u) \int_{-\infty}^{\infty} \tilde{f}(t - \lambda)\tilde{f}^*(u - \lambda)\, d\lambda. \qquad (88)$$

The integral can be written as

$$\int_{-\infty}^{\infty} d\lambda \int_{-\infty}^{\infty} \tilde{F}\{f_1\} e^{j2\pi f_1(t-\lambda)}\, df_1 \int_{-\infty}^{\infty} \tilde{F}^*\{f_2\} e^{-j2\pi f_2(t-\lambda)}\, df_2$$
$$= \int_{-\infty}^{\infty} e^{j2\pi f_1(t-u)} \tilde{S}_{\tilde{f}}\{f_1\}\, df_1, \qquad (89)$$

where

$$\tilde{S}_{\tilde{f}}\{f_1\} \triangleq |\tilde{F}\{f_1\}|^2. \qquad (90)$$

Using (89) in (88) and transforming both sides, we obtain

$$\boxed{\tilde{S}_{\tilde{n}_r}\{f\} = E_t \tilde{S}_{Du}\{f\} \otimes \tilde{S}_{\tilde{f}}\{f\}.} \tag{91}$$

Thus the spectrum of the reverberation return is obtained by convolving the Doppler spectrum with the pulse spectrum. To compute the degradation for a stationary input process, we use (76) and (82) and obtain

$$\boxed{\rho_r = \frac{E_t}{N_0} \int_{-\infty}^{\infty} \tilde{S}_{\tilde{n}_r}\{f\} \tilde{S}_{\tilde{f}}\{f - f_d\}\, df.} \tag{92}$$

The results in (91) and (92) specify the degradation for this special case. A simple example illustrates the application of (92).

**Example.** We assume that the Doppler profile is Gaussian, so that $\tilde{S}_{Du}\{f\}$ is

$$\tilde{S}_{Du}\{f\} = \frac{N_r}{\sqrt{2\pi}\, \sigma_D} e^{-f^2/2\sigma_D^2}, \tag{93}$$

where $\sigma_D$ is the root-mean-square Doppler spread in cycles per second. Assume that the signal is a pulse with a Gaussian envelope and linear FM sweep rate of $2b$ cps/sec. Thus $\tilde{f}(t)$ is

$$\tilde{f}(t) = \left(\frac{1}{\pi T^2}\right)^{1/4} \exp\left[-\left(\frac{1}{2T^2} - jb\right)t^2\right], \qquad -\infty < t < \infty. \tag{94}$$

Then

$$\tilde{S}_{\tilde{f}}\{f\} = \frac{1}{\sqrt{2\pi}\, B_f} e^{-f^2/2B_f^2}, \qquad -\infty < f < \infty, \tag{95}$$

where

$$B_f \triangleq \frac{(\overline{\omega^2})^{1/2}}{2\pi} = \frac{\sigma_w}{2\pi} = \frac{1}{2\pi}\left(\frac{1}{2T^2} + 2b^2T^2\right)^{1/2} \tag{96}$$

is the root-mean-square signal bandwidth in cycles per second [recall (10.48)].

Convolving (93) and (96) gives

$$\tilde{S}_{\tilde{n}_r}\{f\} = \frac{N_r E_t}{\sqrt{2\pi}\, \gamma} e^{-f^2/2\gamma^2}, \tag{97}$$

where

$$\gamma^2 \triangleq \sigma_R^2 + B_f^2. \tag{98}$$

Using (95) and (97) in (92), we have

$$\rho_r = \frac{E_t}{N_0} \frac{N_r}{2\pi\gamma B_f} \int_{-\infty}^{\infty} \exp\left(-\frac{(f - f_d)^2}{2B_f^2} - \frac{f^2}{2\gamma^2}\right) df. \tag{99a}$$

Integrating and rearranging terms gives

$$\rho_r = \frac{E_t N_r}{N_0 \sqrt{2\pi}\, \sigma_R} \frac{1}{\sqrt{1 + 2(B_f/\sigma_R)^2}} \exp\left[-\frac{1}{2}\frac{(f_d/\sigma_R)^2}{(1 + 2(B_f/\sigma_R)^2)}\right]. \tag{99b}$$

We see that the loss depends on the three quantities:

1. $D \triangleq \dfrac{E_t N_r}{N_0 \sqrt{2\pi}\, \sigma_R}$, the ratio of the reverberation power to the noise power in the equivalent rectangular bandwidth of the reverberation.

2. $\dfrac{f_d}{\sigma_R}$, the ratio of the target Doppler to the root-mean-square Doppler spread of the reverberation.

3. $\dfrac{B_f}{\sigma_R}$, the ratio of the effective signal bandwidth to the rms the Doppler spread of the reverberation.

The performance is given by substituting (99*b*) into (81) to obtain

$$\Delta_{wo,n} \triangleq \frac{\Delta_{wo}}{\bar{E}_r/N_0} = \frac{1}{1 + \rho_r}. \tag{100}$$

In Fig. 13.11, $\Delta_{wo,n}$ is plotted for some representative values of $D$.

| $D$ | Physical meaning | Figure |
|---|---|---|
| 0.3 | Reverberation < additive noise | 13.11*a* |
| 1.0 | Reverberation = additive noise | 13.11*b* |
| 10.0 | Reverberation/additive noise = 10 db | 13.11*c* |
| 100.0 | Reverberation/additive noise = 20 db | 13.11*d* |

The parameters on the curves are $f_d/\sigma_R$. This is ratio of the target Doppler shift to the root-mean-square Doppler spread of the reverberation. The horizontal axis is $B_f/\sigma_R$. This is the ratio of the signal bandwidth to the root-mean-square Doppler spread of the reverberation. Two observations may be made with respect to this class of signals:

1. For zero target velocities, we have monotone improvement as the bandwidth increases.

2. For nonzero target velocities, one can use either very small or very large bandwidth signals.

The logic of these results follows easily from the diagrams shown in Figs. 13.12 and 13.13.

In Fig. 13.12, we show a zero-velocity target. As we increase the bandwidth, either by shortening the pulse (decreasing $T$) or by increasing the sweep rate (increasing $b$), the common volume between the ambiguity function and the reverberation decreases monotonically. In Fig. 13.13, we show a non-zero-velocity target. By transmitting a *long* pulse with no frequency modulation, the width of the ambiguity function in the Doppler direction is small and the common volume is negligible. If we shorten the pulse (by decreasing $T$) or widen the bandwidth (by increasing $b$), the result in Fig. 13.12*b* is obtained. We have increased the common volume, and the performance is degraded as shown in Fig. 13.10*c* and *d*. Finally, as $B_f$ continues to increase, as shown in Fig. 13.12*c*, the width of the overlapping part of the ambiguity function decreases (it is $\simeq B_f^{-1}$), and the performance increases again.

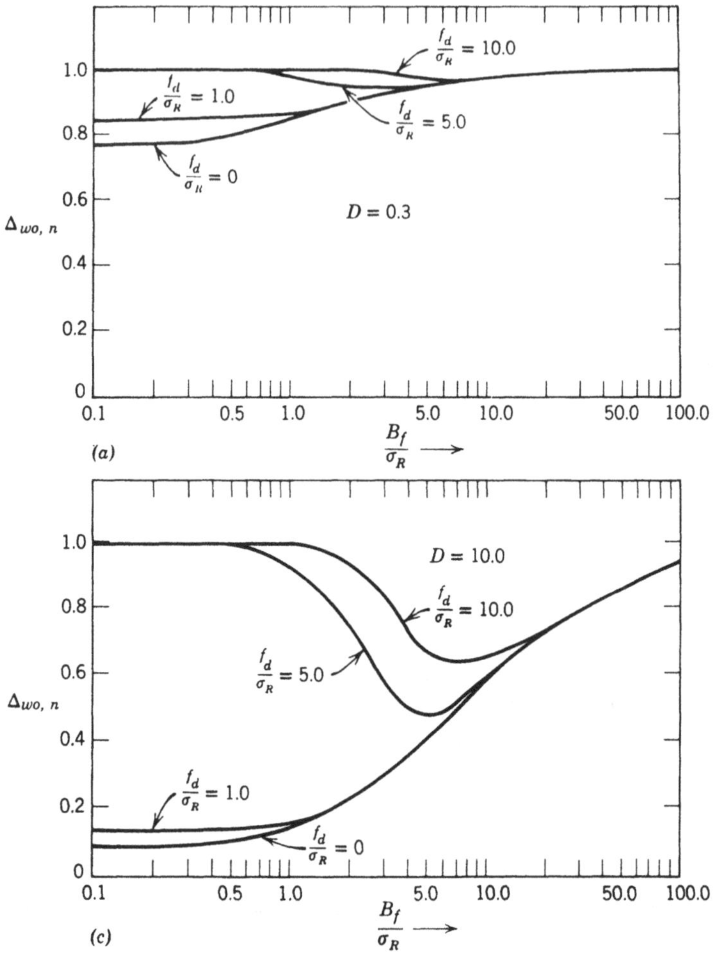


**Fig. 13.11 Performance of conventional receiver in the presence of reverberation.**

This example demonstrates the importance of matching the signal to the environment of interest. In this particular case, one might want to have two types of signals available: a long, unmodulated pulse, which is easy to generate and very effective for moving targets, and a linear FM pulse, which is effective for targets whose velocity was less than the root-mean-square Doppler spread. The example also illustrates an environment in which the signal design problem is relatively insensitive to detailed assumptions in the model. In other words, the basic results depend more on $\sigma_R$, the root-mean-square Doppler spread of the reverberation, than on the detailed shape of $\tilde{S}_{Du}\{f\}$.

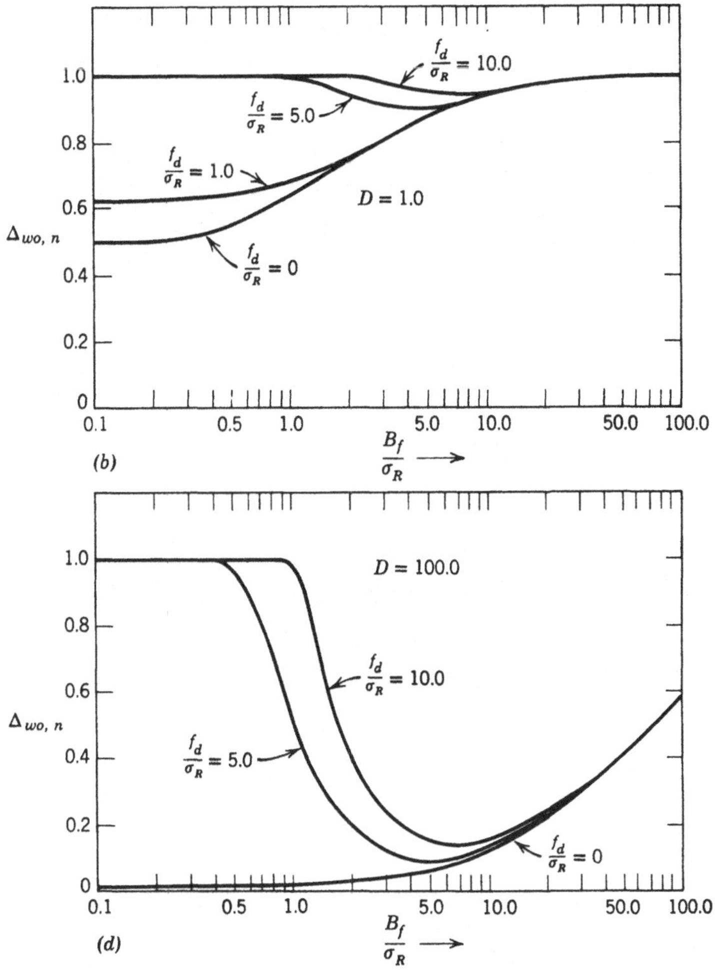


**Fig. 13.11 (continued)**

There are several comments with respect to the effect of reverberation on the performance that apply to both the range-invariant case and the general problem in which $\tilde{S}_{DR}\{f, \lambda\}$ is a function of $\lambda$.

1. The evaluation of receiver performance is always straightforward. At worst, a numerical evaluation of (83), (85), or (92) is required.

2. The problem of designing the optimum signal to minimize $\rho_r$ subject to suitable constraints such as energy, bandwidth, or duration is mathematically difficult and can usually be avoided. Even if we could solve it,

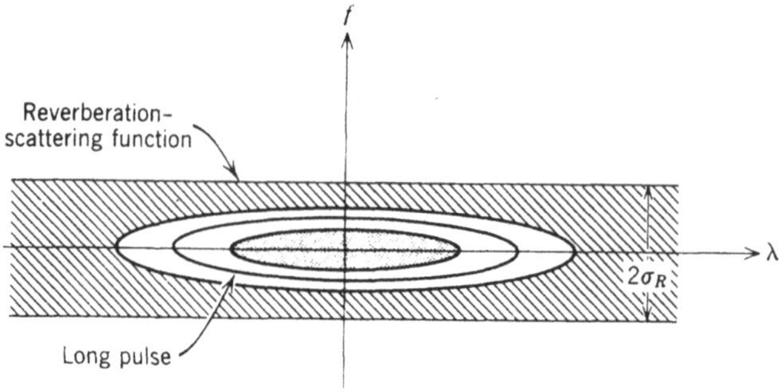


(a)

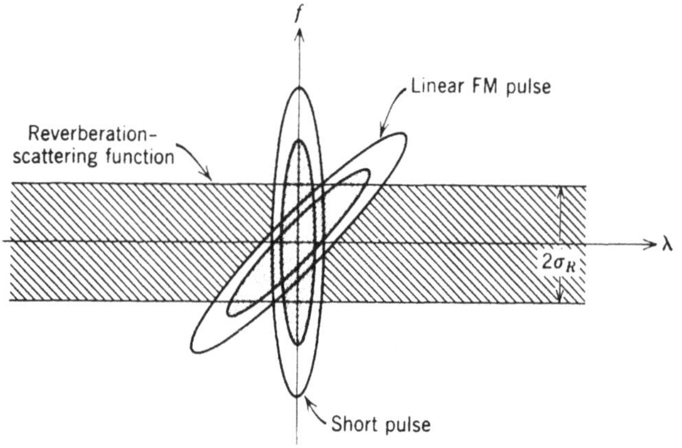


(b)

**Fig. 13.12 Zero-velocity target in reverberation.**

the solution would vary with $\tilde{S}_{DR}\{f, \lambda\}$, $\tau_d$, and $f_d$. A more practical solution is the following:

(a) Choose a class of signals [e.g., the coded pulse sequence in (10.145)]. Maximize their performance by varying the parameters.

(b) Consider a set of allowable scattering functions instead of a specific scattering function. This gives a result that is less sensitive to the detailed environment.

There are a number of references dealing with the signal design problem at various levels of sophistication (e.g., [21]–[28]).

3. The nature of the ambiguity function of a sequence of pulses with complex weightings make it an excellent waveform for reverberation

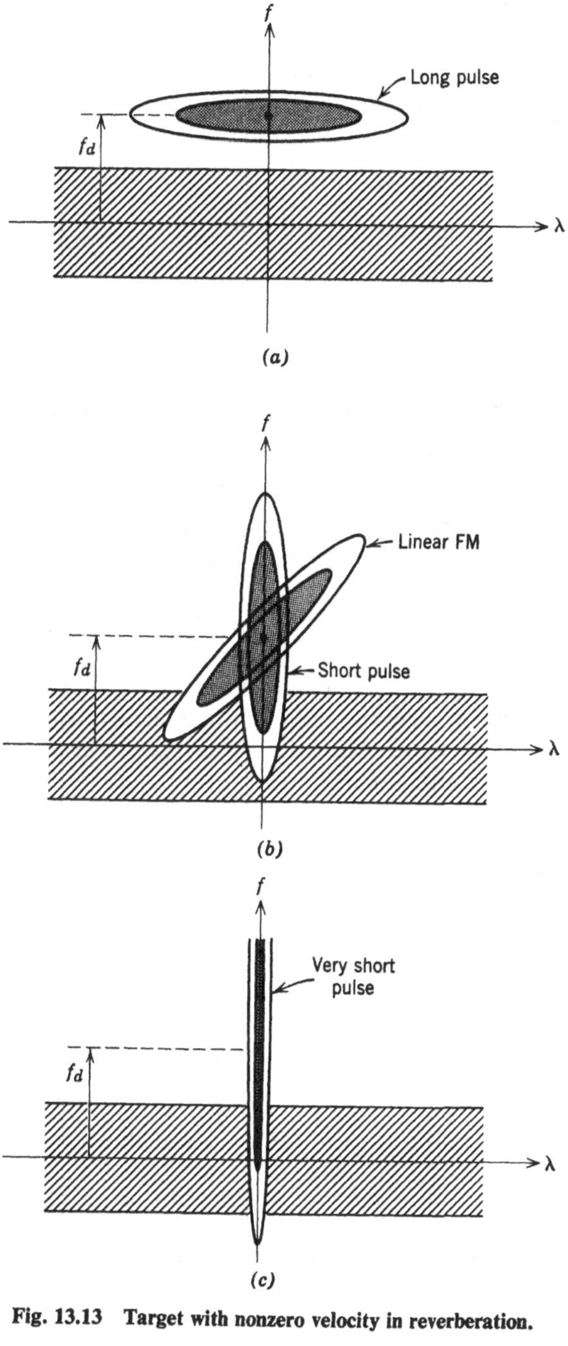


**Fig. 13.13 Target with nonzero velocity in reverberation.**

suppression. Because it is also easy to generate and detect, it is used in many systems.

This completes our discussion of the conventional receiver performance. We now turn to the optimum receiver problem.

### 13.2.2 Optimum Receivers

The problem of interest was given in (72), but we restate it here for convenience. The complex envelopes of the received waveforms on the two hypotheses are

$$\tilde{r}(t) = \tilde{b}\tilde{f}_d(t) + \tilde{n}_r(t) + \tilde{w}(t), \qquad -\infty < t < \infty : H_1 \tag{101}$$

and

$$\tilde{r}(t) = \tilde{n}_r(t) + \tilde{w}(t), \qquad -\infty < t < \infty : H_0, \tag{102}$$

where

$$\tilde{f}_d(t) \triangleq \tilde{f}(t - \tau_d)e^{j\omega_d t}. \tag{103}$$

The reverberation return is a sample function of a complex Gaussian random process whose covariance function is given by (70) as

$$\tilde{K}_{\tilde{n}_r}(t, u) = E_t \int_{-\infty}^{\infty} \tilde{f}(t - \lambda)\tilde{K}_{DR}(t - u, \lambda)\tilde{f}^*(u - \lambda)\, d\lambda. \tag{104}$$

We want to find the optimum receiver to detect $\tilde{f}_d(t)$. This is just the familiar problem of detecting a slowly fluctuating point target in the presence of nonwhite Gaussian noise, which we solved in Section 9.3.

The optimum receiver computes

$$\tilde{l} \triangleq \int_{-\infty}^{\infty} \tilde{g}^*(t)\tilde{r}(t)\, dt, \tag{105}$$

where $\tilde{g}(\cdot)$ satisfies

$$\tilde{f}_d(t) = \int_{-\infty}^{\infty} \tilde{K}_{\tilde{n}_r}(t, u)\tilde{g}(u)\, du + N_0\tilde{g}(t), \qquad -\infty < t < \infty. \tag{106}$$

It then compares $|\tilde{l}|^2$ with a threshold. Using (104) in (106) gives the equation we must solve to find the optimum receiver. It is

$$\tilde{f}_d(t) = E_t \iint_{-\infty}^{\infty} \tilde{f}(t - \lambda)\tilde{K}_{DR}(t - u, \lambda)\tilde{f}^*(u - \lambda)\tilde{g}(u)\, du\, d\lambda + N_0\tilde{g}(t),$$
$$-\infty < t < \infty. \tag{107}$$

For arbitrary $\tilde{K}_{DR}(\cdot, \cdot)$ the solution of (107) is difficult. There are several cases when a solution can be obtained.

**Case 1.** If the functions $\tilde{f}(t)$ and $\tilde{K}_{DR}(t-u, \lambda)$ are of a form so that the eigenvalues and eigenfunctions of (104) can be found, the solution to (107) follows easily. One example of this is the discrete resolution problem of Section 10.5. A second example is when $\tilde{K}_{DR}(t-u, \lambda)$ is a separable kernel. A third example is when $\tilde{f}(t)$ is a Gaussian pulse and $\tilde{K}_{DR}(t-u, \lambda)$ is doubly Gaussian. The basic procedure in this case is familiar, and so we relegate it to the problems.

**Case 2.** If we can describe the channel-scattering function by the distributed differential-equation model of Section 13.1.2, we can find the optimum receiver. We discuss this case in detail in the next subsection.

**Case 3.** If the scattering function is long in the $\lambda$-dimension (as shown in Fig. 13.10) and has a uniform Doppler profile, then $\tilde{n}_r(t)$ is a stationary process and (107) can be solved using Fourier transforms. We studied the conventional receiver for this case on page 466. On page 477, we study the optimum receiver and compare the performance of the two systems.

We now carry out the analysis of Cases 2 and 3.

**Case 2. Optimum Receiver: State-Variable Realization.** In Section 9.4, we developed a realization of the optimum receiver for the detection of a slowly fluctuating target in nonwhite noise. This realization was based on a realizable whitening filter and contained the minimum mean-square error estimate of the nonwhite noise as a basic component. This realization of the optimum receiver is shown in Fig. 13.14 (this is just Fig. 9.8 with modified notation). The test statistic is

$$l_o = \left| \int_{T_i}^{T_f} dt \left[ \int_{T_i}^{t} \tilde{h}_{wr}(t, z)\tilde{r}(z)\, dz \int_{T_i}^{t} \tilde{h}_{wr}^*(t, y)\tilde{f}_d^*(y)\, dy \right] \right|^2, \tag{108}$$

where $\tilde{h}_{wr}(t, z)$ is defined as

$$\tilde{h}_{wr}(t, z) \triangleq \left(\frac{1}{N_0}\right)^{1/2} \{\delta(t-z) - \tilde{h}_{or}(t, z)\}. \tag{109}$$

The filter $\tilde{h}_{or}(t, z)$ is the optimum realizable filter for estimating $\tilde{n}_r(t)$ when the input is

$$\tilde{n}(t) = \tilde{n}_r(t) + \tilde{w}(t). \tag{110}$$

From (9.111) we have the degradation due to the colored noise,

$$\Delta_{dg} = \int_{T_i}^{T_f} [2\tilde{f}_d^*(t)\tilde{f}_r(t) - |\tilde{f}_r(t)|^2]\, dt. \tag{111}$$

The function $\tilde{f}_r(t)$ is the output of $\tilde{h}_{wr}(t, z)$ when the input is $\tilde{f}_d(t)$.

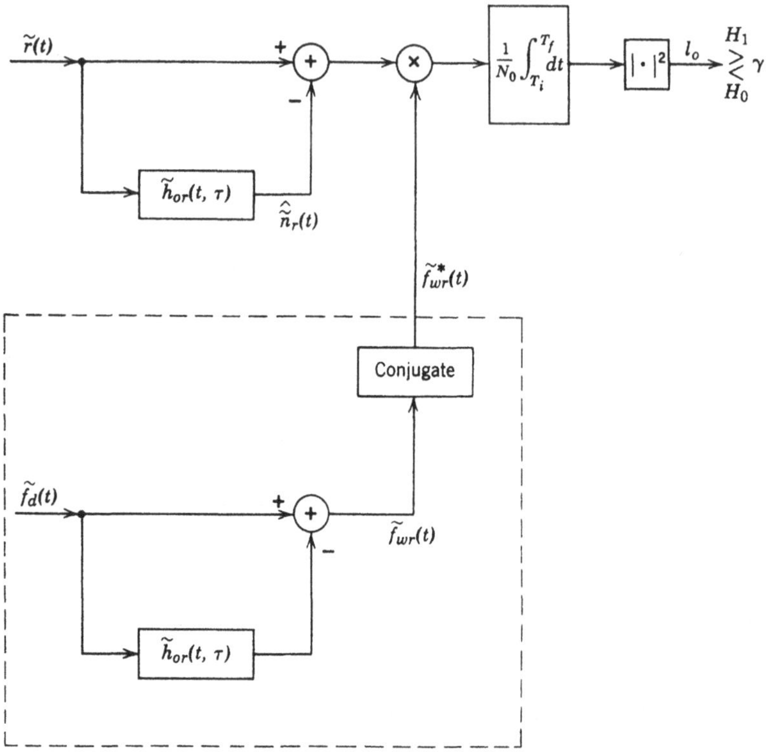


**Fig. 13.14 Optimum receiver**

We see that we can completely design the optimum receiver and analyze the receiver performance if we can find $\tilde{h}_{or}(t, z)$. In this section we derive a set of differential equations that specify $\hat{\tilde{n}}_r(t)$ [and therefore $\tilde{h}_{or}(t, z)$]. Our derivation is based on the distributed state-variable model in Section 13.1.2.

The differential equations describing $\tilde{n}_r(t)$ are analogous to (38)–(41). The state equation is

$$\frac{\partial \tilde{\mathbf{x}}(\lambda, t)}{\partial t} = \tilde{\mathbf{F}}(\lambda)\tilde{\mathbf{x}}(t, \lambda) + \tilde{\mathbf{G}}(\lambda)\tilde{u}(t, \lambda), \qquad t > T_i, \ \lambda \in \Omega_L, \tag{112}$$

where $\Omega_L$ is the range of $\lambda$ where the target-scattering function is nonzero. The covariance function of $\tilde{u}(t, \lambda)$ is

$$E[\tilde{u}(t, \lambda)\tilde{u}^*(\tau, \lambda')] = \tilde{Q}(\lambda)\delta(t - \tau)\,\delta(\lambda - \lambda'). \tag{113}$$

The reverberation process is

$$\tilde{b}(t, \lambda) = \tilde{\mathbf{C}}(\lambda)\tilde{\mathbf{x}}(t, \lambda). \tag{114}$$

The colored noise due to the reverberation is

$$\tilde{n}_r(t) = \int_{-\infty}^{\infty} \sqrt{E_t}\tilde{f}(t-\lambda)\tilde{b}(t,\lambda)\,d\lambda$$

$$\triangleq \tilde{C}(t:\tilde{\mathbf{x}}(t,\lambda)), \tag{115}$$

which we refer to as the *modulation functional.*

The minimum mean-square error estimate is obtained by extending Kalman-Bucy filtering theory to include the spatial integral operator in (115). This extension was done originally by Tzafestas and Nightingale [29], [30]. The results are given in (116–121).†

The estimator equation is

$$\frac{\partial \hat{\tilde{\mathbf{x}}}(t,\lambda)}{\partial t} = \tilde{\mathbf{F}}(\lambda)\hat{\tilde{\mathbf{x}}}(t,\lambda) + \tilde{\mathbf{z}}(t,\lambda)[\tilde{r}(t) - \tilde{C}(t:\hat{\tilde{\mathbf{x}}}(t,\lambda))], \qquad t \geq T_i,\ \lambda \in \Omega_L. \tag{116}$$

$$\hat{\tilde{\mathbf{x}}}(T_i,\lambda) = \mathbf{0}, \qquad \lambda \in \Omega_L. \tag{117}$$

The gain equation is

$$\tilde{\mathbf{z}}(t,\lambda) = \frac{1}{N_0}\int_{\Omega_L} \tilde{\boldsymbol{\xi}}(t:\lambda,\lambda')\tilde{\mathbf{C}}^{\dagger}(\lambda')\sqrt{E_t}\tilde{f}^*(t-\lambda')\,d\lambda'. \tag{118}$$

The function $\tilde{\boldsymbol{\xi}}(t:\lambda,\lambda')$ is the error covariance matrix

$$\tilde{\boldsymbol{\xi}}(t:\lambda,\lambda') \triangleq E[(\tilde{\mathbf{x}}(t,\lambda) - \hat{\tilde{\mathbf{x}}}(t,\lambda))(\tilde{\mathbf{x}}^{\dagger}(t,\lambda') - \hat{\tilde{\mathbf{x}}}^{\dagger}(t,\lambda'))]. \tag{119a}$$

Notice that

$$\tilde{\boldsymbol{\xi}}^{\dagger}(t:\lambda,\lambda') = \tilde{\boldsymbol{\xi}}(t:\lambda',\lambda). \tag{119b}$$

The covariance matrix satisfies the differential equation

$$\frac{\partial \tilde{\boldsymbol{\xi}}(t:\lambda,\lambda')}{\partial t} = \tilde{\mathbf{F}}(\lambda)\tilde{\boldsymbol{\xi}}(t:\lambda,\lambda') + \tilde{\boldsymbol{\xi}}(t:\lambda',\lambda)\tilde{\mathbf{F}}^{\dagger}(\lambda') + \tilde{\mathbf{G}}(\lambda)\tilde{Q}(\lambda)\tilde{\mathbf{G}}^{\dagger}(\lambda')\,\delta(\lambda-\lambda') - \tilde{\mathbf{z}}(t,\lambda)N_0\tilde{\mathbf{z}}^{\dagger}(t,\lambda'), \qquad \lambda,\lambda' \in \Omega_L,\ t \geq T_i. \tag{120}$$

† We have omitted the derivation because, for the particular case described by (112)–(115), the reader should be able to verify that the result is correct (see Problem 13.2.15). The model studied in [29] is much more general than we need. The reader should note the similarity between (116)–(121) and the Kalman-Bucy equations of Section I-6.3. Other references dealing with estimation in distributed parameter systems include [70]–[75].

The reader should notice the similarity between (120) and (A.161). Using (118) in (120) gives

$$\boxed{\begin{aligned}\frac{\partial \tilde{\boldsymbol{\xi}}(t:\lambda,\lambda')}{\partial t} &= \tilde{\mathbf{F}}(\lambda)\tilde{\boldsymbol{\xi}}(t:\lambda,\lambda') + \tilde{\boldsymbol{\xi}}(t:\lambda',\lambda)\tilde{\mathbf{F}}^{\dagger}(\lambda') + \tilde{\mathbf{G}}(\lambda)\tilde{Q}(\lambda)\tilde{\mathbf{G}}^{\dagger}(\lambda)\,\delta(\lambda-\lambda') \\ &\quad - \frac{E_t}{N_0}\left\{\int_{\Omega_L}\tilde{\boldsymbol{\xi}}(t:\lambda,\sigma)\tilde{\mathbf{C}}^{\dagger}(\sigma)\tilde{f}^*(t-\sigma)\,d\sigma\right. \\ &\qquad \left.\times\int_{\Omega_L}\tilde{f}(t-\sigma')\tilde{\mathbf{C}}(\sigma')\tilde{\boldsymbol{\xi}}(t:\sigma',\lambda')\,d\sigma'\right\}, \\ &\qquad\qquad \lambda,\lambda'\in\Omega_L,\ t\geq T_i.\end{aligned}} \tag{121a}$$

The initial condition is

$$\tilde{\boldsymbol{\xi}}(T_i:\lambda,\lambda') = \tilde{\boldsymbol{\xi}}_0(T_i,\lambda)\,\delta(\lambda-\lambda'). \tag{121b}$$

These equations completely specify the filter whose impulse response is $\tilde{h}_{or}(t, z)$. Thus, the receiver is completely specified and its performance can be calculated. The block diagram of the optimum estimator is shown in Fig. 13.15 (the heavy lines represent signals that are functions of both space and time). Using the system in Fig. 13.15 in the diagram of Fig. 13.14 gives the optimum receiver.

Several comments regarding the optimum receiver and the corresponding equations are useful.

1. The optimum filter contains spatial operations. In most cases these will be difficult to implement exactly. We discuss several approximate realizations in Section 13.3.
2. The performance of the optimum receiver provides a bound on the performance of simpler systems. To evaluate the performance we must find $\Delta_{dg}$ as specified in (111). This requires solving the variance equation (120).
3. The equations are algebraically complex. We shall find that we can obtain solutions to them with a reasonable amount of calculation.

We shall defer doing an example to illustrate the solution techniques until Section 13.3. The same filtering problem arises in that section, when we communicate over doubly-spread channels.

This concludes our discussion of the state-variable realization of the optimum receiver in the presence of reverberation. We now consider the third case listed in the introduction on page 473.

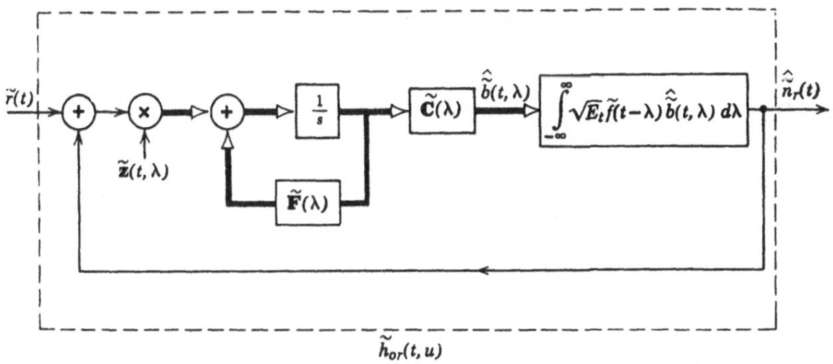


**Fig. 13.15 Distributed state-variable realization of optimum realizable filter.**

**Case 3. Reverberation with Constant Doppler Profile and Infinite Range.** In some cases $\Omega_L$ is long enough compared to the observation time that we can assume it is infinite (see Fig. 13.10). If, in addition,

$$\tilde{K}_{DR}(t-u,\lambda) = \tilde{K}_{Du}(t-u), \tag{122}$$

we can obtain a solution to (106) by using Fourier transform techniques. Using (122) in (104), we obtain

$$\tilde{K}_{\tilde{n}_r}(t,u) = E_t\tilde{K}_{Du}(t-u)\int_{-\infty}^{\infty}\tilde{f}(t-\lambda)\tilde{f}^*(u-\lambda)\,d\lambda, \tag{123}$$

which is identical with (88). From (91),

$$\tilde{S}_{\tilde{n}_r}\{f\} = E_t\tilde{S}_{Du}\{f\} \otimes \tilde{S}_{\tilde{f}}\{f\}, \tag{124}$$

where

$$\tilde{S}_{\tilde{f}}\{f\} \triangleq |\tilde{F}\{f\}|^2. \tag{125}$$

Using (124) in (106), transforming, and solving for $\tilde{G}_\infty\{f\}$ gives

$$\boxed{\tilde{G}_\infty\{f\} = \frac{\tilde{F}\{f-f_d\}}{N_0 + E_t\tilde{S}_{Du}\{f\} \otimes \tilde{S}_{\tilde{f}}\{f\}},} \tag{126a}$$

where $f_d$ is Doppler shift of the desired target. For a zero-velocity target,

$$\boxed{G_\infty\{f\} = \frac{\tilde{F}\{f\}}{N_0 + E_t\tilde{S}_{Du}\{f\} \otimes \tilde{S}_{\tilde{f}}\{f\}}.} \tag{126b}$$

The performance is obtained from (9.77) as

$$\boxed{\Delta_o = \bar{E}_r\int_{-\infty}^{\infty}\frac{|\tilde{F}\{f-f_d\}|^2}{N_0 + \tilde{S}_{\tilde{n}_r}\{f\}}\,df.} \tag{127}$$

We now discuss a particular problem to illustrate the application of these results.

**Example 1.** We consider the same model as in the conventional receiver example on page 466. The signal is specified by (94), and the Doppler profile is given in (93). Using (95) and (97) in (127) gives

$$\Delta_o = \frac{\bar{E}_r}{N_0} \int_{-\infty}^{\infty} \frac{(\sqrt{2\pi}\, B_f)^{-1} e^{-(f-f_d)^2/2B_f^2}}{1 + (N_r E_t/\sqrt{2\pi}\, N_0 \gamma) e^{-f^2/2\gamma^2}}\, df. \tag{128}$$

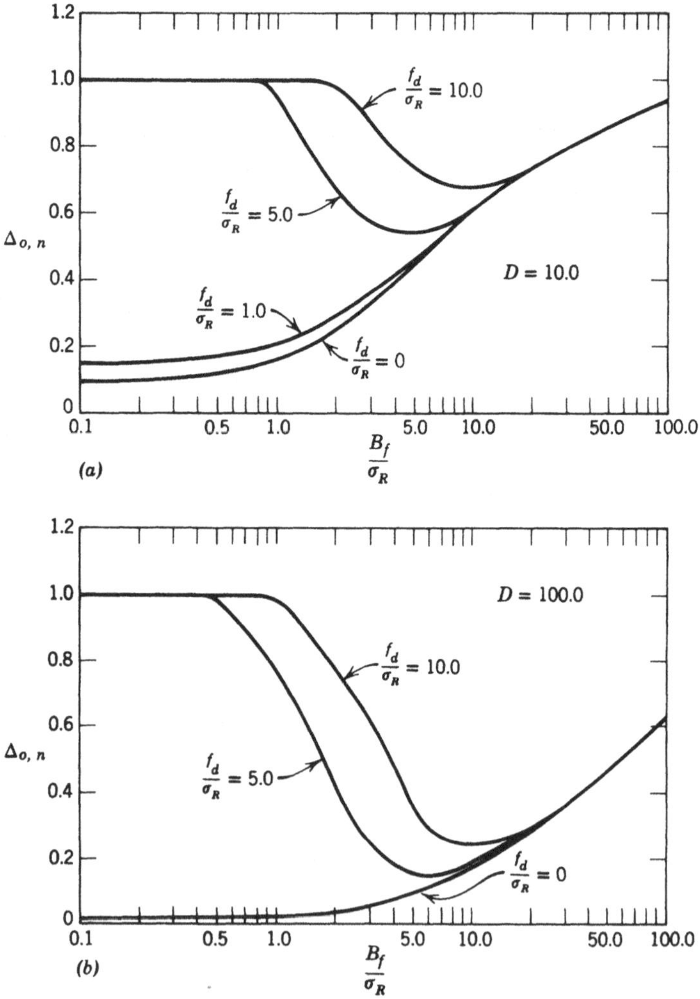


**Fig. 13.16 Performance of optimum receiver in the presence of reverberation.**

As in the conventional receiver case, the performance depends on $D, f_d/\sigma_R$, and $B_f/\sigma_R$, which were defined on page 467. When $D = 0.3$ and $1.0$, the performances of the optimum receiver and the conventional receiver are almost the same. In Fig. 13.16, we have plotted $\Delta_o/(\bar{E}_r/N_0)$ for $D = 10$ and $100$.

It is useful to consider a particular set of values. We assume the following set of parameter values:

$$\begin{aligned} f_d/\sigma_R &= 5.0, \\ B_f/\sigma_R &= 1.0, \\ D &= 100.0. \end{aligned} \tag{129}$$

We see that this set of parameters puts us in the valley on both curves (Figs. 13.16 and 13.11) and corresponds to a poor signal choice.

For this set of parameter values,

$$\Delta_o = 0.762 \frac{\bar{E}_r}{N_0} \tag{130}$$

and

$$\Delta_{wo} = 0.528 \frac{\bar{E}_r}{N_0}. \tag{131}$$

We see that, for a very poor signal choice, the difference is about 1.5 db. For any reasonably good signal choices, the difference in the performance of the two receivers is negligible. This is because a good signal choice decreases the reverberation return to a point where the optimum receiver is not needed.

We consider a second example briefly.

**Example 2.** In some cases, the Doppler spreading of the reverberation (or clutter) is small compared to the bandwidth of the signal energy spectrum, $\tilde{S}_{\tilde{f}}\{f\}$. In these cases it is convenient to approximate $\tilde{S}_{Du}\{f\}$ by an impulse,

$$\tilde{S}_{Du}\{f\} \simeq P_c\, \delta\{f\}. \tag{132}$$

We assume that $f_d = 0$, because a zero-velocity target is the most difficult to detect. Using (132) in (126*b*) gives

$$\tilde{G}_\infty\{f\} = \frac{\tilde{F}\{f\}}{N_0 + E_t P_c |\tilde{F}\{f\}|^2}. \tag{133}$$

For small values of reverberation return, such that

$$\frac{E_t P_c \tilde{S}_{\tilde{f}}\{f\}}{N_0} \ll 1, \tag{134}$$

we see that the optimum filter reduces to a matched filter. However, for large values of reverberation return,

$$\tilde{G}_\infty\{f\} \simeq \frac{1}{E_t P_c \tilde{F}^*\{f\}}, \tag{135}$$

which is an *inverse filter* (this filter was first derived in [31]). Thus, we see that the optimum receiver has an appreciably different character in the frequency ranges where reverberation rather than the additive white noise is the limiting factor.

To evaluate the performance, we use (132) in (127) and obtain

$$\Delta_o = \bar{E}_r \int_{-\infty}^{\infty} \frac{|\tilde{F}\{f\}|^2}{N_0 + E_t P_c |\tilde{F}\{f\}|^2}\, df. \tag{136}$$

It is useful to evaluate $\Delta_o$ for a particular signal set. We consider the Butterworth family whose energy spectrum is

$$\tilde{S}_{\tilde{f}}\{f\} = \frac{(2n/k)\sin(\pi/2n)}{(2\pi f/k)^{2n} + 1}, \qquad -\infty < f < \infty, \quad n = 1, 2, \ldots. \tag{137}$$

Substituting (137) into (136) and integrating gives

$$\Delta_o(n) = \frac{\bar{E}_r}{N_0}\left[1 + \frac{\pi E_t P_c}{k N_0}\,\frac{2n}{\pi}\sin\left(\frac{\pi}{2n}\right)\right]^{1/2n-1}. \tag{138a}$$

For $n = \infty$,

$$\Delta_o(\infty) = \frac{\bar{E}_r}{N_0}\left[1 + \frac{\pi E_t P_c}{k N_0}\right]^{-1}. \tag{138b}$$

Two conclusions follow from (138):

1. Increasing the energy in the transmitted signal is an ineffective way to combat reverberation; we see that, for strong reverberation,

$$\Delta_o(n) \propto E_t^{1/2n} \le E_t^{1/2} \qquad \text{for} \quad E_t \ge 1, \tag{138c}$$

where we obtain equality by using an exponential pulse ($n = 1$).

2. As we would expect, $\Delta_o$ increases monotonically with $k$. (Recall Fig. 13.13.)

Notice that this example has considered a zero-velocity target, which is the most difficult to detect in the reverberation environment specified by (132). The performance increases monotonically with target velocity.

This case completes our discussion of the design and performance of optimum receivers in the presence of reverberation. We now summarize our discussion of the reverberation problem.

### 13.2.3 Summary of the Reverberation Problem

We have investigated the problem of signal and receiver design when the target is a slowly fluctuating point target and the interference consists of reflectors that are spread in range and Doppler and additive white noise. A number of interesting results were obtained.

1. If a conventional matched filter receiver is used, the degradation due to reverberation is

$$\rho_r = \frac{E_t}{N_0} \iint_{-\infty}^{\infty} df\, d\lambda\, \tilde{S}_{DR}\{f, \lambda\}\theta\{\lambda - \tau_d, f_d - f\}. \tag{139}$$

This integral has the simple graphical interpretation shown in Fig. 13.9.

2. The optimal signal design problem for the conventional receiver consists of trying to minimize the common volume of the reverberation-scattering function and the shifted signal ambiguity function. The best signal will depend on the target's location in the range-Doppler plane as well as on $\tilde{S}_{DR}\{f, \lambda\}$.

3. If one can make the two functions essentially disjoint, the conventional and optimum receivers are identical and the performance is limited only by the additive noise.

4. If one is constrained to use a signal that results in an appreciable overlap, the optimum receiver provides an improved performance. It is important to remember that this improved performance requires a more complex receiver and assumes knowledge of the scattering function (including its level) and the additive noise level.

5. For a large number of cases we can find the optimum receiver and evaluate its performance. The techniques for performing this analysis were developed in detail.

We have confined our discussion to conventional and optimum receivers. A third category of receivers is useful in some situations. This receiver computes

$$\tilde{l}_m \triangleq \int_{-\infty}^{\infty} \tilde{r}(t)\tilde{v}^*(t)\, dt \tag{140}$$

and compares $|\tilde{l}_m|^2$ with a threshold. The function $\tilde{v}(t)$ is not necessarily the desired signal, $\tilde{f}_d(t)$, or the optimum correlation function $\tilde{g}(t)$ specified by (106). We might choose a $\tilde{v}(t)$ that is simpler to implement than $\tilde{g}(t)$ but performs better than $\tilde{f}_d(t)$. The performance of the receiver in (140) is given by

$$\Delta_m = \frac{\bar{E}_r \theta_{fv}\{\tau_d, f_d\}}{N_0\left[1 + E_t/N_0 \displaystyle\iint_{-\infty}^{\infty} \tilde{S}_{DR}\{f, \lambda\}\theta_{fv}\{\lambda, -f\}\, df\, d\lambda\right]}, \tag{141}$$

where $\theta_{fv}\{\cdot, \cdot\}$ is the cross-ambiguity function defined in (10.222). We can now choose $\tilde{v}(t)$ to minimize $\Delta_m$. Notice that we must put additional constraints on $\tilde{v}(t)$, or we shall find that the optimum $\tilde{v}(t)$ equals $\tilde{g}(t)$. One possible constraint is to require $\tilde{v}(t)$ to be a piecewise constant function. This would be a logical constraint if $\tilde{f}(t)$ were a sequence of rectangular pulses. Various other constraints are possible.

This particular formulation is attractive because it allows us to design a system that works better than the conventional receiver but can be constrained to be less complicated than the optimum receiver. This problem and variations of it have been studied by Stutt and Spafford [32], Spafford [33], and Rummler [34]. We suggest that the reader consult these references, because they provide an excellent demonstration of how the doubly-spread reverberation model of this section can be used to obtain effective practical systems. Various facets of the question are developed in the problems (e.g., Problems 13.2.17 and 13.2.18).

This completes our discussion of the reverberation and clutter problem. We now turn to a different type of problem.

## 13.3 DETECTION OF DOUBLY-SPREAD TARGETS AND COMMUNICATION OVER DOUBLY-SPREAD CHANNELS

In this section we consider two closely related problems. The first problem arises in the radar and sonar area and consists of trying to detect the return from a doubly-spread target in the presence of additive noise. The second problem consists of communicating digital data over a doubly-spread channel.

The section is divided into four parts. In Section 13.3.1, we formulate the quantitative models for the two problems and derive expressions for the optimum receivers and their performance. The results contain integral equations or differential equations that cannot be solved exactly in most cases. In Section 13.3.2, we develop approximate target and channel models that enable us to obtain a complete solution for the optimum receivers and their performance. In Section 13.3.3, we calculate the performance of a particular binary communication scheme to illustrate the techniques involved. In Section 13.3.4, we discuss some related topics.

### 13.3.1 Problem Formulation

In this section we formulate the detection and binary communication problem quantitatively.

***13.3.1.A. Detection.*** The first problem of interest is the radar or sonar detection problem. We transmit a signal whose complex envelope is $\sqrt{E_t}\tilde{f}(t)$. If a doubly-spread target is present, the complex envelope of the returned signal is

$$\tilde{s}(t) = \int_{-\infty}^{\infty} \sqrt{E_t}\tilde{f}(t-\lambda)\tilde{b}(t,\lambda)\,d\lambda, \tag{142}$$

where $\tilde{b}(t, \lambda)$ is a sample function from a complex Gaussian process whose covariance function is given in (37). We are using the process defined in (36), but the subscript $x$ is omitted. The covariance function of $\tilde{s}(t)$ is given by (22) as

$$\tilde{K}_{\tilde{s}}(t, u) = E_t \int_{-\infty}^{\infty} \tilde{f}(t - \lambda) \tilde{K}_{DR}(t - u, \lambda) \tilde{f}^*(u - \lambda)\, d\lambda. \tag{143}$$

In addition to the signal component, the received waveform contains an additive complex white noise $\tilde{w}(t)$, whose covariance function is

$$E[\tilde{w}(t)\tilde{w}^*(u)] = N_0\, \delta(t - u). \tag{144}$$

The received waveform is just the noise term, $\tilde{w}(t)$, if the target is not present. Thus, we have a binary hypothesis testing problem in which the received complex envelopes on the two hypotheses are

$$\tilde{r}(t) = \tilde{s}(t) + \tilde{w}(t), \qquad T_i \le t \le T_f : H_1, \tag{145}$$

$$\tilde{r}(t) = \tilde{w}(t), \qquad T_i \le t \le T_f : H_0. \tag{146}$$

On both hypotheses, $\tilde{r}(t)$ is a sample function of a complex Gaussian random process. If we compare (145) and (146) with the equations specifying the detection problem in Chapter 11 [(11.30) and (11.31)], we see that the form is identical. The only difference is in the form of the covariance functions of the signal processes. Therefore all of the results in Chapter 11 that contain $\tilde{K}_{\tilde{s}}(t, u)$ as an arbitrary covariance function are valid for the problem of current interest. Specifically, (11.33)–(11.40) and Figs. 11.7–11.9 are valid relations for the receiver structures, and (11.50)–(11.54) are valid expressions for the performance. It is when we evaluate these various formulas that the doubly-spread model becomes important. Specifically, we shall find that the covariance function given in (143) is harder to work with than the covariance functions encountered in the singly-spread cases.

Some of the pertinent results from Chapter 11 are listed for ease of reference. The likelihood ratio test is

$$l_R = \frac{1}{N_0} \int_{T_i}^{T_f} \tilde{r}^*(t) \tilde{h}(t, u) \tilde{r}(u)\, dt\, du \underset{H_0}{\overset{H_1}{\gtrless}} \gamma, \tag{147}$$

where $\tilde{h}(t, u)$ satisfies the integral equation

$$N_0 \tilde{h}(t, u) + \int_{T_i}^{T_f} \tilde{h}(t, z) \tilde{K}_{\tilde{s}}(z, u)\, dz = \tilde{K}_{\tilde{s}}(t, u), \qquad T_i \le t, u \le T_f, \tag{148}$$

and $\tilde{K}_{\tilde{s}}(t, u)$ is given in (143). The estimator-correlator realization is shown in Fig. 11.7.

An alternative expression for the likelihood ratio test is

$$l_R = \frac{1}{N_0} \int_{T_i}^{T_f} \{2 \operatorname{Re} [\tilde{r}^*(t) \hat{\tilde{s}}_r(t)] - |\hat{\tilde{s}}_r(t)|^2\}\, dt \underset{H_0}{\overset{H_1}{\gtrless}} \gamma, \tag{149}$$

where $\hat{\tilde{s}}_r(t)$ is the realizable MMSE estimate of $\tilde{s}(t)$ when $H_1$ is true. An advantage of the implementation in (149) is that whenever $\tilde{s}(t)$ has a distributed state-variable representation we have a set of equations, (116)–(121), that specify $\hat{\tilde{s}}_r(t)$.

The approximate performance expressions that we derived earlier require knowledge of $\tilde{\mu}(s)$, which can be written in three different forms as

$$\tilde{\mu}(s) = \sum_{i=1}^{\infty} \left[ (1 - s) \ln \left(1 + \frac{\tilde{\lambda}_i}{N_0}\right) - \ln \left(1 + (1 - s)\frac{\tilde{\lambda}_i}{N_0}\right)\right]. \tag{150}$$

or

$$\tilde{\mu}(s) = (1 - s) \ln \tilde{D}_{\mathscr{F}}\left(\frac{1}{N_0}\right) - \ln \tilde{D}_{\mathscr{F}}\left(\frac{1 - s}{N_0}\right), \tag{151}$$

or

$$\tilde{\mu}(s) = \frac{(1 - s)}{N_0} \int_{T_i}^{T_f} dt \left[ \xi_P(t, \tilde{s}(t), N_0) - \xi_P\left(t, \tilde{s}(t), \frac{N_0}{1 - s}\right)\right]. \tag{152}$$

We evaluate one of these expressions to find the performance. Before discussing techniques for doing this, we formulate the communications model.

***13.3.1.B. Binary Communication.*** We consider a binary communication system using orthogonal signals. We transmit one of two orthogonal signals,

$$s_t(t) = \text{Re}\,[\sqrt{2E_t}\tilde{f}(t)e^{j\omega_1 t}], \qquad 0 \le t \le T : H_1, \tag{153}$$

$$s_t(t) = \text{Re}\,[\sqrt{2E_t}\tilde{f}(t)e^{j\omega_0 t}], \qquad 0 \le t \le T : H_0, \tag{154}$$

where $\tilde{f}(t)$ has unit energy. Notice that both transmitted signals have the same complex envelope but have different carrier frequencies. We discuss the choice of $\omega_0$ and $\omega_1$ in a moment. The two hypotheses are equally likely.

The received waveforms are

$$r(t) = \text{Re}\,[\sqrt{2}\,\tilde{s}_1(t)e^{j\omega_1 t}] + w(t), \qquad T_i \le t \le T_f : H_1, \tag{155}$$

$$r(t) = \text{Re}\,[\sqrt{2}\,\tilde{s}_0(t)e^{j\omega_0 t}] + w(t), \qquad T_i \le t \le T_f : H_0, \tag{156}$$

where

$$\tilde{s}_i(t) = \sqrt{E_t} \int_{-\infty}^{\infty} \tilde{f}(t - \lambda)\tilde{b}_i(t, \lambda)\, d\lambda, \qquad i = 0, 1. \tag{157}$$

The reflection processes $\tilde{b}_i(t, \lambda)$, $i = 0, 1$, are sample functions from zero-mean complex Gaussian processes, which can be characterized by the same scattering function, $\tilde{S}_{DR}\{f, \lambda\}$.

The channel has two effects on the transmitted signal. The first effect is a delay spread. If the scattering function has a length $L$, there would be a signal component in the received waveform over an interval of length $T + L$. The second effect is a frequency spread. If $\tilde{f}(t)$ is approximately bandlimited to a bandwidth of $W$ cps and the scattering function is approximately bandlimited to $B$ cps the signal portion of the received waveform is approximately bandlimited to $W + B$ cps.

We assume that $\omega_1 - \omega_0$ is large enough so that the signal components at the receiver are in disjoint frequency bands. We see that this separation must take into account both the transmitted signal bandwidth $W$ and the channel-scattering function bandwidth $B$. Thus,

$$\frac{\omega_1 - \omega_0}{2\pi} > W + B. \tag{158}$$

The observation interval is $[T_i, T_f]$, and includes the entire interval in which there is a signal output. This implies

$$T_f - T_i \geq T + L. \tag{159}$$

The receiver must decide between two orthogonal bandpass Gaussian processes in the presence of additive white Gaussian noise. The criterion is minimum probability of error. This is a familiar problem (see Section 11.3). The optimum receiver consists of two parallel branches containing filters centered at $\omega_1$ and $\omega_0$. In the first branch we compute

$$l_1 \triangleq \iint_{T_i}^{T_f} \tilde{r}^*(t)\tilde{h}(t, u)\tilde{r}(u)\, dt\, du, \tag{160}$$

where the complex representation is with respect to $\omega_1$. In the other branch we compute

$$l_0 \triangleq \iint_{T_i}^{T_f} \tilde{r}^*(t)\tilde{h}(t, u)\tilde{r}(u)\, dt\, du, \tag{161}$$

where the complex representation is with respect to $\omega_0$. The complex impulse response is specified by

$$N_0\tilde{h}(t, u) + \int_{T_i}^{T_f} \tilde{h}(t, z)\tilde{K}_{\tilde{s}}(z, u)\, dz = \tilde{K}_{\tilde{s}}(t, u), \qquad T_i \leq t, u \leq T_f, \tag{162a}$$

where

$$\tilde{K}_{\tilde{s}}(t, u) = E_t \int_{-\infty}^{\infty} \tilde{f}(t - \lambda)\tilde{K}_{DR}(t - u, \lambda)\tilde{f}^*(u - \lambda)\, d\lambda. \tag{162b}$$

The optimum test is

$$l_1 \underset{H_0}{\overset{H_1}{\gtrless}} l_0, \tag{163}$$

as shown in Fig. 11.12. We see that (162a) is identical with (148). We can also write $l_1$ and $l_0$ in a form identical with (149). Thus, the equations specifying the optimum receiver are the same for the radar detection problem and the communication problem. Notice that the actual scattering functions will be different in the two problems, because of the different physical environments.

The performance calculation is appreciably simpler in the communication problem, because of the symmetric hypotheses and the zero threshold. Just as for the Doppler-spread case discussed in Section 11.3, we have tight bounds on the error probability. From (11.75),

$$\frac{e^{\tilde{\mu}_{BS}(1/2)}}{2[1 + \sqrt{(\pi/8)\ddot{\tilde{\mu}}_{BS}(\frac{1}{2})}]} \le \Pr(\epsilon) \le \frac{e^{\tilde{\mu}_{BS}(1/2)}}{2[1 + \sqrt{\frac{1}{8}\ddot{\tilde{\mu}}_{BS}(\frac{1}{2})}]} \le \frac{e^{\tilde{\mu}_{BS}(1/2)}}{2}, \tag{164}$$

where $\tilde{\mu}_{BS}(s)$ can be expressed as

$$\tilde{\mu}_{BS}(s) = \tilde{\mu}_{SIB}(s) + \tilde{\mu}_{SIB}(1 - s). \tag{165}$$

The subscript BS denotes binary symmetric, and the subscript SIB denotes simple binary. The formulas for $\tilde{\mu}_{SIB}(s)$ were given in (150)–(152). Substituting (151) into (165) and simplifying gives

$$\tilde{\mu}_{BS}(s) = \ln \left\{\frac{\tilde{D}_{\mathscr{F}}(1/N_0)}{\tilde{D}_{\mathscr{F}}[(1 - s)/N_0]\tilde{D}_{\mathscr{F}}(s/N_0)}\right\}. \tag{166}$$

The exponent in (164) just involves

$$\begin{aligned}\tilde{\mu}_{BS}(\tfrac{1}{2}) &= \ln \left\{\frac{\tilde{D}_{\mathscr{F}}(1/N_0)}{\tilde{D}_{\mathscr{F}}^{\,2}(1/2N_0)}\right\} \\ &= \sum_{i=1}^{\infty} \ln\left(1 + \frac{\tilde{\lambda}_i}{N_0}\right) - 2\sum_{i=1}^{\infty} \ln\left(1 + \frac{\tilde{\lambda}_i}{2N_0}\right).\end{aligned} \tag{167}$$

We can also write $\tilde{\mu}_{BS}(\frac{1}{2})$ in terms of the realizable MMSE filtering error as

$$\tilde{\mu}_{BS}(\tfrac{1}{2}) = \frac{1}{N_0}\int_{T_i}^{T_f} dt[\xi_P(t, \tilde{s}(t), N_0) - \xi_P(t, \tilde{s}(t), 2N_0)]. \tag{168}$$

The basic form of these expressions is familiar from Chapter 11. We must now develop a procedure for finding the required functions.

***13.3.1.C. Summary.*** In this section we have developed the model for the radar detection problem and the binary communication problem. The equations specifying the optimum receivers and their performance were familiar. The new issue that we encountered is that of actually solving these equations when the covariance is given by (143).

There are two cases in which we can solve the equations in a reasonably straightforward manner. We identify them at this point and return to them later in the section. The first case is the low-energy-coherence (LEC) condition that we originally encountered in Chapter 4. We study this case in Section 13.3.4. The second case is a degenerate one in which we choose the transmitted signal so that the target or channel appears to be singly-spread. We discussed this degeneracy in Property 4 (22)–(29) on page 452 and shall study it again in Section 13.3.3. Although these two cases include many of the problems that we encounter in practice, we would like to be able to solve any doubly-spread target (or channel) problem. In the next two sections we develop techniques to deal with the general problem.

### 13.3.2 Approximate Models for Doubly-Spread Targets and Doubly-Spread Channels

In Section 13.3.1 we developed two methods of characterizing a doubly-spread target or channel:

1. The scattering function characterization.
2. The partial differential equation characterization.

These characterizations were easy to visualize and were taken as exact models of the actual physical phenomena. Unfortunately, except for a few special cases, we cannot solve the resulting equations specifying the optimum receiver and its performance.

In this subsection we develop some approximate channel models that allow us to compute the functions needed to specify the optimum receiver and its performance. Our discussion considers three models:

1. The tapped-delay line model.
2. The general orthogonal series model.
3. The approximate differential-equation model.

The tapped-delay line model is intuitively satisfying and relatively easy to implement, and so we present it first. The general orthogonal series model is a logical extension of the tapped-delay line model and leads to simpler computational requirements in many situations. The approximate differential-equation model leads to the general orthogonal series model in a different manner.

In all three cases, the complex envelope of the signal component is

$$\tilde{s}(t) = \sqrt{E_t}\int_{-\infty}^{\infty} \tilde{f}(t-\lambda)\tilde{b}(t,\lambda)\,d\lambda, \qquad -\infty < t < \infty, \tag{169}$$

where we have assumed an infinite observation time for simplicity. The signal $\tilde{s}(t)$ is a sample function from a zero-mean Gaussian random process whose covariance function is given by (143).

The technique that we use in developing our approximate models is straightforward. We expand either $\tilde{f}(t-\lambda)$ or $\tilde{b}(t,\lambda)$ using a complete set of orthonormal functions. This enables us to replace the integral in (169) by an infinite sum. We then truncate the infinite series to obtain an approximate model. The various models differ in their choice of orthogonal functions.

It is important to remember that the "exact" model that we have been working with and the approximate models that we shall develop are both approximations to some physical target or channel. In most cases we have to estimate the target characteristics, and this introduces errors into our model. Thus, in many cases, the approximate models in the next section may represent the physical target or channel as effectively the exact model we have been using.

***13.3.2.A. Tapped-delay Line Model.*** We assume that the transmitted signal $\tilde{f}(t)$ is bandlimited around its carrier frequency. Thus,

$$\tilde{F}\{f\} \stackrel{L}{=} 0, \qquad |f| > \frac{W}{2}. \tag{170}$$

Since $\tilde{f}(t)$ is bandlimited and the interval is infinite, a logical procedure is to expand $f(t-\lambda)$ using the sampling theorem. We write

$$\tilde{f}(t-\lambda) = \sum_{k=-\infty}^{\infty} \tilde{f}\left(t - \frac{k}{W_s}\right)\left(\frac{\sin \pi W_s(\lambda - k/W_s)}{\pi W_s(\lambda - k/W_s)}\right), \tag{171}$$

where $W_s \geq W$. Notice that we could just let $W_s = W$ from the standpoint of the sampling theorem. Introducing $W_s$ gives an additional flexibility in the model, which we shall exploit later.

Observe that we have put the $\lambda$ dependence in the coordinate functions and the $t$ dependence in the coefficients. This separation is the key to the series expansion approach. The sin $x/x$ functions are orthogonal but not normalized. This is for convenience in interpreting the coefficients in (171) as samples. Substituting (171) into (169), we have

$$\tilde{s}(t) = \sqrt{E_t}\sum_{k=-\infty}^{\infty} \tilde{f}\left(t - \frac{k}{W_s}\right)\left[\int_{-\infty}^{\infty} \tilde{b}(t,\lambda)\frac{\sin \pi W_s(\lambda - k/W_s)}{\pi W_s(\lambda - k/W_s)}\right]d\lambda. \tag{172}$$

If we define

$$\tilde{b}_k(t) \triangleq \int_{-\infty}^{\infty} \tilde{b}(t, \lambda) \frac{\sin \pi W_s(\lambda - k/W_s)}{\pi W_s(\lambda - k/W_s)} \, d\lambda, \tag{173}$$

then

$$\tilde{s}(t) = \sum_{k=-\infty}^{\infty} \tilde{f}\left(t - \frac{k}{W_s}\right) \tilde{b}_k(t). \tag{174}$$

Two observations regarding (174) are useful:

1. The functions $\tilde{f}(t - k/W_s)$ can be generated by passing $\tilde{f}(t)$ through a tapped-delay line with taps spaced $1/W_s$ seconds apart.
2. The functions $\tilde{b}_k(t), -\infty < t < \infty$, are defined by (173). This weighted integration is sketched in Fig. 13.17. We see that if the scattering function has length $L$, $\tilde{b}_k(t)$ will be essentially zero for negative values of $k$ and all positive values of $k$ greater than $LW_s$.

These two observations lead us to the target (or channel) model shown in Fig. 13.18.†

The tap gains are sample functions from complex zero-mean Gaussian processes. To specify the model completely, we need their cross-covariance functions

$$E[\tilde{b}_k(t)\tilde{b}_l^*(u)]$$
$$= E\left\{\iint_{-\infty}^{\infty} \tilde{b}(t, \lambda)\tilde{b}^*(u, \lambda_1) \frac{\sin \pi W_s(\lambda - k/W_s)}{\pi W_s(\lambda - k/W_s)} \frac{\sin \pi W_s(\lambda_1 - l/W_s)}{\pi W_s(\lambda_1 - l/W_s)} \, d\lambda \, d\lambda_1\right\}. \tag{175}$$

Bringing the expectation inside the integral, using (37), and performing the integration with respect to $\lambda_1$, we have

$$E[\tilde{b}_k(t)\tilde{b}_l^*(u)]$$
$$= \int_{-\infty}^{\infty} \tilde{K}_{DR}(t - u, \lambda) \left[\frac{\sin \pi W_s(\lambda - k/W_s)}{\pi W_s(\lambda - k/W_s)}\right] \left[\frac{\sin \pi W_s(\lambda - l/W_s)}{\pi W_s(\lambda - l/W_s)}\right] d\lambda. \tag{176}$$

This expression is true for any $\tilde{K}_{DR}(t - u, \lambda)$.

The analysis is somewhat simpler if the tap gains are statistically independent. If $\tilde{K}_{DR}(t - u, \lambda)$ is essentially constant with respect to $\lambda$ over $1/W_s$ units, the integral in (176) is *approximately* zero for $k \neq l$. If $\tilde{K}_{DR}(t - u, \lambda)$ is a smooth function of $\lambda$, we can improve the approximation by increasing $W_s$. Unfortunately, the dimension of the model increases as $W_s$ increases. On page 500 we look at the effect of correlated

† The model in Fig. 13.18 is due to Kailath [35].

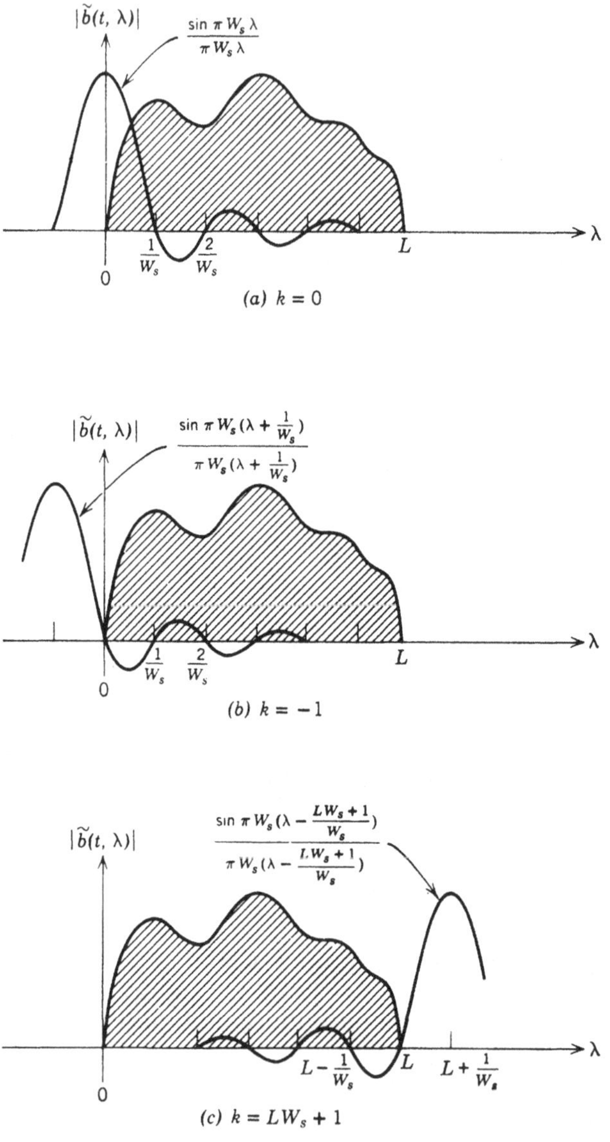


**Fig. 13.17 Location of sin $x/x$ weighting function for various values of $k$.**

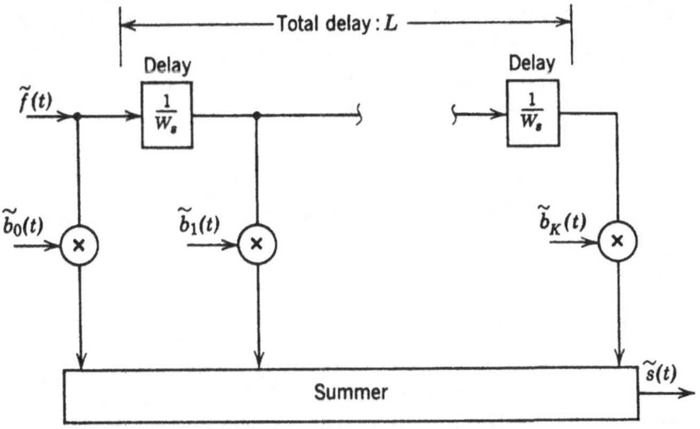


**Fig. 13.18 Tapped-delay line model for doubly-spread target (or channel).**

coefficients. When the independent assumption is valid, we have

$$E[\tilde{b}_k(t)\tilde{b}_l^*(u)] = \begin{cases} \dfrac{1}{W_s}\,\tilde{K}_{DR}\left(t-u, \dfrac{k}{W_s}\right), & k = l, \\ 0, & k \neq l. \end{cases} \tag{177}$$

Because they are stationary, the tap gain processes can also be characterized in terms of their spectra. Transforming (177) gives

$$\tilde{S}_{\tilde{b}_k}\{f\} = \frac{1}{W_s}\,\tilde{S}_{DR}\left\{f, \frac{k}{W_s}\right\}. \tag{178}$$

These spectra are just cross-sections of the scattering function at various values of $\lambda$.

We now have an approximate model for the target (or channel). Looking at (174), we see that we have replaced the doubly-spread channel by a set of $(K + 1)$ singly-spread channels whose signal output is

$$\tilde{s}_K(t) = \sum_{k=0}^{K} \tilde{f}\left(t - \frac{k}{W_s}\right)\tilde{b}_k(t). \tag{179}$$

This is a problem we can solve for a large class of channel processes.

As indicated by (149), the optimum receiver will contain $\hat{\tilde{s}}_{Kr}(t)$ as a waveform. Because $\tilde{f}(t)$ is known,

$$\hat{\tilde{s}}_{Kr}(t) = \sum_{k=0}^{K} \tilde{f}\left(t - \frac{k}{W_s}\right)\hat{\tilde{b}}_{kr}(t). \tag{180}$$

Thus the basic problem in implementing the optimum receiver is to generate the tap gain estimates and weight them with $\tilde{f}(t - k/W_s)$. The tapped-delay model has the advantage that the required functions can be generated in a reasonably straightforward manner. We now discuss the design of the optimum receiver using the tapped-delay model.

If $\tilde{S}_{DR}\{f, \lambda\}$ is a rational function of $f$, each of the tap-gain functions has a finite state representation. When this is true, the optimum receiver and its performance can be evaluated using the techniques that we have already developed. To illustrate this, we set up the state-variable model.†

We assume that the scattering function is such that we need $(K+1)$ taps. Then

$$\tilde{s}_K(t) = \sum_{k=0}^{K} \tilde{f}\left(t - \frac{k}{W_s}\right)\tilde{b}_k(t), \qquad T_i \leq t \leq T_f, \tag{181}$$

where $[T_i, T_f]$ is long enough so that essentially all the output signal energy is contained in the observation interval. The state vector for the $k$th tap gain is $\tilde{\mathbf{x}}_k(t)$, where

$$\dot{\tilde{\mathbf{x}}}_k(t) = \tilde{\mathbf{F}}_k\tilde{\mathbf{x}}_k(t) + \tilde{\mathbf{G}}_k\tilde{u}_k(t), \tag{182}$$

$$\tilde{b}_k(t) = \tilde{\mathbf{C}}_k\tilde{\mathbf{x}}_k(t), \tag{183}$$

$$E[\tilde{u}_k(t)\tilde{u}_k^{\dagger}(\sigma)] = \tilde{Q}_k\,\delta(t-\sigma), \tag{184}$$

and

$$E[\tilde{\mathbf{x}}_k(T_i)\tilde{\mathbf{x}}_k^{\dagger}(T_i)] = \tilde{\mathbf{P}}_k. \tag{185}$$

The dimension of the state vector is $N_k$.

The over-all state vector has the dimension

$$N = \sum_{k=0}^{K} N_k \tag{186}$$

and can be written as

$$\tilde{\mathbf{x}}(t) \triangleq \begin{bmatrix} \tilde{\mathbf{x}}_0(t) \\ \tilde{\mathbf{x}}_1(t) \\ \cdot \\ \cdot \\ \cdot \\ \tilde{\mathbf{x}}_K(t) \end{bmatrix}. \tag{187}$$

Then

$$\tilde{\mathbf{b}}(t) \triangleq \begin{bmatrix} \tilde{b}_0(t) \\ \tilde{b}_1(t) \\ \cdot \\ \cdot \\ \cdot \\ \tilde{b}_K(t) \end{bmatrix} = \begin{bmatrix} \tilde{\mathbf{C}}_0 & & & \\ & \tilde{\mathbf{C}}_1 & & \mathbf{0} \\ & & \ddots & \\ & \mathbf{0} & & \\ & & & \tilde{\mathbf{C}}_K \end{bmatrix} \tilde{\mathbf{x}}(t). \tag{188}$$

† This model is due to Van Trees [36].

The received signal process is just

$$\tilde{s}_K(t) = \left[\tilde{f}(t) \,\vdots\, \tilde{f}\left(t - \frac{1}{W_s}\right) \,\vdots\, \cdots \,\vdots\, \tilde{f}\left(t - \frac{K}{W_s}\right)\right]\tilde{b}(t) \triangleq \tilde{\mathbf{C}}(t)\tilde{\mathbf{x}}(t), \quad (189)$$

where $\tilde{\mathbf{C}}(t)$ is defined by (188) and (189) as

$$\tilde{\mathbf{C}}(t) \triangleq \left[\tilde{f}(t) \,\vdots\, \tilde{f}\left(t - \frac{1}{W_s}\right) \,\vdots\, \cdots \,\vdots\, \tilde{f}\left(t - \frac{K}{W_s}\right)\right] \begin{bmatrix} \tilde{\mathbf{C}}_0 & & & 0 \\ & \tilde{\mathbf{C}}_1 & & \\ & & \ddots & \\ 0 & & & \tilde{\mathbf{C}}_K \end{bmatrix}. \quad (190)$$

We have now reduced the problem to one that we have already solved [see (11.41–11.49)]. The complex receiver structure for the detection problem in (145) and (146) is shown in Fig. 13.19. (This is just the system in Fig. 11.9.) The optimum realizable filter is shown in Fig. 13.20. The only problem is the complexity of this system to generate $\hat{\tilde{s}}_{K_r}(t)$. This complexity is related to the dimension of the variance equation, which is an $N \times N$ matrix equation in this case. As usual, the variance equation can be solved before any data have been received.

To compute the performance, we evaluate $\tilde{\mu}(s)$ by using (152). Recall that $\tilde{\xi}_P(t, \tilde{s}(t), \cdot)$ is the realizable mean-square error in estimating $\tilde{s}(t)$ and is obtained from the solution to the variance equation.

It is important to emphasize that we have the problem in a form in which the optimum receiver and its performance can be calculated using straightforward numerical procedures. Notice that the dimension of the system grows quickly. An analytic solution is not feasible, but this is not important. We defer carrying out the details of an actual example until we complete our development of the various channel models.

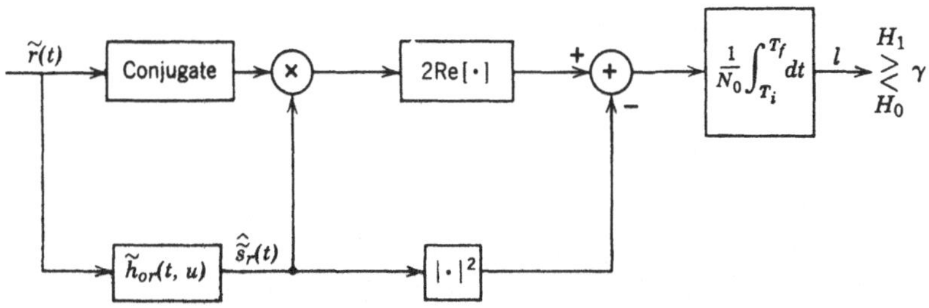


**Fig. 13.19 Optimum receiver for the detection of a doubly-spread target.**

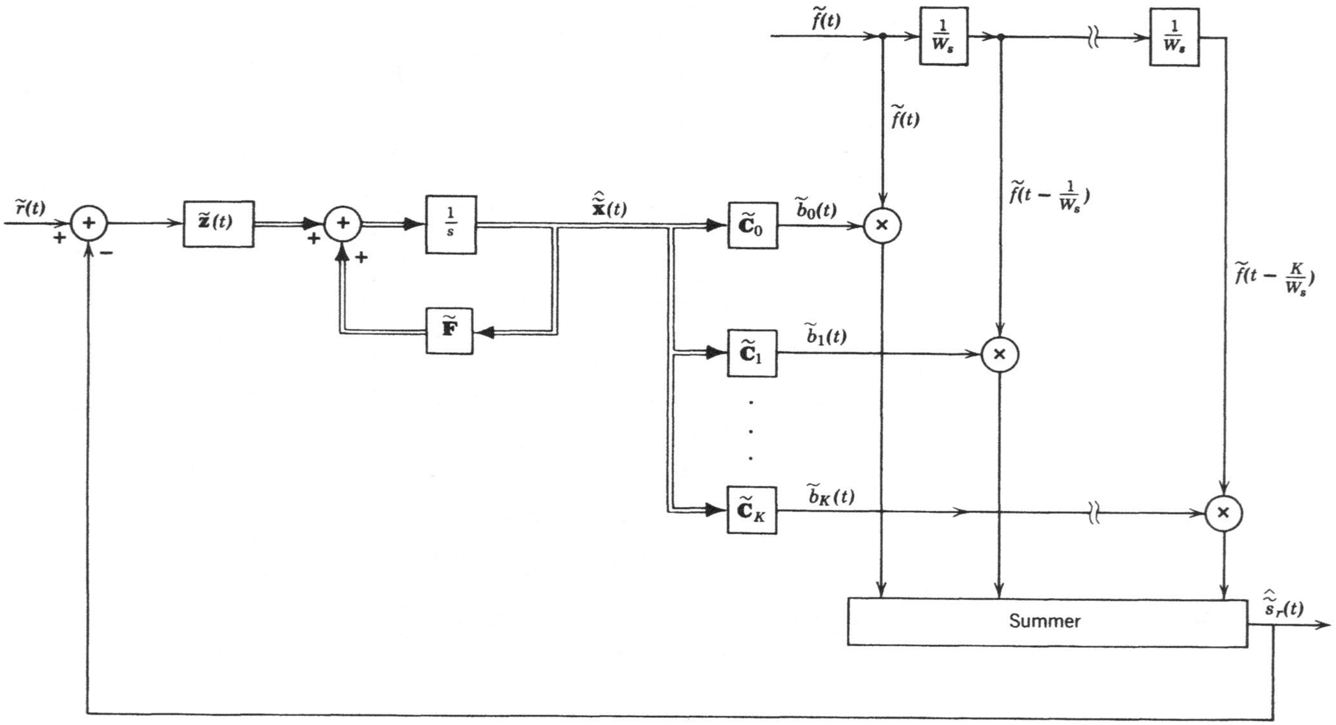


**Fig. 13.20 Optimum realizable filter, $\tilde{h}_{or}(t, z)$, for tapped-delay line model.**

***13.3.2.B. General Orthogonal Series Model.*** The tapped-delay model has a great deal of intuitive appeal and is an adequate model for many physical situations. However, in many problems there are other orthogonal functions that provide a more efficient representation. In this section we develop a general model.

Our starting point is the differential-equation model of the channel that was introduced in Section 13.1.2. The state equation is

$$\frac{\partial \tilde{\mathbf{x}}(t, \lambda)}{\partial t} = \tilde{\mathbf{F}}(\lambda)\tilde{\mathbf{x}}(t, \lambda) + \tilde{\mathbf{G}}(\lambda)\tilde{u}(t, \lambda), \tag{191}$$

where

$$E[\tilde{u}(t, \lambda)\tilde{u}^*(t', \lambda')] = \tilde{Q}(\lambda)\delta(t - t')\,\delta(\lambda - \lambda'). \tag{192}$$

The initial condition of the state vector is

$$E[\tilde{\mathbf{x}}(T_0, \lambda)\tilde{\mathbf{x}}^\dagger(T_0, \lambda')] = \tilde{\mathbf{P}}_0(\lambda)\,\delta(\lambda - \lambda'). \tag{193}$$

The channel process is

$$\tilde{b}(t, \lambda) = \tilde{\mathbf{C}}(\lambda)\tilde{\mathbf{x}}(t, \lambda). \tag{194}$$

The signal component at the channel output is

$$\tilde{s}(t) = \sqrt{E_t}\int_{-\infty}^{\infty} \tilde{f}(t - \lambda)\tilde{b}(t, \lambda)\, d\lambda. \tag{195}$$

In the tapped-delay model we expanded the signal in an orthogonal series. In this case we represent the channel process and its state vector by a series expansion. We assume that the $\tilde{\phi}_i(\lambda)$, $i = 1, 2, \ldots,$ form a complete orthonormal set with

$$\int_{\Omega_L} \tilde{\varphi}_i(\lambda)\tilde{\varphi}_j^*(\lambda)\, d\lambda = \delta_{ij}, \tag{196}$$

where the interval $\Omega_L$ is the target (or channel) length. Notice that the $\tilde{\varphi}_i(\lambda)$ are an arbitrary set of orthonormal functions of $\lambda$ only. We discuss methods of choosing the $\tilde{\varphi}_i(\lambda)$ later.

We first expand the state vector as

$$\tilde{\mathbf{x}}(t, \lambda) = \underset{K\to\infty}{\text{l.i.m.}}\ \tilde{\mathbf{x}}_K(t, \lambda) = \underset{K\to\infty}{\text{l.i.m.}} \sum_{i=1}^{K} \tilde{\mathbf{x}}_i(t)\tilde{\phi}_i(\lambda), \qquad -\infty < t < \infty,\ \lambda \in \Omega_L, \tag{197}$$

where

$$\tilde{\mathbf{x}}_i(t) = \int_{\Omega_L} \tilde{\mathbf{x}}(t, \lambda)\tilde{\phi}_i(\lambda)\, d\lambda, \qquad -\infty < t < \infty. \tag{198}$$

We expand the channel process, using the same set of orthonormal functions, as

$$\tilde{b}(t, \lambda) = \underset{K\to\infty}{\text{l.i.m.}}\ \tilde{b}_K(t, \lambda) = \underset{K\to\infty}{\text{l.i.m.}} \sum_{i=1}^{K} \tilde{b}_i(t)\tilde{\varphi}_i(\lambda), \qquad -\infty < t < \infty,\ \lambda \in \Omega_L, \tag{199}$$

where the $\tilde{b}_i(t)$ are determined by the requirement that

$$\tilde{b}_K(t, \lambda) = \tilde{\mathbf{C}}(\lambda)\tilde{\mathbf{x}}_K(t, \lambda), \qquad -\infty < t < \infty, \quad \lambda \in \Omega_L. \tag{200}$$

We refer to $\tilde{b}_K(t, \lambda)$ as the $K$-term approximation to the channel. We now develop a state-variable representation for $\tilde{b}_K(t, \lambda)$.

From (197),

$$\frac{\partial \tilde{\mathbf{x}}(t, \lambda)}{\partial t} = \sum_{j=1}^{\infty} \frac{d\tilde{\mathbf{x}}_j(t)}{dt}\, \tilde{\phi}_j(\lambda). \tag{201}$$

Substituting (197) and (201) into (191) gives

$$\sum_{j=1}^{\infty} \frac{d\tilde{\mathbf{x}}_j(t)}{dt}\, \tilde{\phi}_j(\lambda) = \tilde{\mathbf{F}}(\lambda) \sum_{j=1}^{\infty} \tilde{\mathbf{x}}_j(t)\tilde{\phi}_j(\lambda) + \tilde{\mathbf{G}}(\lambda)\tilde{u}(t, \lambda). \tag{202}$$

Multiplying both sides of (202) by $\tilde{\phi}_i^*(\lambda)$ and integrating with respect to $\lambda$ over $\Omega_L$, we obtain

$$\frac{d\tilde{\mathbf{x}}_i(t)}{dt} = \sum_{j=1}^{\infty} \left[\int_{\Omega_L} \tilde{\mathbf{F}}(\lambda)\tilde{\phi}_i^*(\lambda)\tilde{\phi}_j(\lambda)\, d\lambda\right] \tilde{\mathbf{x}}_j(t) + \int_{\Omega_L} \tilde{\mathbf{G}}(\lambda)\tilde{u}(t, \lambda)\tilde{\phi}_i^*(\lambda)\, d\lambda. \tag{203}$$

We now define

$$\tilde{\mathbf{F}}_{ij} \triangleq \int_{\Omega_L} \tilde{\mathbf{F}}(\lambda)\tilde{\phi}_i^*(\lambda)\tilde{\phi}_j(\lambda)\, d\lambda, \qquad i, j = 1, 2, \ldots \tag{204}$$

and

$$\tilde{u}_i(t) = \int_{\Omega_L} \mathbf{G}(\lambda)\tilde{u}(t, \lambda)\tilde{\phi}_i^*(\lambda)\, d\lambda, \qquad i = 1, 2, \ldots. \tag{205}$$

Truncating the series gives a $K$-term approximation to the channel state vector. The state equation is

$$\frac{d}{dt}\begin{bmatrix} \tilde{\mathbf{x}}_1(t) \\ \tilde{\mathbf{x}}_2(t) \\ \cdot \\ \cdot \\ \cdot \\ \tilde{\mathbf{x}}_K(t) \end{bmatrix} = \begin{bmatrix} \tilde{\mathbf{F}}_{11} & \tilde{\mathbf{F}}_{12} & \tilde{\mathbf{F}}_{13} & \cdots & \tilde{\mathbf{F}}_{1K} \\ \tilde{\mathbf{F}}_{21} & \tilde{\mathbf{F}}_{22} & & & \\ \tilde{\mathbf{F}}_{31} & & & & \\ \vdots & & & & \\ \tilde{\mathbf{F}}_{K1} & & & & \tilde{\mathbf{F}}_{KK} \end{bmatrix} \begin{bmatrix} \tilde{\mathbf{x}}_1(t) \\ \tilde{\mathbf{x}}_2(t) \\ \cdot \\ \cdot \\ \cdot \\ \tilde{\mathbf{x}}_K(t) \end{bmatrix} + \begin{bmatrix} \tilde{\mathbf{u}}_1(t) \\ \tilde{\mathbf{u}}_2(t) \\ \cdot \\ \cdot \\ \cdot \\ \tilde{\mathbf{u}}_K(t) \end{bmatrix}. \tag{206}$$

If the original distributed state vector is $N$-dimensional, the state vector in (206) has $NK$ dimensions. We can write (206) compactly as

$$\frac{d\tilde{\mathbf{x}}_{\mathbf{M}}(t)}{dt} = \tilde{\mathbf{F}}_{\mathbf{M}}(\lambda)\tilde{\mathbf{x}}_{\mathbf{M}}(t) + \tilde{\mathbf{u}}_{\mathbf{M}}(t). \tag{207}$$

The subscript **M** denotes model. The elements in the covariance function matrix of the driving function are

$$\begin{aligned} E[\mathbf{u}_i(t)\mathbf{u}_j^\dagger(t')] &= E\left\{\int_{\Omega_L} \tilde{\mathbf{G}}(\lambda)\tilde{u}(t, \lambda)\tilde{\phi}_i^*(\lambda)\, d\lambda \int_{\Omega_L} \tilde{\mathbf{G}}^\dagger(\lambda')\tilde{u}^*(t', \lambda')\tilde{\phi}_j(\lambda')\, d\lambda'\right\} \\ &= \left[\int_{\Omega_L} \tilde{\mathbf{G}}(\lambda)\tilde{Q}(\lambda)\tilde{\mathbf{G}}^\dagger(\lambda)\tilde{\phi}_i^*(\lambda)\tilde{\phi}_j(\lambda)\, d\lambda\right] \delta(t - t') \\ &\triangleq \tilde{\mathbf{Q}}_{ij}\delta(t - t'). \end{aligned} \tag{208}$$

The initial conditions are

$$\begin{aligned} E[\tilde{\mathbf{x}}_i(T_0)\tilde{\mathbf{x}}_j^\dagger(T_0)] &= E\left[\int_{\Omega_L} \tilde{\mathbf{x}}(T_0, \lambda)\tilde{\phi}_i^*(\lambda)\, d\lambda \int_{\Omega_L} \tilde{\mathbf{x}}^\dagger(T_0, \lambda')\tilde{\phi}_j(\lambda')\, d\lambda'\right] \\ &= \int_{\Omega_L} \tilde{\mathbf{K}}_{\tilde{\mathbf{x}}}(\lambda)\tilde{\phi}_i^*(\lambda)\tilde{\phi}_j(\lambda)\, d\lambda, \end{aligned} \tag{209}$$

where $\tilde{\mathbf{K}}_{\tilde{\mathbf{x}}}(\lambda)$ is defined in (46).

We must now find the observation matrix relating $\tilde{b}_i(t)$ to $\tilde{\mathbf{x}}_{\mathbf{M}}(t)$. Using (197), (199), and (200), we have

$$\sum_{j=1}^{K} \tilde{b}_j(t)\tilde{\varphi}_j(\lambda) = \tilde{\mathbf{C}}(\lambda)\sum_{j=1}^{K} \tilde{\mathbf{x}}_j(t)\tilde{\varphi}_j(\lambda). \tag{210}$$

Multiplying both sides by $\tilde{\varphi}_i^*(\lambda)$ and integrating over $\Omega_L$ gives

$$\tilde{b}_i(t) = \sum_{j=1}^{K}\left[\int_{\Omega_L} \tilde{\varphi}_i^*(\lambda)\tilde{\mathbf{C}}(\lambda)\tilde{\varphi}_j(\lambda)\, d\lambda\right]\tilde{\mathbf{x}}_j(t). \tag{211}$$

Defining

$$\tilde{\mathbf{C}}_{ij} = \int_{\Omega_L} \tilde{\varphi}_i^*(\lambda)\tilde{\mathbf{C}}(\lambda)\tilde{\varphi}_j(\lambda)\, d\lambda, \tag{212}$$

we obtain

$$\begin{aligned} \tilde{b}_i(t) &= [\tilde{\mathbf{C}}_{i1} \mid \tilde{\mathbf{C}}_{i2} \mid \cdots \tilde{\mathbf{C}}_{iK}]\tilde{\mathbf{x}}_{\mathbf{M}}(t) \\ &\triangleq \tilde{\mathbf{C}}_{i\mathbf{M}}\tilde{\mathbf{x}}_{\mathbf{M}}(t). \end{aligned} \tag{213}$$

The signal component at the output of the channel is

$$\tilde{s}(t) = \sqrt{E_t}\int_{-\infty}^{\infty} \tilde{f}(t - \lambda)\tilde{b}(t, \lambda)\, d\lambda. \tag{214}$$

The $K$-term approximation is

$$\tilde{s}_K(t) = \sqrt{E_t}\int_{-\infty}^{\infty} \tilde{f}(t-\lambda)\sum_{i=1}^{K} \tilde{b}_i(t)\tilde{\varphi}_i(\lambda)\,d\lambda$$

$$= \sum_{i=1}^{K} \tilde{f}_i \tilde{b}_i(t), \tag{215}$$

where

$$\tilde{f}_i \triangleq \sqrt{E_t}\int_{-\infty}^{\infty} \tilde{f}(t-\lambda)\tilde{\varphi}_i(\lambda)\,d\lambda. \tag{216}$$

Using (213) in (215), we obtain

$$\tilde{s}_K(t) = \left(\sum_{i=1}^{K} \tilde{f}_i \tilde{\mathbf{C}}_{i\mathbf{M}}\right)\tilde{\mathbf{x}}_{\mathbf{M}}(t)$$

$$\triangleq \tilde{\mathbf{C}}_{\mathbf{M}}\tilde{\mathbf{x}}_{\mathbf{M}}(t). \tag{217}$$

We now have the $K$-term approximation to the problem completely characterized by a state-variable representation. Once we have this representation, all the results in Section 11.2.2 are immediately applicable. Notice that, although the formulas appear complicated, all the necessary quantities can be calculated in a straightforward manner.

Two comments regarding the model are worthwhile.

1. The tapped-delay line model is a special case of this model (see Problem 13.3.9).
2. The proper choice of the orthogonal set will depend on the scattering function and the signal. A judicious choice will simplify both the structure of the state equation and the value of $K$ required to get a good approximation. It is this simplification in the state equation that has motivated the development of the general orthogonal series model. In the next section we illustrate the choice of the orthogonal set for a typical example.

Up to this point in this section, we have considered various orthogonal series models for doubly-spread channels. The goal was to obtain a finite-dimensional approximation that we could analyze completely. We now consider a direct analysis of the differential-equation model.

***13.3.2.C. Approximate Differential-equation Model.***† The differential-equation model for the doubly-spread channel was described by (38)–(41), which are repeated here for convenience. The state equation is

$$\frac{\partial \tilde{\mathbf{x}}(t,\lambda)}{\partial t} = \tilde{\mathbf{F}}(\lambda)\tilde{\mathbf{x}}(t,\lambda) + \tilde{\mathbf{G}}(\lambda)\tilde{u}(t,\lambda), \tag{218}$$

† The results in this section are due to Kurth [7].

where

$$E[\tilde{u}(t,\lambda)\tilde{u}^*(t',\lambda')] = \tilde{Q}(\lambda)\,\delta(t-t')\,\delta(\lambda-\lambda'). \tag{219}$$

The initial covariance of the state vector is

$$E[\tilde{\mathbf{x}}(T_i,\lambda)\tilde{\mathbf{x}}^\dagger(T_i,\lambda')] = \tilde{\mathbf{P}}_0(\lambda)\,\delta(\lambda-\lambda'). \tag{220}$$

The channel process is

$$\tilde{b}(t,\lambda) = \tilde{\mathbf{C}}(\lambda)\tilde{\mathbf{x}}(t,\lambda). \tag{221}$$

The signal component at the channel output is

$$\tilde{s}(t) = \int_{-\infty}^{\infty} \tilde{f}(t-\lambda)\tilde{b}(t,\lambda)\,d\lambda. \tag{222}$$

The optimum test can be written in terms of the MMSE realizable estimate of $\tilde{s}(t)$. From (149),

$$\boxed{l_R = \frac{1}{N_0}\int_{T_i}^{T_f} \{2\,\text{Re}\,[\tilde{r}^*(t)\hat{\tilde{s}}_r(t)] - |\hat{\tilde{s}}_r(t)|^2\}\,dt.} \tag{223}$$

Notice that $\tilde{s}(t)$ is a function of time only, so that the derivation leading to (149) is applicable without any modification.

To implement the test, we need an expression for $\hat{\tilde{s}}_r(t)$. These equations were encountered previously in Section 13.2.2 (116)–(121). The estimator equation is

$$\frac{\partial\hat{\tilde{\mathbf{x}}}(t,\lambda)}{\partial t} = \tilde{\mathbf{F}}(\lambda)\hat{\tilde{\mathbf{x}}}(t,\lambda) + \tilde{\mathbf{z}}(t,\lambda)[\tilde{r}(t) - \tilde{s}(t:\hat{\tilde{\mathbf{x}}}(t,\lambda))], \qquad t \ge T_i,\ \lambda \in \Omega_L, \tag{224}$$

and

$$\hat{\tilde{\mathbf{x}}}(T_i,\lambda) = \mathbf{0}, \qquad \lambda \in \Omega_L. \tag{225}$$

The gain equation is

$$\tilde{\mathbf{z}}(t,\lambda) = \frac{1}{N_0}\left[\int_{\Omega_L} \tilde{\boldsymbol{\xi}}(t:\lambda,\lambda')\tilde{\mathbf{C}}^\dagger(\lambda')\sqrt{E_t}\tilde{f}^*(t-\lambda')\,d\lambda'\right]. \tag{226}$$

The variance equation is

$$\begin{aligned}\frac{\partial\tilde{\boldsymbol{\xi}}(t:\lambda,\lambda')}{\partial t} &= \tilde{\mathbf{F}}(\lambda)\tilde{\boldsymbol{\xi}}(t:\lambda,\lambda') + \tilde{\boldsymbol{\xi}}^\dagger(t:\lambda',\lambda)\tilde{\mathbf{F}}^\dagger(\lambda') + \tilde{\mathbf{G}}(\lambda)\tilde{Q}(\lambda)\tilde{\mathbf{G}}^\dagger(\lambda')\\ &\quad - \frac{1}{N_0}\left\{\int_{\Omega_L} \tilde{\boldsymbol{\xi}}(t:\lambda,\sigma)\tilde{\mathbf{C}}^\dagger(\sigma)\sqrt{E_t}\tilde{f}^*(t-\sigma)\,d\sigma\right.\\ &\quad \left.\times \int_{\Omega_L} \sqrt{E}\tilde{f}(t-\sigma')\tilde{\mathbf{C}}(\sigma')\tilde{\boldsymbol{\xi}}(t:\sigma',\lambda')\,d\sigma'\right\},\\ &\qquad\qquad \lambda,\lambda' \in \Omega_L,\ t \ge T_i, \end{aligned} \tag{227}$$

with the initial condition

$$\tilde{\boldsymbol{\xi}}(T_i:\lambda, \lambda') = \tilde{\boldsymbol{\xi}}_0(T_i, \lambda)\, \delta(\lambda - \lambda'),$$
$$= \tilde{\mathbf{K}}_{\tilde{\mathbf{x}}}(\lambda)\, \delta(\lambda - \lambda'). \tag{228}$$

The expressions in (224)–(228) characterize the channel estimator. Using these equations, (222) and (223) give the optimum receiver that was shown in Fig. 13.19. We are still faced with the problem of implementing (224)–(228) in order to generate $\hat{\tilde{s}}_r(t)$. A block diagram of a system containing spatial operations that could be used to generate $\hat{\tilde{s}}_r(t)$ was shown in Fig. 13.15 [replace $\hat{\tilde{n}}_r(t)$ with $\hat{\tilde{s}}_r(t)$]. In general we cannot implement this system and must be content with an approximate solution. We consider three procedures for obtaining an approximate solution.

The first procedure is to expand the state vector in an orthonormal expansion and truncate the expansion at $K$ terms. This procedure takes us back to the model on pages 495–498. A second procedure is to sample in $\lambda$. The resulting model would be similar to the tapped-delay line model derived in Section 13.3.2.A, but the tap gains would be correlated. This procedure is generally inefficient from the computational standpoint.

We now develop a third procedure that seems to offer some computational advantages. The first step is to divide the covariance matrix into an impulsive term and bounded term as

$$\tilde{\boldsymbol{\xi}}(t:\lambda, \lambda') = \tilde{\boldsymbol{\xi}}_0(T_i, \lambda)\, \delta(\lambda - \lambda') + \tilde{\mathbf{p}}(t:\lambda, \lambda'), \qquad \lambda, \lambda' \in \Omega_L,\ t \geq T_i. \tag{229}$$

Substituting (229) into (227), we find that $\tilde{\mathbf{p}}(t:\lambda, \lambda')$ must satisfy the differential equation

$$\begin{aligned}\frac{\partial \tilde{\mathbf{p}}(t:\lambda, \lambda')}{\partial t} &= \tilde{\mathbf{F}}(\lambda)\tilde{\mathbf{p}}(t:\lambda, \lambda') + \tilde{\mathbf{p}}(t:\lambda, \lambda')\tilde{\mathbf{F}}^{\dagger}(\lambda') \\ &\quad - \frac{1}{N_0}\Bigg\{\Bigg[\tilde{\mathbf{K}}_{\tilde{\mathbf{x}}}(\lambda)\tilde{\mathbf{C}}^{\dagger}(\lambda)\sqrt{E_t}\tilde{f}(t-\lambda) \\ &\quad + \int_{\Omega_L} \tilde{\mathbf{p}}(t:\lambda, \sigma)\tilde{\mathbf{C}}^{\dagger}(\sigma)\sqrt{E_t}\tilde{f}^*(t-\sigma)\, d\sigma\Bigg] \\ &\quad \times \Bigg[\sqrt{E_t}\tilde{f}(t-\lambda')\tilde{\mathbf{C}}(\lambda')\mathbf{K}_{\tilde{\mathbf{x}}}(\lambda') \\ &\quad + \int_{\Omega_L} \sqrt{E_t}\tilde{f}(t-\sigma')\tilde{\mathbf{C}}(\sigma')\tilde{\mathbf{p}}(t:\sigma', \lambda')\, d\sigma'\Bigg]\Bigg\}, \\ &\qquad t \geq T_i,\ \lambda, \lambda' \in \Omega_L, \end{aligned} \tag{230}$$

with the zero initial condition

$$\tilde{\mathbf{p}}(t:\lambda, \lambda') = \mathbf{0}. \tag{231}$$

We then expand $\tilde{\mathbf{p}}(t:\lambda, \lambda')$ in a series expansion as

$$\tilde{\mathbf{p}}(t:\lambda, \lambda') = \sum_{i=1}^{\infty} \sum_{j=1}^{\infty} \tilde{\mathbf{p}}_{ij}(t) \tilde{\phi}_i(\lambda) \tilde{\phi}_j^*(\lambda'), \qquad \lambda, \lambda' \in \Omega_L, \qquad t \geq T_i, \tag{232}$$

where the $\tilde{\phi}_i(\lambda)$ are an arbitrary set of orthonormal functions and

$$\tilde{\mathbf{p}}_{ij}(t) \triangleq \int_{\Omega_L} d\lambda \int_{\Omega_L} d\lambda' \tilde{\mathbf{p}}(t:\lambda, \lambda') \tilde{\phi}_i(\lambda) \tilde{\phi}_j^*(\lambda'). \tag{233}$$

This procedure is referred to as a *modal expansion technique*.

We truncate the series at $i = j = K$ to obtain an approximate solution. Proceeding as before, we can derive a set of differential equations specifying the $\tilde{\mathbf{p}}_{ij}(t)$ (see Problem 13.3.12). The advantage of separating out the impulse in (229) is that the convergence of the series approximation is usually better. We shall apply this third procedure to a specific problem in Section 13.3.3.

The final step is to compute the performance. We do this by evaluating $\tilde{\mu}(s)$ and using it in our approximate error expressions. We can express $\tilde{\mu}(s)$ in terms of the realizable MMSE signal estimation error, $\xi_P(t, \tilde{s}(t), \cdot)$, by (11.54). Finally, we express $\xi_P(t, \tilde{s}(t), N_0)$ in terms of $\xi_P(t:\lambda, \lambda')$.

$$\xi_P(t, \tilde{s}(t), N_0) \triangleq E[|\tilde{s}(t) - \hat{\tilde{s}}(t)|^2]. \tag{234}$$

Using (221) and (222) in (234) gives

$$\xi_P(t, \tilde{s}(t), N_0) = \int_{\Omega_L} d\sigma \int_{\Omega_L} d\sigma' E_t \tilde{f}(t - \sigma) \tilde{\mathbf{C}}(\sigma) \tilde{\boldsymbol{\xi}}(t:\sigma, \sigma') \tilde{\mathbf{C}}^\dagger(\sigma') \tilde{f}^*(t - \sigma'). \tag{235}$$

Notice that to find $\tilde{\mu}(\frac{1}{2})$ we must solve the variance equation (227) for two values of the additive noise level, $N_0$ and $2N_0$. To find $\tilde{\mu}(s)$, in general, we must solve the variance equation for three values of the additive noise level.

***13.3.2.D. Summary of Approximate Model Discussion.*** In this subsection we have developed various models that we can use tc approximate a doubly-spread target (or channel). The advantage of all these models is that they enable us to obtain a complete solution for the optimum receiver and its performance.

As we pointed out in the introduction to Section 13.3.3, the tapped-delay line model is the simplest to implement and is the only model that has been used in actual systems. At their present state of development, the other two models are most useful in the study of performance limitations.

There are many approximate channel models in addition to those that we have discussed. Suitable references are [35], [61], and [64].

### 13.3.3 Binary Communication over Doubly-Spread Channels

In Section 13.3.1.B we formulated a model for a binary FSK system operating over a doubly-spread channel [see (153)–(168)]. In this section we continue our discussion of the communication problem.

Our discussion is divided into three parts. In Section 13.3.3.A we discuss the performance bounds on binary communication systems and demonstrate some simple signaling schemes that approach these bounds. In Section 13.3.3.B we carry out a detailed performance analysis of a specific system using one of the approximate channel models developed in Section 13.3.2. In Section 13.3.3.C we discuss suboptimum receivers briefly.

***13.3.3.A. Performance Bounds and Efficient Systems.*** As we pointed out in Section 13.3.1.B, the decision problem is that of detecting a complex Gaussian process in complex white Gaussian noise. The covariance function of the signal process, $\tilde{s}(t)$, is given by (5) as

$$\tilde{K}_{\tilde{s}}(t, u) = E_t \int_{-\infty}^{\infty} \tilde{f}(t - \lambda)\tilde{K}_{DR}(t - u, \lambda)\tilde{f}^*(u - \lambda)\, d\lambda. \tag{236}$$

The performance will depend on $E_t$, $N_0$, $\tilde{f}(t)$, and $\tilde{K}_{DR}(t - u, \lambda)$ and may be difficult to evaluate in the general case. However, in Section 11.3 we derived a bound on how well any binary system could perform for a given $E_t$ and $N_0$. Since this bound only depended on the eigenvalues of $\tilde{s}(t)$, it is still valid in this problem.

On page 380 we demonstrated that in order to achieve the bound we would like to design the signal so that the output process has $D_o$ equal eigenvalues, where

$$D_o = \frac{\bar{E}_r/N_0}{3.07} \tag{237}$$

and

$$\bar{E}_r = E_t \iint_{-\infty}^{\infty} \tilde{S}_{DR}\{f, \lambda\}\, df\, d\lambda. \tag{238}$$

For this optimum case,

$$\tilde{\mu}_{BS}(\tfrac{1}{2}) = -0.1488\left(\frac{\bar{E}_r}{N_0}\right). \tag{239}$$

Thus, the probability of error using any signal $\tilde{f}(t)$ is bounded by

$$\Pr(\epsilon) \leq \tfrac{1}{2} \exp\left(-0.1488 \frac{\bar{E}_r}{N_0}\right). \tag{240}$$

This gives us a simple bound on the probability of error for binary orthogonal signals. The difficulty is that there is no guarantee that a signal exists that enables us to achieve this performance. We now discuss two situations in which we can approach the bound with simple signals.

UNDERSPREAD CHANNELS. In (32) we defined an underspread channel as one whose $BL$ product was less than 1. We now discuss the problem of communicating over an underspread channel. (Notice that we allow $B \gg 1$ or $L \gg 1$, as long as $BL \ll 1$.)

In our discussion of communication over Doppler-spread channels in Section 11.3 (specifically pages 384–385), we saw that we could achieve the bound in (240) for any scattering function if there were no peak-power or time-duration constraints. The required signal consisted of a sequence of short pulses, with the number of pulses chosen to achieve the optimum diversity specified in (237) [i.e., $n = D_o$]. The length $T$ of each pulse was much less than $B^{-1}$ (the reciprocal of the bandwidth of the Doppler spread), so that there was no time-selective fading. Here we achieved the desired eigenvalue distribution by reducing the channel to a set of non-fluctuating point channels.

We now consider a similar system for signaling over a doubly-spread channel. The signal is shown in Fig. 13.21. To avoid time-selective fading, we require that

$$T \ll \frac{1}{B}. \tag{241†}$$

To avoid frequency-selective fading, we require that

$$W \ll \frac{1}{L}. \tag{242}$$

Combining (241) and (242), we see that the requirement for flat (nonselective) fading is

$$WT \lll \frac{1}{BL}. \tag{243}$$

However, we know that for any signal

$$WT \geq 1. \tag{244}$$

† Our discussion uses $B$ and $W$ as imprecise bandwidth measures. An exact definition is not needed in the current context.

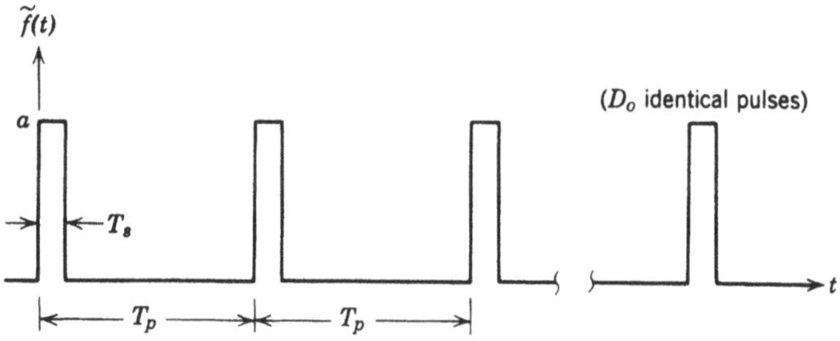


**Fig. 13.21 Signal for communication over an underspread channel.**

Therefore we require that

$$BL \lll 1 \tag{245}$$

in order for us to be able to satisfy (243). The condition in (245) can only be met by underspread channels [see (32)]. The condition in (245) is stronger than the underspread requirement of (32). If the condition in (245) is satisfied and there is no peak-power or time-duration constraint, we can achieve the bound in (240) by using the signal in Fig. 13.21 with its parameters chosen optimally.

We should observe that the requirement in (245) is usually too strict. In many cases we can come close to the performance in (240) with the signal in Fig. 13.21 for $BL$ products approaching unity.

We next consider the case in which $BL$ exceeds unity.

OVERSPREAD CHANNELS. If $BL > 1$, we cannot have fading that is flat in both time and frequency. However, we can choose the signal so that we have either time-selective fading or frequency-selective fading, but not both. We demonstrate this with a simple example.

**Example.** We consider an idealized channel whose scattering function is shown in Fig 13.22. We assume that

$$BL = 5. \tag{246}$$

We transmit a long rectangular pulse

$$\tilde{f}(t) = \begin{cases} \sqrt{\dfrac{1}{T}}, & 0 < t < T, \\ 0, & \text{elsewhere}, \end{cases} \tag{247}$$

We also require that

$$T \geq 10L. \tag{248}$$

Comparing (248) and (242), we see that we can treat the channel as a Doppler-spread channel. From the results in Example 1 of Chapter 11, we know that if

$$2BT = \frac{\bar{E}_r/N_0}{3.07}, \tag{249}$$

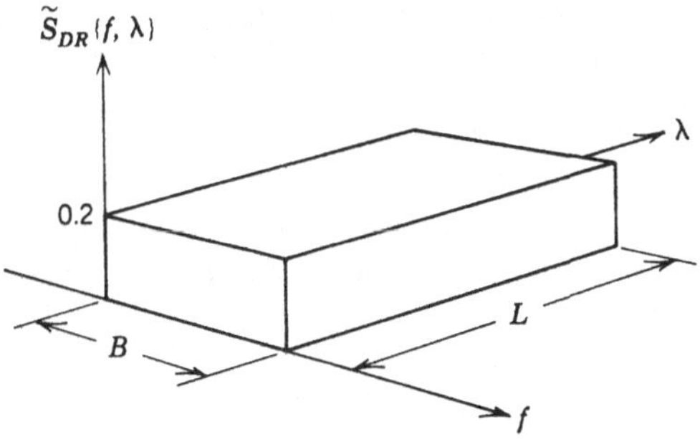


**Fig. 13.22 An idealized scattering function.**

we shall achieve the bound in (240). Using (246) and (248) in (249), we obtain the requirement

$$\frac{\bar{E}_r}{N_0} \geq 307, \tag{250}$$

which is unrealistic [the Pr $(\epsilon) \simeq 10^{-21}$].

We can obtain a more realistic solution by relaxing some of the requirements. For example, if we require that

$$T \geq \tfrac{10}{3} L \tag{251}$$

and

$$2BT = \frac{1}{2}\frac{\bar{E}_r/N_0}{3.07}, \tag{252}$$

then

$$\frac{\bar{E}_r}{N_0} \geq 60 \tag{253}$$

is adequate. The system in (251)–(253) is realistic and will perform close to the bound.

This example illustrates one procedure for signaling efficiently over an overspread channel. The basic principle involved is straightforward. The doubly-spread channel provides a certain amount of implicit diversity in the output signal. If the value of $\bar{E}_r/N_0$ is large enough to make this amount of diversity close to optimum, the system will work close to the bound. On the other hand, if $\bar{E}_r/N_0$ is too small, the performance may be relatively poor.

SUMMARY. In this section we have discussed the performance bounds that apply to any binary system. In addition, we have studied possible signaling schemes for underspread and overspread channels. In the underspread case we could use a signal that reduced the channel to a set of nonfluctuating point channels. By selecting the correct number of subpulses, we could achieve the bound. In the overspread case we could use a signal that reduced the channel to a singly-spread channel. In this case we could approach the bound if the available $\bar{E}_r/N_0$ was large enough.

In both cases we were able to use signals that eliminated the double spreading in the channel. This has several advantages:

1. The optimum receiver is simpler.
2. The performance analysis is simpler.
3. The performance is close enough to the bound that using the channel in a doubly-spread mode could not provide a significant decrease in the error probability.

It appears that in a large number of physical situations we can achieve this simplification, so that the above discussion is relevant. On the other hand, there are at least two reasons why we want to be able to analyze the doubly-spread model directly:

1. There are cases in which we cannot simplify the channel, because of limitations on the signal duration or bandwidth.
2. There is a transitional region between the singly- and doubly-spread cases in which we must check our intuitive arguments; in this region the gross signal and channel characterizations ($W$, $T$, $B$, and $L$) are not adequate.

In Section 13.3.2 we developed the necessary models to carry out this analysis. In the next section we use these models to analyze the binary communication problem.

***13.3.3.B. Performance Analysis for a Specific System.***† In Section 13.3.2 we developed approximate channel models with which we could design the optimum receiver and analyze its performance. We now consider a specific system to illustrate the details of the technique. The discussion has two distinct purposes. The first purpose is to demonstrate with an example the actual steps that one must go through to analyze the system performance. This detailed discussion illustrates the ideas of Section 13.3.2 and enables the reader to analyze any system of interest. The second purpose is to provide an understanding of the important issues in a communication system operating over a doubly-spread channel. The relationship between the signal parameters and the scattering function is explored. The quantitative results apply only to this specific system, but the approach can be used in other problems. This discussion will augment the results in Section 13.3.3.A.

The binary communication problem is described in (153)–(168). The channel-scattering process is described by (38)–(51). We consider a scattering function that is a special case of the scattering function in the example on

† The material in Subsection 13.3.3.B is due to Kurth [7].

page 456. The functions specifying it are

$$\tilde{Q}(\lambda) = \frac{2k}{L}\left[1 - \cos\left(\frac{2\pi\lambda}{L}\right)\right]\sqcap_L(\lambda), \tag{254}$$

where $\sqcap_L(\lambda)$ is a gate function defined as

$$\sqcap_L(\lambda) = \begin{cases} 1, & 0 \le \lambda \le L, \\ 0, & \text{elsewhere.} \end{cases} \tag{255}$$

In addition,

$$\tilde{k}(\lambda) = k \tag{256}$$

and

$$\tilde{C}(\lambda) = 1. \tag{257}$$

Notice that in this simple problem

$$\tilde{b}(t, \lambda) = \tilde{x}(t, \lambda). \tag{258}$$

The scattering function is

$$\tilde{S}_{DR}\{f, \lambda\} = \frac{\tilde{Q}(\lambda)}{(2\pi f)^2 + k^2}. \tag{259}$$

To use (230), we need $\tilde{K}_{\tilde{x}}(\lambda)$. Recalling from (46)–(49) that

$$\tilde{K}_{\tilde{x}}(\lambda) = \tilde{K}_{DR}(0, \lambda), \tag{260}$$

we have

$$\tilde{K}_{\tilde{x}}(\lambda) = \frac{1}{L}\left(1 - \cos\left(\frac{2\pi\lambda}{L}\right)\right)\sqcap_L(\lambda). \tag{261}$$

We assume that the transmitted signal is a rectangular pulse. Thus,

$$\begin{aligned} \tilde{f}(t) &= \begin{cases} \dfrac{1}{\sqrt{T}}, & 0 \le t \le T, \\ 0, & \text{elsewhere,} \end{cases} \\ &= \frac{1}{\sqrt{T}}\sqcap_T(t), \qquad -\infty < t < \infty. \end{aligned} \tag{262}$$

We assume that the propagation time is zero for notational simplicity. (This is equivalent to redefining the time origin.) The endpoints of the observation interval are

$$T_i = 0 \tag{263}$$

and

$$T_f = T + L. \tag{264}$$

We now have the system completely specified and want to determine its performance. To evaluate $\tilde{\mu}_{BS}(\frac{1}{2})$ we must evaluate $\xi_P(t, \tilde{s}(t), \cdot)$ for two noise levels. The function $\xi_P(t, \tilde{s}(t), \cdot)$ is related to $\boldsymbol{\xi}(t{:}\lambda, \lambda')$ by (235).

Using (262) in (235) gives

$$\tilde{\xi}_P(t, \tilde{s}(t), N_0) = \frac{E_t}{T} \int_{-\infty}^{\infty} \mathbb{T}_T(t - \lambda)\mathbb{T}_T(t - \lambda')\boldsymbol{\xi}(t:\lambda, \lambda')\, d\lambda\, d\lambda', \quad (265)$$

where $\boldsymbol{\xi}(t:\lambda, \lambda')$ is specified by (227) and (228). To find $\boldsymbol{\xi}(t:\lambda, \lambda')$, we divide it into two terms as in (229) and solve for $\tilde{\mathbf{p}}(t:\lambda, \lambda')$. Using (254)–(257) and (261) and (262) in (230) gives

$$\begin{aligned}\frac{\partial \tilde{p}(t:\lambda, \lambda')}{\partial t} = -2k\tilde{p}(t:\lambda, \lambda') - \frac{E_t}{N_0 T}\Bigg\{&\left[\frac{1 - \cos(2\pi\lambda/L)}{L}\mathbb{T}_T(t - \lambda)\right. \\ &\left. + \int_{-\infty}^{\infty} \mathbb{T}_T(t - \lambda')\tilde{p}(t:\lambda, \lambda')\, d\lambda'\right]\left[\frac{1 - \cos(2\pi\lambda'/L)}{L}\mathbb{T}_T(t - \lambda')\right. \\ &\left. + \int_{-\infty}^{\infty} \mathbb{T}_T(t - \lambda)\tilde{p}(t:\lambda, \lambda')\, d\lambda\right]^*\Bigg\}, \quad 0 \le \lambda, \lambda' \le L,\ t \ge 0, \quad (266)\end{aligned}$$

with initial conditions

$$\tilde{p}(0:\lambda, \lambda') = 0, \qquad 0 \le \lambda, \lambda' \le L. \quad (267)$$

We now demonstrate how to obtain an approximate solution to (266) by using the modal expansion technique suggested on page 501. We expand $\tilde{p}(t:\lambda, \lambda')$, using (232), as

$$\tilde{p}(t:\lambda, \lambda') = \sum_{i=1}^{\infty}\sum_{j=1}^{\infty} \tilde{p}_{ij}(t)\tilde{\phi}_i(\lambda)\tilde{\phi}_j^*(\lambda'), \qquad 0 \le \lambda, \lambda' \le L, \qquad 0 \le t \le T + L, \quad (268)$$

where the $\phi_i(\lambda)$ are an arbitrary set of orthonormal functions. Proceeding as suggested below (233), we can derive an equation specifying $\tilde{p}_{ij}(t)$. We include the details to guarantee that the actual manipulations are clear.

**Modal Expansion Equations.** Substituting (268) into (266) gives

$$\begin{aligned}\sum_{i=1}^{\infty}\sum_{j=1}^{\infty} \frac{\partial \tilde{p}_{ij}(t)}{\partial t}\tilde{\phi}_i(\lambda)\tilde{\phi}_j^*(\lambda') = &-2k \sum_{i=1}^{\infty}\sum_{j=1}^{\infty} \tilde{p}_{ij}(t)\tilde{\phi}_i(\lambda)\tilde{\phi}_j^*(\lambda') \\ &- \frac{E_t}{N_0 T}\Bigg\{\left[\left(\frac{1 - \cos(2\pi\lambda/L)}{L}\right)\mathbb{T}_T(t - \lambda)\right. \\ &\left. + \int_{-\infty}^{\infty} \mathbb{T}_T(t - \sigma)\sum_{i=1}^{\infty}\sum_{j=1}^{\infty} \tilde{p}_{ij}(t)\tilde{\phi}_i(\lambda)\tilde{\phi}_j^*(\sigma)\, d\sigma\right] \\ &\times\left[\left(\frac{1 - \cos(2\pi\lambda'/L)}{L}\right)\mathbb{T}_T(t - \lambda')\right. \\ &\left. + \int_{-\infty}^{\infty} \mathbb{T}_T(t - \sigma')\sum_{i=1}^{\infty}\sum_{j=1}^{\infty} \tilde{p}_{ij}(t)\tilde{\phi}_i(\sigma')\tilde{\phi}_j^*(\lambda')\, d\sigma'\right]^*\Bigg\}, \\ &0 \le \lambda, \lambda' \le L, \qquad 0 \le t \le T + L. \quad (269)\end{aligned}$$

We now carry out the following steps:

1. Multiply both sides of (269) by $\tilde{\phi}_k^*(\lambda)\tilde{\phi}_l(\lambda')$ and integrate over $\lambda$ and $\lambda'$.
2. Define

$$\tilde{z}_k(t) = \int_{-\infty}^{\infty} \left(\frac{1 - \cos(2\pi\sigma/L)}{L}\right) \Pi_T(t - \sigma)\, \Pi_L(\sigma)\tilde{\phi}_k^*(\sigma)\, d\sigma \tag{270}$$

and

$$\tilde{b}_k(t) = \int_{-\infty}^{\infty} \Pi_T(t - \sigma)\, \Pi_L(\sigma)\tilde{\phi}_k^*(\sigma)\, d\sigma \tag{271}$$

to simplify the equation from step 1.

3. Truncate the equation at $K$ terms to obtain a finite-dimensional Riccati equation.

In the present problem (270) and (271) reduce to

$$\tilde{z}_k(t) = \int_a^b \left(\frac{1 - \cos(2\pi\sigma/L)}{L}\right) \tilde{\phi}_k^*(\sigma)\, d\sigma \tag{272}$$

and

$$\tilde{b}_k(t) = \int_a^b \tilde{\phi}_k^*(\sigma)\, d\sigma, \tag{273}$$

where

$$b \triangleq \min(L, t) \tag{274}$$

and

$$a \triangleq \min(b, \max(0, t - T)). \tag{275}$$

Carrying out the first step and using the definitions in the second step gives the differential equation

$$\frac{d\tilde{p}_{kl}(t)}{dt} = -2k\tilde{p}_{kl}(t) - \frac{E_t}{N_0 T}\left[\tilde{z}_k(t) + \sum_{j=1}^{\infty} \tilde{p}_{kj}(t)\tilde{b}_j(t)\right]\left[\tilde{z}_l^*(t) + \sum_{i=1}^{\infty} \tilde{p}_{il}(t)\tilde{b}_i^*(t)\right]. \tag{276}$$

Truncating the series at $K$, we can put (276) in matrix notation as

$$\frac{d\tilde{\mathbf{p}}(t)}{dt} = -2k\tilde{\mathbf{p}}(t) - \frac{E}{N_0 T}[\tilde{\mathbf{z}}(t) + \tilde{\mathbf{p}}(t)\tilde{\mathbf{b}}(t)][\tilde{\mathbf{z}}(t) + \tilde{\mathbf{p}}(t)\tilde{\mathbf{b}}(t)]^{\dagger}, \tag{277}$$

where the definition of $\tilde{\mathbf{p}}(t)$, $\tilde{\mathbf{z}}(t)$, and $\tilde{\mathbf{b}}(t)$ is clear. The initial condition is

$$\tilde{\mathbf{p}}(t) = \mathbf{0}. \tag{278}$$

We now have reduced the proolem to a finite-dimensional Riccati equation, which we can solve numerically.

The final issue is the choice of the orthogonal functions $\{\tilde{\phi}_i(\lambda)\}$. We want to choose them so that the dimension of the approximating system will be small *and* so that the calculation of the quantities in (272) and (273) will be simple. As pointed out earlier, a judicious choice will reduce the computational problem significantly. In this case, the scattering function is a raised cosine function and the signal is rectangular, so that a conventional Fourier series is a logical choice. We let

$$\tilde{\phi}_1(\lambda) = \frac{1}{\sqrt{L}} \qquad 0 \le \lambda \le L,$$

$$\tilde{\phi}_2(\lambda) = \sqrt{\frac{2}{L}} \cos\left(\frac{2\pi\lambda}{L}\right), \qquad 0 \le \lambda \le L,$$

$$\tilde{\phi}_3(\lambda) = \sqrt{\frac{2}{L}} \sin\left(\frac{2\pi\lambda}{L}\right), \qquad 0 \le \lambda \le L, \tag{279}$$

and so forth. We now have all of the quantities necessary to evaluate the performance.

The performance will depend on $\bar{E}_r/N_0$, $k$, $L$, and $T$. Before carrying out the calculations, we discuss the effect of these parameters.

First, we fix the first three parameters and study the effect of $T$. The length of the input signal affects the number of degrees of freedom in the output waveform. We refer to this as the system "diversity." A crude estimate of this diversity is obtained by multiplying the diversity due to Doppler spreading by the diversity due to range spreading to obtain

$$D = (1 + kT)\left(1 + \frac{L}{T}\right). \tag{280}$$

Three comments regarding (280) are useful:

1. The fading spectrum is a one-pole, and so the best bandwidth measure is not obvious; the equivalent rectangular bandwidth is $k/2$ cps (double-sided) (i.e., one might get a more accurate measure by including a constant before $kT$).
2. More refined diversity measures are discussed by Kennedy [37]; the expression in (280) is adequate for our intuitive discussion.
3. The expression in (280) is for a rectangular transmitted pulse and assumes that $WT = 1$.

The diversity expression in (280) is plotted as a function of $T' \triangleq T\sqrt{k/L}$ in Fig. 13.23. We see that the minimum diversity occurs when

$$T = \sqrt{\frac{L}{k}}, \tag{281}$$

and its value is

$$D_{\min} = (1 + \sqrt{kL})^2. \tag{282}$$

From our earlier work we know that there is an optimum diversity, which we would estimate as

$$D_{\text{opt}} \simeq \frac{1}{3}\frac{\bar{E}_r}{N_0}. \tag{283}$$

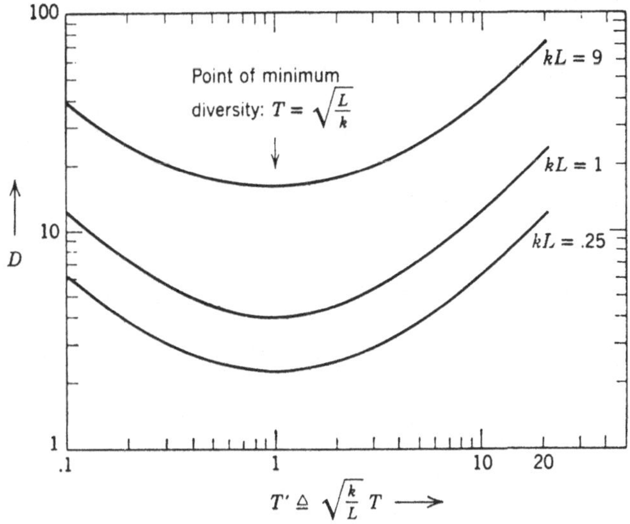


**Fig. 13.23 Diversity of a doubly spread channel ($WT = 1$).**

Comparing (282) and (283), we see that if

$$D_{\min} > D_{\text{opt}}, \tag{284}$$

the optimum value of $T$ will be given by (281) and the performance will decrease for either smaller or larger $T$, as shown in Fig. 13.24*a*. Intuitively, this means that the $kL$ product is such that the channel causes more diversity than we want. On the other hand, if

$$D_{\min} < D_{\text{opt}}, \tag{285}$$

the performance curve will have the general behavior shown in Fig. 13.24*b*. The performance will have a maximum for two different values of $T$.

The minimum diversity increases monotonically with the $kL$ product, while the optimum diversity increases monotonically with $\bar{E}_r/N_0$. Therefore, for a particular $kL$ product, we would expect the behavior in Fig. 13.24*a* for small $\bar{E}_r/N_0$ and the behavior in Fig. 13.24*b* for large $\bar{E}_r/N_0$. From our discussion in (247)–(253), we would expect that increasing the $kL$ product will not decrease the performance significantly if $\bar{E}_r/N_0$ is large enough.

This completes our intuitive discussion. Kurth [7] has carried out the analysis for the system described in (254)–(264), using the modal expansion in (265)–(279). In Figs. 13.25 to 13.27, we show several sets of performance curves. The vertical axis is the efficiency factor,

$$\frac{\tilde{\mu}_{BS}(\frac{1}{2})}{\bar{E}_r/N_0}.$$

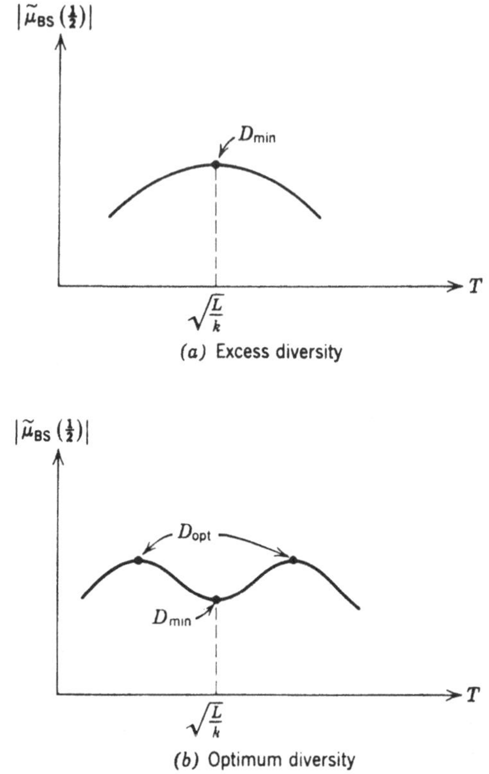


**Fig. 13.24 Qualitative behavior characteristics as a function of the pulse length, $T$.**

The horizontal axis is $T$, the pulse length. In Fig. 13.25, $kL = 0.25$, in Fig. 13.26, $kL = 1.0$, and in Fig. 13.27, $kL = 6.25$. In all cases $k = L$. The different curves correspond to various values of $\bar{E}_r/N_0$. We see that the anticipated behavior occurs. For small $\bar{E}_r/N_0$, $D_{opt} < D_{min}$, and there is a single peak. For larger $\bar{E}_r/N_0$, $D_{opt} > D_{min}$, and there are two peaks, As $kL$ increases, a larger value of $\bar{E}_r/N_0$ is required to obtain the two-peak behavior.

In Figure 13.28, we show the effect of the $kL$ product. To construct these curves, we used the value of $T$ that maximized $|\tilde{\mu}_{BS}(\frac{1}{2})|$ for the particular $kL$ product and $\bar{E}_r/N_0$ ($k = L$ for all curves). The vertical axis is $-\tilde{\mu}_{BS}(\frac{1}{2})$, and the horizontal axis is $\bar{E}_r/N_0$. Each curve corresponds to a different $kL$ product. As the $kL$ product increases, the exponent decreases for a fixed $\bar{E}_r/N_0$, but the change is not drastic.

This example illustrates the performance analysis of a typical system. The reader may be troubled by the seemingly abrupt transition between the

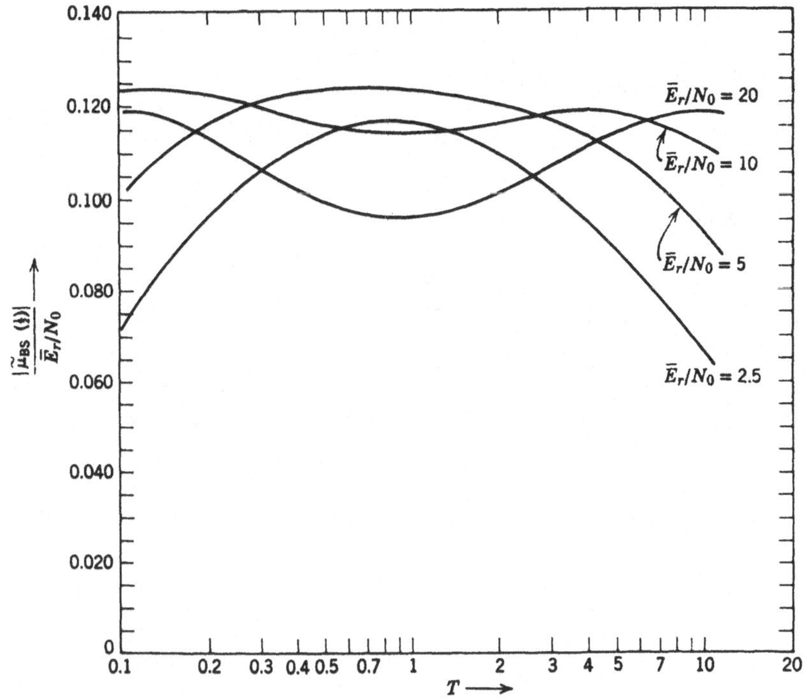


**Fig. 13.25 Optimum receiver performance, binary orthogonal communication, first-order fading, underspread channel; $k = 0.5$, $L = 0.5$, constant $\tilde{f}(t)$. (From [7].)**

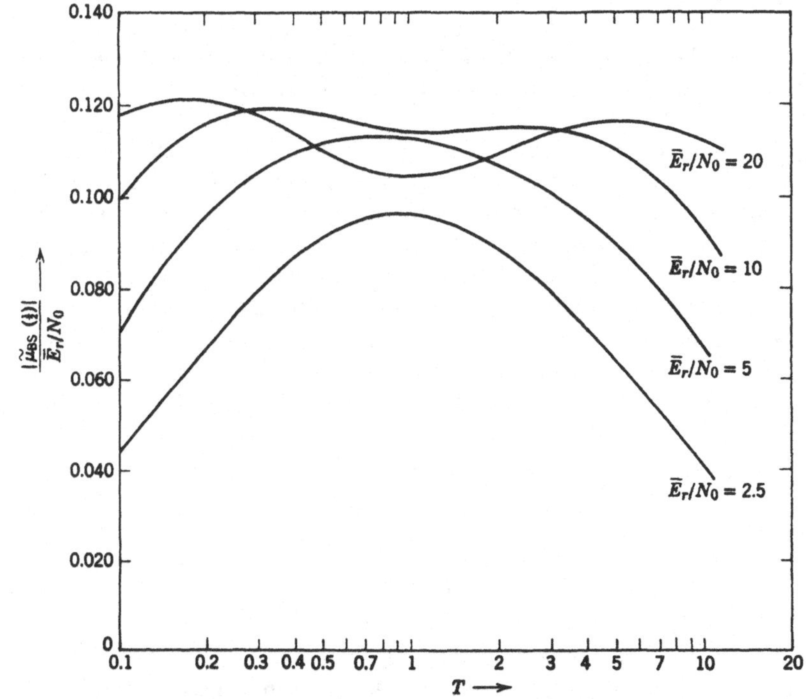


**Fig 13.26 Optimum receiver performance, first-order fading, doubly-spread channel, $k = 1$, $L = 1$, constant $\tilde{f}(t)$. (From [7].)**

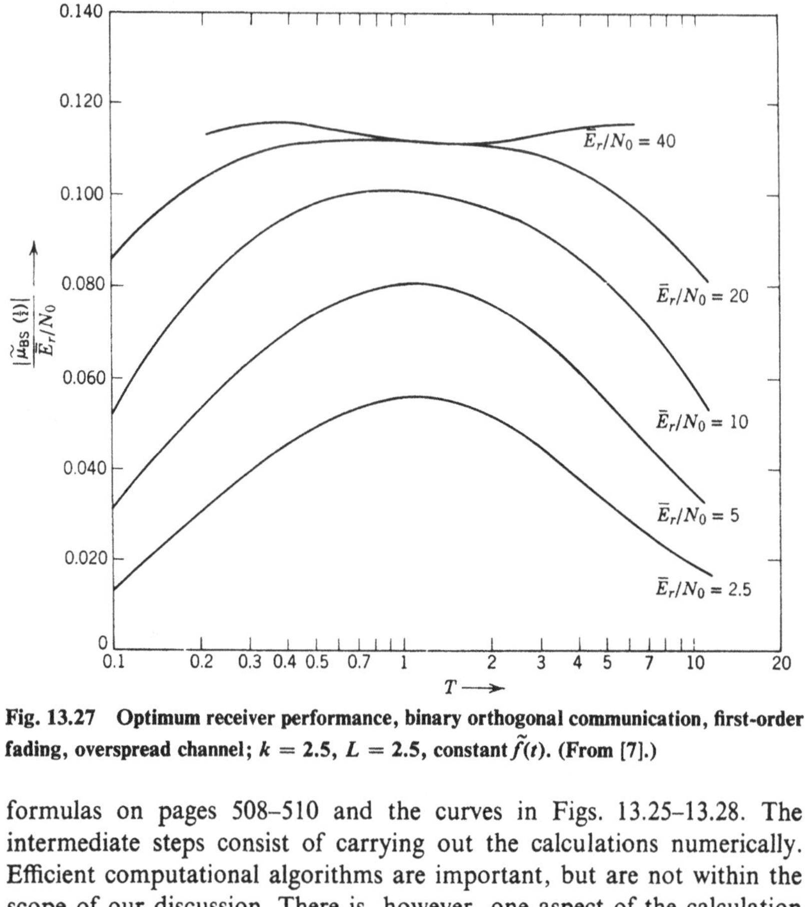


**Fig. 13.27 Optimum receiver performance, binary orthogonal communication, first-order fading, overspread channel; $k = 2.5$, $L = 2.5$, constant $\tilde{f}(t)$. (From [7].)**

formulas on pages 508–510 and the curves in Figs. 13.25–13.28. The intermediate steps consist of carrying out the calculations numerically. Efficient computational algorithms are important, but are not within the scope of our discussion. There is, however, one aspect of the calculation procedure that is of interest. We emphasized that a suitable choice of orthogonal functions reduces the complexity of the calculation. To generate the curves in Figs. 13.25–13.28, we kept increasing $K$ until $\tilde{\mu}_{BS}(\frac{1}{2})$ stabilized. In Table 13.1, we indicate the values of $K$ required to achieve three-place accuracy in $\tilde{\mu}_{BS}(\frac{1}{2})$ as a function of various parameters in the problem. When

$$W = \frac{1}{T} \ll \frac{1}{L} \tag{286}$$

or

$$T \ll \frac{1}{k} \tag{287}$$

and $\bar{E}_r/N_0$ is large, more terms are required. Notice that when (286) is satisfied we can model the channel as a Doppler-spread point channel, and

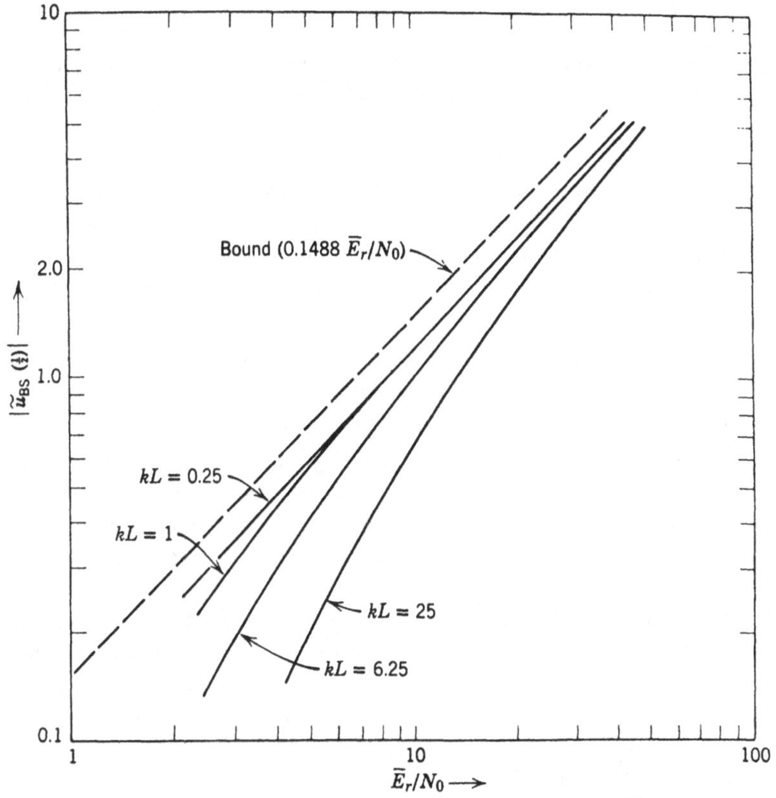


**Fig. 13.28 Optimum receiver performance for optimum $T$, constant $\tilde{f}(t)$, binary orthogonal communication over a doubly-spread channel ($k = L$). (From [7].)**

**Table 13.1 Number of Terms Required to Achieve at Least Three-place Accuracy in the Calculation of $(|\tilde{\mu}_{BS}(1/2)|/\bar{E}_r/N_0)$ (From [7])**

| $\bar{E}_r/N_0$ | $k$ | $L$ | $T$ | $K$ |
|---|---|---|---|---|
| 5 | 0.5 | 0.5 | 0.1 | 17 |
| 5 | 0.5 | 0.5 | 1 | 13 |
| 5 | 0.5 | 0.5 | 10 | 20 |
| 20 | 0.5 | 0.5 | 1 | 13 |
| 5 | 1 | 1 | 1 | 13 |
| 5 | 1 | 1 | 10 | 21 |
| 20 | 1 | 1 | 1 | 17 |
| 5 | 2.5 | 2.5 | 0.1 | 25 |
| 5 | 2.5 | 2.5 | 1 | 17 |
| 20 | 2.5 | 2.5 | 1 | 17 |
| 20 | 2.5 | 2.5 | 10 | 25 |

when (287) is satisfied we can model the channel as a nonfluctuating range-spread channel. Thus, the cases that required the most calculation could be avoided.

In this subsection we have actually carried out the performance analysis for a specific problem. The analysis demonstrates the utility of the channel models developed in Section 13.3.2 for studying problems containing doubly-spread channels or doubly-spread targets. In addition, it demonstrates quantitatively how the various system parameters affect the system performance.

***13.3.3.C. Summary.*** In this section we have studied the problem of binary communication over doubly-spread channels. There are several important points that should be re-emphasized.

1. When the $BL$ product of the channel is small, we can reduce it to a set of nonfluctuating point channels by proper signal design. The resulting system achieves the performance bound. Because the receiver is straightforward, this mode of operation should be used for underspread channels whenever possible.
2. When the channel is overspread, we can reduce it to a singly-spread channel by proper signal design. The efficiency of the resulting system depends on the details of the scattering function and the available $\bar{E}_r/N_0$. Because the singly-spread receiver is simpler than the doubly-spread receiver, the above mode of operation should be used for overspread channels whenever possible.
3. Most scattering functions can be adequately approximated by a distributed state-variable model. For this case, we can analyze the performance using the modal expansion techniques developed in this section. Although the analysis is complicated, it is feasible. The results provide quantitative confirmation of our intuitive arguments in simple cases and enable us to study more complicated systems in which the intuitive arguments would be difficult.

This completes our discussion of binary communication. In Section 13.3.5, we shall discuss briefly the extensions to $M$-ary systems.

### 13.3.4 Detection under LEC Conditions

The model for the detection problem and the binary communication problem were formulated in Section 13.3.1. In the succeeding sections we studied various facets of the general case in detail. There is one special case in which the results are appreciably simpler. This is the lower-energy-coherence (LEC) case that we have encountered several times previously.

In Section 13.3.4.A we study the LEC problem. The discussion suggests suboptimum receivers for the general case, which we discuss briefly in Section 13.3.4.B.

***13.3.4.A. LEC Receivers.*** If we denote the largest eigenvalue of $\tilde{s}(t)$ by $\tilde{\lambda}_{\max}$, the LEC condition is

$$\frac{\tilde{\lambda}_{\max}}{N_0} \ll 1. \tag{288}$$

Paralleling the derivation in Section 11.2.4, we can see that the likelihood ratio test for the simple binary detection problem reduces to

$$l_R = \frac{1}{N_0^2} \iint_{T_i}^{T_f} \tilde{r}^*(t) \tilde{K}_{\tilde{s}}(t, u) \tilde{r}(u)\, dt\, du \underset{H_0}{\overset{H_1}{\gtrless}} \gamma. \tag{289}$$

Substituting (143) into (289) gives

$$l_R = \frac{E_t}{N_0^2} \iint_{T_i}^{T_f} dt\, du \int_{-\infty}^{\infty} d\lambda\, \tilde{r}^*(t) \tilde{f}(t-\lambda) \tilde{K}_{DR}(t-u, \lambda) \tilde{f}^*(u-\lambda) \tilde{r}(u) \underset{H_0}{\overset{H_1}{\gtrless}} \gamma. \tag{290}$$

A particularly simple realization can be obtained when $T_i = -\infty$ and $T_f = \infty$ by factoring $\tilde{K}_{DR}(t-u, \lambda)$ along the time axis as

$$\tilde{K}_{DR}(t-u, \lambda) = \int_{-\infty}^{\infty} \tilde{K}_{DR}^{[\frac{1}{2}]*}(z-t, \lambda) \tilde{K}_{DR}^{[\frac{1}{2}]}(z-u, \lambda)\, dz. \tag{291}$$

Using (291) in (290) gives

$$l_R = \frac{E_t}{N_0^2} \int_{-\infty}^{\infty} dz \int_{-\infty}^{\infty} d\lambda \left| \int_{-\infty}^{\infty} \tilde{K}_{DR}^{[\frac{1}{2}]}(z-u, \lambda) \tilde{f}^*(u-\lambda) \tilde{r}(u)\, du \right|^2. \tag{292}$$

The receiver specified by (292) is shown in Fig. 13.29 (due originally to Price [5]). Because the receiver requires a continuous operation in $\lambda$, it cannot be realized exactly. An approximation to the optimum receiver is obtained by sampling in $\lambda$ and replacing the $\lambda$ integration by a finite sum. This realization is shown in Fig. 13.30. This receiver is also due to Price [5] and is essentially optimum under LEC conditions.

When the LEC condition is valid, (11.65) gives

$$\tilde{\mu}(s) \simeq -\frac{s(1-s)}{2N_0^2} \iint_{T_i}^{T_f} |\tilde{K}_{\tilde{s}}(t, u)|^2\, dt\, du. \tag{293}$$

Using (143) in (293) gives

$$\begin{aligned} \tilde{\mu}(s) \simeq -\frac{s(1-s)}{2} & \\ \times \left\{ \frac{E_t^2}{2N_0^2} \iint_{-\infty}^{\infty} dt\, du \right. & \int_{-\infty}^{\infty} d\lambda_1 \tilde{f}(t-\lambda_1) \tilde{K}_{DR}(t-u, \lambda_1) \tilde{f}^*(u-\lambda_1) \\ & \left. \times \int_{-\infty}^{\infty} d\lambda_2 \tilde{f}^*(t-\lambda_2) \tilde{K}_{DR}^*(t-u, \lambda_2) \tilde{f}(u-\lambda_2) \right\}. \end{aligned} \tag{294}$$

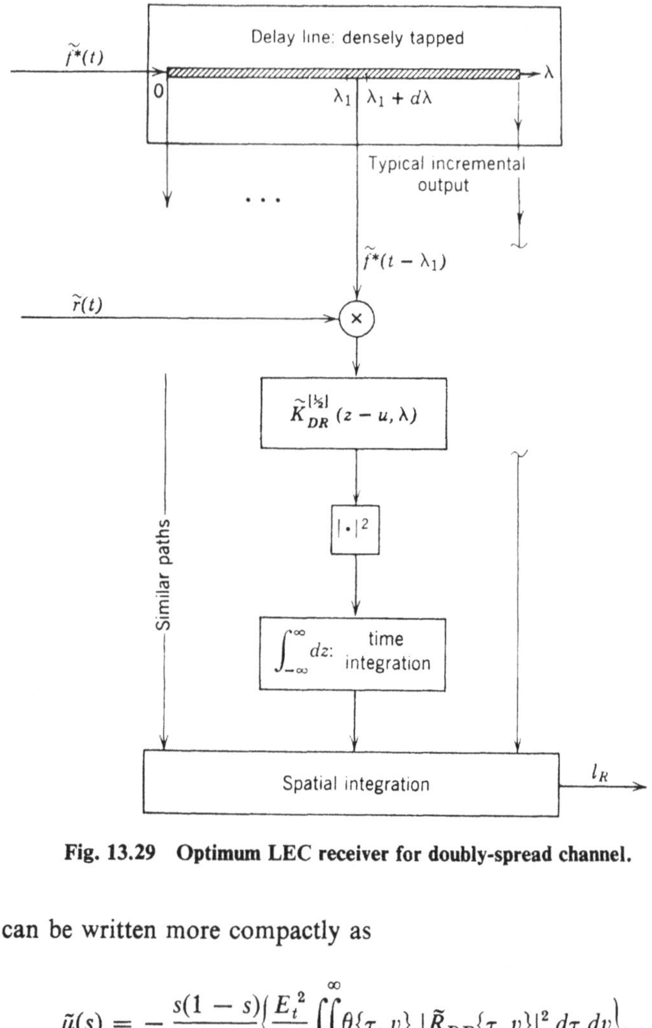


**Fig. 13.29 Optimum LEC receiver for doubly-spread channel.**

This can be written more compactly as

$$\tilde{\mu}(s) = -\frac{s(1-s)}{2}\left\{\frac{E_t^2}{N_0^2}\iint_{-\infty}^{\infty}\theta\{\tau, v\}\,|\tilde{R}_{DR}\{\tau, v\}|^2\,d\tau\,dv\right\}, \tag{295}$$

where $\theta\{\tau, v\}$ is the signal ambiguity function, and $\tilde{R}_{DR}\{\tau, v\}$ is the two-frequency correlation function defined in (21). (See Problem 13.3.21)

Our discussion of the LEC problem has been brief, but the reader should not underestimate its importance. In many cases the system is forced to operate under LEC conditions. Then the results in (292) and (295) are directly applicable. In other cases the LEC condition is not present, but the LEC receiver suggests a suboptimum receiver structure. We explore this problem briefly in the next subsection.

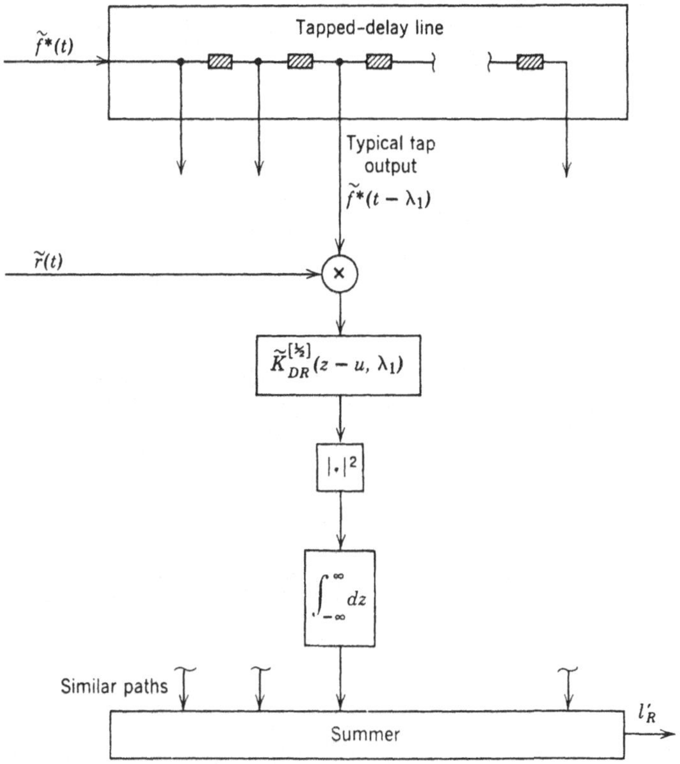


**Fig. 13.30 Approximation to optimum LEC receiver for doubly-spread channel.**

***13.3.4.B. Suboptimum Receivers.*** The first suboptimum receiver follows directly from Fig. 13.30. We retain the structure but allow an arbitrary time-invariant filter in each path. Thus,

$$l_{so} = \sum_{i=1}^{K} \int_{-\infty}^{\infty} dz \left| \int_{-\infty}^{\infty} \tilde{h}(z-u,\lambda_i)\tilde{f}^*(u-\lambda_i)\tilde{r}(u)\,du \right|^2. \tag{296}$$

The performance of this receiver can be analyzed by combining the techniques of Sections 11.3 and 13.3.3. By varying the $\tilde{h}(\cdot,\lambda_i)$, we can optimize the performance within the structural limitations. The actual calculations are complicated but feasible.

The second suboptimum receiver is a generalization of the receivers in Figs. 11.19 and 12.11. This receiver is shown in Fig. 13.31. Notice that there are $N_R$ branches and each branch contains $N_D$ correlation operations.

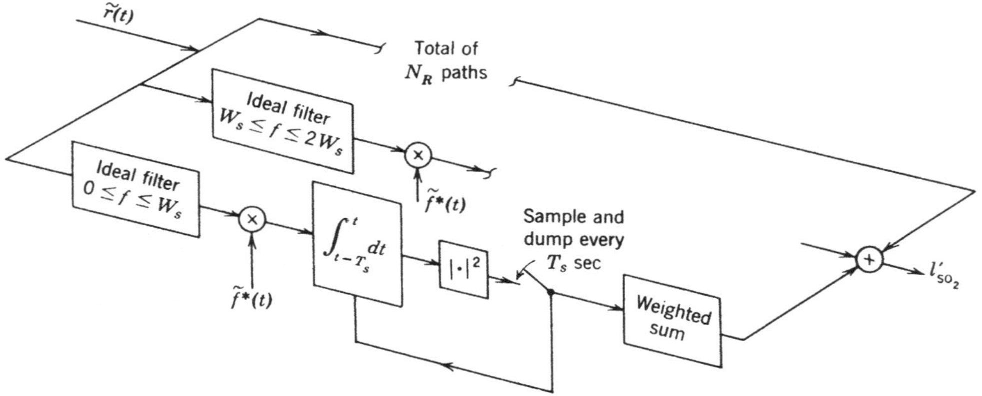


**Fig. 13.31 Suboptimum receiver for detection of doubly-spread signal.**

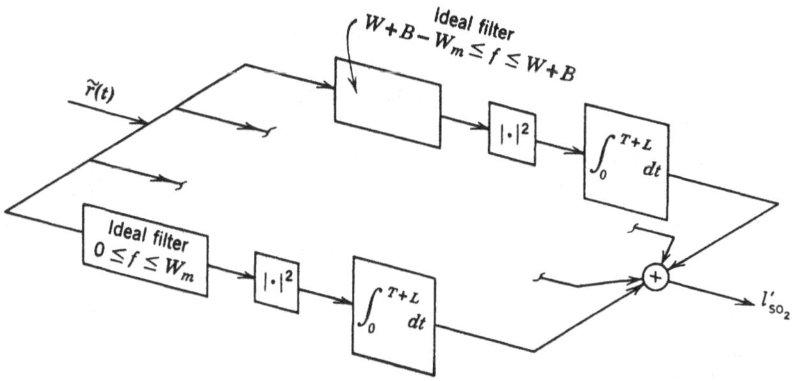


**Fig. 13.32 Approximation to suboptimum receiver No. 2 for rectangular envelope case.**

As a first approximation, we might choose

$$W_s = \frac{1}{T + L} \tag{297a}$$

$$T_s = \frac{1}{B + W}, \tag{297b}$$

so that we have a system with diversity

$$N_R N_D = (T + L)(B + W). \tag{297c}$$

In general, we leave the filter bandwidth, $W_s$ and the correlation time, $T_s$, as parameters. This receiver can be analyzed by using the techniques of Section 11.3.3. Once again, the calculations are complicated but feasible.

When $\tilde{f}(t)$ is a rectangular pulse, a good approximation to the receiver in Fig. 13.31 can be obtained as shown in Fig. 13.32. Here

$$W_m \triangleq \min\left[W_s, \frac{1}{T_s}\right]. \tag{298}$$

This receiver is essentially that suggested by Kennedy and Lebow [38].†

This completes our suboptimum receiver discussion. We now consider some other detection theory topics.

### 13.3.5 Related Topics

In this section we discuss three topics briefly. In Section 13.3.5.A we discuss equivalent channels. In Section 13.3.5.B we comment on $M$-ary

† This reference contains an excellent intuitive discussion of communication over doubly-spread channels, and most engineers will find it worthwhile reading.

communication over doubly-spread channels. In Section 13.3.5.C we re-examine the reverberation problem of Section 13.2.

***13.3.5.A. Equivalent Channels and Systems.*** The idea of an equivalent channel is due to Kennedy [37] and is a generalization of the dual-channel ideas introduced in Section 12.3. Our motivation for developing duality theory was to simplify the design of systems and their analysis. Frequently it was easier to analyze the dual of the system instead of the actual system. The motivation for our discussion of equivalent systems is identical. In many cases, it is easier to analyze an equivalent system instead of the actual system. In addition, the ability to recognize equivalent systems aids our understanding of the general problem.

The first definition is:

**Definition 1. Equivalent Processes.** Consider the two processes $\tilde{r}_1(t)$ and $\tilde{r}_2(t)$ defined over the interval $[T_i, T_f]$. If the eigenvalues of $\tilde{r}_1(t)$ equal the eigenvalues of $\tilde{r}_2(t)$, the processes are equivalent on $[T_i, T_f]$.

For simplicity we now restrict our attention to a simple binary detection problem. The complex envelopes of the received waveforms on the two hypotheses are

$$\tilde{r}(t) = \tilde{s}(t) + \tilde{w}(t), \qquad T_i \le t \le T_f : H_1, \tag{299}$$

$$\tilde{r}(t) = \tilde{w}(t), \qquad T_i \le t \le T_f : H_0. \tag{300}$$

The additive noise $\tilde{w}(t)$ is a sample function from a zero-mean complex white Gaussian noise process with spectral height $N_0$. The signal $\tilde{s}(t)$ is a sample function from a zero-mean complex Gaussian process with covariance function $\tilde{K}_{\tilde{s}}(t, u)$. From our earlier results we know that the performance of the system is completely characterized by the *eigenvalues* of $\tilde{K}_{\tilde{s}}(t, u)$. Notice that the receiver depends on both the eigenfunctions and eigenvalues, but the eigenfunctions do not affect the performance. This observation suggests the following definition.

**Definition 2. Equivalent Detection Problems.** All simple binary detection problems in which the $\tilde{s}(t)$ are equivalent processes are equivalent.

This definition is a generalization of Definition 5 on page 426.

The next idea of interest is that of *equivalent channels*. The covariance function of the signal process at the output of a doubly-spread channel is

$$\tilde{K}_{\tilde{s}_1}(t, u) = \int_{-\infty}^{\infty} \tilde{f}_1(t-\lambda)\tilde{K}_{DR_1}(t-u, \lambda)\tilde{f}_1^*(u-\lambda)\, d\lambda. \tag{301}$$

Now consider the covariance function at the output of a second channel,

$$\tilde{K}_{\tilde{s}_2}(t, u) = \int_{-\infty}^{\infty} \tilde{f}_2(t - \lambda)\tilde{K}_{DR_2}(t - u, \lambda)\tilde{f}_2^*(u - \lambda)\, d\lambda. \quad (302)$$

We can now define equivalent channels.

**Definition 3. Equivalent Channels.** Channel No. 2 is equivalent to Channel No. 1, if, for every $\tilde{f}_1(t)$ with finite energy, there exists a signal $\tilde{f}_2(t)$ with finite energy such that the eigenvalues of $\tilde{K}_{\tilde{s}_2}(t, u)$ are equal to the eigenvalues of $\tilde{K}_{\tilde{s}_1}(t, u)$.

The utility of this concept is that it is frequently easier to analyze an equivalent channel instead of the actual channel.

Some typical equivalent channels are listed in Table 13.2. In columns 1 and 3 we show the relationship between the two-channel scattering functions. Notice that $\tilde{S}_{DR}\{f, \lambda\}$ is an arbitrary scattering function. The complex envelope of the transmitted signal in system 1 is $\tilde{f}(t)$. In column 4, we show the complex envelope that must be transmitted in system 2 to generate an equivalent output signal process. We have assumed an infinite observation interval for simplicity.

We study other equivalent channels and systems in the problems. Once again, we point out that it is a logical extension of the duality theory of Section 12.3 and is useful both as a labor-saving procedure and as an aid to understanding the basic limitations on a system.

***13.3.5.B. M-ary Communications over Doubly-Spread Channels.*** In Section 11.3.4 we discussed communication over Doppler-spread channels using $M$-orthogonal signals. Many of the results were based on the eigenvalues of the output processes. All these results are also applicable to the doubly-spread channel model. In particular, the idea of an optimum eigenvalue distribution is valid. When we try to analyze the performance of a particular system, we must use the new techniques developed in this chapter. The modification of the binary results is straightforward. The reader should consult [37] for a complete discussion of the $M$-ary problem.

***13.3.5.C. Reverberation.*** In Section 13.2 we studied the problem of detecting a *point* target in the presence of doubly-spread interference. One problem of interest was the design of the optimum receiver and an analysis of its performance. The appropriate equations were (116)–(121*b*), and we indicated that we would discuss their solution in this section. We see that all of our discussion in Section 13.3.2 is directly applicable to this problem. The difference is that we want to estimate the reverberation return, $\tilde{n}_r(t)$, in one case, and the reflected signal process in the other. All of the techniques carry over directly.

| | System 1 | | System 2 | |
|---|---|---|---|---|
| | 1 | 2 | 3 | 4 |
| | $\tilde{S}_{DR_1}\{f, \lambda\}$ | $\tilde{f}_1(t)$ | $S_{DR_2}\{f, \lambda\}$ | $\tilde{f}_2(t)$ |
| 1 | $\tilde{S}_{DR}\{-\lambda, f\}$ | $\tilde{f}(t)$ | $\tilde{S}_{DR}\{f, \lambda\}$ | $\tilde{F}(t)$ |
| 2 | $\tilde{S}_{DR}\left\{\frac{f}{a}, a\lambda\right\}$ [scale change] | $\tilde{f}(t)$ | $\tilde{S}_{DR}\{f, \lambda\}$ | $\sqrt{a}\,\tilde{f}(at)$ [see (10.124)] |
| 3 | $\tilde{S}_{DR}\{\lambda \sin\alpha + f\cos\alpha, \lambda\cos\alpha - f\sin\alpha\}$ [rotation] | $\tilde{f}(t)$ | $\tilde{S}_{DR}\{f, \lambda\}$ | $\frac{1}{\sqrt{\cos\alpha}} \exp\left[j\frac{t^2 \cdot \tan\alpha}{2}\right] \int_{-\infty}^{\infty} \tilde{F}\{f\} \times \exp\left[j2\pi\left[\pi t^2 \cdot \tan\alpha + \frac{ft}{\cos\alpha}\right]\right] df$ [see (10.129)] |

**Table 13.2 Typical Equivalent Channels**

### 13.3.6 Summary of Detection of Doubly-Spread Signals

In this section we have studied the detection of doubly-spread targets and communication over doubly-spread channels. In the Section 13.3.1 we formulated the model and specified the optimum receiver and its performance in terms of integral equations and differential equations. Because the problem was one of detecting complex Gaussian processes in complex Gaussian noise, the equations from Section 11.2.1 were directly applicable. The difficulty arose when we tried solve the integral equations that specified the optimum receiver.

In Section 13.3.2 we developed several approximate models. The reason for developing these models is that they reduced the problem to a format that we had encountered previously and could analyze exactly. In particular, we developed a tapped-delay line model, a general orthogonal series model, and an approximate differential-equation model. Each model had advantages and disadvantages, and the choice of which one to use depended on the particular situation.

In Section 13.3.3 we studied a binary communication problem in detail. In addition to obtaining actual results of interest, it provided a concrete example of the techniques involved. Because of the *relative* simplicity of the binary symmetric problem, it is a useful tool for obtaining insight into more complicated problems.

In Section 13.3.4, we studied the LEC problem. In this case the optimum receiver can be completely specified and its performance evaluated. The LEC receiver also suggested suboptimum receiver structures for other problems.

In section 13.3.5, we discussed some related topics briefly. This completes our discussion of the general detection problem. In the next section we consider the parameter estimation problem.

## 13.4 PARAMETER ESTIMATION FOR DOUBLY-SPREAD TARGETS

In this section we consider the problem of estimating the parameters of a doubly-spread target. The model of interest is a straightforward extension of the model of the detection problem in Section 13.1. The complex envelope of the received waveform is

$$\tilde{r}(t) = \tilde{s}(t, \mathbf{A}) + \tilde{w}(t), \qquad T_i \le t \le T_f, \tag{303}$$

where $\tilde{s}(t, \mathbf{A})$, given $\mathbf{A}$, is a sample function of a zero-mean complex Gaussian process whose covariance function is

$$\tilde{K}_{\tilde{s}}(t, u:\mathbf{A}) = E_t \int_{-\infty}^{\infty} \tilde{f}(t-\lambda)\tilde{K}_{DR}(t-u, \lambda:\mathbf{A})\tilde{f}^*(u-\lambda)\, d\lambda. \tag{304}$$

The additive noise $\tilde{w}(t)$ is a sample function of a white Gaussian noise process. The vector parameter **A** is either a nonrandom unknown vector or a value of a random vector that we want to estimate. We consider only the nonrandom unknown parameter problem in the text. Typical parameters of interest are the amplitude of the scattering function or the mean range and mean Doppler of a doubly-spread target.

To find the maximum likelihood estimate of **A**, we construct the likelihood function and choose the value of **A** at which it is maximum. Because the expression for the likelihood function can be derived by a straightforward modification of the analysis in Chapter 6 and Section 11.4, we can just state the pertinent results.

$$l(\mathbf{A}) = l_R(\mathbf{A}) + l_B(\mathbf{A}), \tag{305}$$

where

$$l_R(\mathbf{A}) = \frac{1}{N_0} \iint_{T_i}^{T_f} \tilde{r}^*(t)\tilde{h}_o(t, u:\mathbf{A})\tilde{r}(u)\, dt\, du \tag{306}$$

and

$$l_B(\mathbf{A}) = -\frac{1}{N_0}\int_{T_i}^{T_f} \xi_P(t:\mathbf{A})\, dt. \tag{307}$$

The filter $\tilde{h}_o(t, u:\mathbf{A})$ is specified by the equation

$$N_0\tilde{h}_o(t, u:\mathbf{A}) + \int_{T_i}^{T_f} \tilde{h}_o(t, z:\mathbf{A})\tilde{K}_{\tilde{s}}(z, u:\mathbf{A})\, dz = \tilde{K}_{\tilde{s}}(t, u:\mathbf{A}),$$

$$T_i \le t, u \le T_f. \tag{308}$$

The function $\xi_P(t:\mathbf{A})$ is the realizable minimum mean-square error in estimating $\tilde{s}(t:\mathbf{A})$, assuming that **A** is known. Notice that $l_B(\mathbf{A})$ is usually a function of **A** and cannot be neglected.

A second realization for $l_R(\mathbf{A})$ is obtained by factoring $\tilde{h}_0(t, u:\mathbf{A})$ as

$$\tilde{h}_o(t, u:\mathbf{A}) = \int_{T_i}^{T_f} \tilde{h}^{[1/2]*}(z, t:\mathbf{A})\tilde{h}^{[1/2]}(z, u:\mathbf{A})\, dz, \qquad T_i \le t, u \le T_f. \tag{309}$$

Then

$$l_R(\mathbf{A}) = \frac{1}{N_0}\int_{T_i}^{T_f} dz \left| \int_{T_i}^{T_f} \tilde{h}^{[1/2]}(z, t:\mathbf{A})\tilde{r}(t)\, dt \right|^2. \tag{310}$$

This is the familiar filter-squarer-integrator realization.

A third realization is

$$l_R(\mathbf{A}) = \frac{1}{N_0}\int_{T_i}^{T_f} (2\,\mathrm{Re}\,[\tilde{r}^*(t)\hat{\tilde{s}}_r(t:\mathbf{A})] - |\hat{\tilde{s}}_r(t:\mathbf{A})|^2)\, dt, \tag{311}$$

where $\hat{\tilde{s}}_r(t:\mathbf{A})$ is the realizable minimum mean-square estimate of $\tilde{s}(t:\mathbf{A})$, assuming that $\mathbf{A}$ is known. This is the optimum realizable filter realization.

We see that these realizations are analogous to the realizations encountered in the detection problem. Now we must find the realization for a set of values of $\mathbf{A}$ that span the range of possible parameter values. In the general case, we must use one of the approximate target models (e.g., a tapped delay-line model or a general orthogonal series model) developed in Section 13.2 to find the receiver. The computations are much more lengthy, because we must do them for many values of $\mathbf{A}$, but there are no new concepts involved.

In the following sections we consider a special case in which a more direct solution can be obtained. This is the low-energy-coherence (LEC) case, which we have encountered previously in Section 13.3.4.

There are four sections. In Section 13.4.1, we give the results for the general parameter estimation problem under the LEC condition. In Section 13.4.2, we consider the problem of estimating the amplitude of an otherwise known scattering function. In Section 13.4.3, we consider the problem of estimating the mean range and mean Doppler of a doubly-spread target. Finally, in Section 13.4.4, we summarize our results.

### 13.4.1 Estimation under LEC Conditions

The basic derivation is identical with that in Chapter 6, and so we simply state the results. The LEC condition is

$$\tilde{\lambda}_i(\mathbf{A}) \ll N_0, \tag{312a}$$

for all $\mathbf{A}$ in the parameter space and $i$. The $\tilde{\lambda}_i(\mathbf{A})$ are the eigenvalues of $\tilde{K}_{\tilde{s}}(t, u:\mathbf{A})$. Under these restrictions,

$$l(\mathbf{A}) = \left(\frac{1}{N_0}\right)^2 \iint_{T_i}^{T_f} \tilde{r}^*(t)\tilde{K}_{\tilde{s}}(t, u:\mathbf{A})\tilde{r}(u)\, dt\, du$$
$$- \left(\frac{1}{N_0}\right) \int_{T_i}^{T_f} \tilde{K}_{\tilde{s}}(t, t:\mathbf{A})\, dt - \frac{1}{2}\left(\frac{1}{N_0}\right)^2 \iint_{T_i}^{T_f} |\tilde{K}_{\tilde{s}}(t, u:\mathbf{A})|^2\, dt\, du. \tag{312b}$$

This result is analogous to (7.136).

For simplicity, we assume that $T_i = -\infty$ and $T_f = \infty$ in the remainder of our discussion. Observe that $\tilde{f}(t)$ has unit energy, so that $\tilde{s}(t:\mathbf{A})$ is a nonstationary process whose energy has a finite expected value. Specifically,

$$E\left[\int_{-\infty}^{\infty} |\tilde{s}(t:\mathbf{A})|^2\, dt\right] = E_t \iint_{-\infty}^{\infty} \tilde{S}_{DR}\{f, \lambda:\mathbf{A}\}\, df\, d\lambda. \tag{313a}$$

Thus, the infinite observation interval does not lead to a singular problem, as it would if $\tilde{s}(t:\mathbf{A})$ were stationary. Substituting (304) into (312*b*) gives

$$
\begin{aligned}
l(\mathbf{A}) = {} & \frac{E_t}{N_0^{\,2}} \int_{-\infty}^{\infty} dt \int_{-\infty}^{\infty} du \int_{-\infty}^{\infty} d\lambda\, \tilde{r}^*(t)\tilde{f}(t-\lambda)\tilde{K}_{DR}(t-u, \lambda:\mathbf{A})\tilde{f}^*(u-\lambda)\tilde{r}(u) \\
& - \frac{E_t}{N_0} \int_{-\infty}^{\infty} dt \int_{-\infty}^{\infty} d\lambda\, |\tilde{f}(t-\lambda)|^2\, \tilde{K}_{DR}(0, \lambda:\mathbf{A}) \\
& - \frac{E_t^{\,2}}{2N_0^{\,2}} \int_{-\infty}^{\infty} dt \int_{-\infty}^{\infty} du \int_{-\infty}^{\infty} d\lambda_1 \int_{-\infty}^{\infty} d\lambda_2 \tilde{f}(t-\lambda_1)\tilde{K}_{DR}(t-u, \lambda_1:\mathbf{A}) \\
& \times \tilde{f}^*(u-\lambda_1)\tilde{f}^*(t-\lambda_2)\tilde{K}_{DR}(t-u, \lambda_2:\mathbf{A})\tilde{f}(u-\lambda_2).
\end{aligned} \tag{313b}
$$

The last two terms are bias terms, which can be written in a simpler manner. We can write the second term in (313*b*) as

$$
\begin{aligned}
l_B^{[1]}(\mathbf{A}) &= -\frac{E_t}{N_0} \int_{-\infty}^{\infty} d\lambda \quad {}_{DR}(0, \lambda:\mathbf{A}) \int_{-\infty}^{\infty} |\tilde{f}(t-\lambda)|^2\, dt \\
&= -\frac{E_t}{N_0} \int_{-\infty}^{\infty} d\lambda\, \tilde{K}_{DR}(0, \lambda:\mathbf{A}) = -\frac{\bar{E}_r(\mathbf{A})}{2N_0}.
\end{aligned} \tag{314}
$$

Here, $\bar{E}_r(\mathbf{A})$ is the average received energy written as a function of the unknown parameter $\mathbf{A}$.

To simplify the third term, we use the two-frequency correlation function $\tilde{R}_{DR}(\tau, v:\mathbf{A})$. Recall from (21) that

$$
\tilde{K}_{DR}(\tau, \lambda:\mathbf{A}) = \int_{-\infty}^{\infty} \tilde{R}_{DR}\{\tau, v:\mathbf{A}\} e^{j2\pi v\lambda}\, dv. \tag{315}
$$

Using (315) in the last term of (313*b*) and performing a little manipulation, we find that the third term can be written as

$$
l_B^{[2]}(\mathbf{A}) = -\frac{E_t^{\,2}}{2N_0^{\,2}} \iint_{-\infty}^{\infty} dx\, dy\, \theta\{x, y\}\, |\tilde{R}_{DR}\{x, y:\mathbf{A}\}|^2, \tag{316}
$$

where $\theta\{\cdot, \cdot\}$ is the ambiguity function of $\tilde{f}(t)$. We denote the sum of the last two terms in (313*b*) as $l_B(\mathbf{A})$. Thus,

$$
l_B(\mathbf{A}) = -\frac{\bar{E}_r(\mathbf{A})}{2N_0} - \frac{E_t^{\,2}}{4N_0^{\,2}} \int_{-\infty}^{\infty} dx \int_{-\infty}^{\infty} dy\, \theta\{x, y\}\, |\tilde{R}_{DR}\{x, y:\mathbf{A}\}|^2. \tag{317}
$$

The last step is to find a simpler realization of the first term in (313*b*). The procedure here is identical with that in Section 13.3.4. We factor $\tilde{K}_{DR}(t-u, \lambda:\mathbf{A})$, using the relation

$$
\tilde{K}_{DR}(t-u, \lambda:\mathbf{A}) = \int_{-\infty}^{\infty} \tilde{K}_{DR}^{[1/2]*}(z-t, \lambda:\mathbf{A})\tilde{K}_{DR}^{[1/2]}(z-u, \lambda:\mathbf{A})\, dz. \tag{318}
$$

Since the time interval is infinite and $\tilde{K}_{DR}(\tau, \lambda:\mathbf{A})$ is stationary, we can find a realizable (with respect to $\tau$) $\tilde{K}_{DR}^{[1/2]}(\tau, \lambda:\mathbf{A})$ by spectrum factorization. In the frequency domain

$$\tilde{S}_{DR}^{[1/2]}\{f, \lambda\} = [\tilde{S}_{DR}\{f, \lambda\}]^{+}. \tag{319}$$

Substituting (318) into (313*b*) and denoting the first term by $l_R(\mathbf{A})$, we have

$$l_R(\mathbf{A}) = \frac{E_t}{N_0^2}\int_{-\infty}^{\infty} dz \int_{-\infty}^{\infty} d\lambda \left| \int_{-\infty}^{\infty} \tilde{K}_{DR}^{[1/2]}(z-u, \lambda:\mathbf{A}) \tilde{f}^*(u-\lambda)\tilde{r}(u)\, du \right|^2. \tag{320}$$

Combining (320) and (317) gives an expression for $l(\mathbf{A})$, which is

$$\boxed{\begin{aligned} l(\mathbf{A}) = {} & \frac{E_t}{N_0^2}\int_{-\infty}^{\infty} dz \int_{-\infty}^{\infty} d\lambda \left| \int_{-\infty}^{\infty} \tilde{K}_{DR}^{[1/2]}(z-u, \lambda:\mathbf{A}) \tilde{f}^*(u-\lambda)\tilde{r}(u)\, du \right|^2 \\ & - \frac{\bar{E}_r(\mathbf{A})}{N_0} - \frac{E_t^2}{2N_0^2}\int_{-\infty}^{\infty} d\tau \int_{-\infty}^{\infty} dv\, \theta\{\tau, v\}\, |\tilde{R}_{DR}\{\tau, v:\mathbf{A}\}|^2. \end{aligned}} \tag{321}$$

For *each* value of $\mathbf{A}$ we can realize $l_R(\mathbf{A})$ approximately by sampling in $\lambda$ and replacing the integral in $\lambda$ by a sum. We then add $l_B(\mathbf{A})$ to obtain an approximate likelihood function. This realization is an obvious modification of the structure in Fig. 13.30. Notice that we must carry out this calculation for a set of values of $\mathbf{A}$, so that the entire procedure is quite tedious.

We now have the receiver structure specified. The performance analysis for the general case is difficult. The Cramér-Rao bound gives a bound on the variance of any unbiased estimate. For a *single* parameter, we differentiate (312*b*) to obtain

$$\operatorname{Var}[a - A] \geq \frac{N_0^2}{\displaystyle\iint_{-\infty}^{\infty} dt\, du\, |[\partial \tilde{K}_{\tilde{s}}(t, u:A)]/\partial A|^2}. \tag{322}$$

For multiple parameters, we modify (7.155) to obtain the elements in the information matrix as

$$J_{ij}(\mathbf{A}) = \frac{1}{N_0^2}\operatorname{Re}\iint_{T_i}^{T_f} dt\, du\, \frac{\partial \tilde{K}_{\tilde{s}}(t, u:\mathbf{A})}{\partial A_i} \frac{\partial \tilde{K}_{\tilde{s}}^*(t, u:\mathbf{A})}{\partial A_j}. \tag{323}$$

Substituting (304) into (323) gives the $\tilde{J}_{ij}(\mathbf{A})$ for the doubly-spread target. Using (316), we can write $\tilde{J}_{ij}(\mathbf{A})$ compactly as

$$\boxed{\tilde{J}_{ij}(\mathbf{A}) = \frac{E_t^2}{N_0^2} \operatorname{Re} \iint_{-\infty}^{\infty} \frac{\partial \tilde{R}_{DR}\{x, y:\mathbf{A}\}}{\partial A_i} \theta\{x, y\} \frac{\partial \tilde{R}_{DR}^*\{x, y:\mathbf{A}\}}{\partial A_j} \, dx\, dy.} \tag{324}$$

The principal results of this section are the expressions in (321) and (324). They specify the optimum receiver and the performance bound, respectively. We next consider two typical estimation problems.

### 13.4.2 Amplitude Estimation

In this section we consider the problem of estimating the amplitude of an otherwise known scattering function.† We assume that

$$\tilde{K}_{DR}(t - u, \lambda : A) = A\tilde{K}_{DR}(t - u, \lambda), \tag{325}$$

where $\tilde{K}_{DR}(t - u, \lambda)$ is normalized such that

$$\int_{-\infty}^{\infty} \tilde{K}_{DR}(0, \lambda)\, d\lambda = 1. \tag{326}$$

Thus,

$$A = \frac{\bar{E}_r}{E_t}. \tag{327}$$

The covariance function of the received signal process is

$$\tilde{K}_{\tilde{s}}(t, u : A) = E_t A \int_{-\infty}^{\infty} \tilde{f}(t - \lambda) \tilde{K}_{DR}(t - u, \lambda) \tilde{f}^*(u - \lambda)\, d\lambda \triangleq A\tilde{K}_{\tilde{s}}(t, u). \tag{328}$$

The parameter $A$ is an unknown positive number.

In this case, the likelihood function in (312*b*) has a single maximum, which is located at

$$\hat{a}_0 = \frac{\displaystyle\iint_{-\infty}^{\infty} \tilde{r}^*(t) \tilde{K}_{\tilde{s}}(t, u) \tilde{r}(u)\, dt\, du - N_0 \int_{-\infty}^{\infty} \tilde{K}_{\tilde{s}}(t, t)\, dt}{\displaystyle\iint_{-\infty}^{\infty} |\tilde{K}_{\tilde{s}}(t, u)|^2\, dt\, du}. \tag{329}$$

† This problem was solved originally by Price [39]. Our problem is actually a degenerate case of the problem he considers.

Since $\hat{a}_0$ may be negative, the maximum likelihood estimate is

$$\hat{a}_{ml} = \max\,[0, \hat{a}_0]. \tag{330}$$

We discussed the truncation problem in detail in Section 7.1.2. For simplicity we assume that the parameters are such that the truncation effect is negligible. Using (326), (328), and (315) in (329), we obtain

$$\hat{a}_0 = \frac{\displaystyle\int_{-\infty}^{\infty} dt \int_{-\infty}^{\infty} du \int_{-\infty}^{\infty} d\lambda\, \tilde{r}^*(t)\tilde{f}(t-\lambda)\tilde{K}_{DR}(t-u,\lambda)\tilde{f}^*(u-\lambda)\tilde{r}(u) - N_0}{\displaystyle\int_{-\infty}^{\infty} d\tau \int_{-\infty}^{\infty} dv\, \theta\{\tau, v\}\, |\tilde{R}_{DR}\{\tau, v\}|^2}. \tag{331}$$

It follows easily that $\hat{a}_0$ is unbiased. An approximate receiver realization is shown in Fig. 13.33.

If we neglect the bias on $\hat{a}_{ml}$, we can bound its normalized variance by $J_n^{-1}$, where $J_n$ is obtained from (324) as

$$J_n \triangleq \frac{J(A)}{A^2} = \frac{E_t^{\,2}}{N_0^{\,2}} \iint_{-\infty}^{\infty} \theta\{\tau, v\}\, |\tilde{R}_{DR}\{\tau, v\}|^2\, d\tau\, dv. \tag{332}$$

It is worthwhile observing that we can compute the variance of $\hat{a}_0$ exactly (see Problem 13.4.4). The result is identical with $J_n^{-1}$, except for a term that can be neglected when the LEC condition holds.

To illustrate the ideas involved, we consider an example.

**Example.** We assume that the target has the doubly Gaussian scattering function in Fig. 13.4. Then

$$\tilde{S}_{DR}\{f, \lambda\} = \frac{1}{2\pi BL} \exp\left\{-\frac{f^2}{2B^2} - \frac{\lambda^2}{2L^2}\right\}, \qquad -\infty < f, \lambda < \infty \tag{333}$$

and

$$\tilde{R}_{DR}\{\tau, v\} = \exp\left(-\frac{(2\pi B)^2\tau^2}{2} - \frac{(2\pi L)^2 v^2}{2}\right), \qquad -\infty < \tau, v < \infty. \tag{334}$$

To simplify the algebra, we assume that $\tilde{f}(t)$ is a Gaussian pulse.

$$\tilde{f}(t) = \left(\frac{1}{\pi T^2}\right)^{1/4} \exp\left(-\frac{t^2}{2T^2}\right), \qquad -\infty < t < \infty. \tag{335}$$

Then, from (10.28),

$$\theta\{\tau, f\} = \exp\left[-\frac{1}{2}\left(\frac{\tau^2}{T^2} + T^2(2\pi f)^2\right)\right], \qquad -\infty < \tau, f < \infty. \tag{336}$$

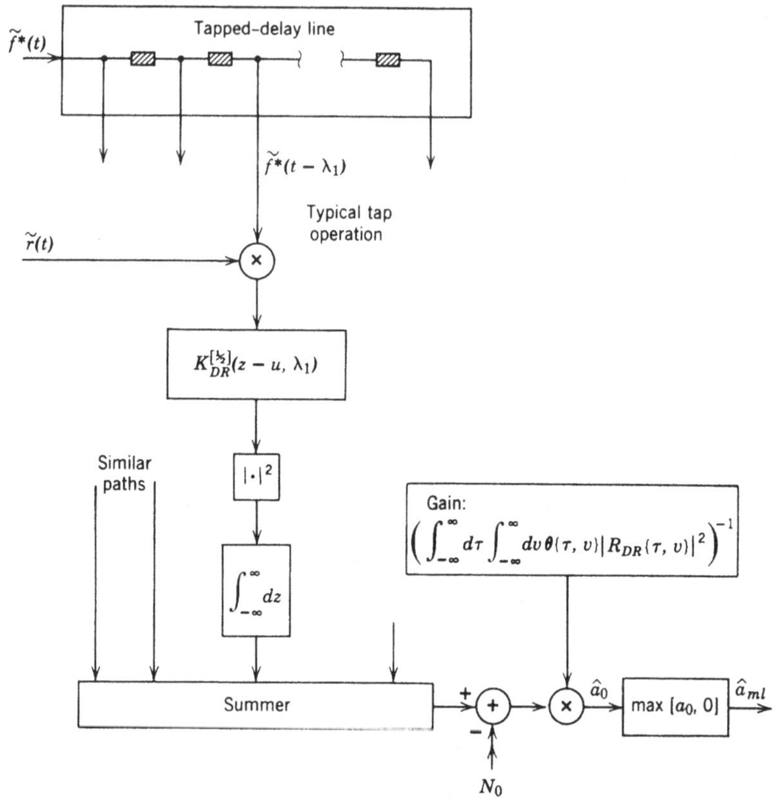


**Fig. 13.33 Maximum likelihood estimator of the amplitude of a scattering function under LEC conditions.**

Substituting in (332), we have

$$J_n = \left(\frac{E_t}{N_0}\right)^2 \iint_{-\infty}^{\infty} \exp\left\{-\frac{1}{2}\left[\tau^2\left(\frac{1}{T^2} + 2(2\pi B)^2\right) + v^2((2\pi T)^2 + 2(2\pi L)^2)\right]\right\} d\tau\, dv. \tag{337}$$

Integrating, we obtain

$$J_n = \left(\frac{E_t}{N_0}\right)^2 \left[\left(1 + 2\left(\frac{L}{T}\right)^2\right)(1 + 2(2\pi BT)^2)\right]^{-1/2}. \tag{338}$$

Looking at (338), we see that $J_n$ will be maximized [and therefore the variance bound in (322) will be minimized] by some intermediate value of $T$. Specifically, the maximum occurs at

$$T = \sqrt{\frac{L}{2\pi B}}. \tag{339}$$

Comparing (339) and (281), we see that if we let

$$k = 2\pi B, \tag{340}$$

then this value of $T$ corresponds to the point of minimum diversity in the output signal process. (Notice that $B$ and $k$ have different meanings in the two scattering functions.)

Intuitively we would expect the minimum diversity point to be optimum, because of the original LEC assumption. This is because there is an optimum "energy per eigenvalue $/N_0$" value in the general amplitude estimation problem (see Problem 13.4.8). The LEC condition in (311) means that we already have the energy distributed among too many eigenvalues. Thus we use the fewest eigenvalues possible. When the LEC condition does not hold, a curve of the form shown in Fig. 13.24*b* would be obtained. If we use the value of $T$ in (339), then

$$J_n = \left(\frac{E_t}{N_0}\right)^2 (1 + 4\pi BL)^{-1} \tag{341}$$

and

$$\text{Var}\,[\hat{a}_{ml} - A] \geq \left(\frac{N_0}{E_t}\right)^2 (1 + 4\pi BL). \tag{342}$$

We see that the variance bound increases linearly with the $BL$ product for $BL > 1$. This linear behavior with $BL$ also depends on the LEC condition and does not hold in general.

This completes our discussion of the amplitude estimation problem. We were able to obtain a closed-form solution for $\hat{a}_{ml}$ because $l(A)$ had a unique maximum. We now consider a different type of estimation problem.

### 13.4.3 Estimation of Mean Range and Doppler

In this subsection we consider the problem of estimating the mean range and mean Doppler of a doubly-spread target. A typical configuration in the $\tau, f$ plane is shown in Fig. 13.34. We denote the mean range by

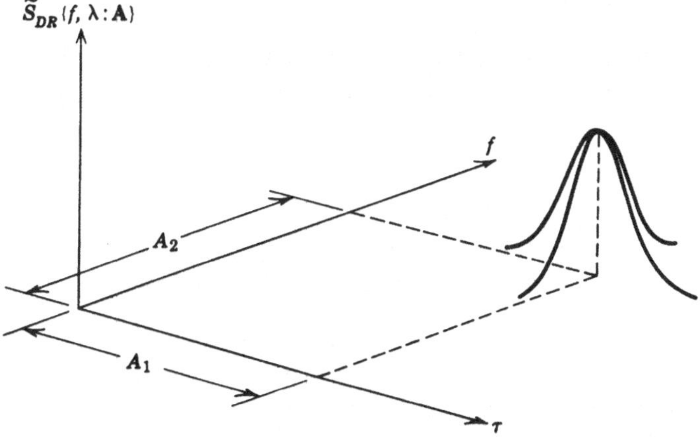


**Fig. 13.34 Target location in $\tau, f$ plane.**

$A_1$, and the mean Doppler by $A_2$. The scattering function is denoted by

$$\tilde{S}_{DR}\{f, \lambda:\mathbf{A}\} = \tilde{S}_{D_0R_0}\{f - A_2, \lambda - A_1\}, \tag{343}$$

where the scattering function on the right side of (343) is defined to have zero mean range and zero mean Doppler. Another function that will be useful is the two-frequency correlation function, which can be written as

$$\tilde{R}_{DR}\{\tau, v:\mathbf{A}\} = \tilde{R}_{D_0R_0}\{\tau, v\} \exp\left[-j2\pi v A_1 + j2\pi\tau A_2\right]. \tag{344}$$

The purpose of (343) and (344) is to express the parametric dependence explicitly.

The returned signal is given by (303). To find the maximum likelihood estimate of $\mathbf{A}$, we first divide the $\tau$, $\omega$ plane into a set of range-Doppler cells. We denote the coordinates of the center of the $i$th cell as $\mathbf{A}_i$. We next construct $l(\mathbf{A}_i)$ for each cell and choose that value of $\mathbf{A}_i$ where $l(\mathbf{A}_i)$ is maximum.

First, we consider the general case and *do not* impose the LEC condition. Then, $l(\mathbf{A}_i)$ is given by (305)–(307). As before, we let $T_i = -\infty$ and $T_f = \infty$. Looking at (307), we see that $\xi_P(t:\mathbf{A})$ does not depend on the mean range or Doppler, so that we do not need to compute $l_B(\mathbf{A})$. Thus,

$$l(\mathbf{A}_i) = l_R(\mathbf{A}_i) = \frac{1}{N_0} \iint_{-\infty}^{\infty} \tilde{r}^*(t)\tilde{h}_o(t, u:\mathbf{A}_i)\tilde{r}(u)\, dt\, du, \tag{345}$$

where $\tilde{h}_o(t, u:\mathbf{A}_i)$ is specified by (308) with $\mathbf{A} = \mathbf{A}_i$. For each cell we must solve (308) [or find one of the equivalent forms given in (309)–(311)]. Actually to carry out the solution, we would normally have to use one of the orthogonal series models in Section 13.3.2.

In analyzing the performance, we must consider both global and local accuracy. To study the global accuracy problem we use the spread ambiguity function that we defined in (11.181). For doubly-spread targets the definition is

$$\theta_{\Omega_{DR}}(\mathbf{A}_a, \mathbf{A}) = \int_{-\infty}^{\infty} dt \int_{-\infty}^{\infty} du \int_{-\infty}^{\infty} d\lambda \tilde{f}^*(t - \lambda) \times \tilde{h}_o(t, u:\mathbf{A})\tilde{K}_{DR}(t - u, \lambda:\mathbf{A}_a)\tilde{f}(u - \lambda), \tag{346}$$

where $\mathbf{A}_a$ corresponds to the actual mean range and mean Doppler of the target. To analyze the local accuracy, we use the Cramér-Rao bound. There is no conceptual difficulty in carrying out these analyses, but the calculations are involved.

When the LEC condition is satisfied, the solution is appreciably simpler. From (320), the likelihood function is

$$l_R(\mathbf{A}_i) = \frac{E_t}{N_0^{\,2}} \int_{-\infty}^{\infty} dz \int_{-\infty}^{\infty} d\lambda \left| \int_{-\infty}^{\infty} \tilde{K}_{DR}^{[1/2]}(z-u, \lambda:\mathbf{A}_i)\tilde{f}^*(u-\lambda)\tilde{r}(u)\, du \right|^2. \tag{347}$$

The spread ambiguity function under LEC conditions is

$$\boxed{\begin{aligned}\theta_{\Omega_{DR},\mathrm{LEC}}(\mathbf{A}_a, \mathbf{A}) &= \frac{E_t^2}{N_0^{\,2}} \int_{-\infty}^{\infty} dt \int_{-\infty}^{\infty} du \int_{-\infty}^{\infty} d\lambda_1 \int_{-\infty}^{\infty} d\lambda_2 \tilde{f}(t-\lambda_1) \\ &\quad \times \tilde{K}_{DR}(t-u, \lambda_1:\mathbf{A})\tilde{f}^*(u-\lambda_1)\tilde{f}^*(t-\lambda_2) \\ &\quad \times \tilde{K}_{DR}^*(t-u, \lambda_2:\mathbf{A}_a)\tilde{f}(u-\lambda_2) \\ &= \frac{E_t^2}{N_0^{\,2}} \iint_{-\infty}^{\infty} dx\, dy \tilde{R}_{DR}\{x, y:\mathbf{A}\}\theta\{x, y\}\tilde{R}_{DR}^*\{x, y:\mathbf{A}_a\}.\end{aligned}} \tag{348}$$

Notice that

$$\tilde{J}_{ij}(\mathbf{A}) = \left.\frac{\partial^2 \theta_{\Omega_{DR},\mathrm{LEC}}(\mathbf{A}_a, \mathbf{A})}{\partial A_i\, \partial A_{aj}}\right|_{\mathbf{A}=\mathbf{A}_a}, \tag{349}$$

which is identical with (324). To evaluate the Cramér-Rao bound, we use (344) in (324) to obtain

$$\tilde{J}_{11}(\mathbf{A}) = \frac{E_t^2}{N_0^{\,2}} \iint_{-\infty}^{\infty} (2\pi v)^2 \theta\{\tau, v\}\, |\tilde{R}_{D_0R_0}\{\tau, v\}|^2\, d\tau\, dv, \tag{350}$$

$$\tilde{J}_{12}(\mathbf{A}) = -\frac{E_t^2}{N_0^{\,2}} \iint_{-\infty}^{\infty} (2\pi)^2 \tau v \theta\{\tau, v\}\, |\tilde{R}_{D_0R_0}\{\tau, v\}|^2\, d\tau\, dv, \tag{351}$$

and

$$\tilde{J}_{22}(\mathbf{A}) = \frac{E_t^2}{N_0^{\,2}} \iint_{-\infty}^{\infty} (2\pi \tau)^2 \theta\{\tau, v\}\, |\tilde{R}_{D_0R_0}\{\tau, v\}|^2\, d\tau\, dv. \tag{352}$$

As we would expect, the error performance depends on both the signal ambiguity function and the target-scattering function. Some typical situations are analyzed in the problems.

### 13.4.4 Summary

In this section we studied parameter estimation for doubly-spread targets. We first formulated the general estimation problem and developed the expressions for the likelihood ratio. The resulting receiver was closely related to those encountered earlier in the detection problem.

In the remainder of the section, we emphasized the low-energy-coherence case. In Section 13.4.1 we developed the expressions for the likelihood function and the Cramér-Rao bound under the LEC assumption. In Section 13.4.2 we found an explicit solution for the estimate of the amplitude of the scattering function. A simple example illustrated the effect of the pulse length and the $BL$ product. In Section 13.4.3, we studied the problem of estimating the mean range and Doppler of a doubly-spread target. This problem is a generalization of the range-Doppler estimation problem that we studied in Chapter 10.

Our goal in this section was to illustrate some of the important issues in the estimation problem. Because of the similarity to the detection problem, a detailed discussion was not necessary.

## 13.5 SUMMARY OF DOUBLY-SPREAD TARGETS AND CHANNELS

In this chapter we have studied targets and channels that are spread in both range and Doppler. The complex envelope of the signal returned from the target is

$$\tilde{s}(t) = \sqrt{E_t} \int_{-\infty}^{\infty} \tilde{f}(t - \lambda)\tilde{b}(t, \lambda)\, d\lambda. \tag{353}$$

The target reflection process is a sample function of zero-mean complex Gaussian random processes, which can be characterized in two ways:

1. By a scattering function $\tilde{S}_{DR}\{f, \lambda\}$ or an equivalent form such as $\tilde{K}_{DR}(\tau, \lambda)$, $\tilde{R}_{DR}\{\tau, v\}$, or $\tilde{P}_{DR}\{f, v\}$.
2. By a distributed state-variable description in which the state equations are ordinary differential equations containing the spatial variable $\lambda$ as a parameter and $\tilde{s}(t)$ is related to the state vector by a modulation functional.

After formulating the model and discussing its general characteristics, we looked at three areas in which we encounter doubly-spread targets.

In Section 13.2 we discussed the problem of resolution in a dense environment. Here, the desired signal was a nonfluctuating point target and the interference was a doubly-spread environment. We examined both conventional and optimum receivers and compared their performance. We found that when a conventional matched filter was used, the spread interference entered through a double convolution of the signal ambiguity function and the target-scattering function. As in the discrete resolution problem, examples indicated that proper signal design is frequently more important than optimum receiver design.

In Section 13.3 we discussed the problem of detecting the return from a doubly-spread target and the problem of digital communication over a doubly-spread channel. After formulating the general problem, we developed several approximate target/channel models using orthogonal series expansions. The purpose of these models was to reduce the problem to a form that we could analyze. The tapped-delay line model was the easiest to implement, but the general orthogonal series model offered some computation advantages. We next studied the binary communication problem. For underspread channels we found signals that enabled us to approach the performance bound for any system. For overspread channels we could only approach the bound for large $\bar{E}_r/N_0$ with the simple signals we considered. To verify our intuitive argument, we carried out a detailed performance analysis for a particular system. The effect of the signal parameters and the scattering function parameters on the performance of a binary communication system was studied. Finally, we indicated the extensions to several related problems.

In Section 13.4 we studied the problem of estimating the parameters of a doubly-spread target. We first formulated the general estimation problem and noted its similarity to the detection problem in Section 13.3. We then restricted our attention to the LEC case. Two particular problems, amplitude estimation and mean range and Doppler estimation, were studied in detail.

There are two important problems which we have not considered that should be mentioned. The first is the problem of measuring the instantaneous behavior of $\tilde{b}(t, \lambda)$. We encountered this issue in the estimator-correlator receiver but did not discuss it fully. The second problem is that of measuring (or estimating) the scattering function of the target or channel. We did not discuss this problem at all. An adequate discussion of these problems would take us too far afield; the interested reader should consult the references (e.g., [39]–[52] and [66]–[69]).

This completes our discussion of doubly-spread targets and channels. In the next chapter we summarize our discussion of the radar-sonar problem.

## 13.6 PROBLEMS

### P.13.1 Target Models

**Problem 13.1.1.** Read the discussion in the Appendix of [5]. Verify that the scattering function of a rough rotating sphere is as shown in Fig. 13.2.

**Problem 13.1.2.** Consider the target shown in Fig. P.13.1. The antenna pattern is constant over the target dimensions. The discs are perpendicular to the axis of propagation (assume plane wave propagation). The carrier is at $f_c$ cps. The dimensions $x_0$, $y_0$, $d_0$, and $d_1$ are in meters. The rates of rotation, $g_0$ and $g_1$, are in revolutions per second.

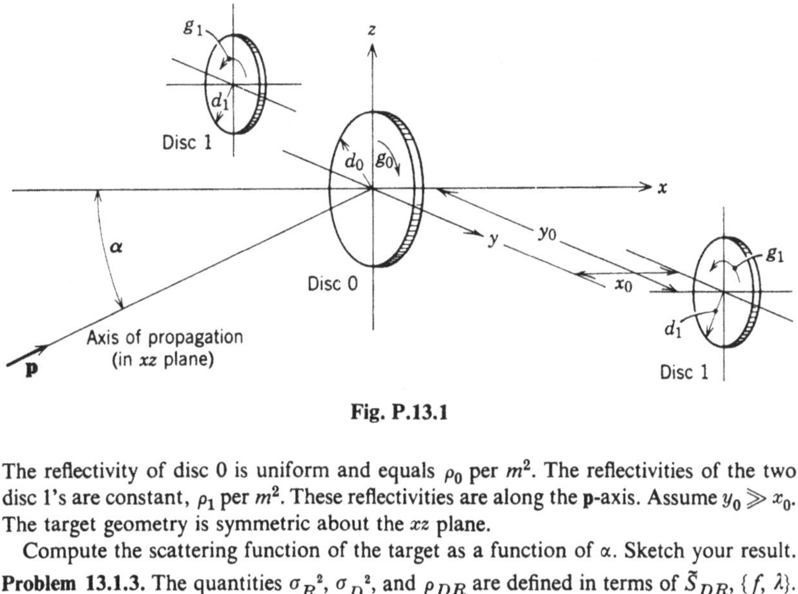


**Fig. P.13.1**

The reflectivity of disc 0 is uniform and equals $\rho_0$ per $m^2$. The reflectivities of the two disc 1's are constant, $\rho_1$ per $m^2$. These reflectivities are along the **p**-axis. Assume $y_0 \gg x_0$. The target geometry is symmetric about the $xz$ plane.

Compute the scattering function of the target as a function of $\alpha$. Sketch your result.

**Problem 13.1.3.** The quantities $\sigma_R^2$, $\sigma_D^2$, and $\rho_{DR}$ are defined in terms of $\tilde{S}_{DR}\{f, \lambda\}$. Find equivalent expressions for them in terms of $\tilde{P}_{DR}\{f, v\}$, $\tilde{R}_{DR}\{\tau, v\}$, and $\tilde{K}_{DR}(\tau, \lambda)$.

**Problem 13.1.4.** Assume that

$$\tilde{S}_{DR}\{f, \lambda\} = \frac{2\sigma_b^2}{2\pi BL} \exp\left[-\frac{f^2}{2B^2} - \frac{\lambda^2}{2L^2}\right], \qquad -\infty < f < \infty, \quad -\infty < \lambda < \infty.$$

1. Find $\tilde{P}_{DR}\{f, v\}$, $\tilde{R}_{DR}\{\tau, v\}$, and $\tilde{K}_{DR}(\tau, \lambda)$.
2. Calculate $\sigma_R^2$, $\sigma_D^2$, and $\rho_{DR}$.

**Problem 13.1.5.** Assume that

$$\tilde{S}_{DR}\{f, \lambda\} = \frac{2\sigma_b^2}{2\pi BL(1 - \rho^2)^{1/2}} \times \exp\left[-\frac{L^2(f - m_D)^2 - 2BL\rho(f - m_D)(\lambda - m_R) + B^2(\lambda - m_R)^2}{2B^2L^2(1 - \rho^2)}\right].$$

1. Find $\tilde{P}_{DR}\{f, v\}$, $\tilde{R}_{DR}\{\tau, v\}$, and $\tilde{K}_{DR}(\tau, \lambda)$.
2. Find the five quantities defined in (14)–(19).

**Problem 13.1.6.** Assume that

$$\tilde{S}_{DR}\{f, \lambda\} = \frac{4k\sigma_b^2}{L[(2\pi f)^2 + k^2]}, \qquad -\infty < f < \infty, \qquad 0 < \lambda < L.$$

Find $\tilde{P}_{DR}\{f, v\}$, $\tilde{R}_{DR}\{\tau, v\}$, and $\tilde{K}_{DR}\{\tau, \lambda\}$.

**Problem 13.1.7.** Assume that

$$\tilde{S}_{DR}\{f, \lambda\} = \begin{cases} \dfrac{2\sigma_b^2}{BL}, & |f| \le \dfrac{B}{2}, \quad |\lambda| \le \dfrac{L}{2}, \\ 0, & \text{elsewhere.} \end{cases}$$

1. Find $\tilde{P}_{DR}\{f, v\}$, $\tilde{R}_{DR}\{\tau, v\}$, and $\tilde{K}_{DR}\{\tau, \lambda\}$.
2. Find $\sigma_R^2$ and $\sigma_D^2$.

**Problem 13.1.8.** Assume that

$$\tilde{S}_{DR}\{f, \lambda\} = \left[\frac{8\sqrt{2}\sigma_b^2 \sin^2(\pi\lambda/L)}{kL[(2\pi f/k)^4 + 1]}\right], \qquad -\infty < f < \infty, \qquad 0 \le \lambda \le L.$$

1. Find $\tilde{P}_{DR}\{f, v\}$, $\tilde{R}_{DR}\{\tau, v\}$, and $\tilde{K}_{DR}(\tau, \lambda)$.
2. Find $\sigma_D^2$ and $\sigma_R^2$.

**Problem 13.1.9.** Consider the target process whose scattering function is given in Problem 13.1.8.

1. Describe this process with a differential-equation model.
2. Describe the received signal process $\tilde{s}(t)$ in terms of the results in part 1.

**Problem 13.1.10.** We frequently use the doubly-Gaussian scattering function in Problem 13.1.4. Construct a differential-equation model to represent it approximately. (*Hint:* Recall Case 2 on page I-505.)

**Problem 13.1.11.** Assume that the scattering function is

$$\tilde{S}_{DR}\{f, \lambda\} = \frac{a(\lambda)}{[(j2\pi f)^2 + \tilde{k}_1^2(\lambda)][(j2\pi f)^2 + \tilde{k}_2^2(\lambda)]}, \qquad -\infty < f < \infty, \quad 0 < \lambda < L.$$

1. Sketch the scattering function for various allowable $\tilde{k}_1(\lambda)$, $\tilde{k}_2(\lambda)$, and $a(\lambda)$.
2. Write out the differential equations that characterize this target. (*Hint:* Recall Example 2 in the Appendix, page 594.)

## P.13.2 Detection in Reverberation

### Conventional Receivers

In Problem 13.2.1–13.2.9, we use the model in (69)–(72) and assume that a conventional receiver is used.

**Problem 13.2.1.** The transmitted signal in given in (10.43). The scattering function is

$$\tilde{S}_{DR}\{f, \lambda\} = \frac{1}{2\pi BL(1-\rho^2)^{1/2}} \exp\left[-\frac{L^2 f^2 - 2BL\rho f\lambda + B^2\lambda^2}{2B^2L^2(1-\rho^2)}\right].$$

Find $\rho_r$ [see (13.83)] as a function of $E_t$, $N_0$, $B$, $L$, $\rho$, and $T$.

**Problem 13.2.2.** Consider the reverberation model in (132). Assume that

$$\tilde{f}(t) = \sqrt{2\alpha}\, e^{-\alpha t} u_{-1}(t).$$

Calculate $\rho_r$ as a function of $E_t$, $P_c$, $f_d$, and $\alpha$.

**Problem 13.2.3.** Consider the reverberation model in (132).

1. Verify that

$$\rho_r = \frac{E_t P_c}{N_0} \int_{-\infty}^{\infty} \tilde{S}_{\tilde{f}}^{\,2}\{f\}\, df$$

for a zero-velocity target.

2. Choose $\tilde{S}_{\tilde{f}}\{f\}$ subject to the energy constant

$$\int_{-\infty}^{\infty} \tilde{S}_{\tilde{f}}\{f\}\, df = 1,$$

so that $\rho_r$ is minimized.

**Problem 13.2.4.** Consider the reverberation model in (132).

1. Verify that

$$\rho_r = \frac{E_t P_c}{N_0} \int_{-\infty}^{\infty} \tilde{S}_{\tilde{f}}\{f\} \tilde{S}_{\tilde{f}}\{f - f_d\}\, df \qquad \text{(P.1)}$$

for a target with Doppler shift $f_d$.

2. What type of constraints must be placed on $\tilde{f}(t)$ in order to obtain a meaningful result when we try to minimize $\rho_r$?

3. Assume that we require

$$\int_{-\infty}^{\infty} f^2 \tilde{S}_{\tilde{f}}\{f\}\, df = \sigma_w^{\,2}. \qquad \text{(P.2)}$$

Minimize $\rho_r$ subject to the constraint in (P.2) and an energy constraint.

**Problem 13.2.5.** Assume that we have the *constant-height* reverberation scattering function shown in Fig. P.13.2. The signal is the pulse train shown in Fig. 10.9.

1. Show how to choose $T_s$, $T_p$, and $n$ to minimize the effect of the reverberation.
2. Calculate $\rho_r$ (13.83) for the signal parameters that you selected.

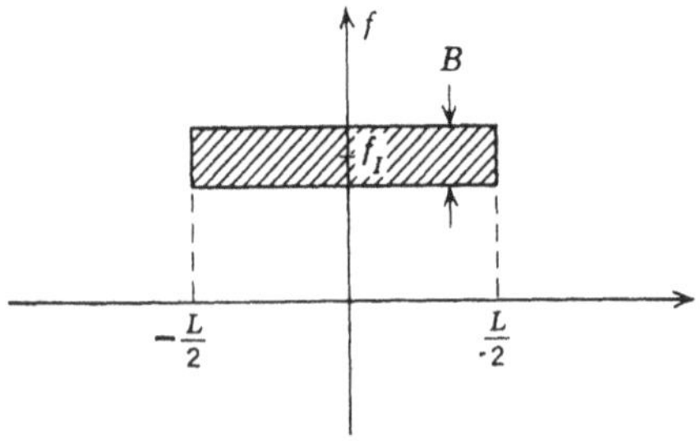


**Fig. P.13.2**

**Problem 13.2.6.** Consider the signal given in (10.145), which has $3N$ parameters to choose. Consider the scattering function in Fig. P.13.2.

1. Write an expression for $\rho_r$ (13.83) in terms of the signal parameters and $B$, $L$, and $f_I$. Assume that $L/T_s$ is an integer for simplicity.
2. Consider the special case of (10.145) in which $\omega_n = 0$, and define

$$\tilde{\mathbf{a}} = \begin{bmatrix} a_1 e^{j\theta_1} \\ \cdot \\ \cdot \\ \cdot \\ a_N e^{j\theta_N} \end{bmatrix}.$$

Express $\rho_r$ in terms of $\tilde{\mathbf{a}}$.

3. We want to minimize $\rho_r$ by choosing $\tilde{\mathbf{a}}$ properly. Formulate the optimization problems and derive the necessary equations.

**Problem 13.2.7.** Repeat parts 2 and 3 of Problem 13.2.6 for the following special cases:

1. We require

$$\omega_n = 0, \qquad n = 1, \dots, N,$$
$$a_n = 1, \qquad n = 1, \dots, N.$$

2. We require

$$\theta_n = 0, \qquad n = 1, \dots, N,$$
$$a_n = 1 \quad \text{or} \quad 0, \qquad n = 1, \dots, N.$$

**Problem 13.2.8.** We want to estimate the range and Doppler of a nonfluctuating point target in the presence of reverberation. The conventional receiver in Section 10.2 is used

Derive a bound on the variance of the range and Doppler estimation errors.

**Problem 13.2.9** In Section 12.3 we developed duality theory. These ideas are also useful in reverberation problems. Assume that a conventional receiver is used.

Derive the dual of the result in (83).

OPTIMUM RECEIVERS

**Problem 13.2.10.** Consider the model in (101)–(107). One procedure for solving (107) is to approximate the integrals with finite sums. Carry out the details of this procedure and obtain a matrix equation specifying $\tilde{g}(t_i)$, $i = 1, \dots, N$. Discuss how you selected the sampling interval and the resulting computational requirements.

**Problem 13.2.11.** Consider the model in (101)–(107). The complex envelope of the transmitted signal is given by (10.25), and the scattering function is given by (13.333). Assume that $\tau_d = \omega_d = 0$.

1. Find a series solution to (107) by using Mehler's expansion (e.g., [53] or [54].)
2. Evaluate $\Delta_o$.

**Problem 13.2.12.** Generalize the result in Problem 13.2.11 to include a nonzero target range and Doppler, the transmitted signal in (10.43), and the scattering function in Problem 13.2.1.

**Problem 13.2.13.** One procedure for obtaining an approximate solution to (107) is to model $\tilde{S}_{DR}\{f, \lambda\}$ as a piecewise constant function and then replace the each piecewise constant segment by an impulse that is located at the center of the segment with the same volume as the segment. This reduces the problem to that in Section 10.5.

1. Discuss how one selects the grid dimensions.
2. Carry out the details of the procedure. Use (10.202) to write an explicit solution to the approximate problem. Identify the various matrices explicitly.
3. The performance is given by (10.203). We would like to choose $\tilde{f}(t)$ to maximize $\Delta_o$. What constraints are necessary? Carry out the optimization.

**Problem 13.2.14.** Assume that $\tilde{K}_{DR}(t-u, \lambda)$ can be factored as

$$\tilde{K}_{DR}(t-u, \lambda) = \tilde{K}_{Du}(t-u)\tilde{K}_{Ru}(\lambda).$$

1. Evaluate $\tilde{K}_{\tilde{n}_r}(t, u)$ in (104) for this case.
2. We want to approximate $\tilde{K}_{\tilde{n}_r}(t, u)$ by a separable kernel. What functions would minimize the approximation error? Discuss other choices that might be more practical. Consider, for example,

$$\tilde{K}_{Ru}(\lambda) = \sum_{i=1}^{M} a_i \varphi_i(\lambda)$$

as a preliminary expansion.

**Problem 13.2.15.** In this problem we derive the optimum estimator equations in (116)–(121).

1. The first step is to derive the generalization of (I-6.55). The linear operation is

$$\hat{\tilde{\mathbf{x}}}(t, \lambda) = \int_{T_i}^{t} \tilde{\mathbf{h}}_o(t, \tau:\lambda)\tilde{r}(\tau)\,d\tau. \tag{P.1}$$

We want to minimize the realizable MMSE error. Show that the optimum impulse response must satisfy

$$E[\tilde{\mathbf{x}}(t, \lambda)\tilde{r}^*(u)] = \int_{T_i}^{t} \tilde{\mathbf{h}}_o(t, \tau:\lambda)\tilde{K}_{\tilde{r}}(\tau, u)\,d\tau, \qquad T_i < u < t. \tag{P.2}$$

2. Using (P.2) as a starting point, carry out an analysis parallel to that in Section 6.3.2 to obtain (116)–(121).

**Problem 13.2.16.** Consider the scattering function given in (53)–(63). Assume that

$$\tilde{f}(t) = \begin{cases} \dfrac{1}{\sqrt{T}}, & 0 \le t \le T, \\ 0, & \text{elsewhere.} \end{cases}$$

Write out the optimum receiver equations (116)–(121) in detail for this case.

**Problem 13.2.17.** Consider the model in (101)–(104) and assume that $\tau_d = \omega_d = 0$. We use a receiver that computes

$$\tilde{l}_m = \int_{-\infty}^{\infty} \tilde{v}^*(t)\tilde{r}(t)\,dt$$

and compares $|\tilde{l}_m|^2$ with a threshold. The function $\tilde{v}(t)$ is an arbitrary function that we want to choose. Do not confuse $\tilde{v}(t)$ and $\tilde{g}(t)$ in (106). The performance of this receiver is a monotonic function of $\Delta_m$, where

$$\Delta_m = \frac{E[|\tilde{l}_m|^2 \mid H_1] - E[|\tilde{l}_m|^2 \mid H_0]}{E[|\tilde{l}_m|^2 \mid H_0]}$$

[see (9.49)].

1. Derive an expression for $\Delta_m$.
2. Find an equation that specifies the $\tilde{v}(t)$ which maximizes $\Delta_m$. Call the solution $\hat{\tilde{v}}_1(t)$.
3. In [32], this problem is studied from a different viewpoint. Stutt and Spafford define

$$J = E[|\tilde{l}_m|^2 \mid \tilde{n}_r(t) \quad \text{only}].$$

Prove that

$$J = \iiint\limits_{-\infty}^{\infty} \tilde{v}^*(t-\lambda)\tilde{f}(t-\lambda)\tilde{K}_{DR}(t-u, \lambda)\tilde{f}^*(u-\lambda)\tilde{v}(u-\lambda)\, dt\, du\, d\lambda$$

$$= \iint\limits_{-\infty}^{\infty} \tilde{S}_{DR}\{f, \lambda\}\theta_{fv}\{\lambda, -f\}\, d\lambda\, df,$$

where $\theta_{fv}\{\cdot, \cdot\}$ is the cross-ambiguity function defined in (10.222).

4. We want to minimize $J$ subject to the constraints

$$\int_{-\infty}^{\infty} |\tilde{v}(t)|^2\, dt = 1$$

and

$$\int_{-\infty}^{\infty} \tilde{f}^*(t)\tilde{v}(t)\, dt = K,$$

where

$$0 \le |K| \le 1.$$

a. Explain these constraints in the context of the result in part 1.

b. Carry out the minimization using two Lagrange multipliers. Call the solution $\hat{\tilde{v}}_2(t)$.

c. Does

$$\hat{\tilde{v}}_2(t) = \hat{\tilde{v}}_1(t)$$

in general?

d. Verify that we can force

$$\hat{\tilde{v}}_2(t) = \hat{\tilde{v}}_1(t)$$

by choosing the two constraints appropriately.

e. Read [32] and discuss why one might want to use $\hat{\tilde{v}}_2(t)$ instead of $\hat{\tilde{v}}_1(t)$.

*Comments:*

1. You should have solved part 2 by inspection, since $\hat{\tilde{v}}_1(t)$ must equal $\tilde{g}(t)$ in (106).
2. The equation in part 3b is solved by a sampling approach in [32]. The same procedures can be used to solve (106).

**Problem 13.2.18.** Consider the model in Problem 13.2.17.

1. Verify that $\Delta_m$ can be written as

$$\Delta_m = \frac{\left|\int_{-\infty}^{\infty} \tilde{f}(t)\tilde{v}^*(t)\, dt\right|^2}{\iiint\limits_{-\infty}^{\infty} \tilde{v}^*(t-\lambda)[N_0\,\delta(t-u) + \tilde{f}(t-\lambda)\tilde{K}_{DR}(t-u, \lambda)\tilde{f}^*(u-\lambda)]\tilde{v}(u-\lambda)\, d\lambda\, dt\, du}. \tag{P.1}$$

2. We require $\tilde{v}(t)$ to be of the form

$$\tilde{v}(t) = \tilde{a}_1\tilde{f}(t) + \tilde{a}_2\tilde{f}(t - T_s),$$

where $T_s$ is a fixed constant and $\tilde{a}_1$ and $\tilde{a}_2$ are complex weightings. Choose $\tilde{a}_1$ and $\tilde{a}_2$ to maximize $\Delta_m$. Call these values $\hat{\tilde{a}}_1$ and $\hat{\tilde{a}}_2$.

3. Now maximize $\Delta_m(\hat{\tilde{a}}_1, \hat{\tilde{a}}_2)$ as a function of $T_s$.

**Problem 13.2.19.** Consider the result in (P.1) in Problem 13.2.18. We would like to optimize $\tilde{f}(t)$ and $\tilde{v}^*(t)$ jointly. We studied this problem for Doppler-spread reverberation in Problem 11.2.14. Our procedure resulted in a set of nonlinear differential equations that we were unable to solve. The basic difficulty was that both the conventional and optimum receivers were related to $\tilde{f}(t)$.

We now try a new procedure. For simplicity we begin with Doppler-spread reverberation,

$$\tilde{K}_{DR}(t - u, \lambda) = \tilde{K}_D(t - u)\,\delta(\lambda).$$

We select an initial $v(t)$ with unity energy, which we denote as $\tilde{v}_1(t)$. Now conduct the following minimization:

(i) Constrain

$$\int_0^T |\tilde{f}(t)|^2\,dt = 1.$$

(ii) Constrain

$$\int_{-\infty}^{\infty} f^2\,|\tilde{F}\{f\}|^2\,df = B^2$$

and

$$\tilde{f}(0) = \tilde{f}(T) = 0.$$

(iii) Constrain

$$\int_0^T \tilde{f}(t)\tilde{v}_1^*(t)\,dt = K.$$

(iv) Minimize

$$J = N_0 + \iint_0^T \tilde{v}_1^*(t)\,\tilde{f}(t)\tilde{K}_D(t - u)\tilde{f}^*(u)\tilde{v}_1(u)\,dt\,du,$$

subject to these constraints.

1. Carry out the required minimization. Verify that the resulting equation is linear. Reduce the problems to a set of differential equations that specify the solution. Observe that these can be solved using Baggeroer's algorithm [55], [56]. Denote the solution as $\tilde{f}_1(t)$.

2. Assume that $\tilde{f}_1(t)$ is transmitted. Choose $\tilde{v}(t)$ to maximize $\Delta_m$. Denote the solution as $\tilde{v}_2(t)$. Is there any difficulty in carrying out this procedure?

3. Repeat part 1, using $\tilde{v}_2(t)$. What is the difficulty with this procedure?

4. Discuss the problems in extending this procedure to the doubly-spread case. Using the distributed state-variable model, derive a set of differential equations that specify the optimum signal as in part 1.

**Problem 13.2.20.** The complex envelope of the transmitted signal is $\sqrt{E_t}\,\tilde{f}(t)$, where

$$\tilde{f}(t) \triangleq a \sum_{n=1}^{N} \tilde{u}(t - nT_p), \tag{P.1}$$

with $\tilde{u}(t)$ defined as in (10.29). The desired target is located at the origin. Instead of correlating with $\tilde{f}^*(t)$, we correlate with $\tilde{x}^*(t)$;

$$\tilde{x}(t) = \sum_{n=1}^{N} W_n T_s \tilde{u}(t - nT_p),$$

where $W_n$ is an arbitrary complex number. Both $\tilde{f}(t)$ and $\tilde{x}(t)$ are normalized to have unit energy. The receiver output is

$$l \triangleq \left| \int_{-\infty}^{\infty} \tilde{f}(t)\tilde{x}^*(t)\, dt \right|^2.$$

1. Calculate $\Delta$ when the complex input is the signal plus complex white noise with spectral height $N_0$.
2. Denote the complex weightings by the vector $\mathbf{W}$. Choose $\mathbf{W}$ to maximize $\Delta$.
3. Calculate $\Delta$ for the case in which there is clutter that has a rectangular scattering function

$$\tilde{S}_{DR}\{f, \lambda\} = \begin{cases} \dfrac{2\sigma_b^2}{BL}, & B_1 \le f \le B_2, \quad L_1 \le \lambda \le L_2, \\ 0, & \text{elsewhere.} \end{cases}$$

Write $\Delta$ in the form

$$\frac{\Delta}{\bar{E}_r/N_0} = \frac{\tilde{\mathbf{W}}^\dagger \mathbf{U} \tilde{\mathbf{W}}}{\tilde{\mathbf{W}}^\dagger [N\mathbf{I} + \lambda \mathbf{C}] \tilde{\mathbf{W}}},$$

where

$$\tilde{u}_{ij} = 1.$$

Specify the other matrices.

4. We want to choose $\tilde{\mathbf{W}}$ to maximize $\Delta$. Carry out the maximization and find the equations specifying the optimum $\tilde{\mathbf{W}}$.

*Comment:* This problem and generalizations of it are studied in detail in [22], [24], and [34].

**Problem 13.2.21.** Consider the reverberation model in (132). From (136).

$$\Delta_o = \bar{E}_r \int_{-\infty}^{\infty} \frac{\tilde{S}_{\tilde{f}}\{f\}}{N_0 + \tilde{S}_{\tilde{n}_r}\{f\}}\, df,$$

where $\tilde{S}_{\tilde{n}_r}\{f\}$ is specified by (124) and (132). We constrain the transmitted signal to be bandlimited with unit energy,

$$\tilde{S}_{\tilde{f}}\{f\} = 0, \qquad |f| \ge W.$$

Find an equation specifying the optimum $S_{\tilde{f}}\{f\}$ to maximize $\Delta_o$.

**Problem 13.2.22.** Consider the reverberation model in (132). If the target is moving, then

$$\Delta_o = \bar{E}_r \int_{-\infty}^{\infty} \frac{\tilde{S}_{\tilde{f}}\{f - f_d\}}{N_0 + \tilde{S}_{\tilde{n}_r}\{f\}}\, df.$$

Repeat Problem 13.2.21 for this case.

**Problem 13.2.23.** Consider the reverberation model in (132). Assume that

$$\tilde{f}(t) = a \sum_{i=1}^{2} \tilde{u}(t - iT_p),$$

where $\tilde{u}(\cdot)$ is defined in (10.29). The desired target has a *known* velocity corresponding to a Doppler shift of $f_d$ cps.

1. Draw a block diagram of the optimum receiver and evaluate its performance.
2. Now assume that we generate two random variables,

$$\tilde{r}_1 \triangleq \int_0^{T_s} \tilde{r}(t)\tilde{u}^*(t)\,dt,$$

$$\tilde{r}_2 \triangleq \int_{T_p}^{T_p+T_s} \tilde{r}(t)\tilde{u}^*(t-T_p)\,dt.$$

Derive a formula for the optimum operations on $\tilde{r}_1$ and $\tilde{r}_2$. Evaluate the performance of the resulting receiver.

3. Consider the receiver shown in Fig. P.13.3. The target is assumed to be at zero-range. Analyze the performance of this receiver as a function of $E_t$, $N_0$, $P_c$, and $f_d$. Compare the results in parts 1, 2, and 3.

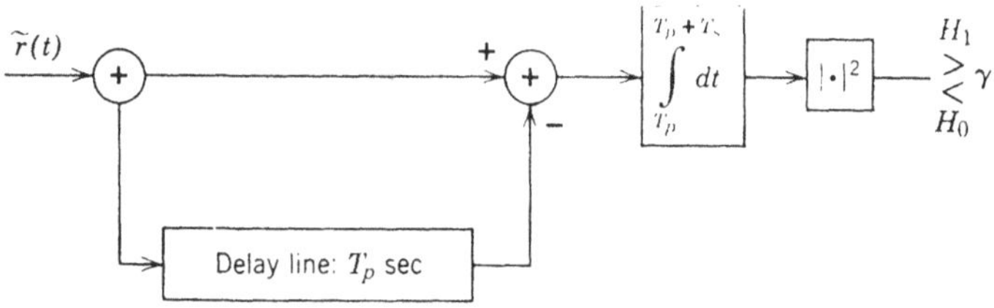


**Fig. P.13.3**

4. Read Steinberg's discussion of MTI (moving target indication) radars [57]. Compare his model and results with our model. Other interesting discussions of MTI systems are given in [58]–[60].

## P.13.3 Detection of Doubly-Spread Targets

### Detection Models

**Problem 13.3.1.** Consider the binary detection problem in which the complex envelopes of the received waveforms on the two hypotheses are

$$\tilde{r}(t) = \tilde{s}_1(t) + \tilde{w}(t), \qquad -\infty < t < \infty : H_1,$$

$$\tilde{r}(t) = \tilde{s}_0(t) + \tilde{w}(t), \qquad -\infty < t < \infty : H_0,$$

where $\tilde{s}_0(t)$, $\tilde{s}_1(t)$, and $\tilde{w}(t)$ are statistically independent complex Gaussian random processes with covariance functions.

$$\tilde{K}_{\tilde{s}_0}(t, u) = E_t \int_{-\infty}^{\infty} \tilde{f}(t-\lambda)\tilde{K}_{DR,0}(t-u, \lambda)\tilde{f}^*(u-\lambda)\,d\lambda,$$

$$\tilde{K}_{\tilde{s}_1}(t, u) = E_t \int_{-\infty}^{\infty} \tilde{f}(t-\lambda)\tilde{K}_{DR\,1}(t-u, \lambda)\tilde{f}^*(u-\lambda)\,d\lambda,$$

and

$$\tilde{K}_{\tilde{w}}(t, u) = N_0\delta(t - u).$$

Derive the equations specifying the optimum receiver.

**Problem 13.3.2.** Consider the model in Problem 13.3.1. Assume that

$$\tilde{K}_{\tilde{s}_0}(t, u) = E_t \int_{-\infty}^{\infty} \tilde{f}_0(t - \lambda)\tilde{K}_{DR}(t - u, \lambda)\tilde{f}_0^*(u - \lambda)\, d\lambda$$

and

$$\tilde{K}_{\tilde{s}_1}(t, u) = E_t \int_{-\infty}^{\infty} \tilde{f}_1(t - \lambda)\tilde{K}_{DR}(t - u, \lambda)\tilde{f}_1^*(u - \lambda)\, d\lambda.$$

Derive the equations specifying the optimum receiver.

**Problem 13.3.3.** The complex envelopes of the received waveforms on the two hypotheses are

$$\tilde{r}(t) = \sqrt{E_t}\, \tilde{b}\tilde{f}(t) + \sqrt{E_t} \int_{-\infty}^{\infty} \tilde{b}(t, \lambda)\tilde{f}(t - \lambda)\, d\lambda + \tilde{w}(t), \qquad -\infty < t < \infty : H_1,$$

$$\tilde{r}(t) = \tilde{w}(t), \qquad -\infty < t < \infty : H_0.$$

The process $\tilde{b}(t, \lambda)$ is characterized in (4). The random variable $\tilde{b}$ is a complex Gaussian variable ($E(|\tilde{b}|^2) = 2\sigma_b^2$) and is statistically independent of $\tilde{b}(t, \lambda)$.

Derive the equations specifying the optimum receiver.

**Problem 13.3.4.** Consider the statement below (176) regarding the statistical independence of the tap gain processes. Investigate the issue quantitatively.

**Problem 13.3.5.** Consider the scattering function in Problem 13.1.6. Assume that we approximate it with the tapped-delay line in Fig. 13.18.

1. Specify the spectrum of the tap gain processes.
2. Find the cross-correlation (or cross-spectrum) of tap gain processes as a function of $W_s$.
3. Assume that we use three taps and that the tap gain processes are statistically independent. Write out the state equations specifying the model.
4. Draw a block diagram of the optimum receiver for the detection model in (142)–(152). Write an expression for $\tilde{\mu}(s)$.

**Problem 13.3.6 [61].** Assume that the transmitted signal is time-limited; that is,

$$\tilde{f}(t) = 0, \qquad |t| > \frac{T}{2}.$$

Develop the dual of the tapped-delay line model.

**Problem 13.3.7.** In the Doppler-spread case the SPLOT condition enabled us to obtain answers reasonably easily. Consider the doubly-spread problem in which $\tilde{f}(t)$ is a time-limited rectangular pulse $[0, T]$ and $\tilde{S}_{DR}\{f, \lambda\}$ is range-limited $[0, L]$. The observation interval is $[-\infty, \infty]$.

1. Is the output signal process stationary?
2. Is any time segment of the output signal process stationary?
3. Consider the following procedure:

(i) Analyze the SPLOT problem for the observation interval $[L, T]$.
(ii) Analyze the SPLOT problem for the observation interval $[0, L + T]$.

a. Will the performance of the system in (i) underbound the performance of the actual system?

b. Will the performance of the system in (ii) overbound the performance of the actual system?

c. For what ranges of parameter values would this procedure be useful?

**Problem 13.3.8.** The scattering function is

$$\tilde{S}_{DR}\{f, \lambda\} = \frac{4k\sigma_b^2[1 - \cos(2\pi\lambda/L)]}{L[(2\pi f)^2 + k^2]}, \qquad -\infty < f < \infty, \qquad 0 < \lambda < L.$$

We expand the channel using the general orthogonal signal model in (196)–(217). The transmitted signal is a rectangular pulse $[0, T]$. The orthogonal functions are

$$\varphi_1(t) = \frac{1}{\sqrt{L}}, \qquad 0 \le \lambda \le L,$$

$$\varphi_2(t) = \sqrt{\frac{2}{L}} \cos\left(\frac{2\pi\lambda}{L}\right), \qquad 0 \le \lambda \le L,$$

$$\varphi_3(t) = \sqrt{\frac{2}{L}} \sin\left(\frac{2\pi\lambda}{L}\right), \qquad 0 \le \lambda \le L,$$

and so forth.

Evaluate the various quantities needed to specify the model completely. Be careful about the intervals.

**Problem 13.3.9.** Prove that the tapped-delay line model is a special case of the general orthogonal signal model.

**Problem 13.3.10.** The scattering function is

$$\tilde{S}_{DR}\{f, \lambda\} = \frac{8k\sigma_b^2}{L} \frac{(1 - (2|\lambda|)/L)}{((2\pi f)^2 + k^2)}, \qquad -\infty < f < \infty, \qquad |\lambda| < L/2.$$

Repeat Problem 13.3.8.

**Problem 13.3.11.** Consider the model in (224)–(228). Show that a direct orthogonal series expansion leads back to the model in Section 13.3.2.B.

**Problem 13.3.12.** Consider the expression for $\tilde{\mathbf{p}}_{ij}(t)$ in (233).

1. Derive a set of differential equations that the $\tilde{\mathbf{p}}_{ij}(t)$ must satisfy.
2. Compare the result in part 1 with that in Problem 13.3.11. Identify the impulsive term in (229).

**Problem 13.3.13.** Consider the decomposition in (229). Verify that the results in (230) is correct. [*Hint:* Recall (47).]

### Binary Communication

**Problem 13.3.14** [38]. Assume that

$$B = 1 \text{ kcps}$$

and

$$L = 250\ \mu\text{sec}.$$

The power-to-noise ratio at the receiver is

$$\frac{P_R}{N_0} = 5 \times 10^5 \qquad (57 \text{ db}).$$

We require a probability of error of $10^{-3}$.

1. Show that the maximum achievable rate using a binary system with the above parameters in 15,000 bits/sec.
2. Design a simple system that achieves this rate.

**Problem 13.3.15.** Consider the scattering function in Fig. 13.1.6 and assume that

$$kL = 10.$$

Design signals for a binary system to communicate effectively over this channel.

**Problem 13.3.16.** Consider the model in (254)–(264) and (279).

1. Find the terms in (277) for the case when $K = 3$.
2. Repeat part 1 for $K = 5$.

LEC CONDITIONS

*Comment:* The next three problems develop some simple tests to verify the LEC condition.

**Problem 13.3.17** [5].

1. Prove that

$$\tilde{\lambda}_{\max} \le E_t \max_t |\tilde{f}(t)|^2 \int_{-\infty}^{\infty} (\max_f \tilde{S}_{DR}\{f, \lambda\}) \, d\lambda. \tag{P.1}$$

2. Consider the special case in which $\tilde{f}(t)$ is a constant. Prove that

$$\tilde{\lambda}_{\max} \le \max_f \tilde{S}_{\tilde{s}}\{f\}. \tag{P.2}$$

**Problem 13.3.18** [5]. Derive the dual of the bound in (P.1) of Problem 13.3.17. Specifically, prove that

$$\tilde{\lambda}_{\max} \le \tfrac{1}{2}\{\max_f |\tilde{F}\{f\}|^2\} \int_{-\infty}^{\infty} (\max_\lambda \tilde{S}_{DR}\{f, \lambda\}) \, df. \tag{P.1}$$

**Problem 13.3.19** [5]. Prove that

$$\tilde{\lambda}_{\max} \le \bar{E}_r \max_{f,\lambda} S_{DR}\{f, \lambda\}.$$

**Problem 13.3.20.** In this problem we develop lower bounds on $\tilde{\lambda}_{\max}$.

1. Prove that

$$\tilde{\lambda}_{\max} \ge \iint_{T_i}^{T_f} \tilde{z}(t)\tilde{K}_{\tilde{s}}(t, u)\tilde{z}^*(u) \, dt \, du \tag{P.1}$$

for any $\tilde{z}(t)$ such that

$$\int_{T_i}^{T_f} |\tilde{z}(t)|^2 \, dt = 1. \tag{P.2}$$

2. Assume that $T_i = -\infty$ and $T_f = \infty$. Prove

$$\tilde{\lambda}_{\max} \ge E_t \iint_{-\infty}^{\infty} \theta\{\lambda, f\}\tilde{S}_{DR}\{f, \lambda\} \, df \, d\lambda. \tag{P.3}$$

3. Give an example of an $\tilde{S}_{DR}\{f, \lambda\}$ in which (P.3) is satisfied with equality.

**Problem 13.3.21.** Consider the model in Section 13.3.4.A.

1. Derive (292).
2. Derive (293)–(295).

**Problem 13.3.22.** Consider the signal in (10.44a) and the scattering function in Problem 13.1.5 with $m_R = m_D = 0$. Evaluate $\tilde{\mu}(s)$ in (295).

**Problem 13.3.23.** Consider a binary communication system operating under LEC conditions and using a rectangular pulse. Assume that $T$ is fixed.

1. Prove that

$$\ln \Pr(\epsilon) \simeq a\left(\frac{P_r}{N_0}\right)^2 + b,$$

where $P_r/N_0$ is the received power-to-noise level ratio. Find $a$ and $b$.

2. Compare this result with (239).

**Problem 13.3.24.** Consider the suboptimum receiver in (296). Set up the equations necessary to analyze its performance.

**Problem 13.3.25.** Consider the suboptimum receiver in Fig. 13.31. Set up the equations necessary to analyze its performance.

**Problem 13.3.26.** Consider the suboptimum receiver in Fig. 13.32.

1. Set up the equations necessary to analyze its performance.
2. Discuss the utility of the SPLOT approach suggested in Problem 13.3.7 for this particular problem.

**Problem 13.3.27.** Consider the equivalent channel definition on page 523. Verify that the relations in Table 13.2 are correct.

**Problem 13.3.28.** Consider the channel whose scattering function is shown in Fig. P.13.4. The height is $2\sigma_b^2/BL$ in the shaded rectangle and zero elsewhere. Assume that

$$BL = 0.01.$$

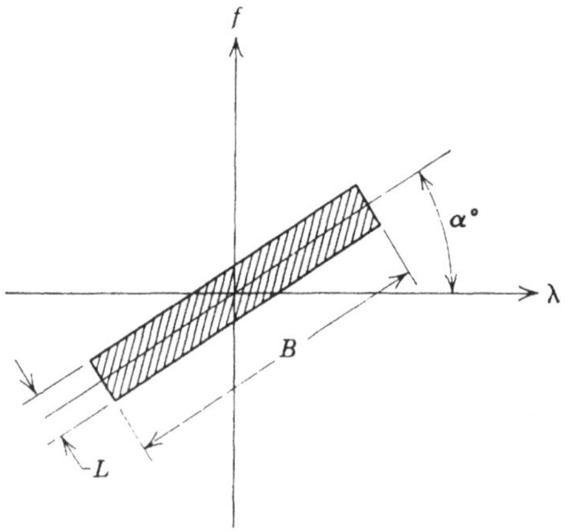


**Fig. P.13.4**

Design a binary communication system that will operate over this channel with

$$\tilde{\mu}_{BS}(\tfrac{1}{2}) \simeq -0.149.$$

Specify both the transmitted signal and the optimum receiver.

**Problem 13.3.29.** Consider the degenerate scattering function

$$S_{DR}\{f, \lambda\} = \sum_{i=1}^{N} \frac{2\sigma_b^2}{N} \delta\{f - f_i\}\, \delta\{\lambda - \lambda_i\}. \tag{P.1}$$

1. Assume that $N = 2$. Prove that all channels with this scattering function are equivalent.
2. Is this result true for $N \geq 3$?

**Problem 13.3.30** [38]. Consider the system in Problem 13.3.15. We require a bit error rate of $10^{-3}$.

1. Show that by using a system with four orthogonal signals we can achieve a rate of 25,000 bits/sec. (*Hint:* Use the results of Problem 5.1.4.)
2. Design a system to achieve this rate.

**Problem 13.3.31** [37]. Prove that all channels whose scattering functions have the form

$$\tilde{S}_{DR}\{f, \lambda{:}a, k, c\} = \tilde{S}_{DR}\left\{ak\lambda + \frac{1 - kc}{a} f,\, a\lambda - \frac{cf}{a}\right\}$$

are equivalent for any values of $c$, $k$, and $a$.

## P.13.4 Parameter Estimation

**Problem 13.4.1.**

1. Derive the expression for $l_R(\mathrm{A})$ in (306).
2. Derive an expression for the elements of the information matrix $\mathbf{J}$. Do not assume LEC conditions.

**Problem 13.4.2.** Derive the expression for $l_B^{[2]}(\mathrm{A})$ given in (317).

**Problem 13.4.3.** Assume that the LEC condition is valid. Derive the result in (324).

**Problem 13.4.4.** Consider the amplitude estimation problem in Section 13.4.2.

1. Verify that $\hat{a}_0$ [defined in (331)] is unbiased under all conditions (i.e., the LEC condition is not required).
2. Find an exact expression for

$$\xi_{\hat{a}_0} \triangleq E[(\hat{a}_0 - A)^2].$$

3. Verify that

$$\xi_{\hat{a}_0} \triangleq J^{-1}(A)$$

when the LEC condition holds.

**Problem 13.4.5.** Express the result in (332) in an alternative form that contains $\tilde{S}_{DR}\{f, \lambda\}$ instead of $\tilde{R}_{DR}\{\tau, \nu\}$.

*Comment:* Notice that the LEC assumption is *not* made in Problems 13.4.6–13.4.8.

**Problem 13.4.6.** Consider the degenerate case of the amplitude estimation problem in which $\tilde{s}(t, A)$ has a finite number of *equal* eigenvalues. Thus,

$$\tilde{K}_{\tilde{s}}(t, u{:}A) = A\tilde{\lambda}_c \sum_{i=1}^{N} \tilde{\varphi}_i(t)\tilde{\varphi}_i^*(u), \qquad -\infty < t, u < \infty.$$

1. Find the receiver to generate $\hat{a}_0$ and $\hat{a}_{ml}$.
2. Evaluate $\xi_{\hat{a}_0}$ and $J(A)$. Verify that $\hat{a}_0$ is an efficient unbiased estimate.
3. Constrain

$$AN\tilde{\lambda}_c = \bar{E}_r.$$

Treat $N$ as a continuous variable that is greater than or equal to 1. Find the value of $N$ that minimizes $\xi_{\hat{a}_0}$. Notice that the answer depends on $A$, the unknown parameter. How would you use this result in an actual system? (*Hint:* Is $\xi_{\hat{a}_0}$ sensitive to the exact choice of $N$?) Plot $\xi_{\hat{a}_0}$ as a function of $N$.

**Problem 13.4.7.** Consider the general amplitude estimation problem. Assume that

$$\tilde{K}_{\tilde{s}}(t, u{:}A) = A\tilde{K}_{\tilde{s}}(t, u), \qquad -\infty < t, u < \infty.$$

1. Express the Cramér-Rao bound of the variance of any unbiased estimate of $A$ in terms of the eigenvalues of $\tilde{K}_{\tilde{s}}(t, u)$.
2. Constrain

$$A \int_{-\infty}^{\infty} \tilde{K}_s(t, t)\, dt = \bar{E}_r.$$

Find the eigenvalue distribution that minimizes the value of the bound in part 1. Compare your result with that in part 3 of Problem 13.4.6.

3. Interpret the result in part 2 in the context of estimating the amplitude of an otherwise known scattering function. Notice that this gives a bound on the variance of an unbiased estimate of the amplitude that does not depend on the scattering function.

**Problem 13.4.8.** Assume that

$$\tilde{S}_{DR}\{f, \lambda{:}A\} = A\tilde{S}_{DR}\{f, \lambda\},$$

where $\tilde{S}_{DR}\{f, \lambda\}$ is known. We know that

$$\frac{A\bar{E}_r}{N_0} = \frac{AE_t}{N_0} \iint_{-\infty}^{\infty} \tilde{S}_{DR}\{f, \lambda\}\, df\, d\lambda \simeq 20,$$

and want to estimate $A$ more exactly.

1. Assume that

$$L = 10$$

and

$$BL = 0.001.$$

Design a signal $\tilde{f}(t)$ that will result in an unbiased estimate whose variance is close to that in part 3 of Problem 13.4.7. Draw a block diagram of the optimum receiver.

2. Repeat part 1 for the case in which

$$B = 10$$

and

$$BL = 0.001.$$

**Problem 13.4.9.** Consider the generalization of the example on page 531, in which the Gaussian pulse has a linear FM [see (10.44a)] and the scattering function has a skewed Gaussian shape.

$$\tilde{S}_{DR}\{f, \lambda\} = \frac{2\sigma_b^2}{2\pi\sigma_D\sigma_R(1 - \rho_{DR}^2)^{1/2}} \exp\left[-\frac{\sigma_R^2 f^2 - 2\sigma_R\sigma_D\rho_{DR} f\lambda + \sigma_D^2\lambda^2}{2\sigma_D^2\sigma_R^2(1 - \rho_{DR}^2)}\right]. \tag{P.1}$$

The LEC condition is assumed.

1. Show that this can be reduced to an equivalent problem with a nonskewed Gaussian scattering function.

2. Evaluate the bound in (332).
3. What linear sweep rate minimizes the variance bound?

**Problem 13.4.10.** Consider the problem of estimating the scale of range axis under LEC conditions,

$$\tilde{K}_{DR}(\tau, \lambda : A) = \tilde{K}_{D_1R_1}\left(\tau, \frac{\lambda}{A}\right),$$

where $\tilde{K}_{D_1R_1}(\cdot, \cdot)$ is known.

1. Derive a lower bound on the variance of an unbiased estimate [62].
2. Consider the special case in which $\tilde{f}(t)$ is given by (335) and $\tilde{S}_{D_1R_1}\{f, \lambda\}$ satisfies (333). Evaluate the bound in part 1.
3. Choose $T$ to minimize the bound.

**Problem 13.4.11.** Consider the problem of estimating the frequency scale under LEC conditions,

$$\tilde{S}_{DR}\{f, \lambda : A\} = \tilde{S}_{D_1R_1}\left\{\frac{f}{A}, \lambda\right\},$$

where $\tilde{S}_{D_1R_1}\{\cdot, \cdot\}$ is known.

1. Repeat Problem 13.4.10.
2. Solve this problem by using duality theory and the results of Problem 13.4.10.

**Problem 13.4.12.** Consider the generalization of the two previous problems in which

$$\tilde{R}_{DR}\{\tau, v : \mathbf{A}\} = A_1 A_2 \tilde{R}_{D_1R_1}\{A_1\tau, A_2 v\}.$$

1. Derive an expression for the element in the bound matrix, $\mathbf{J}(\mathbf{A})$ [62].
2. Evaluate the terms for the case in part 2 of Problem 13.4.10.

## REFERENCES


[1] H. L. Van Trees, "Optimum Signal Design and Processing for Reverberation-Limited Environments," IEEE Trans. Mil. Electronics **MIL-9**, 212–229 (July 1965).

[2] I. S. Reed, "The Power Spectrum of the Returned Echo from a Random Collection of Moving Scatterers," paper presented at IDA Summer Study, July 8, 1963.

[3] E. J. Kelly and E. C. Lerner, "A Mathematical Model for the Radar Echo from a Random Collection of Scatterers," Massachusetts Institute of Technology, Lincoln Laboratory, Technical Report 123, June 15, 1956.

[4] H. L. Van Trees, "Optimum Signal Design and Processing for Reverberation-Limited Environments," Technical Report No. 1501064, Arthur D. Little, Inc., Cambridge, Mass., October 1964.

[5] R. Price and P. E. Green, "Signal Processing in Radar Astronomy—Communication via Fluctuating Multipath Media," Massachusetts Institute of Technology, Lincoln Laboratory, TR 234, October 1960.

[6] P. E. Green, "Radar Astronomy Measurement Techniques," Massachusetts Institute of Technology, Lincoln Laboratory, Technical Report 282, December 12, 1962.

[7] R. R. Kurth, Distributed-Parameter State-Variable Techniques Applied to Communication over Dispersive Channels, Sc.D. Thesis, Department of Electrical Engineering, Massachusetts Institute of Technology, June 1969.

[8] H. Bremmer, "Scattering by a Perturbed Continuum," *in Electromagnetic Theory and Antennas*, E. C. Jorden (Ed.), Pergamon Press, London, 1963.

[9] I. Tolstoy and C. S. Clay, *Ocean Acoustics—Theory and Experiment in Underwater Sound*, McGraw-Hill, New York, 1966.

[10] P. Faure, "Theoretical Model of Reverberation Noise," J. Acoust. Soc. Am. **36,** 259–268 (Feb. 1964).

[11] H. R. Carleton, "Theoretical Development of Volume Reverberation as a First-Order Scattering Phenomenon," J. Acoust. Soc. Am. **33,** 317–323 (March 1961).

[12] V. P. Antonov and V. V. Ol'shevskii, "Space-Time Correlation of Sea Reverberation," Soviet Phys.-Acoust., **11,** 352–355 (Jan.–Mar. 1966).

[13] D. Middleton, "A Statistical Theory of Reverberation and Similar First-Order Scattered Fields. I," IEEE Trans. Information Theory **IT-13,** No. 3, 372–392 (July 1967).

[14] D. Middleton, "A Statistical Theory of Reverberation and Similar First-Order Scattered Fields. II," IEEE Trans. Information Theory **IT-13,** No. 3, 393–414 (July 1967).

[15] C. S. Clay, Jr., "Fluctuations of Sound Reflected from the Sea Surface," J. Accoust. Soc. Am. **32,** 1547–1555 (Dec. 1960).

[16] E. J. Kelly, Jr., "Random Scatter Channels," Massachusetts Institute of Technology, Lincoln Laboratory, Group Report 1964-61, November 4, 1964.

[17] D. Middleton, "Statistical Models of Reverberation and Clutter. I," Litton Systems, Inc., Waltham, Mass., TR 65-2-BF, April 15, 1965.

[18] L. A. Chernov, *Wave Propagation in a Random Medium*, McGraw-Hill, New York, 1960.

[19] J. L. Stewart and E. C. Westerfeld, "A Theory of Active Sonar Detection," Proc. IRE 47, 872–881 (1959).

[20] E. C. Westerfeld, R. H. Prager, and J. L. Stewart, "Processing Gains against Reverberation (Clutter) Using Matched Filters," IRE Trans. Information Theory **IT-6,** 342–348 (June 1960).

[21] D. F. DeLong, Jr., and E. M. Hofstetter, "The Design of Clutter-Resistant Radar Waveforms with Limited Dynamic Range," IEEE Trans. Information Theory **IT-15,** No. 3, 376–385 (May, 1969).

[22] D. F. DeLong, Jr., and E. M. Hofstetter, "On the Design of Optimum Radar Waveforms for Clutter Rejection," IEEE Trans. Information Theory **IT-13,** 454–463 (July 1967).

[23] J. S. Thompson and E. L. Titlebaum, "The Design of Optimal Radar Waveforms for Clutter Rejection Using the Maximum Principle," IEEE Trans. Aerospace Electronic Syst. **AES-3** (Suppl.), No. 6, 581–589 (Nov. 1967).

[24] W. D. Rummler, "A Technique for Improving the Clutter Performance of Coherent Pulse Train Signals," IEEE Trans. Aerospace Electronic Syst. **AES-3,** No. 6, 898–906 (Nov. 1967).

[25] R. Manasse, "The Use of Pulse Coding to Discriminate against Clutter," Massachusetts Institute of Technology, Lincoln Laboratory Report, 312-12, June 1961.

[26a] A. V. Balakrishnan, "Signal Design for a Class of Clutter Channels," IEEE Trans. Information Theory 170–173 (Jan. 1968).

[26b] T. E. Fortmann, "Comments on Signal Design for a Class of Clutter Channels," IEEE Transactions on Information Theory, **IT-16,** No. 1, 90–91. January, 1970.

[27] M. Ares, "Optimum Burst Waveforms for Detection of Targets in Uniform Range-Extended Clutter," General Electric Co., Syracuse, N.Y., Technical Information Series TIS R66EMH16, March 1966.

[28] E. N. Fowle, E. J. Kelly, and J. A. Sheehan, "Radar System Performance in a Dense Target Environment," 1961 IRE Int. Conv. Rec., Pt. 4.

[29] S. G. Tzafestas and J. M. Nightingale, "Optimal Filtering, Smoothing, and Prediction in Linear Distributed-Parameter Systems," Proc. IEE **115,** No. 8, 1207–1212 (Aug. 1968).

[30] S. G. Tzafestas and J. M. Nightingale, "Concerning the Optimal Filtering Theory of Linear Distributed-Parameter Systems," Proc. IEE **115,** No. 11, 1737–1742 (Nov. 1968).

[31] H. Urkowitz, "Filters for Detection of Small Radar Signals in Clutter," J. Appl. (Nov. 1968). Phys. **24,** 1024–1031 (1953).

[32] C. A. Stutt and L. J. Spafford, "A 'Best' Mismatched Filter Response for Radar Clutter Discrimination," IEEE Trans. Information Theory **IT-14,** No. 2, 280–287 (March 1968).

[33] L. J. Spafford, "Optimum Radar Signal Processing in Clutter," IEEE Trans. Information Theory **IT-14,** No. 5, 734–743 (Sept. 1968).

[34] W. D. Rummler, "Clutter Suppression by Complex Weighting of Coherent Pulse Trains," IEEE Trans. Aerospace Electronic Syst. **AES 2,** No. 6, 689–699 (Nov. 1966).

[35] T. Kailath, "Sampling Models for Linear Time-Variant Filters," Massachusetts Institute of Technology, Research Laboratory of Electronics, TR 352, May 25, 1959.

[36] H. L. Van Trees, Printed Class Notes, Course 6.576, Massachusetts Institute of Technology, Cambridge, Mass., 1965.

[37] R. S. Kennedy, *Fading Dispersive Communication Channels*, Wiley, New York, 1969.

[38] R. S. Kennedy and I. L. Lebow, "Signal Design for Dispersive Channels," IEEE Spectrum 231–237 (March 1964).

[39] R. Price, "Maximum-Likelihood Estimation of the Correlation Function of a Threshold Signal and Its Application to the Measurement of the Target Scattering Function in Radar Astronomy," Massachusetts Institute of Technology, Lincoln Laboratory, Group Report 34-G-4, May 1962.

[40] T. Kailath, "Measurements in Time-Variant Communication Channels," IRE Trans. Information Theory **IT-8,** S229–S236 (Sept. 1962).

[41] P. E. Green, Jr., "Radar Measurements of Target Scattering Properties," in *Radar Astronomy*, J. V. Evans and T. Hagfors (Eds.), McGraw-Hill, New York, 1968, Chap. 1.

[42] T. Hagfors, "Some Properties of Radio Waves Reflected from the Moon and Their Relationship to the Lunar Surface," J. Geophys. Res. **66,** 777 (1961).

[43] M. J. Levin, "Estimation of the Second-Order Statistics of Randomly Time-Varying Linear Systems," Massachusetts Institute of Technology, Lincoln Laboratory Report 34-G-7, November 1962.

[44] J. J. Spilker, "On the Characterization and Measurement of Randomly-Varying Filters," Commun. Sci. Dept., Philco Western Div. Labs. **TM 72** (Oct. 1963).

[45] A. Krinitz, "A Radar Theory Applicable to Dense Scatterer Distributions," Massachusetts Institute of Technology, Electronic Systems Laboratory, Report ESL-R-131, January 1962.

[46] P. Bello, "On the Measurement of a Channel Correlation Function," IEEE Trans. Information Theory **IT-10,** No. 4, 381–383 (Oct. 1964).

[47] N. T. Gaarder, "Scattering Function Estimation," IEEE Trans. Information Theory **IT-14,** No. 5, 684–693 (Sept. 1968).

[48] R. G. Gallager, "Characterization and Measurement of Time- and Frequency-Spread Channels," Massachusetts Institute of Technology, Lincoln Laboratory, TR 352, April 1964.

[49] T. Hagfors, "Measurement of Properties of Spread Channels by the Two-Frequency Method with Application to Radar Astronomy," Massachusetts Institute of Technology, Lincoln Laboratory, TR 372, January 1965.

[50] B. Reiffen, "On the Measurement of Atmospheric Multipath and Doppler Spread by Passive Means," Massachusetts Institute of Technology, Lincoln Laboratory, Technical Note 1965-6, Group 66, March 1965.

[51] I. Bar-David, "Radar Models and Measurements," Ph.D. Thesis, Department of Electrical Engineering, Massachusetts Institute of Technology, January 1965.

[52] G. Pettengill, "Measurements of Lunar Reflectivity Using the Millstone Radar," Proc. IRE **48,** No. 5, 933 (May 1960).

[53] J. L. Brown, "On the Expansion of the Bivariate Gaussian Probability Density Using the Results of Nonlinear Theory, IEEE Trans. Information Theory **IT-14,** No. 1,158–159 (Jan. 1968).

[54] N. Wiener, *The Fourier Integral and Certain of Its Applications*, Cambridge University Press, London, 1933.

[55] A. B. Baggeroer, "State Variables, the Fredholm Theory, and Optimal Communication," Sc.D. Thesis, Department of Electrical Engineering, Massachusetts Institute of Technology, 1968.

[56] A. B. Baggeroer, *State Variables and Communication Theory*, Massachusetts Institute of Technology Press, Cambridge, Mass., 1970.

[57] B. D. Steinberg, "MTI Radar Filters," in *Modern Radar*, R. S. Berkowitz (Ed.), Wiley, New York, 1965, Chap. VI-2.

[58] J. Capon, "Optimum Weighting Functions for the Detection of Sampled Signals in Noise," IEEE Trans. Information Theory **IT-10,** No. 2, 152–159 (April 1964).

[59] L. N. Ridenour, *Radar System Engineering*, McGraw-Hill, New York, 1947.

[60] L. A. Wainstein and V. D. Zubakov, *Extraction of Signals from Noise*, Prentice-Hall, Englewood Cliffs, N.J., 1962.

[61] P. A. Bello, "Characterization of Randomly Time-Variant Linear Channels," IEEE Trans. Commun. Syst. **CS-11,** 360–393 (Dec. 1963).

[62] M. J. Levin, "Parameter Estimation for Deterministic and Random Signals," Massachusetts Institute of Technology, Lincoln Laboratory, Group Report 34-G-11 (preliminary draft not generally available).

[63] A. W. Rihaczek, "Optimum Filters for Signal Detection in Clutter," IEEE Trans. Aerospace Electronic Syst. **AES-1,** 297–299 (Dec. 1965).

[64] T. Kailath, "Optimum Receivers for Randomly Varying Channels," in *Fourth London Symp. Information Theory*, C. Cherry (Ed.), Butterworths, Washington, D.C., 1961.

[65] P. E. Green, "Time-Varying Channels with Delay Spread," in *Monograph on Radio Waves and Circuits*, S. Silver (Ed.), Elsevier, New York, 1963.

[66] P. A. Bello, "Measurement of Random Time-Variant Linear Channels," IEEE Trans. Information Theory **IT-15,** No. 4, 469–475 (July 1969).

[67] I. Bar-David, "Estimation of Linear Weighting Functions in Gaussian Noise," IEEE Trans. Information Theory 395–407 (May 1968).

[68] W. L. Root, "On the Measurement and Use of Time-Varying Communications Channels," Information Control 390–422 (Aug. 1965).

[69] P. A. Bello, "Some Techniques for the Instantaneous Real-Time Measurement of Multipath and Doppler Spread," IEEE Trans. Commun. Technol. 285–292 (Sept. 1965).

[70] F. E. Thau, "On Optimum Filtering for a Class of Linear Distributed-parameter Systems," in "Proc. 1968 Joint Automatic Control Conf.," Univ. of Mich., Ann Arbor, Mich., pp. 610–618, 1968.

[71] H. J. Kushner, "Filtering for Linear Distributed Parameter Systems," Center for Dynamical Systems, Brown Univ., Providence, R.I., 1969.

[72] S. G. Tzafestas and J. M. Nightingale, "Maximum-likelihood Approach to Optimal Filtering of Distributed-parameter Systems," *Proc. IEE*, Vol. 116, pp. 1085–1093, 1969.

[73] G. A. Phillipson and S. K. Mitter, "State Identification of a Class of Linear Distributed Systems," in "Proc. Fourth IFAC Congress," Warsaw, Poland, June 1969.

[74] A. V. Balakrishnan and J. L. Lions, "State Estimation for Infinite-dimensional Systems," *J. Computer Syst. Sci.*, Vol. 1, pp. 391–403, 1967.

[75] J. S. Meditch, "On State Estimation for Distributed Parameter Systems," Jour. of Franklin Inst., **290**, No. 1, 49–59 (July 1970).

# 14

# *Discussion*

In this chapter we discuss three topics briefly. In Section 14.1, we summarize some of the major results of our radar-sonar discussion. In Section 14.2, we outline the contents of *Array Processing*, the final volume of this series. In Section 14.3, we make some concluding comments on the over-all sequence.

## 14.1 SUMMARY: SIGNAL PROCESSING IN RADAR AND SONAR SYSTEMS

In Chapter 8 we introduced the radar-sonar problem and discussed the hierarchy of target and channel models of interest. We then detoured to the Appendix and developed a complex representation for narrow-band signals, systems, and processes. For signals,

$$f(t) = \sqrt{2}\,\mathrm{Re}\,[\tilde{f}(t)e^{j\omega_c t}], \tag{1}$$

where $\tilde{f}(t)$ is the complex envelope. For systems,

$$h(t, u) = 2\,\mathrm{Re}\,[\tilde{h}(t, u)e^{j\omega_c t}], \tag{2}$$

where $\tilde{h}(t, u)$ is the complex impulse response. For random processes,

$$n(t) = \sqrt{2}\,\mathrm{Re}\,[\tilde{n}(t)e^{j\omega_c t}], \tag{3}$$

where $\tilde{n}(t)$ is the complex envelope process. By restricting our attention to processes where

$$E[\tilde{n}(t)\tilde{n}(u)] = 0, \tag{4}$$

we have a one-to-one relationship between the covariance function of the complex envelope process $\tilde{K}_{\tilde{n}}(t, u)$ and the covariance function of the actual process,

$$K_n(t, u) = \sqrt{2}\,\mathrm{Re}\,[\tilde{K}_{\tilde{n}}(t, u)e^{j\omega_c(t-u)}]. \tag{5}$$

This class of processes includes all stationary processes and the nonstationary processes that we encounter in practice. We also introduced complex state variables and developed their properties. The complex notation enabled us to see the important features in the problems more clearly. In addition, it simplified all of the analyses, because we could work with one complex quantity instead of two real quantities.

In Chapter 9, we studied the problem of detecting the return from a slowly fluctuating point target in the presence of noise. The likelihood ratio test was

$$|\tilde{l}|^2 \triangleq \left| \int_{T_i}^{T_f} \tilde{r}(t)\tilde{g}^*(t)\,dt \right|^2 \underset{H_0}{\overset{H_1}{\gtrless}} \gamma, \tag{6}$$

where $\tilde{g}(t)$ satisfies the integral equation

$$\tilde{f}(t) = \int_{T_i}^{T_f} \tilde{K}_{\tilde{n}}(t, u)\tilde{g}(u)\,du, \qquad T_i \le t \le T_f. \tag{7}$$

The performance was completely characterized by the quantity

$$\Delta \triangleq \frac{E\{|\tilde{l}|^2 \mid H_1\} - E\{|\tilde{l}|^2 \mid H_0\}}{E\{|\tilde{l}|^2 \mid H_0\}}. \tag{8}$$

This quantity could be used in (9.50) to determine the error probabilities. In addition, we specified the receiver and its performance in terms of a set of differential equations that could be readily solved using numerical techniques. Although we formulated the optimal signal design problem, we did not study it is detail.

In Chapter 10 we discussed the problem of estimating the range and velocity of a slowly fluctuating point target in the presence of additive white noise. We found that the time-frequency correlation function,

$$\phi\{\tau, f\} = \int_{-\infty}^{\infty} \tilde{f}\left(t - \frac{\tau}{2}\right) \tilde{f}^*\left(t + \frac{\tau}{2}\right) e^{j2\pi f t}\,dt, \tag{9}$$

and the ambiguity function,

$$\theta\{\tau, f\} = |\phi\{\tau, f\}|^2, \tag{10}$$

played a key part in most of our discussion. When the estimation errors were small, the accuracy was directly related to the shape of the ambiguity function at the origin. However, if the ambiguity function had subsidiary peaks whose heights were close to unity, the probability of making a large error was increased. These two issues were related by the radar uncertainty principle, which said that the total volume under the ambiguity function was unity for *any* transmitted signal,

$$\iint_{-\infty}^{\infty} \theta\{\tau, f\}\, d\tau\, df = 1. \tag{11}$$

It is important to re-emphasize that the ambiguity function is important because the receiver has been designed to be optimum in the presence of additive white Gaussian noise. We found that, in some environments, we want to use a different filter [e.g., $\tilde{v}^*(t)$]. This function, $\tilde{v}^*(t)$, could correspond to the $\tilde{g}^*(t)$ specified by (7), or it could be a function chosen for ease in receiver implementation. Now the cross-ambiguity function

$$\theta_{fv}\{\tau, f\} = \left| \int_{-\infty}^{\infty} \tilde{f}\left(t - \frac{\tau}{2}\right) \tilde{v}^*\left(t + \frac{\tau}{2}\right) e^{j2\pi f t}\, dt \right|^2 \tag{12}$$

played the central role in our analyses.

A particularly important problem is the resolution problem. In Section 10.5, we considered resolution in a discrete environment. A typical situation in which this type of problem arises is when we try to detect a target in the presence of decoys. Although we could always find the optimum receiver, the conventional matched-filter receiver was frequently used because of its simplicity. In this case, the degradation due to the interference was

$$\rho_r = \sum_{i=1}^{K} \frac{\bar{E}_i}{N_0}\, \theta(\tau_i - \tau_d, \omega_i - \omega_d). \tag{13}$$

Thus, if we could make the ambiguity function zero at those points in the $\tau$, $\omega$ plane where the interfering targets were located, there would be no degradation. In general, this was not a practical solution, but it did provide some insight into the selection of good signals. Whenever $\rho_r$ was appreciable, we could improve the performance by using an optimum receiver. If there were no white noise, the optimum receiver would simply tune out the interference (this eliminates some of the signal energy also). In the presence of white noise the optimum receiver cannot eliminate all of the interference without affecting the detectability, and so the resulting filter is a compromise that maximizes $\Delta$ in (8).

We continued our discussion of resolution in Section 13.2. The reverberation (or clutter) return was modeled as a dense, doubly-spread target. Once again we considered both conventional and optimum receivers. In the conventional matched-filter receiver the degradation due to the reverberation was given by the expression

$$\rho_r = \frac{E_t}{N_0} \iint_{-\infty}^{\infty} df\, d\lambda\, \tilde{S}_{DR}\{f, \lambda\} \theta\{\lambda - \tau_d, f_d - f\}. \tag{14}$$

Now the signal design problem consisted of minimizing the common volume of the signal ambiguity function and target-scattering function. When $\rho_r$ was appreciable, some improvement was possible using an optimum receiver. In the general case we had to approximate the target by some orthogonal series model, such as the tapped-delay line of Fig. 13.18, in order actually to find the optimum receiver. Several suboptimum configurations for operation in a reverberation environment were developed in the problems.

In Chapter 11 we discussed Doppler-spread point targets. The basic assumption in our model was that the reflection process was a stationary, zero-mean Gaussian process. The covariance function of the complex envelope of the received signal process was

$$\tilde{K}_{\tilde{s}}(t, u) = E_t\tilde{f}(t-\lambda)\tilde{K}_D(t-u)\tilde{f}^*(u-\lambda), \tag{15}$$

where $\tilde{K}_D(\tau)$ was the covariance function of the reflection process. Equivalently, we could characterize the reflection process by the Doppler scattering function,

$$\tilde{S}_D\{f\} = \int_{-\infty}^{\infty} \tilde{K}_D(\tau)e^{-j2\pi f\tau}\, d\tau. \tag{16}$$

We saw that whenever the pulse length $T$ was greater than the correlation time of the reflection process ($\simeq B^{-1}$), the target or channel caused time-selective fading. The optimum receiver problem was simply the bandpass version of the Gaussian signal in noise problem that we had studied in Chapters 2–4. Several classes of reflection processes allowed us to obtain complete solutions. In particular, whenever $\tilde{S}_D\{f\}$ was rational or could be approximated by a rational function, we could obtain a complete solution for the optimum receiver and a good approximation to its performance. This rational-spectrum approximation includes most cases of interest. We also studied binary communication over Doppler-spread channels. We found that there is a bound on the probability of error,

$$\Pr(\epsilon) \leq \tfrac{1}{2}\exp\left(-0.1488\frac{\bar{E}_r}{N_0}\right), \tag{17}$$

that is independent of the shape of the scattering function. In addition, we were able to demonstrate systems using simple signals and receivers whose performance approached this bound. We found that the key to efficient performance was the use of either implicit or explicit diversity. In addition to being important in its own right, the communication problem gave us further insight into the general detection problem.

In Chapter 12 we discussed dispersive (or range-spread) targets and channels. The basic assumptions in our model were that the return from disjoint intervals in range were statistically independent and that the received signal was a sample function of a zero-mean Gaussian random process. The covariance function was

$$\tilde{K}_{\tilde{s}}(t, u) = E_t \int_{-\infty}^{\infty} \tilde{f}(t - \lambda)\tilde{S}_R(\lambda)\tilde{f}^*(u - \lambda)\, d\lambda, \tag{18}$$

where $\tilde{S}_R(\lambda)$ was the range-scattering function. Equivalently, we could characterize the target in terms of a two-frequency correlation function,

$$\tilde{K}_R\{v\} = \int_{-\infty}^{\infty} \tilde{S}_R(\lambda)e^{j2\pi v\lambda}\, d\lambda. \tag{19}$$

Whenever the bandwidth of the transmitted signal was greater than the reciprocal of the target length ($L^{-1}$), we saw that the target caused frequency-selective fading. We next introduced the concept of time-frequency duality. Because the performance of a system is completely determined by the eigenvalues of the received process, we could analyze either a system or its dual. The duality theory enabled us to deal with a large class of range-spread targets that would be difficult to analyze directly. In addition, it offered new insights into the problem. The availability of efficient Fourier transform algorithms makes the synthesis of dual receivers practical.

In Chapter 13 we discussed the final class of targets in our hierarchy, doubly-spread targets. Here we assumed that the reflection process from each incremental range element was a sample function from a stationary Gaussian process and that the reflections from disjoint intervals were statistically independent. The covariance function of the received signal was given by

$$\tilde{K}_{\tilde{s}}(t, u) = E_t \int_{-\infty}^{\infty} \tilde{f}(t - \lambda)\tilde{K}_{DR}(t - u, \lambda)\tilde{f}^*(u - \lambda)\, d\lambda, \tag{20}$$

where $\tilde{K}_{DR}(t - u, \lambda)$ is the covariance function of the received process as a function of $\lambda$. Equivalently, we could characterize the target by a range-Doppler scattering function,

$$\tilde{S}_{DR}\{f, \lambda\} = \int_{-\infty}^{\infty} \tilde{K}_{DR}(\tau, \lambda)e^{-j2\pi f\tau}\, d\tau. \tag{21}$$

If $BL < 1$, we could obtain flat fading by a suitable signal choice. On the other hand, for $BL > 1$, the target was overspread and the received signal had to exhibit either time-selective or frequency-selective fading (or both).

After discussing the reverberation problem, we considered the detection problem for doubly-spread targets. For the low-energy-coherence case the results were straightforward. For the general case we used an orthogonal series model for the channel. The most common model for this type is the tapped-delay line model. In this case, if we could approximate the spectrum of the tap gain processes by rational functions, we could find a complex state-variable model for the entire system. This enabled us to specify the optimum receiver completely and obtain a good approximation to its performance. A second method of solving the doubly-spread channel problem relied on a differential-equation characterization of the channel. This method also led to a set of equations that could be solved numerically. Although the optimum receivers were complicated, we could obtain a good approximation to them in most situations.

The final topic was the discussion of parameter estimation for doubly-spread targets. After deriving the likelihood function, we introduced the *generalized spread ambiguity function* in order to study the performance. Several specific estimation problems were studied in detail.

This concludes our discussion of signal processing in radar and sonar systems. In the next section we briefly discuss the contents of *Array Processing*.

## 14.2 OPTIMUM ARRAY PROCESSING

In the subsequent volume [1] we study the array-processing problem for sonar and seismic systems. The first topic is detection of known signals in noise. The basic derivation is just a special case of the results in Chapter I-4. The important problem is a study of the various issues that arise in a particular physical situation. To explore these issues, we first develop a model for spatially distributed noise fields. We then introduce the ideas of array gain, beam patterns, and distortionless filters, and demonstrate their utility in the signal-processing problem.

The next topic is the detection of unknown signals in noise. This model is appropriate for the passive sonar and seismic problem. By exploiting the central role of the distortionless filter, we are able to develop a receiver whose basic structure does not depend on the detailed assumptions of the model.

The final topic in [1] is the study of multivariable processes as encountered in continuous receiving apertures. Although the basic results are a straightforward extension of the multidimensional results, we shall find that both new insight and simplified computational procedures can be obtained from this general approach.

Just as in Parts II and III, we present a large number of new research results in the book. As in this volume, the result is a mixture of a research monograph and a graduate-level text.

## 14.3 EPILOGUE

Because of the specialized nature of the material in *Array Processing*, many readers will stop at this point, and so a few comments about the over-all development are worthwhile.

We hope that the reader appreciates the close relationships among the various problems that we have considered. A brief glance at the table of contents of the books indicates the wide range of physical situations that we have studied. By exploiting a few fundamental concepts, we were able to analyze them efficiently. An understanding of the relationships among the various areas is important, because it enables one to use results from other problems to solve the problem of current interest.

A second point that the reader should appreciate is the utility of various techniques for solving problems. A standard course in communications theory is no longer adequate. One should understand the techniques and concepts used in control theory, information theory, and other disciplines in order to be an effective analyst. We may have "oversold" the use of state-variable techniques, because it is an item of current research interest to us. We do feel that it is certain to have an important influence on many sophisticated systems in the future.

The reader should remember that we have been working with mathematical models of physical situations. More specifically, we have emphasized Gaussian process models throughout our discussion. In many cases, they are adequate to describe the actual situation, and our predicted performance results can be confirmed experimentally. In other cases, more complicated models employing non-Gaussian processes must be used. There are still other cases in which experimental (or simulation) procedures provide the only feasible approach. These comments do not negate the value of a thorough study of the Gaussian problem, but serve to remind us of its limitations.

Although it is not conventional, we feel that an appropriate final comment is to thank those readers who have followed us through this lengthy development. We hope that you have obtained an appreciation of *Detection, Estimation, and Modulation Theory*.

## REFERENCES

[1] H. L. Van Trees, *Array Processing*, Wiley, New York, 1971.

# Appendix: *Complex Representation of Bandpass Signals, Systems, and Processes*

In this appendix we develop complex representations for narrow-band signals, systems, and random processes. The idea of representing an actual signal as the real part of a complex signal is familiar to most electrical engineers. Specifically, the signal,

$$\cos \omega_1 t = \text{Re}\,[e^{j\omega_1 t}], \tag{A.1}$$

and the associated phasor diagram in Fig. A.1 are encountered in most introductory circuit courses. The actual signal is just the projection of the complex signal on the horizontal axis. The ideas in this appendix are generalizations of this familiar notation.

In Section A.1, we consider bandpass deterministic signals. In Section A.2, we consider bandpass linear systems. In Section A.3, we study bandpass random processes. In Section A.4, we summarize the major results. Section A.5 contains some problems to demonstrate the application of the ideas discussed.

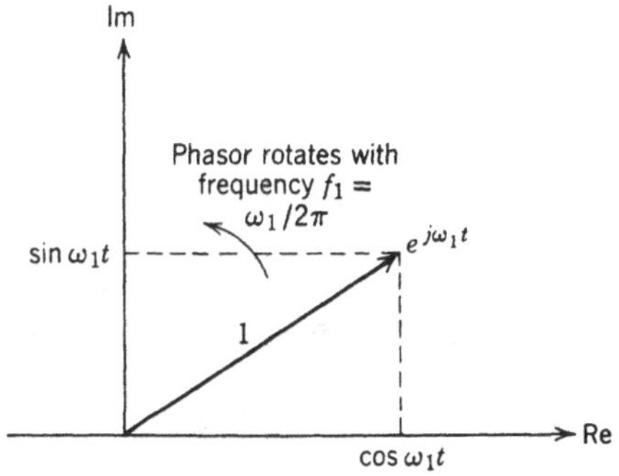


**Fig. A.1 Phasor diagram.**

The treatment through Section A.3.1 is reasonably standard (e.g., [1]–[8]), and readers who are familiar with complex representation can skim these sections in order to learn our notation. The material in Section A.3.2 is less well known but not new. The material in Section A.3.3 is original [9], and is probably not familiar to most readers. With the exception of Sections A.2.3 and A.3.3, the results are needed in order to understand Chapters 9–14.

## A.1 DETERMINISTIC SIGNALS

In this section we consider deterministic finite-energy signals. We denote the signal by $f(t)$, and its Fourier transform by $F(j\omega)$.

$$F(j\omega) = \int_{-\infty}^{\infty} f(t)e^{-j\omega t}\,dt. \qquad \text{(A.2)†}$$

For simplicity, we assume that $f(t)$ has unit energy. A typical signal of interest might have the Fourier transform shown in Fig. A.2. We see that

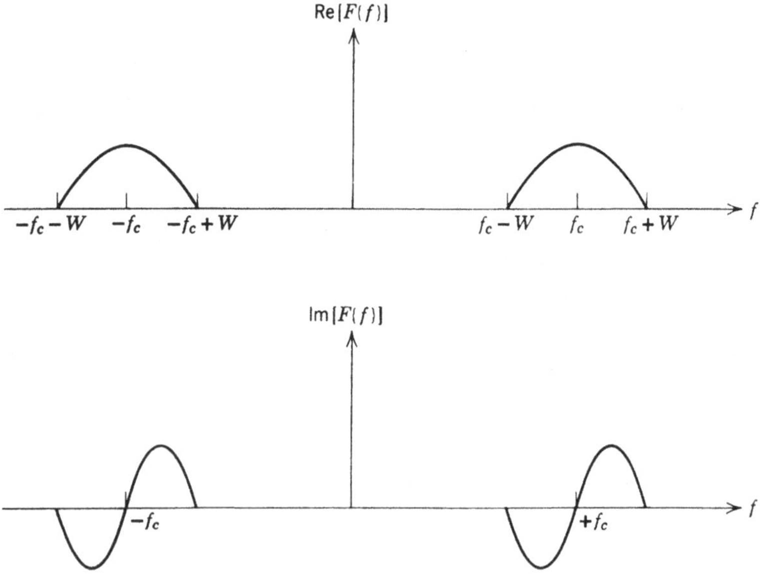


**Fig. A.2 Fourier transform of a bandpass signal.**

† We also use the transform $F\{f\} = \int_{-\infty}^{\infty} f(t)e^{-j2\pi ft}\,dt$. The braces $\{\cdot\}$ imply this definition.

the transform is bandlimited to frequencies within $\pm W$ cps of the carrier $f_c$. In practice, very few signals are strictly bandlimited. If the energy outside the band is negligible, it is convenient to neglect it. We refer to a signal that is essentially bandlimited around a carrier as a *bandpass* signal. Normally the bandwidth around the carrier is small compared to $\omega_c$, and so we also refer to the signal as a *narrow-band* signal. A precise statement about how small the bandwidth must be in order for the signal to be considered narrow-band is not necessary for our present discussion.

It is convenient to represent the signal in terms of two low-pass quadrature components, $f_c(t)$ and $f_s(t)$,

$$f_c(t) \triangleq [(\sqrt{2}\cos \omega_c t) f(t)]_{LP}, \tag{A.3}$$

$$f_s(t) \triangleq [(\sqrt{2}\sin \omega_c t) f(t)]_{LP}. \tag{A.4}$$

The symbol $[\cdot]_{LP}$ denotes the operation of passing the argument through an ideal low-pass filter with unity gain. The low-pass waveforms $f_c(t)$ and $f_s(t)$ can be generated physically, as shown in the left of Fig. A.3. The transfer function of the ideal low-pass filters is shown in Fig. A.4. Given $f_c(t)$ and $f_s(t)$, we could reconstruct $f(t)$ by multiplying $\cos \omega_c t$ and $\sin \omega_c t$ and adding the result as shown in the right side of Fig. A.3. Thus,

$$f(t) = \sqrt{2}\,[f_c(t)\cos(\omega_c t) + f_s(t)\sin(\omega_c t)]. \tag{A.5}$$

One method of verifying that the representation in (A.5) is valid is to follow the Fourier transforms through the various operations. A much easier procedure is to verify that the entire system in Fig. A.3 is identical

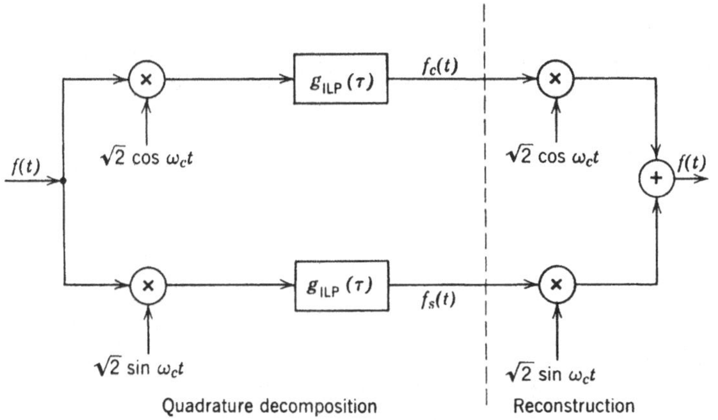


**Fig. A.3 Generation of quadrature components and reconstruction of bandpass signal.**

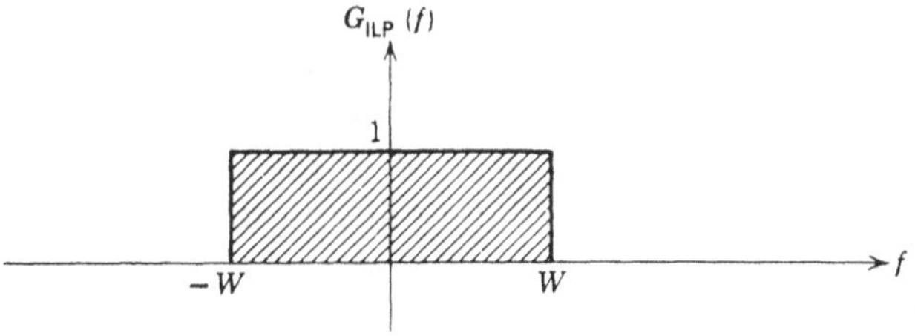


**Fig. A.4 Transfer function of ideal low-pass filter.**

with the ideal bandpass filter whose transfer function is shown in Fig. A.5.† To do this we put an impulse into the system and calculate its output. We denote the output due to an impulse at time $t = \tau$ as $g_\tau(t)$.

$$\begin{aligned} g_\tau(t) &= \sqrt{2}\cos(\omega_c t)\int_{-\infty}^{\infty} \delta(t-\tau)\sqrt{2}\cos(\omega_c u_1) g_{\text{ILP}}(t-u_1)\,du_1 \\ &\quad + \sqrt{2}\sin(\omega_c t)\int_{-\infty}^{\infty} \delta(t-\tau)\sqrt{2}\sin(\omega_2 u_2) g_{\text{ILP}}(t-u_2)\,du_2 \\ &= 2g_{\text{ILP}}(t-\tau)[\cos(\omega_c t)\cos(\omega_c\tau) + \sin(\omega_c t)\sin(\omega_c\tau)] \\ &= 2g_{\text{ILP}}(t-\tau)\cos[\omega_c(t-\tau)]. \end{aligned} \tag{A.6}$$

The transfer function is the Fourier transform of the impulse response,

$$\begin{aligned} &\int_{-\infty}^{\infty} 2g_{\text{ILP}}(\sigma)\cos(\omega_c\sigma)e^{-j\omega\sigma}\,d\sigma \\ &\qquad = \int_{-\infty}^{\infty} g_{\text{ILP}}(\sigma)e^{-j(\omega+\omega_c)\sigma}\,d\sigma + \int_{-\infty}^{\infty} g_{\text{ILP}}(\sigma)e^{-j(\omega-\omega_c)\sigma}\,d\sigma \\ &\qquad = G_{\text{ILP}}\{f+f_c\} + G_{\text{ILP}}\{f-f_c\}. \end{aligned} \tag{A.7}$$

The right side of (A.7) is just the transfer function shown in Fig. A.5, which is the desired result. Therefore the system in Fig. A.3 is just an ideal bandpass filter and any bandlimited input will pass through it undistorted. This verifies that our representation is valid. Notice that the

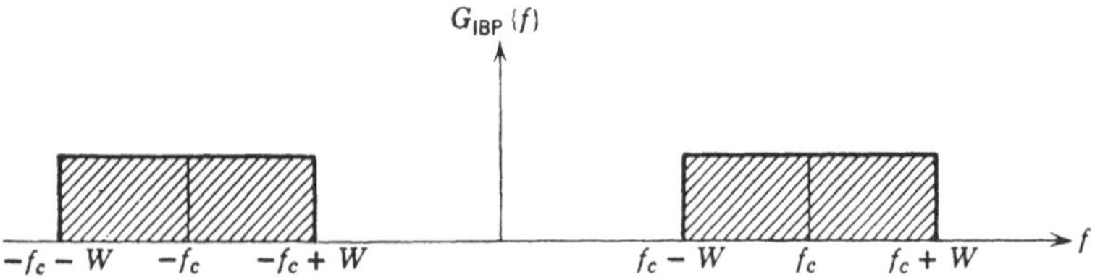


**Fig. A.5 Transfer function of overall system in Fig. A.3.**

† This procedure is due to [10, page 497].

assumption that $f(t)$ has unit energy implies

$$\int_{-\infty}^{\infty} (f_c^2(t) + f_s^2(t))\, dt = \int_{-\infty}^{\infty} f^2(t)\, dt = 1. \tag{A.8}$$

We can represent the low pass waveforms more compactly by defining a *complex signal*,

$$\boxed{\tilde{f}(t) \triangleq f_c(t) - jf_s(t).} \tag{A.9}$$

Equivalently,

$$\tilde{f}(t) = |\tilde{f}(t)|\, e^{j\phi_{\tilde{f}}(t)}, \tag{A.10}$$

where

$$|\tilde{f}(t)| = \sqrt{f_c^2(t) + f_s^2(t)} \tag{A.11}$$

and

$$\phi_{\tilde{f}}(t) = \tan^{-1}\left(\frac{f_s(t)}{f_c(t)}\right). \tag{A.12}$$

Notice that we can also write

$$\tilde{f}(t) = [f(t)\sqrt{2}\, e^{-j\omega_c t}]_{\text{LP}}. \tag{A.13}$$

The actual bandpass signal is

$$\begin{aligned} f(t) &= \sqrt{2E_t}\, \text{Re}\, [\tilde{f}(t)e^{j\omega_c t}] \\ &= \sqrt{2E_t}\, |\tilde{f}(t)|\, e^{j(\omega_c t + \phi_{\tilde{f}}(t))}. \end{aligned} \tag{A.14}$$

Some typical signals are shown in Fig. A.6. Notice that the signals in Fig. A.6*a–c* are not strictly bandlimited but do have negligible energy outside a certain frequency band. We see that $|\tilde{f}(t)|$ is the actual envelope of the narrow-band signal and $\phi_{\tilde{f}}(t) + \omega_c t$ is the *instantaneous phase*. The function $\tilde{f}(t)$ is commonly referred to as the *complex envelope*.

The utility of the complex representation for a bandpass signal will become more apparent as we proceed. We shall find that the results of interest can be derived and evaluated more easily in terms of the complex envelope.

There are several properties and definitions that we shall find useful in the sequel. All of the properties are straightforward to verify.

**Property 1.** Since the transmitted energy is unity, it follows from (A.8) that

$$\int_{-\infty}^{\infty} |\tilde{f}(t)|^2\, dt = 1. \tag{A.15}$$

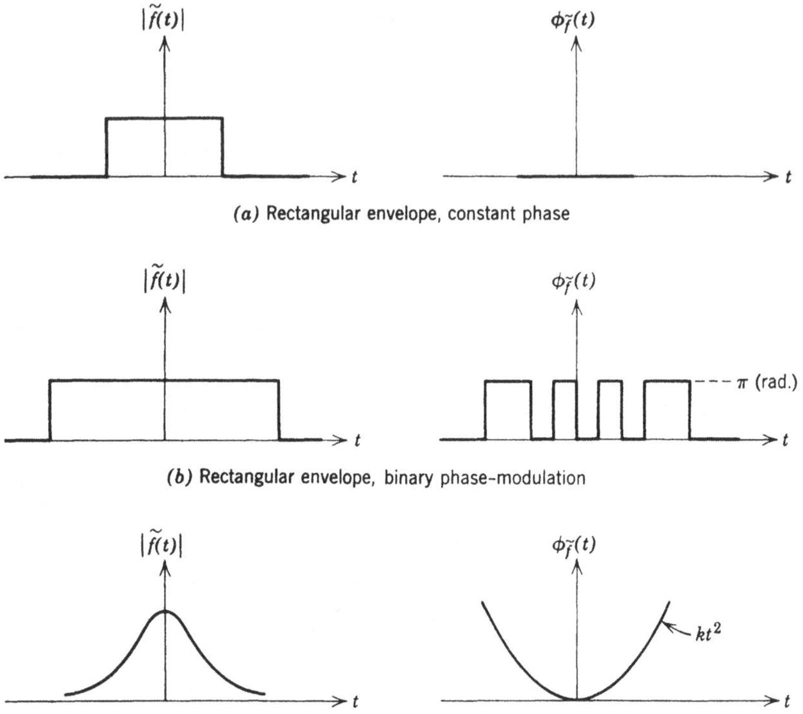


*(a)* Rectangular envelope, constant phase

*(b)* Rectangular envelope, binary phase-modulation

*(c)* Gaussian envelope, linear frequency sweep

**Fig. A.6 Typical signals.**

**Property 2.** The mean frequency of the envelope is defined as the first moment of the *energy* spectrum of the complex envelope,

$$\bar{\omega} \triangleq \int_{-\infty}^{\infty} \omega\, |\tilde{F}(j\omega)|^2 \frac{d\omega}{2\pi} = \int_{-\infty}^{\infty} \omega \tilde{S}_{\tilde{f}}(\omega) \frac{d\omega}{2\pi}, \tag{A.16}$$

where $\tilde{F}(j\omega)$ is the Fourier transform of $\tilde{f}(t)$,

$$\tilde{F}(j\omega) = \int_{-\infty}^{\infty} \tilde{f}(t) e^{-j\omega t}\, dt. \tag{A.17}$$

In our model the actual signal is $f(t)$ and is fixed. The complex envelope $\tilde{f}(t)$ depends on what frequency we denote as the carrier. Since the carrier frequency is at our disposal, we may always choose it so that

$$\bar{\omega} = 0 \tag{A.18}$$

(see Problem A.1.1). Later we shall see that other considerations enter into the choice of the carrier frequency, so that (A.18) may not apply in all cases.

**Property 3.** The *mean time* of the envelope is defined as the first moment of the squared magnitude of the complex envelope,

$$\bar{t} \triangleq \int_{-\infty}^{\infty} t\, |\tilde{f}(t)|^2\, dt = 0. \tag{A.19}$$

Since the time origin is arbitrary, we may always choose it so that

$$\bar{t} = \int_{-\infty}^{\infty} t\, |\tilde{f}(t)|^2\, dt = 0. \tag{A.20}$$

The assumptions in (A.18) and (A.20) will lead to algebraic simplifications in some cases.

**Property 4.** There are several quadratic quantities that are useful in describing the signal. The first two are

$$\overline{\omega^2} \triangleq \int_{-\infty}^{\infty} \omega^2\, |\tilde{F}(j\omega)|^2\, \frac{d\omega}{2\pi} \tag{A.21a}$$

and

$$\sigma_w^{\,2} \triangleq \overline{\omega^2} - (\bar{\omega})^2 \tag{A.21b}$$

The latter quantity is called the *mean-square bandwidth*. It is an approximate measure of the frequency spread of the signal.

Similarly, we define

$$\overline{t^2} = \int_{-\infty}^{\infty} t^2\, |\tilde{f}(t)|^2\, dt \tag{A.22a}$$

and

$$\sigma_t^{\,2} \triangleq \overline{t^2} - (\bar{t})^2. \tag{A.22b}$$

The latter quantity is called the *mean-square duration* and is an approximate measure of the time spread of the signal.

The definitions of the final quantities are

$$\overline{\omega t} = \operatorname{Im} \int_{-\infty}^{\infty} t\tilde{f}(t)\, \frac{d\tilde{f}^*(t)}{dt}\, dt \tag{A.23a}$$

and

$$\rho_{\omega t} = \frac{\overline{\omega t} - \bar{\omega}\bar{t}}{\sigma_\omega \sigma_t}. \tag{A.23b}$$

These definitions are less obvious. Later we shall see that $\rho_{\omega t}$ is a measure of the frequency modulation in $\tilde{f}(t)$.

The relations in (21)–(23) can be expressed in alternative ways using Fourier transform properties (see Problem A.1.2). Other useful interpretations are also developed in the problems.

**Property 5.** Consider two unit-energy bandpass signals $f_1(t)$ and $f_2(t)$. The correlation between the two signals is

$$\rho = \int_{-\infty}^{\infty} f_1(t) f_2(t)\, dt. \tag{A.24}$$

We now represent the two signals in terms of complex envelopes and the *same* carrier, $\omega_c$.

$$\tilde{f}_i(t) = [\sqrt{2}\, f_i(t) e^{-j\omega_c t}]_{\text{LP}}, \qquad i = 1, 2. \tag{A.25}$$

The complex correlation $\tilde{\rho}$ is defined to be

$$\tilde{\rho} = \int_{-\infty}^{\infty} \tilde{f}_1(t) \tilde{f}_2^*(t)\, dt. \tag{A.26a}$$

Then

$$\rho = \text{Re}\, \tilde{\rho}. \tag{A.26b}$$

To verify this, we write $\rho$ in terms of complex envelopes, perform the integration, and observe that the double frequency terms can be neglected.

This concludes our discussion of the complex envelope of a deterministic bandpass signal. We now consider bandpass systems.

## A.2 BANDPASS LINEAR SYSTEMS

We now develop a complex representation for bandpass linear systems. We first consider time-invariant systems and define a bandpass system in that context.

### A.2.1 Time-Invariant Systems

Consider a time-invariant linear system with impulse response $h(\sigma)$ and transfer function $H(j\omega)$.

$$H(j\omega) = \int_{-\infty}^{\infty} h(\sigma) e^{-j\omega\sigma}\, d\sigma. \tag{A.27}$$

A typical transform of interest has the magnitude shown in Fig. A.7. We see that it is bandlimited to a region about the carrier $\omega_c$. We want to represent the bandpass impulse response in terms of two quadrature components. Because $h(\sigma)$ is deterministic, we may use the results of Section A.1 directly. We define two low-pass functions,

$$h_c(\sigma) \triangleq [h(\sigma) \cos \omega_c \sigma]_{\text{LP}} \tag{A.28}$$

and

$$h_s(\sigma) \triangleq [h(\sigma) \sin \omega_c \sigma]_{\text{LP}}. \tag{A.29}$$

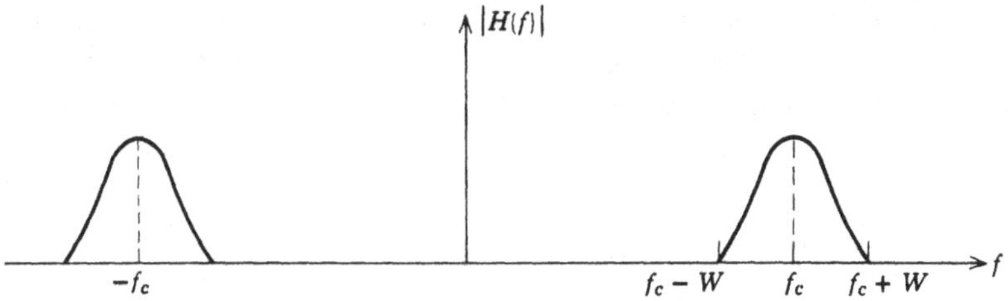


**Fig. A.7 Magnitude of the transfer function of a bandpass linear system.**

Then

$$h(\sigma) = 2h_c(\sigma) \cos (\omega_c\sigma) + 2h_s(\sigma) \sin (\omega_c\sigma). \tag{A.30}$$

Defining a *complex impulse response* as

$$\tilde{h}(\sigma) = h_c(\sigma) - jh_s(\sigma), \tag{A.31}$$

we have the complex representation

$$h(\sigma) \triangleq \text{Re}\,[2\tilde{h}(\sigma)e^{j\omega_c\sigma}]. \tag{A.32}$$

The introduction of the factor $\sqrt{2}$ is for convenience only.

We now derive an expression for the output of a bandpass linear system $h(t)$ when the input to the system is a bandpass waveform $f(t)$,

$$f(t) = \sqrt{2}\,\text{Re}\,[\tilde{f}(t)e^{j\omega_c t}]. \tag{A.33}$$

Notice that the carrier frequencies of the input signal and the system are identical. This common carrier frequency is implied in all our subsequent discussions. The output $y(t)$ is obtained by convolving $f(t)$ and $h(t)$.

$$\begin{aligned} y(t) &= \int_{-\infty}^{\infty} h(t-\sigma)f(\sigma)\,d\sigma \\ &= \int_{-\infty}^{\infty} [\tilde{h}(t-\sigma)e^{j\omega_c(t-\sigma)} + \tilde{h}^*(t-\sigma)e^{-j\omega_c(t-\sigma)}] \\ &\quad \times \left[\frac{\tilde{f}(\sigma)e^{j\omega_c\sigma} + \tilde{f}^*(\sigma)e^{-j\omega_c\sigma}}{\sqrt{2}}\right] d\sigma. \end{aligned} \tag{A.34}$$

Now we define

$$\boxed{\tilde{y}(t) \triangleq \int_{-\infty}^{\infty} \tilde{h}(t-\sigma)\tilde{f}(\sigma)\,d\sigma, \qquad -\infty < t < \infty.} \tag{A.35}$$

Because $\tilde{h}(t)$ and $\tilde{f}(t)$ are low-pass, the two terms in (A.34) containing $e^{\pm 2j\omega_c\sigma}$ integrate to approximately zero and can be neglected. Using (A.35) in the other two terms, we have

$$y(t) = \sqrt{2}\,\text{Re}\,[\tilde{y}(t)e^{j\omega_c t}]. \tag{A.36}$$

This result shows that the complex envelope of the output of a bandpass system is obtained by convolving the complex envelope of the input with the complex impulse response. [The $\sqrt{2}$ was introduced in (A.21) so that (A.35) would have a familiar form.]

### A.2.2 Time-Varying Systems

For time-varying bandpass systems, the complex impulse response is $\tilde{h}(t, \tau)$ and

$$h(t, u) = \text{Re}\,[2\tilde{h}(t, u)e^{j\omega_c(t-u)}]. \tag{A.37}$$

The complex envelope of the output is

$$\tilde{y}(t) = \int_{-\infty}^{\infty} \tilde{h}(t, u)\tilde{f}(u)\,du. \tag{A.38}$$

The actual bandpass output is given by (A.36).

### A.2.3 State-Variable Systems

In our work in Chapters I-6, II-2, and II-3, we encountered a number of problems in which a state-variable characterization of the system led to an efficient solution procedure. This is also true in the radar-sonar area. We now develop a procedure for characterizing bandpass systems using complex state variables.† The complex input is $\tilde{f}(t)$. The complex state equation is

$$\frac{d\tilde{\mathbf{x}}(t)}{dt} = \tilde{\mathbf{F}}(t)\tilde{\mathbf{x}}(t) + \tilde{\mathbf{G}}(t)\tilde{f}(t), \qquad T_i \leq t, \tag{A.39}$$

with initial condition $\tilde{\mathbf{x}}(T_i)$. The observation equation is

$$\tilde{y}(t) = \tilde{\mathbf{C}}(t)\tilde{\mathbf{x}}(t). \tag{A.40}$$

The matrices $\tilde{\mathbf{F}}(t)$, $\tilde{\mathbf{G}}(t)$, and $\tilde{\mathbf{C}}(t)$ are complex matrices. The complex state vector $\tilde{\mathbf{x}}(t)$ and complex output $\tilde{y}(t)$ are low-pass compared to $\omega_c$. The complex block diagram is shown in Fig. A.8.

We define a complex state transition matrix $\tilde{\boldsymbol{\phi}}(t, \tau)$ such that

$$\frac{d}{dt}\tilde{\boldsymbol{\phi}}(t, \tau) = \tilde{\mathbf{F}}(t)\tilde{\boldsymbol{\phi}}(t, \tau), \tag{A.41}$$

$$\tilde{\boldsymbol{\phi}}(t, t) = \mathbf{I}. \tag{A.42}$$

† This discussion is based on our work, which originally appeared in [9]. It is worthwhile emphasizing that most of the complex state-variable results are logical extensions of the real state-variable results.

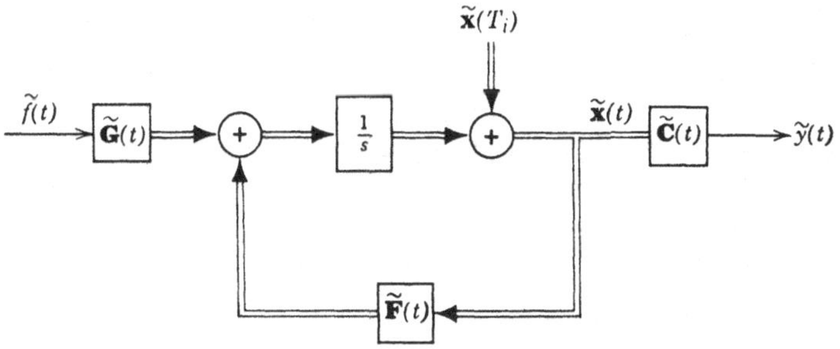


**Fig. A.8 State-variable model for a complex linear system.**

Then it is readily verified that

$$\tilde{y}(t) = \tilde{\mathbf{C}}(t)\left[\tilde{\boldsymbol{\phi}}(t, T_i)\tilde{\mathbf{x}}(T_i) + \int_{T_i}^{t} \tilde{\boldsymbol{\phi}}(t, \tau)\tilde{\mathbf{G}}(\tau)\tilde{f}(\tau)\, d\tau\right]. \tag{A.43}$$

The first term is the output due to the initial conditions, and the second term is the output due to $\tilde{f}(t)$. The complex impulse response is obtained by letting the initial condition $\tilde{\mathbf{x}}(T_i)$ equal zero and $T_i = -\infty$. Then

$$\tilde{y}(t) = \tilde{\mathbf{C}}(t)\int_{-\infty}^{t} \tilde{\boldsymbol{\phi}}(t, \tau)\tilde{\mathbf{G}}(\tau)\tilde{f}(\tau)\, d\tau, \qquad -\infty < t. \tag{A.44}$$

Recall from (A.38) that

$$\tilde{y}(t) = \int_{-\infty}^{\infty} \tilde{h}(t, \tau)\tilde{f}(\tau)\, d\tau, \qquad -\infty < t. \tag{A.45}$$

Thus,

$$\tilde{h}(t, \tau) = \begin{cases} \tilde{\mathbf{C}}(t)\tilde{\boldsymbol{\phi}}(t, \tau)\tilde{\mathbf{G}}(\tau), & -\infty < \tau \le t, \\ 0, & \text{elsewhere.} \end{cases} \tag{A.46}$$

Notice that this is a realizable impulse response. Using (A.46) in (A.37) gives the actual bandpass impulse response,

$$\begin{aligned} h(t, \tau) &= \operatorname{Re}\,[2\tilde{h}(t, \tau)e^{j\omega_c(t-\tau)}] \\ &= \begin{cases} \operatorname{Re}\,[2\tilde{\mathbf{C}}(t)\tilde{\boldsymbol{\phi}}(t, \tau)\tilde{\mathbf{G}}(\tau)e^{j\omega_c(t-\tau)}], & -\infty < \tau \le t, \\ 0, & \text{elsewhere.} \end{cases} \end{aligned} \tag{A.47}$$

There are two alternative procedures that we can use actually to implement the system shown in Fig. A.8. The first procedure is to construct a circuit that is essentially a bandpass analog computer. The second procedure is to perform the operations digitally using complex arithmetic.

We now consider the problem of representing bandpass random processes.

## A.3 BANDPASS RANDOM PROCESSES

We would expect that an analogous complex representation could be obtained for bandpass random processes. In this section we discuss three classes of random processes:

1. Stationary processes.
2. Nonstationary processes.
3. Finite state processes.

Throughout our discussion we assume that the random processes have zero means. We begin our discussion with stationary processes.

### A.3.1 Stationary Processes

A typical bandpass spectrum is shown in Fig. A.9. It is bandlimited to $\pm W$ cps about $\omega_c$. We want to represent $n(t)$ as

$$n(t) = \sqrt{2}\, n_c(t) \cos(\omega_c t) + \sqrt{2}\, n_s(t) \sin(\omega_c t), \qquad \text{(A.48)}$$

where $n_c(t)$ and $n_s(t)$ are low-pass functions that are generated as shown in the left side of Fig. A.10.

$$n_c(t) = [(\sqrt{2}\cos(\omega_c t))n(t)]_{\text{LP}} \qquad \text{(A.49)}$$

and

$$n_s(t) = [(\sqrt{2}\sin(\omega_c t))n(t)]_{\text{LP}}. \qquad \text{(A.50)}$$

The over-all system, showing both the decomposition and reconstruction, is depicted in Fig. A.10. In Section A.1 we showed that this system was just an ideal bandpass filter (see Fig. A.5). Thus, the representation is valid for all processes that are bandlimited to $\pm W$ cps around $\omega_c$.

Alternatively, in complex notation, we define

$$\boxed{\tilde{n}(t) \triangleq n_c(t) - jn_s(t)} \qquad \text{(A.51)}$$

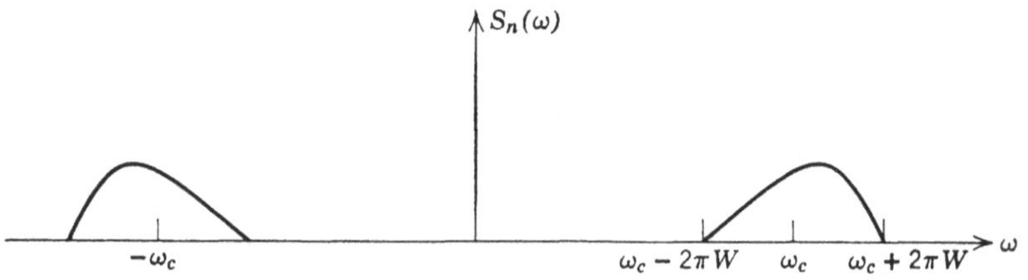


**Fig. A.9 Typical spectrum for bandpass process.**

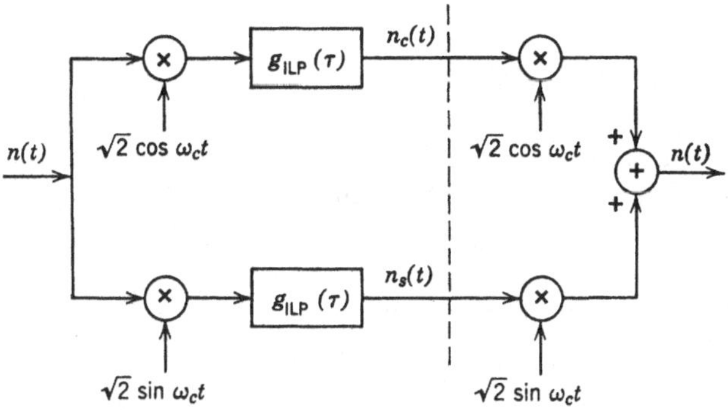


**Fig. A.10 Generation of quadrature components and reconstruction of bandpass process.**

or

$$\boxed{\tilde{n}(t) = [\sqrt{2}\, n(t)e^{-j\omega_c t}]_{\text{LP}}} \tag{A.52}$$

and write

$$n(t) = \sqrt{2}\,\text{Re}\,[\tilde{n}(t)e^{j\omega_c t}]. \tag{A.53}$$

We now derive the statistical properties of $\tilde{n}(t)$.

The operation indicated in (A.52) corresponds to the block diagram in Fig. A.11. We first compute the covariance function of $\tilde{z}(t)$,

$$\begin{aligned}
\tilde{K}_{\tilde{z}}(t, t-\tau) \triangleq E[\tilde{z}(t)\tilde{z}^*(t-\tau)] &= E[\sqrt{2}\, n(t)e^{-j\omega_c t}\cdot \sqrt{2}\, n(t-\tau)e^{j\omega_c(t-\tau)}] \\
&= 2(E[n(t)n(t-\tau)])e^{-j\omega_c\tau} \\
&= 2K_n(\tau)e^{-j\omega_c\tau}.
\end{aligned} \tag{A.54}$$

The spectrum of $\tilde{z}(t)$ is

$$\begin{aligned}
\tilde{S}_{\tilde{z}}(\omega) &= \int_{-\infty}^{\infty} K_{\tilde{z}}(\tau)e^{-j\omega\tau}\, d\tau = 2\int_{-\infty}^{\infty} K_n(\tau)e^{-j(\omega+\omega_c)\tau}\, d\tau \\
&= 2S_n(\omega+\omega_c).
\end{aligned} \tag{A.55}$$

Now $\tilde{n}(t)$ is related to $\tilde{z}(t)$ by an ideal low-pass filter. Thus,

$$\boxed{\tilde{S}_{\tilde{n}}(\omega) = 2[S_n(\omega+\omega_c)]_{\text{LP}}} \tag{A.56}$$

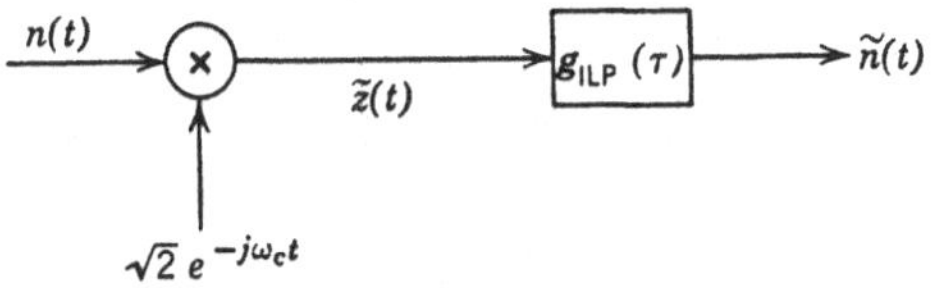


**Fig. A.11 Generation of $\tilde{n}(t)$.**

and

$$E[\tilde{n}(t)\tilde{n}^*(t-\tau)] \triangleq \tilde{K}_{\tilde{n}}(\tau) = 2\int_{-2\pi W}^{2\pi W} S_n(\omega + \omega_c)e^{j\omega\tau}\,\frac{d\omega}{2\pi}. \tag{A.57}$$

The next result of interest concerns the expectation without a conjugate on the second term. We shall prove that

$$\boxed{E[\tilde{n}(t_1)\tilde{n}(t_2)] = 0, \qquad \text{for all} \quad t_1, t_2.} \tag{A.58}$$

$$E[\tilde{n}(t_1)\tilde{n}(t_2)] =$$

$$2E\left\{\int_{-\infty}^{\infty} n(x_1)e^{-j\omega_c x_1} g_{\text{ILP}}(t_1 - x_1)\,dx_1 \int_{-\infty}^{\infty} n(x_2)e^{-j\omega_c x_2} g_{\text{ILP}}(t_2 - x_2)\,dx_2\right\}$$

$$= 2\iint_{-\infty}^{\infty} K_n(x_1 - x_2)e^{-j\omega_c(x_1+x_2)} g_{\text{ILP}}(t_1 - x_1) g_{\text{ILP}}(t_2 - x_2)\,dx_1\,dx_2$$

$$= 2\iiint_{-\infty}^{\infty} S_n(f)e^{j2\pi f(x_1-x_2)-j2\pi f_c(x_1+x_2)} g_{\text{ILP}}(t_1 - x_1) g_{\text{ILP}}(t_2 - x_2)\,dx_1\,dx_2\,df$$

$$= 2\int_{-\infty}^{\infty} S_n(f)\,df \int_{-\infty}^{\infty} g_{\text{ILP}}(t_1 - x_1)e^{j2\pi x_1(f-f_c)}\,dx_1$$

$$\times \int_{-\infty}^{\infty} g_{\text{ILP}}(t_2 - x_2)e^{-j2\pi x_2(f+f_c)}\,dx_2$$

$$= 2\int_{-\infty}^{\infty} S_n(f)\{G_{\text{ILP}}(f - f_c)G_{\text{ILP}}^*(f + f_c)\}\,e^{j2\pi t_1(f-f_c)-j2\pi t_2(f+f_c)}\,df. \tag{A.59}$$

Now the term in the braces is identically zero for all $f$ if $f_c > W$. Thus the integral is zero and (A.58) is valid. The property in (A.58) is important because it enables us to characterize the complex process in terms of a single covariance function. We can obtain the correlation function of the actual process from this single covariance function.

$$K_n(t, t-\tau) = E[n(t)n(t-\tau)]$$

$$= E\left\{\left[\frac{\sqrt{2}\,\tilde{n}(t)e^{j\omega_c t} + \sqrt{2}\,\tilde{n}^*(t)e^{-j\omega_c t}}{2}\right]\right.$$

$$\left.\times \left[\frac{\sqrt{2}\,\tilde{n}(t-\tau)e^{j\omega_c(t-\tau)} + \sqrt{2}\,\tilde{n}^*(t-\tau)e^{-j\omega_c(t-\tau)}}{2}\right]\right\}$$

$$= \text{Re}\,[\tilde{K}_{\tilde{n}}(\tau)e^{j\omega_c\tau}] + \text{Re}\,\{E[\tilde{n}(t)\tilde{n}(t-\tau)]e^{j\omega_c(2t-\tau)}\}. \tag{A.60}$$

Using (A.58) gives

$$\boxed{K_n(\tau) = \text{Re}\,[\tilde{K}_{\tilde{n}}(\tau)e^{j\omega_c\tau}].} \tag{A.61}$$

In terms of spectra,

$$S_n(\omega) = \int_{-\infty}^{\infty} \frac{\tilde{K}_{\tilde{n}}(\tau)e^{j\omega_c\tau} + \tilde{K}_{\tilde{n}}^*(\tau)e^{-j\omega_c\tau}}{2}\, e^{-j\omega\tau}\, d\tau, \tag{A.62}$$

or

$$\boxed{S_n(\omega) = \frac{\tilde{S}_{\tilde{n}}(\omega - \omega_c) + \tilde{S}_{\tilde{n}}(-\omega - \omega_c)}{2},} \tag{A.63}$$

where we have used the fact that $\tilde{S}_{\tilde{n}}(\omega)$ is a real function of $\omega$.

The relations in (A.61) and (A.63) enable us to obtain the statistics of the bandpass process from the complex process, and vice versa. In Fig. A.12, we indicate this for some typical spectra. Notice that the spectrum of the complex process is even if and only if the bandpass process is symmetric about the carrier $\omega_c$. In Fig. A.13, we indicate the behavior for some typical pole-zero plots. We see that the pole-zero plots are always symmetric about the $j\omega$-axis. This is because $\tilde{S}_{\tilde{n}}(\omega)$ is real. The plots are not necessarily symmetric about the $\sigma$-axis, because $\tilde{S}_{\tilde{n}}(\omega)$ is not necessarily even.

Although we shall normally work with the complex process, it is instructive to discuss the statistics of the quadrature components briefly. These follow directly from (A.57) and (A.58).

$$\begin{aligned} E[\tilde{n}(t)\tilde{n}^*(t-\tau)] &= E[(n_c(t) + jn_s(t))(n_c(t-\tau) - jn_s(t-\tau))] \\ &= K_c(\tau) + K_s(\tau) + j[K_{sc}(\tau) - K_{cs}(\tau)] \\ &= \tilde{K}_{\tilde{n}}(\tau) \end{aligned} \tag{A.64}$$

and

$$\begin{aligned} E[\tilde{n}(t)\tilde{n}(t-\tau)] &= E[(n_c(t) + jn_s(t))(n_c(t-\tau) + jn_s(t-\tau))] \\ &= K_c(\tau) - K_s(\tau) + j[K_{sc}(\tau) + K_{cs}(\tau)] \\ &= 0. \end{aligned} \tag{A.65}$$

Therefore,

$$\boxed{K_c(\tau) = K_s(\tau) = \tfrac{1}{2}\text{Re}\,[\tilde{K}_{\tilde{n}}(\tau)]} \tag{A.66}$$

and

$$\boxed{K_{sc}(\tau) = -K_{cs}(\tau) = -K_{sc}(-\tau) = \tfrac{1}{2}\,\text{Im}\,[\tilde{K}_{\tilde{n}}(\tau)].} \tag{A.67}$$

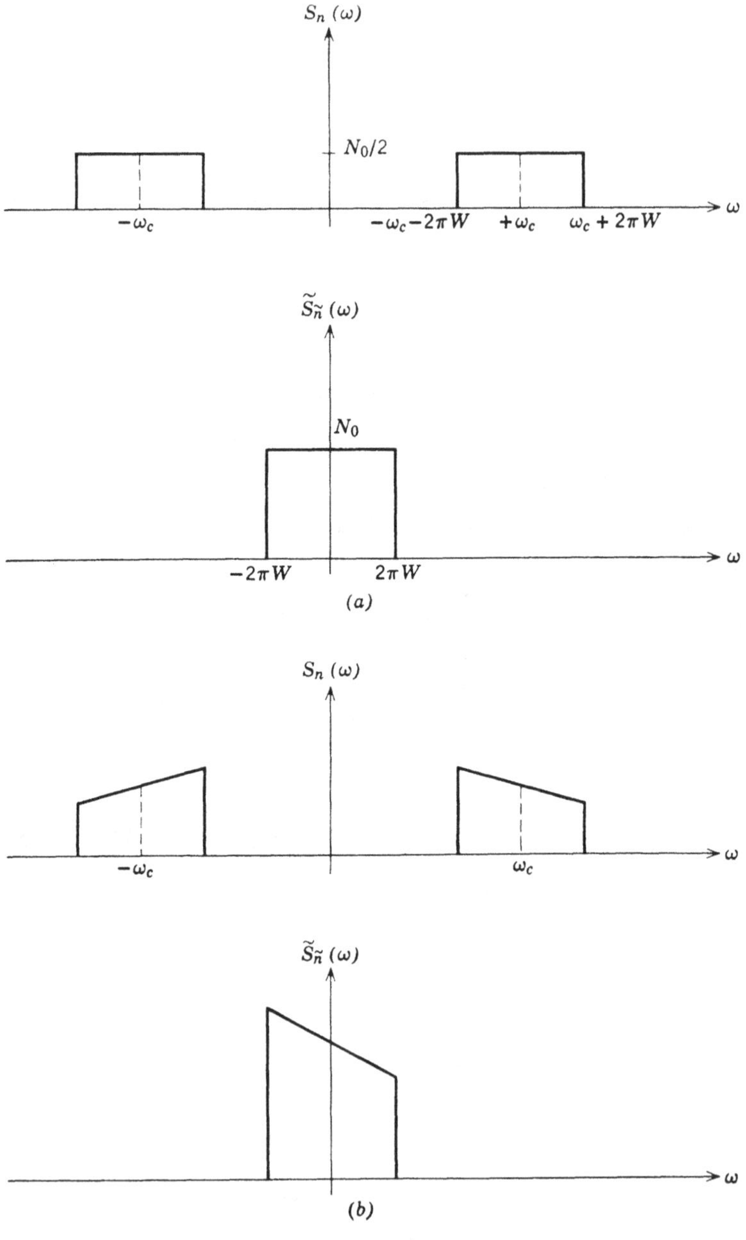


**Fig. A.12 Representative spectra.**

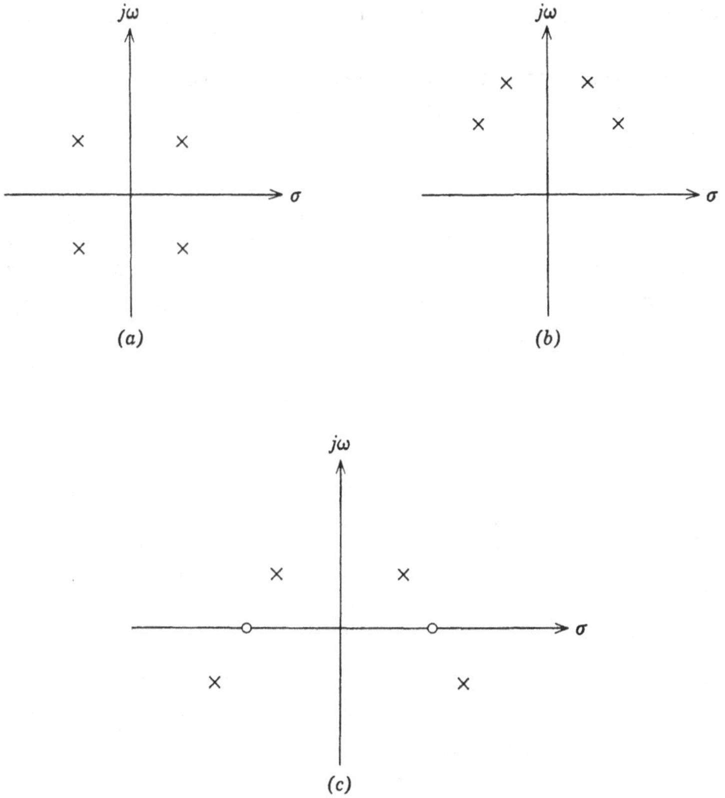


**Fig. A.13 Possible pole-zero plots for spectra of complex processes.**

In terms of spectra,

$$S_c(\omega) = S_s(\omega) = \frac{1}{2}\int_{-\infty}^{\infty} (\mathrm{Re}\,[\tilde{K}_{\tilde{n}}(\tau)])e^{-j\omega\tau}\,d\tau$$

$$= \frac{1}{2}\left[\frac{\tilde{S}_{\tilde{n}}(\omega) + \tilde{S}_{\tilde{n}}(-\omega)}{2}\right], \tag{A.68}$$

or

$$S_c(\omega) = \tfrac{1}{2}[\tilde{S}_{\tilde{n}}(\omega)]_{\mathrm{EV}}$$

$$= [[S_n(\omega + \omega_c)]_{\mathrm{LP}}]_{\mathrm{EV}}, \tag{A.69}$$

where $[\cdot]_{\mathrm{EV}}$ denotes the operation of taking the even part. Similarly,

$$S_{cs}(\omega) = -\frac{1}{2}\int_{-\infty}^{\infty} \mathrm{Im}\,[\tilde{K}_{\tilde{n}}(\tau)]e^{-j\omega\tau}\,d\tau$$

$$= \frac{j}{2}\,[\tilde{S}_{\tilde{n}}(\omega)]_{\mathrm{ODD}}$$

$$= j[[S_n(\omega + \omega_c)]_{\mathrm{LP}}]_{\mathrm{ODD}}. \tag{A.70}$$

Notice that $S_{cs}(\omega)$ is imaginary. [This is obvious from the asymmetry in (A.67).] From (A.70) we see that the quadrature processes are correlated unless the spectrum is even around the carrier. Notice that, at any single time instant, $n_c(t_1)$ and $n_s(t_1)$ are uncorrelated. This is because (A.67) implies that

$$K_{cs}(0) = 0. \tag{A.71}$$

**Complex White Processes.** Before leaving our discussion of the second-moment characterization of complex processes, we define a particular process of interest. Consider the process $w(t)$ whose spectrum is shown in Fig. A.14. The complex envelope is

$$\tilde{w}(t) = w_c(t) - jw_s(t). \tag{A.72}$$

Using (A.69) and (A.70) gives

$$S_{n_c}\{f\} = S_{n_s}\{f\} = \begin{cases} \dfrac{N_0}{2}, & |f| \leq W, \\ 0, & \text{elsewhere,} \end{cases} \tag{A.73}$$

and

$$S_{n_c n_s}\{f\} = 0. \tag{A.74}$$

The covariance function of $\tilde{w}(t)$ is

$$\tilde{K}_{\tilde{w}}(t, u) \triangleq \tilde{K}_{\tilde{w}}(\tau) = 2K_{w_c}(\tau)$$

$$= N_0\left(\frac{\sin(2\pi W\tau)}{\pi\tau}\right), \qquad -\infty < \tau < \infty. \tag{A.75}$$

Now, if $W$ is larger than the other bandwidths in the system of interest, we can approximate (A.75) with an impulse. Letting $W \to \infty$ in (A.75) gives

$$\boxed{\tilde{K}_{\tilde{w}}(\tau) = N_0\, \delta(\tau).} \tag{A.76}$$

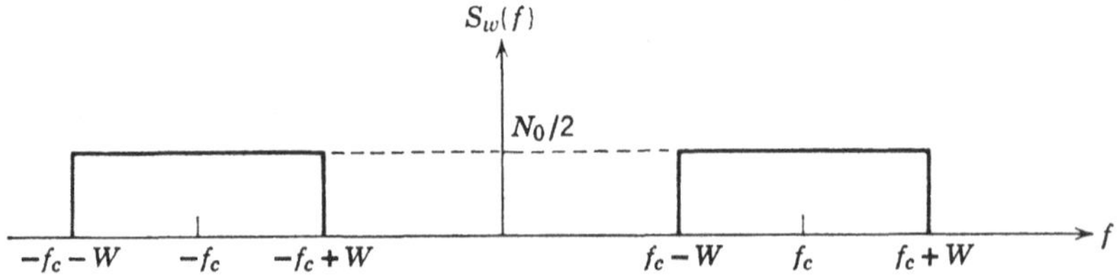


**Fig. A.14 Spectrum of bandpass white noise.**

We refer to $\tilde{w}(t)$ as a *complex white noise process*. We refer to the actual process $w(t)$ as a bandpass white noise process. Notice that, just as in the case of white noise, they are convenient approximations to the actual physical process.

**Complex Gaussian Processes.** In many cases of interest to us the processes are *Gaussian* random processes. If $n(t)$ is a stationary Gaussian process, then $n_c(t)$ and $n_s(t)$ are stationary jointly Gaussian processes, because they are obtained by linear operations on $n(t)$. The complex envelope is

$$\tilde{n}(t) = n_c(t) - jn_s(t), \tag{A.77}$$

so that we might logically refer to it as a *stationary complex Gaussian random process*. Since we shall use this idea frequently, an exact definition is worthwhile.

**Definition.** Let $n_c(t)$ and $n_s(t)$ be two zero-mean stationary jointly Gaussian random processes with identical covariance functions. The process $\tilde{n}(t)$ is defined by (A.77). The relation

$$E[\tilde{n}(t)\tilde{n}(t - \tau)] = 0, \qquad \text{for all} \quad t \text{ and } \tau, \tag{A.78}$$

is satisfied. The process $\tilde{n}(t)$ is a *zero-mean stationary complex Gaussian random process*.

The modification to include a mean value function is straightforward. Notice that a complex process whose real and imaginary parts are both Gaussian processes is not necessarily Gaussian. The condition in (A.78) must be satisfied. This implies that the real part and imaginary part are Gaussian processes with identical characteristics that are related by the covariance function in (A.67). They are statistically independent *if and only if* the original spectrum is symmetric around the carrier. We also note that a real Gaussian process is *not* a special case of a complex Gaussian process.

If we sample the complex envelope at time $t_1$, we get a complex random variable $\tilde{n}(t_1)$. To specify the density of a complex random variable, we need the joint density of the real part, $n_c(t_1)$, and the imaginary part, $n_s(t_1)$. Since $n_c(t_1)$ and $n_s(t_1)$ are samples of a jointly Gaussian process, they are jointly Gaussian random variables. Since (A.71) implies that they are uncorrelated, we know that they are statistically independent. Therefore,

$$p_{n_{c_{t_1}}, n_{s_{t_1}}}(N_c, N_s) = \frac{1}{2\pi\sigma_n^2} \exp\left\{-\frac{N_c^2 + N_s^2}{2\sigma_n^2}\right\}, \qquad -\infty < N_c, N_s < \infty, \tag{A.79}$$

where

$$\sigma_{\tilde{n}}^2 = K_c(0) = K_s(0) = \tfrac{1}{2}K_{\tilde{n}}(0). \tag{A.80}$$

Equivalently,

$$\boxed{p_{\tilde{n}_{t_1}}(\tilde{N}) = \frac{1}{2\pi\sigma_n^{\,2}} \exp\left\{-\frac{|\tilde{N}|^2}{2\sigma_n^{\,2}}\right\}, \qquad -\infty < \text{Re}\,[\tilde{N}],\ \text{Im}\,[\tilde{N}] < \infty.} \tag{A.81}$$

We define a *complex Gaussian random variable* as a random variable whose probability density has the form in (A.81). Notice that

$$E(|\tilde{n}_{t_1}|^2) = 2\sigma_n^{\,2}. \tag{A.82}$$

Properties analogous to those for real Gaussian random variables and real Gaussian random processes follow easily (see Problems A.3.1–A.3.7). One property that we need corresponds to the definitions on page I-183. Define

$$\tilde{y} = \int_{T_\alpha}^{T_\beta} \tilde{g}(u)\tilde{x}(u)\,du, \tag{A.83}$$

where $\tilde{g}(u)$ is a function such that $E[|\tilde{y}|^2] < \infty$. If $\tilde{x}(u)$ is a complex Gaussian process, then $\tilde{y}$ is a complex Gaussian random variable. This result follows immediately from the above definitions.

A particular complex Gaussian process that we shall use frequently is the complex Gaussian white noise process $\tilde{w}(t)$. It is a complex Gaussian process whose covariance function is given by (A.76).

Two other probability densities are of interest. We can write $\tilde{n}(t)$ in terms of a magnitude and phase angle.

$$\tilde{n}(t) = |\tilde{n}(t)|\, e^{j\phi_{\tilde{n}}(t)}. \tag{A.84}$$

The magnitude corresponds to the envelope of the actual random process. It is easy to demonstrate that it is a Rayleigh random variable at any given time. The phase angle corresponds to the instantaneous phase of the actual random process minus $\omega_c t$, and is a uniform random variable that is independent of the envelope variable. Notice that the envelope and phase processes are not independent processes.

We now turn our attention to nonstationary processes.

### A.3.2 Nonstationary Processes

A physical situation in which we encounter nonstationary processes is the reflection of a deterministic signal from a fluctuating point target. We shall see that an appropriate model for the complex envelope of the return is

$$\tilde{s}(t) = \tilde{f}(t)\tilde{b}(t), \tag{A.85}$$

where $\tilde{f}(t)$ is a complex deterministic signal and $\tilde{b}(t)$ is a zero-mean stationary complex Gaussian process. We see that $\tilde{s}(t)$ is a zero-mean nonstationary process whose second-moment characteristics are

$$E[\tilde{s}(t)\tilde{s}^*(u)] = \tilde{f}(t)\tilde{K}_{\tilde{b}}(t-u)\tilde{f}^*(u) \tag{A.86}$$

and

$$E[\tilde{s}(t)\tilde{s}(u)] = \tilde{f}(t)E[(\tilde{b}(t)\tilde{b}(u))]\tilde{f}^*(u) = 0. \tag{A.87}$$

The condition in (A.87) corresponds to the result for stationary processes in (A.58) and enables us to characterize the complex process in terms of a single covariance function. Without this condition, the complex notation is less useful, and so we include it as a condition on the nonstationary processes that we study. Specifically, we consider processes that can be represented as

$$n(t) \triangleq \sqrt{2}\,\text{Re}\,[\tilde{n}(t)e^{j\omega_c t}], \tag{A.88}$$

where $n(t)$ is a complex low-pass process such that

$$E[\tilde{n}(t)\tilde{n}^*(\tau)] = \tilde{K}_{\tilde{n}}(t, \tau) \tag{A.89}$$

and

$$E[\tilde{n}(t)\tilde{n}^*(\tau)] = 0, \qquad \text{for all} \quad t \text{ and } \tau. \tag{A.90†}$$

For a nonstationary process to be low-pass, all of its eigenfunctions with non-negligible eigenvalues must be low-pass compared to $\omega_c$. This requirement is analogous to the spectrum requirement for stationary processes.

The covariance of the actual bandpass process is

$$\begin{aligned} K_n(t, u) &= E[n(t)n(u)] \\ &= E\left\{\left[\frac{\tilde{n}(t)e^{j\omega_c t} + \tilde{n}^*(t)e^{-j\omega_c t}}{\sqrt{2}}\right]\left[\frac{\tilde{n}(u)e^{j\omega_c u} + \tilde{n}^*(u)e^{-j\omega_c u}}{\sqrt{2}}\right]\right\} \\ &= \text{Re}\left\{\tilde{K}_{\tilde{n}}(t, u)e^{j\omega_c(t-u)}\right\} + \text{Re}\left\{E[\tilde{n}(t)\tilde{n}(u)]e^{j\omega_c(t+u)}\right\}. \end{aligned} \tag{A.91}$$

The second term on the right-hand side is zero because of the assumption in (A.90). Thus, we have the desired one-to-one correspondence between the second-moment characteristics of the two processes $n(t)$ and $\tilde{n}(t)$. The assumption in (A.90) is not particularly restrictive, because most of the processes that we encounter in practice satisfy it.

As before, the eigenvalues and eigenfunctions of a random process play an important role in many of our discussions. All of our discussion in

† It is worthwhile emphasizing that (A.90) has to be true for the complex envelope of a *stationary* bandpass process. For nonstationary processes it is an additional assumption. Examples of nonstationary processes that do not satisfy (A.90) are given in [11] and [12].

Chapter I-3 (page I-166) carries over to complex processes. The equation specifying the eigenvalues and eigenfunctions is

$$\tilde{\lambda}_i\tilde{\phi}_i(t) = \int_{T_i}^{T_f} \tilde{K}_{\tilde{n}}(t, u)\tilde{\phi}_i(u)\, du, \qquad T_i \le t \le T_f. \tag{A.92}$$

We assume that the kernel is Hermitian,

$$\tilde{K}_{\tilde{n}}(t, u) = \tilde{K}_{\tilde{n}}^*(u, t). \tag{A.93}$$

This is analogous to the symmetry requirement in the real case and is satisfied by all complex covariance functions. The eigenvalues of a Hermitian kernel are all real. We would expect this because the spectrum is real in the stationary case. We now look at the complex envelope process and the actual bandpass process and show how their eigenfunctions and eigenvalues are related. We first write the pertinent equations for the two processes and then show their relationship.

For the bandpass random process, we have from Chapter I-3 that

$$n(t) = \underset{K\to\infty}{\text{l.i.m.}} \sum_{i=1}^{K} n_i\phi_i(t), \qquad T_i \le t \le T_f, \tag{A.94}$$

where the $\phi_i(t)$ satisfy

$$\lambda_i\phi_i(t) = \int_{T_i}^{T_f} K_n(t, u)\phi_i(u)\, du, \qquad T_i \le t \le T_f \tag{A.95}$$

and the coefficients are

$$n_i = \int_{T_i}^{T_f} n(t)\phi_i(t)\, dt. \tag{A.96}$$

This implies that

$$E[n_i n_j] = \lambda_i\delta_{ij} \tag{A.97}$$

and

$$K_n(t, u) = \sum_{i=1}^{\infty} \lambda_i\phi_i(t)\phi_i(u), \qquad T_i \le t, u \le T_f. \tag{A.98}$$

Similarly, for the complex envelope random process,

$$\tilde{n}(t) = \underset{K\to\infty}{\text{l.i.m.}} \sum_{i=1}^{K} \tilde{n}_i\tilde{\phi}_i(t), \qquad T_i \le t \le T_f, \tag{A.99}$$

where the $\tilde{\phi}_i(t)$ satisfy the equation

$$\tilde{\lambda}_i\tilde{\phi}_i(t) = \int_{T_i}^{T_f} \tilde{K}_{\tilde{n}}(t, u)\tilde{\phi}_i(u)\, du, \qquad T_i \le t \le T_f. \tag{A.100}$$

The complex eigenfunctions are orthonormal,

$$\int_{T_i}^{T_f} \tilde{\phi}_i(t)\tilde{\phi}_j^*(t)\,dt = \delta_{ij}. \tag{A.101}$$

The coefficients are

$$\tilde{n}_i = \int_{T_i}^{T_f} \tilde{n}(t)\tilde{\phi}_i^*(t)\,dt. \tag{A.102}$$

We can then show that

$$E[\tilde{n}_i\tilde{n}_j^*] = \tilde{\lambda}_i\delta_{ij}, \tag{A.103}$$

$$E[n_i n_j] = 0, \qquad \text{for all} \quad i \text{ and } j, \tag{A.104}$$

and

$$\tilde{K}_n(t, u) = \sum_{i=1}^{\infty} \tilde{\lambda}_i\tilde{\phi}_i(t)\tilde{\phi}_i^*(u), \qquad T_i \le t, u \le T_f. \tag{A.105}$$

The processes are related by (A.88) and (A.91),

$$n(t) = \sqrt{2}\,\mathrm{Re}\,[\tilde{n}(t)e^{j\omega_c t}], \tag{A.106}$$

$$K_n(t, u) = \mathrm{Re}\,[\tilde{K}_n(t, u)e^{j\omega_c(t-u)}]. \tag{A.107}$$

To find how the eigenfunctions are related, we substitute

$$\phi_i(t) = \sqrt{2}\,\mathrm{Re}\,[\tilde{\phi}_i(t)e^{j(\omega_c t+\theta)}], \qquad T_i \le t \le T_f \tag{A.108}$$

into (A.95) and use (A.107). The result is

$$\lambda_i[\tilde{\phi}_i(t)e^{j(\omega_c t+\theta)} + \tilde{\phi}_i^*(t)e^{-j(\omega_c t+\theta)}]$$
$$= \frac{1}{2}\left\{e^{j(\omega_c t+\theta)}\int_{T_i}^{T_f} \tilde{K}_{\tilde{n}}(t, u)\tilde{\phi}_i(u)\,du + e^{-j(\omega_c t+\theta)}\int_{T_i}^{T_f} \tilde{K}_{\tilde{n}}^*(t, u)\tilde{\phi}_i^*(u)\,du\right\}. \tag{A.109}$$

Equivalently,

$$\mathrm{Re}\left\{\left(\lambda_i\tilde{\phi}_i(t) - \frac{1}{2}\int_{T_i}^{T_f} \tilde{K}_{\tilde{n}}(t, u)\tilde{\phi}_i(u)\,du\right)e^{j\theta+j\omega_c t}\right\} = 0. \tag{A.110}$$

If we require that

$$\lambda_i = \frac{\tilde{\lambda}_i}{2}, \tag{A.111}$$

then (A.109) will be satisfied for any $\theta$. Because (A.109) is valid for any $\theta$, each eigenvalue and eigenfunction of the complex process corresponds to an eigenvalue and a family of eigenfunctions of the bandpass process. Clearly, not more than two of these can be algebraically linearly independent. These can be chosen to be orthogonal by using $\theta = 0$ and $\theta = -\pi/2$. (Any two values of $\theta$ that differ by 90° are also satisfactory.) Thus,

**Table A.1**

| Complex process | Actual bandpass process | |
|---|---|---|
| $\tilde{\lambda}_1 \tilde{\phi}_1(t)$ | $\lambda_1 = \dfrac{\tilde{\lambda}_1}{2}$ | $\phi_1(t) = \sqrt{2}\ \text{Re}\ [\tilde{\phi}_1(t) e^{j\omega_c t}]$ |
| | $\lambda_2 = \dfrac{\tilde{\lambda}_1}{2}$ | $\phi_2(t) = \sqrt{2}\ \text{Re}\ [\tilde{\phi}_1(t) e^{j(\omega_c t - \pi/2)}] = \sqrt{2}\ \text{Im}\ [\tilde{\phi}_1(t) e^{j\omega_c t}]$ |
| $\tilde{\lambda}_2 \tilde{\phi}_2(t)$ | $\lambda_3 = \dfrac{\tilde{\lambda}_2}{2}$ | $\phi_3(t) = \sqrt{2}\ \text{Re}\ [\tilde{\phi}_2(t) e^{j\omega_c t}]$ |
| | $\lambda_4 = \dfrac{\tilde{\lambda}_2}{2}$ | $\phi_4(t) = \sqrt{2}\ \text{Im}\ [\tilde{\phi}_2(t) e^{j\omega_c t}]$ |

we can index the eigenvalues and eigenfunctions as shown in Table A.1. The result that the eigenvalues of the actual process occur in pairs is important in our succeeding work. It leads to a significant simplification in our analyses.

The relationship between the coefficients in the Karhunen-Loève expansion follows by direct substitution:

$$\begin{aligned} n_1 &= \text{Re}\ [\tilde{n}_1], \\ n_2 &= \text{Im}\ [\tilde{n}_1], \\ n_3 &= \text{Re}\ [\tilde{n}_2], \\ n_4 &= \text{Im}\ [\tilde{n}_2], \end{aligned} \tag{A.112}$$

and so forth.

From (A.89) and (A.90) we know that $n_c(t)$ and $n_s(t)$ have identical covariance functions. When they are uncorrelated processes, the eigenvalues of $\tilde{n}(t)$ are just twice the eigenvalues of $n_c(t)$. In the general case, there is no simple relationship between the eigenvalues of the complex envelope process and the eigenvalues of the quadrature process.

Up to this point we have considered only second-moment characteristics. We frequently are interested in Gaussian processes. If $n(t)$ is a nonstationary Gaussian process and (A.88)–(A.90) are true, we could define $\tilde{n}(t)$ to be a complex Gaussian random process. It is easier to define a complex Gaussian process directly.

**Definition.** Let $\tilde{n}(t)$ be a random process defined over some interval $[T_\alpha, T_\beta]$ with a mean value $\tilde{m}_{\tilde{n}}(t)$ and covariance function

$$E[(\tilde{n}(t) - \tilde{m}_{\tilde{n}}(t))(\tilde{n}^*(u) - \tilde{m}_{\tilde{n}}^*(u))] = \tilde{K}_{\tilde{n}}(t, u), \tag{A.113}$$

which has the property that

$$E[(\tilde{n}(t) - \tilde{m}_{\tilde{n}}(t))(\tilde{n}(u) - \tilde{m}_{\tilde{n}}(u))] = 0, \qquad \text{for all} \quad t \text{ and } u. \tag{A.114}$$

If every complex linear functional of $\tilde{n}(t)$ is a complex Gaussian random variable, $\tilde{n}(t)$ is a complex Gaussian random process. In other words, assume that

$$\tilde{y} = \int_{T_\alpha}^{T_\beta} \tilde{g}(u)\tilde{x}(u)\, du, \tag{A.115}$$

where $\tilde{g}(u)$ is any function such that $E[|\tilde{y}|^2] < \infty$. Then, in order for $\tilde{x}(u)$ to be a complex Gaussian random process, $\tilde{y}$ must be a complex Gaussian random variable for every $\tilde{g}(u)$ in the above class.

Notice that this definition is exactly parallel to the definition of a real Gaussian process on page I-183. Various properties of nonstationary Gaussian processes are derived in the problems. Since stationary complex Gaussian processes are a special case, they must satisfy the above definition. It is straightforward to show that the definition on page 583 is equivalent to the above definition when the processes are stationary.

Returning to the Karhunen-Loève expansion, we observe that if $\tilde{n}(t)$ is a complex Gaussian random process, $\tilde{n}_i$ is a complex Gaussian random variable whose density is given by (A.81), with $\sigma_n^2 = \tilde{\lambda}_i/2$,

$$p_{\tilde{n}_i}(\tilde{N}_i) = \frac{1}{\pi\tilde{\lambda}_i} \exp\left(-\frac{|\tilde{N}_i|^2}{\tilde{\lambda}_i}\right), \qquad -\infty < \text{Re}\,[\tilde{N}_i], \text{Im}\,[\tilde{N}_i] < \infty. \tag{A.116}$$

The complex Gaussian white noise process has the property that a series expansion using any set of orthonormal functions has statistically independent coefficients. Denoting the $i$th coefficient as $\tilde{w}_i$, we have

$$p_{\tilde{w}_i}(\tilde{W}_i) = \frac{1}{\pi N_0} \exp\left(-\frac{|\tilde{W}_i|^2}{N_0}\right), \qquad -\infty < \text{Re}\,[W_i], \text{Im}\,[W_i] < \infty. \tag{A.117}$$

This completes our general discussion of nonstationary processes. We now consider complex processes with a finite state representation.

### A.3.3 Complex Finite-State Processes†

In our previous work we found that an important class of random processes consists of those which can be generated by exciting a finite-dimensional linear dynamic system with a white noise process. Instead of

† This section is based on [9].

working with the bandpass process, we shall work with the complex envelope process. We want to develop a class of complex processes that we can generate by exciting a finite-dimensional complex system with complex white noise. We define it in such a manner that its properties will be consistent with the properties of stationary processes when appropriate. The complex state equation of interest is

$$\dot{\tilde{\mathbf{x}}}(t) = \tilde{\mathbf{F}}(t)\tilde{\mathbf{x}}(t) + \tilde{\mathbf{G}}(t)\tilde{\mathbf{u}}(t). \tag{A.118}$$

This is just a generalization of (A.39) to include a vector-driving function $\tilde{\mathbf{u}}(t)$. The observation equation is

$$\tilde{\mathbf{y}}(t) = \tilde{\mathbf{C}}(t)\tilde{\mathbf{x}}(t). \tag{A.119}$$

A block diagram of the system is shown in Fig. A.15. We assume that $\tilde{\mathbf{u}}(t)$ is a complex vector white noise process with zero mean and covariance matrix

$$E[\tilde{\mathbf{u}}(t)\tilde{\mathbf{u}}^{\dagger}(\sigma)] = \tilde{\mathbf{K}}_{\tilde{\mathbf{u}}}(t, \sigma) = \tilde{\mathbf{Q}}\, \delta(t - \sigma), \tag{A.120}$$

where

$$\tilde{\mathbf{u}}^{\dagger}(t) \triangleq [\tilde{\mathbf{u}}(t)^*]^T. \tag{A.121}$$

We further assume that

$$E[\tilde{\mathbf{u}}(t)\tilde{\mathbf{u}}^T(\sigma)] = \mathbf{0}, \qquad \text{for all} \quad t \text{ and } \sigma. \tag{A.122}$$

This is just the vector analog to the assumption in (A.90). In terms of the quadrature components,

$$\begin{aligned} \tilde{\mathbf{K}}_{\tilde{\mathbf{u}}}(t, \sigma) &= E[(\mathbf{u}_c(t) - j\mathbf{u}_s(t))(\mathbf{u}_c^T(\sigma) + j\mathbf{u}_s^T(\sigma))] \\ &= \mathbf{K}_{\mathbf{u}_c}(t, \sigma) + \mathbf{K}_{\mathbf{u}_s}(t, \sigma) + j\mathbf{K}_{\mathbf{u}_c\mathbf{u}_s}(t, \sigma) - j\mathbf{K}_{\mathbf{u}_s\mathbf{u}_c}(t, \sigma) \\ &= \tilde{\mathbf{Q}}\, \delta(t - \sigma). \end{aligned} \tag{A.123}$$

The requirement in (A.122) implies that

$$\mathbf{K}_{\mathbf{u}_c}(t, \sigma) = \mathbf{K}_{\mathbf{u}_s}(t, \sigma) = \tfrac{1}{2}\, \text{Re}\, [\tilde{\mathbf{Q}}]\, \delta(t - \sigma), \tag{A.124}$$

$$\mathbf{K}_{\mathbf{u}_c\mathbf{u}_s}(t, \sigma) = -\mathbf{K}_{\mathbf{u}_s\mathbf{u}_c}(t, \sigma) = \tfrac{1}{2}\, \text{Im}\, [\tilde{\mathbf{Q}}]\, \delta(t - \sigma). \tag{A.125}$$

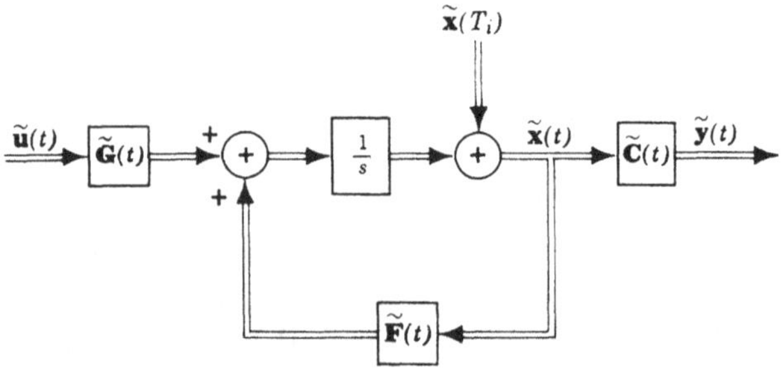


**Fig. A.15 Generation of a finite-state complex process.**

The covariance matrices for the two quadrature components are identical non-negative-definite matrices, and the cross-covariance matrix is a skew symmetric matrix (i.e., $a_{ij} = -a_{ji}$). This implies that $\tilde{\mathbf{Q}}$ is a Hermitian matrix with a non-negative-definite real part.

Usually we do not need a correlation between the components of $\tilde{\mathbf{u}}(t)$ (i.e., we can let $E[\mathbf{u}_c(t)\mathbf{u}_s^T(t)] = \mathbf{0}$), since any correlation between the components of the state vector may be represented in the coefficient matrices $\tilde{\mathbf{F}}(t)$ and $\tilde{\mathbf{G}}(t)$. In this case $\tilde{\mathbf{Q}}$ is a real non-negative-definite symmetric matrix.

The next issue that we want to consider is the initial conditions. In order that we be consistent with the concept of state, whatever symmetry assumptions we make regarding the state vector at the initial time $T_i$ should be satisfied at an arbitrary time $t$ ($t \geq T_i$).

First, we shall assume that $\tilde{\mathbf{x}}(T_i)$ is a complex random vector (we assume zero mean for simplicity). The complex covariance matrix for this random vector is

$$\begin{aligned}\tilde{\mathbf{P}}_i &\triangleq \tilde{\mathbf{K}}_{\tilde{\mathbf{x}}}(T_i, T_i) = E[\tilde{\mathbf{x}}(T_i)\tilde{\mathbf{x}}^\dagger(T_i)] \\ &= \mathbf{K}_{\mathbf{x}_c}(T_i, T_i) + \mathbf{K}_{\mathbf{x}_s}(T_i, T_i) + j\mathbf{K}_{\mathbf{x}_c\mathbf{x}_s}(T_i, T_i) - j\mathbf{K}_{\mathbf{x}_s\mathbf{x}_c}(T_i, T_i). \qquad \text{(A.126)}\end{aligned}$$

We assume that

$$E[\tilde{\mathbf{x}}(T_i)\tilde{\mathbf{x}}^T(T_i)] = \mathbf{0}. \qquad \text{(A.127)}$$

Notice that (A.126) and (A.127) are consistent with our earlier ideas. They imply that

$$\mathbf{K}_{\mathbf{x}_c}(T_i, T_i) = \mathbf{K}_{\mathbf{x}_s}(T_i, T_i) = \tfrac{1}{2}\,\mathrm{Re}\,[\tilde{\mathbf{P}}_i], \qquad \text{(A.128)}$$

$$\mathbf{K}_{\mathbf{x}_c\mathbf{x}_s}(T_i, T_i) = -\mathbf{K}_{\mathbf{x}_s\mathbf{x}_c}(T_i, T_i) = \tfrac{1}{2}\,\mathrm{Im}\,(\tilde{\mathbf{P}}_i). \qquad \text{(A.129)}$$

Consequently, the complex covariance matrix of the initial condition is a Hermitian matrix with a non-negative-definite real part.

Let us now consider what these assumptions imply about the covariance of the state vector $\tilde{\mathbf{x}}(t)$ and the observed signal $\tilde{\mathbf{y}}(t)$. Since we can relate the covariance of $\tilde{\mathbf{y}}(t)$ directly to that of the state vector, we shall consider $\tilde{\mathbf{K}}_{\tilde{\mathbf{x}}}(t, u)$ first.

For real state-variable random processes, we can determine $\mathbf{K}_{\mathbf{x}}(t, \sigma)$ in terms of the state equation matrices, the matrix $\mathbf{Q}$ associated with the covariance of the excitation noise $\mathbf{u}(t)$, and the covariance $\mathbf{K}_{\mathbf{x}}(T_i, T_i)$ of the initial state vector, $\mathbf{x}(T_i)$. The results for complex state variables are parallel. The only change is that the transpose operation is replaced by a conjugate transpose operation. Because of the similarity of the derivations, we shall only state the results (see Problem A.3.19).

The matrix $\tilde{\mathbf{K}}_{\tilde{\mathbf{x}}}(t, t)$ is a Hermitian matrix that satisfies the linear matrix differential equation

$$\frac{d\tilde{\mathbf{K}}_{\tilde{\mathbf{x}}}(t, t)}{dt} = \tilde{\mathbf{F}}(t)\tilde{\mathbf{K}}_{\tilde{\mathbf{x}}}(t, t) + \tilde{\mathbf{K}}_{\tilde{\mathbf{x}}}(t, t)\tilde{\mathbf{F}}^{\dagger}(t) + \tilde{\mathbf{G}}(t)\tilde{\mathbf{Q}}\tilde{\mathbf{G}}^{\dagger}(t), \tag{A.130}$$

where the initial condition $\tilde{\mathbf{K}}_{\tilde{\mathbf{x}}}(T_i, T_i)$ is given as part of the system description. [This result is analogous to (I-6.279).] $\tilde{\mathbf{K}}_{\tilde{\mathbf{x}}}(t, \sigma)$ is given by

$$\tilde{\mathbf{K}}_{\tilde{\mathbf{x}}}(t, \sigma) = \begin{cases} \tilde{\boldsymbol{\phi}}(t, \sigma)\tilde{\mathbf{K}}_{\tilde{\mathbf{x}}}(\sigma, \sigma), & t > u, \\ \tilde{\mathbf{K}}_{\tilde{\mathbf{x}}}(t, t)\tilde{\boldsymbol{\phi}}^{\dagger}(\sigma, t), & u > t, \end{cases} \tag{A.131}$$

where $\tilde{\boldsymbol{\phi}}(t, \sigma)$ is the complex transition matrix associated with $\tilde{\mathbf{F}}(t)$. (This result is analogous to that in Problem I-6.3.16.) In addition,

$$\tilde{\mathbf{K}}_{\tilde{\mathbf{x}}}(t, \sigma) = \tilde{\mathbf{K}}_{\tilde{\mathbf{x}}}^{\dagger}(\sigma, t) \tag{A.132}$$

and

$$E[\tilde{\mathbf{x}}(t)\tilde{\mathbf{x}}^T(\sigma)] = \mathbf{0}, \qquad \text{for all} \quad t \text{ and } \sigma. \tag{A.133}$$

Therefore the assumptions that we have made on the covariance of the initial state vector $\tilde{\mathbf{x}}(T_i)$ are satisfied by the covariance of the state vector $\tilde{\mathbf{x}}(t)$ for all $t \geq T_i$.

Usually we are not concerned directly with the state vector of a system. The vector of interest is the observed signal, $\tilde{\mathbf{y}}(t)$, which is related to the state vector by (A.119). We can simply indicate the properties of the covariance $\tilde{\mathbf{K}}_{\tilde{\mathbf{y}}}(t, \sigma)$, since it is related directly to the covariance of the state vector by

$$\tilde{\mathbf{K}}_{\tilde{\mathbf{y}}}(t, \sigma) = \tilde{\mathbf{C}}(t)\tilde{\mathbf{K}}_{\tilde{\mathbf{x}}}(t, \sigma)\tilde{\mathbf{C}}^{\dagger}(\sigma). \tag{A.134}$$

Consequently, it is clear that $\tilde{\mathbf{K}}_{\tilde{\mathbf{y}}}(t, t)$ is Hermitian. Similarly, from (A.133) we have the result that $E[\tilde{\mathbf{y}}(t)\tilde{\mathbf{y}}^T(\sigma)]$ is zero.

The properties of the quadrature components follow easily:

$$E[\mathbf{y}_c(t)\mathbf{y}_c^T(\sigma)] = E[\mathbf{y}_s(t)\mathbf{y}_s^T(\sigma)] = \tfrac{1}{2}\,\mathrm{Re}\,[\tilde{\mathbf{K}}_{\tilde{\mathbf{y}}}(t, \sigma)] \tag{A.135}$$

and

$$E[\mathbf{y}_c(t)\mathbf{y}_s^T(\sigma)] = \tfrac{1}{2}\,\mathrm{Im}\,[\tilde{\mathbf{K}}_{\tilde{\mathbf{y}}}(t, \sigma)]. \tag{A.136}$$

In this section we have introduced the idea of generating a complex random process by exciting a linear system having a complex state variable description with a complex white noise. We then showed how we could describe the second-order statistics of this process in terms of a complex covariance function, and we discussed how we could determine this function from the state-variable description of the system. The only assumptions that we made were on the second-order statistics of $\tilde{\mathbf{u}}(t)$ and $\tilde{\mathbf{x}}(T_i)$. Our results were independent of the form of the coefficient matrices $\tilde{\mathbf{F}}(t)$, $\mathbf{G}(t)$, and $\tilde{\mathbf{C}}(t)$. Our methods were exactly parallel to those for real state variables. It is easy to verify that all of the results are consistent with those derived in Sections A.1 and A.2 for stationary and nonstationary random processes.

We now consider a simple example to illustrate some of the manipulations involved.

**Example 1.** In this example we consider a first-order (scalar) state equation. We shall find the covariance function for the nonstationary case and then look at the special case in which the process is stationary, and find the spectrum. The equations that describe this system are

$$\frac{d\tilde{x}(t)}{dt} = -\tilde{k}\tilde{x}(t) + \tilde{u}(t), \qquad T_i \le t \tag{A.137}$$

and

$$\tilde{y}(t) = \tilde{x}(t). \tag{A.138}$$

The assumptions on $\tilde{u}(t)$ and $\tilde{x}(T_i)$ are

$$E[\tilde{u}(t)\tilde{u}^*(\sigma)] = 2\,\mathrm{Re}[\tilde{k}]P\,\delta(t-\sigma) \tag{A.139}$$

and

$$E[|\tilde{x}(T_i)|^2] = P_i. \tag{A.140}$$

Because we have a scalar process, both $P$ and $P_i$ must be real. In addition, we have again assumed zero means.

First, we shall find $K_x(t, t)$. The differential equation (A.130) that it satisfies is

$$\begin{aligned}\frac{d\tilde{K}_{\tilde{x}}(t,t)}{dt} &= -\tilde{k}\tilde{K}_{\tilde{x}}(t,t) - \tilde{k}^*\tilde{K}_{\tilde{x}}(t,t) + 2\,\mathrm{Re}\,[\tilde{k}]P \\ &= -2\,\mathrm{Re}\,[\tilde{k}]\tilde{K}_{\tilde{x}}(t,t) + 2\,\mathrm{Re}\,[\tilde{k}]P, \qquad t \ge T_i.\end{aligned} \tag{A.141}$$

The solution to (A.141) is

$$\tilde{K}_{\tilde{x}}(t,t) = P - (P - P_i)e^{-2\,\mathrm{Re}\,[\tilde{k}](t-T_i)}, \qquad t \ge T_i. \tag{A.142}$$

In order to find $\tilde{K}_{\tilde{x}}(t, \sigma)$ by using (A.131), we need to find $\tilde{\phi}(t, \sigma)$, the transition matrix for this system. This is

$$\tilde{\phi}(t,\sigma) = e^{-\tilde{k}(t-\sigma)}, \qquad t > \sigma. \tag{A.143}$$

By substituting (A.142) and (A.143) into (A.131), we can find $\tilde{K}_{\tilde{x}}(t, \sigma)$, which is also $\tilde{K}_{\tilde{y}}(t, \sigma)$ for this particular example.

Let us now consider the stationary problem in more detail. This case arises when we observe a segment of a stationary process. To make $\tilde{x}(t)$ stationary, we let

$$P_i = P. \tag{A.144}$$

If we perform the indicated substitutions and define $\tau = t - u$, we obtain

$$\tilde{K}_{\tilde{x}}(\tau) = \begin{cases} Pe^{-\tilde{k}\tau}, & \tau \ge 0, \\ Pe^{\tilde{k}^*\tau}, & \tau \le 0. \end{cases} \tag{A.145}$$

This may be written as

$$\tilde{K}_{\tilde{x}}(\tau) = Pe^{-\mathrm{Re}\,[\tilde{k}]|\tau|}e^{-j\,\mathrm{Im}\,[\tilde{k}]\tau}. \tag{A.146}$$

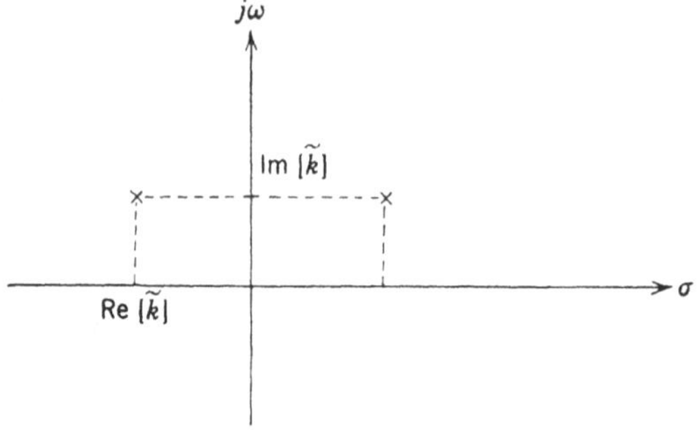


**Fig. A.16 Pole location for stationary process generated by first-order system.**

The spectrum of the complex process is

$$\tilde{S}_{\tilde{y}}(\omega) = \tilde{S}_{\tilde{x}}(\omega) = \frac{2\,\mathrm{Re}\,[\tilde{k}]P}{(\omega + \mathrm{Im}\,[\tilde{k}])^2 + (\mathrm{Re}\,[\tilde{k}])^2}. \tag{A.147}$$

From (A.147), we see that in the stationary case, the net effect of the complex pole $\tilde{k}$ is that the complex spectrum has a frequency shift equal to the imaginary part of $\tilde{k}$. In the actual bandpass process, this corresponds to shifting the carrier frequency. This is obvious if we look at the pole-zero plot of $\tilde{S}_{\tilde{x}}(\omega)$ as shown in Fig. A.16.

**Example 2.** Consider the pole-zero plot shown in Fig. A.17. The spectrum is

$$\tilde{S}_{\tilde{y}}(\omega) = \frac{C}{(j\omega - \tilde{k}_1)(j\omega - \tilde{k}_2)(-j\omega - \tilde{k}_1^*)(-j\omega - \tilde{k}_2^*)}. \tag{A.148}$$

We can generate this spectrum by driving a two-state system with complex white noise. The eigenvalues of the $\tilde{\mathbf{F}}$ matrix must equal $-\tilde{k}_1$ and $-\tilde{k}_2$. If we use the state representation

$$\tilde{x}_1(t) = \tilde{y}(t) \tag{A.149}$$

$$\tilde{x}_2(t) = \dot{\tilde{x}}_1(t), \tag{A.150}$$

then the equations are

$$\frac{d}{dt}\begin{bmatrix}\tilde{x}_1(t)\\ \tilde{x}_2(t)\end{bmatrix} = \begin{bmatrix}0 & 1\\ -\tilde{k}_1\tilde{k}_2 & -(\tilde{k}_1 + \tilde{k}_2)\end{bmatrix}\begin{bmatrix}\tilde{x}_1(t)\\ \tilde{x}_2(t)\end{bmatrix} + \begin{bmatrix}0\\ 1\end{bmatrix}u(t), \tag{A.151}$$

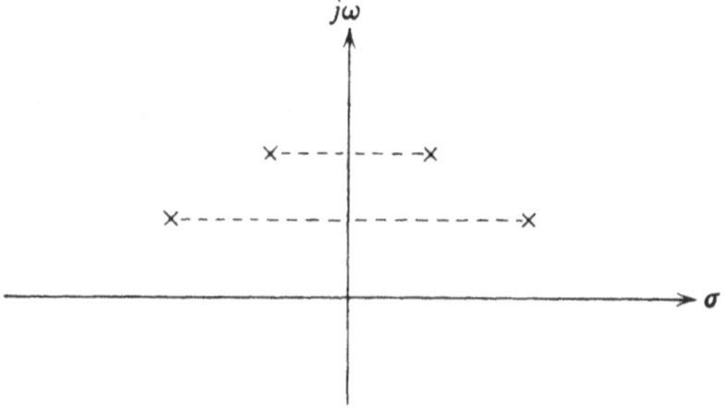


**Fig. A.17 Pole location for a particular second-order spectrum.**

and

$$\tilde{y}(t) = [1 \quad 0]\begin{bmatrix} \tilde{x}_1(t) \\ \tilde{x}_2(t) \end{bmatrix}, \tag{A.152}$$

where we assume that

$$\text{Re}\,[\tilde{k}_1] > 0, \tag{A.153a}$$

$$\text{Re}\,[\tilde{k}_2] > 0, \tag{A.153b}$$

and

$$\tilde{k}_1 \neq \tilde{k}_2. \tag{A.153c}$$

We can carry out the same type of analysis as in Example 1 (see Problem A.3.14).

Our two examples emphasized stationary processes. The use of complex state variables is even more important when we must deal with nonstationary processes. Just as with real state variables, they enable us to obtain complete solutions to a large number of important problems in the areas of detection, estimation, and filtering theory. Many of these applications arise logically in Chapters 9–13. There is one application that is easy to formulate, and so we include it here.

***Optimal Linear Filtering Theory.*** In many communication problems, we want to estimate the complex envelope of a narrow-band process. The efficiency of real state-variable techniques in finding estimator structures suggests that we can use our complex state variables to find estimates of the complex envelopes of narrow-band processes. In this section we shall indicate the structure of the realizable complex filter for estimating the complex envelope of a narrow-band process. We shall only quote the results of our derivation, since the methods used are exactly parallel to those for real state variables. The major difference is that the transpose operations are replaced by conjugate transpose operations.

We consider complex random processes that have a finite-dimensional state representation. In the state-variable formulation of an optimal linear filtering problem, we want to estimate the state vector $\tilde{\mathbf{x}}(t)$ of a linear system when we observe its output $\tilde{\mathbf{y}}(t)$ corrupted by additive white noise, $\tilde{\mathbf{w}}(t)$. Therefore, our received signal $\tilde{\mathbf{r}}(t)$ is given by

$$\begin{aligned}\tilde{\mathbf{r}}(t) &= \tilde{\mathbf{y}}(t) + \tilde{\mathbf{w}}(t) \\ &= \tilde{\mathbf{C}}(t)\tilde{\mathbf{x}}(t) + \tilde{\mathbf{w}}(t), \qquad T_i \le t \le T_f,\end{aligned} \tag{A.154}$$

where

$$E[\tilde{\mathbf{w}}(t)\tilde{\mathbf{w}}^\dagger(\tau)] = \tilde{\mathbf{R}}(t)\,\delta(t-\tau). \tag{A.155}$$

We assume that $\tilde{\mathbf{R}}(t)$ is a positive-definite Hermitian matrix.

In the realizable filtering problem, we estimate the state vector at the endpoint time of the observation interval, i.e., at $T_f$. This endpoint time, however, is usually a variable that increases as the data are received. Consequently, we want to have our estimate $\tilde{\mathbf{x}}(t)$ evolve as a function of the endpoint time of the observation interval $[T_i, t]$. We choose the estimate $\hat{\tilde{\mathbf{x}}}(t)$ to minimize the mean-square error,

$$\tilde{\boldsymbol{\xi}}_P(t) \triangleq E\{[\tilde{\mathbf{x}}(t) - \hat{\tilde{\mathbf{x}}}(t)][\tilde{\mathbf{x}}(t) - (\hat{\tilde{\mathbf{x}}}(t)]^\dagger\}. \tag{A.156}$$

We assume that $\hat{\tilde{\mathbf{x}}}(t)$ is obtained by a linear filter. For complex Gaussian processes, this gives the best MMSE estimate without a linearity assumption.

We can characterize the optimum realizable filter in terms of its impulse response $\tilde{\mathbf{h}}_o(t, \tau)$, so that the optimal estimate is given by

$$\hat{\tilde{\mathbf{x}}}(t) = \int_{T_i}^{t} \tilde{\mathbf{h}}_o(t, \tau)\tilde{\mathbf{r}}(\tau)\, d\tau, \qquad t > T_i. \tag{A.157}$$

It is easy to show that this impulse response $\tilde{\mathbf{h}}_o(t, \tau)$ is the solution of the complex Wiener-Hopf integral equation,

$$\tilde{\mathbf{K}}_{\tilde{\mathbf{x}}}(t, \tau)\tilde{\mathbf{C}}^\dagger(\tau) = \int_{T_i}^{t} \tilde{\mathbf{h}}_o(t, \sigma)\tilde{\mathbf{K}}_{\tilde{\mathbf{r}}}(\sigma, \tau)\, d\sigma, \qquad T_i \le \tau < t \tag{A.158}$$

(see Problem A.3.15). In the state-variable formulation we find $\hat{\tilde{\mathbf{x}}}(t)$ directly without finding the optimum impulse response explicitly. By paralleling the development for real state variables, we can implicitly specify $\hat{\tilde{\mathbf{x}}}(t)$ as the solution of the differential equation

$$\frac{d\hat{\tilde{\mathbf{x}}}(t)}{dt} = \tilde{\mathbf{F}}(t)\hat{\tilde{\mathbf{x}}}(t) + \tilde{\mathbf{z}}(t)[\tilde{\mathbf{r}}(t) - \tilde{\mathbf{C}}(t)\tilde{\mathbf{x}}(t)], \qquad T_i \le t, \tag{A.159}$$

where

$$\tilde{\mathbf{z}}(t) = \tilde{\mathbf{h}}_o(t, t) = \tilde{\boldsymbol{\xi}}_P(t)\tilde{\mathbf{C}}^\dagger(t)\tilde{\mathbf{R}}^{-1}(t). \tag{A.160}$$

The covariance matrix $\tilde{\boldsymbol{\xi}}_P(t)$ is given by the nonlinear equation

$$\frac{d\tilde{\boldsymbol{\xi}}_P(t)}{dt} = \tilde{\mathbf{F}}(t)\tilde{\boldsymbol{\xi}}_P(t) + \tilde{\boldsymbol{\xi}}_P(t)\tilde{\mathbf{F}}^\dagger(t) - \tilde{\mathbf{z}}(t)\tilde{\mathbf{R}}(t)\tilde{\mathbf{z}}^\dagger(t) + \tilde{\mathbf{G}}(t)\tilde{\mathbf{Q}}\tilde{\mathbf{G}}^\dagger(t), \qquad T_i \le t, \tag{A.161}$$

which can also be written as

$$\frac{d\tilde{\boldsymbol{\xi}}_P(t)}{dt} = \tilde{\mathbf{F}}(t)\tilde{\boldsymbol{\xi}}_P(t) + \tilde{\boldsymbol{\xi}}_P(t)\tilde{\mathbf{F}}^\dagger(t) - \tilde{\boldsymbol{\xi}}_P(t)\tilde{\mathbf{C}}^\dagger(t)\tilde{\mathbf{R}}^{-1}(t)\tilde{\mathbf{C}}(t)\tilde{\boldsymbol{\xi}}_P(t) + \tilde{\mathbf{G}}(t)\tilde{\mathbf{Q}}\tilde{\mathbf{G}}(t), \qquad T_i \le t. \tag{A.162}$$

The initial conditions reflect our a-priori information about the initial state of the system.

$$\hat{\tilde{\mathbf{x}}}(T_i) = E[\tilde{\mathbf{x}}(T_i)], \tag{A.163}$$

$$\tilde{\xi}_P(T_i) = \tilde{\mathbf{P}}_i. \tag{A.164}$$

$\hat{\tilde{\mathbf{x}}}(T_i)$ is an a-priori estimate of the initial state. (Often it is assumed to be zero for zero-mean processes.) $\tilde{\mathbf{P}}_i$ is the covariance of this a-priori estimate.

As in the case of real variables, the variance equation may be computed independently of the estimator equation. In order to obtain solutions, it may be integrated numerically or the solution may be computed in terms of the transition matrix of an associated set of linear equations. Several interesting examples are discussed in the problems.

A particular case of interest corresponds to a scalar received waveform. Then we can write

$$\tilde{\mathbf{R}}(t) = N_0. \tag{A.165}$$

We also observe that $\tilde{\mathbf{C}}(t)$ is a $1 \times n$ matrix. In Fig. A.18 we show two pole-zero plots for the spectrum of $\tilde{y}(t)$. We denote the modulations matrix of the two systems as $\tilde{\mathbf{C}}_a(t)$ and $\tilde{\mathbf{C}}_b(t)$, respectively. Clearly, we can

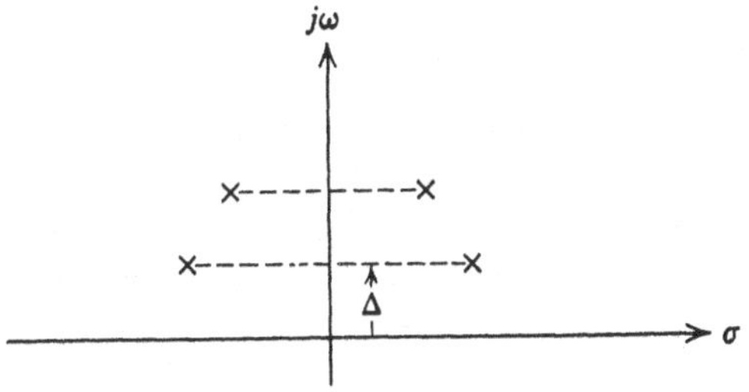


*(a)* Pole locations in system *A*

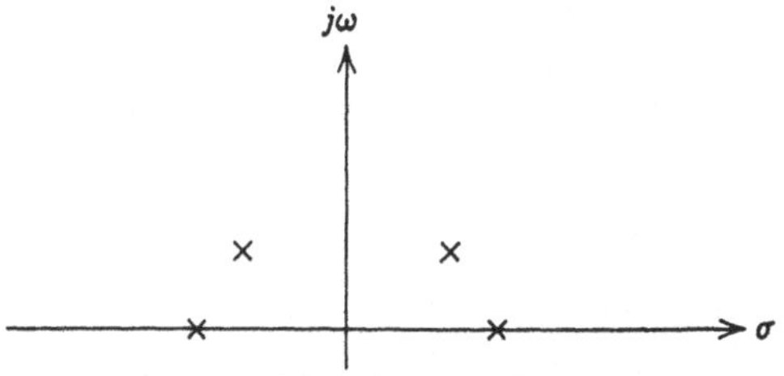


*(b)* Pole locations in system *B*

**Fig. A.18 Effect of carrier frequency shift on pole location.**

use the same state equations for the two systems and let

$$\tilde{\mathbf{C}}_b(t) = e^{-j\Delta t}\tilde{\mathbf{C}}_a(t). \tag{A.166}$$

Using (A.165) and (A.166) in (A.162), we see that $\tilde{\xi}_P(t)$ is not a function of $\Delta$. Since $\Delta$ corresponds to a carrier frequency shift in the actual bandpass problem, this result is just what we would expect. Notice that $\tilde{\xi}_P(t)$ is also invariant to an arbitrary phase modulation on $\tilde{\mathbf{C}}_a(t)$,

$$\tilde{\mathbf{C}}_n(t) = e^{-j\phi(t)}\tilde{\mathbf{C}}_a(t). \tag{A.167}$$

This result is less obvious intuitively, but follows easily from (A.162).

The results in (A.157)–(A.164) are valid for nonstationary processes and arbitrary observation intervals. For stationary processes and semi-infinite observation intervals, the problem is equivalent to the complex version of the Wiener filtering problem. All of the techniques carry over with obvious modifications (see Problem A.3.15).

Our discussion has considered the MMSE estimate of a complex random process. As we would expect from our work in Chapters 2 and 3, we shall encounter the problem of estimating a complex Gaussian random process in the detection problem.

## A.4 SUMMARY

In this appendix we have developed a complex representation for bandpass signals, systems, and processes. Several important ideas should be re-emphasized at this point. The first idea is that of a *complex envelope*, $\tilde{f}(t)$. It is a low-pass function whose magnitude is the actual envelope and whose phase is the phase modulation of the carrier. We shall find that the complex envelope plays the same role as the signal itself did in our earlier discussions. The second idea is that of a *complex Gaussian random process*. It plays the same role in the bandpass problem that the real Gaussian random process played previously. The third idea is that of *complex state variables*. They play the same role as real state variables did earlier.

We have spent a fair amount of time developing the complex notation. As we proceed through Chapters 9–14, we shall find that it was time well spent, because of the efficiency and insight it adds to the development.

## A.5 PROBLEMS

### P.A.1 Complex Signals

**Problem A.1.1.** The mean frequency $\bar{\omega}$ is defined in (A.16). Prove that we can always choose the carrier frequency so that

$$\bar{\omega} = 0.$$

**Problem A.1.2.** Derive the following expressions:

$$\overline{\omega^2} = \int_{-\infty}^{\infty} \left| \frac{d\tilde{f}(t)}{dt} \right|^2 dt,$$

$$\bar{\omega} = -j \int_{-\infty}^{\infty} \frac{d\tilde{f}(t)}{dt} \tilde{f}^*(t)\, dt,$$

$$\overline{t^2} = \int_{-\infty}^{\infty} \left| \frac{d\tilde{F}(j\omega)}{d\omega} \right|^2 \frac{d\omega}{2\pi},$$

$$\bar{t} = j \int_{-\infty}^{\infty} \frac{d\tilde{F}(j\omega)}{d\omega} \tilde{F}^*(j\omega) \frac{d\omega}{2\pi},$$

$$\overline{\omega t} = \operatorname{Im} \int_{-\infty}^{\infty} \omega \frac{d\tilde{F}(j\omega)}{d\omega} \tilde{F}^*(j\omega) \frac{d\omega}{2\pi}.$$

**Problem A.1.3** [18]. Write

$$\tilde{f}(t) \triangleq A(t) e^{j\varphi(t)},$$

where

$$A(t) \triangleq |\tilde{f}(t)|$$

is the signal envelope. Assume that

$$\bar{\omega} = \bar{t} = 0.$$

1. Prove that

$$\overline{t^2} = \int_{-\infty}^{\infty} t^2 A^2(t)\, dt, \tag{P.1}$$

$$\overline{\omega^2} = \int_{-\infty}^{\infty} \left(\frac{dA(t)}{dt}\right)^2 dt + \int_{-\infty}^{\infty} \left(\frac{d\varphi(t)}{dt}\right)^2 A^2(t)\, dt. \tag{P.2}$$

Notice that the first term in (P.2) is the frequency spread due to amplitude modulation and the second term is the frequency spread due to frequency modulation.

2. Derive an expression for $\overline{\omega t}$ in terms of $A(t)$ and $\varphi(t)$. Interpret the result.

**Problem A.1.4** [2].

1. Prove that

$$\operatorname{Re}\left[\int_{-\infty}^{\infty} t\tilde{f}(t) \frac{d\tilde{f}^*(t)}{dt}\, dt\right] = \frac{1}{2}, \tag{P.1}$$

and therefore

$$\int_{-\infty}^{\infty} t\tilde{f}(t) \frac{d\tilde{f}^*(t)}{dt}\, dt = \frac{1}{2} + j\overline{\omega t}. \tag{P.2}$$

2. Use the Schwarz inequality on (P.2) to prove

$$\overline{\omega^2}\,\overline{t^2} - (\overline{\omega t})^2 \geq \tfrac{1}{4}, \tag{P.3}$$

assuming

$$\bar{\omega} = \bar{t} = 0.$$

This can also be written as

$$\sigma_\omega \sigma_t [1 - \rho_{\omega t}^2]^{1/2} \geq \tfrac{1}{2}. \tag{P.4}$$

3. Prove that

$$\sigma_\omega \sigma_t \geq \tfrac{1}{2}. \tag{P.5}$$

An alternative way of stating (P.5) is to define

$$\sigma_f \triangleq \frac{\sigma_\omega}{2\pi}.$$

Then

$$\sigma_f \sigma_t \geq \pi. \tag{P.6}$$

The relation in (P.5) and (P.6) is called the *uncertainty relation*.

**Problem A.1.5.** Assume that

$$\tilde{f}(t) = \left(\frac{1}{\pi T^2}\right)^{1/4} \exp\left[-\left(\frac{1}{2T^2} - jb\right)t^2\right].$$

Find $\overline{\omega^2}$, $\overline{t^2}$, and $\overline{\omega t}$.

**Problem A.1.6.** In Chapter 10 we define a function

$$\phi(\tau, \omega) \triangleq \int_{-\infty}^{\infty} \tilde{f}\left(t - \frac{\tau}{2}\right) \tilde{f}^*\left(t + \frac{\tau}{2}\right) e^{-j\omega t}\, dt.$$

Evaluate $\sigma_\omega^2$, $\sigma_t^2$, and $\overline{\omega t} - \bar{\omega}\bar{t}$ in terms of derivatives of $\phi(\tau, \omega)$ evaluated at $\tau = \omega = 0$.

## P.A.3 Complex Processes

**Problem A.3.1.** Consider a complex Gaussian random variable whose density is given by (A.81). The characteristic function of a complex random variable is defined as

$$\tilde{M}_{\tilde{y}}(j\tilde{v}) = E[e^{j \operatorname{Re}[\tilde{v}^* \tilde{y}]}].$$

1. Find $\tilde{M}_{\tilde{y}}(j\tilde{v})$ for a complex Gaussian random variable.
2. How are the moments of $\tilde{y}$ related to $\tilde{M}_{\tilde{y}}(j\tilde{v})$ in general (i.e., $\tilde{y}$ is not necessarily complex Gaussian)?

**Problem A.3.2.** Consider the $N$-dimensional complex random vector $\tilde{\mathbf{x}}$, where

$$E[\tilde{\mathbf{x}}] = \mathbf{0},$$

$$E[\tilde{\mathbf{x}}\tilde{\mathbf{x}}^\dagger] \triangleq 2\tilde{\mathbf{\Lambda}}_{\tilde{\mathbf{x}}},$$

and

$$E[\tilde{\mathbf{x}}\tilde{\mathbf{x}}^T] = \mathbf{0}.$$

We define $\tilde{\mathbf{x}}$ to be a complex Gaussian vector if

$$p_{\tilde{\mathbf{x}}}(\tilde{\mathbf{X}}) = \frac{1}{(2\pi)^N |\tilde{\mathbf{\Lambda}}_{\tilde{\mathbf{x}}}|} \exp\left(-\tfrac{1}{2}\tilde{\mathbf{X}}^\dagger \tilde{\mathbf{\Lambda}}_{\tilde{\mathbf{x}}}^{-1}\tilde{\mathbf{X}}\right),$$

$$-\infty < \operatorname{Re}[\tilde{\mathbf{X}}] < \infty, \quad -\infty < \operatorname{Im}[\tilde{\mathbf{X}}] < \infty.$$

We refer to the components of $\tilde{\mathbf{x}}$ as *joint complex Gaussian random variables*.

The characteristic function of a complex random vector is defined as

$$\tilde{M}_{\tilde{\mathbf{x}}}(\tilde{\mathbf{v}}) \triangleq E[e^{j \operatorname{Re}[\tilde{\mathbf{v}}^\dagger \tilde{\mathbf{x}}]}].$$

Prove that

$$\tilde{M}_{\tilde{x}}(\tilde{v}) = \exp(-\tfrac{1}{4}\tilde{v}^{\dagger}\tilde{\Lambda}_{\tilde{x}}\tilde{v})$$

for a complex Gaussian random vector.

**Problem A.3.3.** A complex Gaussian random variable is defined in (A.81). Define

$$\tilde{y} = \tilde{\mathbf{g}}^{\dagger}\tilde{\mathbf{x}}.$$

If $\tilde{y}$ is a complex Gaussian random variable for every finite **g**, we say that $\tilde{\mathbf{x}}$ is a complex Gaussian random vector. Prove that this definition is equivalent to the one in Problem A.3.2.

**Problem A.3.4.** Assume that $\tilde{y}$ is a complex Gaussian random variable. Prove that

$$E[|\tilde{y}|^{2n}] = n!\,(E(|\tilde{y}|^2))^n.$$

**Problem A.3.5.** Assume that $\tilde{y}_1$ and $\tilde{y}_2$ are joint complex Gaussian random variables. Prove that

$$E[(\tilde{y}_1\tilde{y}_2^*)^n] = n!\,[E(\tilde{y}_1\tilde{y}_2^*)]^n.$$

**Problem A.3.6.** Assume that $\tilde{y}_1$, $\tilde{y}_2$, $\tilde{y}_3$, and $\tilde{y}_4$ are joint complex Gaussian random variables. Prove that

$$E[\tilde{y}_1^*\tilde{y}_2^*\tilde{y}_3\tilde{y}_4] = E[\tilde{y}_1^*\tilde{y}_3]E[\tilde{y}_2^*\tilde{y}_4] + E[\tilde{y}_2^*\tilde{y}_3]E[\tilde{y}_1^*\tilde{y}_4].$$

(This result is given in [16].)

**Problem A.3.7** [8]. Derive the "factoring-of-moments" property for complex Gaussian random processes. (Recall Problem I-3.3.12.)

**Problem A.3.8.** Consider the problem outlined on pages 161–165 of [15]. Reformulate this problem using complex notation and solve it. Compare the efficiency of the two procedures.

**Problem A.3.9.** In Problem I-6.2.1, we developed the properties of power density spectra of real random processes.

Let $n(t)$ be a stationary narrow-band process with a rational spectra $S_n(\omega)$. Denote the complex envelope process by $\tilde{n}(t)$, and its spectrum by $\tilde{S}_{\tilde{n}}(\omega)$.

1. Derive properties similar to those in Problem I-6.2.1.
2. Sketch the pole-zero plots of some typical complex spectra.

**Problem A.3.10.** The definition of a complex Gaussian process is given on page 588. Derive the complex versions of Properties 1 through 4 on pages I-183–I-185.

**Problem A.3.11.** Prove that the eigenvalues of a complex envelope process are invariant to the choice of the carrier frequency.

**Problem A.3.12.** Consider the results in (A.99)–(A.105).

1. Verify that one gets identical results by working with a real vector process,

$$\mathbf{n}(t) = \begin{bmatrix} n_c(t) \\ n_s(t) \end{bmatrix}.$$

[Review Section 3.7 and observe that $\mathbf{K}_{\mathbf{n}}(t, u)$ has certain properties because of the assumption in (A.90).]

2. What is the advantage of working with the complex process instead of $\mathbf{n}(t)$?

**Problem A.3.13** [14]. Consider the process described by (A.118)–(A.122) with

$$\tilde{\mathbf{G}}(t) = 1,$$

$$\tilde{\mathbf{F}}(t) = a - \frac{jbt^2}{2},$$

$$\tilde{\mathbf{C}}(t) = 1.$$

1. Find the covariance function of $\tilde{y}(t)$.
2. Demonstrate that $\tilde{y}(t)$ has the same covariance as the output of a Doppler-spread channel with a one-pole fading spectrum and input signal given by (10.52).

**Problem A.3.14.** Consider the process described by (A.148)–(A.153).

1. Find the covariance function of $\tilde{y}(t)$.
2. Calculate $E[|\tilde{y}|^2]$.

**Problem A.3.15.** Consider the linear filtering model in (A.154)–(A.156).

1. Derive the Wiener-Hopf equation in (A.158).
2. Derive the complex Kalman-Bucy equations in (A.159)–(A.164).
3. Assume that $T_i = -\infty$ and $\tilde{y}(t)$ is stationary. Give an explicit solution to (A.158) by using spectrum factorization.
4. Prove that, with a complex Gaussian assumption, a linear filter is the optimum MMSE processor.

**Problem A.3.16.** The complex envelope of the received waveform is

$$\tilde{r}(u) = \tilde{s}(u) + \tilde{w}(u), \qquad -\infty < u \le t,$$

where $\tilde{s}(u)$ and $\tilde{w}(u)$ are statistically independent complex Gaussian processes with spectra

$$\tilde{S}_{\tilde{s}}(\omega) = \frac{\text{Re}\,[\tilde{k}_1]P}{(\omega + \text{Im}\,[\tilde{k}_1])^2 + \text{Re}\,[\tilde{k}_1])^2} + \frac{k_2 P}{\omega^2 + k_2{}^2},$$

and

$$\tilde{S}_{\tilde{w}}(\omega) = N_0.$$

1. Find the minimum mean-square realizable estimate of $\tilde{s}(t)$.
2. Evaluate the minimum mean-square error.

**Problem A.3.17.** The complex envelope of the received waveform is

$$\tilde{r}(u) = \tilde{s}(u) + \tilde{n}_c(u) + \tilde{w}(u), \qquad -\infty < u \le t,$$

where $\tilde{s}(u)$, $\tilde{n}_c(u)$, and $\tilde{w}(u)$ are statistically independent Gaussian processes with spectra

$$\tilde{S}_{\tilde{s}}(\omega) = \frac{2\,\text{Re}\,[\tilde{k}_1]P_s}{(\omega + \text{Im}\,[\tilde{k}_1])^2 + (\text{Re}\,[\tilde{k}_1]^2)},$$

$$\tilde{S}_{\tilde{n}_c}(\omega) = \frac{2\,\text{Re}\,[\tilde{k}_1]P_c}{\omega^2 + (\text{Re}\,[\tilde{k}_1])^2},$$

and

$$\tilde{S}_{\tilde{w}}(\omega) = N_0,$$

respectively.

1. Find the minimum mean-square realizable estimate of $\tilde{s}(t)$.
2. Evaluate the minimum mean-square error. Do your results behave correctly as $\text{Im}\,[\tilde{k}_1] \to \infty$?

**Problem A.3.18.** Consider the system shown in Fig. P.A.1. The input $u(t)$ is a sample function of a real white Gaussian process.

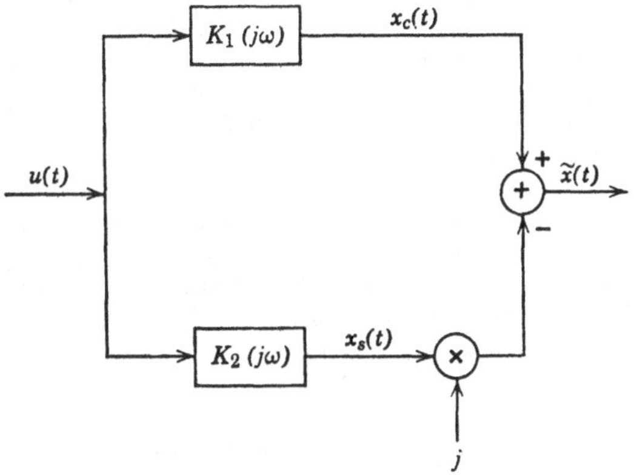


**Fig. P.A.1**

1. Compute $S_{x_c}(\omega)$, $S_{x_s}(\omega)$, and $S_{x_c x_s}(\omega)$.
2. Under what conditions is $\tilde{x}(t)$ a complex Gaussian process (according to our definition)?
3. Let $K_1(j\omega)$ and $K_2(j\omega)$ be arbitrary transfer functions. We observe that

$$\tilde{r}(t) = \tilde{x}(t) + \tilde{w}(t),$$

where $\tilde{w}(t)$ is a sample function of a complex white Gaussian process with spectral height $N_0$. Find the minimum mean-square *unrealizable* estimate of $\tilde{x}(t)$.

(This type of problem is discussed in [12] and [17].)

**Problem A.3.19.** Verify the results in (A.130)–(A.133).

## REFERENCES


[1] P. M. Woodward, *Probability and Information Theory, with Applications to Radar*, Pergamon Press, Oxford, 1953.

[2] D. Gabor, "Theory of Communications," J. IEE **93,** 429–457 (1946).

[3] R. Arens, "Complex Envelopes for Envelopes of Normal Noise," IRE Trans. Information Theory **IT-3,** 204–207 (Sept. 1957).

[4] E. J. Kelly and I. S. Reed, "Some Properties of Stationary Gaussian Processes," Massachusetts Institute of Technology, Lincoln Laboratory, TR-157, June 5, 1957.

[5] E. J. Kelly, I. S. Reed, and W. L. Root, "The Detection of Radar Echoes in Noise. I," J. SIAM **8,** 309–341 (June 1960).

[6] M. Schwartz, W. R. Bennett, and S. Stein, *Communication Systems and Techniques*, McGraw-Hill, New York, 1966.

[7] J. Dugundji, "Envelopes and Pre-Envelopes of Real Waveforms," IRE Trans. Information Theory **IT-4,** 53–57 (March 1958).

[8] I. S. Reed, "On a Moment Theorem for Complex Gaussian Processes," IRE Trans. Information Theory **IT-8,** 194–195 (April 1962).
[9] H. L. Van Trees, A. B. Baggeroer, and L. D. Collins, "Complex State Variables: Theory and Application," WESCON, Los Angeles, August 1968.
[10] J. M. Wozencraft and I. M. Jacobs, *Principles of Communication Engineering*, Wiley, New York, 1965.
[11] P. Bello, "On the Approach of a Filtered Pulse Train to a Narrowband Gaussian Process," IRE Trans. Information Theory, **IT-7,** 144–150 (July 1961).
[12] W. M. Brown and R. B. Crane, "Conjugate Linear Filtering," IEEE Trans. Information Theory **IT-15,** No. 4, 462–465 (July 1969).
[13] A. B. Baggeroer, "State Variables, the Fredholm Theory, and Optimal Communication," Sc.D. Thesis, Department of Electrical Engineering, Massachusetts Institute of Technology, 1968.
[14] L. D. Collins, Internal Memo, Detection and Estimation Theory Group, Massachusetts Institute of Technology, 1968.
[15] W. B. Davenport and W. L. Root, *An Introduction to the Theory of Random Signals and Noise*, McGraw-Hill, New York, 1958.
[16] J. L. Doob, *Stochastic Processes*, Wiley, New York, 1953.
[17] W. M. Brown and C. J. Palermo, *Random Processes, Communications and Radar*, McGraw-Hill, New York, 1969.
[18] C. W. Helstrom, *Statistical Theory of Signal Detection*, Pergamon Press, New York, 1960.

# *Glossary*

In this section we discuss the conventions, abbreviations, and symbols used in the book.

## CONVENTIONS

The following conventions have been used:

1. Boldface roman denotes a vector or matrix.
2. The symbol $|\quad|$ means the magnitude of the vector or scalar contained within.
3. The determinant of a square matrix $\mathbf{A}$ is denoted by $|\mathbf{A}|$ or det $\mathbf{A}$.
4. The script letters $\mathscr{F}(\cdot)$ and $\mathscr{L}(\cdot)$ denote the Fourier transform and Laplace transform respectively.
5. Multiple integrals are frequently written as,

$$\int d\tau\, f(\tau) \int dt\, g(t, \tau) \triangleq \int f(\tau) \left\{ \int dt\, g(t) \right\} d\tau,$$

that is, an integral is inside all integrals to its left unless a multiplication is specifically indicated by parentheses.

6. $E[\cdot]$ denotes the statistical expectation of the quantity in the bracket. The overbar $\bar{x}$ is also used infrequently to denote expectation.
7. The symbol $\otimes$ denotes convolution.

$$x(t) \otimes y(t) \triangleq \int_{-\infty}^{\infty} x(t - \tau) y(\tau)\, d\tau$$

8. Random variables are lower case (e.g., $x$ and $\mathbf{x}$). Values of random variables and nonrandom parameters are capital (e.g., $X$ and $\mathbf{X}$). In some estimation theory problems much of the discussion is valid for both random and nonrandom parameters. Here we depart from the above conventions to avoid repeating each equation.

9. The probability density of $x$ is denoted by $p_x(\cdot)$ and the probability distribution by $P_x(\cdot)$. The probability of an event $A$ is denoted by Pr $[A]$. The probability density of $x$, given that the random variable $a$ has a value $A$, is denoted by $P_{x|a}(X \mid A)$. When a probability density depends on non-random parameter $A$ we also use the notation $p_{x|a}(X \mid A)$. (This is non-standard but convenient for the same reasons as 8.)

10. A vertical line in an expression means "such that" or "given that"; that is Pr $[A \mid x \leq X]$ is the probability that event $A$ occurs given that the random variable $x$ is less than or equal to the value of $X$.

11. Fourier transforms are denoted by both $F(j\omega)$ and $F(\omega)$. The latter is used when we want to emphasize that the transform is a real-valued function of $\omega$. The form used should always be clear from the context.

12. Some common mathematical symbols used include,

| | |
|---|---|
| (i) $\propto$ | proportional to |
| (ii) $t \to T^-$ | $t$ approaches $T$ from below |
| (iii) $A + B \triangleq A \cup B$ | $A$ or $B$ or both |
| (iv) l.i.m. | limit in the mean |
| (v) $\int_{-\infty}^{\infty} d\mathbf{R}$ | an integral over the same dimension as the vector |
| (vi) $\mathbf{A}^T$ | transpose of $\mathbf{A}$ |
| (vii) $\mathbf{A}^{-1}$ | inverse of $\mathbf{A}$ |
| (viii) $\mathbf{0}$ | matrix with all zero elements |
| (ix) $\binom{N}{k}$ | binomial coefficient $\left(= \frac{N!}{k!\,(N-k)!}\right)$ |
| (x) $\triangleq$ | defined as |
| (xi) $\int_{\Omega} d\mathbf{R}$ | integral over the set $\Omega$ |

## ABBREVIATIONS

Some abbreviations used in the text are:

| | |
|---|---|
| ML | maximum likelihood |
| MAP | maximum a posteriori probability |
| PFM | pulse frequency modulation |
| PAM | pulse amplitude modulation |
| FM | frequency modulation |
| DSB-SC-AM | double-sideband–suppressed carrier–amplitude modulation |

| | |
|---|---|
| DSB-AM | double sideband-amplitude modulation |
| PM | phase modulation |
| NLNM | nonlinear no-memory |
| FM/FM | two-level frequency modulation |
| MMSE | minimum mean-square error |
| ERB | equivalent rectangular bandwidth |
| UMP | uniformly most powerful |
| ROC | receiver operating characteristic |
| LRT | likelihood ratio test |
| LEC | low energy coherence |
| SPLOT | stationary process–long observation time |
| SK | separable kernel |

**SYMBOLS**

The principal symbols used are defined below. In many cases the vector symbol is an obvious modification of the scalar symbol and is not included. Similarly, if the complex symbol is an obvious modification of the real symbol, it may be omitted.

| | |
|---|---|
| $A$ | class of detection problem |
| $A_a$ | actual value of parameter |
| $A_i$ | sample at $t_i$ |
| $A_w$ | class of detection problem, white noise present |
| $\hat{a}_0$ | solution to likelihood equation |
| $\hat{a}_{\text{abs}}$ | minimum absolute error estimate of $a$ |
| $\hat{a}_{\text{map}}$ | maximum a posteriori probability estimate of $a$ |
| $\hat{a}_{ml}$ | maximum likelihood estimate of $A$ |
| $\hat{a}_{\text{ms}}$ | minimum mean-square estimate of $a$ |
| $\alpha$ | amplitude weighting of specular component in Rician channel |
| $B$ | constant bias |
| $B$ | Bhattacharyya distance (equals $-\mu(1/2)$) |
| $B$ | class of detection problem |
| $B$ | signal bandwidth |
| $B(A)$ | bias that is a function of $A$ |
| $B_w$ | class of detection problem, white noise present |
| $\tilde{b}$ | random variable describing target or channel reflection |
| $\tilde{b}_D(t)$ | complex Gaussian process describing reflection from Doppler-spread target |
| $\tilde{b}_R(\lambda)$ | complex Gaussian process describing reflection from range-spread target |

| | |
|---|---|
| $\mathbf{B}_d(t)$ | matrix in state equation for desired signal |
| $\beta$ | parameter in PFM and angle modulation |
| $C$ | channel capacity |
| $C(a_\epsilon)$ | cost of an estimation error, $a_\epsilon$ |
| $C_F$ | cost of a false alarm (say $H_1$ when $H_0$ is true) |
| $C_{ij}$ | cost of saying $H_i$ is true when $H_j$ is true |
| $C_M$ | cost of a miss (say $H_0$ when $H_1$ is true) |
| $C_\infty$ | channel capacity, infinite bandwidth |
| $\tilde{C}(t:\tilde{\mathbf{x}}(t, \lambda))$ | modulation functional |
| $c$ | velocity of propagation |
| $\mathbf{C}(t)$ | modulation (or observation) matrix |
| $\mathbf{C}_d(t)$ | observation matrix, desired signal |
| $\mathbf{C}_M(t)$ | message modulation matrix |
| $\mathbf{C}_N(t)$ | noise modulation matrix |
| $\chi$ | parameter space |
| $\chi_a$ | parameter space for $a$ |
| $\chi_\theta$ | parameter space for $\theta$ |
| $\chi^2$ | chi-square (description of a probability density) |
| $D(\omega^2)$ | denominator of spectrum |
| $D_{\mathscr{F}}(\cdot)$ | Fredholm determinant |
| $D_{\min}$ | minimum diversity |
| $D_{\text{opt}}$ | optimum diversity |
| $D_o$ | optimum diversity |
| $d$ | desired function of parameter |
| $d$ | performance index parameter on ROC for Gaussian problems |
| $d(t)$ | desired signal |
| $\hat{d}(t)$ | estimate of desired signal |
| $\hat{d}_o(t)$ | optimum MMSE estimate |
| $d_\epsilon(t)$ | error in desired point estimate |
| $\delta$ | phase of specular component (Rician channel) |
| $\Delta$ | performance measure (9.49) |
| $\Delta_{dg}$ | performance degradation due to colored noise |
| $\Delta_o$ | performance measure in optimum receiver |
| $\Delta_\omega$ | width of Doppler cell |
| $\Delta_\tau$ | length of range cell |
| $\Delta_v$ | performance measure in suboptimum test |
| $\Delta_{wo}$ | performance measure in "white-optimum" receiver |
| $\mathbf{\Delta m}$ | mean difference vector (i.e., vector denoting the difference between two mean vectors) |
| $\mathbf{\Delta Q}$ | matrix denoting difference between two inverse covariance matrices |

| | |
|---|---|
| $E$ | energy (no subscript when there is only one energy in the problem) |
| $E_a$ | expectation over the random variable $a$ only |
| $E_I$ | energy in interfering signal |
| $E_i$ | energy on $i$th hypothesis |
| $\bar{E}_r$ | expected value of received energy |
| $E_t$ | transmitted energy |
| $E_y$ | energy in $y(t)$ |
| $E_1, E_0$ | energy of signals on $H_1$ and $H_0$ respectively |
| $E(R)$ | exponent in M-ary error bound |
| $E_\epsilon$ | energy in error signal (sensitivity context) |
| $e_N(t)$ | error waveform |
| $\epsilon_I$ | interval error |
| $\epsilon_{T'}$ | total error |
| $\text{erf}\,(\cdot)$ | error function (conventional) |
| $\text{erf}_*\,(\cdot)$ | error function (as defined in text) |
| $\text{erfc}\,(\cdot)$ | complement of error function (conventional) |
| $\text{erfc}_*\,(\cdot)$ | complement of error function (as defined in text) |
| $\eta$ | (eta) threshold in likelihood ratio test |
| $E(\cdot)$ | expectation operation (also denoted by $\overline{(\cdot)}$ infrequently) |
| $F$ | function to minimize or maximize that includes Lagrange multiplier |
| $\tilde{f}(t)$ | complex envelope of signal |
| $\tilde{f}_d(t)$ | complex envelope of signal returned from desired target |
| $f_c$ | oscillator frequency ($\omega_c = 2\pi f_c$) |
| $\mathbf{F}$ | matrix in differential equation |
| $\mathbf{F}(t)$ | time-varying matrix in differential equation |
| $G^+(j\omega)$ | factor of $S_r(\omega)$ that has all of the poles and zeros in LHP (and $\frac{1}{2}$ of the zeros on $j\omega$-axis). Its transform is zero for negative time. |
| $GB$ | general binary detection problem |
| $g(t)$ | function in colored noise correlator |
| $g(t, A), g(t, \mathbf{A})$ | function in problem of estimating $A$ (or $\mathbf{A}$) in colored noise |
| $g(\lambda_i)$ | a function of an eigenvalue |
| $g_{ILP}(\tau)$ | impulse response of ideal low-pass filter |
| $g_d(\lambda)$ | efficiency factor for diversity system |
| $g_h(t)$ | homogeneous solution |
| $g_l(\tau)$ | filter in loop |
| $g_{lo}(\tau), G_{lo}(j\omega)$ | impulse response and transfer function optimum loop filter |

| | |
|---|---|
| $g_{pu}(\tau)$ | unrealizable post-loop filter |
| $g_{puo}(\tau)$, $G_{puo}(j\omega)$ | optimum unrealizable post-loop filter |
| $g_\delta(t)$ | impulse solution |
| $g_\Delta(t)$ | difference function in colored noise correlator |
| $g_\lambda$ | a weighted sum of $g(\lambda_i)$ |
| $g_\infty(t)$, $G_\infty(j\omega)$ | infinite interval solution |
| $\tilde{g}(t)$ | complex function for optimum colored noise correlator |
| $\mathbf{G}$ | matrix in differential equation |
| $\mathbf{G}(t)$ | time-varying matrix in differential equation |
| $\mathbf{G}_d$ | linear transformation describing desired vector **d** |
| $\mathbf{G}_d(t)$ | matrix in differential equation for desired signal |
| $\mathbf{g}(t)$ | function for vector correlator |
| $\mathbf{g_d}(\mathbf{A})$ | nonlinear transformation describing desired vector **d** |
| $\Gamma(x)$ | Gamma function |
| $\gamma$ | parameter ($\gamma = k\sqrt{1 + \Lambda}$) |
| $\gamma$ | threshold for arbitrary test (frequently various constants absorbed in $\gamma$) |
| $\gamma_a$ | factor in nonlinear modulation problem which controls the error variance |
| $\sqcap_L(\lambda)$ | gate function |
| $H_0, H_1, \ldots, H_i$ | hypotheses in decision problem |
| $h_i$ | $i$th coefficient in orthogonal expansion of $h(t, u)$ |
| $H_n(t)$ | $n$th order Hermite polynomial |
| $h(t, u)$ | impulse response of time-varying filter (output at $t$ due to impulse input at $u$) |
| $h_1(\tau, u \mid z)$ | optimum unrealizable filter when white noise spectral height is $z$ |
| $h_1(\tau, u{:}t)$ | optimum filter for $[0, t]$ interval |
| $h_1^{[1/2]}(t, z)$ | functional square root of $h_1(t, z)$ |
| $h_{1\infty}(\tau)$, $H_{1\infty}(j\omega)$ | filter using asymptotic approximation |
| $h_{1d}(t, u)$ | filter to give delayed unrealizable MMSE estimate |
| $h_{\text{ch}}(t, u)$ | channel impulse response |
| $h_f(t, z)$ | filter in Canonical Realization No. 3 |
| $h_{fr}(t, z)$ | realizable filter in Canonical Realization No. 3 |
| $h_{fu}(t, z)$ | unrealizable filter in Canonical Realization No. 3 |
| $h_o(t, u)$ | optimum linear filter |
| $h_o'(\tau)$, $H_o'(j\omega)$ | optimum processor on whitened signal: impulse response and transfer function, respectively |
| $h_{or}(t, u)$ | optimum realizable linear filter for estimating $s(t)$ |
| $h_{ou}(\tau)$, $H_{ou}(j\omega)$ | optimum unrealizable filter (impulse response and transfer function) |
| $h_{\text{sub}}(\tau)$ | suboptimum filter |

| | |
|---|---|
| $h_w(t, u)$ | whitening filter |
| $h_{w_0}(t, u)$ | filter whose output is white on $H_0$ |
| $h_\Delta(t, u)$ | filter corresponding to difference between inverse kernels on two hypotheses (3.31) |
| $\tilde{h}(t, u)$ | complex envelope of impulse response of bandpass filter |
| $\tilde{h}_{wr}(t, z)$ | complex realizable whitening filter |
| $\mathrm{H}$ | linear matrix transformation |
| $\mathbf{h}_o(t, u)$ | optimum linear matrix filter |
| $I_o(\cdot)$ | modified Bessel function of 1st kind and order zero |
| $I_k(\cdot)$ | integrals involved in Edgeworth series expansion (defined by (2.160)) |
| $I_1, I_2$ | integrals |
| $I_\Gamma$ | incomplete Gamma function |
| $\mathbf{I}$ | identity matrix |
| $J(A)$ | function in variance bound |
| $J^{ij}$ | elements in $\mathbf{J}^{-1}$ |
| $J^{-1}(t, u)$ | inverse information kernel |
| $J_{ij}$ | elements in information matrix |
| $J_k(t, u)$ | $k$th term approximation to information kernel |
| $\mathbf{J}$ | information matrix (Fisher's) |
| $K_{\text{com}}(t, u{:}s)$ | covariance function of composite signal (3.59) |
| $K_s(t, u)$ | covariance function of signal |
| $K_{H_i}(t, u)$ | covariance of $r(t)$ on $i$th hypothesis |
| $K_{H_0}^{[-1/2]}(t, u)$ | functional square root of $K_{H_0}^{[-1]}(t, u)$ |
| $K_x(t, u)$ | covariance function of $x(t)$ |
| $\tilde{K}_D(\tau)$ | correlation function of Doppler process |
| $\tilde{K}_{DR}(\tau, \lambda)$ | target correlation function |
| $\tilde{K}_R\{v\}$ | two-frequency correlation function |
| $\mathbf{K}$ | covariance matrix |
| $\tilde{\mathbf{K}}_{\tilde{\mathbf{x}}}(t, u)$ | covariance function of $\tilde{\mathbf{x}}(t)$ |
| $\mathbf{k}_d(t)$ | linear transformation of $\mathbf{x}(t)$ |
| $L_n(x)$ | $n$th order Laguerre polynomial |
| $l(\mathbf{R}), l$ | sufficient statistic |
| $l(A)$ | likelihood function |
| $l_B$ | bias term in log likelihood ratio |
| $l_D$ | term in log likelihood ratio due to deterministic input |
| $l_R$ | term on log likelihood ratio due to random input |
| $l_a$ | actual sufficient statistic (sensitivity problem) |
| $l_c, l_s$ | sufficient statistics corresponds to cosine and sine components |
| $\tilde{l}_v$ | correlator output in suboptimum test |

| | |
|---|---|
| $l_{wo}$ | output of correlator in "white-optimum" receiver |
| $\Lambda$ | a parameter which frequently corresponds to a signal-to-noise ratio in message ERB |
| $\Lambda(\mathbf{R})$ | likelihood ratio |
| $\Lambda(r_k(t))$ | likelihood ratio |
| $\Lambda(r_k(t), A)$ | likelihood function |
| $\Lambda_B$ | signal-to-noise ratio in reference bandwidth for Butterworth spectra |
| $\Lambda_{\text{ef}}$ | effective signal-to-noise ratio |
| $\Lambda_g$ | generalized likelihood ratio |
| $\Lambda_m$ | parameter in phase probability density |
| $\Lambda_{3db}$ | signal-to-noise ratio in 3-db bandwidth |
| $\mathbf{\Lambda}_{\mathbf{x}}$ | covariance matrix of vector $\mathbf{x}$ |
| $\mathbf{\Lambda}_{\mathbf{x}}(t)$ | covariance matrix of state vector ($= \mathbf{K}_{\mathbf{x}}(t, t)$) |
| $\lambda$ | Lagrange multiplier |
| $\lambda_{\text{max}}$ | maximum eigenvalue |
| $\lambda_i$ | eigenvalue of matrix or integral equation |
| $\lambda_i(A)$ | $i$th eigenvalue, given $A$ |
| $\lambda_i^{\text{ch}}$ | eigenvalues of channel quadratic form |
| $\lambda_i^s$ | eigenvalue of signal process |
| $\lambda_i^T$ | total eigenvalue |
| $\lambda_i^*$ | eigenvalues of $r_*(t)$ |
| ln | natural logarithm |
| $\ln \Lambda(A)$ | log likelihood function |
| $\log_a$ | logarithm to the base $a$ |
| $M_x(jv)$, $M_{\mathbf{x}}(j\mathbf{v})$ | characteristic function of random variable $x$ (or $\mathbf{x}$) |
| $M_{l\|H_i}(s)$ | generating function of $l$ on $H_i$ |
| $m_D$ | mean Doppler shift |
| $m_i$ | $i$th coefficient in expansion of $m(t)$ |
| $m_R$ | mean delay |
| $m_x(t)$ | mean-value function of process |
| $m_\Delta(t)$ | difference between mean-value functions |
| $\mathbf{M}$ | matrix used in colored noise derivation |
| $\mathbf{m}$ | mean vector |
| $\mu(s)$ | logarithm of $\phi_{l(\mathbf{R})\|H_0}(s)$ |
| $\mu_{BP}(s)$ | $\mu(s)$ for bandpass problem |
| $\mu_{BS}(s)$ | $\mu(s)$ for binary symmetric problem |
| $\mu_D(s)$ | component of $\mu(s)$ due to deterministic signal |
| $\mu_{LEC}(s)$ | $\mu(s)$ for low energy coherence case |
| $\mu_{LP}(s)$ | $\mu(s)$ for low-pass problem |
| $\mu_R(s)$ | component of $\mu(s)$ due to random signal |
| $\mu_{SIB}(s)$ | $\mu(s)$ for simple binary problem |

| | |
|---|---|
| $\mu_{SK}(s)$ | $\mu(s)$ for separable kernel case |
| $\mu_\infty(s)$ | asymptotic form of $\mu(s)$ |
| $\tilde{\mu}(s)$ | complex version of $\mu(s)$ |
| $N$ | dimension of observation space |
| $N$ | number of coefficients in series expansion |
| $N(m, \sigma)$ | Gaussian (or Normal) density with mean $m$ and standard deviation $\sigma$ |
| $N(\omega^2)$ | numerator of spectrum |
| $N_0$ | spectral height (joules) |
| $n(t)$ | noise random process |
| $n_c(t)$ | colored noise (does not contain white noise) |
| $n_i$ | $i$th noise component |
| $n_*(t)$ | noise component at output of whitening filter |
| $\hat{n}_{c_r}(t)$ | MMSE realizable estimate of colored noise component |
| $\hat{n}_{c_u}(t)$ | MMSE unrealizable estimate of colored noise component |
| $\tilde{n}(t)$ | complex envelope of noise process |
| $\mathbf{N}$ | noise correlation (matrix numbers) |
| $n$, $\mathbf{n}$ | noise random variable (or vector variable) |
| $\xi_{CR}$ | Cramér-Rao bound |
| $\xi_{ij}(t)$ | elements in error covariance matrix |
| $\xi_{ml}$ | variance of ML interval estimate |
| $\xi_P(t)$ | expected value of *realizable* point estimation error |
| $\xi_P\left(t \mid s(\cdot), \frac{N_0}{2}\right)$ | minimum mean-square realizable filtering error of $s(t)$ in the presence of white noise with spectral height $N_0/2$ |
| $\xi_{Pi}(t)$ | variance of error of point estimate of $i$th signal |
| $\xi_{Pn}(t)$ | normalized realizable point estimation error |
| $\xi_{P\infty}$ | expected value of point estimation error, statistical steady state |
| $\xi_u$ | optimum unrealizable error |
| $\xi_{un}$ | normalized optimum unrealizable error |
| $\boldsymbol{\xi}_d(t)$ | covariance matrix in estimating $d(t)$ |
| $\boldsymbol{\xi}_{P\infty}$ | steady-state error covariance matrix |
| $\tilde{\boldsymbol{\xi}}(t)$ | function in optimum receiver equations (9.90) |
| $\tilde{\boldsymbol{\xi}}(t:\lambda, \lambda')$ | distributed error covariance function matrix |
| $P$ | power |
| $Pr(\epsilon)$ | probability of error |
| $Pr_{FSK}(\varepsilon)$ | probability of error for binary FSK system |
| $Pr_{PSK}(\varepsilon)$ | probability of error for binary PSK system |
| $P_{BP}$ | power in bandpass problem |
| $P_D$ | probability of detection (a conditional probability) |

| | |
|---|---|
| $P_{ef}$ | effective power |
| $P_F$ | probability of false alarm (a conditional probability) |
| $P_i$ | a priori probability of $i$th hypothesis |
| $P_{LP}$ | power in low-pass problem |
| $P_M$ | probability of a miss (a conditional probability) |
| $P_M^{[1]}$ | one-term approximation to $P_M$ |
| $P_r$ | received power |
| $P_t$ | transmitted power |
| $\tilde{P}_{DR}\{f, v\}$ | transform of $\tilde{S}_{DR}\{f, \lambda\}$ |
| $p_{\mathbf{r}\mid H_i}(\mathbf{R} \mid H_i)$ | probability density of $\mathbf{r}$, given that $H_i$ is true |
| $\phi(t)$ | eigenfunction |
| $\phi(Y)$ | Gaussian density, $N(0, 1)$ |
| $\phi_i(t)$ | $i$th coordinate function, $i$th eigenfunction |
| $\phi_{l(\mathbf{R})\mid \mathbf{H}_0}(s)$ | moment generating function of $l(\mathbf{R})$, given $H_0$ |
| $\phi_x(s)$ | moment generating function of random variable $x$ |
| $\phi(t)$ | phase of signal |
| $\phi(\tau, \omega)$, $\phi\{\tau, f\}$ | time-frequency correlation function |
| $\phi_{fg}(\tau, \omega)$ | time-frequency cross-correlation function |
| $\psi_L(t)$ | low pass phase function |
| $\psi_\Omega\{\lambda, f\}$ | spread cross-ambiguity function |
| $\mathbf{P}(t)$ | cross-correlation matrix between input to message generator and additive channel noise |
| $\boldsymbol{\phi}(t, \tau)$ | state transition matrix, time-varying system |
| $\boldsymbol{\phi}(t - t_0) \triangleq \boldsymbol{\phi}(\tau)$ | state transition matrix, time-invariant system |
| Pr [·], Pr (·) | probability of event in brackets or parentheses |
| $\Omega_B$ | bandwidth constraint |
| $\omega_c$ | carrier frequency (radians/second) |
| $\omega_D$ | Doppler shift |
| $\bar{\omega}$ | mean frequency |
| $Q(\alpha, \beta)$ | Marcum's $Q$ function |
| $Q_{H_i}(t, u)$ | inverse kernel on $i$th hypothesis |
| $Q_n(t, u)$ | inverse kernel |
| $q$ | height of scalar white noise drive |
| $\mathbf{Q}$ | covariance matrix of vector white noise drive |
| $\mathbf{Q}_n(u, z)$ | inverse matrix kernel |
| $R$ | transmission rate |
| $R(t)$ | target range |
| $R_x(t, u)$ | correlation function |
| $\tilde{R}_{DR}\{\tau, v\}$ | two-frequency correlation function |
| $\mathcal{R}_B$ | Bayes risk |
| $r(t)$ | received waveform (denotes both the random process and a sample function of the process) |

| | |
|---|---|
| $r_c(t)$ | combined received signal |
| $r_g(t)$ | output when inverse kernel filter operates on $r(t)$ |
| $r_K(t)$ | $K$ term approximation |
| $r_*(t)$ | output of whitening filter |
| $r_{**}(t)$ | output of $S_Q(\omega)$ filter (equivalent to cascading two whitening filters) |
| $\tilde{r}(t)$ | complex envelope of signal process |
| $\rho_{ij}$ | normalized correlation $s_i(t)$ and $s_j(t)$ (normalized signals) |
| $\rho_{12}$ | normalized covariance between two random variables |
| $\rho_{DR}$ | target skewness |
| $\rho_r$ | degradation due to interference |
| $\mathbf{R}(t)$ | covariance matrix of vector white noise $\mathbf{w}(t)$ |
| $\mathbf{r}$, $\mathbf{R}$ | observation vector |
| $S(j\omega)$ | Fourier transform of $s(t)$ |
| $S_c(\omega)$ | spectrum of colored noise |
| $S_Q(\omega)$ | Fourier transform of $Q(\tau)$ |
| $S_r(\omega)$ | power density spectrum of received signal |
| $S_x(\omega)$ | power density spectrum |
| $S_{\epsilon_o}(j\omega)$ | transform of optimum error signal |
| $\tilde{S}_D\{f\}$ | Doppler scattering function |
| $\tilde{S}_{DR}\{f, \lambda\}$ | scattering function |
| $\tilde{S}_{Du}\{f\}$ | uniform Doppler profile |
| $\tilde{S}_{\tilde{n}_r}\{f\}$ | spectrum of reverberation return |
| $\tilde{S}_R(\lambda)$ | range scattering function |
| $s(t)$ | signal component in $r(t)$, no subscript when only one signal |
| $s(t, A)$ | signal depending on $A$ |
| $s(t, a(t))$ | modulated signal |
| $s_a(t)$ | actual $s(t)$ (sensitivity context) |
| $s_{\text{com}}(t, s)$ | composite signal process (3.58) |
| $s_i$ | coefficient in expansion of $s(t)$ |
| $s_i$ | $i$th signal component |
| $s_R(t)$ | random component of signal |
| $s_r(t)$ | received signal |
| $\hat{s}_r(t)$ | realizable MMSE estimate of $s(t)$ |
| $s_t(t)$ | signal transmitted |
| $s_0(t)$ | signal on $H_0$ |
| $s_1(t)$ | signal on $H_1$ |
| $s_1(t, \boldsymbol{\theta})$, $s_0(t, \boldsymbol{\theta})$ | signal with unwanted parameters |
| $s_\Omega(t)$ | random signal |
| $s_*(t)$ | signal component at output of whitening filter |

| | |
|---|---|
| $\tilde{\Sigma}(t)$ | complex covariance matrix $(=\tilde{\xi}_P(t))$ |
| $\sigma^2$ | variance |
| $\sigma_1^2, \sigma_0^2$ | variance on $H_1$, $H_0$ |
| $\sigma_D^2$ | mean-square Doppler spread |
| $\sigma_\omega^2$ | mean-square bandwidth |
| $\sigma_R^2$ | mean-square delay spread |
| $\sigma_t^2$ | mean-square duration |
| $\mathbf{s}(t)$ | vector signal |
| $T$ | pulse duration |
| $T_b$ | initial observation time (same as $T_i$) |
| $T_d$ | duration of pulse sequence |
| $T_f$ | final observation time |
| $T_i$ | initial observation time |
| $T_p$ | pulse repetition interval |
| $\bar{t}$ | mean (arrival) time |
| $\tau$ | round-trip delay time |
| $\theta$, $\boldsymbol{\theta}$ | unwanted parameter |
| $\theta(\mathbf{A}, \mathbf{A}_a)$ | generalized ambiguity function |
| $\theta(\tau, \omega)$, $\theta\{\tau, f\}$ | signal ambiguity function |
| $\theta_{fg}(\tau, \omega)$ | cross-ambiguity function |
| $\theta_\Omega(\mathbf{A}_a, \mathbf{A})$ | generalized spread ambiguity function |
| $\theta_{\Omega_D}\{\lambda_a, \lambda : m_a, m\}$ | Doppler-spread ambiguity function |
| $\hat{\theta}$ | phase estimate |
| $\theta_{\text{ch}}(t)$ | phase of channel response |
| $\hat{\boldsymbol{\theta}}_1$ | estimate of $\boldsymbol{\theta}_1$ |
| $\mathbf{T}(t, \tau)$ | transition matrix |
| $[\ ]^T$ | transpose of matrix |
| $[\ ]^\dagger$ | conjugate transpose of matrix |
| $u_{-1}(t)$ | unit step function |
| $u(t)$, $\mathbf{u}(t)$ | input to system |
| $\tilde{u}(t)$ | elementary rectangular signal |
| $V$ | variable in piecewise approximation to $V_{\text{ch}}(t)$ |
| $V_{\text{ch}}(t)$ | envelope of channel response |
| $v$ | target velocity |
| $W$ | bandwidth parameter (cps) |
| $W(j\omega)$ | transfer function of whitening filter |
| $W_{\text{ch}}$ | channel bandwidth (cps) single-sided |
| $W^{-1}(j\omega)$ | transform of inverse of whitening filter |
| $w(t)$ | white noise process |
| $w(\tau)$ | impulse response of whitening filter |
| $\tilde{w}(t)$ | complex white noise process |

| | |
|---|---|
| $\mathbf{W}$ | a matrix operation whose output vector has a diagonal covariance matrix |
| $x(t)$ | input to modulator |
| $x(t)$ | random process |
| $\hat{x}(t)$ | estimate of random process |
| $\mathbf{x}$ | random vector |
| $\mathbf{x}(t)$ | state vector |
| $\mathbf{x}_a(t)$ | augmented state vector |
| $\mathbf{x}_d(t)$ | state vector for desired operation |
| $\mathbf{x}_f(t)$ | prefiltered state vector |
| $\mathbf{x}_M(t)$ | state vector, message |
| $\mathbf{x}_N(t)$ | state vector, noise |
| $\tilde{\mathbf{x}}(t, \lambda)$ | distributed complex state variable |
| $Y(t, u)$ | kernel in singularity discussion (3.151) |
| $y(t)$ | output of differential equation |
| $y(t)$ | transmitted signal |
| $Z$ | observation space |
| $Z_1, Z_2$ | subspace of observation space |
| $z(t)$ | output of whitening filter |
| $\mathbf{z}(t)$ | gain matrix in state-variable filter ($\triangleq \mathbf{h}_o(t, t)$) |

# Author Index

# Subject Index

9 780471 10793